A Field Guide
to the Families and Genera
of

Woody Plants of Northwest South America

(Colombia, Ecuador, Peru)

with supplementary notes on herbaceous taxa

A Field Guide to the Families and Genera of Woody Plants of Northwest South America (Colombia, Ecuador, Peru)

with supplementary notes on herbaceous taxa

Alwyn H. Gentry

Illustrations by Rodolfo Vasquez

Published in Association with Conservation International

THE UNIVERSITY OF CHICAGO PRESS

CHICAGO AND LONDON

THE UNIVERSITY OF CHICAGO PRESS, CHICAGO 60637
THE UNIVERSITY OF CHICAGO PRESS. LTD. LONDON

University of Chicago Press edition 1996
Printed in the United States of America

15 14 13 12 11 10 09 08 07 06 3 4 5 6 7

ISBN: 0-226-28944-3 (paper)

This printing contains every word of text and every illustration that appeared in the first printing. For this printing, the size of the book's pages has been reduced photographically, and the book has been printed on special paper to make it less bulky so that it may be carried and used more conveniently in the field.

Library of Congress Cataloging-in-Publication Data

Gentry, Alwyn H.

A field guide to the families and genera of woody plants of northwest South America (Colombia, Ecuador, Peru), with supplementary notes on herbaceous taxa / Alwyn H. Gentry ; illustrations by Rodolfo Vasquez.

p. cm.

Originally published: Washington, DC : Conservation International, 1993 (Conservation biology series ; contribution no. 1).

"Published in association with Conservation International."

Includes indexes.

1. Woody plants—Colombia—Identification. 2. Woody plants—Ecuador—Identification. 3. Woody plants—Peru—Identification. I. Conservation International. II. Title. III. Title: Woody plants of northwest South America (Colombia, Ecuador, Peru), with supplementary notes on herbaceous taxa

[QK241.G45 1996] 95-44911
582.1′5′0986—dc20 CIP

♾ The paper used in this publication meets the minimum requirements of the American National Standard for Information Sciences—Permanence of Paper for Printed Library Materials, ANSI Z39.48-1992.

CONTENTS

FERNS AND GYMNOSPERMS

MONOCOTS

DICOTS

Abbreviations used in this book:

C. Am.	Central America
incl.	includes
N. Am.	North America
n. temperate	north temperate
S. Am.	South America
s. temperate	south temperate
sp. (spp.)	species
subspp.	subspecies

FOREWORD

Biologists and conservationists who work in tropical countries face formidable scientific obstacles. Just knowing the biota is a Herculean task. While a scientist working in a temperate country might easily learn to identify many of the woody plant species in a single year, a tropical biologist could easily spend a lifetime in a single small country without mastering the flora. Tiny Ecuador, for example, has an estimated flora of 13,000 species, more than all of Europe combined. Neighboring Colombia supports an estimated 50,000 plants species, virtually a fifth of the entire global flora. The problem of identifying this floristic abundance is compounded by the dearth of taxonomic identification materials available to tropical field biologists. Al Gentry makes a vital contribution to tropical biology and conservation with this field guide. In tackling northwestern South America, a region with the most diverse flora on earth, he has created an exhaustive reference work tailored to meet the needs of field researchers and conservationists alike.

Gentry's near legendary expertise makes him uniquely qualified to write this ambitious work. For the past quarter of a century, Gentry has combed the Neotropics collecting and studying plants. His intense devotion to field work, great stamina, and willingness to botanize under the most difficult conditions, have enabled him to amass larger collections of plant specimens from more countries than any other living botanist. Moreover, he has helped build the plant collections of dozens of national museums and research institutions around the world. His writings — a body of some 200 scientific papers — are no less impressive a contribution to scientific study.

Gentry has incorporated much of his pragmatic field abilities into this book and gives us a sorely needed new approach to the identification of tropical plants. Traditionally, taxonomic determinations have been made using flower and fruit characteristics. This is a source of considerable frustration in the field where it is often impossible to find a flowering or fruiting individual of the most common trees. Instead, Gentry uses sterile vegetative characters such as leaves, bark, and odor to lead us through the maze of plant families and genera.

At Conservation International we work at developing the local capacity to do conservation in the field. The work is urgent. Gentry's encyclopedic knowledge of the neotropical flora has made him a vital member of Conservation International's Rapid Assessment Program. He is one of only a handful of people who can be dropped in the most remote, unexplored areas of the Neotropics and be able to quickly assess the conservation significance of the area in terms of species richness and endemism.

It is with great pleasure then, that we publish this innovative, authoritative and, above all, highly useful, volume. Gentry's contribution is the first in a series of field guides to support conservation biology in the tropics.

Adrian B. Forsyth, Ph.D.
Director of Conservation Biology

ACKNOWLEDGEMENTS

Publication of this book has been made possible by the support of the Pew Scholars Program in Conservation and the Environment. I also thank Adrian Forsyth and Conservation International for serving as publisher and distributor for it. Special thanks go to Jane Koplow of Jane Koplow Designs who took a rough manuscript replete with inconsistencies and eased it towards camera-ready copy complete with indices, a process far more demanding in time and editorial skill than either of us envisioned originally.

My greatest thanks are reserved for Rodolfo Vasquez, friend and colleague as well as artist, who prepared the illustrations. His unique combination of skill as both botanist and illustrator has allowed him to capture salient taxonomic characters, often exquisitely. In addition to providing the illustrations, he is the source of many of the Peruvian vernacular names. Moreover, some of the suggested field recognition characters indicated here are ideas that stem from his observations or that we have worked out together over the course of years of shared field work.

Many colleagues have found time to review the familial treatments of their taxonomic specialties, although not all of them agree with all of my taxonomic decisions. Taxonomists who have reviewed familial treatments include G. Schatz (MO) and P. Maas (U), Annonaceae, T. Croat (MO), Araceae, D. Stevens (MO), Asclepiadaceae, J. Miller (MO), Boraginaceae, D. Daly (NY), Burseraceae, D. Hunt (K), Cactaceae, R. Faden (US), Commelinaceae, M. Dillon (F), Compositae, E. Forero (NY), Connaraceae, B. Hammel (MO), Cyclanthaceae and Guttiferae, M. Huft (F), Euphorbiaceae, R. Moran (MO), ferns, L. Skog (US), Gesneriaceae, G. Davidse (MO), Gramineae, P. Goldblatt (MO), Iridaceae, R. Harley (K), Labiatae, H. van der Werf (MO), Lauraceae, S. Mori (NY), Lecythidaceae, G. Lewis (K), Leguminosae, J. Kuijt (LEA), Loranthaceae, W. Anderson (MICH), Malpighiaceae, P. Fryxell (TAES), Malvaceae, S. Renner (AAU) and J. Wurdack (US), Melastomataceae, C. Berg (ARBOHA), Moraceae, J. Pipoly (MO), Myrsinaceae, B. Holst (MO), Myrtaceae, C. Dodson and G. Carnevali (MO), Orchidaceae, A. Henderson (NY), and W. Hahn (WIS), Palmae, C. Taylor (MO), Rubiaceae, R. Gereau (MO), Rutaceae and Sapindaceae, W. Thomas (NY), Simaroubaceae, M. Nee (NY), Solanaceae, A. Weitzman (US), Theaceae.

I also thank M. Mathias and L. Holdridge for early stimulation of my interest in identifying tropical plants by vegetative characters; H. Cuadros, C. Dodson, C. Díaz, C. Grandes, R. Foster, P. Nuñez, R. Ortíz, O. Phillips, C. Reynel, A. Repizzo, and J. Zarucchi for contributions of common names or other observations; and graduate students at the University of Turku, San Marcos University, Washington University, St. Louis, and the University of Missouri-St. Louis, for helping to test the familial keys.

Finally, thanks are due to Richard C. Schaefer and Joel R. Manesberg for generously making available computer time and equipment during the production phase of this book.

LIST OF FIGURES

Figure		Page

Figure **Page**

Figure		Page

Figure		Page

Figure **Page**

Figure **Page**

Figure **Page**

Figure **Page**

INTRODUCTION

This book aims to make possible the identification to genus, usually even when only sterile material is available, of all the woody plants of Colombia, Ecuador, and Peru. I have generally adopted traditional concepts of families and genera. At the family level, this means that the families under which specimens are filed in the Missouri Botanical Garden herbarium, essentially the old Englerian concepts, are adopted here, with very few exceptions. This does not imply any value judgements as to the status of various groups sometimes recognized as families, but only what I judge to be the most convenient technique for a majority of users. I have made a value judgement in rejecting all of the proposed substitutions of new familial names like Poaceae for Gramineae or Asteraceae for Compositae.

At the generic level, on the other hand, I have made numerous value judgements as to what genera to recognize. When applicable, I have parenthetically indicated segregate generic names that are here included within a more broadly circumscribed genus. Although these are often somewhat subjective decisions, I have been guided by the principal that, unless compellingly different morphologically, segregate genera should only be recognized when demonstrably polyphyletic. Philosophically my taxonomic premise is that the generic limits proposed by a particular monographer constitute a hypothesis, to be accepted or rejected like any other in science. Thus, the generic concept that seems to make the most sense to the taxonomic consumer is likely to be the best one, no matter what the current specialist in a particular group may happen to say. In this connection it is useful for the nontaxonomic reader to remember that generic concepts have been more or less at the whim of particular taxonomists. In my opinion it should be the responsibility of the taxonomic consumer to evaluate the evidence and make his/her own decisions as to what generic delimitations most accurately reflect the real world, rather than blindly following the most recent monographer, especially if that monographer has been misled in formulating his taxonomic decisions by concern about avoiding paraphyletic groups.

All families that contain woody, epiphytic, or scandent species in northwest South America are included, even if most of their species are herbaceous. For each family, the characteristics for recognizing that family are indicated along with an indication of how to recognize its constituent genera. In addition, for most families each genus is listed along with a few of its most outstanding characteristics and hints for its differentiation from related taxa. A few important cultivated or naturalized genera are included parenthetically. For each genus the number of neotropical species is indicated parenthetically, where such an estimate is available; for some wide-ranging genera only a worldwide estimate of species numbers is available as indicated. Also included parenthetically, without indication of species numbers, is a second listing for many genera that are polymorphic with respect to characters (e.g., tree vs. liana) that are used to subdivide various families. Most woody genera, and a few herbaceous ones, are illustrated. The illustrations reflect the largely Amazonian material available to the Iquitos-based artist.

Common names are indicated for many genera, along with their geographic usage, with C referring to Colombia, E to Ecuador, and P to Peru. No attempt has been made at a complete catalogue of vernacular names. Instead I have listed only those names that I have personally encountered to be used by reliable informants in the respective areas (supplemented by additional similar observations of R. Vasquez for Amazonian Peru and C. Dodson for coastal Ecuador). Thus, the indicated vernacular names are strongly biased by my field experience, mostly in the coastal lowlands in Colombia and Ecuador, mostly in northern Amazonia in Peru. On the other hand, these are emphatically not the uncritical compilations that are found in many monographs or lists of local names. It should also be noted that most of the Peruvian vernacular names are Quechua-derived while most of the Colombian and Ecuadorian names are based on Spanish. No indigenous names are indicated unless they have been taken up by the mestizo populations of the respective regions. Most vernacular names are specific only to genus; in those relatively few cases where distinct names are applied to different species of the same genus, that is also indicated.

Treatment of the woody and scandent taxa is straight forward, and these are the only groups included in the key to families. It will be noted that the keys to family (as well as those to genus in the larger families) are not the traditional dichotomous ones of classical taxonomy. Rather I have attempted to talk the user through the relevant character states and possible taxonomic outcomes much as I would in a field course, with different-sized groups of taxa sharing a useful set of characters considered together along with their individual defining characters.

In general, I have included herbaceous genera only when they belong to a family with a significant complement of woody, scandent, or epiphytic species. Moreover, herbs are not considered in the introductory keys (although, of course, herbaceous members of a family generally share the features of the family's woody representatives that are treated in the keys). However, I have included at least synopses of the nonaquatic herbaceous families which occur in our area. In addition, for the very large, mostly herbaceous, families Gramineae and Compositae, I have included a listing only of woody taxa, plus some of the most important herbaceous genera. For predominantly epiphytic Orchidaceae, an excellent key to genera has been published in Spanish by C. Dodson and a previously unpublished English version of that key, kindly provided by Dr. Dodson, is included here. Although the key is mostly based on technical floral characters, those are generally unavoidable in generic delimitation in this family. Although the huge family Compositae is mostly herbaceous, the family is diverse and also includes trees and lianas. Like the orchids, its taxonomy depends inordinantly on technical floral characters that most nonspecialists, myself included, find abstruse and difficult to use. For the Compositae, Dr. M. Dillon has been kind enough to provide a tribal key as well as supplementary notes on many of the woody genera of each tribe. For both orchids and Compositae, generic identifications by the nonspecialist will rarely be possible without flowers.

Families that are exclusively or essentially aquatic or semiaquatic in our area are omitted entirely from this treatment. These include Alismataceae, Butomaceae, Callitrichaceae, Ceratophyllaceae, Elatinaceae, Hydrocharitaceae, Juncaceae, Lemnaceae, Mayacaceae, Najadaceae (and various segregates), Nymphaeaceae, Podostemaceae, Pontederiaceae, Potamogetonaceae, Taccaceae, Typhaceae, and Xyridaceae. For a treatment of these taxa see B. Leon (*Catalogo Anotado de las Fanerogamas Acuaticas del Peru.* In F. Kahn, B. Leon and K. Young (eds.) *Plantas Acuaticas del Peru.* ORSTOM and Inst. Frances de Estudios Andinos, Lima, 1992).

Most neotropical plants are surprisingly easy to identify to family, even in sterile condition. Indeed, in many ways it is probably *easier* to identify woody tropical plants in sterile condition to family than it is to identify the fertile material to which many systematic botanists tend to restrict themselves. This is true both because of strong convergence by many different families that share a common disperser or pollinator and because the technical characters on which most plant families are defined are so obscure and esoteric (typically involving a determination of ovule number, placement, and orientation) that they are of limited practical use. Moreover, the plants one wants to identify are more often encountered sterile than in fertile condition. Vegetative characters, on the other hand, are always available, mostly macroscopically obvious, and at least in the rain forest, apparently have been subjected to much less of the kind of convergence-inducing selection on taxonomically useful characters than have flowers or fruits. This book stresses identifications based on leaves. In some other vegetation types like deserts or Mediterranean-climate areas, the leaves, too, have converged to the extent that they have lost many of the obvious characters for identification, at least to my tropically experienced eye. The common plaint of the temperate botanist who encounters his first tropical forest is "The leaves all look just alike." However, I view this "lauraceous" look as a blessing in disguise. I suspect that it is precisely this generalized aspect that facilitates full expression of the taxonomically useful vegetative characters by tropical woody plants.

In addition to leaf characters, many woody families and genera have taxonomically useful bark, wood, and habit characters. Wood anatomists can often identify a piece of wood to genus or family. Anyone who has ever observed a good "matero" effortlessly identifying trees with nothing more than a machete slash of the bark and a sniff of his nose can begin to appreciate some of these additional characters. I am much less adept at such techniques, but have included some of the more obvious bark and slash characters here, generally as a supplementary aid to identification.

It has been my experience that any neophyte tropical botanist can learn to recognize the great majority of the plants he encounters to family by learning to recognize a particular combination of only two or three characters, although most families have a few nasty exceptions that have to be learned individually. Thus, opposite compound leaves are enough to suggest Bignoniaceae, and opposite compound leaves with tendrils are definitively Bignoniaceae. Similarly, primitive odor and strong bark is diagnostic for most Annonaceae, and lianas with simple, opposite, nonaromatic leaves having serrate or serrulate margins are found almost exclusively in Hippocrateaceae. Opposite leaves with punctations strongly suggest Myrtaceae, and if the leaves have a spicy odor, the identification is positive. I have tried to indicate this level of simple characters for recognizing each of the families included in this book.

A special word of encouragement on the sense of smell may prove useful to the neophyte. Virtually everyone whom I have instructed in field techniques for tropical plant identification has complained at first that his/her nose is below average and unable to detect the sometimes rather faint odors that may be critical for recognizing tropical plants vegetatively. But with a bit of practice, essentially everyone discovers that they really are able to pick up rather subtle nuances of plant odors (the main exception being herpetologists whose long exposure to formaldehyde has apparently "pickled" their sense of smell). It is useful to break or crush both leaves and young twigs to check for odor. Also, remember that slash and leaf odors are not always the same; for example, the sweetish odor of most Meliaceae seems restricted to the trunk slash.

Finally, beware the odor of anise. There is a common epiphyllous leafy liverwort that has a strong anise odor that is easily confused with some of the variants of what I here call the "primitive" or Ranalean odor (i.e., the odor produced by the ethereal oil cells of the leaves of woody Ranalean families). Whenever you encounter an anise odor, wash your hands and look for a different twig, from a different part of the plant. There is also a faint nondescript "leafy" aroma that emanates from the crushed leaves of many nonaromatic plants and this "leafy" odor (not to be confused with the typical green-bean odor of most legumes) similarly should be disregarded when doing vegetative identifications.

Latex characters can also be tricky. Some species or individuals of some families or genera that are supposed to be characterized by latex have exceedingly minute amounts; and these latex traces may be apparent only in the trunk or only in the leaves. Usually the petiole and midvein are the best places to look for latex. Frequently a trunk slash must be allowed several minutes for latex to ooze out or to turn to its characteristic color. Also the line between latex and sap is a very fine one and some saps, e.g., the bright red exudate of a number of papilionate legumes, are here informally called latex. For convenience, an exudate, even a weak one, that is white or colored may be referred to as latex, while an exudate that is clear may be generally considered as sap rather than latex, unless it is very copious. Thus, the red latex of Myristicaceae is in reality sap, while the legitimate latex or resin of Burseraceae and Anacardiaceae may be faint and often tends to exude clear, turning whitish or blackish, respectively, only on drying.

Obviously such characters as latex or odor must be observed or recorded in the field. Although this book is intended primarily as a field manual, I have also noted, coincidentally, some characteristics of dried specimens, especially the colors of dried leaves, that may be extremely useful taxonomically. Contrary to popular supposition, these color characters, which may be taxonomically definitive, even at the familial level (e.g., the characteristic black or olive color of dried Olacaceae, Loranthaceae, and their relatives), are not significantly affected by preservatives or other drying techniques.

In some families it is very easy to recognize all or most genera by vegetative characters. In others, of which Myrtaceae and Lauraceae are the most infamous examples, the family is easy to recognize vegetatively but generic differentiation tends to require fertile material. In such cases I have been forced to emphasize the traditional taxonomic differentiations, based on fertile material, while merely suggesting some of the vegetative trends that tend to characterize individual genera. At the opposite extreme are whole groups of families, e.g., those of the Malvales, that can easily be recognized to order and genus but often can only be differentiated to family by using fertile characters or by recognizing the individual genera involved. In addition, there are a few families like Euphorbiaceae that are so vegetatively heterogeneous that field identification to family may only be possible with recognition of the constituent genera.

In addition to those vegetative characters that are most generally useful in family recognition and emphasized in the keys, a number of unusual morphological features that may be useful for recognizing particular taxa are noted in the lists and figures that preceed the familial treatments. These traits include parasitic and saprophytic growth forms, leafless plants with chlorophyllous stems, tendrillate lianas, anomalous liana stem cross sections, branching patterns, ant domatia, stilt roots, spines, strongly fenestrated trunks, and strangling habit.

KEY OUTLINE

I. COMPOUND AND OPPOSITE

IA. Bipinnate to biternate (page 11)
Bignoniaceae, Ranunculaceae, Compositae, Leguminosae

IB. Simply pinnate (page 11)
Bignoniaceae, Staphyleaceae, Brunelliaceae, Cunoniaceae, Caprifoliaceae, Quiinaceae, Zygophyllaceae, Leguminosae, Juglandaceae, Rutaceae, Sapindaceae, Ranunculaceae, Valerianaceae

IC. 3-foliate (page 13)
Bignoniaceae, Ranunculaceae, Compositae, Caryocaraceae, Hippocastanaceae, Verbenaceae, Rutaceae

ID. Palmately compound (page 13)
Bignoniaceae, Verbenaceae

II. ALTERNATE AND COMPOUND

IIA. Bipinnate (page 14)
Leguminosae, Sapindaceae, Araliaceae, Rutaceae, Vitaceae, Meliaceae

IIB. Simply Pinnate (page 14)

IIBa. Parallel venation
Palmae, Cycadaceae

IIBb. Rank odor
Juglandaceae, Meliaceae, Proteaceae, Leguminosae

IIBc. Odor of essential oils or turpentine
Rutaceae, Anacardiaceae, Burseraceae

IIBd. Sweetish odor in trunk
Meliaceae

IIBe. Punctations
Rutaceae, Leguminosae

IIBf. Bitter taste
Simaroubaceae

IIBg. Spines
Leguminosae, Rutaceae, Sapindaceae, Palmae, Cycadaceae

IIBh. Latex
Sapindaceae, Anacardiaceae, Burseraceae, Julianaceae, Leguminosae

IIBi. Even-pinnate
Leguminosae, Meliaceae, Sapindaceae, Palmae, Cycadaceae

IIBj. Miscellaneous useful features
Winged rachis (various families), Terminal "bud" on rachis (Meliaceae), cylindrical pulvinuli (Connaraceae, Leguminosae,

Lepidobotryaceae, Simaroubaceae), aborted rachis tip (Sapindaceae, *Euplassa*), expanded petiole base (*Polylepis*), apical tendril (Polemoniaceae, Compositae, Leguminosae)

IIBk. Nondescript

Meliaceae (*Trichilia*), Staphyleaceae (*Huertea*), Sabiaceae (*Ophiocaryon*), Anacardiaceae (*Tapirira*), Simaroubaceae

IIC. 3-foliolate and alternate (page 20)

IICa. Trees

Anthodiscus, Allophylus, Erythrina, Hevea, Rutaceae

IICb. Vines

Leguminosae, Connaraceae, Sapindaceae, Cucurbitaceae, Vitaceae, (Euphorbiaceae), (Dioscoreaceae), (Menispermaceae)

IID. Palmately compound and alternate (page 21)

IIDa. Trees

Palmae, Rutaceae, Araliaceae, Bombacaceae, Caricaceae, Moraceae, Sterculiaceae, Cochlospermaceae

IIDb. Vines

Cucurbitaceae, Passifloraceae, Convolvulaceae, Dioscoreaceae, Sapindaceae, Araceae, Cyclanthaceae

---

III. SIMPLE AND OPPOSITE

IIIA. Lianas (page 24)

IIIAa. T-shaped trichomes

Malpighiaceae

IIIAa. 3(–7)-veined leaves

Loganiaceae, Coriariaceae, Melastomataceae, Compositae, Valerianaceae

IIIAc. Milky latex

Apocynaceae, Asclepiadaceae, Guttiferae, (Compositae)

IIIAd. Miscellaneous

Hippocrateaceae, Combretaceae, Rubiaceae, Bignoniaceae, Acanthaceae, Gesneriaceae, Verbenaceae, Nyctaginaceae, Amaranthaceae, Gnetaceae, Trigoniaceae, Saxifragaceae

IIIB. Trees and shrubs (page 29)

IIIBa. Stipules (or stipule scars)

Rubiaceae, Quiinaceae, Dialypetalanthaceae, Malpighiaceae, Chloranthaceae, Vochysiaceae, Rhizophoraceae

IIIBb. Latex

Apocynaceae, Guttiferae

IIIBc. Punctations

Myrtaceae, Melastomataceae, Lythraceae, Rutaceae, Guttiferae, Loranthaceae, (Rhizophoraceae)

IIIBd. 3(–7)-veined leaves with parallel cross veins

Melastomataceae

IIIBe. Odor of essential oils

Monimiaceae, Chloranthaceae, Lauraceae, Verbenaceae, Labiatae, Compositae

IIIBf. Glands on twig at petiole base

Vochysiaceae

IIIBg. Serrate margins

Hippocrateaceae, Violaceae, Elaeocarpaceae, Rhizophoraceae, Loganiaceae, Flacourtiaceae, Verbenaceae, Monimiaceae, Caprifoliaceae, Brunelliaceae, Chloranthaceae, Cunoniaceae, Columelliaceae, Acanthaceae, Euphorbiaceae

IIIBh. Miscellaneous

Nyctaginaceae, Myrtaceae, Melastomataceae, Lythraceae, Alzateaceae, Polygalaceae, Loganiaceae, Verbenaceae, Hippocrateaceae, Rhamnaceae, Acanthaceae, Gesneriaceae, Caprifoliaceae, Cornaceae, Columelliaceae, Oleaceae, (Buxaceae), Malpighiaceae, (Guttiferae), Elaeocarpaceae

---

IV. SIMPLE AND ALTERNATE

IVA. Trees (page 37)

IVAa. Latex

Sapotaceae, Moraceae, Euphorbiaceae, (Olacaceae), Apocynaceae, Papaveraceae, Myristicaceae, Campanulaceae, Anacardiaceae, (Chrysobalanaceae), (Annonaceae)

IVAb. Conical terminal stipule

Moraceae, (Magnoliaceae), Polygonaceae, (Winteraceae)

IVAc. Odor of essential oils

Piperaceae, Winteraceae, Magnoliaceae, Hernandiaceae, Annonaceae, Myristicaceae, Lauraceae, (Anacardiaceae), (Burseraceae), (Leguminosae), (Araliaceae), (Icacinaceae)

IVAd. Palmately veined (or 3-veined)

(1) Malvalean Pulvinus

Tiliaceae, Sterculiaceae, Bombacaceae, Malvaceae, Elaeocarpaceae, Bixaceae

(2) Without swollen pulvinus

Ulmaceae, Urticaceae, Euphorbiaceae, Buxaceae, (Caricaceae), Begoniaceae, Cochlospermaceae, Flacourtiaceae, Hamamelidaceae, (Hernandiaceae), Araliaceae, Rhamnaceae, Rhizophoraceae, Olacaceae, (Leguminosae), (Menispermaceae)

IVAe. Strong bark

(Annonaceae), Lecythidaceae, Thymelaeaceae, (Leguminosae), (Malvales), (Urticales)

IVAf. Unequal petioles

Araliaceae, Capparidaceae, Euphorbiaceae, (Sterculiaceae)

IVAg. Petiole glands

Chrysobalanaceae, Combretaceae, (Euphorbiaceae), (Flacourtiaceae), (Rhamnaceae)

IVAh. Serrate margins

Actinidiaceae, Aquifoliaceae, Betulaceae, Boraginaceae, Celastraceae, Clethraceae, Compositae, Dilleniaceae, (Elaeocarpaceae), Euphorbiaceae, Fagaceae, Flacourtiaceae, Humiriaceae, Icacinaceae, Lacistemataceae, Leguminosae, Myricaceae, (Myrsinaceae), Ochnaceae, Rhamnaceae, Rosaceae, Sabiaceae, Saxifragaceae, (Solanaceae), Symplocaceae, Theaceae, Theophrastaceae, Violaceae

IVAi. Thickened or flexed petiole apex

Dipterocarpaceae, Elaeocarpaceae, (Euphorbiaceae), Flacourtiaceae, (Meliaceae)

IVAj. Punctations

Flacourtiaceae, (Rutaceae), Myrsinaceae, Theaceae

IVAk. Stipules

Celastraceae, Chrysobalanaceae, Dichapetalaceae, Erythroxylaceae, Euphorbiaceae, (Flacourtiaceae), Lacistemataceae, Rosaceae, (Violaceae)

IVAl. Lepidote or stellate trichomes

(Annonaceae), Capparidaceae, Clethraceae, (Compositae), (Dilleniaceae), Euphorbiaceae, Icacinaceae, Fagaceae, (Malvales), Solanaceae, Styracaceae

IVAm. Leaves parallel-veined or lacking secondary veins

Podocarpaceae, Theaceae, Goodeniaceae, (Theophrastaceae), (monocots), Gramineae, (Palmae)

IVAn. Parallel tertiary venation

Opiliaceae, Guttiferae, Icacinaceae, Lacistemataceae, Lecythidaceae, Linaceae, (Myristicaceae), (Sapotaceae), Ochnaceae, Olacaceae

IVAo. Spines or spine-tipped leaves

Berberidaceae, Boraginaceae, Compositae, Cactaceae, Celastraceae, Euphorbiaceae, Flacourtiaceae, (Moraceae), Nyctaginaceae), Olacaceae, Phytolaccaceae, Rhamnaceae, (Rosaceae), (Solanaceae), Theophrastaceae, (Urticaceae)

IVAp. None of above

Amaranthaceae (*Pleuropetalum*), Aquifoliaceae (few *Ilex*), Bignoniaceae (*Crescentia, Amphitecna*), Boraginaceae, Capparidaceae, Celastraceae (*Gymnosporia,* few *Maytenus*), (Chrysobalanaceae), Combretaceae, Compositae, Cyrillaceae, Dichapetalaceae, Ebenaceae, Euphorbiaceae, Flacourtiaceae, Humiriaceae, Icacinaceae, (Leguminosae), (Moraceae), (Myricaceae), Olacaceae, Onagraceae, (Passifloraceae), Phytolaccaceae, (Polygonaceae), Rhamnaceae, (Sabiaceae), Solanaceae, Violaceae

IVB. Lianas (page 60)

IVBa. Tendrils

Passifloraceae, Cucurbitaceae, Vitaceae, Smilacaceae, Rhamnaceae, (Polygonaceae), Leguminosae

IVBb. Tendrils absent

(1) Parallel veins

Araceae, Cyclanthaceae, Gramineae, (other monocots)

(2) Serrate leaves

Dilleniaceae, Ulmaceae, Urticaceae, Euphorbiaceae, Celastraceae, Violaceae, (Malvaceae), Loasaceae, Compositae, (Basellaceae), (Boraginaceae)

(3) Deeply lobed and/or peltate

Tropaeolaceae, Euphorbiaceae, (Solanaceae), (Caricaceae), (Menispermaceae), (Aristolochiaceae)

(4) Primitive odor

Aristolochiaceae, Hernandiaceae, (Annonaceae), Lauraceae

(5) Petiolar or lamina base glands

Euphorbiaceae

(6) Palmately 3(–5)-veined

Sterculiaceae, Menispermaceae, Dioscoreaceae, Ericaceae, (Urticaceae), Rhamnaceae, Compositae, Basellaceae, (Olacaceae), (Leguminosae), Euphorbiaceae

(7) Latex

Convolvulaceae, Moraceae, Campanulaceae, (Euphorbiaceae), (Caricaceae), (Olacaceae)

(8) Spines

(Ulmaceae), Cactaceae, Phytolaccaceae, Polygalaceae, Solanaceae

(9) None of the above

Polygonaceae, Marcgraviaceae, Ericaceae, Dichapetalaceae, Icacinaceae, (Basellaceae), Compositae, Plumbaginaceae, Solanaceae, Convolvulaceae, Polygalaceae, (Phytolaccaceae), Boraginaceae, Amaranthaceae, (Olacaceae), (Dilleniaceae), (Onagraceae)

KEY TO FAMILIES

The families may be divided into four main vegetative groups:

I. Leaves compound and opposite

II. Leaves compound and alternate

III. Leaves simple and opposite

IV. Leaves simple and alternate

Hints for compound leaves: If in doubt, look for the axillary bud to determine whether the leaf is really compound or not. In deciduous species, thick twigs tend to indicate compound leaves. In fallen leaflets, asymmetric leaf bases often suggest origin from a compound leaf.

KEY I

LEAVES OPPOSITE OR WHORLED (AND COMPOUND) — the easiest category. Only one important family (Bignoniaceae) plus a few other small families and miscellaneous genera or species are characterized by opposite compound leaves.

IA. Leaves bipinnate to biternate (and opposite)

Bignoniaceae — *Jacaranda, Memora, Pleonotoma,* few *Arrabidaea* species.

Ranunculaceae — *Clematis:* lianas with deeply multicostate stems and sensitive petiole.

Compositae — Bipinnate only in herbs: leaflets serrate; plants aromatic.

Leguminosae — A few *Parkia* species with glandular area on top of petiole and characteristic legume pulvinus and pulvinuli.

IB. Leaves simply pinnately compound (and opposite)

Useful differentiating characters include: the rachis (winged in Cunoniaceae; angled and more or less grooved above in Bignoniaceae, Zygophyllaceae, Brunelliaceae, and Juglandaceae; conspicuously jointed, especially in Staphyleaceae); presence or absence of interpetiolar line or ridge (absent only in three families that are usually alternate-leaved: Rutaceae, Juglandaceae, Sapindaceae); type of marginal serrations and pubescence of the leaflets; and presence and type of stipules.

Bignoniaceae (few genera, *Tecoma, Digomphia* [1 sp.]) — Very weak interpetiolar line (V-shaped in *Digomphia*); leaflets petiolulate and sharply serrate in *Tecoma,* entire and subsessile in *Digomphia.*

Staphyleaceae — Only *Turpinia* with glabrous leaves on conspicuously jointed rachis, and with closely finely serrate or serrulate leaflet margins.

Brunelliaceae — Montane except in Chocó; almost always more or less densely pubescent leaves, usually closely serrate and with prominulously reticulate venation below and numerous secondary veins making obtuse angle with midvein; interpetiolar ridge usually with small subulate stipulelike projections.

Cunoniaceae — Montane; rachis usually winged, the sessile leaflets with coarsely serrate to remotely dentate margins; distinctive small leafy caducous stipules.

Caprifoliaceae — Only *Sambucus:* montane; finely and rather unevenly serrate, always petiolulate leaflets with stipel-like glands between upper leaflets.

Quiinaceae — Only *Touroulia:* leaflets often incompletely separated with bases usually decurrent on rachis, glabrous, with secondary veins rather close together and ending in spinose teeth.

Zygophyllaceae — Very dry areas; leaflets entire, round-tipped, oblong (sometimes very small) to broadly oval, sessile on strongly angled rachis more or less grooved above.

Leguminosae — Very few genera (*Platymiscium, Taralea*), typical legume pulvinus and pulvinuli; green-bean odor.

Juglandaceae — Only *Alfaroa:* montane; lower leaflets smaller, in some species becoming stipulelike, twigs and petiole hispid or the twigs with conspicuous round white lenticels; typical rank walnutlike odor.

Rutaceae — Only *Amyris:* with conspicuously punctate leaflets; acute to acuminate, petiolulate.

Sapindaceae — Few species, e.g., *Matayba apetala* with margins entire, petiolule short, thick-based.

Ranunculaceae — A few *Clematis* species are our only lianas with opposite simply pinnate leaves.

Valerianaceae — A few *Valeriana* species, all montane, are vines with opposite simply pinnate leaves, distinguished by drying blackish, and by opposite petiole bases joined to form an ochrea-like sheath.

IC. Leaves 3-foliolate

All groups listed as simply pinnate above and some of those listed as palmately compound below, may have occasional variants (or species) which are 3-foliolate, at least in part. The first three families listed below are mostly vines when 3-foliolate (also herbs in Compositae and trees with lepidote scales in *Tabebuia* of Bignoniaceae); the rest are trees.

Bignoniaceae (most lianas of the family are basically 3-foliolate) — Unique in the terminal leaflet of some leaves replaced by a tendril (or tendril scar)

Ranunculaceae (only a few *Clematis*) — Vines or lianas; similar to bignons but very different deeply costate-ribbed stems and evenly costate-striate twigs; lacking tendrils but petioles or petiolules sometimes twining.

Compositae (*Hidalgoa* [vine] and several herbs, especially *Bidens*) — Characteristic composite odor, serrate margins; *Hidalgoa* has undeveloped terminal leaflets forming kind of tenuous grappling hooks.

Caryocaraceae (only *Caryocar:* lowland) — Trees; many species are characterized by distinctive gland pair at apex of petiole (unique in opposite-leaved taxa); branchlets with a conspicuous rubiac-like terminal stipule; leaflets often serrate or serrulate).

Hippocastanaceae (only *Billia:* montane) — Vegetatively very similar to *Caryocar* but lacks petiolar glands and elongate terminal stipules; completely entire leaflets.

Verbenaceae (*Vitex*) — Trees; leaf bases usually gradually tapering into indistinct petioles is unique among opposite, 3-foliolate taxa.

Rutaceae (*Metrodorea*) — Trees; very characteristic in hollowed petiole base with ligulelike dorsal projection.

ID. Leaves palmately compound (and opposite)

Bignoniaceae (most neotropical trees of the family) — Leaflet bases rounded to cuneate, not tapering into petiolule (except *Godmania* with characteristic rank vegetative odor [cf., horse urine]); pubescence usually of stellate trichomes or lepidote scales.

Verbenaceae (*Vitex*) — Leaflet bases taper into indistinct petiolule unlike most Bignoniaceae; also differs from almost all area *Tabebuia* species in pubescence of simple trichomes.

KEY II

LEAVES ALTERNATE (AND COMPOUND) — Several families in this group are very difficult to separate on the basis of sterile characters, especially some Sapindaceae (*Cupania, Matayba, Talisia,* etc.) from some Meliaceae (*Trichilia*) and a few nonaromatic Anacardiaceae (*Tapirira*).

IIA. Leaves bipinnate (sometimes ternately so) (and alternate)

Leguminosae — Vines, trees or herbs. Bipinnate leaves and spines on trunk, branches, or rachises are unique to mimosoid legumes and some species of *Caesalpinia;* like nonspiny bipinnate legumes these taxa are characterized by the typical cylindrical legume pulvinus and pulvinulus and (in mimosoids) by development of an often elaborate gland on the dorsal side of the lower petiole or sometimes between the lower rachises.

Sapindaceae — Vines and lianas (usually basically ternate and usually with milky latex and sometimes characteristic compound stem anatomy) or trees (*Dilodendron* with many small serrate leaflets and the typical aborted rachis apex of the family).

Araliaceae (*Sciadodendron*) — Trees; thick branches and very large leaves with a ligulelike dorsal projection on petiole base.

Rutaceae (*Dictyoloma*) — Small second-growth tree of dry areas; leaflets with marginal punctations; charcteristic large flat-topped inflorescence.

(Vitaceae) (a few *Cissus* spp.) — Vines with basically ternate leaves; leaf-opposed tendrils and enlarged often reddish nodes are characteristic.

(Meliaceae) (*Melia*) — A tree cultivated and also escaped, characterized by thin serrate leaflets; young growth often conspicuously whitish; looks much more like Araliaceae than Meliaceae but twigs much thinner than *Sciadodendron* and lacking characteristic ligulelike base.

IIB. Leaves simply pinnate (and alternate)

IIBa. Leaflets with parallel venation

Palmae — The only woody monocots with pinnately compound leaves, the leaf segments unmistakable in their parallel venation.

Cycadaceae — Similar only to palms from which the leaflets differ in being more coriaceous and having the parallel leaflet veins all equal; often +/- spiny with spines shorter and thicker-based than in pinnate-leaved spiny palms.

IIBb. Rank odor

Juglandaceae (montane) — Usually serrate leaflets with characteristic walnut odor.

Meliaceae — *Cedrela* typically has a somewhat garliclike odor but always entire leaflets.

Proteaceae — Canned-meat odor is typical of *Roupala,* mostly with compound juvenile leaves; the trunk slash has a characteristic odor resembling low-quality canned beef; the leaflets are extremely asymmetrical, usually with one side flat (even concave), the other margin serrate.

Leguminosae — Most legumes have a characteristic more or less rank green-bean-like odor. Easy to recognize by distinctively round, swollen petiolules (whole length of petiolule uniformly cylindric) and petiole base; Connaraceae (lianas and occasionally treelets) lack the odor but are otherwise indistinguishable vegetatively from Leguminosae). Some *Picramnia* species (Simaroubaceae) have similarly pulvinate petioles and petiolules, but often with bitter taste and the strongly alternate leaflets becoming much smaller toward leaf base.

IIBc. Odor of essential oils or turpentine

Rutaceae — *Zanthoxylum;* look for spines on trunk or stems (unique in simply pinnate taxa) and typically punctate leaflets, at least along margin below.

Anacardiaceae and Burseraceae — These two families usually have a fairly strong turpentine-like or mangolike odor but are very difficult to tell apart vegetatively; check both families.

Anacardiaceae — Usually more weakly turpentine-odored or with a somewhat sweetish mangolike odor (only with a very faintly mango-like odor in the common *Tapirira guianensis* which might be taken for a *Trichilia* with too many leaflets); sometimes with a watery latex which dries black (wounded trunks often stained with black); may be considered a derived version of burseracs differentiated by the technical character of one anatropous ovule per locule or a single ovule in ovary.

Burseraceae — Usually strongly turpentine-odored, even in bark; almost always with milky latex either in the twigs or as few widely scattered droplets from the bark slash, the latex drying whitish around trunk wounds; technically separated from Anacardiaceae by ovules pendulous, epitropous, and two per locule.

IIBd. Sweetish odor — *(In trunk only; excluding mangolike odors)*

Meliaceae — Most Meliaceae are characterized by a faint but distinctly sweetish odor from the trunk slash (but *Cedrela* has a very different rank garliclike odor).

IIBe. Punctations — *(Look both against strong light and out of it; punctations are often restricted to the sinuses of marginal teeth or serrulations.)*

Rutaceae — Most species of Rutaceae are punctate at least along leaflet margin; most pinnate-leaved rutacs are species of *Zanthoxylum* and most of these have spines on the trunks, branchlets, or leaves.

Leguminosae — A few genera of legume have punctate leaves, the punctations often rather linear, at least in part.

IIBf. Very bitter taste

Simaroubaceae — Most simaroubacs are characterized by a bitter taste when the twig is chewed. (Apparently a few people don't taste the bitter principal of Simaroubaceae; check your taste buds on a known specimen. Another helpful hint: If you suspect Simaroubaceae, get someone else to taste for you or be sure you have water handy!).

IIBg. Spines — (Rare in simply pinnate taxa)

Leguminosae — *Machaerium,* mostly lianas, usually only with paired stipular spines and usually with red latex.

Rutaceae — *Zanthoxylum,* always trees in Neotropics; thick spines on trunk typical, also often with spines on petioles and leaflets.

(Sapindaceae) — A very few *Paullinia* species, all lianas, have branch-derived spines or short spines on angles of stem; characterized by milky latex.

(Burseraceae) — *Bursera orinocensis,* distinctive in its strongly pungent aroma, has branch spines.

Palmae — Several pinnate-leaved palm genera have spiny trunks and/or leaves.

Cycadaceae — The majority of *Zamia* species have short spines on the petiole and/or rachis.

IIBh. Latex (rare in compound-leaved taxa)

Sapindaceae — Latex present in most lianas (characterized also by bifurcating inflorescence-derived tendril and frequently compound wood) but never in trees.

Anacardiaceae — Subwatery latex in *Toxicodendron* (montane; highly allergenic) and few other genera, tends to dry blackish and sometimes visible around old wounds; (black resinous latex in many Old World species).

Burseraceae — Latex sometimes present in twigs, almost always in exceedingly inconspicuous scattered droplets in stem slash, these typically continuing to exude and forming cloudy white drippings below trunk wounds.

Julianaceae — Essentially an anacard reduced to wind-pollination; in our area occurring only in very dry parts of the western Andean slopes of Peru. Distinctive in the few thick branches with terminally clustered leaves and serrate leaflets.

Leguminosae — Red latex in a number of papilionate genera (e.g., *Dussia, Machaerium, Pterocarpus, Swartzia,* sometimes very faintly in caesalpinioid *Dialium*).

IIBi. Even-pinnate leaves

Leguminosae — *Inga* (unique in being even-pinnate and with glands between all leaflets); *Cassia* (often with glands between basal pair or pairs of leaflets), and several other caesalpinioid genera, all with typical pulvinus and pulvinuli and often with typical legume odor.

Meliaceae — Most meliacs except *Trichilia* are even-pinnate, especially *Guarea* with typical apical "bud".

Sapindaceae — Most Sapindaceae are basically even-pinnate but with alternate leaflets and a very characteristic aborted rachis apex at base of what often appears to be a terminal leaflet.

Palmae and Cycadaceae — Even-pinnate but very distinctive in their parallel-veined leaflets (see above).

IIBj. Miscellaneous useful characters for genera or common species with pinnately compound alternate leaves

(1) Winged rachis —Individual species of many genera and families: **Leguminosae** — Several unrelated genera have winged rachis: *Inga,* even-pinnate with glands between all leaflet pairs; *Dipteryx* (only subwinged); *Swartzia* (few species, most obvious in juveniles, odd-pinnate).

Meliaceae — *Guarea pterorachis* is even-pinnate with many thick leaflets and broad coriaceous rachis wings.

Sapindaceae — One tree with even-pinnate leaves with distinctively subalternate leaflets (*Sapindus saponaria*) and many lianas (*Paullinia* and *Serjania*), mostly with compound wood and/or bicompound leaves.

Solanaceae — Some herbaceous and vine *Solanum* species have winged rachises, often with incompletely divided leaflets.

Simaroubaceae — *Quassia amara* is characterized by mostly 5-foliolate leaves and bitter taste.

Burseraceae — A few *Bursera* species, distinctive in their strongly aromatic odor, have winged rachises, usually in conjunction with strongly serrate leaflets.

(2) Terminal "bud" of unfolding leaflet pair at tip of rachis (see illustration)—*Guarea*(Meliaceae).

(3) Uniformly cylindrical pulvinuli and pulvinus — Typical of nearly all legumes and connaracs which are extremely hard to distinguish vegetatively (and to some extent of *Picramnia*). Legumes can be either trees or lianas; in our area connaracs are nearly always lianas (rarely treelets). Most legumes have the typical legume green-bean odor and many once-pinnate legume lianas have red latex; Connaraceae generally lack a noticeable vegetative odor and never have red latex. Lepidobotryaceae have a unifoliolate leaf with legumelike pulvinulus.

(4) Naked rachis apex — Tree Sapindaceae (especially *Cupania* and *Matayba*) with pinnate leaves almost always have a very characteristic aborted rachis apex extended as a small projection at base of what appears to be a terminal leaflet. *Euplassa* (Proteaceae) has a similar rachis apex.

(5) Petiole base expanded to form sheath surrounding twig— *Polylepis* (Rosaceae), restricted to the highest altitude Andean forests has a very characteristic growth form with leaves borne on short shoots and having an enlarged petiole base that surrounds the twig; the small sessile oblong leaflets, usually conspicuously pubescent below, are also characteristic.

(6) Tendril from apex of rachis

Polemoniaceae — *Cobaea* is a cloud-forest vine with tenuous leaflets and a much-branched terminal leaf-tendril.

Compositae — *Mutisia* has the leaf rachis ending in a tendril.

Leguminosae — Herbaceous *Vicia* (and some species of bipinnate *Entada*) has the leaf rachis ending in tendril.

IIBk. Nondescript — *(Odd-pinnately compound, alternate, no spines, odor, punctations, latex, etc.)*

Meliaceae — *Trichilia,* unfortunately the commonest Meliaceae genus, is atypical in the family in odd-pinnate leaves. The leaflets are entire, and there is usually a sweetish odor from the trunk slash; one species distinctive in glandular area on petiole (cf., mimosoid legumes).

Staphyleaceae — *Huertea* has membranaceous asymmetric leaflets with distinctive serrulate-glandular margins.

Sabiaceae — *Ophiocaryon* has smooth, coriaceous, olive-grayish-drying leaflets, very irregular leaflet numbers (some leaves often 1(–2)-foliolate), and thickened flexed petiolules and often subwoody petiole bases.

Anacardiaceae — Odorless anacards (e.g., *Tapirira*) that also lack obvious latex are very nondescript and especially easy to confuse with *Trichilia.* Often there is at least a faint trace of a mangolike odor. The commonest species dries with a characteristic reddish tint.

Simaroubaceae — Nonbitter simaroubs are often characterized by legumelike cylindrical pulvinuli. *Picramnia* can be distinguished from legumes by the typical alternate leaflets, progressively smaller toward base of the rachis.

IIC. Leaves 3-foliolate — Alternate consistently 3-foliolate leaves are not very common although they may occur as variants in basically pinnately compound-leaved individuals (or species or genera).

IICa. Trees — *Four fairly common tree genera plus miscellaneous genera and species of Rutaceae have consistently 3-foliolate alternate leaves.*

Anthodiscus **(Caryocaraceae)** — Usually blunt apex and crenate leaflet margins.

Allophylus **(Sapindaceae)** — Usually acute or acuminate apex and toothed (or entire) margins; a few species have simple leaves.

Erythrina **(Leguminosae)** — Usually with spiny trunks and branchlets; with the typical legume cylindrical pulvinus and pulvinuli; margin always entire.

Hevea **(Euphorbiaceae)** — Easily recognized by the latex; a second 3-foliolate euphorb genus, *Piranhea* is common in central Amazonia in seasonally inundated forest.

Rutaceae — Several rutac genera have 3-foliolate leaves, at least in some species; all are characterized by the pellucid punctations and most have a more or less citruslike vegetative odor.

- -

IICb. Vines — *The great majority of 3-foliolate climbers are legumes (the leaflets uniformly entire; very rarely with very broad entire lobes) and Sapindaceae (the leaflets nearly always somewhat serrate or dentate).*

Leguminosae — Most 3-foliolate climbers are papilionate legumes characterized by typical legume odor, uniformly cylindrical pulvinuli and pulvinus and often with (unique) stipels at base of lateral pulvinuli, and red latex.

Connaraceae — Very like legumes in the cylindrical pulvinuli and pulvinus but without a green-bean odor and always lacking stipels and red latex. Most 3-foliolate Connaraceae have alternate basal leaflets (rare in legumes). Connaraceae leaflets mostly have finely prominulous venation and a characteristic chartaceous texture that is subtley different from those of legume climbers.

Sapindaceae — *Urvillea, Thinouia, Lophostigma,* and a few species of *Serjania* and *Paullinia* have uniformly 3-foliolate leaves; they are characterized by the bifurcating inflorescence-derived Sapindaceae tendril and usually irregularly coarsely toothed margin.

A few other vine genera (all but *Psiguria* usually with pinnate or bipinnate or simple and palmately lobed leaves) have some individual species with alternate 3-foliolate leaves:

Cucurbitaceae — 3-foliolate cucurbits (*Gurania, Psiguria, Cayaponia,* few *Fevillea* species) are recognizable by the divided spirally coiling tendrils that make a right angle with base of petiole, and by tendency for remotely toothed margins and/or scabrous surface and/or large glands near apex of petiole (*Fevillea*), and/or cucurbit odor.

Vitaceae — Several *Cissus* species have 3-foliolate leaves. The commonest of these have a characteristic 4-angled subwinged branchlet; all differ from cucurbits in having the tendril arising opposite the petiole base and are also usually distinctive in swollen nodes.

(Euphorbiaceae) — Only a few species of *Dalechampia,* all more or less herbaceous, are 3-foliolate vines.

(Dioscoreaceae) (*Dioscorea*) — A few species of *Dioscorea,* characterized by the rather thickish and usually curved and angled base of the long petiole, are 3-foliolate.

Menispermaceae — At least one menisperm has a 3-foliolate leaf; the anomalous stem cross section with concentric rings is distinctive.

IID. Leaves palmately compound (and alternate)

IIDa. Trees

Palmae — Fan palms are our only arborescent plants with palmately compound leaves with parallel-veined segments.

Rutaceae — Some species of *Angostura* and *Raputia* (and extralimital species of *Esenbeckia* and *Casimiroa*) have palmately compound leaves, characterized by the punctations and more or less pungent or citruslike odor.

Araliaceae — Most area araliacs have palmately compound leaves, characterized by the rank or medicinal odor and the thickly triangular ligule projecting up from the more or less clasping petiole base; only a few of species become large trees (characterized by tan leaf undersurface).

Bombacaceae — Most bombacs have palmately compound leaves, always with a Malvalean pulvinus at petiole apex. Several genera have spines on the trunk (at least when young) (a unique combination except for *Jacaratia*); whether with or without spines, bombacs are often unusually large emergents with distinctively swollen thick trunks. One spineless genus has the leaflets continuous with the digitately parted petiole apex (unique in Malvales).

Caricaceae — One genus (*Jacaratia*), is the only palmately compound-leaved tree with milky latex in our area; several species have spiny trunks and resemble Bombacaceae except for the latex and lack of a pulvinus.

Moraceae — One common *Cecropia* and one common *Pourouma* have the leaves completely split into separate leaflets, the latter with dark brown latex in the young twigs and both with stilt roots and the conspicuous conical Moraceae stipule.

Sterculiaceae — *Herrania* always has palmately compound leaves, usually large, conspicuously hairy and borne on pachycaul treelet with maroon cauliflorous flowers and cacao-like fruit; two extralimital neotropical species of *Sterculia* (one Mexican, one Bolivian) also have Bombacaceae-like compound leaves.

Cochlospermaceae — One *Cochlospermum* has very bombaclike palmately compound leaves, the petiole apex expanded as in *Pseudobombax* but without the Malvalean pulvinus and leaflets thinner than in *Pseudobombax*.

***IIDb. Vines** — (Most vines with palmately compound leaves are tendrillate and most are atypical members of predominantly simple-leaved taxa).*

Cucurbitaceae — A few species of various genera (e.g., *Cyclanthera*) have palmately compound leaves, usually irregularly divided and with distinctive rank cucurbit odor; lacking petiolar glands except in *Siolmatra.* The tendril arises at 90° angle from petiole base.

Passifloraceae — Two area *Passiflora* species have palmately compound leaves; characterized by distinctive petiolar glands. The tendril is axillary.

Convolvulaceae — Most species of *Merremia* have palmately compound leaves; tendrils are absent in Convolvulaceae.

Dioscoreaceae — One or two rare *Dioscorea* species have palmately compound leaves, without tendrils and with distinctively angulate petiole with thickened curved base.

Sapindaceae — One species of *Paullinia* has palmately 5-foliolate leaves; characterized by compound stem anatomy and milky latex.

Araceae — *Syngonium* and a few atypical species of *Philodendron* and *Anthurium* are hemiepiphytic climbers with palmately compound leaves, very different from the above taxa in succulent stems, adventitious attachment roots, and usually in finely and closely parallel secondary venation.

Cyclanthaceae — Nearly all climbing Cyclanthaceae have deeply bifid, rather than truly compound, leaves very distinctive in their parallel venation.

KEY III

LEAVES SIMPLE AND OPPOSITE (OR WHORLED)

IIIA. Lianas

A majority of lianas have simple opposite leaves. Look for: serrate or serrulate margins (Hippocrateaceae); petiolar glands and/or sericeous petiole (and/or other parts) with T-shaped trichomes (Malpighiaceae); 3-veined from base (*Strychnos, Coriaria, Melastomataceae*), or above base (Valerianaceae and Compositae [usually aromatic]); milky latex (Apocynaceae and Asclepiadaceae, the former usually woody, the latter mainly herbaceous); spines (several rubiacs mostly with paired spines from leaf-axils); *Combretum,* sometimes with branch spines on stem; *Pisonia* with leaves mostly on short-side branches; swollen jointed nodes (acanths, amaranths, *Gnetum,* the latter with [slow] resin from cut stem); asperous surface (*Petrea* [Verbenaceae]), *Prionostemma* (Hippocrateaceae, some Compositae).

The four main liana families with opposite simple leaves are apocs, malpighs, hippocrats, and combretacs. Apocynaceae lianas always have milky latex, at least in the leaves and young stems and center of stem; the others never do (though a kind of thin watery latex may be present). Only Hippocrateaceae (plus a very few comps and a few montane forest oddballs of miscellaneous families) are ever serrate or serrulate; they also usually have green petioles and twigs, a good indicator for most entire-leaved hippocrats. Some malpigh genera have petiolar glands which may be stipulelike enations on the petiole (*Hiraea*) or secretory glands at the petiole apex (especially *Stigmaphyllon*); when present these are a sure familial indicator. Most malpighs have tannish petioles (from the T-shaped trichomes) and brownish or tannish twigs. Malpighs with nonglandular petioles might be confused with Combretaceae but the petioles and twigs of the latter are usually brown, lack sericeous indumentum and may have an exfoliating cortex, even on the petioles. Combretaceae mostly have conspicuous rigidly parallel tertiary connecting veins adjacent secondaries, a feature rare in malpighs and nonexistent in hippocrats (which may have parallel tertiary venation but then almost parallel to the secondaries); the young branchlets of Combretaceae are generally hollow-centered and even the large stems tend to have one or more small mucilage-secreting channels in extreme center.

Some unusual features of opposite-leaved lianas:
Spines — *Pisonia, Combretum,* Rubiaceae (*Uncaria, Chomelia, Randia*).

Asperous leaves — *Petrea, Prionostemma,* several Compositae.

Three-veined leaves — *Strychnos, Coriaria,* Compositae, Melastomataceae, Valerianaceae

Serrate or serrulate margins — Hippocrateaceae; some Gesneriaceae, Compositae, Valerianaceae, *Hydrangea.*

Petiolar enations or "glands" — Malpighiaceae.

IIIAa. T-shaped trichomes

Malpighiaceae — T-shaped trichomes (= malpighiaceous hairs) give rise to a macroscopically sericeous look, especially on petioles (and buds) and are definitive among plants with opposite leaves; not always obvious to naked eye, if in doubt use a hand lens. Several genera have pair of stipulelike appendages (glands) on petiole or pair of thicker glands near petiole apex. Stems sometimes fragment into cables, unlike combretacs and hippocrats. Sometimes with watery submilky latex approaching that of some ascleps in texture.

IIIAb. 3-veined leaves

Loganiaceae — The only neotropical liana genus is *Strychnos,* easy to recognize by the opposite 3-veined leaves; rather sparse but very characteristic tendrils are usually present, thick and coiled in a single plane.

Coriariaceae — The only neotropical species is *Coriaria ruscifolia,* a very distinctive and common weedy Andean species with small, sessile, palmately veined leaves. The entire leaves are arranged along thin opposite branchlets giving the effect of a multifoliolate pinnately compound leaf. Grows at higher altitudes than *Strychnos* and the leaf arrangement is very different.

Melastomataceae — Easy to recognize by the leaves with one to four pairs of longitudinal veins arcuately subparallel to the midvein and perpendicularly connected by finer cross veins, these lacking in other 3-veined taxa.

Compositae — Only a few opposite-leaved genera (e.g., some *Mikania, Wulffia*) are actually lianas; most other scandent comps with opposite leaves tend to be clambering, more or less herbaceous vines. The combination of opposite leaves 3-veined *above* the base and aromatic odor (also frequently more or less asperous) is unique to comps; some *Mikania* species have the leaves 3-veined from base but the leaves then cordate; the margins are usually entire but may be (usually more or less irregularly) toothed, and the surface often distinctively asperous. Liabeae climbers lack odor but have milky leaves.

Valerianaceae (*Valeriana*) — The leaves of climbing species are 3-veined above the base just as in Compositae; distinctive in having the petiolar bases united to form a kind of nodal sheath, also in usually drying black; most scandent species are more or less toothed but a few are quite entire. The stems are usually very flexible and smooth-barked except for conspicuously raised lenticels.

IIIAc. Milky latex

Apocynaceae — Abundant milky latex is unique to this and the following family among lianas or vines with opposite leaves. The two families are very difficult to distinguish vegetatively and are sometimes united into the same family. Leaves usually with characteristic glands on midrib above, at least at base of midrib. In general woody lianas are apocynacs while herbaceous climbers are mostly asclepiads.

Asclepiadaceae — Essentially a herbaceous version of Apocynaceae and often indistinguishable vegetatively. A few species have rather watery latex, a phenomenon which also occurs in a few malpighs.

Compositae — Tribe Liabeae climbers have milky latex but usually triangular or serrate leaves and/or winged petioles.

Guttiferae — Guttiferae climbers rarely have conspicuous latex; all have rather strongly coriaceous leaves with *Clusia*-like venation; and they tend to be hemiepiphytic.

IIIAd. Miscellaneous

Hippocrateaceae — Leaves usually serrate or at least +/- serrulate; the only lowland liana family (except a few subwoody Gesneriaceae and aromatic sub-3-veined Compositae) with opposite serrate leaves.

Also characterized by typical bent tendril-like lateral branches which twist around support (but sometimes occur in other families). Leaves of nonserrate genera are coriaceous and often dry with a characteristic grayish-olive color and are typically coriaceous, glabrous, and smooth-surfaced with immersed fine venation, this especially pronounced in entire-leaved genera. One genus has reddish or pinkish watery latex. When the tertiary venation is more or less parallel it tends to also be parallel to the secondary veins unlike Combretaceae. Stems of some genera strongly anomalous, typically with a few irregular reddish concentric circles interconnected by spokes.

Combretaceae — The best vegetative character may be the usually hollow or secretory (discolored when dry) stem center of young branches and the tendency to have even older stems with 1 or 3 mucilage-oozing canals in extreme center. Fibrous bark and leaves typically with rigidly parallel often somewhat raised tertiary venation and brown petioles (sometimes with more or less exfoliating cortex) are typical; the leaves tend to be somewhat subopposite and a few species actually have alternate leaves (these species with stem spines); an interpetiolar line or ridge is never present. Pubescence (if present) of simple trichomes.

Rubiaceae — Vine genera have the typical interpetiolar rubiac stipules but these are not always obvious in many liana taxa. Commonest liana genus is *Uncaria* with paired curved axillary spines.

Bignoniaceae — A few compound-leaved genera (e.g., *Arrabidaea,* but usually also with some tendrillate 2-foliolate leaves) have simple-leaved species or variants which are 3-veined from base; one hemiepiphytic genus (*Schlegelia,* which may be a scroph) has uniformly simple leaves, strikingly coriaceous and with a characteristic appressed-conical "pseudostipular" axillary bud, and adventitious roots along stem.

Acanthaceae — One liana genus (*Mendoncia*) and several genera which include clambering vines. Thickened, more or less swollen nodes and rather fragile branchlets are typical of all acanth climbers. *Mendoncia* has a soft easily broken stem with an amazingly dissected cross section and often with corky bark; also, more or less membranaceous leaves, often with conspicuous simple trichomes.

Gesneriaceae — Nearly all climbing Gesneriaceae are subwoody hemiepiphytes growing appressed to a tree trunk. The leaves are usually membranaceous or succulent and are often either serrate or strongly anisophyllous (sometimes strikingly red-tipped or reddish below or with red apical "eyespots").

Verbenaceae — Two genera of climbers, *Petrea* with asperous leaves, and *Aegiphila.* Both often have squarish stems or twigs; the bark is usually smoothish and light colored; prominent subulate axillary bud scales usually present; the leaves are often membranaceous and/or noticeably simple-pubescent in *Aegiphila.* The only lianas other than *Petrea* with asperous opposite leaves are *Prionostemma aspera* (hippocrat) and some comps (3-veined above base).

Nyctaginaceae — Two opposite-leaved genera become lianas in our area, one (*Pisonia*) characterized by spines and rather clustered, not clearly opposite leaves, the other (*Colignonia*) woody only in cloud forests and characterized by the tendency to verticillate leaves (and the very characteristic *Hydrangea*-like inflorescence bracts); both have concentric rings of anomalous growth.

Amaranthaceae — Two genera of more or less woody opposite-leaved climbers in our area, *Iresine* and *Pfaffia* (other genera are usually prostrate or scrambling herbs; see also alternate-leaved *Chamissoa*), characterized by acanthlike swollen nodes (sometimes shrunken above node when dried) and by the evenly striate-ribbed branchlets; *Iresine* has hollow twigs.

Gnetaceae (*Gnetum*) — Swollen, more or less jointed nodes, coriaceous leaves, and resin oozing from cut stem with concentric rings in cross section are diagnostic characters. (Note: there is no way the uninitiated will recognize this as a gymnosperm; it looks much like *Salacia* and related hippocrats.)

Trigoniaceae — Rather similar to Malpighiaceae because of the usually whitish more or less sericeous puberulous leaf undersides, but with a dense mat of spiderweb-like trichomes and/or simple trichomes, never T-shaped ones. Nonpubescent species tend to have hollow or discolored twig centers like Combretaceae but differ from that family in having an interpetiolar line.

Saxifragaceae — In our area *Hydrangea* is a thick woody liana (sometimes hemiepiphytic), with entire or serrate leaves; entire-leaved species similar to Rubiaceae when sterile but with petiolar bases connected across nodes to form conspicuous ochrea-like joint-sheath.

IIIB. Trees and shrubs (with leaves simple and opposite [or whorled])

IIIBa. Stipules (or stipule scars!)

Rubiaceae — Interpetiolar stipules are present at least 99% of the time; if stipules are not readily apparent check terminal bud to see if it is enclosed by caducous stipules. These should be in plane at right angle to two uppermost leaves and leave an interpetiolar line when they fall.

Quiinaceae — The other main family with interpetiolar stipules, these always separate (usually fused in Rubiaceae) and often rather long and subfoliaceous; differs from Rubiaceae in the usually serrate or serrulate leaf margin (deeply incised in juveniles of few species).

Dialypetalanthaceae — The single species similar to Quiinaceae in the completely separate, large triangular pair of interpetiolar stipules, but the margin entire; the fluted-based trunk with thick reddish fibrous bark also distinctive.

Malpighiaceae — The tree genera have *intra*petiolar stipules in the axil between the petiole and twig (looking like ligular dorsal projection from petiole base), these differing from the few rubiacs with similar stipules by being fused (usually bifid in Rubiaceae except *Capirona* with conspicuously smooth red bark); interpetiolar lines are also usually present. The main familial vegetative characteristic is the presence of malpighiaceous or T-shaped hairs, these almost always visible at least on the petioles and young twigs.

Chloranthaceae — The more or less swollen node has a stipulelike sheath; the plants are easily distinguished by the strong Ranalean odor and the serrate leaves.

Vochysiaceae — *Vochysia* usually has whorled leaves and/or rather thick-based stipules (other genera have stipule apex broken off to leave characteristic gland, see below).

Rhizophoraceae — *Rhizophora,* restricted to coastal mangroves, is utterly distinctive as the only mangrove with stilt roots; the other two opposite-leaved genera are less striking, with *Sterigmapetalum* having mostly whorled leaves with tannish-puberulous petioles and caducous narrowly triangular-pubescent stipules between them and *Cassipourea* in the leaves usually obscurely and remotely denticulate or serrate (unlike Rubiaceae) and the small, narrowly triangular, early-caducous stipule usually sericeous.

IIIBb. Latex

Apocynaceae — (Some genera alternate-leaved and many are climbers). Latex white and free-flowing (red in some species with alternate leaves). Lacks the typical guttifer terminal bud (i.e., the petioles of terminal leaf pair not hollow-based with terminal bud growing from within cavity except in a few species with very profuse latex).

Guttiferae — Very distinctive in typical terminal bud and colored latex; latex commonly yellow, cream, or orange, usually slower-flowing than in apocs. Terminal bud characteristically from between the hollowed-out leaf bases; typical terminal bud not developed only when latex strongly colored; latex white only when the leaf bases form conspicuous chamber. The latex may not be very obvious; try breaking a leaf and twig as well as the trunk slash; stilt roots are rather common.

IIIBc. Punctations

Myrtaceae — Usually further characterized by more or less parallel and close-together secondary and tertiary venation ending in a submarginal vein. Many have smooth, white, peeling bark; some have aromatic leaves. The only possible confusion comes from a very few guttifers that have punctations (but also latex, unknown in Myrtaceae) or from some myrtacs that are not obviously punctate (also beware *Mouriri* [see below]).

Melastomataceae — *Mouriri* completely lacks the ascending veins of other melastomes, looks almost exactly like Myrtaceae and may have punctations: it differs from myrtacs in the somewhat jointed nodes.

Lythraceae (*Adenaria* and *Pehria*) — Leaves thinner than in most myrtacs and with more ascending secondary veins and absence of collecting vein.

Rutaceae — *Ravenia* has opposite simple leaves with punctations, characterized by sheathing guttifer-like petiole bases in which the apical bud is protected; differs from Guttiferae in lacking latex and the small glandular punctations. One upland *Amyris* species has small, mostly opposite, but unifoliolate, leaves.

Guttiferae — Shrubby *Hypericum* and some *Vismia* species have punctate leaves, the former characterized by mostly upland habitat and the small coriaceous ericoid leaflets, the latter usually by orange latex (or a moist orangish area just inside bark where latex should be.

Loranthaceae — *Gaiadendron* is an exclusively montane free-standing tree with conspicuous punctations, differing from Myrtaceae in secondary and tertiary venation of the thick-coriaceous leaves completely invisible.

(Rhizophoraceae) — *Rhizophora,* keyed out above on account of the terminal stipule, can have leaf punctations.

IIIBd. 3(–7)-veined leaves with parallel cross veins more or less perpendicular to main veins

Melastomataceae — The very characteristic leaf venation makes this one of the easiest families to identify. A few other families have 3-veined bases but lack the typical cross veins (*Delostoma* [Bignoniaceae; Andean]), *Strychnos* (few shrubby species). Also beware of *Mouriri* which lacks the typical venation and looks almost exactly like Myrtaceae.

IIIBe. Odor of essential oils — *(Only two Ranalean families are characterized by aromatic opposite leaves.)*

Monimiaceae — *Siparuna,* usually puberulous or with lepidote scales; the second genus (*Mollinedia*) often lacks obvious odor but is characterized by very distinctive leaf with few very separated marginal teeth.

Chloranthaceae — *Hedyosmum* (our only genus) has very characteristic swollen nodes with stipulelike sheath.

Lauraceae — A very few atypical Lauraceae, most notably peculiar 3-veined *Caryodaphnopsis,* have opposite leaves.

Verbenaceae and Labiatae — These two families have opposite leaves and are usually aromatic but the odor is clearer and sweeter (often somewhat minty) and less pungent than in the Ranales; aromatic members of these families usually have tetragonal branchlets unlike the Ranales. Strongly aromatic species of both families are mostly herbs or subshrubs.

Compositae — Opposite-leaved Compositae are nearly always 3-veined *above* the base (unlike any of the above families) and have more pungent odors.

IIIBf. Glands on twig at base of petiole

Vochysiaceae — Very characteristic glands from the fallen stipules or stipule bases characterize most Vochysiaceae (except *Vochysia*).

IIIBg. Serrate (serrulate) margins — The combination of opposite simple leaves and serrate margins is a rare one and found only in eight woody neotropical lowland families besides the Quiinaceae, Melastomataceae, and Chloranthaceae which are easily recognized (cf., above). A few additional herbaceous families (e.g., Gesneriaceae) have some viny (see above) or subwoody members with opposite serrate leaves.

Hippocrateaceae — Mostly lianas but a few are trees and *Cheiloclinium* can be both a tree and have serrate leaves; it is characterized by tertiary venation more or less parallel and perpendicular to midvein.

Violaceae — *Rinorea* usually has opposite leaves and is one of the commonest understory-tree genera of many forests. Characterized by the nodes noticeably jointed, the typically short petioles, and the tendency to have a small acute stipule-enclosed apical bud immediately subtended by oblique, whitish-margined, interpetiolar ridge.

(Eleaocarpaceae) — *Sloanea* is characterized by a mixture of alternate and opposite leaves, even on the same branch, but the leaves are almost never uniformly opposite; also very distinctive in the flexed, but nonpulvinate petiole apex and strictly pinnate venation. The margins vary from quite entire to rather shallowly and coarsely subdentate; species with more serrate leaves tend to be more pubescent and some of the pubescent species have conspicuous persistent leafy stipules.

Rhizophoraceae — *Cassipourea* (secondary veins few and brochidodromous strikingly far from margin; margin mostly remotely serrulate (resembling similarly few-toothed *Mollinedia* but teeth blunter); caducous triangular terminal stipule pair leaving interpetiolar line.

Loganiaceae — *Buddleja* (mostly Andean upland except for one weedy species, *B. americana*) characterized by leaves white or tan stellate-tomentose below and usually narrow or somewhat rhombic shape. *Desfontainia* has spinose margins like holly. *Peltanthera* and *Gomara* have +/- serrulate margins.

Flacourtiaceae — *Abatia* (always pubescent, at least the leaf below and twigs, usually with floccose trichomes very like *Callicarpa*) rather sharply serrate; when fertile distinguished by narrow terminal raceme or spike (cf., *Clethra*).

Verbenaceae — Typically with more or less tetragonal stem and aromatic odor. Most woody verbenacs are entire but usually serrate-leaved *Callicarpa*, with conspicuously floccose indument on leaf undersides and twigs, is small tree. Interpetiolar lines lacking.

Monimiaceae — *Mollinedia* has the teeth usually very widely separated (typically only one or two per side) and rather sharp; *Siparuna* also is frequently toothed but easy to recognize by the Ranalean odor.

Several other mostly exclusively montane families also have taxa with opposite serrate leaves, at least on occasion:
Viburnum **(Caprifoliaceae)** —Leaves with few strongly ascending veins, puberulous at least below, sparsely and bluntly serrulate or more or less bluntly few-toothed toward apex.

(*Brunellia* [Brunelliaceae]) — The occasional simple-leaved species, like compound-leaved congeners, are closely serrate and with prominulously reticulate venation below and numerous secondary veins making obtuse angle with midvein; strong interpetiolar line.

Hedyosmum **(Chloranthaceae)** — As noted above unmistakable in the sheathing node and strong Ranalean odor.

(*Weinmannia* [Cunoniaceae]) — A few species are simple-leaved but otherwise remarkably similar to the pinnate species in the rather coarsely toothed margins and distinctive small leafy caducous stipules.

Columellia **(Columelliaceae)** — Mostly entire-leaved (with apical spine or apicule) but sometimes with a few thickened subterminal teeth; distinctive in the small grayish-sericeous leaves with opposite petioles strongly connected by line or flap of tissue.

(Aphelandra [Acanthaceae]) — A few montane species have conspicuously spinose leaf margins.

(*Alchornea pearcei* ([Euphorbiaceae]) — A montane cloud-forest taxon with mostly opposite leaves, easily recognized by the strongly 3-veined leaves with glands in the lower vein axils, thus, looking almost exactly like typical alternate-leaved species of the genus.

IIIBh. Miscellaneous opposite simple-leaved trees — *(Lacking latex, essential oils, serrate margins)*

Nyctaginaceae — Rather nondescript and might be confused with *Psychotria* or similar rubiacs, even when in flower or fruit, except for lacking stipules. The best sterile character is the reddish-brown pubescent terminal bud. The combination of somewhat succulent, often different sized and/or subopposite blackish-drying leaves and rufescent terminal bud immediately indicates Nyctaginaceae.

Myrtaceae — In some Myrtaceae the punctations are not very evident. They are usually characterized (as are the punctate-leaved taxa) by the straight often rather close-together secondary and intersecondary veins that end almost perpendicular to a well-developed submarginal collecting vein.

Melastomataceae — *Mouriri* looks much more like Myrtaceae in vegetative condition than like typical 3–7-veined melastomes. It differs from myrtacs, most notably, in the jointed nodes.

Lythraceae — Usually with tetragonally angled young twigs and/or longitudinally exfoliating, often reddish twig bark in older branchlets; interpetiolar lines or ridges lacking except in *Lafoensia* (with close-together secondary veins prominulous above and below, each adjacent pair separated by a well-developed intersecondary). *Physocalymma* is very distinctive in entire-margined strongly scabrous leaves.

Alzateaceae (close to Lythraceae) — Easily distinguished vegetatively by very strongly tetragonal thickish twigs and jointed stems with strong interpetiolar ridges; leaves thick-coriaceous, oval with rounded apex and base and very short petiole, the secondary veins immersed or slightly prominulous below.

(Polygalaceae) — *Polygala scleroxylon,* which vegetatively has nothing whatsoever to suggest this family, is a large Amazonian lowland tree with opposite leaves having a legumelike odor and cylindrical legumelike petioles.

Loganiaceae — Vegetatively heterogeneous, several genera have serrate leaves, these always pubescent, sometimes with stellate trichomes (*Buddleja*); one genus glabrous with spinose hollylike margins, another a pachycaul treelet with large coriaceous entire, oblanceolate leaves; a few genera are herbs, the commonest of which has +a terminal whorl of 4 leaves subtending the inflorescence.

Verbenaceae — Usually with tendency to tetragonal branchlets and raised petiole attachments; leaf base typically attenuate onto petiole and in many species (most *Citherexylon*) with an elongate gland in the laminar attenuation on either side of petiole apex. Leaves, at least of forest taxa, usually rather membranaceous and somewhat aromatic.

Hippocrateaceae — Two genera are sometimes trees, one (*Cheiloclinium*) usually with finely crenate-serrate margins (also distinguished by conspicuously parallel tertiary venation more or less perpendicular to midvein), the other (*Salacia*) with large very thick-coriaceous entire leaves with immersed fine venation and drying a characteristic dull olive.

Rhamnaceae — *Rhamnidium* and some *Colubrina* have opposite leaves, the former characterized by the close, straight, parallel secondary veins of the family and the pale leaf undersurface, the latter by conspicuous glands at base of lamina (typically in basal auricles).

Acanthaceae — Only two real tree genera (*Bravaisia* and *Trichanthera*, both with very weak wood) although several genera include shrubby or small tree species. Most acanths characterized by conspicuously jointed nodes, swollen when fresh and contracted when dried. Most acanths have an obvious interpetiolar line; several of the taxa lacking this line are spinose-margined (cf., holly) upland *Aphelandra* species. Except for the spiny-margined species, our acanths all have entire or merely serrulate, but never truly serrate, leaves. Cystoliths (look like short black lines) often present on upper leaf surface (also in Urticaceae).

Gesneriaceae — Only *Besleria* become small soft-wooded trees in our area (these usually with entire leaves smaller than in acanths); although several serrate-leaved taxa can be shrubs or treelets. Woody generiads mostly have noticeably membranaceous, long-petioled, often pubescent leaves and differ from most acanths in lacking cystoliths and interpetiolar lines.

Caprifoliaceae — *Viburnum*, exclusively montane, has the leaves sometimes subentire (though usually with at least a few inconspicuous teeth toward apex), characterized by the few strongly ascending secondary veins and tendency to be +/- puberulous at least on veins; always with conspicuous interpetiolar line.

Cornaceae — *Cornus*, exclusively montane and extremely similar to *Viburnum*, even when fertile, but completely entire margins, more strongly ascending secondary veins and petiole bases decurrent onto the somewhat angled tannish puberulous twig.

Columelliaceae — *Columellia,* exclusively montane, has a conspicuous interpetiolar line (sometimes accentuated into an actual flap of tissue) like *Viburnum* but is very distinctive in the small grayish-sericeous always apiculate or spine-tipped leaves.

Oleaceae — *Chionanthus,* has petioles usually somewhat thickened at base (cf., Sapindaceae petiolules), the leaf blade either pubescent or else rather narrow and oblong; twigs lacking interpetiolar lines, often with scattered round raised white lenticels. In flower unmistakable in only 2 anthers and very narrow petals.

(Buxaceae) — *Buxus,* with very characteristic coriaceous, sub-3-veined leaves and the petiole bases attenuating into strong ridges on the thus irregularly 6-angled twig, has one species on limestone outcrops in northern Columbia.

Malpighiaceae — A few genera of shrubs and small trees lack obvious intrapetiolar stipules (as do the lianas); *Malpighia* and many species of *Bunchosia* have neither obvious stipules nor interpetiolar lines and are often characterized instead by pair of ocellar glands near base of lamina below. Like the species with stipules, they are also vegetatively characterized by the typical T-shaped trichomes at least on petioles and young branchlets.

(Guttiferae) — Occasionally lacks apparent latex (*Tovomitopsis,* some *Chrysochlamys*), but then with the typical hollowed-out Guttiferae petiole bases that form a protective chamber for the developing bud; a few *Vismia* species (which lack the typical hollowed petiole base) may not always show the orange latex but there is always a hint of orange color under the bark where the latex should be.

(Elaeocarpaceae) — *Sloanea* is characterized by a mixture of alternate and opposite leaves, even on same branch, almost never uniformly opposite; also very distinctive in the flexed but nonpulvinate petiole apex and strictly pinnate venation. Margins sometimes bluntly irregularly toothed (see also above).

KEY IV

LEAVES SIMPLE AND ALTERNATE — This "grab bag" category constitutes by far the largest and generally the most nondescript group. In preceding groups any sterile woody plant should be identifiable to family; in this group there will be many plants which end up as family indets, unless they are fertile and technical characters are used.

In trees —Look for (in approximate order of importance): latex, odor of essential oils (Ranalean odor), conical terminal stipules (usually = Moraceae), 3-veined base (frequently suggests Malvales), punctations (and the undersurface texture which accompanies punctations in Myrsinaceae), serrate margins (uncommon in tropical forest species), strong bark (pull a leaf off a twig to see if a strip of bark comes with it; also check the twig bark itself), petiole length and flexion, glands at tip of petiole (usually Euphorbiaceae or Flacourtiaceae), whether petioles are thickened at base or apex or of unequal lengths, spines.

In vines — Look for (in approximate order of importance): tendrils (only ten families have true tendrils and the type of tendril is usually specific to a given family), glands on petiole (especially common in Passifloraceae), latex (a few Convolvulaceae lianas have latex).

IVA. Trees

IVAa. ***Latex*** *(and alternate simple leaves) — (Look carefully, breaking the midveins or petioles of several leaves as well as young twigs; be sure to check both trunk and leaves since sometimes obvious latex is apparent only in one or the other; note whether the trunk slash has discrete latex droplets, how these are arranged, and what color the latex is.)*

Sapotaceae — Latex (in Neotropics) always white and milky; sometimes not very apparent but almost always visible in either trunk slash or leaves (if not both). Leaves typically with base of petiole enlarged (petiole more or less pop-bottle shaped) and with numerous parallel secondary veins. Some genera lack the typical petiole but these mostly have finely parallel tertiary and secondary veins (the extremes with leaves similar to *Clusia*); margins always entire and latex of slash usually emerging in discrete droplets. Never with conical terminal stipule or glands on petiole.

Moraceae — Latex of over half of species a unique tan shade (exactly the color of "cafe con leche"), but many other species with milky white latex (usually only watery in *Trophis*), and in a few varying to tannish yellow (some *Naucleopsis*) or dark brown (*Pourouma*). Conical terminal stipules and the scar from these stipules usually obvious (and definitive for Moraceae). Leaf venation very characteristic with the brochidodromous lower secondary veins closer together and/or joining midvein at different angle from others.

Euphorbiaceae — White to cream milky latex typically present, often caustic and harmful to eyes. *Pausandra* and some *Croton* have bright red latex; *Omphalea* (liana in our area) has a cloudy latex that turns rather purplish. Note: Although latex is considered characteristic of Euphorbiaceae, *many* species have no latex at all. Serrate (or serrulate) leaf margins, long petioles with flexed apices and often of different lengths, and a pair of glands near petiole apex are good indicators of Euphorbiaceae and are unique to this family among species with alternate simple leaves and latex.

Olacaceae — Latex present only in few species and usually present only in leaves and petioles, white and milky (usually) to somewhat watery (*Minquartia*). Look for a slightly longish, distinctively curved (putatively U-shaped) and somewhat apically thickened petiole; leaves of most species of this family have a characteristic grayish or tannish-green color when dry. The margins are always entire (at least in our area).

Apocynaceae — Relatively few apocs have alternate leaves. Alternate-leaved apocs usually have white and milky free-flowing latex but this may be bright red in some species of *Aspidosperma* and orangish or pinkish in others (one *Aspidosperma* has both white and bright red latex in the same twigs!). *Aspidosperma* often has little or no latex in the trunk; other genera with conspicuous trunk latex. Many species of *Aspidosperma* are extremely difficult to distinguish from Sapotaceae vegetatively; a good character for some of these species (series *Nitida*) is the conspicuously fenestrated trunk, a feature never found in Sapotaceae.

Papaveraceae — Exclusively high-altitude *Bocconia* is unique among alternate-leaved taxa in its orange latex, also characterized by the large irregularly pinnatifidly lobed leaves.

Myristicaceae — Usually with red latex (only in trunk, this sometimes rather watery at first but almost always soon becoming obviously red, especially when drying [except *Osteophloeum* which is persistently straw-colored but still dries reddish]). Very easy to distinguish from other families with occasional species or genera with red latex by the typical myristicaceous branching, lack of petiolar glands, and presence of Ranalean odor.

Campanulaceae — Always with white and milky latex. Mostly herbs but a few higher-altitude shrubby trees (actually overgrown herbs).

Anacardiaceae — Most species have compound leaves, at least in Neotropics. Resinous black-drying latex is sometimes present in simple-leaved species, this most obvious around old trunk wounds and not usually evident in fresh slashes of the leaves or twigs.

(Chrysobalanaceae) — Very rarely with a distinct trace of reddish latex, this not always visible in individual trees. Look for stipules on the young twigs or their scars; lack of Ranalean odor separates from Myristicaceae, the only potential confusion.

(Annonaceae) — At least one species of Annonaceae may have a faint trace of red latex (*Unonopsis floribunda*). It can be distinguished from Myristicaceae by its strong bark.

IVAb. Conical terminal stipule *(+/- definitive for Moraceae) — Again look carefully; the stipule is not always obvious and in* ***Trophis*** *can only be considered present by stretching the imagination; also note that other families may have young leaves which are superficially somewhat similar to the Moraceae stipule.*

Moraceae — The combination of milky latex and conical terminal stipule (that falls to leave a distinct scar) is definitive for and almost universal in Moraceae. The exception is *Trophis* where neither latex nor stipule may be discernible, where recognizable as Moraceae only by the typical leaf venation.

Magnoliaceae — *Talauma* has a Moraceae-like terminal stipule that falls to leave a conspicuous nodal ring, but is aromatic and nonlactiferous.

Winteraceae — *Drimys,* lacking latex and strongly aromatic, has an inconspicuous terminal stipule.

Polygonaceae — Some Polygonaceae have conical terminal stipules but these rupture to form an ochrea rather than falling cleanly as in Moraceae.

(Theaceae and Myrsinaceae) — A number of genera of Theaceae and Myrsinaceae and related families have young leaves rolled at branch apex and are superficially similar to the conical terminal stipule of Moraceae.

IVAc. Odor of essential oils (Ranalean odor) *— Most Ranalean plants have alternate simple leaves and most have more or less conspicuous rank or turpentiny odors. As a group these are easy to recognize by their "primitive" odor and many of these families are very common and important in neotropical forests. The beginner almost always complains of either not having an adequately developed sense of smell or of being unable to discriminate nuances of different vegetative odors. Don't despair — with a little bit of practice you really can (usually) pick out the Ranalean families by the combination of simple alternate leaves and their odor. Also, an important warning: There is a common epiphyllous leafy liverwort with a rather licorice-like smell. Learn that anise odor well and eliminate it from consideration. Frequently a twig split longitudinally will give off a more easily detectable odor than the leaves themselves. In some Lauraceae, with little or no leaf odor, the bark slash is aromatic; in others the reverse may be true.*

Families with Ranalean odors *— All with completely entire margins except for a very few somewhat lobed-leaved (but never serrate) species.*

Piperaceae — Swollen nodes with shoot proceeding from leaf axil. Distinctive spicate inflorescence; odor tends to be peppery (not surprisingly since pepper comes from this family); leaf base often strikingly asymmetric. Usually shrubs; when trees (usually small), typically with prominent stilt roots.

Winteraceae — Only *Drimys* (montane) which has the leaves strongly whitish below and lacking noticeable secondary veins.

Magnoliaceae — Mostly montane; the petiole is conspicuously grooved above in *Talauma,* the only significant South American genus; note complete rings around twig at nodes from the distinctive caducous stipule that completely covers terminal bud.

Hernandiaceae — Three-veined (occasionally peltate or subpeltate and rarely somewhat 3-lobed in part) leaves are unique in aromatic Ranalean taxa except for a very few atypical Lauraceae. Also distinctive among Ranales in long often somewhat different-length petioles. The vegetative odor is ranker than in most Lauraceae in which 3-veined taxa also differ in shorter petioles. Differ from similar Araliaceae in lacking conspicuously smaller short-petioled leaves and in basal lateral vein pair curving upward rather than being straight or curving outward.

This leaves three very large and very important Ranalean families — **Annonaceae, Myristicaceae, and Lauraceae** — which are easy to tell apart when fertile but can be confusing when sterile.

Typically, Myristicaceae have relatively long oblong leaves with dull surfaces, short petioles, and many close-together parallel secondary veins. Typically Lauraceae have short elliptic leaves with glossy shiny surfaces, relatively long petioles, and relatively glossy shiny surfaces, relatively long petioles, and relatively few, often strongly ascending and not strictly parallel secondary veins. Although there is little room for confusion between Myristicaceae and Lauraceae, Annonaceae are intermediate and overlap with both vegetatively. Both Myristicaceae and Annonaceae (but not Lauraceae) are characterized by myristicaceous branching with the lateral branches at right angles to the trunk and the evenly spaced leaves 2-ranked (except *Tetrameranthus*) along these or along their lateral branches; in Myristicaceae, especially, the lateral branches tend to be clustered and appear to have an almost whorled arrangement ("myristicaceous branching" [Fig. 4]). Lauraceae never have such a phyllotaxy and their leaves are often irregularly spaced along the branches with a definite clustering towards the branchlet apex. Very many Lauraceae are distinguished by the way the leaf blade gradually merges with the petiole apex, typically with at least the hint of an involution of the margin and sometimes with a distinctly involute auricle on each side. Lauraceae leaves typically have shinier surfaces than do the other two families, and the pubescence, when present is usually sericeous with appressed simple trichomes or softly rufescent. Myristicaceae trichomes are either stellate or 2-branched (T-shaped), frequently very conspicuous, and usually rufescent (to whitish on the leaf undersurface). Annonaceae, as usual, are intermediate but stellate (or lepidote) trichomes are rare (mostly *Duguetia*) and sericeous pubescence is common only in *Xylopia.* Most Annonaceae have strong bark ("cargadero" = useful for tying cargo) a feature not found in Myristicaceae or Lauraceae. Lauraceae twigs are typically green while those of Myristicaceae are brownish; Annonaceae commonly have either green or brown twigs; all the green-twigged annonacs have strong bark but only some of the brown-twigged ones. The odor of Lauraceae is usually either clear, spicy, and almost sweetish or foetid and unpleasant; that of Annonaceae tends to be slightly rank; that of Myristicaceae is usually more pungently turpentiny and typically not very strong.

At least five other taxa with simple alternate leaves have odors that might be confused with the Ranalean group: Most simple-leaved species of **Anacardiaceae** (e.g., *Anacardium*) have a more strongly turpentiny odor, as do the few simple-leaved species of **Burseraceae**. The latter also have prominently flexed petiole apices indicating their compound-leaved affinities. Simple-leaved **Leguminosae** (*Bocoa, Lecointea,* some *Swartzia*) have the distinctive green-bean odor typical of their family as well as a petiole that assumes the round pulvinate-cylindrical form typical of the pulvinuli of compound-leaved legumes. **Araliaceae** have aromatic leaves and some are reminiscent of some Lauraceae; they differ prominently in their varying petiole lengths. *Dendrobangia* (**Icacinaceae**) has a more medicinal odor than typical of Ranalean families and is also characterized by a grooved petiole, appressed-stellate indumentum, and black-drying color. Alternate-leaved weedy **Compositae** are mostly not strongly aromatic.

IVAd. Leaves palmately 3(–9)-veined at base *(and alternate and simple) — The majority of taxa with palmate basal veins (here referred to as "3-veined") belong to one of two quite unrelated main groups: Malvales (Malvaceae, Tiliaceae, Bombacaceae, Elaeocarpaceae, Sterculiaceae) or Hamamelidae (especially Ulmaceae, Urticaceae). The Malvalean woody taxa have petioles with a distinctive swollen apical pulvinus; the Hamamelidae and other three-veined families do not.*

IVAd(1). Petioles with apical pulvinar thickening (or with leafy stipules) *(= Malvales) — Perhaps the main palmately veined group of plants, as an order also characterized by strong bark fibers, by stellate (or lepidote) trichomes and the very distinctive petiole apex which is more or less swollen and pulvinar. Only Elaeocarpaceae and most Malvaceae lack the typical pulvinus, the fomer distinctive in their foliaceous stipules, the latter in their mostly herbaceous habit.* ***Bixa*** *is not usually included in Malvales but has a similar, though shorter, pulvinus. The bark slashes of tree Malvales all tend to have a mucilaginous secretion which can be felt when fresh or seen as globules after a few hours. Although recognition to order on vegetative characters is easy, separation of the individual Malvalean families without flowers is frequently problematic. When sterile, Tiliaceae, Sterculiaceae, and simple-leaved Bombacaceae are reliably differentiated only by first knowing the genera. Bombacaceae are all trees and the simple-leaved ones are entire (very weakly sublobed in* ***Ochroma****); Malvaceae are mostly herbs and subshrubs, with the woody species in our area serrate-margined; Tiliaceae (mostly serrate) and Sterculiaceae (trees mostly entire except* ***Guazuma****) include both large trees and small weedy shrubs.*

Tiliaceae — Most serrate Malvalean trees are Tiliaceae (see also sterculiaceous *Guazuma*); entire-leaved tiliacs (except a few genera rare in our area) have the lower leaf surface canescent, a character combination not found in simple-leaved bombacs and only in a few *Theobroma* species in Sterculiaceae (from which entire-leaved *Apeiba* species can be differentiated by longer more slender petioles). Shrub genera *Corchorus* and *Triumfetta,* respectively, differ from stercul shrubs by more crenate marginal serrations and a tendency to 3-lobed leaves. Flowers characterized by multiple stamens arranged in single whorl and with free filaments.

Sterculiaceae — Tree genera differ from most tiliacs in being entire-leaved (or palmately lobed or compound), except *Guazuma* which has leaves more jaggedly serrate than in any Tiliaceae. Shrub sterculs (i.e., most Malvalean shrubs) have serrate leaves (see Tiliaceae above for distinguishing characters). The flowers can have fused or distinct filaments, the former differing from Malvaceae and most Bombacaceae in having 2-celled anthers.

Bombacaceae — The relatively few simple-leaved genera, exclusively large trees, are best characterized by fused filaments, a feature shared in Malvales only with Malvaceae, which differ in being mostly herbs and shrubs, and with some Sterculiaceae. The only definitive difference from Malvaceae is the absence of spinulose pollen, although the stamen tube often differs from Malvaceae in being fused only at base.

Malvaceae — Essentially the herbaceous counterpart of Bombacaceae with which they share the distinctive feature of fused filaments. The most definitive difference is spinulose pollen, a feature never found in Bombacaceae. Mostly differing from sterculs and tiliacs in combination of more broadly ovate leaves with serrate or lobed margins and from most other Malvales in less developed pulvinus. Flowers distinctive by numerous stamens with filaments fused around style into staminal column and/or an epicalyx.

Elaeocarpaceae — The three genera with 3-veined leaves are distinctive in the order in persistent foliaceous stipules and in lacking the typical Malvalean pulvinus.

Bixaceae — *Bixa,* closer to Flacourtiaceae than Malvales, has a distinct apical pulvinus similar to that of the Malvales, but shorter. It is also characterized by scattered, reddish, peltate scales below (but lacks the typical Malvalean stellate trichomes).

IVAd(2). Petioles lacking apical pulvinus

Ulmaceae — Most taxa with pinnate venation, but *Trema* and *Celtis* have 3-veined alternate leaves, the petioles always of equal lengths; the common *Trema* has asperous leaves with fine close-together teeth; *Celtis* is often spiny and has leaves with coarse rather irregular teeth, but the commonest erect species has entire leaves recognizable by the noticeably asymmetric base that characterizes most Ulmaceae.

Urticaceae — Close to Ulmaceae but leaves usually with cystoliths in upper surface and/or with stinging hairs, in tree taxa always serrate.

Euphorbiaceae — Conspicuously 3-veined euphorbs mostly have glands at apex of petiole (sometimes also with latex, see above) or at base of lamina (usually in axils of basal vein pair below) or have stellate or peltate trichomes (*Croton*) or are deeply palmately lobed.

Buxaceae — *Styloceras;* coriaceous with smooth surface, sub-3-veined from above base, drying olive.

(Caricaceae) — Some milky latex usually present.

Begoniaceae — *Begonia parviflora* is our only erect woody *Begonia.* Trunk with swollen nodes; leaves large, very asymmetric, shallowly jaggedly serrate.

Cochlospermaceae (*Cochlospermum*) — Palmately lobed with serrate margins, a combination unique among area trees (although also in some extralimital *Oreopanax*).

Flacourtiaceae — Two of the commonest 3-veined flacourt genera have a very characteristic pair of glands at petiole apex (euphorbs with similar glands differ in having latex or leaves larger and more broadly ovate). *Neosprucea* and *Lunania* have the leaves more strongly 3-veined to near apex than do most other taxa with alternate–3-veined leaves; the common *Prockia* has distinctive semicircular foliaceous stipules, differing from elaeocarps with similar stipules by the combination of longer slender petioles and serrate margin.

Hamamelidaceae — Our only genus has distinctively oblong-ovate leaves with somewhat asymmetric base and a short dorsally grooved petiole.

Hernandiaceae — Leaves long-petioled and entire (in part 3-lobed in *Gyrocarpus*), usually rank-smelling (but the odor not clearly Ranalean). Most like Araliaceae but usually either with two main lateral veins arising slightly below base of lamina (*Gyrocarpus*) or the base subpeltate (many *Hernandia*); if 3-veined from exact base, the main lateral vein pair curving upward unlike similar araliacs.

Araliaceae — Characterized by leaves with rank odor and of dramatically different sizes and with petioles of different length. Three-veined species of *Dendropanax* differ from nonpeltate *Hernandia* in +/- wrinkled usually tannish-drying twig bark and main lateral vein pair straight or curving slightly outward rather than upward. *Oreopanax* usually either epiphytic or with leaves palmately lobed and conspicuously tannish-pubescent below.

Rhamnaceae — *Zizyphus* has conspicuously 3-veined leaves, the species mostly in dry areas where also characterized by spines; moist-forest *Z. cinnamomum* lacks spines but has oblong leaves 3-veined all the way to apex. Some *Colubrina* have 3-veined leaves, but in our area only when opposite.

Rhizophoraceae — *Anisophyllea* of Amazonian sandy soil areas has distinctive oblong 3–5-plinerved leaves.

(Olacaceae) — Extralimital *Curupira* has 3-veined leaves.

(Leguminosae [*Bauhinia*]) — A few species of *Bauhinia* have the two leaflets completely fused (e.g., *B. brachycalyx*).

(Menispermaceae) — A few *Abuta* species are trees with 3-veined leaves and longish petioles with wiry flexed apex.

IVAe. Strong bark — *(Pull off a leaf and see if a long strip of bark comes off with it.) All neotropical species with strong bark fibers have alternate, mostly simple, leaves, and this is a very useful character for several families, some of them (e.g., Thymelaeaceae) otherwise nondescript.*

(Annonaceae) — Keyed out above under plants with primitive odors; if odor not apparent, can be identified by the strong, often greenish, twig bark, entire leaf margins, and vertical fiber lines in a very shallow bark slash.

Lecythidaceae — Differs from other strong-barked families in bark of trunk peeling off in layers rather than as single unit. Faint but characteristic "huasca" odor. Leaves nearly always with serrate or serrulate margins and distinctive secondary (and usually intersecondary) veins that turn up and fade out at margins.

Thymelaeaceae — Very distinctive in the thick homogeneous bark that strips as a unit from entire twig; the only family with thick strong nonlayered homogeneous bark.

(Leguminosae [*Bauhinia*]) — Erect *Bauhinia brachycalyx* with completely fused leaflets has surprisingly strong Lecythidaceae-like bark.

(Malvales and Urticales) — The Malvalean and Urticalean families, keyed out above on account of 3-veined leaves, are also characterized by strong bark fibers, the entire trunk bark peeling off when pulled (as opposed to peeling in layers in Lecythidaceae).

IVAf. Unequal petioles

Araliaceae — Leaves with rank vegetative aroma.

Capparidaceae — Petioles unequal only when leaves terminally clustered; leaves more oblong and/or petioles more wiry than in other taxa with unequal petioles.

Euphorbiaceae — The combination of serrate leaf margins with conspicuously different-length petioles having flexed apices is definitive for Euphorbiaceae; nonserrate taxa with unequal petioles also have the flexed petiole apex (*Nealchornea, Senefeldera, Sagotia, Pogonophora, Dodecastigma, Caryodendron, Didymocistus, Garcia, Gavarretia*).

(Sterculiaceae [*Sterculia*]) — Although *Sterculia* petioles are conspicuously unequal, the genus is keyed out above on account of the Malvalean pulvinus.

IVAg. Petiole glands

Chrysobalanaceae — Some Chrysobalanaceae species have a pair of lateral glands at extreme apex of petiole or at extreme base of leaf blade below; they can usually be recognized by the red gritty-textured inner bark and/or small inconspicuous stipules on young twigs.

Combretaceae — Most tree combretacs (except most *Terminalia*) have a distinctive pair of glands on upper petiole surface, also characterized by leaves clustered at tips of ascending short-shoot branchlets or branch tips.

(Euphorbiaceae) — All taxa with pair of glands near petiole apex have latex and/or are conspicuously 3-veined (see above).

(Flacourtiaceae) — The flacourt genera with glands at apex of petiole have conspicuously 3-veined leaves (see above).

(Rhamnaceae) — Some *Colubrina* species have pair of large glands at extreme base of lamina, typically in basal auricles, and thus, in effect at petiole apex.

IVAh. Serrate (or serrulate) margins

Actinidiaceae — Numerous straight parallel secondary veins; surface frequently rough-pubescent; petiole base not enlarged, unlike Sabiaceae; trichomes simple, unlike *Clethra* or *Curatella.*

Aquifoliaceae — Usually conspicuously coriaceous with faint blackish tracing of tertiary venation and/or blackish dots below on light green undersurface. Characteristic green outer layer in bark slash.

Betulaceae — Doubly toothed leaf margin (with teeth over secondary vein endings slightly larger).

Boraginaceae — Although most tree Boraginaceae have entire leaves, *Saccellium* (with ascending close-together secondary veins, cf., Rhamnaceae) and a few arborescent *Cordia* and *Tournefortia* species have serrate or serrulate margins.

Celastraceae — Twig usually irregularly angled from decurrent petiole base and often zigzag and/or greenish when fresh. *Maytenus,* the main lowland genus, usually has coriaceous olive-drying leaves, but lowland species often entire.

Clethraceae — Distinctive in the densely tannish-stellate tomentum of the leaf undersurface, usually rather remotely serrate or serrulate, sometimes only toward apex.

Compositae — Rather few arborescent Compositae have alternate serrate leaves, *Baccharis* (usually shrubby and characterized by resinous coriaceous leaves, *Tessaria* (with shallowly remotely serrate, narrow, gray leaves), and *Verbesina* (typically deeply pinnately lobed) being the most frequently encountered. Like most other comps these can be recognized by their rather pungent aroma.

Dilleniaceae (*Curatella*) — Leaves asperous, stellate-pubescent, very coriaceous; restricted to open savannas.

(Elaeocarpaceae) — Some *Sloanea* species have remotely serrate or serrulate margins; they are recognizable by the flexed petiole apex and tendency to have both opposite and alternate leaves.

Euphorbiaceae — Only a few arborescent euphorbs have pinnately veined leaves with eglandular equal petioles, lack latex, and have serrate margins. These very nondescript taxa include *Cleidion* (=*Alchornea?*), *Richeria* (the margin only slightly crenulate, leaves cuneate and petiole base slightly enlarged), and several shrubs (*Acidoton, Adenophaedra, Sebastiana*).

Fagaceae — Characterized by clustered terminal buds with scales; round white lenticels. Inconspicuously serrate or serrulate, cuneate to short petiole.

Flacourtiaceae — Many serrate(-serrulate)-leaved pinnate-veined flacourts are characterized by very small pellucid punctations; stipules are always present but usually early caducous and leaving inconspicuous scar. Slightly zigzag twigs are another frequent character. *Banara* lacks punctations but has a conspicuous marginal gland pair near base of lamina from glandular basal teeth; *Xylosma* lacks punctations but is frequently spiny.

Humiriaceae — Most genera (except entire *Vantanea*) have *crenate* margins and festooned-brochidodromous venation. Young leaves at shoot apex rolled into narrow cone; inner bark red or dark red.

Icacinaceae — Most species have groove on top of the often somewhat twisted petiole. Only two genera with serrate margins: *Calatola* (black-drying) and *Citronella* (olive-drying coriaceous and usually with conspicuous axillary domatia; when serrate, typically with spinose teeth (cf., *Ilex opaca*).

Lacistemataceae — Membranaceous to chartaceous. Tertiary venation parallel and perpendicular to midvein; finely serrate or remotely serrulate.

(Lecythidaceae) — Nearly all Lecythidaceae (keyed out above on account of their strong bark) have serrate or serrulate margins.

Leguminosae — The vanishingly few truly simple-leaved legume genera are generally characterized by olive-drying leaves with asymmetric bases and serrulate margins (*Zollernia, Lecointea*) (related *Etaballia* has entire margins).

Myricaceae — Unique leaves strongly yellow gland-dotted, always +/- coriaceous, mostly oblanceolate and +/- marginally toothed, young growth densely lepidote-glandular and macroscopically yellowish or tannish; twigs strongly ridged from decurrent petiole base. High Andes only.

(Myrsinaceae) — A very few mostly shrubby myrsinacs have finely serrate leaf margins; like other members of the family they are characterized by the typical, usually nonpellucid punctations (see below). Serrate myrsinacs can be differentiated from similarly punctate Theaceae (*Ternstroemia*) by the more elongate punctations that are pellucid in bud.

Ochnaceae — Always serrate or serrulate, usually with caducous stipules leaving annular scar; three leaf types — one with secondary veins marginally curved and becoming almost submarginal (with several of these marginal extensions paralleling each other at a given point), or with close, rigidly parallel, secondary veins and finely parallel tertiary veins perpendicular to secondaries; occasionally *Clusia*-type venation (*Blastomanthus*).

Rhamnaceae — *Rhamnus* has leaves with pinnate venation and finely crenate-serrulate on margins; it is recognizable by the typical Rhamnaceae leaf venation with rather close-together, straight, strongly ascending secondary veins.

Rosaceae — The majority of Andean trees with alternate coriaceous leaves and serrate margins are probably Rosaceae. In theory they should have stipules but these are usually not very obvious; *Kageneckia* and *Quillaja* have resinous leaves, like *Escallonia* but narrower, with more strongly ascending secondary veins, tapering gradually at base to a less-defined petiole; most *Prunus* leaves have distinctive large ocellate glands near base of lamina below. Rosaceae trunks lack the green inner bark layer of *Ilex*, the petiole lacks the Sapotaceae-like woody thickened base of *Meliosma*, the leaves are less festooned-brochidodromous than *Symplocos*.

Sabiaceae — Most *Meliosma* species have conspicuously serrate, or at least serrulate margins, usually with numerous, fairly straight, secondary veins. Similar to *Saurauia*, but more coriaceous and the petiole base thickened and often woody; trichomes simple unlike *Clethra* or *Curatella*.

Saxifragaceae — *Escallonia*, exclusively montane, has finely serrate, more or less resinous leaves with a characteristic undersurface from the immersed tertiary venation.

(Solanaceae) — Leaves of many *Solanum* species distinctively irregularly, broadly, and shallowly toothed, usually also with stellate or dendroid trichomes and/or prickles.

Symplocaceae — Characterized by festooned-brochidodromous venation, the leaves usually loosely and rather irregularly reticulate below with nonprominulous venation, the surface between the secondary veins usually rather smooth. Vegetatively very similar to *Ilex* but lacks a green inner bark layer. Species with small coriaceous leaves could be confused with some *Ternstroemia* (Theaceae) but lack black punctations.

Theaceae — Most cloud-forest taxa (*Gordonia, Freziera, Symplococarpon, Ternstroemia*) have more or less serrate leaves (although this can vary even within a species), at least inconspicuously near apex. Leaves characteristically markedly asymmetric at least basally, typically coriaceous and oblanceolately tapering to sessile or subsessile base, secondary venation often immersed and nonapparent. *Freziera* can have long petioles but is easy to recognize by the unusually numerous nearly parallel secondary (and intersecondary) veins and dorsally grooved petiole. *Ternstroemia* (usually only inconspicuously serrate near apex) has well-developed petiole but is distinctively punctate with blackish glands.

Theophrastaceae — *Clavija,* consisting mostly of pachycaul treelets, always has narrowly obovate to oblanceolate leaves, typically with strongly spiny-serrate margins; when not obviously serrate-margined the margin usually distinctively cartilaginous or the plant reduced to a small erect subshrub.

Violaceae — Very nondescript and often impossible to differentiate from Flacourtiaceae vegetatively. Stipules present but usually caducous; leaves usually membranaceous; *Gloeospermum* leaves often dry light green with a paler central area.

IVAi. Thickened and/or flexed petiole apices

Diptercarpaceae — Our only species recognizable by the broadly ovate leaf with close-together, straight, secondary veins and Malvalean pulvinus.

Elaeocarpaceae (*Sloanea*) — Usually recognizable by the highly unusual mixture of opposite and alternate leaves, most species also distinctive in the large, unusually thin buttresses; a few large-leaved species have distinctive leafy stipules.

(Euphorbiaceae) — Many euphorbs have flexed petiole apices but most are 3-veined and/or serrate and/or have latex and/or petiolar glands. A few entire-margined nonlactiferous euphorbs that lack petiolar glands have thickened or flexed petioles; taxa that do not always have conspicuously different-lengthed petioles include *Caryodendron* and *Sagotia.*

Flacourtiaceae—A few flacourts (*Lindackeria, Mayna, Carpotroche*) have flexed petiole apices and +/- entire margins; their petioles tend to be shorter and/or less variable in length than the above euphorbs.

(Meliaceae) — *Trichilia acuminata* has unifoliolate leaves that appear when fresh to be simple leaves with short petioles having flexed apices. Unifoliolate Leguminosae and Lepidobotyraceae have cylindrically thickening at petiole apices but are more clearly unifoliolate.

IVAj. Punctations

Flacourtiaceae — Many genera (*Casearia, Homalium, Banara, Xylosma, Neoptychocarpus*) pellucid-punctate, sometimes with almost linear punctations, but the majority of their species serrate-margined. Usually have stipules or stipule scars, unlike Myrsinaceae which also differ in usually dark nonpellucid punctations.

Rutaceae — Only a few (mostly nonaromatic) genera have simple leaves, these almost always narrowly obovate to oblanceolate with cuneate bases and clustered at branch apices or at apex of pachycaul treelet.

Myrsinaceae — Punctations usually nonpellucid (except in bud), usually elongate, and often reddish or blackish; associated with distinctive pale green "matte" undersurface; stipules completely absent unlike Flacourtiaceae. Trichomes, when present, nearly always +/- branched, unlike other punctate taxa.

Theaceae — Some punctate Theaceae have entire leaves; these differ from Myrsinaceae in having rounder punctations, which are blackish even in juvenile leaves and buds.

IVAk. Stipules

Celastraceae (*Goupia*) — Usually serrate or serrulate but occasionally subentire; leaves characteristically asymmetric-based, blackish-drying, and with strongly ascending lateral veins and finely prominulous parallel tertiary venation. Stipules very conspicuous in juveniles, only.

Chrysobalanaceae — The stipules are typically small and inconspicuous and are usually visible only on young twigs. Many Chrysobalanaceae have very characteristic leaves with close-together, rigidly parallel, secondary veins and a whitish undersurface, but *Hirtella* and some *Licania* species are very nondescript. The whole family is usually recognizable by having red inner bark with a gritty-sandy texture.

Dichapetalaceae (*Dichapetalum*) — Most tree species of *Dichapetalum* have conspicuous stipules, sometimes with a very unusual serrate or fimbriate margin.

Erythroxylaceae (*Erythroxylon*) — When present, faint venation lines paralleling the midvein below are very typical; stipules triangular and brownish or tannish, often longitudinally striate.

Euphorbiaceae — Several entire-margined nondescript euphorb genera have distinctive +/- caducous stipules (e.g., *Chaetocarpus* with subpersistent thick-foliaceous stipule, *Margaritaria* with reddish slightly zigzag puberulous twigs and conspicuous stipule scar, *Sagotia* with moraclike terminal stipule falling to leave conspicuous scar).

(Flacourtiaceae [*Casearia*]) — A very few nondescript *Casearia* species are both entire and lack punctations; stipule scars are about their only useful character.

Lacistemataceae (*Lacistema*) — Leaves usually subentire with faint tendency to marginal serrulation; characterized by stipule caducous to leave conspicuous scar, with both stipule and young twigs tending to dry blackish, contrasting with the whitish stipule scar.

Rosaceae — Entire-leaved *Prunus* species have early-caducous inconspicuous stipules; they are usually recognizable by the pair of large dark-drying glandular ocelli near base of lamina below.

(Violaceae) — A few nondescript Violaceae with entire or subentire leaves are characterized by very inconspicuous stipules (see *Leonia* and *Paypayrola* below).

IVAl. Lepidote scales and/or stellate trichomes

(Annonaceae) — *Duguetia* and some *Annona* species have stellate trichomes or lepidote scales but should key out above under primitive odor.

Capparidaceae (*Capparis*) — Conspicuous tannish scales in many species including some that have uniform petioles; a characteristic patelliform gland just above the leaf axil on young twigs is frequently apparent.

Clethraceae — Distinctive in the densely white-stellate leaf undersurface, but most species +/- serrate or serrulate (see above). Its margin entire differs from area *Styrax* in longer, laxer arms on twig trichomes, and in lacking the scattered rufescent trichomes of the more strongly reticulate leaf undersurface.

(Compositae) — A few arborescent Compositae have stellate trichomes, usually distinctive in a blackish inner bark layer.

(Dilleniaceae [*Curatella*]) — Stellate-pubescent but keyed out above on account of the serrate margins.

Euphorbiaceae — Pinnate-veined euphorbs characterized by lepidote scales and/or stellate trichomes include *Hieronyma, Pera,* some *Croton* (also *Gavarettia* with petiole glands and the dry-area shrubs *Chiropetalum* and *Argythamnia,* both with +/- sericeous leaves, sometimes in part with malpighiaceous trichomes).

Icacinaceae (*Dendrobangia*) — Leaves characteristically membranaceous and black-drying.

Fagaceae — Margins usually +/- serrulate; when entire the mostly stellate trichomes (*Trigonobalanus* only with, in part, 2-branched trichomes) a useful indicator.

(Malvales) — Most Malvales have stellate trichomes or scales but are keyed out above by the pulvinar petiole apex.

Solanaceae — Several genera (especially many *Solanum* species) have stellate to variously dendroid trichomes; the family is usually recognizable by the rank tomato-like odor of crushed leaves; and many *Solanum* species are spiny.

Styracaceae — Characterized by densely white-stellate or lepidote leaf undersurface, usually also rufescent with reddish-stellate hairs, especially on twigs; similar densely white-below Solanaceae lack the rufous-stellate twig pubescence and are usually spiny and/or with slightly lobed leaf margins and/or have asymmetric leaf bases.

IVAm. Leaves parallel-veined or lacking secondary veins

Podocarpaceae — Leaves very coriaceous, linear-oblong with a strong midvein, completely lacking secondary veins or with a few faint longitudinal veins paralleling midvein.

Theaceae — *Bonnetia* has very faint longitudinal secondary veins paralleling the midvein; it is recognizable by the terminally clustered spiral leaf arrangement.

Goodeniaceae — *Scaevola* is a beach shrub with very succulent leaves with invisible secondary veins, characterized by petiole base expanded and subclasping.

(Theophrastaceae) — A few *Jacquinia* species have such thick-coriaceous leaves that the secondary veins are invisible; always distinctive in the sharply spinose leaf apex (see below).

(Monocots) — Most monocots have parallel-veined leaves, but very few of these are woody enough to include in this key.

Gramineae — Bamboos have parallel-veined leaves, the plants distinctive in the segmented often hollow stems with characteristic swollen nodes.

(Palmae) — A few understory palms have simple leaves.

IVAn. Parallel tertiary venation

Opiliaceae (*Agonandra*) — Moist-forest species have the tertiary veins finely parallel and perpendicular to midvein. Very *Heisteria*-like in olive-drying leaves and olive-colored fresh twigs. Distinctive in the leaf blade decurrent on the poorly differentiated petiole and the few often poorly defined secondary veins.

Guttiferae — *Caraipa* has the tertiary venation finely parallel and +/- perpendicular to the secondary veins; it is otherwise very nondescript and not very obviously a guttifer although there is usually a very faint trace of latex.

Icacinaceae — A rather nondescript family, but most genera have grooved petioles and the tertiary veins conspicuously finely parallel and arranged perpendicular to the midvein; in addition *Pouraqueiba* and *Emmotum* are distinctively sericeous below and *Discophora* has a characteristic smooth "matte" undersurface.

Lacistemataceae — Tertiary venation perpendicular to midvein; usually finely serrate to serrulate; when entire (some *Lacistema*) with conspicuous stipule scars.

Lecythidaceae — Strong bark (the only combination of strong bark and parallel tertiary venation). Brochidodromous genera lacking the family's typical upcurved veins and serrate margins have tertiary veins closely parallel and perpendicular to midvein.

Linaceae — Most species of *Roucheria* have *Clusia*-type venation with the secondaries straight and parallel and well-developed intersecondaries; margins usually crenate (at least when *Clusia*-venation lacking).

(Ixonanthaceae) — Whether to recognize Guayana Shield genera *Cyrillopsis* and *Ochthocosmus* as members of this African Linaceae segregate is debatable.

(Myristicaceae [*Compsoneura*]) — The tertiary veins are conspicuously finely parallel and perpendicular to the midvein and the primitive odor is not always apparent.

(Sapotaceae) — Many Sapotaceae have conspicuously parallel tertiary venation or *Clusia*-type venation. They have latex (and are keyed out above), but the latex sometimes is not very conspicuous, especially during periods of water stress.

Ochnaceae (except *Ouratea*) — *Blastomanthus* has *Clusia*-type venation; most other area genera have the parallel tertiary veins perpendicular to the secondary veins. They are serrate or serrulate and keyed out above.

Olacaceae — Most genera have finely parallel tertiary venation +/- perpendicular to the midvein or secondary veins; usually there is a slight bit of latex in petiole.

- -

IVAo. Spines or spine-tipped leaves

(Annonaceae) — One Llanos species of *Annona* has branch-spines.

Berberidaceae — Branches usually armed with trifurcate spines (unique to family). Leaves spine-tipped or spinose-margined and clustered on bracteate short shoots.

Boraginaceae (*Rochefortia*) — Leaves clustered on short shoots in spine axils; secondary veins more conspicuously raised than in spiny phytolacs and *Pisonia.*

Compositae — Most Mutisieae trees in our area have spines on the branches or in leaf axils or spinose leaf apices or teeth. Spiny Compositae differ from *Pereskia* in the spines (or at least the primary spines) at each node arranged in equal-length pairs (sometimes also with additional shorter spines).

Cactaceae (*Pereskia*) — Although normal leaves are present, *Pereskia* is recognizable as Cactaceae by the numerous long clustered spines on the branches. Differs from spiny Compositae trees in several spines of different lengths clustered together at each node.

Celastraceae (*Schaefferia*) — Branch apices spine-tipped; leaves small with strongly ascending very inconspicuous secondary venation, often clustered in short-shoots. Twigs distinctively green and strongly angled from decurrent petiole base.

Euphorbiaceae — The family that can have virtually any characteristic has only a few spiny members, including spiny-trunked *Hura* (with latex) and *Adelia* (with inconspicuously spine-tipped branches).

Flacourtiaceae — *Casearia* and *Xylosma* sometimes have branch-spines, the latter sometimes with very striking branched spines covering trunk.

(Moraceae) — The only spiny Moraceae in our area are *Maclura* (*Chlorophora*) and *Poulsenia*, both with milky latex (see above).

(Nyctaginaceae) — *Pisonia* actually has opposite leaves but in spiny taxa they are mostly clustered on short shoots and the disposition is not evident.

Olacaceae — *Ximenia* has branch spines and leaves clustered at lateral branch tips, drying olive to blackish, usually retuse at apex.

Phytolaccaceae — *Achatocarpus,* often spiny, is restricted to seasonally dry areas and usually multitrunked. It is characterized by small blackish-drying leaves that are usually at least in part clustered several per node.

Rhamnaceae — Several Rhamnaceae genera have spines; these include a few taxa with nondescript entire leaves but most of these are easy to recognize by the accentuated photosynthetic, densely branched, spiny branches.

(Rosaceae) — *Hesperomeles,* with serrate leaves (see above), usually has spine-tipped branches.

Simaroubaceae — *Castela,* mostly only shrubs but occasionally small trees, always has strongly spiny twigs; characterized by the bitter simaroub taste.

(Solanaceae) — The small thick-based spines that characterize many species of *Solanum* are actually prickles and may be present on leaves as well as on twigs and branches; spiny members of the family have stellate trichomes (and are keyed out above), in addition usually recognizable by the rank tomato-like odor of crushed leaves.

Theophrastaceae — *Jacquinia* has small thick-coriaceous strongly spine-tipped leaves.

(Urticaceae) — The few spiny-trunked Urticaceae have serrate or incised leaf margins.

IVAp. None of the above

Amaranthaceae(*Pleuropetalum*) — Black-drying, membranaceous, narrowly elliptic, long-petiolate leaves.

Aquifoliaceae — The few entire-leaved lowland *Ilex* species (e.g., common tahuampa species *I. inundata*) mostly have the coriaceous leaves distinctively black-drying; a green layer in trunk slash is characteristic of most Aquifoliaceae.

Bignoniaceae — *Crescentia* and *Amphitecna* are totally unbignoniaceous vegetatively in simple alternate leaves. *Crescentia* has the narrowly obovate leaves in characteristic fascicles alternating along thick branches; *Amphitecna* has elliptic to obovate coriaceous leaves, poorly demarcated from woody based petioles, in South America drying grayish with pale secondary veins below.

Boraginaceae — Leaves and stem often stiff-pubescent and asperous. Most tree *Cordia* species have distinctive node with a leaf arising from each branch dichotomy and held parallel to the dichotomy (Fig. 4); *Saccellium* has ascending close-together secondary veins; (cf., Rhamnaceae).

Capparidaceae — Capparids that lack different-length petioles have the leaves 2-ranked, usually with a raised patelliform axillary gland.

Celastraceae — A few lowland *Maytenus* species and its larger-leaved segregate *Gymnosporia* have entire leaves, these mostly drying olive with paler inconspicuous (*Maytenus*) or strongly arcuate (*Gymnosporia*) secondary veins.

Chrysobalanaceae — Although small stipules are present, they are usually inconspicuous and early-caducous; even if stipules not apparent, recognizable by the gritty-textured, red inner bark.

Combretaceae — Leaves usually apically clustered; except for *Terminalia,* usually with petiole glands. Alternate-leaved taxa typically with leaves clustered and pagoda branching form or bark very smooth and white.

Compositae — Tree composites with alternate entire leaves mostly have the leaves +/- conspicuously whitish- or grayish-pubescent below. Some (especially Vernonieae) have a distinctive blackish layer in inner bark.

Cyrillaceae — Leaves narrowly obovate, blunt-tipped, glabrous, with poorly developed secondary veins and tertiary venation intricately prominulous above and below.

Dichapetalaceae — Tree dichapetalacs usually have serrate stipules (*Dichapetalum*) or uniformly terete tannish-puberulous thickish petiole, usually in part with distinctive scars from fallen petiole-borne inflorescence.

Ebenaceae — Tropical species with trunk slash characterized by black bark ring; leaves typically with large darkish glands on lower surface usually scattered along (but removed from) midvein (rarely in *Prunus*-like basal pair).

Euphorbiaceae — A notoriously heterogeneous and difficult to recognize family. "Left-over" genera include *Drypetes* (asymmetric base and prominulous tertiary venation), *Margaritaria* (caducous stipules and a characteristic twig apex), *Croizatia, Tacarcuna,*

Discocarpus, Maprounea, and *Phyllanthus* (sometimes resembling a compound-leaved legume.

Flacourtiaceae — While flacourts have stipules, these are often not very evident; a few *Casearia* species with nonobvious stipules have entire margins and lack punctations.

Humiriaceae — *Vantanea* has coriaceous, +/- obovate leaves, usually drying a dark reddish color; young leaves rolled at shoot apex.

Icacinaceae — Groove on top of the often twisted petiole; *Citronella* and *Calatola,* both usually with at least a few serrations, lack conspicuously parallel tertiary venation. When subentire, the former is black-drying, the latter usually with conspicuous domatia.

(Leguminosae) — Rare simple-leaved legumes have the entire petiole cylindrically pulvinate (cf., pulvinulus of leaflets of compound leaves). They usually have asymmetric bases, frequently serrulate margins, and dry a distinctive light olive-green; of these *Zollernia* and *Lecointea* are often remotely serrate but *Etaballia* is uniformly entire. Unifoliolate legumes (e.g., *Cyclolobium, Bocoa,* some *Swartzia,* and *Poecilanthe*) are more common and easy to recognize by the apical and basal pulvinular area of the 2-parted "petiole" with typical cylindrical pulvinus of the single leaflet forming its apical part.

(Moraceae) — *Trophis* usually lacks conical stipule and the latex is watery, not milky, but the leaf venation is typically moraceous.

(Myricaceae) — Usually +/- marginally toothed; if subentire recognizable by the yellow gland dots, especially on young growth.

Olacaceae — Usually characterized by curved (sometimes almost U-shaped) green petiole, slightly thicker toward apex; most taxa have slight latex in petiole and/or finely parallel tertiary venation.

(Onagraceae) — A few *Ludwigia* species are subarborescent; they are restricted to swampy areas and have exfoliating reddish bark. (Tree species of *Fuchsia* have opposite leaves.)

(Passifloraceae) — A few *Passiflora* species are arborescent; they have broad succulent leaves with abaxial glands at base of midrib; erect *Dilkea* is a wandlike shrub with narrowly obvovate or oblanceolate, usually clustered leaves with prominulous venation.

Phytolaccaceae — Tree phytolacs have entire, thin or somewhat succulent leaves, soft wood, and few obvious recognizing characters; *Gallesia,* the commonest large-tree phytolac has a conspicuous garlic odor.

Polygonaceae — The whole family very easy to recognize by presence of an ochrea, an irregular broken sleeve of stipular tissue that covers the node above petiole base.

Rhamnaceae — *Colubrina* usually has nondescript pinnate-veined entire leaves, but can be recognized to family by the typical Rhamnaceae venation with parallel rather straight strongly ascending secondary veins, also frequently with pair of large glands at base of lamina.

(Sabiaceae) — A few *Meliosma* species are essentially entire; like their more numerous serrate-leaved congeners they are recognizable by the thickened sometimes subwoody petiole base (cf., Sapotaceae).

Solanaceae — Some species of *Cestrum* and *Solanum* are glabrous or have simple trichomes and are very nondescript; the best vegetative character is the rank tomato-like vegetative odor; some species of *Solanum* also distinctive in the peculiar nodes, apparently with very small leaves opposite the regular ones.

Violaceae — *Paypayrola* and the commonest *Leonia* species have coriaceous leaves with entire or subentire margins, nonobvious stipules, and are very nondescript. The leaves are rather Lauraceae-like in appearance, tend to have prominulous tertiary venation, and have a peculiarly dryish texture when fresh.

IVB. Lianas with alternate simple leaves

IVBa. Tendrils present *— In our area seven families include taxa with alternate simple leaves and tendrils. They are very easy to distinguish by the form and position of the tendril — axillary in Passifloraceae, arising at 90° angle from petiole in Cucurbitaceae, arising opposite petiole in Vitaceae, arising in pair from lower part of petiole in Smilacaceae, coiled in one plane like a butterfly's tongue in Rhamnaceae, axillary with a straight basal portion and a sharp median bend or zigzag below twining part in* ***Antigonum*** *(originally terminating inflorescence and usually branching). (**Bauhinia** can have tendrils similar to those of Rhamnaceae but apparently not in the few nonbifid-leaved species.) See Figure 2.*

Passifloraceae — Tendril axillary; also the leaves distinctive in having either striking petiolar glands or large glands in the axils of basal vein pair, or both, and in many species in their peculiar shapes, often broader than long. Stems of lianas usually woody, often somewhat irregularly lobed or with an inconspicuous 4–5-armed cross in cross section.

Cucurbitaceae — Tendril arising at 90° angle from petiole base, often bifurcate; also distinctive in palmate-veined (usually palmately lobed) leaves usually with remote teeth on the margin; stems soft and variously (often complexly) anomalous in cross section.

Vitaceae — Tendril arising at 90° angle from petiole base; also recognizable by the nodes distinctively jointed and or swollen and the leaves palmately veined or 3–5-foliolate; stems soft and flexible, with large vessels and characteristically differentiated outer layer but no obvious anomalous structure; often with papery reddish bark and/or pendent stemlike aerial roots.

Smilacaceae — Unique tendrils arising in pair from near base of petiole, actually representing modified stipules partially fused to base of petiole; leaves entire, palmately veined (with the lateral veins continuing into leaf apex: actually a monocot despite the net venation); stems smooth green and spiny (could be confused only with *Dioscorea* with usually thicker spines).

Rhamnaceae — The main scandent genus has a distinctive tendril coiled in one plane exactly like a butterfly's tongue, this usually terminal on young branch; leaves with rather straight, parallel, strongly ascending secondary veins and tertiary veins perpendicular to the secondaries and parallel to each other, with gland pair at base of lamina above and/or serrate.

(Polygonaceae) — *Antigonum*, in our area only in cultivation or seminaturalized, has an axillary tendril with a zigzag basal part below twining part (originally terminating inflorescence and usually branching [cf., compound-leaved Sapindaceae]); it is also characterized by the typical ochrea of the family.

(Leguminosae) — Several compound-leaved legume vines have tendrils, and one scandent tendrillate genus, *Bauhinia*, has incompletely 2-foliolate leaves that are sometimes barely split only at extreme apex and rarely not at all. The tendrils of *Bauhinia* are rather woody and hooklike and not strongly twining, but may not occur in nonbifid species.

***IVBb. Tendrils absent** (and plants scandent with alternate simple leaves), athough young branch apex can be coiling and tendril-like in a few taxa (e.g., **Omphalea**). Note that the categories below are not mutually exclusive.*

(1) Parallel veins — (All veins parallel to midvein or with finely parallel secondary and tertiary venation more or less perpendicular to midvein (cf., *Clusia*-venation).

Araceae — Climbing Araceae are mostly parallel-veined (unlike mostly nonscandent *Anthurium*) with the venation arising +/- perpendicular to the strong midvein; only *Heteropsis* is truly woody, the others more or less succulent.

Cyclanthaceae — *Ludovia* is the only scandent cyclanth with simple leaves, the other taxa all being more or less deeply bifid-leaved. Like other cyclanths it is characterized by leaves with long thin petioles having sheathing bases, by weakly defined midveins and thin secondary vines all strictly parallel to each other and nearly parallel to the midvein.

Gramineae — Climbing bamboos are easily recognized by their parallel leaf venation and the distinctive segmented stems with characteristic swollen nodes.

(Other Monocots) — Commelinaceae, Liliaceae (*s.s.*), Amaryllidaceae, and Orchidaceae all have a few scandent herbaceous genera or species with strictly parallel-veined leaves.

***(2) Serrate leaves** (without tendrils)* — Except for Dilleniaceae, our climbers with serrate alternate leaves are all habitally atypical genera (or species) of predominantly erect families.

Dilleniaceae — Characterized by the combination of usually sharply serrate margins (at least when young), and usually asperous surface. (*Doliocarpus* is nonasperous and can also be subentire.) The leaf shape is also distinctive with strictly pinnate secondary veins and the leaf base typically narrowly cuneate, at least when young. Bark usually distinctive: papery peeling and more or less reddish; stem uniformly woody, the section always showing concentric growth rings crossed by paler rays radiating out from center.

Ulmaceae — *Celtis iguanea,* the only area ulmac liana, has spiny branchlets and is the only area vine with alternate serrate leaves and spines; the leaves are 3-veined.

Urticaceae — Leaves distinctively membranaceous, also differing from other alternate-leaved climbers in cystoliths on upper surface; usually long-petiolate, the petioles lacking any hint of apical pulvinulus or flexion. In our area three mostly erect genera have a few scandent (usually merely clambering) species, mostly restricted to cloud forests: *Urera* (the scandent species nonurticating), *Pouzolzia,* and *Boehmeria.*

Euphorbiaceae — The family is mostly erect but we have three genera that are lianas (but with entire leaves) and two that are +/- herbaceous vines (sometimes with urticating hairs) with a few erect genera also having a scandent species or two. All euphorb climbers in our area have the leaves 3-veined (rarely 3-foliolate or deeply 3-lobed in *Dalechampia* and *Manihot*) at the base, usually broadly ovate in shape, and usually with a pair of glands at apex of petiole or near base of blade. Most of our euphorb climbers lack milky latex, although entire-leaved *Omphalea* has a cloudy-watery sap that sometimes turns pinkish or bluish with oxidation.

Celastraceae — Our only scandent genus is *Celastrus,* restricted to montane cloud forest, and easily characterized by the pinnate-veined, crenate-serrate, oblong-elliptic leaves and the prominent white lenticels on the twigs.

Violaceae — The most distinctive character is the slightly but distinctively raised petiole attachment (leaving raised scar on branchlets); the branchlets are longitudinally striate, lacking the large raised lenticels of *Celastrus.* One of our scandent genera (*Anchietea*) is restricted to Andean cloud forests in our area, the other (*Corynostylis:* the leaves are often barely subserrate) to swamps.

(Malvaceae) — The only species I know to become a true liana is *Malvaviscus arboreus* and even that species is more often erect or subclambering. It looks vegetatively much like *Urera* but lacks cystoliths and usually has a more broadly ovate sub-3-lobed leaf.

Loasaceae — Our taxa mostly +/- scandent but usually only herbaceous vines; leaves mostly coarsely or doubly serrate and or deeply toothed or lobed. Mostly readily recognizable by the strongly urticating trichomes (see also *Tragia* with more finely serrate leaves); if not urticating the leaves more or less asperous. A few genera have opposite leaves; others very from alternate to opposite.

Compositae — Most Compositae are erect and most composite climbers are opposite-leaved and/or entire-leaved, but a few scandent genera of Senecioneae (*Pentacalia, Pseudogynoxys*) mostly with young stems hollow, can have alternate, +/- remotely serrate leaves, and *Jungia,* with distinctive leaves broader than long, is usually more or less marginally lobed.

(Basellaceae) — *Tournonia,* a cloud-forest vine, has broadly ovate, palmately veined, cucurbit-like leaves with an irregularly serrate margin; other Basellaceae have entire margins.

(Boraginaceae) — Most Boraginaceae are erect and most of the climbers are entire-leaved species of *Tournefortia.* Only *Cordia spinescens* is a serrate-leaved, more or less clambering liana, distinctive in the inflorescence base fused to the base of subtending leaves with the remnants of this structure forming a blunt, subwoody spine on older stems.

*(3) **Deeply lobed and/or peltate leaves** (and lacking tendrils)*

Tropaeolaceae — The only scandent family in our area characterized by peltate leaves; slender herbaceous vines nearly restricted to cloud forests and nearly always with roundish, very shallowly, broadly lobed leaves, unique in the frequently twining petiole.

Euphorbiaceae — *Manihot* is the only euphorb genus with deeply palmately 3–5-lobed leaves that is mostly scandent in our area; a few species of scandent *Dalechampia* have deeply lobed leaves but are subwoody weedy vines. *Manihot* differs from *Carica horowitzii* in having watery sap.

(Solanaceae) — The only family in our area with climbers having deeply pinnately lobed leaves; the lobe-leaved climbers all belong to *Solanum* and most have spiny prickles on the branchlets.

(Caricaceae) — One unusual species of *Carica* (*C. horowitzii* of western Ecuador) is a vine with deeply palmately lobed leaves, vegetatively very similar to *Manihot* but with milky latex.

(Menispermaceae) — A few species of *Cissampelos* have slightly peltate leaves differing from *Tropaeolum* in woodier stem, non-twining petiole and not at all lobed or angled margins; a few species of *Disciphania* have deeply 3-lobed leaves, these differing from scandent *Manihot* and *Carica* in pubescent undersurfaces.

(Aristolochiaceae) — A few *Aristolochia* species have deeply 3-lobed leaves, these with leafy stipules.

(4) ***Primitive odor*** — Only one Ranalean family in our area is primarily scandent — Aristolochiaceae. In addition Lauraceae (*Cassytha*), Hernandiaceae (*Sparattanthelium*) and Piperaceae (*Sarcorhachis*) each have a scandent genus and there are miscellaneous climbing species in mostly erect genera of Annonaceae (*Annona*), Lauraceae (*Ocotea*), and Piperaceae (*Piper, Peperomia*).

Aristolochiaceae — Besides the vegetative odor, characterized by the distinctly palmately veined leaves, usually with three ascending main veins and two laterals forming *margin* of sinus and then branching to send main vein to sinus, the frequent presence of foliaceous stipules, and the petiole with base almost always extended into thickened nodal ridge or decurrent as raised striation on opposite side of branchlet.

Hernandiaceae — *Sparattanthelium,* the only scandent genus in our area, is a lowland forest liana, characterized by the 3-veined entire leaves with no hint of an apical pulvinulus or flexion and by the reflexed petiole bases which tend to become woody and form hook-like climbing organs on older branches.

(Annonaceae) — Curiously, although most paleotropical annonacs are lianas, the few scandent species of Annonaceae in our area all belong to mostly erect *Annona.* They are characterized, besides the odor, by the pinnate leaf venation (unlike *Sparattanthelium*) and a typical dark bark which shows tiny interconnecting fibers when shallowly slashed.

Piperaceae — Liana Piperaceae, like the erect ones, are easily recognized by the combination of Ranalean odor and swollen jointed nodes.

Lauraceae — The only truly scandent Lauraceae genus in our area is the bizarre leafless achlorophyllous parasitic vine *Cassytha* (a close look-alike for *Cuscuta* except for the flower). A few species of *Ocotea* are more or less scandent, these differing from scandent *Annona* in smaller more delicate leaves and smoother gray bark.

(5) ***Petiolar or lamina-base glands*** *(and no tendrils)* — Petiolar glands are most typical of the tendrillate family Passifloraceae and also occur in some similarly tendrillate Cucurbitaceae (especially *Fevillea*). However, they also characterize the nonserrate scandent euphorb genera *Omphalea* and *Plukenetia,* the latter with characteristic narrow glands at extreme base of lamina.

Euphorbiaceae — Our two true liana genera of Euphorbaceae both have a pair of glands near petiole apex. In *Plukenetia,* the commonest species also characterized by squarish basal "corners" on the ovate leaf, the glands are at extreme base of the lamina and tend to be narrow or on basal auricles. In *Omphalea* the pair of thick glands are at the petiole apex.

***(6) Palmately 3–5-veined entire leaves** (and no tendrils)* — In addition to the 3-veined Ranalean taxa (Piperaceae, Aristolochiaceae, Hernandiaceae, see above), milky latexed Moraceae and peltate-leaved *Tropaeolum,* there are a number of nontendrillate families with scandent taxa characterized by otherwise unremarkable 3-veined or palmately veined entire leaves.

Sterculiaceae — *Byttneria* is our only scandent Malvalean genus, thus easy to recognize by the Malvalean characters of pulvinate petiole apex and usual presence of stellate trichomes; some species have recurved prickles on the branchlets.

Menispermaceae — The main neotropical liana family with entire 3-veined alternate leaves (only *Telitoxicum* has strictly pinnate venation, although a few other species are rather inconspicuously 3-veined). Most of the truly woody genera have a characteristic wiry or subwoody pulvinal flexion at the petiole apex; genera lacking the apical thickening have the petiole base conspicuously flexuous and usually also have relatively soft flexuous stems and corky bark. A useful vegetative character for most of the woody lianas is the contrasting concentric rings of xylem that are usually strikingly asymmetric and/or flattened with the center of the stem usually far to one side.

Dioscoreaceae — Despite the conspicuously reticulate venation, a monocot; characterized by several lateral veins running evenly and uniformly all the way to leaf apex and by the base of the well-developed petiole which is usually flexed and/or twisted above the extreme base which tends to be rigid and is usually somewhat decurrent onto the slender evenly striate branchlet (+/- hinting at the typical sheathing petiole base of many monocots). Stems never becoming >2.5 cm diameter, except at nodes, often green and spiny (with the spines thicker than in similar *Smilax*). Many species have conspicuous enlarged tubers.

Ericaceae — The leaves are typically the most coriaceous of any of the 3-veined scandent taxa and are generally 3–5-plinerved rather than truly 3-nerved, usually 2-ranked and borne on a short petiole that becomes subwoody. Most of the scandent Ericaceae are restricted to cloud forests and most are hemiepiphytic, often with adventitious roots.

(Urticaceae) — *Pouzolzia* has a few scandent species mostly with entire leaves, including the common tahuampa species *P. formicaria.* Distinctive in cystoliths in upper leaf surface. Most species also characterized by small but conspicuous stipules.

Rhamnaceae — *Ampelozizyphus* has the leaves entire, coriaceous, and elliptic, strongly 3-nerved to apex, usually also with much weaker pair of nearly marginal veins; tertiary venation tends to be finely and prominulously parallel on upper surface (only); best character is the dorsally narrowly grooved petiole.

Compositae — The simple-leaved Mutisieae liana genera mostly have 3–5-veined leaves (in *Jungia,* 3–5-veined and broader than long; in *Lycoseris,* densely white-tomentose below and usually 3-veined) as does *Baccharis trinervis* of the Astereae.

Basellaceae — Leaves succulent with pinnate venation, but usually more or less distinctly plinerved at base and +/- decurrent onto apically winged petiole; usually rounded or obtuse at apex; *Anredera* is the only genus with acute to acuminate leaves.

(Olacaceae) — *Heisteria scandens,* the only scandent species of its family is a canopy liana, characterized, like most other species of the family, by the curved greenish petiole slightly thicker toward the apex, and usually with a bit of latex in it. The scandent species has distinctly sub-3-veined leaves unlike most of its relatives.

(Leguminosae) — A few scandent *Bauhinia* species have the two leaflets fused together all the way to the apex; apparently the non-bifid species do not have the *Gouania*-like butterfly-tongue tendrils that characterize most scandent *Bauhinia.* Sometimes simple-leaved *Bauhinia* can be recognized by the bifid drip tip of some leaves; otherwise only by the cylindrical pulvinus at both base and apex of petiole.

Euphorbiaceae — Although the euphorb lianas are a mixed bag, most of them have 3-veined entire leaves. Most are also characterized by petiolar or lamina-base glands (*Omphalea, Plukenetia*), milky latex (*Mabea eximia*), or deep lobing (*Manihot*). However, a few weedy *Dalechampia* species lack all these characters, and might not be recognizable when not fertile.

(7) Latex — While presence or absence of latex is usually an important taxonomic character, in alternate simple-leaved climbers it is of relatively little importance, characterizing only part of Convolvulaceae plus miscellaneous species of other families.

Convolvulaceae — Our only climbing family with essentially pinnately veined leaves having conspicuously cordate bases. Only *Ipomoea* and its relatives, most of which are slender vines, have latex. The only woody Convolvulaceae lianas with latex are the unusually woody species of the *Ipomoea* group (e.g., the common *I. phillomega*).

Moraceae — Most Moraceae species are trees or stranglers, although a few strangling figs (e.g., *F. schippii*) remain vinelike at maturity. While most Moraceae, including the strangler figs, have the family's distinctive conical terminal stipule, our only true Moraceae liana, *Maclura brasiliensis,* does not, although it can be recognized by the combination of distinctive Moraceae leaf venation and milky latex.

Campanulaceae — Characterized by alternate leaves and milky latex, but almost entirely herbaceous, with only a few cloud-forest species of *Burmeistera* becoming scandent or hemiepiphytic. These are distinct from milky latexed Convolvulaceae in their more oblong leaves completely lacking a cordate base.

(Euphorbiaceae) — *Mabea eximia* may be the only scandent Euphorbiaceae in our area with milky latex, although *Omphalea* (characterized by the pair of thick glands at the petiole apex) has a cloudy-watery latex.

(Caricaceae) — One *Carica* species, *C. horowitzii* of western Ecuador, is scandent.

(Olacaceae) — The only scandent species of Olacaceae, *Heisteria scandens,* sometimes has a trace of latex in its petioles.

(8) Spines — While spines are taxonomically useful for recognizing a few genera of climbers, they are generally variable from species to species within a genus, and thus, not very useful for familial recognition. The climbing families that may be more or less defined by the presence of spines are Cactaceae, tendrillate Smilacaceae, Ulmaceae, Phytolaccaceae, Solanaceae, and Polygalaceae, the latter four only because an individual scandent genus or species happens to have spines. Otherwise, spines occur in miscellaneous scandent taxa

including *Passiflora* (*P. spinosa* and relatives), *Combretum* (alternate-leaved *C. decandra* and relatives), *Cordia spinosa,* and a few *Byttneria* and *Dioscorea* species. Another spiny liana, *Pisonia aculeata* of the Nyctaginaceae, might be confused with this group; it has opposite leaves but these tend to be clustered and might be taken for alternate.

(Ulmaceae) — *Celtis iguanea,* our only scandent Ulmaceae, has spines (see "serrate leaves" above).

Cactaceae — Most scandent Cactaceae, like their erect relatives, are unmistakable on account of their succulent photosynthetic stems and lack of leaves. However *Pereskia* has normal leaves and a few species are scandent, these very reminiscent of *Seguieria* except for the more numerous longer straight spines on the stem.

Combretaceae — *Combretum decandra* and its relatives are atypical for the genus both in the usually alternate leaves and in having straight rather blunt spines on the stem.

(Dioscoreaceae) — Many *Dioscorea* species (see 3-veined taxa above) have irregularly scattered thickish spines, the stems easy to recognize on account of their smooth green surface.

Phytolaccaceae — *Seguieria* is a canopy liana with distinctive slightly raised nodes having a pair of recurved spines. The entire olive-drying leaves are also distinctive.

Polygalaceae — *Moutabea,* one of the main Polygalaceae liana genera, has inconspicuous prickles scattered on its branches; it is distinctive in the rather narrow, thick-coriaceous, olive-drying, leaves with immersed and hardly visible secondary and tertiary venation.

(Smilacaceae) — The smooth green stems of *Smilax* (see tendrillate taxa above) could be confused only with *Dioscorea* from which the stem spines differ in being small and evenly recurved or (when larger) in having the remnants of dried tendril bases at their apices.

Solanaceae — Many scandent Solanaceae have recurved prickles on the branchlets. They can be identified by the stellate trichomes and the leaves with completely pinnate venation, usually also with spines on main veins below, and often with irregularly remotely dentate-lobed margin.

(9) Scandent plants with alternate simple leaves and lacking any of the above characters — The taxa belonging to this most nondescript group of climbers are listed individually with their individual distinguishing characters, if any.

Polygonaceae — Andean *Muehlenbeckia* plus about half our species of *Coccoloba* are scandent; like erect members of the family they can be readily distinguished by the sheathlike ochrea at the nodes. Prior to its rupture to form the ochrea, woody Polygonaceae have a distinctive Moraceae-like conical terminal stipule (this especially useful in a few *Coccoloba* species where the ochrea is poorly developed).

Marcgraviaceae — Characterized by hemiepiphytic habit and the usually succulent-coriaceous leaves, frequently dark-punctate and/or with the secondary venation reduced or suppressed; these often rolled around branchlet apex prior to expansion and resembling the terminal stipule of Moraceae. The distinctive juvenile form of *Marcgravia* has overlapping leaves appressed against a tree trunk which it climbs by adventitious roots (cf., juvenile *Monstera*).

Ericaceae — Some Ericaceae climbers are only slightly or not at all 3-nerved; they are mostly strongly coriaceous and the vines are hemiepiphytic, usually with loose fibrous bark.

Dichapetalaceae — Most area *Dichapetalum* climbers have nondescript pinnately veined leaves (usually with characteristic but inconspicuous early caducous stipules), the best vegetative character being the smooth usually light-colored trunk with conspicuously raised darker, round lenticels (cf., *Valeriana,* but restricted to lowland moist forest).

Icacinaceae — Two of the three vine genera have pubescent sometimes rather asperous leaves that are strongly raised-reticulate below with more scalariform-parallel tertiary veins than in *Dichapetalum;* the third is very common *Leretia cordata* with nondescript grayish-drying leaves with conspicuously reticulate ultimate venation and, at least, a few T-shaped trichomes on the lower surface, less coriaceous and the petiole shorter than in similar Convolvulaceae.

(Basellaceae) — Succulent-leaved vines, the broadly ovate round-tipped leaves usually more or less plinerved at base (see 3-veined taxa above), but *Anredera* has acute to acuminate leaves with pinnate venation.

Compositae — Alternate-leaved scandent composites are mostly pinnately veined, lack latex, and tend to be rather nondescript. Many have the typical composite odor, but *Senecio* relatives often do not. Senecioneae climbers often have hollow branchlets with fine longitudinal striations; Vernonieae climbers usually have a thin blackish layer in the inner bark similar to trees of that tribe, and the leaves tend to be grayish-pubescent below.

Plumbaginaceae (*Plumbago*) — Herbaceous vines restricted to dry forest or cloud forest, characterized by the blade attenuate onto a poorly differentiated petiole which is more or less expanded and clasping at base.

Solanaceae — Scandent species of *Solanum* recognizable by stellate trichomes and frequent presence of prickles on leaves and branchlets; other climbing genera are mostly +/- hemiepiphytic, typically with elliptic to obovate +/- coriaceous clustered leaves, and mostly occur in cloud forest. Except for hemiepiphytic tendency, climbing Solanaceae are mostly nondescript and difficult to recognize although the rather few widely separated secondary veins and sometimes presence of branched trichomes or tomato-like odor can be useful characters.

Convolvulaceae — The woody lianas mostly lack latex and are characterized by anomalous stem cross sections with interrupted rings of secondary xylem. Most likely to be confused with Solanaceae on one hand and Icacinaceae on the other. The leaves of the nonlactiferous genera are more or less recognizable from the unusually long eglandular, epulvinate, often sericeous petiole and the oblong leaf, either +/- sericeous below or very coriaceous.

Polygalaceae — Leaves coriaceous, usually with prominulous venation (except *Moutabea* which has secondary and tertiary veins immersed but is distinctive in the inconspicuous prickles on the branchlets). Polygalac lianas have anomalous stem cross sections with more or less interrupted concentric rings of secondary xylem, similar to Convolvulaceae, but the leaves have shorter petioles; the secondary veins are less strongly ascending and the apex less acute than in *Leretia. Securidaca* is distinctive in the sensitive greenish-olive twigs that make tendril-like twists (cf., opposite-leaved Hippocrateaceae).

(Phytolaccaceae) — *Trichostigma octandra* is a nondescript liana with membranaceous black-drying leaves; stem with anomalous broken rings of secondary xylem like Convolvulaceae and Polygalaceae, but *Trichostigma* differs from liana members of both in either thinner more membranaceous leaves and/or nonprominulous tertiary venation.

Boraginaceae — Most of the liana species of *Tournefortia* are utterly nondescript and especially likely to be confused with *Trichostigma* and *Chamissoa.* A few common species have distinctive long dark-brownish trichomes. The leaves of lianas are noncoriaceous but often subsucculent, differing from *Chamissoa* in shorter petioles with nonflexed bases; from Phytolaccaceae lianas in lacking anomalous stem section, and unlike thin-leaved Polygalaceae in having prominulous tertiary venation.

Amaranthaceae — *Chamissoa,* the only amaranth climber with alternate leaves, is vegetatively nondescript with membranacous narrowly ovate leaves. Most likely to be confused with *Tournefortia* from which it differs in longitudinally finely striate, usually hollow twigs and especially in the long slender petiole with flexion at base.

(Olacaceae) — *Heisteria scandens,* our only liana Olacaceae has an indistinct tendency to 3-veined leaf bases and also usually has a faint trace of latex in the petioles. Otherwise it can be recognized by the curved greenish petiole slightly thickened toward the apex. Frequently with scattered linear glandular area on midvein above.

(Dilleniaceae) — A few dilleniac species (mostly *Doliocarpus*) have virtually entire mature leaf margins. Unfortunately, most of these same species also have nonasperous leaves; however, they are obviously Dilleniaceae in the reddish, papery outer bark, and the typical leaf shape with more or less cuneate base and usually numerous straight secondary veins.

(Onagraceae) — A few more or less scandent *Fuchsia,* typically characterized by +/- papery reddish bark and/or red petioles when fresh, have alternate leaves.

SPECIAL HABIT LISTS AND FIGURES

1. Leafless achlorophyllous parasites and saprophytes (Figure 1)

Small terrestrial herbs

Triuridaceae (**A**, *Sciaphila;* **B**, *Triuris*)
Burmanniaceae (**C-D**, *Burmannia;* **E**, *Dictyostega;* **F**, *Thismia;* **G**, *Gymnosiphon*)
Orchidaceae (very few) (e.g., **H**, *Wullschlaegelia*)
Gentianaceae (**J**, *Voyria*)
Pyrolaceae (**I**, *Monotropa*)
Balanophoraceae (e.g., **L**, *Corynaea;* see also Figure 62)

Vines

Cuscuta (Convolvulaceae)
Cassytha (Lauraceae)

Apparently epiphytic

Rafflesiaceae (e.g., **K**, *Apodanthes*)
(Loranthaceae) — Most have leaves and all have chlorophyll but are true parasites nevertheless.

2. Leafless plants with chlorophyllous stems (excluding parasites) (sometimes with small early caducous leaflets)

Ephedraceae
Cactaceae (except *Pereskia*)
Baccharis trimera (Compositae)
Muehlenbeckia platyclada (Polygonaceae)
Rhamnaceae (*Scutia, Colletia*)
Leguminosae (introduced *Spartium junceum,* extralimital *Prosopis kuntzei,* some extralimital *Cassia;* some *Parkinsonia* with leaves reduced to rachises)
Zygophyllaceae (*Bulnesia retama*)
Koeberlinia (Koeberliniaceae; Bolivia only)

Figure 1

Parasites and Saprophytes

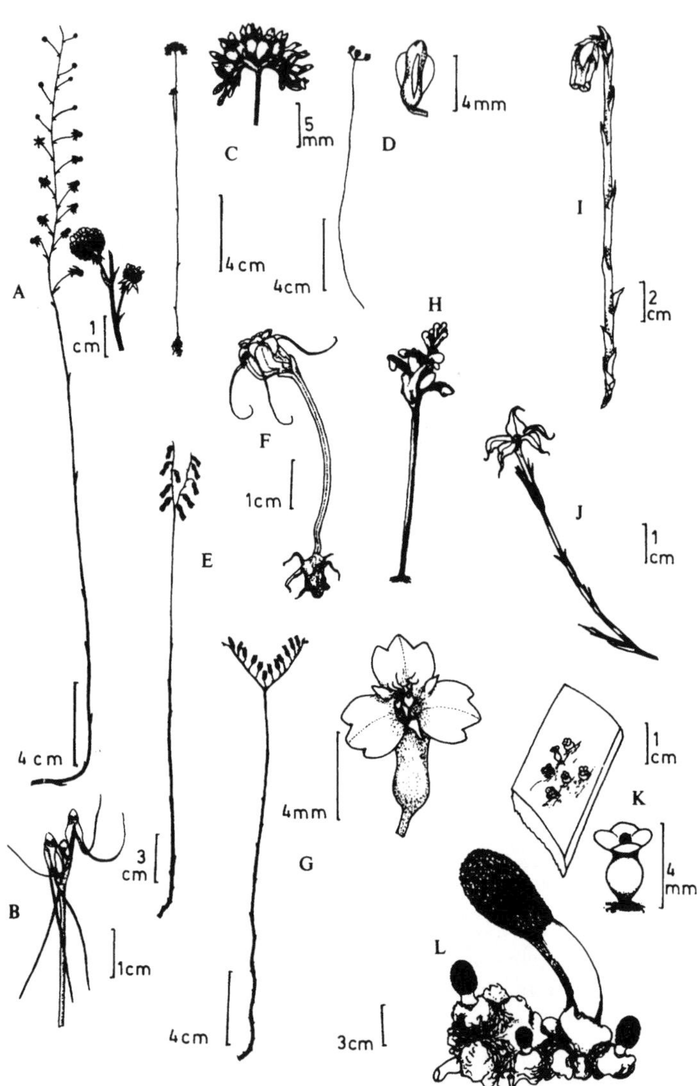

A - *Sciaphila* **C** - *Burmannia* **D** - *Burmannia* **I** - *Monotropa*

E - *Dictyostegia* **F** - *Thismia* **H** - *Wullschlaegelia* **J** - *Voyria*

G - *Gymnosiphon* **K** - *Apodanthes*

B - *Triuris* **L** - *Cornynaea*

3. Tendrillate Lianas — Only ten families in our area (plus extralimital but escaped *Antigonum*) include tendrillate lianas, although several other families have tendril-like climbing mechanisms. Lianas with tendrils are extremely easy to identify to family, as follows. **(Figure 2)**

I. Compound-leaved

A. Terminal leaflet or rachis apex converted into tendril

a. Opposite leaves — Bignoniaceae **(F)**

b. Alternate leaves

Leguminosae (*Vicia*, simply pinnate; *Entada*, bipinnate), *Cobaea* (Polemoniaceae), *Mutisia* (Compositae)

B. Inflorescence-derived tendril arising from nodes of branchlets

a. Forked tendril, stem often with latex — Sapindaceae **(B)**

b. Much-coiled variously branching tendril; stem without latex — *Psiguria* and few *Gurania* and *Siolmatra* (Cucurbitaceae), few *Cissus* (Vitaceae).

II. Simple-leaved

Gouania **(D)** (Rhamnaceae) — butterfly-tongue tendril coiled in one plane

Passifloraceae **(H)** — axillary tendril

Cucurbitaceae **(E)** — tendril at right angles to petiole base

Vitaceae **(I)** — tendril opposite petiole base

Smilacaceae **(C)** — pair of stipule-derived tendrils arising from petiole

Antigonum (Polygonaceae) (not illustrated) — inflorescence-derived tendril with coiling apical part arising from zigzag straight-segmented basal part

Tendril-like climbing mechanisms

Tendril-like branchlet apex — e.g., *Omphalea* **(A)** (Euphorbiaceae).

Tendril-like sensitive lateral branchlets — Hippocrateaceae, Malpighiaceae, Connaraceae, *Securidaca* (Polygalaceae)

Sensitive petiole — Tropaeolaceae

Woody hooklike tendrils — *Bauhinia* (Leguminosae), *Strychnos* **(G)** (Loganiaceae)

Figure 2

Tendrillate Lianas

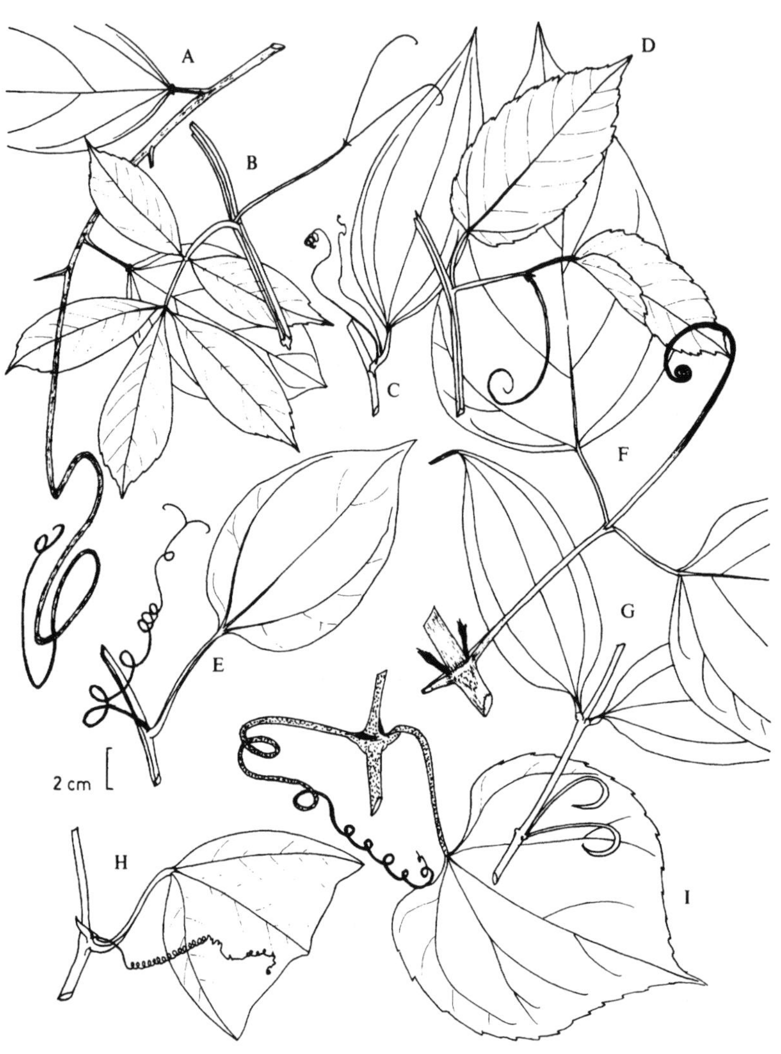

A - *Omphalea* (Euphorbiaceae)

B - Sapindaceae **C** - Smilacaceae **D** - *Gouania* (Rhamnaceae)

F - Bignoniaceae

E - Cucurbitaceae **G** - *Strychnos* (Loganiaceae)

H - Passifloraceae **I** - Vitaceae

4. Anomalous liana stem cross sections (Figure 3)

(1) Regular structure, no obvious growth rings
Many families, e.g., *Dilkea* **(V)** (Passifloraceae), some *Stigmaphyllon* **(S)** and related genera (Malpighiaceae), *Ampelozizyphus* **(P)** (Rhamnaceae), some *Paullinia* **(AA)** (Sapindaceae)

(2) Regular structure with stem center hollow and/or with exudate
Combretaceae (**G**, *Combretum*)

(3) "Islets" of secondary phloem scattered in xylem
Strychnos **(Q)**

(4) Regular growth rings with strong radial rays
Hippocratea **(I)** (wind-dispersed Hippocrateaceae), *Doliocarpus* **(K)** (Dilleniaceae)

(5) Very large vessels and soft thick inner bark layer
Cissus **(CC)** (Vitaceae)

(6) Regular concentric cylinders of successive cambia
Odontadenia **(A)** (Apocynaceae) (with white latex), *Machaerium* **(M)** (Leguminosae) (with red latex), Menispermaceae **(T)** (often eccentric), Phytolaccaceae (e.g., **X**, *Seguieria*), Polygalaceae (e.g., **Y**, *Moutabea*)

(7) Broken concentric cylinders of successive cambia
Convolvulaceae (e.g., **H**, *Maripa*), indehiscent-fruited Hippocrateaceae (e.g., **N**, *Salacia*)

(8) Xylem cylinder dissected but complete

- **a.** Irregularly slightly dissected — *Passiflora* **(W)** (some Malpighiaceae)
- **b.** Irregularly deeply dissected — *Plukenetia* **(L)** (Euphorbiaceae) (and some Malpighiaceae and *Passiflora*)
- **c.** With 6 phloem arms, soft stem and thick bark
  Aristolochia **(B)** (Aristolochiaceae)
- **d.** With 4 (or multiples of 4) phloem arms
  Bignoniaceae (e.g., **D**, *Callichlamys;* **E**, *Clytostoma;* **C**, *Parabignonia*

(9) Stem strongly flattened
Bauhinia **(O)** (Leguminosae), the classical "monkey's ladder", also in some other legumes including *Dalbergia, Machaerium,* and some mimosoid climbers, *Coccoloba* **(Z)** (2-parted, also in some *Solanum*), Menispermaceae **(U)** (eccentric rings)

Figure 3

Liana Cross Sections

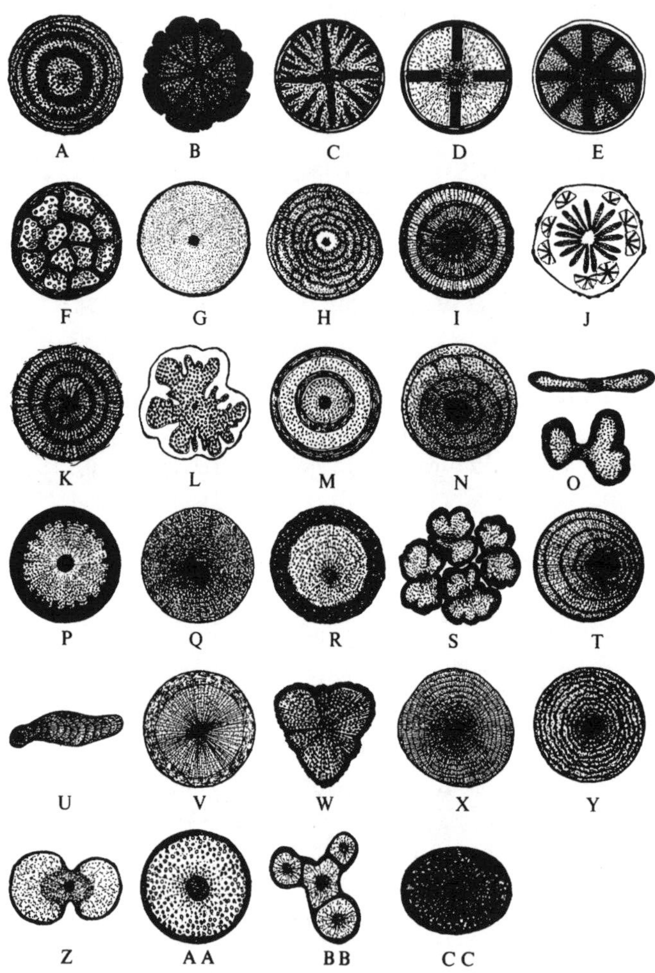

Figure 3 Legend

A - *Odontadenia* **B** - *Aristolochia* **C** - *Parabignonia* **D** - *Callichlamys* **E** - *Clytostoma*

F - *Mendoncia* **G** - *Combretum* **H** - *Maripa* **I** - *Hippocratea* **J** - *Gurania*

K - *Doliocarpus* **L** - *Plukenetia* **M** - *Machaerium* **N** - *Salacia* **O** - *Bauhinia*

P - *Apelozizyphus* **Q** - *Strychnos* **R** - *Stigmaphyllon* **S** - *Mascagnia* **T** - *Menispermaceae*

U - *Menispermaceae* **V** - *Dilkea* **W** - *Passiflora* **X** - *Seguiera* **Y** - *Moutabea*

Z - *Coccoloba* **AA** - *Paullinia* **BB** - *Paullinia* **CC** - *Cissus*

(10) Intact stem extremely irregularly dissected internally

- **a.** Soft stem with combination of radial and peripheral dissection Cucurbitaceae (e.g., **J**, *Gurania*)
- **b.** Soft stem with multiple small separate irregular cylinder *Mendoncia* **(F)** (Acanthaceae)

(11) Xylem cylinder completely divided into individual subunits

- **a.** Triangular stem with 3 accessory xylem cylinders Sapindaceae (many *Paullinia* **[BB]** and *Serjania*)
- **b.** Irregular dissection Malpighiaceae (e.g., **S**, *Mascagnia*), few Bignoniaceae (e.g., some *Mansoa*)

5. Some common whole-tree branching patterns (Figure 4)

(1) Branches more or less in whorls — (branching monopodial with rhythmic growth); leaves 2-ranked (= myristicaceous branching), (**B**, *Iryanthera* [Myristicaceae])

(2) Pagoda-style (or candelabra) branching — e.g., Combretaceae (**F**, *Buchenavia;* **G**, *Terminalia catappa*)

(3) Dichotomous branching with trunk elongation from new leaders — (branching sympodial by substitution); leaves (or flowers) in fork of bifurcation (**C**, *Cordia* [Boraginaceae]).

(4) Whorled branching with trunk elongation from new leaders — (branching sympodial by apposition)

- **a.** Branching of lateral branches alternate or lacking (**A**, *Mabea* [Euphorbiaceae])
- **b.** First branching of lateral branches bifurcating (**D**, *Theobroma* [Sterculiaceae])

(5) Unbranched pachycaul growth form — (**E**, *Clavija* [Theophrastaceae])

Tree Habits

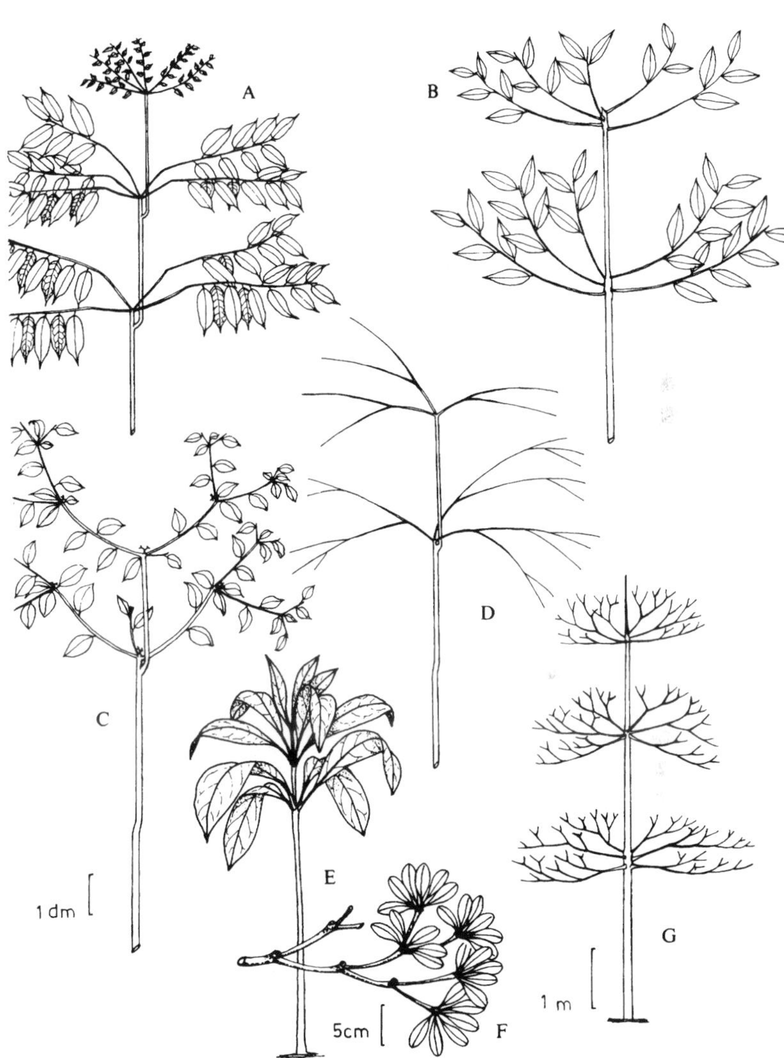

A - *Mabea* **B** - *Iryanthera*

C - *Cordia* **D** - *Theobroma*

E - *Clavija* **F** - *Buchenavia* **G** - *Terminalia catappa*

6. Ant domatia (Figure 5)

(1) On petioles or leaf base
Melastomataceae: *Tococa* **(B)**, *Maieta* **(C)**, some *Clidemia*
Chrysobalanaceae: *Hirtella* **(A)**
Leguminosae: *Sclerolobium, Tachigali* **(F)**
Rubiaceae: *Duroia saccifera* (not illustrated)

(2) Swollen twigs or nodes
Rubiaceae: *Duroia hirsuta* **(E)**
Boraginaceae: *Cordia nodosa* **(G)**

(3) Swollen thorns
Leguminosae: *Acacia* **(D)**

7. Stilt roots

(1) Genera regularly characterized by stilt roots
Rhizophoraceae (*Rhizophora*)
Guttiferae (*Clusia, Symphonia, Dystovomita, Tovomita*)
Piperaceae (*Piper*)
Moraceae (*Pourouma, Cecropia, Ficus* [stranglers])
Palmae (tribe Iriarteae: *Socratea, Iriartea, Dictyocaryum, Wettinia, Iriartella*)
Euphorbiaceae (*Micrandra*)
Alzateaceae (*Alzatea*)

(2) Genera with occasional stilt-rooted species
Chrysobalanaceae (*Licania heteromorpha* and allies)
Elaeocarpaceae (*Sloanea* aff. *latifolia*)
Melastomataceae (few *Miconia*?)
Annonaceae (*Oxandra espintana*)

Ant Domatia

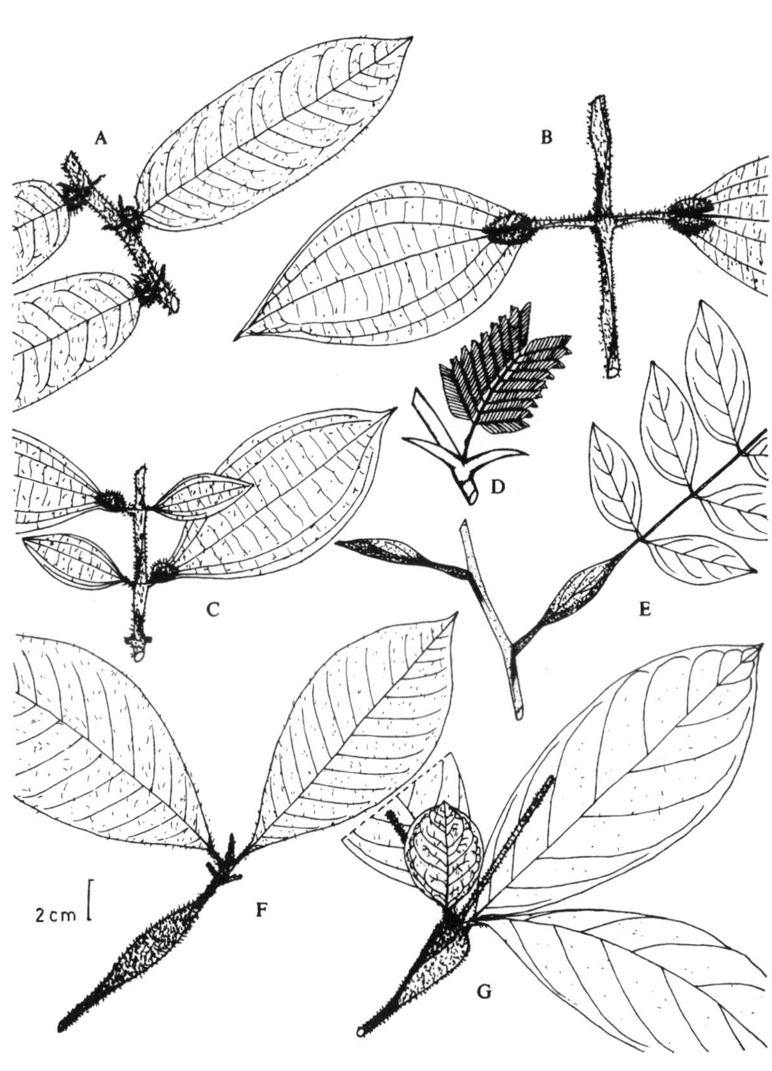

A - *Hirtella* **B** - *Tococa*

C - *Maieta* **D** - *Acacia* **E** - *Tachigali*

F - *Duroia hirsuta* **G** - *Cordia nodosa*

8. Spines — Mostly useful at level of individual species except for the genera indicated with * and the whole family Cactaceae.

(1) Lianas

a. Trunk spines

*Smilax**
Dioscorea
*Desmoncus** (actually on sheathing petiole base)
Paullinia (few)
Strychnos panamensis (and few others)
Passiflora spinosa
Combretum decandrum
Leguminosae: *Acacia**, *Piptadenia**, *Machaerium**
*Pereskia** (and other climbing Cactaceae)
*Seguieria**
Pisonia

b. Twig and/or leaf spines

Leguminosae: scandent *Caesalpinia, Machaerium,* most mimosoid climbers
Solanum
*Moutabea**
*Smilax**
*Seguieria**
Sagaretia
Pisonia aculeata
Ophellantha
Byttneria
Celtis iguanea
Rubiaceae: *Uncaria**, *Randia**, *Chomelia*
Lantana
*Rubus**

c. "Grappling Hook" apices of compound leaves

*Desmoncus**

(2) Trees

a. Trunk spines

Compositae (*Barnadesia** and relatives)
Bombacaceae (*Ceiba**, *Chorisia**, *Spirotheca**, few *Pochota*)
Caricaceae (*Jacaratia*)
Solanaceae (*Solanum*)
Apocynaceae (*Lacmellea* [varies within species])
Rutaceae (*Zanthoxylum**)
Flacourtiaceae (*Xylosma,* the spines strongly branched in *X. benthamii*)
Urticaceae (*Urera laciniata, U. baccifera*)
Palmae (*Bactris**, *Astrocaryum**, *Aiphanes**, *Acrocomia**, *Crysophila**, *Mauritiella**)

Tree ferns (*Cyathea, Alsophila**)
Leguminosae (*Acacia**, *Piptadenia**, *Erythrina** few tree *Machaerium,* extralimital *Gleditsia*)
Cactaceae*
Araceae (*Montrichardia*)
Malvaceae (*Wercklea*)
Euphorbiaceae (*Hura*)

b. Branch-tip spines at ends of twigs or short-shoots

Euphorbiaceae (*Adelia*)
Rosaceae (*Hesperomeles**)
Burseraceae (*Bursera orinocensis*)
Annonaceae (*Annona punicifolia*)
Celastraceae (*Schaefferia**)
Rhamnaceae (*Colletia**, *Scutia**, *Condalia*)
Sapotaceae (*Sideroxylon*)
Solanaceae (*Grabowskia**, *Lycium**)
Scrophulariaceae (*Basistemon*)
Anacardiaceae (*Schinus*)
Geraniaceae (*Rhynchotheca*)
Zygophyllaceae (*Porliera*)

c. Twig and/or leaf spines

Berberidaceae (*Berberis*)
Boraginaceae (*Rochefortia**)
Compositae (Mutisieae: *Barnadesia**, *Arnaldoa, Chuquiraga, Dasyphyllum, Fulcaldea*)
Cycadaceae (*Zamia**)
Ferns (Tree ferns: (*Cyathea**, *Alsophila**)
Flacourtiaceae (*Xylosma,* few *Casearia*)
Gramineae (*Guadua**)
Hydrophyllaceae (*Hydrolea*)
Leguminosae (*Acacia**, *Piptadenia**, *Pithecellobium, Prosopis**, *Caesalpinia**, *Machaerium**, *Bauhinia, Adesmia,* etc.)
Moraceae (*Maclura, Poulsenia**)
Nyctaginaceae (*Pisonia macranthocarpa*)
Olacaceae (*Ximenia**)
Phytolaccaceae (*Achatocarpus**)
Rhamnaceae (*Zizyphus**, *Colubrina*)
Rubiaceae (*Chomelia**, *Randia**)
Santalaceae (*Acanthosyris*)
Simaroubaceae (*Castela**)
Solanaceae (*Solanum**)
Ulmaceae (*Celtis*)
Verbenaceae (*Duranta*)

9. Strongly fenestrated trunks

Apocynaceae — *Aspidosperma* series *Nitida*
Celastraceae — *Perrottetia sessiliflora*
Leguminosae — *Inga neblinensis, Caesalpinia, Platypodium*
Olacaceae — *Minquartia* (one form)
Rubiaceae — *Amaioua, Macrocnemum*
Sapotaceae — few *Pouteria* spp.

10. Stranglers

Moraceae — *Ficus, Coussapoa*
Guttiferae — *Clusia* (?)
Marcgraviaceae — *Souroubea* (?)
Cunoniaceae — *Weinmannia* (1 sp.)
Bombacaceae — *Spirotheca*
Alzateaceae — *Alzatea*(?)

FERNS (MOSTLY CYATHEACEAE)

Tree ferns are unmistakable in their large much-divided leaves which unfurl from fiddleheads and usually have scales and/or prickles at base of petiole. Although easy to recognize as ferns, generic taxonomy of the tree ferns is in a state of flux, with nearly all of the species at one time or another included in a *sensu lato Cyathea.* The various attempts at generic segregation depend heavily on the nature of the scales of the petiole base and young fiddleheads and on whether the sporangia are borne in covered (= indusiate) sori. One taxonomic circumscription treats the tree ferns with petiole scales uniform in texture throughout (i.e., without distinctive margins) as *Sphaeropteris,* those with distinctly marginate scales which have a dark apical seta as *Alsophila* and *Nephelea,* and those with distinctly marginate scales lacking a dark apical seta as *Trichipteris* (lacking an indusium), *Cyathea* (indusiate and the veins of the leaf segments free) and *Cnemidaria* (indusiate and the veins of the leaf segments forming areolae or fused to veins of adjacent segment). Most authors currently combine *Nephelea* with *Alsophila* and *Trichipteris* with *Cyathea.*

In addition to the true tree ferns, there are a number of climbing ferns some of which are more or less woody. The most important of these is *Polybotrya,* climbing appressed to tree trunks, which has dimorphic fertile fronds. Other climbing genera, like *Lygodium* and *Salpichlaena,* are slender vines which are actually single straggling leaves with a twining rachis).

Although ferns lack secondary growth and therefore are not truly woody, they often have well-developed sclerenchymatous support tissue and tree ferns can reach 20 cm or more in trunk diameter and heights of 10 to 15 meters. It is worth noting that tree ferns can often be identified to species in sterile condition, especially when complete descriptions of leaf size, pinnae number, and stem characteristics are available, since the scales and spines of petiole base and young fiddleheads provide many of the specific recognition characters; for such identification, the petiole base, frond apex, and at least one medial pinna are usually necessary.

Although there are also very many epiphytic (as well as terrestrial) fern genera, these are not treated here. The largest epiphytic genera are *Asplenium* (400/650 spp. epiphytic, worldwide), *Grammitis* (400/400 spp. worldwide), *Elaphoglossum* (250/500 spp. worldwide), *Ctenopteris* (200/200 spp. worldwide), *Polypodium* (140/150 spp. worldwide), and the filmy ferns *Hymenophyllum* (250/300 spp. worldwide) and *Trichomanes* (140/150 spp. worldwide).

1. Tree Ferns

Sphaeropteris (15 spp.) — Distinctive in having nonmarginate scales, uniform in texture throughout.

Alsophila (incl. *Nephelea*) (30 spp.) — Like *Cyathea* in marginate scales but the scales with a dark apical seta. Other distinctive characters are the often dark pinnae rhachises and the lower surface of the leaf lamina lighter green than the upper. Differing from sometime segregate *Nephelea* in the unfurling young leaves lacking spines and the petiole bases nonspiny or with blunt or scale-tipped spines rather than having large black spines tapering to slender apex.

Cyathea (incl. *Trichipteris*) (60 spp., plus 200 Old World) — The main genus of tree ferns. Vegetatively characterized by marginate scales lacking a dark apical seta; differs from sometime segregate *Trichipteris* in having an indusium or "lid" over the sori and from *Cnemidaria* in having free veins not forming areoles. *Trichipteris* cannot be maintained since its defining character, loss of the indusium, has occurred independently in different evolutionary lines.

Cnemidaria (27 spp.) — Scales similar to *Cyathea* in lacking a dark terminal seta but distinctive in having a dark central stripe and wide contrasting whitish margins; also differs in the veins of the leaf segments joining to form areoles or free but with basal veins of adjacent segments fused to sinus. It is also the only tree fern genus with the upper surface of the rachises glabrous and usually has less-divided fronds than *Cyathea* and *Alsophila* (the lamina only once-pinnate [though with pinnatifid lamina segments]); the stem is usually short and the plant thus less arborescent than in the other genera.

Blechnum (ca. 50 spp., plus 100 Old World) — Only distantly related to the other tree ferns. However, some species of this mostly herbaceous genus become definitely arborescent, especially in moist paramos, where their conspicuous cycadlike growth-form makes them one of the most characteristic elements. The leaves are once-pinnate unlike Cyatheaceae and the treelike species all have dimorphic fertile fronds.

Dicksonia (1 sp., plus 20 Old World) — Differs from *Cyathea* relatives by having dense hairs instead of scales and in margined sori.

2. Climbing Ferns

2A. Woody hemiepiphytes with straight rachis, climbing appressed to supporting tree

Polybotrya (35 spp.) — Leaves usually more divided than in other hemiepiphytic climbing ferns, the lamina 1–4-pinnate with tapered pinnatifid apex; also distinctive, even when simply pinnate, in the veins in pinnate groups.

Lomariopsis (25 spp.) — Differs from *Polybotrya* in the lamina uniformly 1-pinnate and with an undivided apex; the veins close together and nearly parallel, rather than pinnately arranged.

Lomagramma (1 spp., plus 20 Old World) — Lamina uniformly 1-pinnate and with an undivided apex as in *Lomariopsis,* but differing in the veins areolate rather than long-parallel.

2B. Climbing by slender twining leaf rachises

Lygodium (7 spp., plus 23 Old World) — Differs from *Salpichlaena* in having the sori borne on modified lobes at the tips of the leaflet segments.

Salpichlaena (3 spp.) — Differs from *Lygodium* in having the sori borne in lines along both sides of the costa.

Cycadaceae

Palmlike understory (one Chocó species epiphytic) gymnosperms, poorly represented in our area by a single genus, usually essentially stemless or with short thick stem crowned by ring of pinnately compound leaves. The pinnately compound leaves are somewhat similar only to palms (and a few ferns), differing from the former in being more coriaceous and having the parallel leaflet veins all equal. The petioles and rachises are frequently more or less spiny but the spines are shorter than in pinnate-leaved palms and have more conspicuously thickened bases. Most species rarely encountered in fertile condition, but the large thick strobili, when present, are utterly distinctive.

Zamia (30 spp.)

Chigua (2 spp.) — Western Colombia. Differs from *Zamia* in the leaflets having a distinct midrib.

(Several other genera occur in Mexico and Cuba.)

Podocarpaceae and Ephedraceae

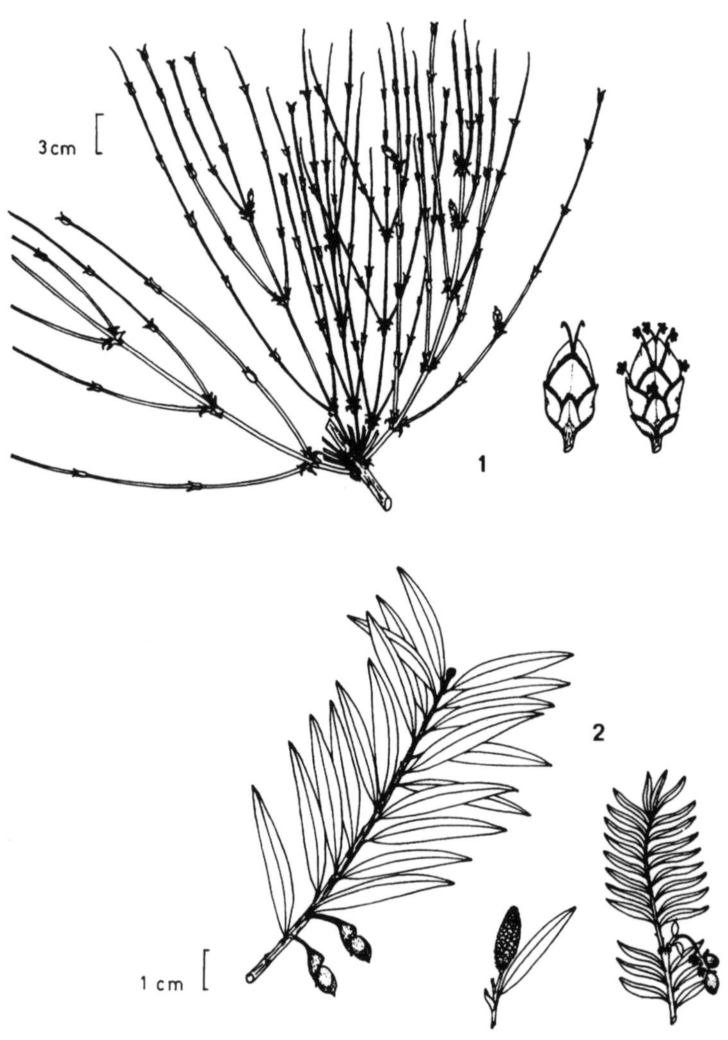

1 - *Ephedra*

2 - *Podocarpus*

Cycadaceae and Gnetaceae

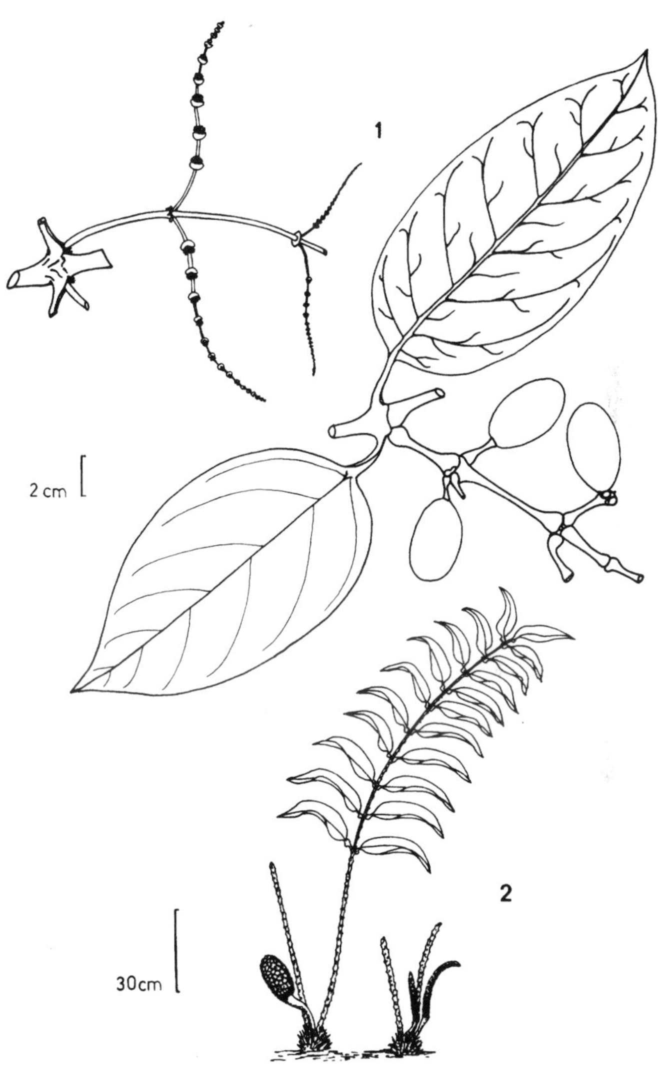

1 - *Gnetum*

2 - *Zamia*

Ephedraceae

Leafless shrubs or subshrubs (puna), in our area restricted to the Andes, mostly in dry, inter-Andean valleys. The segmented greenish stems are conspicuously jointed with opposite or verticillate branching, each node with a sheath formed from the fused bases of the opposite or whorled scales. When fertile, the small sessile or subsessile ovoid or subglobose strobili borne in axils of the scales. None of the leafless angiosperms in our area has opposite scales.

Ephedra (40 spp., incl. Old World and temperate N. Am.)

Gnetaceae

There is nothing in the large opposite leaves of this canopy liana to suggest to the uninitiated that it is a gymnosperm. In fact the very coriaceous, uniformly entire leaves are quite like those of *Salacia* (Hippocrateaceae), and also dry a similar olive color. The best distinguishing character is the conspicuously jointed branchlet; the opposite petiole bases are continuous across the more or less thickened node, with the next segment of the branchlet emerging from the node to leave a conspicuously articulate joint. The stem is also distinctive in having concentric rings in cross section (cf., Menispermaceae) and exudes a characteristic jellylike resin when cut. "Fruits" oblong-ellipsoid, 3–5 cm long, orangish or reddish when mature, not at all gymnosperm-like. Male inflorescence very distinctive, resembling a short string of beads, the "beads" immediately adjacent or widely spaced, depending on the species.

Gnetum (6 spp., plus 24 Old World).
P: palo huayo, hambre huayo

Podocarpaceae

One of the most characteristic tree genera of upland Andean forests, easy to recognize by the small, thick, coriaceous, linear-oblong leaves with a strong midvein but lacking secondary venation (or with several more or less faint longitudinal veins). The only possibility for confusion would be that the few species with longer strap-shaped leaves somewhat resemble a few *Xylopia* species with narrow leaves and indistinct secondary venation, but the leaves of these are much less coriaceous. The "fruit" of most species is a naked seed borne at the end of a fleshy receptacle that turns red at maturity; in a few species (the segregates *Decussocarpus* and *Prumnopitys*) the receptacle expands to cover the seed and the "fruit" is drupaceous.

Podocarpus (37 spp., plus many in Old World) — Mostly very large trees, formerly often forming nearly pure stands in montane forest, but most species now nearly eliminated; a few species occasionally reaching sea level as widely scattered individuals in broad-leafed forest (in Amazonia only on white sand). One group (with a drupaceous fruit and sometimes segregated as *Decussocarpus*) has very short, relatively broad, opposite, sharp-pointed leaves all in the same plane and with several +/- parallel longitudinal veins; the majority of species have spirally arranged leaves, these small (<3 cm long and 3 mm wide), usually relatively obtuse apically, and lacking hypodermis in *P. montanus* and allies (with a drupaceous fruit and sometimes segregated as *Prumnopitys*), usually larger (and with the fruit naked at end of a fleshy receptacle) in *Podocarpus sensu stricto.*

C: pino hayuelo, romerón, chaquiro; P: diablo fuerte, ulcumamo, romerillo

AGAVACEAE

A family closely related to Liliaceae and Amaryllidaceae from both of which it is differentiated by the stiff, thick, fibrous, usually succulent leaves, typically with spiny apices or prickles along margin. Our two genera, *Furcraea,* with superior-ovaried flowers borne singly on a loose open inflorescence, and *Agave,* with inferior-ovaried flowers borne in characteristic flat-topped clusters aggregated near apex of huge inflorescence, have traditionally been placed, respectively, in Liliaceae and Amaryllidaceae, on account of their respective ovary position. Vegetatively, they could only be confused with terrestrial Bromeliaceae, but their leaves are far broader and more succulent than in any Bromeliaceae. In our area Agavaceae, much better represented in Mexico and the Antilles, occurs only in dry, inter-Andean valleys.

Furcraea (20 spp.) — Differs from *Agave* in the openly paniculate inflorescence of singly borne flowers. The succulent spiny-margined leaves are less bluish than in *Agave.*

Agave (300 spp.) — Differs from *Furcraea* in the very typical inflorescence with the flowers densely clustered together into flat-topped bunches which are borne near apex of the giant inflorescence. Vegetatively, the bluish color of the leaves differentiates our species from *Furcraea.*

C: motua, penca

(*Dracaena*) — In the Paleotropics speciose and sometimes becoming large trees, but very poorly represented in the Neotropics by a single slender-pachycaul Central American species and occurring in our area only in cultivation.

AMARYLLIDACEAE

An entirely herbaceous family, mostly lacking an above-ground stem, the strictly parallel-veined strap-shaped leaves borne from an underground bulb from the apex of which is borne a rather succulent inflorescence stalk which produces a terminal umbel (or reduced to 1–few flowers) of often spectacular lilylike flowers. The flowers differ from Liliaceae primarily in having inferior ovaries; in addition a corona (which may arise from fused filaments or from petal outgrowths) is often present in Amaryllidaceae but never in Liliaceae. Inclusion of this family in this volume is mandated by *Bomarea,* the largest Amaryllidaceae genus in our region, which is mostly scandent. In addition to a few *Bomarea* species, one

Agavaceae and Amaryllidaceae

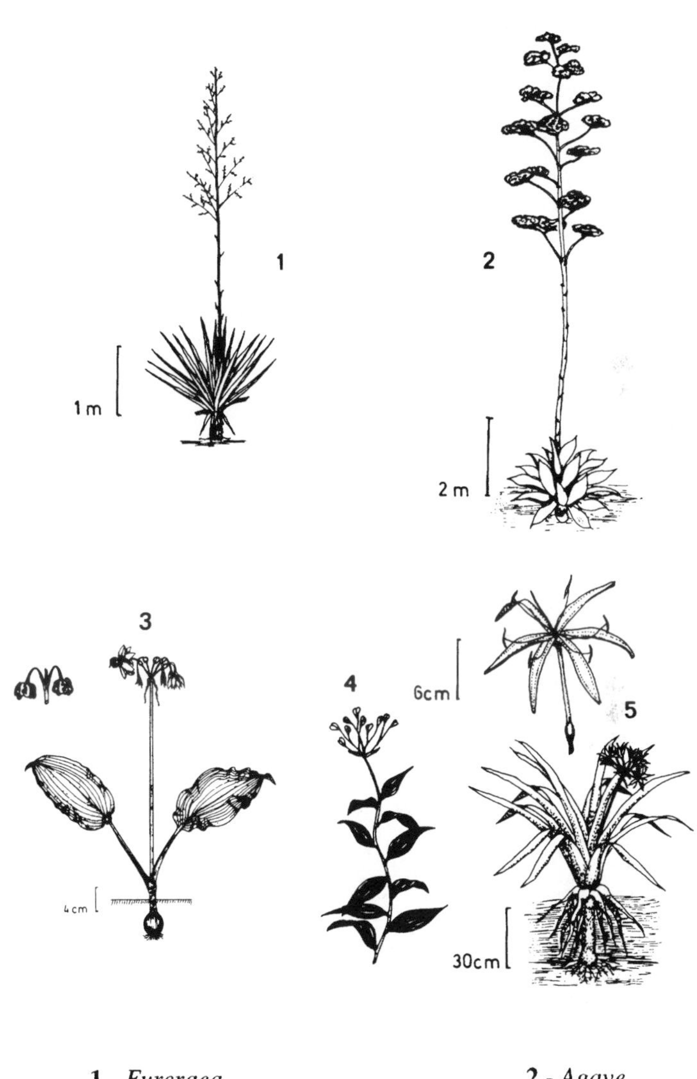

1 - *Furcraea* **2** - *Agave*

3 - *Eucharis* **4** - *Bomarea* **5** - *Crinum*

other genus, *Eucharis,* occurs inside lowland tropical forest, several genera occur in open (usually swampy) lowland areas outside the forest, and many genera occur in dry and/or upland Andean areas.

Bomarea (150 spp.) — A very characteristic cloud-forest and high-Andean scandent genus with the separate conspicuous orange or red tepals (usually conspicuously maroon-dotted inside) forming a more or less flaring tube. The stem is twining, the strictly parallel-veined leaves vary from linear to narrowly elliptic, and the 3-valved fruit is a round dark brown or blackish capsule that dehisces incompletely to reveal conspicuous bright red or red-orange seeds.

Eucharis (10 spp.) — The only Amaryllidaceae, except for a few ecologically atypical species of *Bomarea,* to occur in lowland tropical forest (a few other genera may occur in open areas outside the actual forest). Very characteristic in its large white daffodil-like flower with well-developed corona.

Araceae

Araceae is the predominant family of herbaceous hemiepiphytic climbers; many are also terrestrial or swamp herbs, one genus (*Pistia*) is a floating aquatic (extralimital *Jasarum* is a submerged aquatic with linear leaves!), and one genus (*Heteropsis*) is a slender distinctly subwoody climber. Vegetatively the family can usually be recognized by the large often succulent leaves borne on +/- succulent petioles with +/- sheathing bases. The inflorescence, which consists of a fleshy cylindrical spadix subtended by a single leaflike spathe, is unmistakable.

Vegetative differentiation of the genera is usually possible. The leaves of the largest genus, *Anthurium*, have a marginal collecting vein and +/- reticulate venation while those of *Philodendron,* the other very large genus, lack a collecting vein and have all lateral veins parallel. Scandent *Heteropsis,* the genus least obviously an Araceae when sterile, has the woodiest (though slender) stems and clearly alternate distichous elliptic leaves subsessile on short petioles and with *Clusia*-like straight parallel secondary veins and a marginal collecting vein. *Monstera* usually has the oblong leaves with naturally occurring holes (unique); if lacking these, the leaf blade tends to be deeply pinnately laciniate (the few similarly laciniate *Philodendron* species have deep basal lobes unlike *Monstera*). The leaves are deeply divided (frequently even compound) in epiphytic *Syngonium,* a few epiphytic species of *Philodendron* and *Anthurium,* and terrestrial *Dracontium, Xanthosoma* (some species), *Chlorospatha* (some species),

Asterostigma, and *Taccarum.* Among compound-leaved epiphytic taxa, only *Anthurium* (and one *Philodendron: P. goeldii*) can have multisegmented leaves; *Syngonium* usually has 5-foliolate leaves, these with net veins and marginal collecting veins like *Anthurium* but distinctive in milky latex and the lateral leaflets with strong basal lobes; the few 3-foliolate *Philodendron* species have completely parallel veins and no collecting vein. *Stenospermation* has oblong-elliptic leaves with hardly visible, strongly ascending, closely parallel secondary veins, whereas, other genera have obvious secondary veins. *Rhodospatha* is vegetatively very like *Philodendron* but has the petiole uniformly grooved or winged above right to base of lamina while *Philodendron* (except section *Pteromischum*) has the petiolar wings ending below base of lamina; the secondary veins are usually closer together and more numerous in *Rhodospatha* than in *Philodendron.* Two terrestrial genera have common species with spines or prickles: *Homalomena peltata* (stemless with inconspicuous prickles on petiole bases) and *Montrichardia* (treelike with spines on trunk). Other terrestrial genera include: *Spathiphyllum,* differing from *Anthurium* in closely parallel rather than reticulate leaf venation and a more conspicuously sheathing petiole base (also in the broad open spathe); *Dieffenbachia* (with succulent stems and petioles and a submilky, caustic, often foul-smelling sap) having oblong or elliptic parallel-veined leaves without basal lobes; *Xanthosoma,* very like *Syngonium* except for being terrestrial and less frequently compound-leaved, having net venation and strong basal lobes, often on +/- triangular leaves. Genera that usually grow in swamps include *Montrichardia* (almost treelike), *Urospatha* (sagittate-triangular leaves with mottled petioles), and *Spathiphyllum* (elliptic leaves with completely parallel veins); one genus, *Pistia,* is a floating aquatic which resembles a small head of lettuce.

When fertile, aroid genera are easy to distinguish. Most of the epiphytic genera (except *Philodendron* and *Syngonium*) have a uniform-thickness spadix of perfect flowers (in *Anthurium* this more or less free from the usually rather narrow spathe; in other genera, the spathe is broader and more or less enclosing spadix); *Philodendron* and *Syngonium* have a 2-parted spadix (bottom half with female flowers and cylindric, top half with male flowers and club-shaped) almost completely enclosed by the broad curved spathe; the spathe is caducous in *Syngonium* and *Xanthosoma,* persistent in *Philodendron, Dieffenbachia,* and *Homalomena. Stenospermation, Rhodospatha,* and *Monstera* have uniform spadices like *Anthurium,* differing from *Anthurium* in a large enclosing spathe like *Philodendron,* but this caducous, unlike *Philodendron. Monstera* and to a lesser extent *Stenospermation* have the pistils united into a rather pineapple-like syncarpous fruit very different from the individual berries of *Anthurium; Syngonium* and terrestrial *Montrichardia* and *Xanthosoma,* with inflorescence divided into separate male and female portions, also have the pistils fused into a syncarpous fruit. Among terrestrial genera, only three — *Dracontium* (with a single huge treeletlike deeply dissected leaf), *Spathiphyllum* (with elliptic cuneate-based leaves and a broad white leaflike spathe), and *Urospatha* (semiaquatic with triangular-sagittate leaves having well-developed basal lobes), plus some *Anthurium* species, have a uniform spadix of perfect flowers. *Dieffenbachia,* with parallel venation and no marginal vein, has the spathe entirely enclosing the spadix except during a very brief anthesis, with the fruiting inflorescence curling to split irregularly and reveal the rather separated orange-arillate seeds. *Xanthosoma,* with net venation and a marginal vein, has an infructescence-like *Philodendron; Caladium* and *Chlorospatha* are essentially reduced versions of *Xanthosoma.*

1. Floating Aquatic

Pistia (1 sp.) — A common and very distinctive free-floating lettuce-like rosette plant.

P: huama

2. Epiphytes

2. Epiphytes — Usually hemiepiphytic climbers with adventitious roots; most epiphytic aroids have the spadix uniform throughout (first five genera), only *Philodendron* and *Syngonium* having a 2-parted spadix.

2A. Spadix uniform throughout

Anthurium (750 spp.) — A very large and variable genus, mostly epiphytic (but seldom truly climbing), but also including numerous terrestrial species. Leaves distinctive in marginal collecting vein (except few species of section *Pachyneurium*) and more or less net venation. Spadix (at least in flower) and spathe both narrower than in other genera; fruits are individual berries, usually bright red or white.

C, E, P: anturio; P: jergón quiro (smallest spp.)

Heteropsis (12 spp.) — Slender rather woody climbers with very distinctive subsessile leaves having straight, close-together, prominulous, parallel secondary veins and a strong marginal collecting vein; spadix uniform; spathe caducous.

P: tamshi

Monstera (50 spp.) — More or less succulent hemiepiphytic climbers, the mature leaves usually with holes (or these reaching margin and the leaf deeply laciniate), the dimorphic juvenile leaves small, overlapping and often appressed against tree trunks. Spathe caducous; fruit +/- syncarpous, often edible, before maturity covered by the fused pistil apices.

E: camachillo, chirrivaca (*M. delicosa*); P: costilla de adán

Araceae
(Floating and Epiphytic)

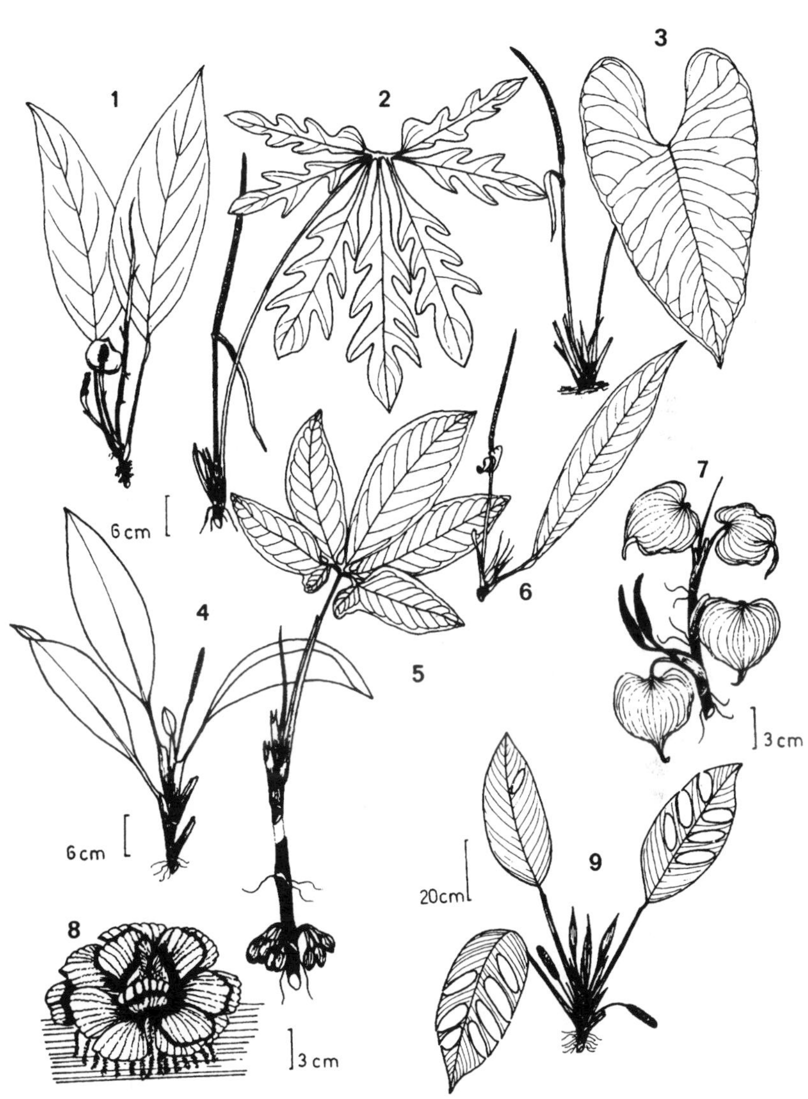

1 - *Rhodospatha* **2** - *Anthurium* **3** - *Anthurium*

4 - *Stenospermation* **5** - *Syngonium* **6** - *Anthurium* **7** - *Philodendron*

8 - *Pistia* **9** - *Monstera*

Rhodospatha (25 spp.) — Mostly hemiepiphytic climbers but a few species terrestrial. Vegetatively very similar to those *Philodendron* species that have oblong leaves with numerous secondary veins and without basal lobes, but the petiole winged to near apex (some *Philodendron* species also with fully winged petiole but these non-oblong-leaved).

P: huasca bijao

Stenospermation (25 spp.) — Epiphytic climbers with distinctive elliptic leaf having a strong midvein but reduced, hardly discernible secondary veins very strongly ascending and almost parallel to midvein.

2B. Spadix with thick cylindrical basal part (female flowers) and slender apical part (male flowers)

Philodendron (275 spp.) — The second largest genus of the family, easily differentiated vegetatively from *Anthurium* by the completely parallel secondary and tertiary venation and lack of a marginal collecting vein.

P: itininga, huambé (large spp.)

Syngonium (33 spp.) — Rather succulent hemiepiphytic climbers, mostly distinctive in palmately 3–5-parted leaves with basal lobes on the lateral leaflets; species with undivided leaves have strong basal lobes. The parallel-reticulate leaf venation is rather intermediate between reticulate-veined *Anthurium* and strictly parallel-veined *Philodendron;* in addition the marginal collecting vein is similar to *Anthurium* but the succulent leaf texture more like *Philodendron.* The flowering inflorescence is similar to *Philodendron* but the spathe is caducous to reveal a syncarpous fruit. Except for being epiphytic, very similar to *Xanthosoma.*

P: patiquina

3. Terrestrial (or Erect, Rooted, Swamp Plants)

3A. Spadix homogeneous throughout

Dracontium (13 spp.) — Forest-understory plants consisting of a single giant, pinnately deeply divided leaf borne from an underground tuber on a long thick petiole that resembles the trunk of a treelet. When fertile, the inflorescence, emerging from the ground near base of the petiole and with the spadix enclosed by a large acuminate maroon spathe, is unmistakable.

P: jergón sacha

Spathiphyllum (35 spp.) — Most commonly growing colonially in swampy places. Leaves always elliptic with a cuneate base, completely parallel rather strongly ascending secondary and intersecondary veins, and no marginal vein. The inflorescence is very characteristic with a rather leaf-like, large, open, white or greenish spathe.

Araceae
(Miscellaneous Epiphytic and Terrestrial)

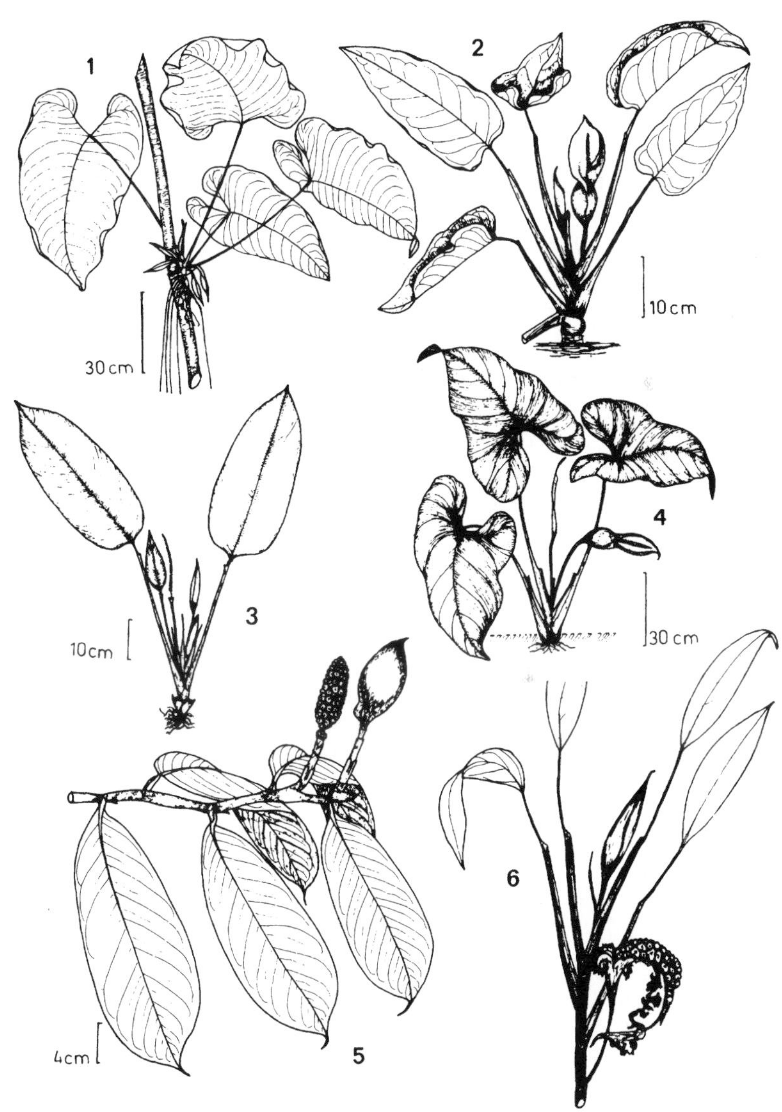

1 - *Philodendron* **2** - *Xanthosoma*

3 - *Rhodospatha* **4** - *Homalonema*

5 - *Heteropsis* **6** - *Dieffenbachia*

Urospatha (20 spp.) — Stemless swamp herb with strongly triangular leaves with the elongate sagittate basal lobes longer than anterior portion (unique). Inflorescence with a brownish-magenta acuminate spathe similar to *Dracontium* but with the spathe spirally twisted.

P: jergón sacha

(*Anthurium*) — Some *Anthurium* species are terrestrial or grow on rocks; many of these are bird's nest forms with short petioles and large clustered oblanceolate leaves. Terrestrial *Anthurium* leaves are mostly undivided, mostly lack strong basal lobes, and differ from *Spathiphyllum* in reticulate venation and a marginal vein.

(*Rhodospatha*) — A few *Rhodospatha* species are terrestrial; except for the inflorescence they look like *Philodendron* with oblong parallel-veined leaves without marginal vein.

3B. Spadix 2-parted—Basal (female) part cylindrical and usually thinner, apical (male) part club-shaped, narrower where it emerges from spathe.

Montrichardia (2 spp.) — Tall almost treelike plants forming dense stands in swampy areas, the commonest species with spines on trunk. Leaves large, broadly triangular-ovate, with strong basal lobes, the venation parallel-reticulate as in *Syngonium*. Fruit (edible) somewhat resembling *Syngonium*, but the closely appressed fleshy berries not truly syncarpous.

P: raya balsa

Dieffenbachia (30 spp.) — Succulent forest-understory herbs with well-developed succulent erect stems and uniformly oblong or elliptic leaves without basal lobes. Venation uniformly parallel and marginal vein lacking as in *Philodendron*. Usually with rather unpleasant pungent odor (sometimes almost skunklike) and sometimes with the leaves splotched white above. Spathe entirely enclosing spadix except during a very brief anthesis, the fruiting inflorescence curling to split irregularly and reveal the rather separated orange-arillate seeds.

P: patiquina

Xanthosoma (45 spp.) — The main terrestrial aroid genus, characterized by palmately dissected or triangular succulent leaves with strong basal lobes, reticulate venation, and usually both marginal and submarginal collecting veins. Typically rather coarse herbs, occasionally with soft "trunk" to 2 m high. Spathe strongly constricted at throat and with upper part of spathe more open than in *Philodendron* and caducous immediately after anthesis.

E: camacho; P: oreja de elefante

Araceae
(Terrestrial and Hemiaquatic)

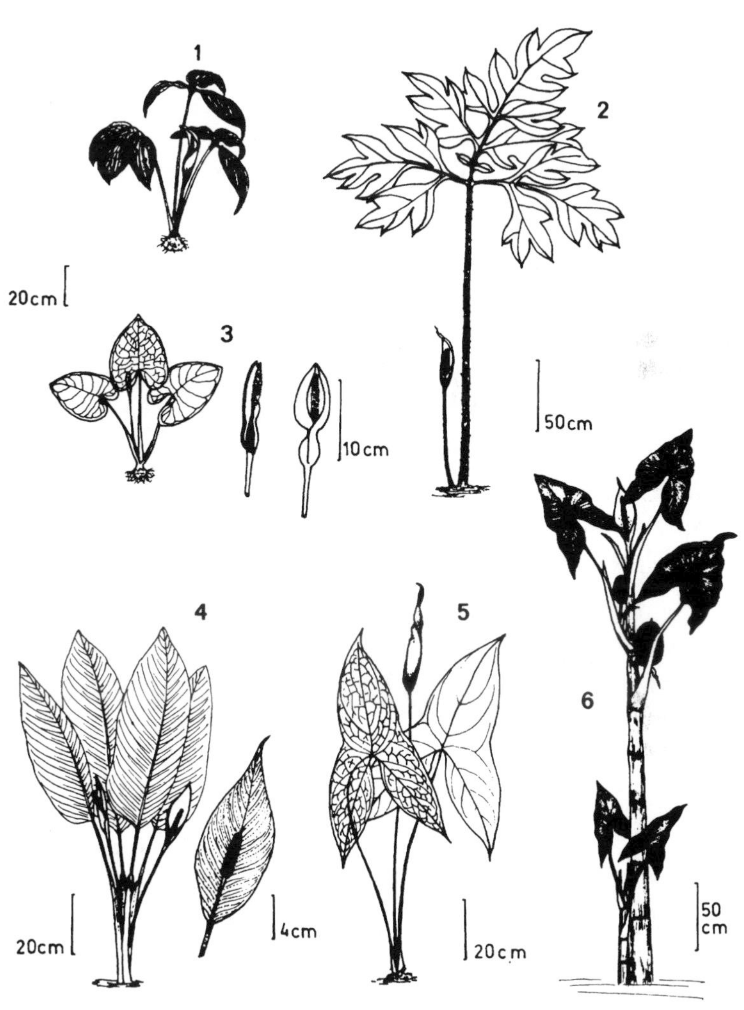

1 - *Chlorospatha* **2** - *Dracontium*

3 - *Caladium*

4 - *Spathiphyllum* **5** - *Urospatha* **6** - *Montricardia*

Caladium (15 spp.) — Very close to *Xanthosoma* but smaller and usually with strongly peltate leaves, often conspicuously splotched with red or white and/or with purple undersurface. The technical distinguishing characters are ovary with 1–2 (vs. 3–4) placentae and <20 (vs. >20) ovules, style nondiscoid, and pollen grains solitary (vs. tetrads).

C, E, P: corazón de jesús

Chlorospatha (3 spp.) — Small, cloud-forest, understory herbs, similar to *Caladium* but the leaves sometimes palmately divided and when undivided nonpeltate, noncolored, and with more pronounced sagittate basal lobes. Most easily distinguished by the several long-pedunculate inflorescences with the female portion of spadix mostly fused to spathe which is relatively narrow and elongate.

Homalomena (8 spp., plus 130 in Old World) — Terrestrial understory herbs with underground rhizomes, usually distinguished by pubescent petioles (and sometimes blades), pubescence otherwise rare in the family. Inflorescence like *Philodendron* but with staminodia scattered among female flowers. Usually with anise-scented sap.

Taccarum (4 spp.) — Mostly subtropical dry areas, probably reaching southern Peru. Essentially a dwarf *Dracontium* with deeply pinnatifidly divided leaves and a green and maroon mottled spathe; spadix unusual in having the individual flowers not strongly congested.

Asterostigma (6 spp.) — Mostly Brazilian; in our area (Peru) rare understory herbs characterized by pinnately laciniate-serrate leaves (one species from Loja with entire leaves).

Ulearum (1 sp.) — Rare upper Amazonian understory herb with +/ - hastate thin leaves with reticulate veins and inflorescence overtopping the leaves; spathe free and spreading (cf., *Anthurium*), the spadix distinct in having a large sterile segment between the male and female portions.

Zomicarpella (1 sp.) — Rare Amazonian understory herb. Leaves like *Xanthosoma* but blooming precociously when no leaves present.

(*Philodendron*) — A few *Philodendron* species are terrestrial.

Bromeliaceae

Typically a very distinctive family, characterized by the sessile rosette of narrow basally imbricated leaves, often forming a water-containing tank and/or having spiny margins. The typically conspicuously bracteate terminal inflorescsence is also characteristic. This is the second largest epiphytic family and most species are epiphytic, but some (especially the primitive Pitcairnioideae with capsular fruit and entire-winged seeds) are terrestrial. A few of the terrestrial *Puya* species of the high Andes have well-developed trunks several meters tall below the rosette of long spiny-margined leaves.

There are three subfamilies, generally vegetatively distinguishable. Two mostly have spinose-serrate leaves, often narrowed to a petiole-like base (Pitcairnioideae with capsular fruits and Bromelioideae with berry-fruits) and the third (Tillandsioideae) never does. The berry-fruited Bromelioideae are mostly epiphytic (except in some lowland taxa) and typically have strongly spiny leaf margins; the Pitcairnioideae are mostly terrestrial and, except for completely terrestrial high-Andean *Puya,* have weakly spinose leaf margins (often with spines only on the petiole-like basal constriction, especially when epiphytic).

1. Leaves (Usually) Spinose-Serrate; Fruit a Berry (= Bromelioideae)

Aechmea (172 spp.) — Epiphytic (usually) or terrestrial. The main genus (and only common lowland genus) with spiny leaf margins; in flower characterized by spiny-tipped sepals.

Billbergia (54 spp.) — Differs from *Aechmea* in zygomorphic petals and nonspiny sepals. Flowers borne in openly spicate inflorescence and with long-exserted stamens whereas *Aechmea* has densely spicate or branched inflorescence.

Bromelia (47 spp.) — Terrestrial, in our area only at low altitudes in northern Colombian dry forests. Differs from terrestrial *Aechmea* species in connate filaments, a more open inflorescence, and, vegetatively, in the leaf margin spines more strongly and wickedly recurved.

Ananas (8 spp.) — Terrestrial, mostly in extralimital dry areas. Vegetatively, much like some *Aechmea* species but distinguished by the fleshy compound fruit (pineapple) formed from the fused ovaries.

Neoregelia (71 spp.) — Differs from *Aechmea* in the dense sessile inflorescence which is shortened and included inside leaf cup.

Ronnbergia (8 spp.) — Mostly in Chocó region. Tank epiphytes, usually with leaves narrowed to basal "petiole" with fine reduced leaf margin serrations (or these entirely absent). Differs technically from *Aechmea* in naked (not appendaged) petals. The distinctive inflorescence is a rather few-flowered spike with reduced bracts, and oblong-ovoid sessile fruits which are often rather zygomorphic.

Streptocalyx (14 spp.) — Tank epiphyte (or on rocks) with strongly spiny-margined leaves, the commonest species with long narrow leaves abruptly expanded at base; inflorescence rather short and dense, the ovate bracts rather large and often spiny-margined. Technically differs from *Aechmea* in petals lacking appendages.

Araeococcus (5 spp.) — Mostly on Guayana Shield, ours a peculiar epiphyte with very narrow leaves abruptly broadening to wide base and an open almost ebracteate racemose-branched inflorescence with tiny flowers.

2. Leaves (Usually) Spinose-Serrate; Fruit a Capsule with Entire-Winged Seeds (= Pitcairnioideae)

Two main genera: *Puya* and *Pitcairnia*

Puya (168 spp.) — High-altitude terrestrial plants, sometimes with distinct stems. Leaf bases enlarged and triangular. Technical characters are usually superior ovary and the petals spiralled together.

Pitcairnia (262 spp.) — Epiphytic or terrestrial, mostly in cloud forests. Leaves usually contracted at base (often only spiny on contracted basal part). Technical characters are large, conspicuous, separate petals and a +/- inferior ovary.

Fosterella (12 spp.) — Terrestrial and not very obviously a Bromeliaceae, the inflorescence slender-branched and the very small flowers with only minute inconspicuous bracts; leaves thin and weakly or not at all spiny. In our area only at middle elevations in Peru.

3. Leaves Entire; Fruit a Capsule with Plumose Seeds (= Tillandsioideae)

Catopsis (19 spp.) — Mostly Central American; ours are tank epiphytes with thin broad leaves, strikingly obtuse except for an apiculate point. Inflorescence sparsely openly branched with reduced branches and with widely spaced short sessile flowers subtended by small inconspicuous bracteoles. Technical character is the seed appendage (hairs) apical and folded (unique).

Bromeliaceae
(Spiny Leaf Margins)

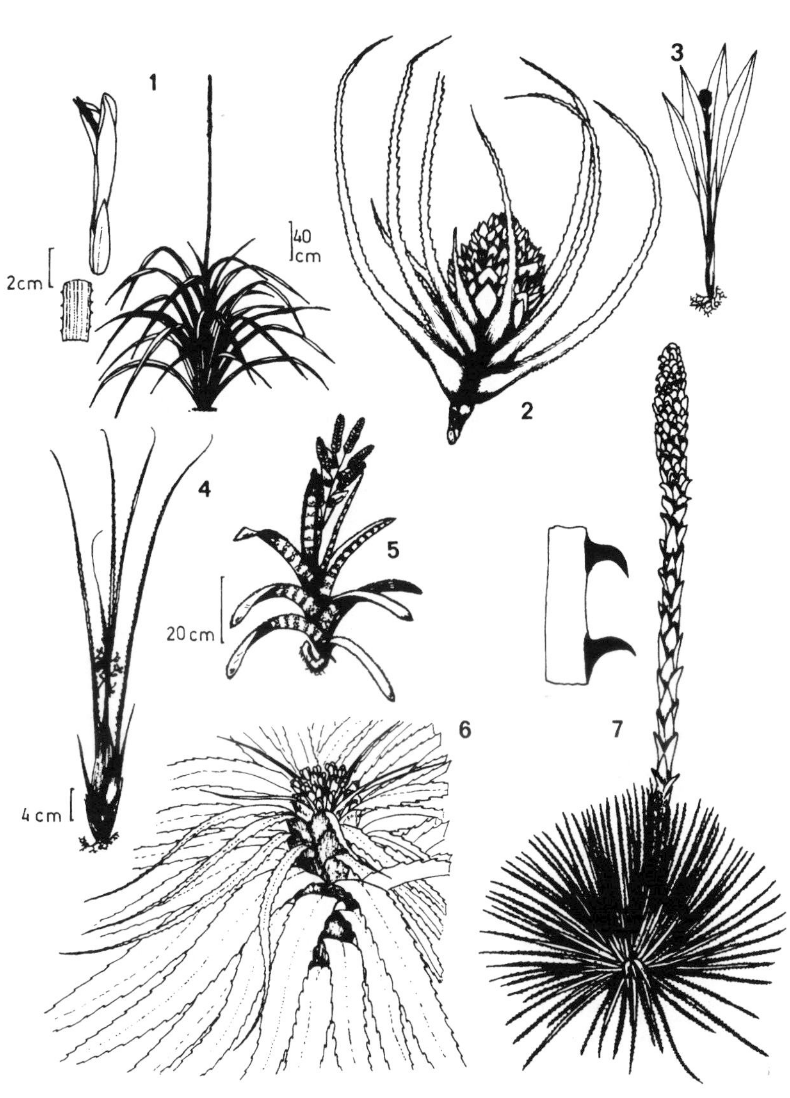

1 - *Pitcairnia* **2** - *Streptocalyx* **3** - *Ronnbergia*

4 - *Araeococcus* **5** - *Aechmea*

6 - *Bromelia* **7** - *Puya*

Tillandsia (410 spp.) — Inflorescence laterally compressed (= distichous), usually branched. The main genus of this group (and of the family). Petals without scales.

C: quiche, huayocoma (*T. flexuosa*), barba de viejo (*T. usneoides*)

Vriesia (257 spp.) — Inflorescence laterally compressed (often unbranched). Very close to *Tillandsia* and separable with certainty only by the technical character of petal scales, a character whose significance in generic delimitation is questioned by some taxonomists.

Guzmania (127 spp.) — Mostly cloud-forest epiphytes. Inflorescence with spiralled bracts unlike *Tillandsia* and *Vriesia*. Flowers distinctive in the petals with claws fused into tube.

C: quiche

Mezobromelia (2 spp.) — Mostly epiphytic in the wettest northern Andean cloud forests. Looks just like a *Guzmania* with rather open inflorescence of bright red spiralled bracts, but differs in petal claws with scales; essentially to *Guzmania* what *Vriesea* is to *Tillandsia*.

There are many other genera of Bromeliaceae, especially in eastern Brazil.

BURMANNIACEAE

Small mostly achlorophyllous (except most *Burmannia*) saprophytic herbs, usually with very reduced leaves and radially symmetric uniformly 3- or 6-parted small flowers with inferior ovaries. Similar only to saprophytic orchids, from which they differ in the radially symmetric flower, to Triuridaceae, which have apocarpous superior ovaries and separate tepals, and to a few saprophytic Gentianaceae (*Voyria*) which have 5 fused petals and a superior ovary. Most of our species have white flowers, typically with 3-lobed tepals, the tepals of several genera falling immediately after anthesis. The leafy species (*Burmannia*) are very similar to orchids except for the radially symmetric flowers, and one especially orchidlike species is actually epiphytic.

The first four achlorophyllous genera below have salverform flowers either with 6 corolla lobes (*Hexapterella, Campylosiphon*) or conspicuously 3-lobed tepals (*Cymbocarpa, Gymnosiphon*), these often caducous. *Apteria* is distinctive in the relatively openly campanulate 6-lobed flower and *Thismia* in the urceolate flower with three filiform appendages. The last two genera (*Dictyostega, Miersiella*) have tubular flowers, like unwinged species of *Burmannia*.

Bromeliaceae (Nonspiny Leaf Margins) and Cannaceae

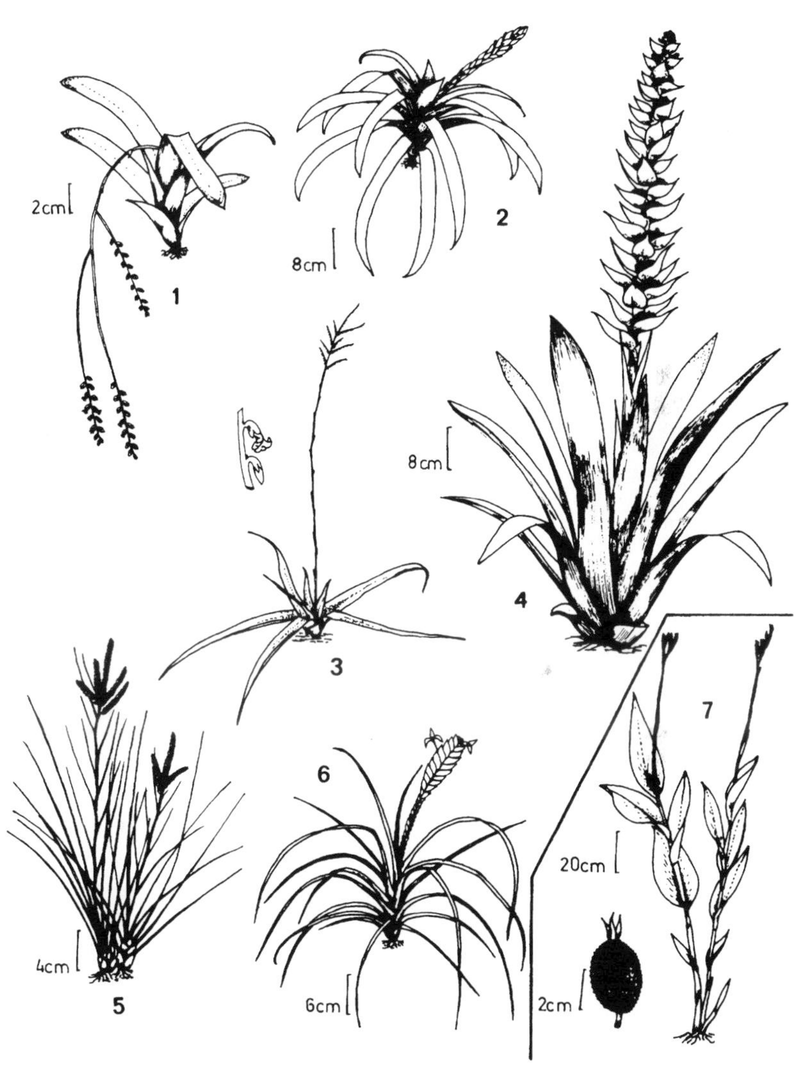

1- *Catopsis* **2** - *Guzmania*

3 - *Fosterella* **4** - *Vriesia*

5 - *Tillandsia* **6** - *Tillandsia* **7** - *Canna*

1. Species with Chlorophyll (and Nearly Always with Normal Leaves)

Burmannia (19 spp., plus 11 African, 30 Asian) — Nonsaprophytic with green stems and nearly always with well-developed green leaves, at least at base. One species is epiphytic and looks exceedingly like an orchid except for the nonzygomorphic flower. Flowers typically light blue and relatively attractive. Many species (including the only saprophyte) have a strongly 3-winged floral tube and most have +/- capitate inflorescences. Mostly occurs in moist, sandy savannas, especially in the Guayana area; only two Andean species, one epiphytic, the other a high-altitude terrestrial.

2. Without Chlorophyll; Leafless or the Leaves Reduced and Scalelike — All are species of the leaf litter of mature forest floor. See Figure 1.

Gymnosiphon (14 spp., plus 3 African, 7 Asian) — The small white flowers are distinctive in having only 3 apparent tepals which are conspicuously 3-lobed and caducous soon after anthesis. Most species have 2-branched inflorescences, a few capitate.

Cymbocarpa (2 spp.) — Virtually the same as *Gymnosiphon,* from which it differs by the fruit held at angle to pedicel and dehiscing on only one side. One species has very distinctive bulging pouches near middle of corolla tube

Hexapterella (1 sp.) — Flowers white to purple with three large entire tepals alternating with three very small ones. Very like *Gymnosiphon* except for the nontrifid tepals.

Campylosiphon (1 sp.) — The largest-flowered burmanniac (flowers 16–28 mm long). Similar to *Gymnosiphon* and allies in salverform corolla but the usually blue flower larger and with 6 subequal tepals. Habitally distinct in stem with larger, more conspicuous, achlorophyllous leaves, and plant from tuberous rhizome. This is the most *Voyria*-like burmanniac, but differs from *Voyria* in having 6 instead of 5 tepals.

Apteria (1 sp.) — Unique in the relatively openly campanulate, 6-lobed flowers; the flowers are larger than in most burmanniacs (7–20 mm) and usually purple. The few (–1) flowers widely spaced along inflorescence on long pedicels.

Thismia (12 spp., plus 1 N. Am. and many Old World) — Tiny mushroomlike fleshy forest-floor saprophytes, more or less hyaline in color, the peculiar flower always solitary, with a more or less cupular tube terminated by six tepals, three of which are usually elongate into filamentous appendages. Could easily be confused with a fungus, but otherwise only with *Triuris,* which has similar appendages, but the openly cupular flower very different from the caplike *Triuris* flower.

Dictyostega (1 sp.) — Characterized by the pendent, white, tubular 6-lobed flowers on a forked inflorescence; mostly in middle-elevation cloud forests.

Miersiella (1 sp.) — Looks like a leafless saprophytic *Burmannia* on account of the tubular 3-lobed corolla but differs in the more umbellate inflorescence. Flowers white to purple, unlike *Dictyostega* in being erect.

There is only one other neotropical genus of Burmanniaceae, monotypic *Marthella,* known only from Trinidad, which is similar to *Miersiella* but with a capitate inflorescence.

CANNACEAE

Large-leaved scitaminous herbs, mostly occurring in disturbed or swampy vegetation at middle elevations. Also widely cultivated. The distichous leaves with well-developed midvein and finely parallel lateral veins are similar to those of *Renealmia* and *Heliconia,* lacking the pulvinulus and minute cross veins of Marantaceae. Cannaceae lacks the Zingiberaceae vegetative odor of *Renealmia* and usually has more ascending lateral veins than *Heliconia.* Differs from all other Scitamineae in the asymmetric flower structure without even a clear bilateral symmetry. The red or yellow flowers are large and conspicuous, with an inferior ovary and three unequal petals united into a basal tube, these equalled or exceeded by the single petaloid stamen and ca. four (1 in *C. paniculata*) large petaloid staminodes (the innermost of which is usually reflexed to form a kind of labellum). The fruit, also very distinctive, is a large, conspicuously rough-surfaced or tuberculate, 3-parted capsule.

Canna (10 spp.)

COMMELINACEAE

A uniformly herbaceous, usually rather succulent, family, vegetatively characterized by the well-developed completely closed cylindrical sheathing petiole base (and resultant jointed stem) and usually by involute leaf vernation. No other monocot family has involute vernation. The only other taxon with a similar petiole base is *Costus* of the Zingiberaceae which is habitally very different from any Commelinaceae in its "spiral staircase" vegetative growth-form; grasses also have sheathing leaf bases but these are usually not completely closed and the plants are less succulent. The small flowers have 3 sepals and 3 nearly always free delicate petals (usually blue or bluish (to white or pink), sometimes the lowermost reduced) that deliquesce as the flower ages, and the inflorescence is often subtended by a spathelike bract.

Generic taxonomy, especially of the *Tradescantia* relatives, has been in a constant state of flux, but may now be stabilizing, the divisions, unfortunately, based largely on anatomical characters. Several genera are distinctive, however. *Dichorisandra* has a few free-climbing scandent species and *Cochliostema* (restricted to Chocó area) is a bromeliad-like tank epiphyte. *Geogenanthus* is remarkable for bearing the flowers at ground level from the base of the stem. *Floscopa* is small and insignificant, but common in wet-forest understory and characterized by the minute flowers in a small much-branched conspicuously pubescent panicle. *Zebrina,* now often treated as part of *Tradescantia,* cultivated and naturalized in our area, is unique in the petals fused into tube and also characterized by the often patterned leaf upper surface and purple leaf undersurface. *Commelina* is weedy and characterized by the well-developed spathelike bract that immediately subtends the flowers. *Aneilema,* with an unusually long petiole-like contraction at base of leaf blade, has white flowers with 2 large petals (cf., the ears of Mickey Mouse) along with a small inconspicuous one. *Tinantia* is vegetatively similar in the well-developed petiole-like blade base but has larger blue flowers. *Campelia,* common in wet-forest, has a few small white flowers subtended by pair of leaflike bracts and borne at end of a +/- elongate lateral inflorescence; it differs from *Tradescantia* principally in the fleshy, black, berrylike fruits.

1. Epiphyte

Cochliostema (2 spp.) — A curious bromeliad-like tank epiphyte with very long leaves, more or less narrowed toward base, and then expanded into the sheathing base. The inflorescence is openly paniculate with conspicuous pinkish bracts and large flowers with white to pinkish sepals and fringed blue petals both to ca. 2 cm long.

2. Terrestrial Herbs or Twining Vines

2A. The next seven genera share a tendency to distinctly zygomorphic flowers with elaborated stamens and staminodes (*Geogenanthus* actinomorphic with conspicuously fringed petals). The first four are related to *Commelina* (tribe Commelineae); but the last four are closer to actinomorphic-flowered *Tradescantia* (tribe Tradescantieae) based on anatomical characters.

Commelina (ca. 10 spp., plus 150 Old World and n. temperate) — Prostrate mostly weedy rather succulent herbs. Leaves mostly distichous, unlike *Floscopa* and *Aneilema.* Leaf blades usually sessile, sometimes

Commelinaceae

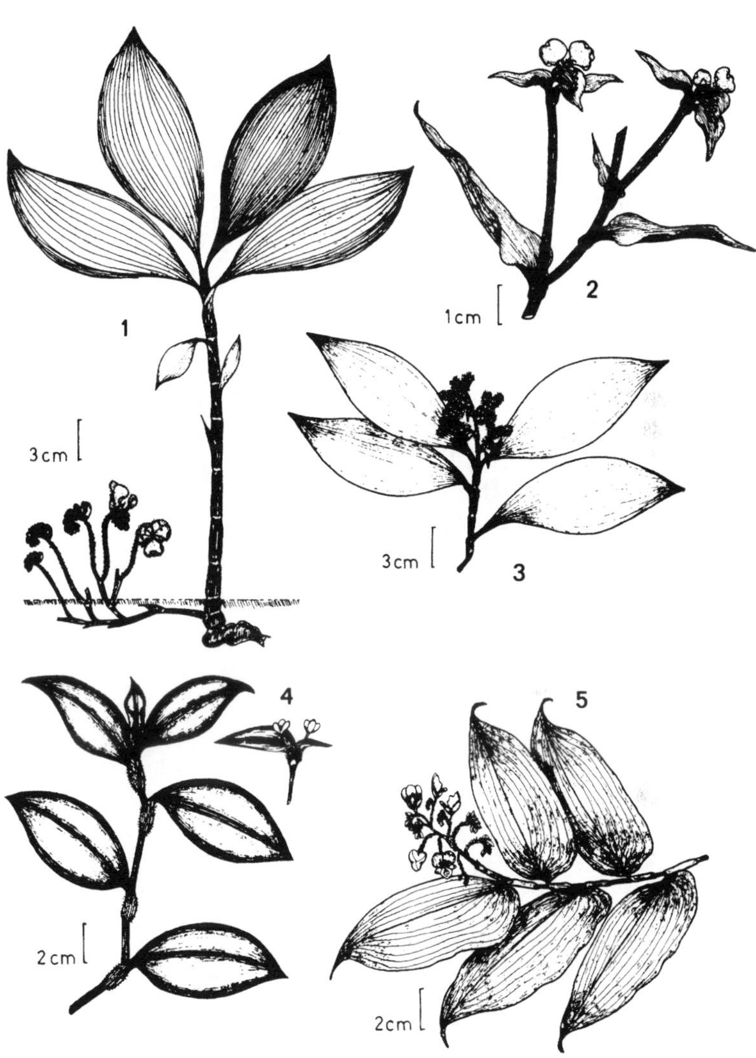

2 - *Commelina*

1 - *Geogenanthus*

3 - *Floscopa*

4 - *Tradescantia (Zebrina)*

5 - *Dichorisandra*

more or less clasping at base, but sometimes with very short petiole-like constriction (e.g., in sometimes segregated *Phaeosphaeron*). Inflorescence leaf-opposed, with several small usually blue flowers with 2 upper petals larger, subtended by a conspicuous leaflike spathe.

Murdannia (5 spp., plus ca. 45 Old World) — Prostrate herb with linear leaves, mostly in swampy areas on sandy soil. Inflorescence composed of cymes, frequently contracted and umbel-like, without leafy spathe, the blue to pinkish flowers small, with stamens sometimes reduced to two.

Floscopa (4 spp., plus 15 Old World) — Small forest-understory herbs with +/- erect flowering stems terminated by densely paniculate, conspicuously pubescent, shortly pedunculate inflorescence of small white or light bluish flowers. The elliptic leaf blades are gradually contracted to narrow petiole-like base.

Aneilema (1 sp., plus 60 Old World) — Subprostrate herbs of disturbed wet forests. Characterized vegetatively by the leaf base narrowed to a distinct petiole above the sheath. Inflorescence terminal, open, with small cup-shaped bracteoles, the flowers strongly zygomorphic with upper 2 petals larger and white, the lowermost small and greenish, and spirally arranged, unlike area *Commelina* and *Tripogandra.*

Tripogandra (22 spp.) — Characterized by lack of leaflike inflorescence bracts and dimorphic stamens, the latter unique in *Tradescantia* relatives. The moderately zygomorphic flowers are unusual among *Tradescantia* relatives. The ovate or oblong-ovate leaves are mostly sessile or subsessile and distinctive in distichous arrangement.

Tinantia (13 spp.) — Forest-understory herbs with broad distinctly petiolate leaves. Inflorescence of 1–several cymes, usually appearing umbel-like or racemose with umbel-like branches, the flowers relatively (at least to *Aneilema*) large (1–2 cm wide) and zygomorphic, with 6 stamens.

Dichorisandra (25 spp.) — Some species are twining vines, others erect herbs, all with blue or blue and white flowers in terminal inflorescences without spathelike bracts. The leaf blade is subsessile but usually has a very short petiole-like contraction above the sheath. The taxonomy is very confused with several distinct species passing under the name *D. hexandra.*

Geogenanthus (4 spp.) — A distinctive forest-floor herb characterized by the short leafless inflorescence arising at ground level from leafless base of the stem and with several relatively conspicuous fringed blue actinomorphic flowers. Vegetatively distinctive in the rather broad (sometimes orbicular) leaves with a well-developed petiolar contraction and tending to cluster near stem apex.

2B. The last five genera are related to *Tradescantia* on account of technical characters of the inflorescence and pollen. All have radially symmetric flowers.

Tradescantia (40 spp., plus 19 in USA) — A poorly defined, largely Central and North American genus now interpreted to include such segregates as *Cymbispatha, Rhoeo, Setcreasea,* and *Zebrina.* The most important character is the inflorescence cymes fused in pairs subtended by spathaceous bracts. The flowers are mostly radially symmetrical unlike *Commelina* and its relatives.

Gibasis (11 spp.) — Mostly Mexican herbs, characterized by the conspicuously stipitate paired or umbellate cymes which are simple rather than fused in pairs as in *Tradescantia;* leaflike bracts absent and flowers always regular and 6-staminate.

Callisia (incl. *Phyodina*) (20 spp.) — Essentially represents a trend in floral reduction (toward anemophily?) from *Tradescantia.* The very small flowers regular, white to pink or greenish, with 1–6 (typically 3) stamens. Inflorescences variable in form but always lacking spathelike subtending bracts.

Campelia (3 spp.) — Forest-floor herb with narrowly elliptic leaf blade gradually tapering to narrow base. Flowers small, white or purplish, borne on ends of long erect peduncle from well down the stem, the flower clusters subtended by folded leaflike spathes. Sometimes included in *Tradescantia* from which it differs in having a fleshy black berrylike fruit.

Elasis (1 sp.) — A high-Andean subparamo herb with actinomorphic flowers, known only from Ecuador. Described as, and very similar to, *Tradescantia,* except for the inflorescence.

CYCLANTHACEAE

A very distinctive family that could only be confused with palms. The characteristic thick-spicate unbranched inflorescence of Cyclanthaceae with its appressed thick four-sided female flowers and fruits is unlike that of any palm. Most cyclanths are bifid-leaved epiphytic root climbers while palms (except the spiny, multifoliolate liana *Desmoncus*) are erect and terrestrial. Terrestrial cyclanths are always essentially stemless and generally more obviously herbaceous than similar-appearing palms. While a few palms do have simple leaves with bifid apices, (especially as seedlings) these are always associated with an elongate midvein and are obviously derived from a pinnate condition as opposed to the more obviously palmate venation of bifid cyclanths. Cyclanths with palmately divided leaves (*Carludovica*) can be distinguished from many fan palms (but not from trunkless juveniles of the scaly-fruited lepidocaryoid ones predominating in Amazonia), by the absence of a raised triangular projection (hastula) at the intersection between petiole apex and leaf blade. The distinction between *Cyclanthus* with its screwlike inflorescence composed of alternating cycles of male and female flowers (and with each segment of the bifid leaf having a strongly developed midvein) and the rest of the genera with the flowers in groups of single pistillate flowers surrounded by four male flowers (and leaves palmately divided to entire or bifid but with the segments lacking a strong midvein) is a fundamental one. The differences between the groups with palmately divided leaves (*Carludovica*), bifid leaves (most genera) and undivided leaves (*Ludovia*) are clearly apparent if not as indicative of natural relationships.

1. Leaves Palmately Divided (Usually into Four Main Segments with Less Deeply Divided Lobes)

Carludovica (3 pp.)— Completely unmistakable in its palmately divided palmlike leaves from an underground stem; always terrestrial. The mature fruiting inflorescence is also unique in the bright orange-red pulp which is revealed as the outer seed-containing layer of the inflorescence separates and falls off; presumably this is the only bird-dispersed cyclanth. Thin slices of the fibrous petioles are much used in basketry and to make the famous "Panama" hats of Ecuador.

C: palmicha; P: bombonaje

2. Leaves Undivided (and with Only a Single (or No) Well-Developed Midrib) — At least, the commonest species tends to be more lianescent and free-climbing than bifid-leaved genera.

Ludovia (2 spp., incl. *Pseudoludovia,* based on a mixture) — Two species, the common one with narrowly obovate to oblanceolate leaves, the second with shorter wider leaves, both with irregularly crenulate apices. The inflorescence is rather thin and reduced with the flowers completely lacking tepals.

3. Leaves Bifid (Undivided but with Two Strongly Developed Parallel "Midribs" in One Rare *Cyclanthus*)

Cyclanthus (2 spp.) — Always terrestrial; unique in thickened midrib of each segment of the bifid leaf. The spiralled inflorescence is also unique. A rare second species of Chocó has the two segments fused at maturity but the pair of "midveins" is still clearly evident.

E: hoja de lapa

Cyclanthaceae
(Terrestrial)

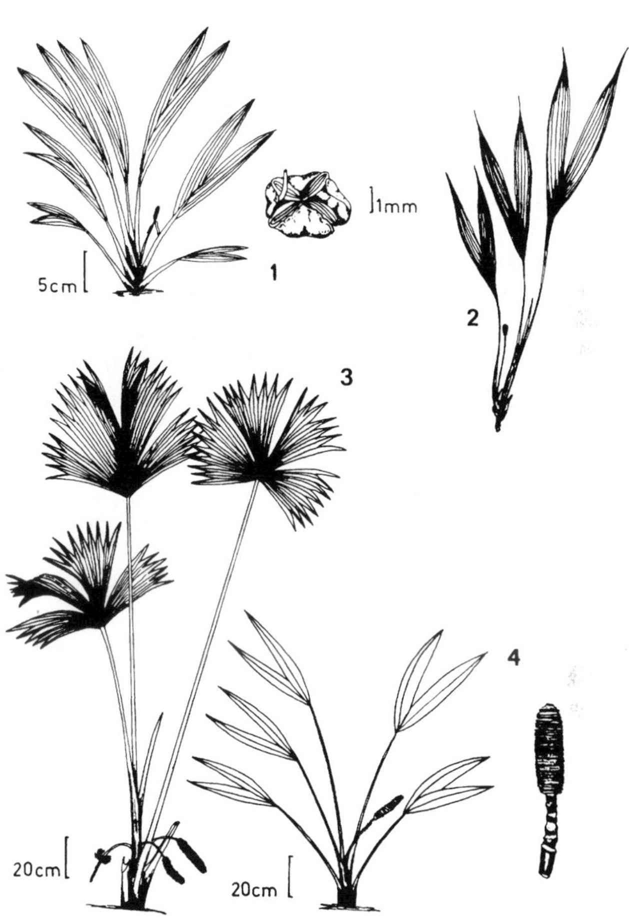

1 - *Sphaeradenia* **2** - *Dicranopygium*

3 - *Carludovica* **4** - *Cyclanthus*

Asplundia (82 spp.) — The main genus of the family; includes both terrestrial species and epiphytic climbers. The leaves are always bifid, arranged spirally on the stem, usually relatively (to *Sphaeradenia*) dull-surfaced, and thin-textured. The inflorescences have the usually 3–5 spathes (or their scars) scattered along at least the upper half of the peduncle.

E: totora; P: tamshi

Dicranopygium (44 spp.) — Usually terrestrial, especially on rocks near rapidly flowing streams; occasionally +/- epiphytic in Chocó. Differs from *Asplundia* in the smaller (usually tiny) few-flowered inflorescence with more broadly separated individual flowers having reduced perianth lobes, and the spathes clustered at base of spadix.

Evodianthus (1 sp.) — Epiphytic climber; monotypic but wide-spread and common. Unique in the family (along with newly discovered *Dianthoveus* in the leaves scabrous (but this apparent only when dry). Differs from *Asplundia* in having the three spathes (or scars) densely crowded at base of spadix; also in having the pistillate flowers free from each other, the obviously separate adjacent narrowly columnar thin-tepaled floral units being especially distinct in fruit.

E: jarre; P: tamshi

Dianthoveus (1 sp.) — A newly discovered Chocó area endemic, similar to *Evodianthus* in the scabrous leaves but terrestrial.

Thoracocarpus (1 sp.) — Epiphytic root climber, monotypic but common and widespread. Vegetatively distinctive in the unusually wide, thin, concave petiole. Differs from *Asplundia* in 8–11 spathes (or spathe scars) instead of 3–5(–8), and the typically rather shiny surface of the thick fruit epidermis and the characteristically hardened but brittle surrounding tepals.

P: tamshi

Sphaeradenia (38 spp.) — Mostly terrestrial but also many epiphytes. Differs from *Asplundia* in the stiff distichous 2-ranked leaves (cf., miniature *Ravenala*); in most species the leaves are more shiny and coriaceous than in *Asplundia* and also differ in having only one main vein in each segment.

Schultesiophytum (1 sp.) — Terrestrial in northwestern Amazonia. Characterized by very long peduncles and the leaves drying black.

Cyclanthaceae
(Hemiepiphytic)

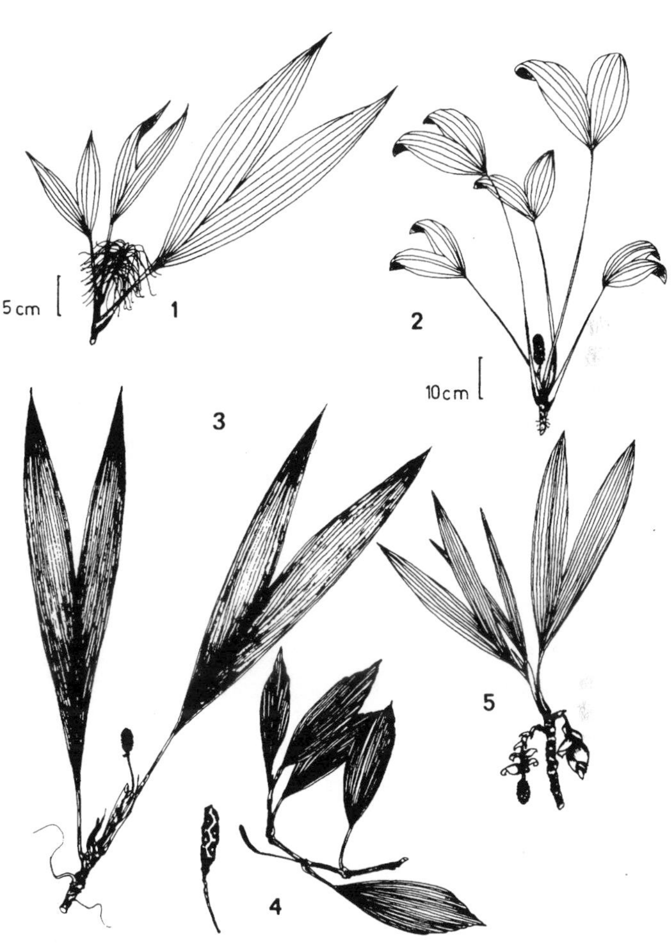

1 - *Asplundia*

2 - *Asplundia*

3 - *Evodianthus*

5 - *Thoracocarpus*

4 - *Ludovia*

CYPERACEAE

Mostly grasslike herbs with linear leaves (sometimes leafless) and reduced wind-pollinated flowers in complexly bracteate inflorescences, differentiated from grasses by 3-ranked leaves, usually triangular stems, and fused margins of the sheathing leaf base. A few forest-floor species have broader petiolate leaves (*Mapania*). A few species are scrambling vines (*Scleria*). Sedges are especially prevalent in swampy situations.

Taxonomic relationships within the family are largely based on technical characters of the inflorescence which are subject to different interpretations. Worldwide there are only six large genera plus many small ones. *Carex* and *Uncinia* (Caricideae), mostly of temperate areas, are characterized by unisexual flowers, the female below the male (or the spikelets unisexual). *Cyperus,* the main tropical lowland genus, and its relatives (first seven genera) have several-flowered spikes, the inflorescence axes lacking terminal flowers and with each flower in a glume axil. *Rhynchospora,* another important neotropical taxon, and its relatives (next four genera) differ from *Cyperus* in 2-flowered spikelets with a perfect basal flower and male upper one. *Mapania* and its relatives (Mapanioideae) are technically characterized by having the spikelets reduced to a pair of unisexual flowers these arranged into pseudospikelets. *Scleria, Diplacrum, Becquerelia,* and *Calyptrocarya* (Sclerioideae) have unisexual spikelets with a terminal solitary female flower and lateral male flowers.

Cyperus (600 spp., incl. Old World) — The most prevalent neotropical sedge genus, generally characterized macroscopically by typical grasslike leaves and the umbellately branched inflorescence subtended by a whorl of 3 or more leaflike bracts. Relatives have sessile, unbranched inflorescences or lack the subtending whorl of leaflike bracts.

P: piri piri

Kyllingia (60 spp., mostly African) — A small herb, close to *Cyperus* but with the inflorescence more contracted (of several "confluent" heads) and sessile.

Fimbristylis (300 spp., mostly Australasian) — Weedy herbs with discrete ovoid spikelets. Similar to some *Cyperus* species and with a similarly whorled inflorescence but with the subtending whorl of bracts inconspicuous or at least not leaflike.

Fuirena (40 spp., incl. Old World) — Large succulent marsh herbs with characteristic leaf sheath forming a segmented 5-angled stem. Upper leaves gradually reduced and each subtending a pedunculate inflorescence branch with several more or less clustered ovoid spikelets.

Cyperaceae
(A – C)

1 - *Bisboecklera* **2** - *Calyptrocarya*

3 - *Becquerelia* **4** - *Diplasia* **5** - *Carex*

6 - *Cyperus*

7 - *Cyperus*

8 - *Cyperus* (fungus-infected)

Scirpus (200 spp., incl. Old World and N. Am.) — Mostly in Andean paramos and puna, especially in marshes (e.g., the Lake Titicaca totora reed); also on or near beaches. Stems rather weakly triangular, often with rounded angles. Inflorescence more or less congested, distinctive in being subterminal, from side of stem apex; small paramo species resemble *Eleocharis* except for the subterminal inflorescence.

P: totora

Bulbostylis (100 spp., incl. Old World) — Densely tussock-forming with very fine linear, erect leaves, mostly occurring in dry, rocky grasslands. Inflorescence branched and *Fimbristylis*-like or reduced to a single terminal spikelet, the latter resembling some *Eleocharis* species but differing in the denser tussock habit and nonmarshy habitat.

Eleocharis (150 spp., incl. Old World) — Leafless with round hollow stems, occurring in swampy or marshy areas. Inflorescence a single terminal spike. With two growth-forms, larger species with conspicuously hollow stems are unmistakable but smaller species are similar to *Bulbostylis* except for the swampy habitat.

Rhynchospora (200 spp., incl. Old World) — A large and variable genus in both upland and lowland Neotropics. Differs technically from *Cyperus* in spikelets with only 2 flowers. The inflorescence is variable but differs from *Cyperus* in either being dense and sessile (when with subtending whorl of leaflike bracts) or more elongate and with separated leaflike bracts below each of lower inflorescence branches.

Pleurostachys (50 spp.) — Mostly Coastal Brazilian. Related to *Rhynchospora*. The inflorescence often small or tenuous, borne in axils of several normal-looking upper leaves.

Cladium (2 spp.) — Coarse 2–3 m tall herb of swamps and salt marshes. Distinctive in round stem. Inflorescences scattered along upper leaf axils.

Oreobolus (10 spp., mostly s. temperate) — A distinctive pincushion-like paramo plant with a tuft of very narrow erect leaves and hard-to-see solitary bisexual flowers.

Scleria (200 spp., incl. Old World) — Variable in habit and sometimes viny. Stem sharply triangular, the leaves often with sharp, nastily cutting edges. Inflorescence with several lower branches individually subtended by reduced leaves or the inflorescence in axils of normal-looking leaves. Spikelets often rather tenuous and less defined than in many sedges; female spikelets with prolonged rachilla, unlike *Calyptrocarya*. Fruits round and shiny, usually white.

Cyperaceae
(E – U)

1 - *Uncinia* **2** - *Fimbristylis* **3** - *Kyllingia* **4** - *Elaeocharis*

5 - *Scleria*

6 - *Hypolytrum* **7** - *Scirpus* **8–10** - *Rhynchospora*

Calyptrocarya (6 spp.) — Related to *Scleria* but with the rachillae of inflorescence not prolonged; so spikelets congested into small irregularly globose heads. The inflorescence rather tenuous and few-branched (like *Pleurostachys*), borne in axils of several normal-looking upper leaves.

Becquerelia (2 spp.) — Essentially an overgrown *Calyptrocarya,* to 1.5 m tall and occurring in understory of poor-soil forest.

Diplacrum (7 spp.) — Mostly in moist Guayana-Shield savannahs. Looks like *Kyllingia* with congested sessile heads but the inflorescence bracts more acuminate and the head, thus, rougher-looking.

Mapania (50 spp., incl. Old World) — A forest-floor herb with very distinctive broad leaves contracted into basal petiole. Inflorescence sessile and congested, subtended by whorl of 3 broad bracts.

Bisboecklera (8 spp.) — A narrow-leaved version of *Mapania,* the leaves sublinear but contracted to a petiole-like base.

Diplasia (1 sp.) — A coarse forest-understory herb, superficially resembling an overgrown clump-forming *Cyperus,* 2–3 m tall. Related to *Mapania* but the leaves epetiolate, longer and with finely toothed cutting margins.

Hypolytrum (50 spp.) — Exclusively swamp or stream-side herbs, rather *Cyperus*-like but with only 2 leaflike inflorescence bracts, these separated and subtending each of lowermost inflorescence branches. Spikelets tending to be more open than in *Cyperus* and with fuzzy appearance at anthesis from exserted anthers of the more numerous flowers. Technically related to *Mapania* from which it differs in having 2 (rather than 3) stigmas.

Carex (2000 spp., incl. n. temperate and Old World) — Mostly temperate; in our region restricted to Andean uplands where it replaces *Cyperus* as the dominant sedge. Inflorescences variable and superficially can resemble *Cyperus,* but very different in the unisexual flowers with female below the male (or unisexual spikelets).

Uncinia (30 spp., incl. Old World, mostly s. temperate) — Restricted to Andean uplands in our area. Related to *Carex* but very distinctive in the narrow, elongate, dense inflorescence with very elongated, protruding, hook-tipped spikelet rachillae.

Dioscoreaceae

Mostly slender twining vines with smooth green stems arising from well-developed rhizomes or basal tubers (these sometimes large and conspicuous with the tops projecting above ground surface); sometimes with rather large, irregular, laterally flattened recurved spines on stem. Leaves not very monocot-like, palmately veined, mostly broadly ovate and basally cordate, rarely 3-lobed or 3(–5)-foliolate, always with entire margin. Differs from *Smilax* in lacking petiolar tendrils (as well as the jaggedly sheathed base of petiole), instead with petiole base thickened and twisted or bent, sometimes also with petiole apex slightly thicker. Other vegetative differences from *Smilax* include the sharply defined, not-at-all decurrent leaf base and the well-defined transversely parallel secondary cross veins (the vein network of *Smilax* is much less regular). The flowers are small and greenish, borne in axillary spikes or racemes or panicles with well-developed central rachises; the inflorescence and characteristic 3-winged capsular fruit are thus totally unlike *Smilax.*

Dioscorea (600 spp., incl. Old World)
C: ñame; P: sacha papa

Eriocaulaceae

Herbs, mostly characterized by the Bromeliaceae-like or grasslike tufts of long, narrow leaves and dense heads of small whitish flowers borne at the end of a long naked stalk. Leaves usually in basal rosettes but sometimes the leaf clusters borne at end of a thick subwoody stem. Mostly grow on exposed rock outcrops or open sandy places, especially in moist situations. Our commonest species, monotypic *Tonina,* the most widespread taxon in our area, is a prostrate or floating weed of open swampy areas with short sheathing-based leaves and short-peduncled axillary inflorescences. The other most important genera are *Eriocaulon* (ca. 100 spp., plus 300 paleotropical), characterized by 4 or 6 stamens (vs. 2–3 in other genera) and uniquely diaphragmed aerenchymous roots, *Paepalanthus* (485 spp.), with unisexual trimerous flowers and pistillate flowers with free petals, and *Syngonanthus* (192 spp., plus 4 African) with unisexual trimerous flowers and pistillate flowers with connate petals. There are numerous additional genera and species in the Guayana area and on the Brazilian Shield, mostly looking about alike and differentiated by technical characters of the minute flowers.

Dioscoreaceae

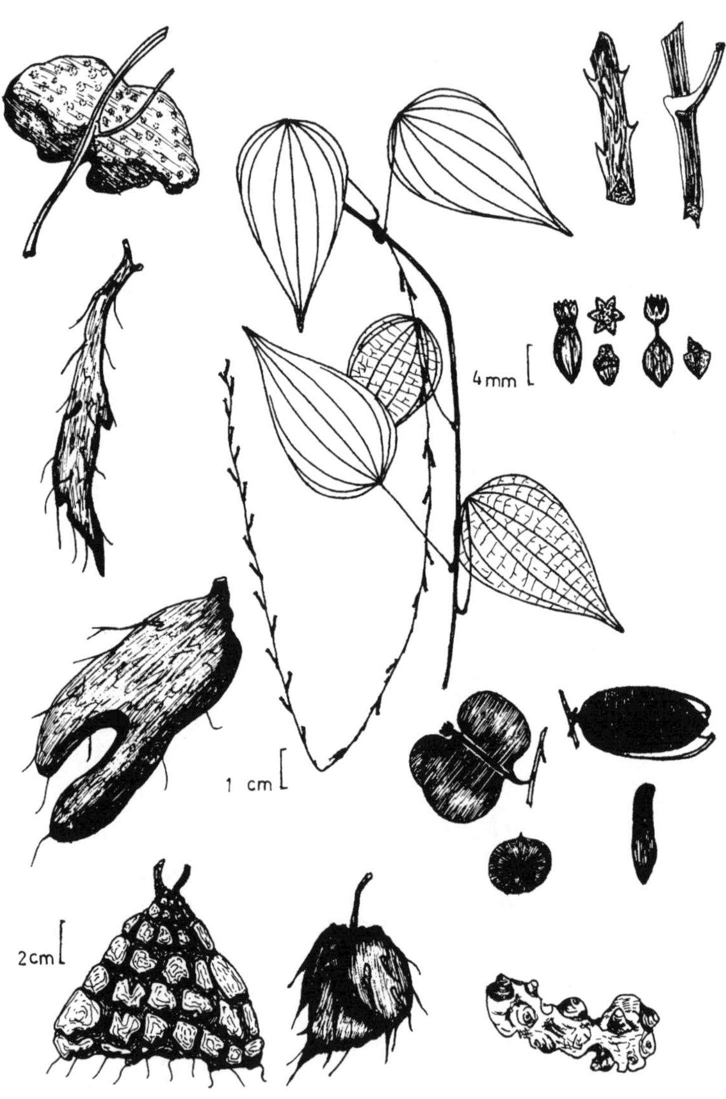

Dioscorea

Eriocaulaceae and Gramineae (Spikelet Details)

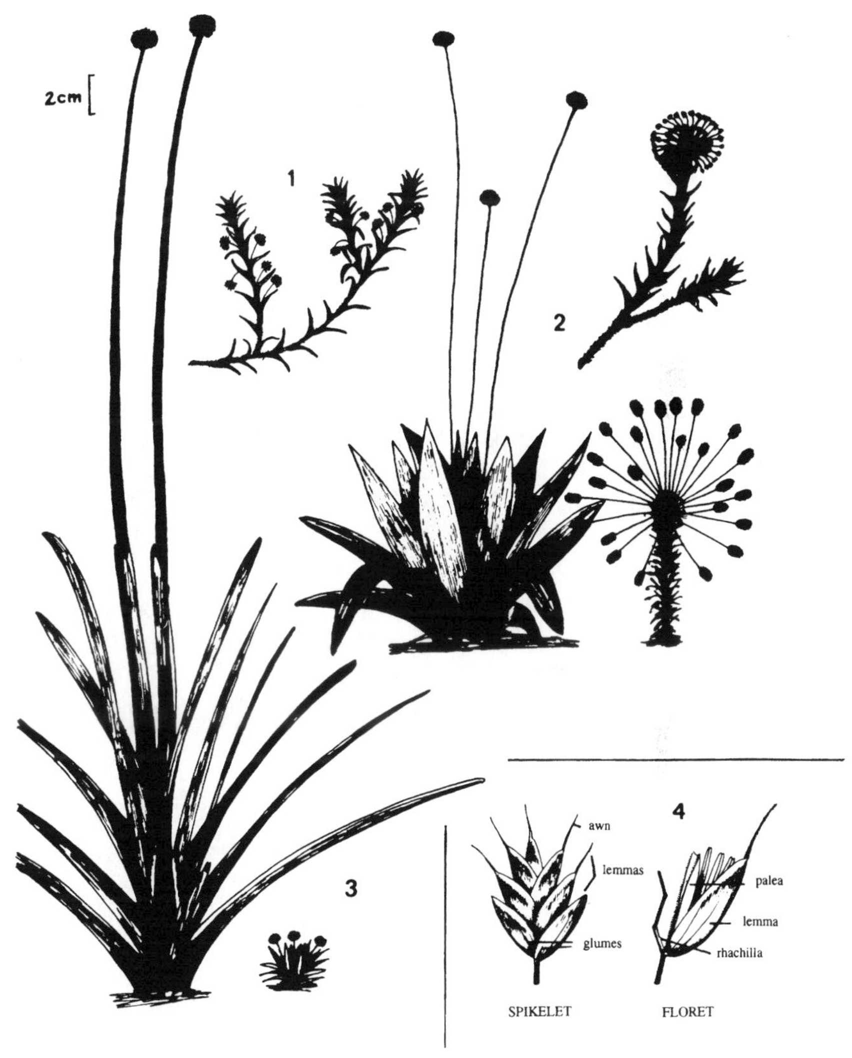

1 - *Tonina*

2 - *Paepalanthus*

3 - *Eriocaulon*

4 - *Gramineae* spikelet details

Gramineae

Although the vast majority of grasses are herbaceous, many bamboos are woody and there are also a number of climbing or subwoody taxa in other subfamilies. Grasses, including the woody ones, are generally very distinctive in their jointed stems, mostly with hollow internodes) and the characteristic leaf which includes a basal sheath surrounding the stem, in addition to the typically narrow lamina; there is usually a flap or projection at the junction of blade and sheath called a ligule. Nevertheless, forest-floor grasses can be tricky to recognize to family, since they typically have much broader leaves than a temperate botanist is accustomed to encountering in this family. Most bamboo leaves are distinctive in the family in having the base of the leaf lamina narrowed above the summit of the sheath into a petiole-like structure called a pseudopetiole.

Grass flowers are highly reduced, with adaptations for wind-pollination, and lack normal sepals and petals. Instead the sexual parts are protected by a series of bracts and bracteoles. The family's taxonomy is largely based in the complex inflorescences which have evolved their own complicated descriptive terminology. The basic inflorescence unit is the spikelet which consists of a short axis with several distichous bracts; the two lowermost bracts are called glumes (rarely only one is present); each pair of glumes subtends one or more florets, each with an outer bract called the lemma (which often has its midrib extended as an awn) and an inner bract called the palea. (Figure 20).

This is a huge family, with only the primitive subfamily Bambusoideae (the bamboos) containing truly woody taxa. Besides being woody, the bambusoids are characterized by a narrowed pseudopetiole between blade base and sheath apex. A few additional bamboo genera are herbaceous (*Cryptochloa, Lithachne, Pariana, Raddia, Pharus, Streptochaeta, Streptogyna,* and some *Olyra* species). A few members of two other subfamilies are coarse and subwoody. These include *Arundo* (introduced and escaped), *Gynerium, Phragmites,* and barely subwoody *Cortaderia* of the Arundinoideae, large reedlike grasses with large plumose panicles, and *Saccharum* (cultivated), *Zea* (cultivated), and some viny *Lasiacis* species of the Panicoideae.

See Pohl's treatment in the *Flora Costaricensis* or *Genera Graminum* (Clayton and Renvoize, 1986) for more complete recent treatments. Pohl's key to the woody Costa Rican taxa is duplicated here (with the addition of Andean *Neurolepis*) in lieu of a complete treatment of grasses. In addition, notes on tribal affiliation and distinguishing characteristics of most of the more important genera in our area are included.

The traditional taxonomy divided grasses into two main subfamilies based on technical characters of the spikelets. Pooideae (= Festucoideae) have generally laterally compressed spikelets, each 1–many-flowered with reduced florets (if any) above the perfect florets, and articulation above the basal bract pair (glumes); Panicoideae have spikelets more or less dorsally compressed each with a single perfect terminal floret subtended by a sterile or staminate floret, articulate below the spikelet (sometimes on the rachis or at base of spikelet cluster). The pooids include ten traditional tribes, defined by whether the spikelets are perfect or unisexual, 1(or more)-flowered, articulate above or below (Oryzeae) the glume, presence and size of glumes, and spikelet arrangement on the inflorescence (pedicellate in open inflorescence = Agrostideae (1-flowered), Aveneae (2–many-flowered and glumes as long as spikelet), or Poeae (Festuceae) (2–many-flowered and glumes mostly shorter than first floret); sessile in spike (or 2–several spikes) = Triticeae (Hordeae) (spikes solitary with spikelets on opposite sides or rachis) or Chlorideae (spikes usually more than one, spikelets only on one side of rachis). The panicoids traditionally included four tribes, two with thick hard glumes (Andropogoneae with characteristic pairs of spikelets, one sessile and perfect, the other pedicellate and either staminate or neuter), and Tripsaceae with unisexual spikelets, the pistillate below and the staminate above, sometimes on separate inflorescences), and two with thin membranaceous glumes (mostly Paniceae with fertile lemma and palea +/- coriaceous).

Nowadays agrostologists mostly recognize six subfamilies with the old festucoids split into five subfamilies, the Bambusoideae (vegetatively defined by the leaf blade contracted at base into a pseudopetiole and including nearly all the woody grasses), Arundinoideae (sometimes included in bambusoids, mostly large and reedlike with large plumelike panicles and including several more or less woody taxa), Oryzoideae (more or less aquatic plants characterized by very reduced glumes and a single fertile floret), Pooideae (Festucoideae), mostly high-altitude and restricted to the Andes in our area, the lemmas mostly with five or more faint veins), and Chloridoideae (the bulk of the lowland tropical nonbambusoid members of the old Festucoideae, typically with lemmas with three strong veins). The Panicoideae are essentially unchanged.

Gramineae
(A – *Chusquea* Plus *Dissanthelium* and *Guadua*)

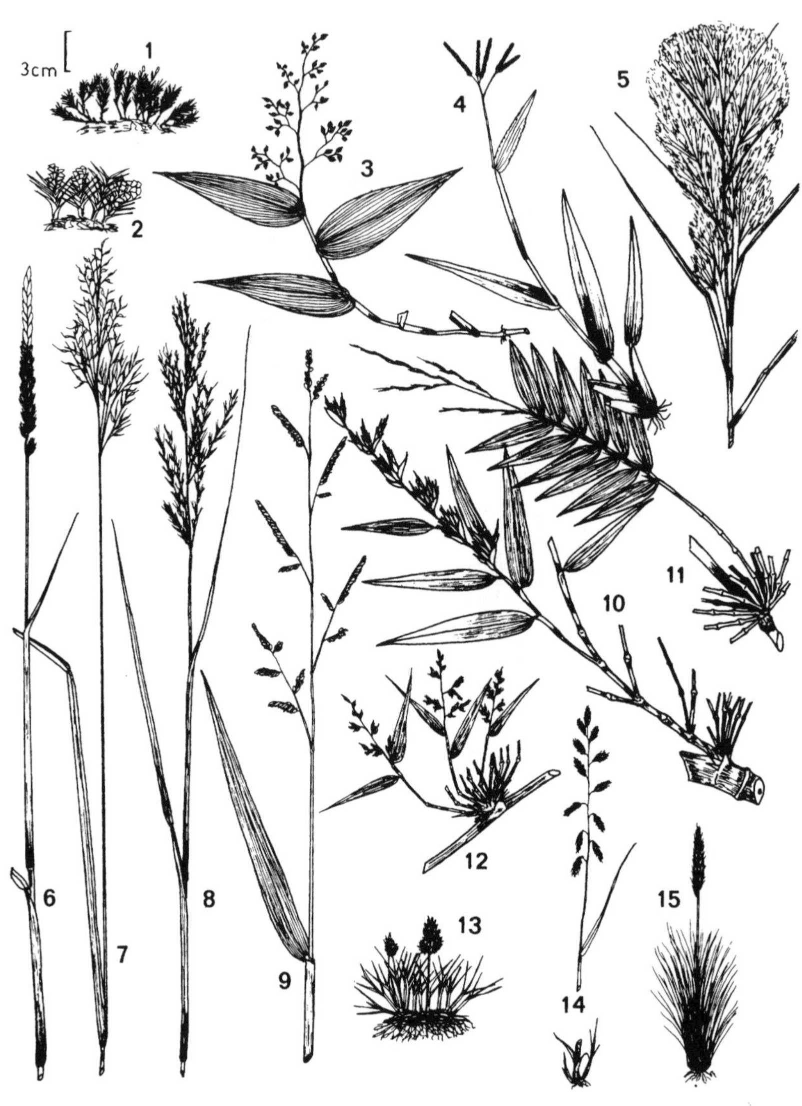

1 - *Aciachne*
3 - *Acroceras*
4 - *Axonopus*
5 - *Andropogon*
2 - *Anthochloa*
11 - *Arthrostylidium*
10 - *Guadua*
12 - *Chusquea*
6 - *Agropyron*
8 - *Aristida*
7 - *Agrostis*
9 - *Brachiaria*
14 - *Bouteloua*
13 - *Dissanthelium*
15 - *Calamagrotis*

Gramineae
(*Bromus* and *Chloris* – H)

1 - *Bromus* **3** - *Coix* **5** - *Echinochloa* **7** - *Eleusine*
2 - *Chloris* **4** - *Digitaria* **6** - *Eriochloa*

8 - *Eragrostis* **10** - *Gynerium* **12** - *Hordeum* **14** - *Hymenachne*
9 - *Festuca* **11** - *Hemarthria* **13** - *Homolepis* **15** - *Hyparrhenia*

Gramineae
(I – O)

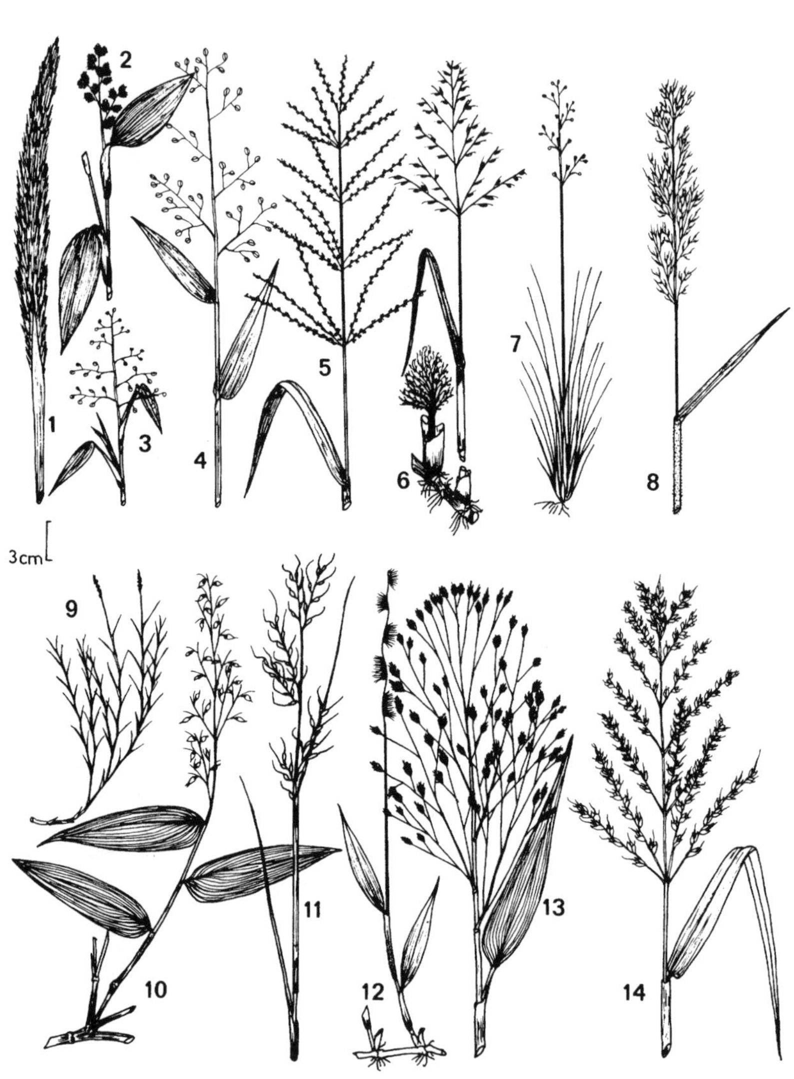

2 - *Ichnanthus*

3 - *Isachne* **5** - *Leptochloa* **7** - *Lorenzochloa*

1 - *Imperata* **4** - *Lasiacis* **6** - *Luziola* **8** - *Melinis*

9 - *Muhlenbergia* **11** - *Nasella* **13** - *Orthochloa*

10 - *Olyra* **12** - *Oplismenus* **14** - *Oryza*

Gramineae
(P – T)

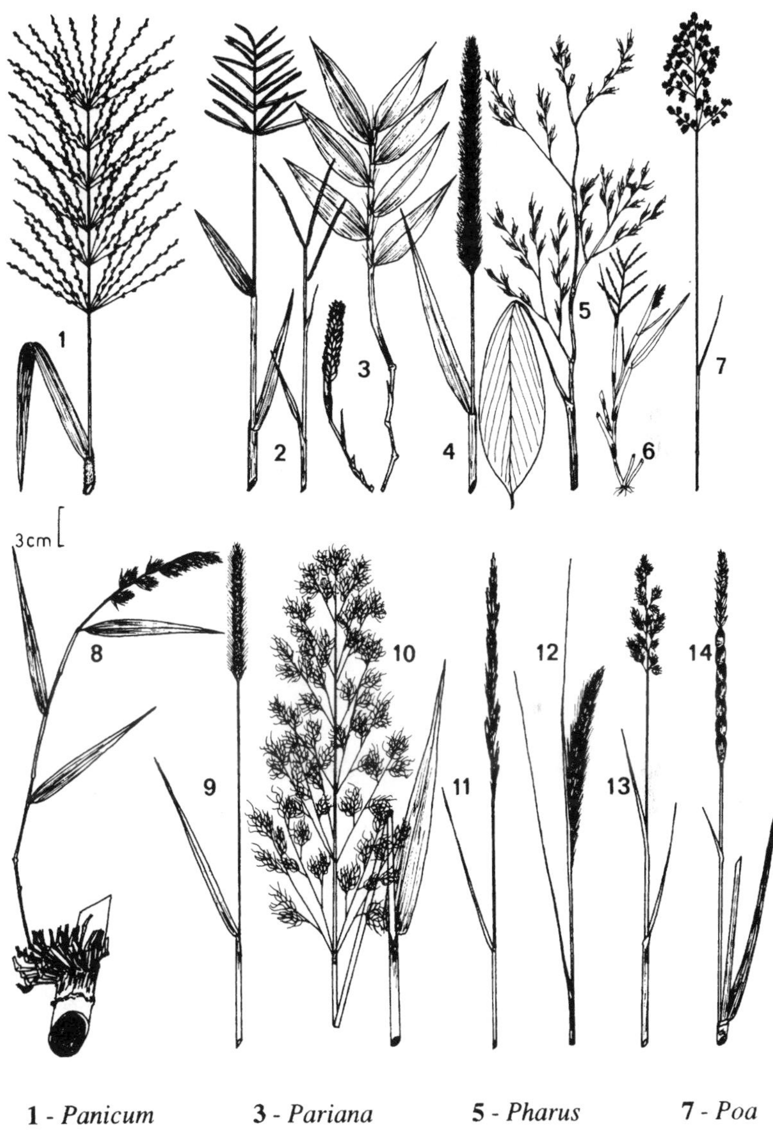

1 - *Panicum* **3** - *Pariana* **5** - *Pharus* **7** - *Poa*

2 - *Paspalum* **4** - *Pennisetum* **6** - *Reimarochloa*

8 - *Rhipidocladum* **10** - *Sorghum* **12** - *Stipa* **14** - *Tripsacum*

9 - *Setaria* **11** - *Sporobolus* **13** - *Trisetum*

Key to Woody Grasses

1. Culm internodes solid; without hollow center.

2. Leaf blades 1–several m long, borne on main stem; stems pithy, not extremely woody.

3. Leaf blades evenly distributed along stems; monoecious; spikelets each with single fertile flower, disarticulating below glumes; cultivated sugar cane..***Saccharum officinarum***

3. Leaf blades borne in a large, fan-shaped cluster at apex of stem; dioecious; solitary spikelets 2-flowered, disarticulating above the glumes; wild plants, usually along river banks...***Gynerium sagittatum***

2. Leaf blades usually less than 20 cm long, mostly borne on small lateral branches or from near base of unbranched subwoody stem; main stems leafless or bearing nearly bladeless sheaths.

4. Stems woody and branching..***Chusquea***

4. Stems nearly herbaceous, unbranched.........................***Neurolepis***

1. Culm internodes hollow.

5. Leaf blades 1–2 m long, cordate-clasping at base, borne on main stem; culms thin-walled; cultivated or escaped....................***Arundo donax***

5. Leaf blades usually <20 cm long, not cordate-clasping, mostly with short pseudopetioles; wild or cultivated.

6. Stems with thorny branches...........................***Bambusa (Guadua)***

6. Stems lacking thorny branches.

7. Stems barely woody, internodes slender (<1 cm wide), green, soft (can be crushed with fingers).

8. Stems with glistening viscid band at nodes; plants rarely blooming..***Aulonemia***

8. Stems not viscid; plants blooming annually.

9. Spikelets alike, black and shiny when mature, placed at angle to pedicel; disarticulation below glumes; leaf blades sessile, without pseudopetioles............................***Lasiacis***

9. Spikelets unisexual, of two kinds, not black, aligned with pedicel; disarticulation above glumes; leaf blades with short pseudopetiole..***Olyra***

7. Stems definitely woody, of various diameters, sometimes very large.

10. Small shrubby bamboos of paramos above 3000 m elevation; hollow stem center small and without definite boundary membrane; branches stiff and erect, usually 3–5 per node..... ....***Swallenochloa***

10. Bamboos of various habits, occurring in forest or savanna below 3000 m elevation; hollow stem center usually large and with definite boundary membrane; branches 2–many per node.

11. Branches at midstem nodes numerous, arising from edges of flat, triangular platelike meristem that is closely appressed to main stem.

12. Midstem sheaths with a narrow reflexed blade that is constricted at its base and much narrower than sheath apex.. ***Merostachys***

12. Midstem sheaths with an erect blade that is as wide as the sheath apex and not constricted at base.......... ....***Rhipidocladum***

11. Branches at midstem nodes not arising from flat plate, of varying number.

13. Primary midstem branches solitary, soon branched near base; auricular bristles very prominent, to 8 cm long on main stem sheaths.................***Elytrostachys***

13. Primary midstem branches 2–several per node; auricular bristles short.

14. Branches 3–many per node, arising above the node at the apex of a prominent bulge that continues down to the node; internodes cylindrical in cross-section; native................***Arthrostylidium***

14. Primary branches usually 2 per node; internodes D-shaped in cross-section; cultivated bamboos ....***Phyllostachys***

1. Woody Grasses

1A. Woody bambusoids with pseudopetiole between sheath and blade

1Aa. Solid stem

Chusquea (120 spp.) — Mostly scandent or at least strongly arching slender montane bamboos. Vegetatively distinguished by the solid stems and the distinctive branching often with one large branch and numerous small leafy branches together at each node.

C: chusque; E: carrizo

1Ab. Hollow stem

Guadua (30 spp.) — Our only native large bamboo. Distinctive in the spiny branches. Generic segregation of the neotropical species from paleotropical *Bambusa* is moot.

E: caña brava

Merostachys (40 spp.) — Clump-forming more or less arching (or scandent in upper portion) bamboos, related to *Rhipidocladum* by the otherwise unique feature of the multiple branches at each node arising from flat triangular plate; characteristic in the midstem blades reflexed and so strongly basally constricted that pseudopetiole base is narrower than sheath apex.

Rhipidocladum (10 spp.) — Slender viny bamboos, mostly of lowland forest, characterized by the fan-shaped branch clusters arising from a flattened triangular meristem just above node; differs from *Merostachys* in the blades stiff and erect with base of blade as wide as sheath apex.

Elytrostachys (2 spp.) — Tall nonspiny lowland bamboos, becoming subviny. Distinctive in the combination of solitary branching and strongly reflexed leaf blades; also in the very long bristles around apex of sheath.

Arthrostylidium (20 spp.) — Viny or subviny cloud-forest bamboos differing from *Elytrostachys* in 3–many branches per node and short bristles around sheath margin; from *Rhipidocladum* and *Merostachys* in lacking the triangular meristem above node.

(*Swallenochloa*) (7 spp.) — Short densely clump-forming paramo bamboos with stiffly erect branches. A high-altitude nonscandent version of *Chusquea* from which it differs, perhaps inadequately, in the (slightly!) hollow internodes, equal primary-branching, and denser inflorescences.

Aulonemia (24 spp.) — Small clump-forming montane nonthorny bamboos with rather succulent barely woody stems and viscid bands at nodes; often on cliffs.

Neurolepis (9 spp.) — Barely subwoody high-Andean grasses with very large broad-bladed leaves mostly >2 cm wide. Differs from *Chusquea* and *Arthrostylidium* which occur in the same habitat by broader leaves with blade continuous with the sheath, in the erect nonbranching habit (even the individual leaves +/- erect), and the less woody stem.

Olyra (22 spp., also 1 African) — Barely subwoody, usually scrambling grasses often with broad blades. Differing from true bamboos in having leaves on the main stem as well as branches. Inflorescence paniculate with separate narrowly ovoid male (basal) and female (terminal) spikelets, floret structure confusing with a single floret per spikelet and formerly considered related to *Panicum.*

E: gramalote

1B. Nonbambusoid +/- woody grasses lacking pseudopetioles

1Ba. Panicoideae— (Cultivated sugar cane and berry-fruited scandent *Lasiacis*)

Lasiacis (20 spp.) — Subwoody clambering vine, very distinctive in the round, black, shiny-fruiting spikelets (highly unusual in grasses in being oily and berrylike at maturity). Leaf blade usually lacks pseudopetiole, unlike the climbing bamboos.

E: carricillo trepador

(Saccharum) (cultivated) — One widely cultivated Asian species, vegetatively characterized by the tall subwoody canes and in flower by the typical Andropogoneae spikelets.

C, E, P: caña de azúcar

1Bb. Arundinoideae

Gynerium (1(–2?) sp.) — Large erect canes to 10 m tall and 4–5 cm thick, barely subwoody, very common in early succession along low elevation river banks. Leaves in a more strongly fan-shaped cluster than in cultivated *Saccharum.* Inflorescence large and rather plumose.

P: caña brava

Phragmites (4 spp.) — Large subwoody marsh canes to several meters tall. Differs from *Gynerium* in perfect spikelets and ecological restriction to true standing-water swamps.

Cortaderia (19 spp., plus 4 in New Zealand and 1 in New Guinea) — The montane equivalent of *Gynerium,* which it resembles in the large paniculate plumose inflorescence; differs vegetatively from *Gynerium* in leaves mostly clustered near base of stem.

2. Herbaceous Grasses (Very Incomplete)

2A. Herbaceous bambusoids — Broad-leaved genera of forest understory; all with pseudopetiole.

Cryptochloa (15 spp.) — Forest-floor grass, characterized by the exceptionally short, narrowly oblong 2-ranked leaves. Inflorescence inconspicuous few-flowered and axillary.

Lithachne (4 spp.) — Forest-understory broad-leaved grass related to *Olyra* and similarly once placed in Panicoideae. Differs from *Olyra* (and *Cryptochloa*) in the asymmetric, apically truncate fertile florets. Leaf blades characteristically asymmetric and truncate-based.

Pariana (30 spp.) — One of the most distinctive forest-floor grasses, characterized by broad leaves with poorly developed pseudopetioles, and especially by the unique spicate inflorescence (probably insect-pollinated which in many species arises straight out of ground at some distance from plant base.

Pharus (5 spp.) — A broad-leaved forest-floor grass that vegetatively resembles *Streptochaeta* but differs in the veins of leaf blade running obliquely to margin and the blades twisted so that anatomical lower surface is upper surface. Inflorescence very different from *Streptochaeta,* paniculate, and with sticky awnless lemmas for exozoochoric dispersal.

Streptochaeta (3 spp.) — Lowland forest-floor grass. Vegetatively differs from *Pharus* in the veins of leaf blade completely parallel and running to tip of blade. Inflorescence spicate; spikelets distinctive in the long coiled awn, often dangling together in large groups.

Streptogyna (1 sp., plus 1 Old World) — Very similar to *Streptochaeta,* the inflorescence like a *Streptochaeta* twisted up with string from the greatly elongate stigmatic branches (rather than an elongate awn) which twist and entangle together so spikelets fall together as group.

(*Olyra*) — Some species are completely herbaceous.

Orthoclada (1 sp., plus 1 African) — Technically in its own small subfamily (Centothecoideae) but vegetatively looks like bambusoid on account of well-developed pseudopetiole. Paniculate inflorescence like *Pharus* but differs in 2-flowered spikelets that are not sticky.

2B. Herbaceous nonbambusoid grasses — Without pseudopetiole

2Ba. Oryzoideae—Marsh plants with glumes absent or very reduced; sometimes included in Bambusoideae despite different flowers and absence of pseudopetiole.

Leersia (18 spp.) — Laterally compressed spikelets, unlike *Luziola* from which it also differs in acuminate perfect florets and distinctive scabrous nodes.

Luziola (11 spp., incl. USA) — Spikelets terete, male and female spikelets in separate inflorescence (= old Zizanieae).

Oryza (20 spp., mostly Old World) — Differs from *Leersia* in the presence of vestigial glumes and the usually awned lemma.

2Bb. Arundinoideae — Most of our genera woody or subwoody (*Cortaderia, Gynerium, and Phragmites,* see above); mostly characterized by large plumelike panicles.

Aristida (250 spp., incl. n. temperate) — A savanna bunchgrass, formerly placed near *Stipa* in Pooideae. Characterized by 3-awned lemma, the awns upwardly scabrous for exozoochorous dispersal.

2Bc. Pooideae (Festucoideae) — Old Pooeae (*Bromus, Poa, Festuca*); plus Hordeae (*Agropyron, Hordeum*), Aveneae (*Dissanthelium, Trisetum*), and Agrostideae (*Aciachne, Agrostis, Calamogrostis, Nasella, Stipa*)

Aciachne (1 sp.) — Low cushion-plant of puna and paramo, characterized by the stiff narrow leaves with spiny tips.

Agrostis (220 spp., incl. n. temperate and Old World) — Paramo bunch grasses with narrow leaves. Inflorescence an open terminal panicle with numerous small spikelets with the pair of glumes exceeding the single floret (= old Agrostideae).

Anthochloa (1 sp.) — Dwarf grass of high-Andean punas. Vegetatively characterized by the sheath completely fused (= tribe Meliceae).

Bromus (150 spp., incl. n. temperate) — Similar to *Poa* (but awned and with larger spikelets) and very similar to *Festuca,* from which it differs in the awn from between two short teeth. Vegetatively distinctive in the closed sheaths.

Calamagrostis (270 spp., incl. Old World) — Paramo and puna bunch grasses, with narrowly paniculate inflorescence often rather woolly-looking from the plumose rachilla extended beyond the single floret.

Dissanthelium (16 spp.) — Dwarf high-Andean puna and paramo bunch grasses with small few-flowered panicles. Distinctive in glumes longer than the awnless 2-flowered spikelets.

Elymus (incl. Andean spp. of *Agropyron*) (150 spp., incl. n. temperate) — Inflorescence an unbranched terminal spike (= old Hordeae), differing from *Hordeum* in the spikelets several-flowered and solitary at each node.

Festuca (450 spp., incl. Old World) — Close to *Poa* (but awned) and *Bromus,* from which it differs in terminal awn.

Hordeum (40 spp., incl. n. temperate) — Like *Elymus* in the unbranched spicate inflorescence but the spikelets single-flowered and three instead of one at each inflorescence node.

Nasella (15 spp.) — An Andean segregate of *Stipa* from which it differs in the often eccentric, readily deciduous awn.

Ortachne (incl. *Lorenzochloa*) (3 spp.) — Paramo bunch grass with rigid, erect, terete, rather spiny-tipped leaves. Inflorescence a narrow few-flowered panicle, the 1-flowered spikelets with large lemma gradually narrowing to thick stiff awn (Stipeae).

Poa (500 spp., incl. n. temperate and Old World) — Close to *Bromus* and *Festuca* but lemmas awnless and acute.

Stipa (300 spp., incl. Old World) — Paramo and puna bunch grasses with paniculate inflorescence, the glumes larger than the single floret, the lemma tip prolonged into stiff, twisted, and usually bent awn (sharp floret base is drilled into the ground by the hydroscopically twisting awn) (Stipeae).

Trisetum (75 spp., incl. Old World) — Bunch grasses with paniculate inflorescence of 2–several-flowered spikelets. Differs from *Festuca* and relatives in glumes as long as lowest floret and lemma awned from near middle.

The next three genera have paniculate inflorescences like many Pooideae (where they were formerly placed) but are now considered Chlorideae on anatomical grounds. These differ from *Calamagrostis* in rachilla not being elongated and from *Stipa* in lacking sharp fruit base.

Eragrostis (300 spp., incl. Old World) — Some species weedy, others in dry areas. Mostly bunch grasses with open panicles of several-flowered spikelets. Characterized by awnless spikelets with 3-nerved lemmas (*Festuca* and relatives have 5–many-nerved lemmas) and by glumes early caducous.

Muhlenbergia (160 spp., incl. Old World) — Inflorescence paniculate (though often narrow- or few-branched) the spikelets 1-flowered (unlike *Eragrostis*), disarticulating above the peculiarly different-sized glumes. Differs from *Sporobolus* in 3-nerved often awned lemma and a membranaceous ligule.

Sporobolus (160 spp., incl. Old World) — Inflorescence paniculate (though frequently narrow), the small spikelets 1-flowered, often with unequal glumes, typically not disarticulating with the adhesive seed released naked from the split ovary wall. Differs from *Muhlenbergia* in awnless 1-nerved lemma, and hairy ligule.

2Bd. Chloridoideae — Characterized by spicate-branched inflorescence with laterally compressed spikelets sessile along one side only (but similar inflorescences found in Panicoideae). (Excluding the paniculate-inflorescenced Eragrosteae — *Eragrostis, Muhlenbergia, Sporobolus* [see above]).

Bouteloua (24 spp.) — Characteristic inflorescence with the short spikes racemosely arranged and several to many rather crowded spikelets per spike, each with 1 fertile basal floret and 1–2 modified sterile florets above.

Chloris (55 spp.) — Inflorescence of whorled spikes, the sessile spikelets appressed along the lower side of rachis, each with one perfect basal floret and 1–several sterile florets above.

Cynodon (introduced) — Stoloniferous and rhizomatous weeds or forage grass, like *Chloris* in whorled spikes of 1-flowered spikelets. Vegetatively distinctive in the short leaves which are +/- opposite on the stolons.
C, E, P: zacate de Bermuda

Eleusine (2 sp., plus 7 African) — Our species a weed. Like *Chloris* and *Cynodon* in whorled spikes with the spikelets borne densely along lower side of rachis, but spikelets longer and with several florets.

Leptochloa (40 spp.) — Mostly weedy bunch grasses, characterized by the unusually slender, racemosely arranged, elongate branches of the paniculate inflorescence, the small several-flowered spikelets not very densely arranged and disarticulating above the glumes which remain on old inflorescences.

2Be. Panicoideae — Spikelets with one (except *Isachne* with two perfect) terminal floret above a sterile or staminate floret, disarticulating below the glumes or in the rachis; spikelets in fruit more less dorsally compressed.

(i) Andropogoneae — Glumes hard and coriaceous in contrast to the hyaline or membranaceous lemma and palea; the fertile spikelets perfect, each paired with a sterile spikelet, in most genera the rachis disarticulating and spikelet pairs fall together. For monoecious taxa (= old Triticeae), see (ii) below.

Andropogon (113 spp., incl. Old World) — Bunch grasses, the inflorescence paniculate with usually plumose, more or less erect, racemose branches, these disarticulating at maturity, each internode with a sessile awned fertile spikelet and a pedicel bearing a reduced or abortive spikelet.

Hyparrhenia (1 sp., plus 54 Old World) — In our area only one native species and a common weedy one originally from Africa. The weedy species vegetatively distinctive in the colorful contrasting reddish and tan segmented mature stems. Technically differs from *Andropogon* in oblique callus, in the first glume of sessile spikelet rounded and incurved rather than marginally keeled, and in each branch having several basal pairs of sessile, sterile, or staminate spikelets.

Imperata (8 spp., incl. Old World) — Weedy rhizomatous grasses. Inflorescence a densely cylindrical spikelike panicle with the spikelets hidden by the very characteristic long silky hairs of the inflorescence. Differs from *Andropogon* in sessile and pedicellate spikelets alike.

Ischaemum (65 spp., incl. Old World) — Sprawling weedy grasses. Inflorescence of 2–many digitately borne racemes which fragment at internodes; unusual in tribe in staminate fertile lower floret.

Sorghastrum (16 spp., incl. African and n. temperate) — American equivalent of *Sorghum;* similar to *Sorghum* in the cylindrically paniculate inflorescence, each branch of which has 1–few nodes each with a sessile perfect-flowered awned spikelet and a slender hairy pedicel (unlike *Sorghum,* this lacks a terminal spikelet), the rachis disarticulating.

Sorghum (1 Mexican sp., plus ca. 19 Old World) — Several originally African species occur in cultivation or as weeds. Differs from *Sorghastrum* in the pedicellate second spikelet of each node being a fertile staminate spikelet.

Trachypogon (2 spp., plus 3 African) — Savanna bunch grasses frequently dominant in well-drained soils of the Llanos. Characterized by usually solitary terminal racemes with pairs of persistent short-pedicellate staminate and disarticulating long-pedicellate perfect spikelets, the latter awned. Also differing from *Andropogon* and relatives by the nondisarticulating rachis and by the base of the perfect spikelet sharp-pointed and hairy (for exozoochorous dispersal?).

(ii) Monoecious Andropogoneae (= old Tripsaceae, now subtribes Tripsacinae and Coicinae) — Characterized by being monoecious with separate male and female flowers, the male above and the female below on the same inflorescence.

Tripsacum (13 spp., incl. USA) — Inflorescence of solitary or whorled spikes, lower portion with the solitary pistillate spikelets sunken into recesses in the thickened rachis joints; upper portion with paired male spikelets.

Coix (introduced) — Cultivated and escaped. Very distinctive in the pistillate spikelets and fruits enclosed in a white, bony, round, beadlike structure, with short staminate racemes extruded from bead.

(iii) Paniceae — Glumes and lower sterile lemma thin and membranaceous (or the first glume absent), contrasting with coriaceous fertile lemma and palea; spikelets disarticulating below the glumes.

Acroceras (1 sp., plus several Old World) — A coarse stoloniferous semiaquatic. Looks much like *Lasiacis* but nonwoody and fruits not black and fleshy at maturity.

Brachiaria (100 spp., mostly Old World) — Sprawling weedy grass. Inflorescence a panicle with dense racemosely arranged racemes, the subsessile spikelets usually in rows. Inflorescence like *Paspalum* but both glumes present as in *Panicum.*

Cenchrus (22 spp., incl. Old World) — Weedy grass, especially in sandy areas. Very distinctive in the several-spined involucre connate into a burlike fruit (for exozoochorous dispersal); closest to *Pennisetum* from which it differs in the bristles thicker and united at base (= Cenchrinae).

Digitaria (230 spp., incl. Old World) — Usually weedy, often stoloniferous grasses. Characterized by the inflorescence with slender unbranched lateral branches arranged either racemosely or umbellately and with spikelets in pairs or triads, more pointed than in most *Paspalum,* though similarly densely arranged along the flattened rachis. Technically differs from *Paspalum* in the soft fertile floret with thin exposed (not inrolled) lemma edge; (= Digitariinae).

Echinochloa (30 spp., incl. Old World) — Mostly weedy or in swamps. Similar to *Panicum* but differing in the spikelets cuspidate or awned, the fertile lemma margin not inrolled, and racemosely arranged branches. Leaves always linear, unlike similarly awned *Oplismenus.*

Eriochloa (30 spp., incl. Old World) — *Paspalum*-like with one-sided mostly short and racemose lateral branches, one to many along central axis. The main character separating from *Paspalum* is the presence of a thickened bead of tissue forming stalk at base of spikelet.

Homolepis (3 spp.) — Stoloniferous weedy grass. Spikelets in open panicles as in many *Panicum* species but the spikelets longer and more acute than in *Panicum* and entirely enclosed by the pair of glumes.

Hymenachne (5 spp., incl. Old World) — Swamp grasses with spongy stem centers. Inflorescence a dense spikelike narrowly cylindric panicle with the individual spikelets more narrowly acuminate than in *Panicum,* and differing from *Pennisetum* in lacking bristles.

Ichnanthus (33 spp.) — Mostly stoloniferous, creeping, forest-margin grasses, usually with short broadly ovate leaves. Closely related to *Panicum,* but differs in the usually more acuminate glumes (and narrower spikelet) and the fertile lemma with wings or broad scars at base.

Isachne (100 spp., mostly Old World) — Unique in Panicoideae in spikelets frequently with 2 fertile florets, the spikelet thus nearly globose but otherwise superficially similar to *Panicum.*

(*Lasiacis*) — A few *Lasiacis* species are only subwoody. They are readily recognized by the fleshy, black, berrylike mature fruit.

Oplismenus (5 spp., incl. Old World) — Creeping grasses with narrowly paniculate inflorescence of conspicuously awned spikelike racemes; unlike similarly awned *Echinochloa*, the plants are stoloniferous and have very characteristic short, broadly lanceolate leaves.

Panicum (500 spp., incl. Old World) — A huge genus characterized by awnless dorsally compressed spikelets, usually borne in open panicles (unlike *Paspalum*); but a few species have racemose *Paspalum*-like inflorescences. Technically differs from *Paspalum* in having a well-developed short first glume and subequal second glume and sterile lemma together enclosing the fertile floret.

Paspalum (330 spp., incl. Old World) — A very large genus differing from *Panicum* in the racemosely branched rather than paniculate inflorescences (sometimes reduced to only 1–2 racemes); the racemes one-sided with subsessile nonawned spikelets, at least in part, with first glume absent.

Pennisetum (80 spp., incl. Old World) — Mostly either weedy or middle-altitude grasses characterized by the combination of a bristly cylindrical inflorescence with panicoid spikelet structure (although one introduced high-altitude species has the inflorescence reduced to a few spikelets hidden within the subtending sheaths). Differs from *Setaria* in the subtending bristles falling with spikelets at maturity; (= Cenchrinae).
C, E, P: paja elefante

Setaria (100 spp.) — Mostly weedy grasses characterized by the combination of bristly cylindrical inflorescence with panicoid spikelet structure. Differs from closely related *Pennisetum* in disarticulation of the spikelets above the bristle attachments.

(iv) Melinideae and Arundinelleae — Paniceae genera distinct from *Panicum* and relatives in disarticulating above the glumes, a reduced lower glume, and the fertile lemma and palea of about same texture as glumes and either the fertile or sterile lemma awned.

Melinis (introduced; 11 African) — Sprawling weed, vegetatively characterized by the viscid pubescence of the plant and characteristic odor of molasses; the inflorescence a panicle, the solitary spikelets pedicellate with an awned sterile lemma below the unawned fertile floret.
(molasses grass)

Arundinella (5 spp., plus 42 Old World) — Characterized by the spikelets in triads and by the peculiar 2-flowered spikelet structure, the lower floret sterile, the fertile upper floret with an awned lemma. Differs from *Melinis* in the fertile, rather than sterile, lemma awned.

There are many additional genera, even within our area but most are infrequently encountered.

Haemodoraceae, Iridaceae, and Liliaceae

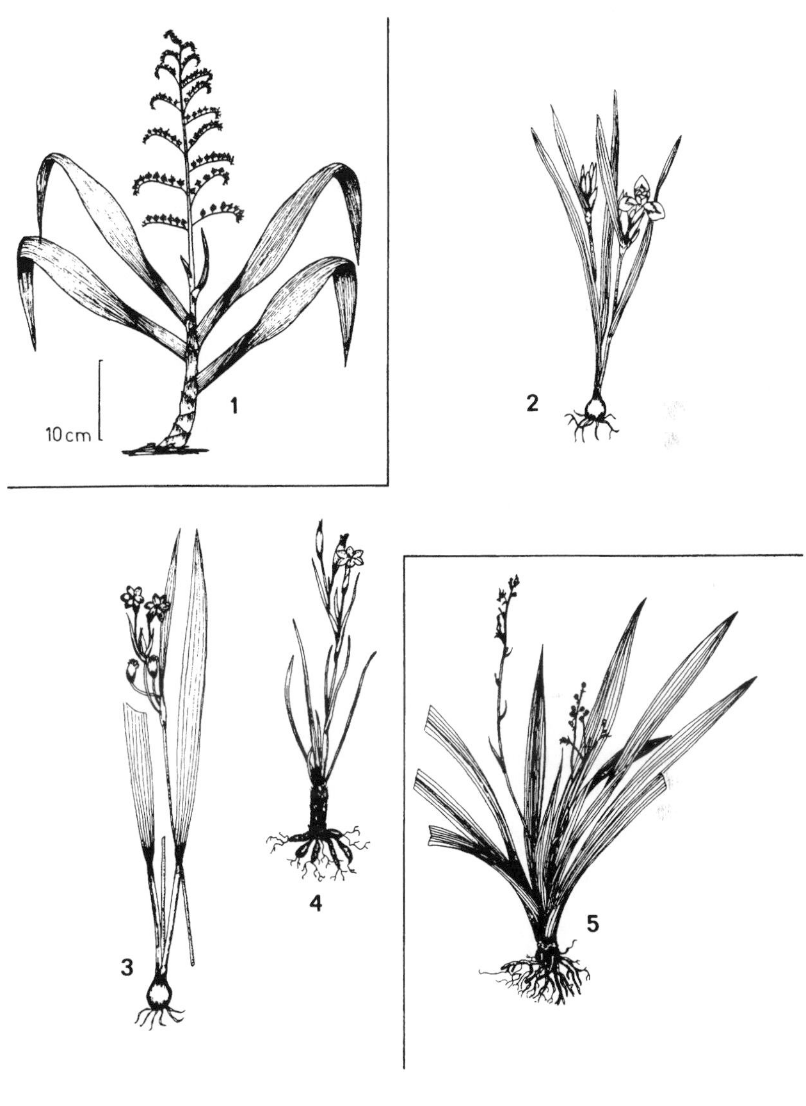

1 - *Xiphidium* (Haemodoraceae)

2 - *Cipura* (Iridaceae)

4 - *Sisyrinchium* (Iridaceae)

3 - *Eleutherine* (Iridaceae)

5 - *Anthericum* (Liliaceae)

HAEMODORACEAE

Differs from all other monocots except Iridaceae and some Rapateaceae in the equitant leaves which are held +/- vertically with the top and bottom surfaces not differentiated. The rather succulent leaves are linear with closely parallel veins and are softer and less fibrous than in Iridaceae. The leaves are distichous and the whole plant is laterally compressed with shallowly sheathing leaf bases. The flowers of our taxa, similar to those of many Liliaceae, are small and regular with three white petals (these sometimes orange at base) fused at base. Only three small genera are represented in South America, two of these in our area.

Xiphidium (2 spp.) — Widespread in moist and wet lowland forest. Leaves mostly >2 cm wide. Inflorescence pyramidal-paniculate with an erect rachis and strongly spreading lateral branches.

Schiekia (1 sp.) — In our area mostly restricted to laja outcrops in the Orinoco area. Vegetatively distinctive in the blood-red rhizomes. Leaves (<1 cm wide) and inflorescence both narrower than in *Xiphidium,* the inflorescence pubescent and racemose-paniculate with reduced lateral branches.

A monotypic third genus *Pyrrhorhiza* is endemic to Cerro Neblina.

IRIDACEAE

In our area poorly represented and uniformly herbaceous, vegetatively characterized by the narrow more or less linear leaves which are very distinctive in being 2-ranked and equitant, with the upper and lower surfaces of the vertically held leaf not differentiated. The leaves of stemmed taxa are usually stiffer and tougher in texture than in the other equitant-leafed family Haemodoraceae. The regular flowers are often very showy, have inferior ovaries, and differ from related Liliaceae and Amaryllidaceae in having only three stamens (and from the former in the inferior ovary). Our only significant genus is *Sisyrinchium* (ca. 60 spp., plus 20 in temperate N. Am.), restricted to high-Andean paramos and puna and characterized by small blue or yellow flowers. Also common in moist disturbed Andean habitats is introduced but widely naturalized orange-flowered *Crocosmia.* Many species, especially in drier areas, have bulbs and may lack above-ground stems. Bulbous genera represented in our area include *Hesperioxiphion* (incl. *Cypella*) (20 spp.), *Mastigostyla* (incl. *Nemastylis*) (16 spp.), *Cardenanthus* (8 spp.), and *Tigridia* (35 spp.), all confined to the Peruvian coastal lomas or open grassy or rocky middle-elevation Andean slopes. Only three genera of bulbous Iridaceae occur in lowland Amazonia, each with a single species in our area — *Cipura* (6 spp.), *Eleutherine* (5 spp.), and *Ennealophus* (2 spp.). Besides *Sisyrinchium*, nonbulbous taxa in our area are *Orthrosanthus* (4 spp.) and *Olsynium* (incl. *Phaiophleps*) (11 spp.), both restricted to high-Andean paramos and puna and both very similar to *Sisyrinchium*.

LILIACEAE

A large diverse herb family with many extralimital genera but relatively poorly represented in our area. Traditionally differentiated from closely related Amaryllidaceae primarily on account of superior rather than inferior ovary, but both families are nowadays often more finely split using different taxonomic criteria. Exclusion of *Smilax* as Smilacaceae (and recognizing the succulent relatives of *Yucca* and *Agave* as Agavaceae) means that in our area Liliaceae are represented entirely by terrestrial herbs. There are climbing liliaceous genera in Chile and subwoody lianascent *Herreria* with characteristic whorls of leaves occurs in the Bolivian chaco and dry forests but none of these reaches as far north as Peru.

MARANTACEAE

Large-leaved monocots with ovate or elliptic to oblong leaves with well-developed midveins and numerous very fine closely parallel lateral veins curving from midrib to margin. Most species are coarse herbs but some species of *Ischnosiphon* are subwoody vines. Differs vegetatively from Musaceae in having a cylindrical pulvinar area at apex of petiole and the presence of minute cross veinlets in the leaves (visible as fine enations along the edge of a torn leaf section). Inflorescence usually conspicuously bracteate with complex but ephermeral, somewhat irregular flowers having 3 usually relatively inconspicuous petals basally united into tube and only a single fertile stamen (with single anther, only one side of which is fertile), also with 3–4 petaloid staminodes, these usually more conspicuous than the actual petals.

Calathea and its closest relatives have a 3-locular ovary while *Ischnosiphon, Maranta* and their relatives have a unilocular ovary. The genera are here arranged in an informal sequence from densest and least-branched to least dense and most-branched inflorescences.

Calathea (250 spp.) — The main neotropical Marantaceae genus; erect lowland-forest herbs mostly ca. 1–2 m tall. Characterized by the dense spikelike (or very rarely with spikelike branches: *C. nodosa*) inflorescence (only 1–few per shoot) with strongly overlapping spirally arranged +/- coriaceous bracts. The inflorescence distinct from *Ischnosiphon* in being broader, nearly always unbranched, and in the wider less woody bracts ("spathes"). Differs technically from *Ischnosiphon* relatives in 3 fertile locules per ovary.

C, E, P: bijao, platanillo, bijao macho (*C. insignis*)

Hylaeanthe (5 spp.) — Forest-floor herb with distinctively obovate leaves with cuneate bases and the petiole margins rotting to leave a netlike fiber network. The broad obtuse overlapping inflorescence bracts arranged along one side of inflorescence: looks like a one-sided *Calathea.* Technical characters are soft herbaceous bracts, the presence of 2 outer staminodes, and the ridged capsules.

Myrosma (3 spp.) — Small dry-forest herb, leaves dying back during dry season, probably reaching our area in the Llanos. Characterized by thin overlapping bracts similar to *Ctenanthe* but white or cream. Inflorescence single or two together, smaller and more tenuous than in most *Calathea,* the flowers white.

Ctenanthe (15 spp.) — Small forest-floor herbs with 2–several inflorescences borne together and having slightly overlapping green bracts and white flowers. Inflorescence +/- intermediate between *Ischnosiphon* and *Calathea* in the bracts narrow but not very close together and arranged in definite spikelike inflorescence branches, but with dorsoventral bract arrangement (radial in *Calathea* and *Ischnosiphon*).

Ischnosiphon (35 spp.) — Mostly erect lowland-forest herbs, but a few species subwoody and distinctly scandent (our only vine marantacs). Easily recognized by the elongate narrow spikelike inflorescence with overlapping subwoody bracts rolled tightly around inflorescence axis, the whole remarkably pencil-like in size and appearance. A useful vegetative character (shared with *Pleiostachya* and *Monotagma*) is the usually obliquely asymmetric leaf apex.

P: bijao

Pleiostachya (3 spp.) — A segregate from *Ischnosiphon,* differing primarily in the broader flattened (rather than cylindrical) inflorescence with strongly overlapping chartaceous bracts.

Monotagma (40 spp.) — Forest-floor herbs, mostly of poor-soil forests. Closely related to *Ischnosiphon* but with the inflorescence more branched, the bracts (spathes) rolled individually and not enclosing inflo-

Marantaceae

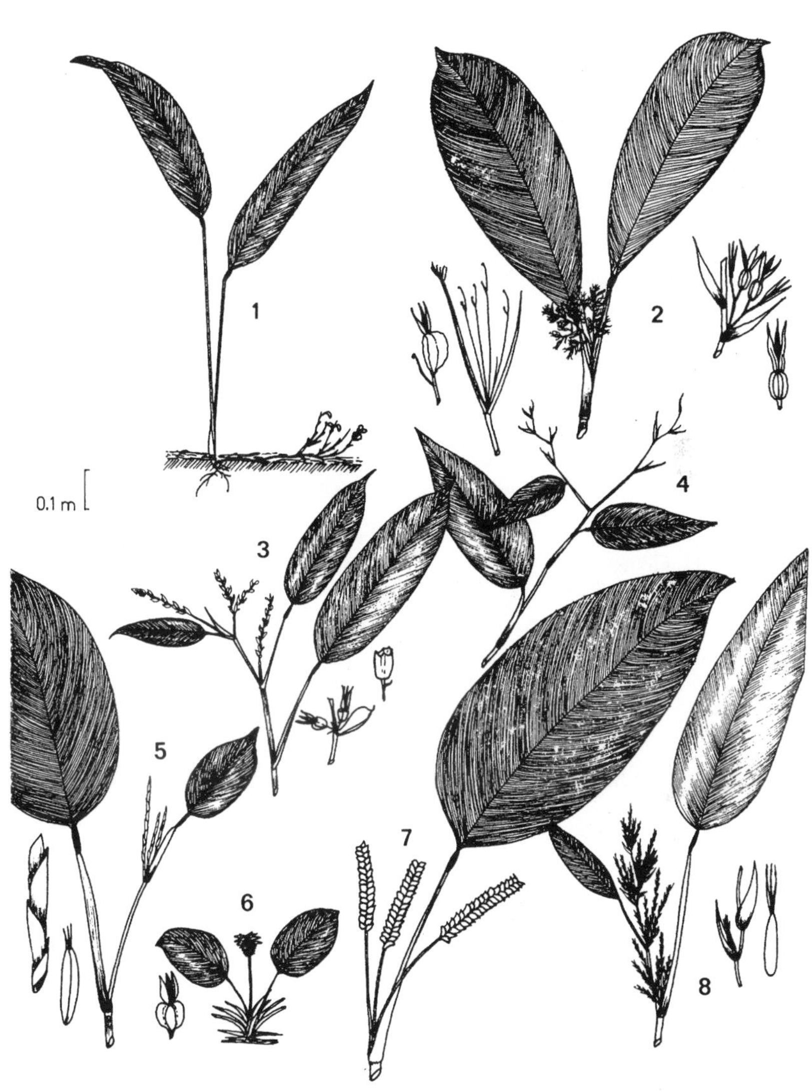

1 - *Hylaeanthe* **2** - *Myrosma*

3 - *Stromanthe* **4** - *Maranta*

5 - *Ischnosiphon* **6** - *Calathea* **7** - *Calathea* **8** - *Monotagma*

rescence axis, less tightly overlapping than in *Ischnosiphon* and the inflorescence somewhat lax and not as rigidly pencil-like. The main technical character is a series of single flowers (rather than paired flowers) per bract. Flowers are often orange, rare elsewhere in family.

Stromanthe (15 spp.) — Plant more branching and canelike and inflorescence more openly branched than in above taxa. The bracts obtuse and slightly overlapping, frequently orange. Closely related to and not clearly separated from *Ctenanthe* with some species moved back and forth.

Maranta (ca. 30 spp.) — Mostly a forest-understory herb; plant smaller and with smaller leaves than *Thalia*. Inflorescence open as in *Thalia* but few-bracted and few-flowered with white flowers.

Thalia (11 spp., incl. African) — A coarse diffusely branched swamp herb with the most open diffuse, least conspicuously bracteate inflorescence of any neotropical marantac. Flowers usually lavender or purple, unlike *Maranta*. A technical character is possession of 1 outer staminode which has 2 long appendages.

Musaceae

Banana-like plants with well-developed stems and large oblong or oblong-elliptic leaf blades with well-developed midvein and numerous very fine parallel veinlets running more or less perpendicularly from midvein to margin. Differs vegetatively from Marantaceae in lacking both a cylindrical pulvinar area at petiole apex and cross-veinlets (the leaf blade, thus, tending to tear easily). Inflorescence very characteristic with large conspicuous brightly colored bracts subtending the flowers. We have only one native genus, *Heliconia*, sometimes segregated as a distinct family (Heliconiaceae). Closely related *Phenakospermum* is here treated under the segregate family Strelitziaceae, differing from *Heliconia* in its tree habit and from cultivated *Musa* in 2-ranked leaves.

Heliconia (80 spp., plus few in Australasia) — The main neotropical Musaceae and one of the most characteristic elements of neotropical forests, on account of its ubiquity, large conspicuous leaves, and especially the large bracteate, bright-colored (usually red) inflorescences with hummingbird-pollinated flowers. Some species have erect inflorescences, others pendent; all have 3-lobed 3-seeded berrylike fruits. Most are coarse herbs, ca. 1–2 m tall, but a few species can be 5 m or more tall.

C, E: platanillo; P: situlli

(*Musa*) — The cultivated banana and its close relative the platano are widely cultivated throughout lowland tropical America.

Musaceae

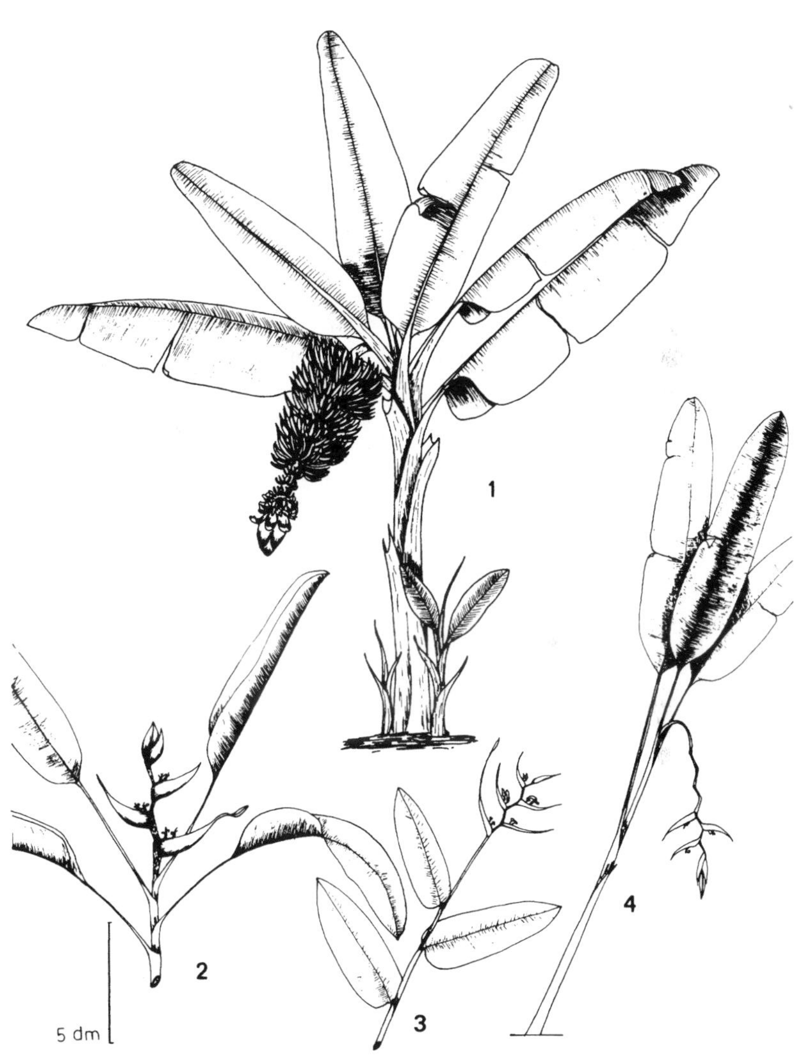

1 - *Musa*

2 - *Heliconia* **3** - *Heliconia* **4** - *Heliconia*

Orchidaceae

An entirely herbaceous family, mostly of epiphytes, but including some hemiepiphytic climbers (*Vanilla*), and some fairly large canelike terrestrial species. The best vegetative character of (most) orchids is the pseudobulb, a solid swollen bulblike stem section from which the leaves arise; pseudobulbs (or underground corms in some terrestrial taxa) are present in the majority of the orchids, which belong to the advanced subfamily Epidendroideae. Orchids without pseudobulbs (essentially the more primitive and mostly terrestrial subfamilies [Cypripedioideae, Spiranthoideae, and Orchidoideae]) can be vegetatively recognized by the white often rather grublike roots (= root-stem tuberoids) (most orchid roots with a distinctive whitish velamen), or by the thick fleshy leaves. The family is unmistakable in flower on account of its highly elaborate flowers (see Fig. 28) with specialized labellum (= lip, the enlarged lower petal), column (fleshy central structure produced by fusion of stamen and pistil), and pollinia (borne at stigma apex and often with a sticky attachment pad called a viscidium), and is famous for its complex coevolutionary relationships, especially with the euglossine bees which are pollinators of many orchids. The fruits are also completely unmistakable, each more or less cylindrical and 3-parted, dehiscent to release huge numbers of tiny dustlike seeds.

Subfamilial arrangement (following Dressler: *The Orchids, Natural History and Classification*) is largely based on characters of the column and anther. The three most primitive subfamilies (in our area), as well as the anomalous achlorophyllous saprophytic *Wullschlaegelia* (Fig. 1), are mostly terrestrial, and lack pseudobulbs. Cypripedioideae (with few genera and species in our area) tend to have elongate stems and conduplicate leaves and are distinctive in their lady-slipper flowers (from the saccate lip). Spiranthoideae (mainly *Erythrodes* and relatives with roots along elongate rhizome and *Cranichis* and relatives with clustered roots) have a distinctive column structure with the rostellum and anther about equal in size, and the viscidium borne at apex of anther and apically attached to pollinia. Orchidoideae (mostly African and European, in our area mostly *Habenaria* and relatives) have clustered fleshy roots or distinctive white grublike "root-stem tuberoids" at base of stem and are often rosette plants, usually with spiral leaves; the erect anther projects beyond (and is longer than) the rostellum, and the viscidia are at base or middle of pollinia. The large advanced subfamilies Epidendroideae and Vandoideae, mostly epiphytic and mostly with pseudobulbs (or corms in some terrestrial species) are most definitively differentiated by anther position, erect and opening basally (and with a stipitate viscidium) in Vandoideae, erect in bud but bending down to make right angle with column apex in Epidendroideae (or the flexion lost in some genera with viscidia (but the epidendroid viscidium nonstipitate).

This is the largest family of plants, and our area is the world's center for orchid diversity, making a complete treatment here out of the question. However, since this is such an important epiphytic family, it seems useful to include a concise generic summary here, even though truly woody members are lacking in the family. The key below has kindly been provided by Calaway Dodson and is largely a translation of his Spanish language key to the orchid genera of the Andean countries (Rev. Mus. Ecuat. Cienc. Natur. 5: 5–35. 1986).

1. Flowers with 2 anthers; lateral sepals forming a synsepal; lip slipperlike (with the exception of *Phragmipedium lindenii*).
2. Plants with canelike stems; leaves plicate .... ***Selenipedium***
2. Plants caespitose; leaves smooth .... ***Phragmipedium***
1. Flowers with one fertile anther; lateral sepals not forming a synsepal or if so the lip not slipperlike.
3. Pollinia soft and mealy.
4. Anthers bending down to become more or less operculate on the apex of the column.
5. Stems woody and canelike or cormous or pseudobulbous.
6. Plants cormous or pseudobulbous.
7. Plants terrestrial, cormous .... ***Bletia***
7. Plants epiphytic; pseudobulbs of several leafy nodes.......... .... ***Chysis***
6. Plants with canelike stems.
8. Roots wiry; basal portion of the lip united to the column...... .... ***Palmorchis***
8. Roots relatively soft; basal portion of the lip free from the column.
9. Leaves and bracts similar; column with a pair of falcate appendages below the stigma..................... ***Xerorchis***
9. Leaves and bracts dissimilar, column without appendages below the stigma.
10. Flowers less than 1.5 cm long.
11. Inflorescence lateral; lip flattened laterally with a transverse platelike ridge in the middle of the cavity .... ***Sertifera***
11. Inflorescence terminal; lip saccate at the base with 1 or 2 pairs of fleshy calli.

Orchidaceae Floral Details

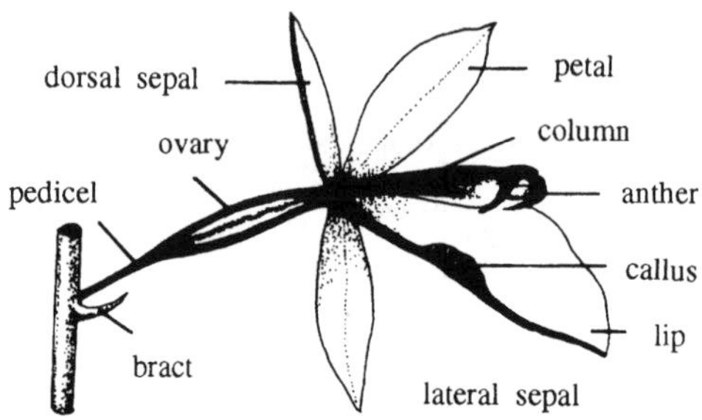

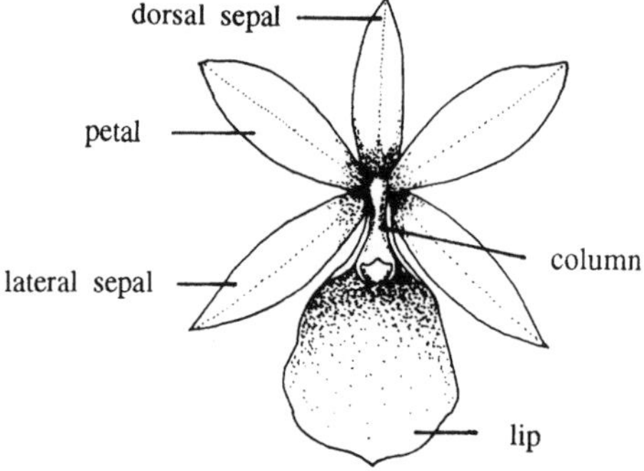

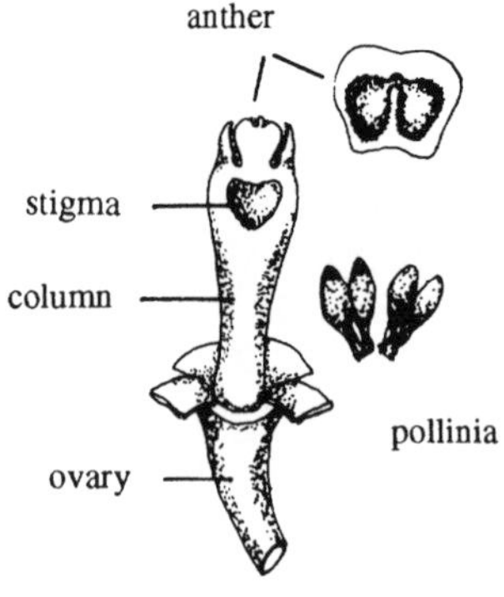

Orchidaceae

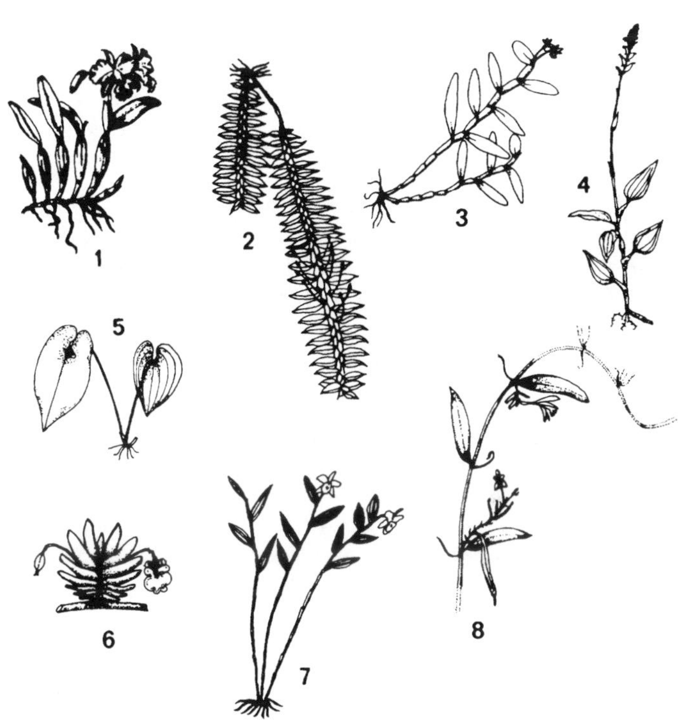

1 - *Cattleya* **2** - *Dichaea* **3** - *Epidendrum* **4** - *Erythrodes*

5 - *Pleurothallis*

6 - *Psygmorchis* **7** - *Sobralia* **8** - *Vanilla*

12. Leaves nonarticulate; lip with a pair of kidney-shaped calli at the base (sometimes separated into 2 pairs) ....***Epilyna***

12. Leaves articulated with the sheath, lip with 1 pair of ovoid calli at the base ....***Elleanthus***

10. Flowers more than 2.5 cm long ....***Sobralia***

5. Stems herbaceous.

13. Plants fleshy vines (sometimes leafless) ....***Vanilla***

13. Plants herbaceous.

14. Pollinia 4..***Duckeella***

14. Pollinia 2.

15. Base of lip with a pair of fleshy, clavate glands; flower without a calyculus beneath the perianth; leaves parallel-veined (sometimes included in *Pogonia*).............***Cleistes***

15. Base of lip without glands; flowers with a calyculus beneath the perianth; leaves reticulate-veined.................... ....***Epistephium***

4. Anther remaining erect or bending back, not short and operculate on the apex of the column.

16. Anther more or less embedded in the column; saprophytes with sectile pollinia of many slender massulae; roots ellipsoid; plants with erect leafless stems; terrestrial..............................***Wullschlaegelia***

16. Anther not surrounded by columnar tissue; plants leafy terrestrials or leafy epiphytes.

17. Sepals and petals united to near their apices, open between the lateral sepals; column-foot short ....***Uleiorchis***

17. Sepals and petals free to the base; column-foot prominent.

18. Anther usually projecting beyond the rostellum; viscidium, if present, usually at the base or middle of the pollinia; plants often with root-stem tuberoids.

19. Lip with a spur .... ***Habenaria***

19. Lip without a spur.

20. Leaves spirally arranged, conduplicate; often rosette plants ....***Chloraea***

20. Leaves scattered and subdistichous or plicate.

21. Plants 1-leafed; leaf subplicate, cordate........... ....***Monophyllorchis***

21. Plants with a leafy stem; leaves not cordate.

22. Pollinia 2; column winged; leaves and bracts distinctive...............................***Psilochilus***

22. Pollinia 4; column terete; leaves and bracts similar.. ***Triphora***

18. Anther subequal to the rostellum; viscidium at the apex of the anther and attached to the apex of the pollinia or caudical; plants without root-stem tuberoids.

23. Plants with a fleshy, recumbent stem or erect, woody, canelike stems, not rosulate; pollinia sectile.

24. Stem woody; leaves plicate; inflorescence terminal or lateral.

25. Base of lip narrow .... ***Corymborkis***

25. Base of lip concave .... ***Tropidia***

24. Stem herbaceous, recumbent; leaves conduplicate; inflorescence terminal .... ***Erythrodes***

23. Plants with an herbaceous, erect stem, rosulate; pollinia not sectile.

26. Flowers resupinate.

27. Plants small, rosulate epiphytes with pendent inflorescences.

28. Inflorescences densely capitate.................. ***Eurystylis***

28. Inflorescence lax.................................. ***Lankesterella***

27. Plants usually large and terrestrial, if rosulate and epiphytic then with an erect inflorescence.

29. Rostellum transversely 3-parted, median lobe spathulate to anchor-shaped................................. ***Buchtienia***

29. Rostellum vertical, entire or bifid.

30. Stigmata at the apex of a truncate or subtruncate column and at a right angle to the rostellum.

31. Plants with a single, sessile, cordate-subrotund leaf surrounding the stem .... ***Discyphus***

31. Plants with well-developed leaves either cuneate or petiolate at the base.

32. Rostellum soft, short, broadly triangular with an apical fovea; lip fleshy, apex cochleate................................. ***Sauroglossum***

32. Rostellum rigid, more or less cartilaginous, sharply pointed; lip membranaceous, apex not cochleate.

33. Lateral sepals with the column foot do not form an observable mentum; column-foot very short, subequal in length to the column; lip pandurate..... .... ***Lyroglossa***

33. Lateral sepals with the free part of the column-foot form either a mentum or a tubular, spurlike extension; column-foot much longer than the column; lip not pandurate.

34. Stigmata bilobed, more or less separated by the terminal edge of a distinct fold running the full length of the face of the column

35. Lateral sepals together with the column-foot form a short, protruding mentum, not spurlike......................***Pteroglossa***

35. Lateral sepals together with the free part of the column-foot form a pendulous, spurlike process.........***Eltroplectris***

34. Stigmata confluent, semicircular at the apex of a terete column .... ***Sacoila***

30. Stigmata beneath or on both sides of the rostellum.

36. Column free from the sepaline tube.

37. Flowers very large; rostellum acicular in the center of a large, deeply bilobed, blunt plate; stigmata confluent...........***Cybebus***

37. Flower medium to small; rostellum laminar, without a deeply bilobed plate...***Helonoma***

36. Column partially adnate to the dorsal sepal; rostellum short, triangular in outline; lip either sessile or with a long claw which is adnate to the sepaline tube.

38. Column ballooned out in front due to an inflated clinandrium..... .... ***Beloglottis***

38. Column not ballooned out in front.

39. Rostellum divided into two distinct segments....***Spiranthes***

39. Rostellum undivided with an entire, acute, truncate, emarginate or denticulate apex.

40. Rostellum rigid, more or less cartilaginous.

41. Rostellum broadly triangular, acute at the more or less obscurely 3-lobulate or 3-dentate apex........... ....***Odontorrhynchus***

41. Rostellum linear-lanceolate to nearly aciculate, sharply acute or with a small, lateral tooth on each side.

42. Column footless, sometimes with an oblique base at the top of ovary.

43. Inflorescence subcorymbose, slightly drooping; lip unguiculate with large thickened auricles; column elongate, slender.... ....***Coccineorchis***

43. Inflorescence spicate, erect; lip sessile, calliferous along the margins at the base; column short, stout.........***Stenorrhynchos***

42. Column with a distinct, decurrent foot attached to the side of the ovary.

44. Flowers small in a loosely-flowered, spirally twisted rachis; stigmata approximate; lip sagittate at the base; rostellum linear-subulate........................***Mesadenella***

44. Flowers of medium size in a densely flowered rachis; stigmata confluent or V-shaped; lip conduplicate at the base, more or less sigmoid; rostellum with a small lateral tooth on each side at the base....***Skeptrostachys***

40. Rostellum soft, pliable, laminar to filiform.

45. Rostellum very short..***Brachystele***

45. Rostellum prominent.

46. Column with an obliquely extended base on top of the ovary.

47. Lip unguiculate with a sagittate, auriculate or cordate base..***Cyclopogon***

47. Lip sessile, either excavate or conduplicate-channeled at the base.

48. Stem ascending from a rhizomatous or subrhizomatous base with fleshy tuberlike roots originating from distant to more or less approximate nodes rostellum oblong to ligulate, excised at the apex.. ***Hapalorchis***

48. Stem caespitose with fasciculate roots or tubers; rostellum not oblong or ligulate, or excised at the apex..***Stalkya***

46. Column with a distinct foot attached to the side of the ovary, either visible externally or included within the ovary.

49. Column foot embedded internally in the ovary with the connate lateral sepals forming an internal nectary with out a line of adnation on the outside.........***Sarcoglottis***

49. Column-foot attached to the outside of the ovary with a line of adnation; lateral sepals basally connate into a ventricose, saccate, or tubular spur..................***Pelexia***

26. Flowers not resupinate.

50. Petals and lip adnate to the column at a distance from the base.

51. Lateral sepals connate to connivent into a tube or a more or less globose sac in which the channeled, 3-lobed lip is completely enclosed..***Pseudocentrum***

51. Lateral sepals free or rarely connate into a concave synsepal, not tubular or saccate; lip not enclosed in the lateral sepals.

52. Petals adnate to the column dorsally, commonly reflexed, their apices free from the dorsal sepal; lip saccate, 3-lobed, with replicate lateral lobes..............................***Baskervilla***

52. Petals adnate to the column laterally, their apices connivent with the tip of the dorsal sepal; lip not saccate, without replicate lobes..***Ponthieva***

50. Petals and lip free from the column.

53. Lateral sepals wider than long, parallel with the lip; lip with a distinct spur.. ***Solenocentrum***

53. Lateral sepals longer than wide; lip without a spur.

54. Dorsal sepal connate at the base with the lateral sepals to form a tube or cup of various lengths; column erect.

55. Column short, inconspicuous, glabrous; petals and lip firmly adnate to or fused with the sepaline cup; leaves usually petioled... ***Prescottia***

55. Column elongate, prominent, papillose or puberulent; petals and lip free from the sepals; leaves tapered toward the base.

56. Sepals erect, appressed with spreading apices, together forming a tube; lip helmet-shaped; column footless.. ***Stenoptera***

56. Sepals diverging from different levels with recurved apices; lateral sepals decurrent on the ovary; lip concave, adnate to the column foot and together with the lateral sepals forming a tube; column with a long, decurrent foot................. ***Porphyrostachys***

54. Dorsal sepal free from lateral sepals; column erect or sharply bent back.

57. Column sharply bent back at a right angle to the ovary; lip parallel to and covering but not surrounding the column... ***Gomphicis***

57. Column erect, perpendicular to the ovary; lip erect, more or less surrounding the column.

58. Rostellum erect, pointed.

59. Scape terminal, slender; perianth thin; lip cochleate; clinandrium cup-shaped... ***Cranichis***

59. Scape lateral, rigid, perianth fleshy; lip conduplicate; clinandrium with inrolled margins.... .... ***Pterichis***

58. Rostellum transverse, truncate or obtusely lobulate.

60. Column pubescent; stigma small; scape terminal; anthesis occurs after leaf development...... .... ***Altensteinia***

60. Column glabrous; stigma large; scape lateral; anthesis occurs before full leaf development.

61. Peduncle enveloped by cylindric, hyaline-diaphanous sheaths; dorsal sepal and petals free from the column; lip calceolate with an involute margin around the orifice .... ***Aa***

61. Peduncle with imbricating to spreading, infundibuliform, scarious sheaths; dorsal sepal and petals adnate to the back of the column above the base; lip tubular or flared without involute margins..***Myrosmodes***

3. Pollinia hard or waxy.

62. Flowers with a joint between the ovary and the pedicel; pedicel persisting; without pseudobulbs (the Pleurothallidinae).

63. Sheaths of the stems leaflike; pollinia 2, spherical.....***Frondaria***

63. Sheaths of the stems not leaflike; pollinia 2 to 8.

64. Pollinia 4 to 8.

65. Pollinia 6 or 8.

66. Pollinia 6 or 8; stigma bilobulate; flowers nonresupinate...................................***Brachionidium***

66. Pollinia 8; stigma entire; flowers resupinate...***Octomeria***

65. Pollinia 4.

67. Leaves pubescent..............................***Dresslerella***

67. Leaves glabrous.

68. Leaves narrow with abbreviated stems with a few basal, tubular sheaths..............***Barbosella***

68. Leaves ovate with well-developed stems with a series of sheaths.

69. Lip with a hypochile with a pair of hairlike appendages, connate by an immobile rod to the pedistal-like column foot...***Restrepia***

69. Lip without a hypochile, hinged to the column foot.......................***Restrepiopsis***

64. Pollinia 2 (except in the monotypic *Pleurothallis,* subgenus *Chamelophyton,* with 6).

70. Sheath of the stems lepanthiform (tubular, ribbed, more or less imbricating, with oblique, margined ostia, the ribs and margins of the ostia usually ciliate or scabrous).

71. Column more or less cylindrical, footless...***Lepanthes***

71. Column short and broad with a rudimentary foot, or short to elongate with a well-developed foot.

72. Column short and broad with a rudimentary foot; the stigma apical and transversely bilobed ....***Lepanthopsis***

72. Column short to elongate with a short or well-developed foot; the stigma not transversely bilobed ....***Trichosalpinx***

70. Sheaths of the stem not lepanthiform.
73. Lip motile, sensitively hinged to the column foot.
74. Lip broadly hinged under tension to the column foot...***Acostaea***
74. Lip attached to the column foot by a slender strap.
75. Base of the blade of the lip with a callus appressed to the slender, curved column foot.........................***Porroglossum***
75. Base of the blade of the lip with a pair of armlike calli appressed to the bulbous base of the column foot...***Condylago***
73. Lip not sensitively hinged to the column foot.
76. Lateral sepals with a transverse callus above the base...***Dryadella***
76. Lateral sepals without a transverse callus above the base.
77. Apex of the column terminated by a more or less flat, collarlike disk surrounding the anther and stigma...***Salpistele***
77. Apex of the column not terminated by a collarlike disk.
78. Petals callous on the labellar margin (rarely in very small species the callus may be lacking)...***Masdevallia***
78. Petals not callous on the labellar margin, often callous at the apex.
79. Lip divided into an obvious hypochile and epichile.
80. Sepals free at the apex.......................***Dracula***
80. Sepals united at the apex................***Ophidion***
79. Lip not divided into an obvious hypochile and epichile.
81. Lateral sepals with a callous pad at the apex...***Scaphosepalum***
81. Lateral sepals without a callous pad at the apex.
82. Base of the lip cordate or cleft, articulated with a laterally compressed column foot...***Trisetella***
82. Base of the lip not divided and articulated with a laterally compressed column foot.
83. Column membranous and hooded at the apex.............................***Platystele***
83. Column stout, not membranous-hooded at the apex.
84. Petals transverse and callous at the apex...................................***Stelis***
84. Petals not transverse and callous at the apex.
85. Sepals united at the apex.................................***Zootrophion***

85. Sepals free at the apex (with the exception of one or two members of *Pleurothallis* subgenus *Acianthera* and subgenus *Specklinia*).

86. Inflorescence 1-flowered, borne laterally with out an annulus; cauline sheaths usually scurfy; petals thickened at the apex..***Myoxanthus***

86. Not with above combination of characters; cauline sheaths not scurfy; inflorescence 1–many-flowered; lateral sepals free or connate; petals membranous, or sometimes thickened at the apex...***Pleurothallis***

62. Flowers without a joint between the ovary and the pedicel; pedicel dehiscing with the ovary; with or without pseudobulbs.

87. Stems successionally produced from the apex of the previous stem, therefore, stem segments superimposed.

88. Leaves terete or subulate, very elongate (to 70 x 0.3 cm); flowers dull colored, usually green with white petals and lip...***Reichenbachanthus***

88. Leaves flat, relatively short (to 20 x 1 cm); flowers brightly colored, either red, red-orange yellow or white...***Scaphyglottis***

87. Stems produced from the base of the plant or near the apex of the stem but segments not superimposed.

89. Pollinia naked or with caudicles, with or without viscidia, without a stipe.

90. Pollinia quite naked, without caudicles.

91. Leaves articulated.

92. Plants with obvious pseudobulbs...***Bulbophyllum***

92. Plants without pseudobulbs.................***Vargasiella***

91. Leaves not articulate.

93. Column elongate, terete.............................***Liparis***

93. Column very short, thick...........................***Malaxis***

90. Pollinia with caudicles.

94. Pollinia superposed with cylindrical, translucent caudicles attached to an obvious viscidium; lip with an elongate crested claw at the base..................***Cryptarrhena***

94. Pollinia laterally flattened or ovoid with prominent opaque caudicles, sometimes with an indistinct viscidium.

95. Pollinia 2...***Epidanthus***

95. Pollinia 4 to 8.

96. Pollinia 4 (in some cases with a rudimentary pair each, or with a single pair).

97. Basal portion of the lip with a pair of hollow calli................................***Caularthron***

97. Basal portion without hollow calli; calli either lacking or solid.

98. Lip united to the column to its apex.

99. Inflorescences usually lateral, emerging from the sheath opposite a leaf; sheaths rugose............................***Oerstedella***

99. Inflorescence terminal (lateral only in section *Pleuranthium* and then sheaths not rugose)...........................***Epidendrum***

98. Lip free from the column for at least the upper half.

100. Flowers large, exceeding 5 cm in diameter........***Cattleya***

100. Flowers medium to small, not exceeding 4 cm in diameter.

101. Inflorescences paniculate or racemose, elongate except in sections *Osmophytum* and *Hormidium* of ....***Encyclia***

102. Lip articulated to the apex of the column foot... ....***Orleanesia***

102. Lip united to the base of the column; column without a foot....................................***Encyclia***

101. Inflorescences solitary, fasciculate or abbreviated racemes.

103. Leaves terete or substerete.........***Jacquiniella***

103. Leaves flat.

104. Column with an elongate spreading foot; lip bilobed at the apex................***Ponera***

104. Column with a short, erect foot; lip acute at the apex......................***Isochilus***

96. Pollinia 8.

105. Leaves terete; pseudobulbs usually inconspicuous......***Brassavola***

105. Leaves flat.

106. Flowers large, diameter more than 5 cm.......***Schomburgkia***

106. Flowers small, diameter less than 4 cm.

107. Inflorescence racemose.

108. Flowers red-orange..........................***Sophronitis***

108. Flowers yellow-green or pink.

109. Inflorescence loose, to 5-flowered; flowers resupinate.................................... ***Nidema***

109. Inflorescence dense, many-flowered; flowers nonresupinate, pink.........***Arpophyllum***

107. Inflorescence uniflorous.

110. Inflorescence elongate; flowers green................. ....***Homalopetalum***

110. Inflorescence sessile.......................***Dimerandra***

89. Pollinia with caudicles (often reduced), viscidium *and* stipe.

111. Plants always monopodial; flowers with a conspicuous spur developed from the base of the lip; inflorescences multiflorous; flowers strictly distichous..***Campylocentron***

111. Plants usually sympodial; flowers with or without a spur developed from the united lateral sepals or from the base of the lip; inflorescences 1–many-flowered; flowers loosely alternate.

112. Pollinia 4 (except in *Anthosiphon roseus*).

113. Plants with pseudobulbs of several internodes; inflorescence terminal; flowers nonresupinate; column with a foot..***Polystachya***

113. Plants various; inflorescence usually lateral; flowers usually resupinate; column with or without a foot.

114. Plants cormous, terrestrial; lip thin; leaves 2, subopposite, with sheaths surrounding the petiole; column with a foot................................***Govenia***

114. Plants various, usually epiphytic or without pseupseudobulbs; lip usually with a prominent callus; leaves usually several or one, alternate, sheaths if present short; column with or without a foot.

115. Plants small, monopodial; leaves strongly distichous; the one-lowered inflorescence arising opposite the leaf axil; column generally with a 'ligule' beneath the stigma.... ....***Dichaea***

115. Plants usually medium to large, usually sympodial, inflorescences from the leaf axil; column without a 'ligule' beneath the stigma.

116. Pollinia flattened and superposed.

117. Stipes long and narrow; leaves clearly plicate.

118. Flowers erect, subglobose; lip articulated to the column at the base... ***Anguloa***

118. Flowers nodding, open; lip firmly fixed to the column at the base............***Lycaste***

117. Stipes short and wide or lacking; leaves either plicate or conduplicate.

119. Leaves conduplicate; viscidium wide, usually semilunate; inflorescence one-flowered.

120. Lip with a spur.

121. Plant with unifoliate pseudobulbs; lip and sepals with a short spur at base..***Anthosiphon***

121. Plant usually without pseudobulbs (if present, e.g., *Cryptocentrum pseudobulbosum,* with 3–4 apical leaves); lip and sepals with a conspicuous spur at the base............ ....***Cryptocentrum***

120. Lip without a spur.

122. Leaves terete...***Scuticaria***

122. Leaves flattened, at times thick and subterete.

123. Pseudobulbs with several leaves at the apex; leaves less than 1 cm long................................***Pityphyllum***

123. Pseudobulbs, if present, with 1–4 leaves at the apex; leaves longer than 4 cm.

124. Sepals spreading; lip and petals without glossy pad at their apices.

125. Column without a distinct foot.

126. Plants with caespitose pseudobulbs. ....***Mormolyca***

126. Plants with basal pseudobulbs and elongate, canelike stems, often with additional pseudobulbs at the branches.

127. Inflorescence in the axil of distichous leaves on canelike stems......................***Cyrtidium***

127. Inflorescence from appressed sheaths either at the base of the pseudobulbs or along leafless stems................***Chrysocycnis***

125. Column with a distinct foot......***Maxillaria***

124. Sepals forming an erect funnel; lip and petal apices with a glossy pink pad........***Trigonidium***

119. Leaves plicate or conduplicate; viscidium flattened, usually longer than wide; inflorescence various.

128. Callus usually prominent and with longitudinal ridges or keels; with or without pseudobulbs, may have pseudobulbs of several internodes.

129. Inflorescences uniflorous.

130. Plant with conspicuous, leafless pseudobulbs........... ....***Chaubardia***

130. Plant without pseudobulbs.

131. Column with a longitudinal keel on the underside.............................***Kefersteinia***

131. Column without a keel on the underside.

132. Callus of the lip long-fimbriate....................................***Huntleya***

132. Callus of the lip radiate, lobed, entire or dentate.

133. Lip deeply saccate or calceolate, either for its entirety or at the base only.

134. Column short, thick, with variously developed fleshy wings on each side of the stigma, the column foot very short...***Chaubardiella***

134. Column relatively elongate, slender, usually without fleshy wings on each side of the stigma, the column foot elongate.

135. Lip deeply saccate or calceolate, the flower stem erect, the flowers resupinate.

136. Apical portion of the lip acute to truncate ....***Stenia***

136. Apical portion of the lip bilobed, the lobes falcate, incurved........................***Dodsonia***

135. Lip shallowly concave at the base, flower pendent or the ovary geniculate to produce the lip uppermost.......................................***Benzingia***

133. Lip not deeply saccate, often tubular or concave to broadly expanded.

137. Sides of the lip surrounding the column at the base.

138. Callus of the lip tongue- or keel-like in the mid-portion of the lip...................***Chondrorhyncha***

138. Callus of the lip fanlike at the base of the lip, composed of radiating ribs divided to the surface of the lip............................***Cochleanthes***

137. Sides of the lip not surrounding the column at the base, the base of the lip may touch the base of the column.

139. Column broad, concave on the underside surrounding the callus................................***Bollea***

139. Column narrow, not surrounding the callus..... ....***Pescatorea***

129. Inflorescences multiflorous.

140. Rhizome elongate between the pseudobulbs.

141. Anther cap with an elongate hornlike projection from its apex...***Zygosepalum***

141. Anther cap without a hornlike projection.

142. Lateral lobes of the lip narrow, hornlike; stipe of the pollinarium rectangular...............................***Acacallis***

142. Lateral lobes of the lip broad, rounded; stipe of the pollinarium elongate, oblong.......................***Aganisia***

140. Rhizome short, pseudobulbs approximate.

143. Lip simple without lobes, callus of 3 elongate, longitudinal ribs... ***Warrea***

143. Lip 3-lobed; callus basal, bilobed or fan-shaped.

144. Callus erect, bilobed or bipartite........***Koellensteinia***

144. Callus fan-shaped, ribbed or a flattened transverse ridge.

145. Flowers blue; callus of 3–5 parallel, radiating ribs.. ***Warreella***

145. Flowers of various colors but not blue; callus fan-shaped.

146. Plants very small, adult leaves less than 7 cm long................................***Chieradenia***

146. Plants large, adult leaves more than 15 cm long.

147. Lateral lobes of the lip fimbriate... ***Galeottia***

147. Lateral lobes of the lip not fimbriate.

148. Callus fan-shaped... ***Zygopetalum***

148. Callus a flattened transverse ridge.

149. Column with a longitudinal keel on the underside, wingless...... ***Warreopsis***

149. Column without a keel on the underside, winged or wingless.

150. Column without wings...........................***Batemannia***

150. Column winged...................***Otostylis***

128. Callus usually low, smooth or without keels; always with pseudobulbs of a single internode.

151. Inflorescences 1-flowered.

152. Pseudobulbs covered by a scarious sheath; inflorescences produced from the base of the pseudobulb; flowers with a spur produced at the base of the united lateral sepals, the base of the lip united to the elongate column-foot and inserted in the sepaline spur............................***Teuscheria***

152. Pseudobulbs not surrounded by a scarious sheath; inflorescence produced at the midpoint of an elongate rhizome; flowers spurless...***Horvatia***

151. Inflorescences multiflowered.
153. Pseudobulbs terete to ovoid; lip entire to 3-lobed................ ....***Xylobium***
153. Pseudobulbs quadrangular or flattened; lip 3-4-lobed.
154. Pseudobulbs quadrangular; lip with a single, elongate callus...***Bifrenaria***
154. Pseudobulbs flattened; lip with a callus on the isthmus and another between the lateral lobes.....***Rudolfiella***
116. Pollinia usually ovoid or clavate, not markedly superposed.
155. Viscidium hooklike; column and callus usually bristly (insect-like); stigma at the apex of the column.
156. Column enlarged into a pair of parallel lobes at the base, sepals and petals spreading, curved upward...***Hofmeisterella***
156. Column terete or very short and immersed in the callus; sepals and petals spreading equally.
157. Pseudobulbs conspicuous, rhizome usually elongate between the pseudobulbs.......................***Trichoceros***
157. Pseudobulbs, if present, not conspicuous; stems short or caulescent.
158. Flowers large, more than 1.5 cm in diameter... ....***Telipogon***
158. Flowers small, less than 1 cm in diameter....... ....***Stellilabium***
155. Viscidium not hooklike; column not bristly; stigma under the apex or at the base of the column; anther often dorsal on the column, often long-beaked.
159. Flowers with a triangular spur from the base of the lip..... ....***Dunstervillea***
159. Flowers without a spur.
160. Inflorescence capitate with the flowers in a whorl of four...***Caluera***
160. Inflorescence not with flowers in a whorl of 4.
161. Plants with pseudobulbs and conduplicate leaves ....***Dipteranthus***
161. Plants fan-shaped (psygmoid) laterally flattened leaves, without pseudobulbs.
162. Column winged on each side at the base, with a curled ligule at the junction with the lip..***Eloyella***
162. Column wingless, without a ligule at the junction with the lip.
163. Stem elongate; flowers yellow........ ....***Sphyrastylis***
163. Stem short; flowers white................ ....***Ornithocephalus***

112. Pollinia 2.

164. Plants generally with pseudobulbs of several internodes; leaves plicate.

165. Male flowers with sensitive rostellum, which discharges the viscidia when triggered; flowers unisexual or bisexual.

166. Flowers bisexual, both stigma and anther functional.

167. Flowers pink, red, purple, brown, yellow or combinations, but not green............. ***Mormodes***

167. Flowers green.

168. Inflorescence erect; margin of the lip entire; leaves foul-smelling......... ***Dressleria***

168. Inflorescence pendent; margin of the lip fimbriate; leaves normal-smelling.. ***Clowesia***

166. Flowers normally unisexual, either stigma or anther functional but not both (except rarely perfect).

169. Column with elongate antennae at base of male flowers; lip sessile in both sexes, the lip of female flowers hood-shaped............ ***Catasetum***

169. Column in male flowers very elongate, without antennae; lip clawed in both sexes... ***Cycnoches***

165. Rostellum not sensitive; flowers bisexual.

170. Flowers with a spur at the base of the lip.

171. Leaves thick, dark green spotted with white, not articulated with the sheaths.......... ***Oeceoclades***

171. Leaves thin, green, articulated with the sheaths... ***Galeandra***

170. Flowers without a spur at base of lip.

172. Lip deeply concave at the base.......... ***Eulophia***

172. Lip not concave at the base.

173. Lip obcordiform with a callus at apex; foot with a forward projecting fingerlike projection..................................... ***Grobya***

173. Lip conspicuously trilobed.

174. Leaves thin; pseudobulbs of 8–10 internodes, elongate; inflorescence paniculate................... ***Cyrtopodium***

174. Leaves thick; pseudobulbs ovoid; of 2–3 internodes; inflorescence racemose................................... ***Eriopsis***

164. Plants usually with pseudobulbs of one internode, or without pseudobulbs; leaves plicate or conduplicate.

175. Leaves plicate.

176. Lateral sepals forming a saccate hood around the petals, lip and column.................................. ***Schlimia***

176. Lateral sepals free or if united not saccate.

177. Dorsal sepal and petals united to the middle of the column...***Gongora***
177. Dorsal sepal and petals not united to the column.
178. Lip forming a bucket with water faucets dripping from the base of the column..***Coryanthes***
178. Lip not bucketlike, water glands not present on the column.
179. Margin of the apical lobe of the lip (epichile) papillate-dactylate...***Paphinia***
179. Margin of the apex of the lip not papillate (fimbriate in *Sievekingia reichenbachiana*).
180. Claw of the lip with a toothlike laterally flattened callus...***Lueddemannia***
180. Claw of the lip without a callus.
181. Pseudobulbs with more than 1 leaf.
182. Rachis of the inflorescence densely black-pilose.....................................***Kegeliella***
182. Rachis of inflorescence glabrous or finely pubescent.
183. Epichile of the lip articulated and hinged to the hypochile which has well-developed lateral lobes......... ....***Peristeria***
183. Epichile of the lip firmly fixed to the hypochile which is simple, without lateral lobes.
184. Inflorescence erect.................. ....***Soterosanthus***
184. Inflorescence pendent.
185. Pseudobulbs of several internodes...................... ....***Lycomormium***
185. Pseudobulbs of a single internode.
186. Pseudobulbs pyriform; column densely pubescent. ....***Vasqueziella***
186. Pseudobulbs spherical, cylindrical or ovate-flattened; column glabrous.

187. Leaves petiolate; pseudobulbs spherical or cylindrical...... ....***Trevoria***

187. Leaves narrowing to a subpetiolate base; pseudobulbs ovate, flattened and longitudinally ribbed............***Acineta***

181. Pseudobulbs with a single leaf.

188. Hypochile of the lip superimposed over the epichile.

189. Lip of the flower not articulated to the column foot; inflorescence arching to subpendent........***Polycycnis***

189. Lip of the flower articulated to the column foot; inflorescence stiffly erect.............................***Braemia***

188. Hypochile of the lip, when present, connected directly to the mesochile or epichile.

190. Lip concave or excavate at the base; inflorescence pendent.

191. Lip excavate at the base, usually divided into an hypochile, mesochile and epichile.

192. Inflorescence uniflorous; mesochile wings T-shaped....................................***Embreea***

192. Inflorescence 2–many-flowered; mesochile wings, if present, hornlike.......***Stanhopea***

191. Lip concave, not divided into an hypochile, mesochile and epichile..................***Sievekingia***

190. Lip with a solid claw at the base; inflorescence pendent or erect...***Houlletia***

175. Leaves conduplicate.

193. Plants monopodial, with (or without) 2 cylindrical, translucent caudicles longer than the stipe.

194. Flowers yellow-spotted with brown or bright red (sometimes with yellow lip); caudicles elongate.

195. Flowers red or red and yellow; callus glabrous......... ....***Fernandezia***

195. Flowers yellow-spotted with brown; callus lobes pubescent..***Raycadenco***

194. Flowers yellow-green; caudicle short..........***Pachyphyllum***

193. Plants sympodial or pseudomonopodial (branching from the base only); pollinarium with short caudicles.

196. Stems elongate, surrounded by flattened, distichous, imbricating, bractlike leaves to give a braided aspect..***Lockhartia***

196. Stems forming pseudobulbs or fan-shaped and often without pseudobulbs; growths not elongate (although rhizome sometimes elongate).

197. Flowers with a spurlike structure or a well-developed gibbous cavity in the rear.

198. Spurlike structure or gibbous cavity without horns produced by the lip and column base.

199. Rear of the flower with a gibbous cavity formed by the bases of the fused lateral sepals............***Ionopsis***

199. Rear of the flower with a spurlike structure formed by the base and free lateral sepals.

200. Plants not psygmoid, without pseudobulbs; leaves thick, leathery, conduplicate; column apex with broad, forward-projecting column wings. ....***Trichocentrum***

200. Plants psygmoid with reduced pseudobulbs; leaves laterally flattened (rarely conduplicate); column apex without appendages (although base may have 2 expanded ridges that fuse with lip)..***Plectrophora***

198. Spurlike structure or gibbous cavity with one or two horns produced by the lip and column base.

201. Spurlike structure formed by the full length of the lateral sepals; column arms or stigmatic arms present; anther dorsal...***Rodriguezia***

201. Spurlike structure formed by only the bases of the lateral sepals; column without appendages; anther terminal.

202. Spurlike structure with one horn.

203. Spurlike structure long; spathulate horn broadly attached to the column base...... ....***Diadenium***

203. Spurlike structure short; acute horn shortly attached to the column base................. ....***Scelochilopsis***

202. Spurlike structure with two horns.

204. Lip much larger than the other perianth parts; spurlike structure equal to or longer than the column....................***Comparettia***

204. Lip equal to or only slightly larger than the other perianth parts; spurlike structure shorter than the column.........***Scelochilus***

197. Flowers without a spurlike or gibbous structure at the rear.

205. Petals and dorsal sepal alike and very narrow; lateral sepals broad and petal-like...***Psychopsis***

205. Petals and dorsal sepal usually dissimilar; sepals similar.

206. Petals united with the dorsal sepal, free from the lateral sepals..***Sanderella***

206. Petals not united with the dorsal sepal.

207. Column with a prominent column-foot to which the lateral sepals and hinged lip are attached..***Systeloglossum***

207. Column without a prominent column-foot; lateral sepals and unhinged lip attached directly to the column.

208. Stigmatic cavities 2.

209. Column with a pair of falcate column wings projecting forward beyond the anther cap; column-hood not present... ....***Solenidiopsis***

209. Column with short, obtuse stigmatic arms; column-hood present, extending beyond the anther cap.

210. Plant caespitose; apical lobe of the 3-lobed lip well-developed and not fingerlike; flowers brilliantly colored..***Cochlioda***

210. Plant with an elongate rhizome; apical lobe of the lip reduced to a fingerlike process or an apicle; flowers green..***Oliveriana***

208. Stigmatic cavities 1.

211. Stigmatic cavity a slit longitudinally oriented on the column.

212. Column without paired appendages...............***Notylia***

212. Column with paired appendages.

213. Apex of the column with two elongate, narrow column arms..................................***Cypholoron***

213. Apex of the column with two short, broad stigmatic arms.

214. Column with a hood exceeding the anther cap; lip 3-lobed, with the apical lobe much narrower than the basal lobes; plants with only apical, conduplicate leaves on pseudobulbs..........................***Macradenia***

214. Column without a hood; lip either unlobed or pandurate with the two portions equal in width; plants psygmoid with pseudobulbs very reduced or absent.

215. Column with the paired stigmatic arms twisted under the column ....***Hirtzia***

215. Column with the paired stigmatic arms lateral on the column.

216. Inflorescence much exceeding the leaves with the flowers produced in succession; lip with a prominent toothlike callus near the apex.......................***Iquitoa***

216. Inflorescence subequal to the leaves with the flowers produced nearly simultaneously; lip with a gently rounded callus, bilobed near the lip apex...***Pterostemma***

211. Stigmatic cavity oval to round (rarely a transversely oriented slit).

217. Column apex with a dorsal hood or fringe extending beyond the anther cap (the lip callus not digitate-tuberculate).

218. Lip fused to the base of the column by a median keel... ....***Trichopilia***

218. Lip free from the column.

219. Lip enfolding the column throughout its length.

220. Leaves narrow (nearly terete); column-hood bent ventrally and covering a large portion of the anther cap.................................***Leucohyle***

220. Leaves not narrow; column-hood projecting straight forward...........................***Cischweinfia***

219. Lip not enfolding the column or only the basal lobes of the lip enfolding the column.

221. Lateral sepals fused..................***Osmoglossum***

221. Lateral sepals free.

222. Lip 3-lobed with the apical lobes smaller than the basal lobes................***Warmingia***

222. Lip 3-lobed with the apical lobe larger than the basal lobes.......................***Helcia***

217. Column apex without a dorsal hood or fringe (but may have dorsal structure if the lip callus is digitate-tuberculate).

223. Flowers tiny, less than 0.4 cm in diameter.

224. Perianth campanulate.................................***Buesiella***

224. Perianth parts spreading, at least at apex.

225. Flowers arranged in dense, terminal clusters..... ....***Trizeuxis***

225. Flowers loosely arranged.

226. Apical leaf aborted; basal leaves well-developed and conduplicate......***Konanzia***

226. Apical leaf present; basal leaves present or not, all leaves terete to subterete.

227. Column 2/3 or more the length of the lip, with 2 introrse acute stigmatic arms................................***Quekettia***

227. Column 1/2 or less the length of the lip, with 2 introrse obtuse stigmatic arms...................***Stictophyllorchis***

223. Flowers small to large, more than 0.8 cm in diameter.

228. Flowers campanulate................................***Neodryas***

228. Flowers not campanulate.

229. Leaves semiterete; anther dorsal, the anther cap wider near the viscidium..***Determannia***

229. Leaves conduplicate or psygmoid, not semiterete; anther terminal; anther cap either equidimensional or narrower near the viscidium (beaked).

230. Leaves psygmoid (fan-shaped)..............................***Psygmorchis***

230. Leaves conduplicate.

231. Flowers with the combination of the column without appendages and the lip fused to the column for 1/3-1/2 its length (appearing to emerge above the point of attachment of the other perianth parts).

232. Lateral sepals free...***Aspasia***

232. Lateral sepals fused 3/4 or more of their length...***Rusbyella***

231. Flowers with the column generally with prominent appendages and the lip free from or fused to the column for 1/4 or less of its length (appearing from the side to emerge from the same point as the other perianth parts).

233. Lip base with a cavity, often formed by a pair of fleshy, raised lamellae.

234. Column without paired appendages, but often greatly swollen at its base.

235. Column with unlobed stigmatic arms perpendicular to the column..........***Leochilus***

235. Column with extrorse bilobed stigmatic arms.....................................***Polyotidium***

234. Column with 2, obtuse, stigmatic arms.

236. Lateral sepals fused 1/2-3/4 of their length ....***Mesospinidium***

236. Lateral sepals free or only basally fused.

237. Floral bracts less than 1/2 the length of the ovary..........................***Brassia***

237. Floral bracts 2/3 or more the length of the ovary.

238. Floral bracts exceeding the flowers; flowers arranged on only one side of the inflorescence; cavity on the lip base embedded in the ovary...... ....***Brachtia***

238. Floral bracts not exceeding the flowers; flowers arranged distichously; cavity on the lip base not embedded in the ovary ....***Ada***

233. Lip base without a cavity.

239. Inflorescences with several bracts per flower-bearing node; flowers generally borne in coordinated succession; column generally elongate-arcuate..***Sigmatostalix***

239. Inflorescence with a single bract per node; flowers rarely borne in coordinated succession; column when elongate not arcuate.

240. Foliage gray-green; lip broad and flat with two acute basal projections; column long................................***Miltoniopsis***

240. Foliage green; lip generally not broad, if broad, then with a short column.

241. Plant small, 9 cm or less in diameter, with narrow (0.3–0.4 cm) grassy leaves; leaves subtending the pseudobulb as well-developed as the apical leaf; lip without a callus and wider toward the middle....... ....***Suarezia***

241. Plant generally larger than 8 cm in diameter; leaves 0.8 cm or more wide, generally with the apical leaves larger than basal ones; lips generally broadly attached to the column, if narrowly so, then with a prominent callus and wider to either the base or apex than in the middle.

242. Column without appendages and a swollen base; plant with only an apical leaf as an adult (often flowering as a psygmoid juvenile)..................... ....***Scelochiloides***

242. Column rarely without appendages; if absent, then the plant with several basal leaves in addition to the apical leaves.

243. Flowers with the combination of the lip callus inserted into the swollen basal lobes of the column and the fused lateral sepals wider than the lip, which is wider at the base..***Caucaea***

243. Flower with the lip callus not inserted into the swollen basal lobes of the column; if so, then the fused lateral sepals narrower than the lip, which is wider at apex.

244. Flower with the combination of the column with only 2 short, obtuse stigmatic arms and the lip with two prominent raised lamellae that extend to the midpoint of the lip.

245. Anther cap with the portion nearest the viscidium revolute; lip clawed and with several pubescent mounds on the erect lamellae or with the erect lamellae entirely pubescent............***Solenidium***

245. Anther cap with the portion nearest the viscidium not revolute; lip not clawed and the lamellae glabrous..........***Symphyglossum***

244. Flowers with the column generally without stigmatic arms, if so, then the callus not composed of erect lamellae.

246. Column entirely and densely pubescent; stigmatic arms, if present, short and obtuse..***Cymbiglossum***

246. Column generally glabrous, pubescent only on the dorsal side of the tabula infrastigmatica, if entirely glabrous, then the stigmatic arms elongate and acute (includes *Odontoglossum*)............ ....***Oncidium***

Habitally, the most distinctive orchid genus is *Vanilla* (65 spp., incl. Old World), which is a succulent-stemmed root-climbing hemiepiphytic vine. Our largest genera, all predominantly epiphytic, are five genera of small (mostly miniature), +/- small-flowered plants characterized by lack of pseudobulbs and a single leaf per stem with the inflorescence produced from stem apex at leaf base — *Pleurothallis* (1600 spp.), with free or nearly free sepals held +/- erect to form a cup, *Stelis* (600 spp.), differing in the three obtuse sepals (much larger than the reduced petals and column) spreading to form a small flat flower, *Masdevallia* (400 spp.), with the greatly enlarged sepals of the single flowers basally joined into a tube and usually apically extended as long tails, *Lepanthes* (600 spp.), with flowers similar to *Pleurothallis* but with lip fused to base of column and vegetatively characterized by the conspicuous ovate lepanthoid scales along the stem, and *Octomeria* (130 spp.), differing from the above in having the sepals and petals similar (and not very elaborated) — and six genera of usually larger, +/- large-flowered plants either with pseudobulbs or with >1 leaf per stem (or both) — *Epidendrum* (800 spp.), with apical inflorescence, flowers with a characteristic lip, the base strongly fused to column and the apex widely spreading, and vegetatively characterized by thick narrowly oblong leaves, in most species borne on reed-like stems, in others one or two together at tip of a pseudobulb, *Maxillaria* (600 spp.), with typical flowers, having the lip smaller than rest of perianth and the base of column produced into a foot and borne singly from base of the pseudobulbs (which may be reduced and inconspicuous) typically each pseudobulb with a single narrow thick distinctly petiolate leaf from apex, *Oncidium* (500 spp.), called "dancing lady" orchids because of the distinctive flower typically with a large flat medially contracted lip (the dancer's arms and skirt) held below the smaller sepals and lateral petals (the dancer's elaborate headress), usually with the multiflowered inflorescences borne from the base of an often small pseudobulb, *Odontoglossum* (300 spp., but often included in *Oncidium*), the prominent pseudobulbs with pair of thick narrow leaves at top and basally borne inflorescences with flowers mostly strikingly spotted or mottled and having a flat *Oncidium*-like lip but the sepals and lateral petals at least as large as lip (unlike most *Oncidium*), *Encyclia* (130 spp.), an *Epidendrum* segregate specialized for pollination by bees (instead of lepidoptera or hummingbirds) and differentiated by the lip free from base of column (and the much more frequent occurrence of pseudobulbs), and *Catasetum* (100 spp.), with unisexual flowers, usually thick and greenish, the males with very characteristic crossed ("wishbonelike") antennae that forcibly release the pollinia, vegetatively characterized by the large thick pseudobulbs a few of which have large, unusually thin, plicate leaves at apex.

The largest neotropical terrestrial genera are *Habenaria* (500 spp., incl. Old World), which has the fleshy white root-stem tuberoids and poorly developed pollinia of Orchidoideae, and *Sobralia* (100 spp., rarely epiphytic), with large *Cattleya*-like flowers and reedlike stems with strong-veined leaves.

Several other neotropical orchid genera, all epiphytic, have 50 to 100 species, including *Elleanthus* (70 spp.), with small flowers clustered into heads, vegetatively like a miniature *Sobralia* with small strong-veined leaves on a canelike stem, *Cattleya* (65 spp.), the "queen of orchids" on account of its spectacular flowers, with one or two narrowly oblong leaves at apex of prominent pseudobulb, *Dichaea* (40+ spp.), with a very characteristic growth-form lacking pseudobulbs and with the short narrowly oblong leaves with strongly overlapping bases distichous and close together along the stem and with solitary rather small flowers from the leaf axils, *Telipogon* (80 spp., mostly high-Andean), similar to *Dichaea* in growth-form but smaller, the 1–few flat flowers appearing too large for plant and characterized by reduced sepals and three rather similar large yellowish petals with maroon veins or shading, the column and petal bases often obviously resembling a perched fly or bee (pollination by pseudocopulation), *Scaphyglottis* (50 spp.) with small inconspicuous flowers and a peculiar spindly growth-form with long narrow pseudobulbs borne on top of each other and each with a pair of narrow apical leaves, and *Brassia* (50 spp.), called "spider orchids" on account of the narrrow elongate sepals (resembling spindly spider legs), the several-flowered racemose inflorescence borne from base of well-developed pseudobulb.

Palmae

Palms are so distinctive that familial identification is usually self-evident. They are well-characterized by their arborescent (often colonial) habit, the parallel-veined segmented leaves (each segment also with a strong midvein [except some Iriarteae]), when pinnate with a very strong rachis or midrib, when palmate either costapalmate (= midrib extended into blade) or with a hastula or erect dorsal projection at apex of petiole (this lacking in the only similar family, Cyclanthaceae which are also never arborescent). Many palms have underground stems and appear acaulescent with a rosette of leaves emerging from the ground; in our area acaulescent palms are mostly pinnately segmented, never palmate (except *Chelyocarpus repens*), whereas cyclanths are palmately divided or bifid. One genus (*Desmoncus*) is climbing and characterized by spiny stems and pinnate leaves with the reduced terminal leaflets thickened and hooklike; climbing cyclanths all have bifid or subentire leaves and are spineless.

The palm inflorescence is also very characteristic with the fertile part enclosed in bud by a series of peduncular bracts (sometimes called spathes) and subtended by these bracts (or their scars) at anthesis. The flowers are usually unisexual (the male and female sometimes borne on separate inflorescences and sometimes dioeceous) and tend to be borne in clusters. They are almost always sessile (or even in pits) and borne more or less densely along the inflorescence branches. Although the flowers are rather small and reduced, the large inflorescences are often spectacular. Most palm fruits are single-seeded with a fleshy (to fibrous) mesocarp, in some taxa covered by reptilelike scales (Calamoideae [= Lepidocaryoideae]) or prickles (some *Bactris* and *Astrocaryum*); some palm fruits are spectacularly large.

1. Leaves Palmate (or "Costapalmate" with Petiole Apex Extended into Lamina)

1A. Fruits not scaly; leaves with induplicate segments (V-shaped in cross section)

Chelyocarpus (3 spp.) (incl. *Tessmanniophoenix*) — Amazonia and Colombian Chocó. Petiole base closed, not split. Fruit round, the surface somewhat rough and corky but not covered with scales; stamens 6–8, not connate.

Itaya (1 sp.) — Exclusively upper Amazonian; close to *Chelyocarpus*, but the stamens 18–24 and connate at base; fruit somewhat bean-shaped. Vegetatively different in sheath and petiole base split.

P: sacha bombonaje

Palmae
(Palmate Leaves)

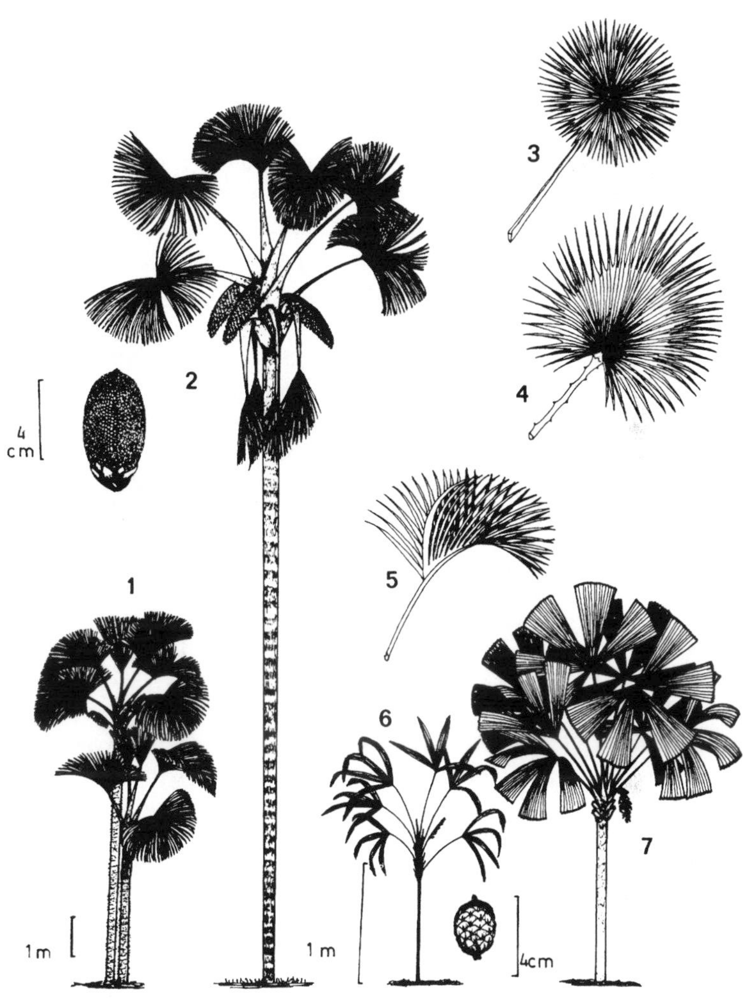

3 - *Crysophila*

2 - *Mauritia*

4 - *Copernicia*

5 - *Sabal*

1 - *Mauritiella*

7 - *Chelyocarpus*

6 - *Lepidocaryum*

Crysophila (8 spp.) — Mostly Central America, barely into northwestern Colombia. Distinctive in spines on trunk (unique in nonscaly fruited palmate-leaved species); Differs from *Chelyocarpus* and *Itaya* in the stamens 6 and with filaments united half their length.

C: palma barbasco, palma escoba

Sabal (14 spp.) — Mostly Antillean, in our area only reaching northern Colombia. Distinctive among fan palms of its area in being strongly costapalmate with the petiole apex continued into leaf blade. Differs technically from above genera by having the 3 carpels connate. Our only species is a tall tree with nonspiny trunk and nonspiny petiole margin (unlike *Crysophila* and *Copernicia,* respectively, the only other fan palms in its range).

C: palma amarga

Copernicia (25 spp., mostly Antillean) — Solitary palms growing in swampy areas in strongly seasonal part of the northern Colombia coastal plain. Unique among our strictly palmate fan palms in short recurved spines along petiole angles (a trait shared with costapalmate *Mauritia* and *Mauritiella*). At a distance distinguishable from *Sabal* by the more stiffly horizontal leaf blade. A technical floral character unique in our area is having 3 carpels united only by their styles.

C: palma sara

1B. Fruits covered with scales; leaf segments reduplicate (= shaped like inverted V in cross section)

Mauritia (3 spp.) — Large nearly always solitary trees with unarmed trunks, forming pure stands in palm swamps (aguajales). Leaves strongly costapalmate with many leaf segments. Inflorescence with many more or less parallel pendulous branches more or less forming a vertically oriented fan.

C: canangucha, aguaje; P: aguaje

Mauritiella (ca. 8 spp.) — Differs from *Mauritia,* with which it is often lumped, in spiny trunks, colonial growth-form, and generally smaller fruits.

C: quitasol; P: aguajillo

Lepidocaryum (2 spp.) — Small understory Amazonian treelet of rather sandy soils, distinct in having only 4 leaf segments. Inflorescence with many fewer branches than other fan palms.

P: irapay

Palmae
(Pinnate-Leaved Taxa with Distinctive Fruit)

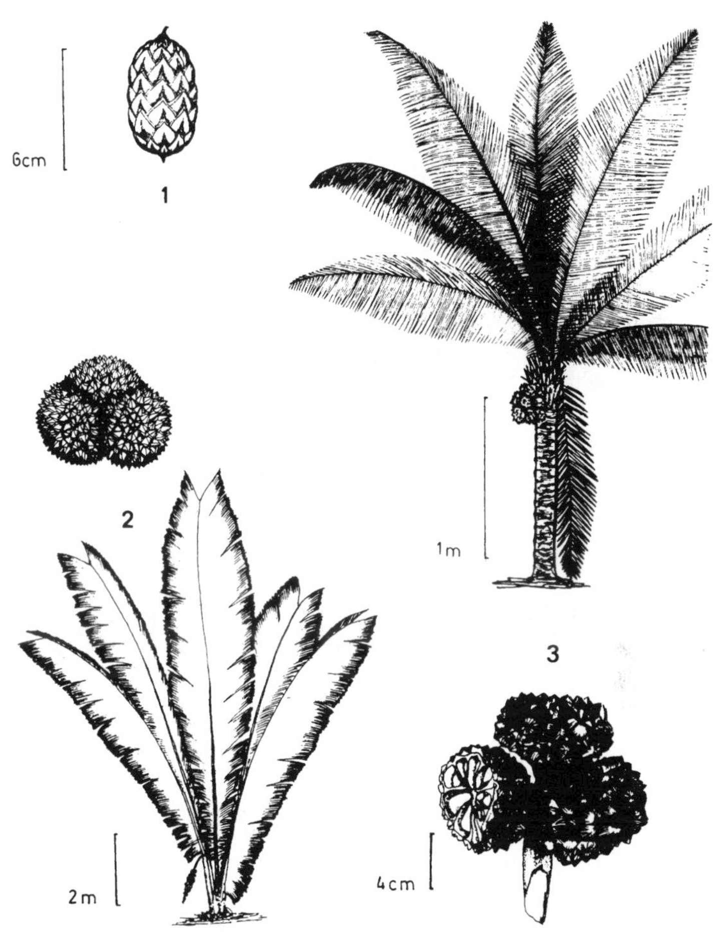

1 - *Raphia*

3 - *Phytelephas*

2 - *Manicaria*

2. Leaves Pinnately Compound (or Occasionally Undivided)

2A. Miscellaneous distinctive genera

Phytelephas (15 spp.) (incl. *Ammandra, Aphandra, Palandra*) — Mostly short-stemmed or stemless; when trunk well-developed, with irregular surface from subpersisting leaf bases. Leaflets mostly all in one plane (grouped and in two planes in *Palandra*). Fruits very large, sessile, crowded into dense more or less tuberculate masses in leaf axils; male flowers in sessile clusters forming dense catkins (*Phytelephas*) or with the flower clusters on long stalks which may be partly fused (*Palandra*) or completely fused (*Ammandra*).

C: tagua; E: tagua, cadi; P: yarina

Raphia (1 sp., plus reputedly 20 in Africa) — Forming monospecific palm swamps along Caribbean coast as in the Atrato Delta. The only pinnately leaved neotropical palm with scaly fruits.

C: panga

Manicaria (1 sp.) — Palm of wet areas (either edaphically or climatically), distinctive in the characteristic fibrous sacklike bract which completely encloses inflorescence at anthesis (and can be removed and worn as a kind of stocking cap). Fruits also very distinctive, large and warty in small dense cluster at base of leaves. Vegetatively easy to recognize on account of the large, rather erect, incompletely, and irregularly separated (but all in the same plane) pinnae.

C: cabecinegro

(*Phoenix*) — Sometimes cultivated, especially in coastal Peru. Easily distinguished vegetatively by having the basal leaflets modified into spines.

2B. Cocosoid palms — By definition cocosoid palms have the fruit with a thick "bony" endocarp penetrated by 3 pores. However, in the first several genera below (all spiny) the fruits are (sometimes) small and berrylike and the endocarp relatively weakly developed. All truly spiny palms and climbing palms belong here as do the palms with strongly woody spathes.

2Ba. Vines

Desmoncus (60 spp.) — The only neotropical "rattan" genus is completely unrelated to the Old World rattans which are lepidocaryoid with scaled fruits. The terminal leaflets are modified into thickened spine-like hooks; rachis and stem (actually leaf sheath) usually also spiny.

C: matamba, bejuco de alcalde; P: varacasha

Palmae
(Spiny - A)

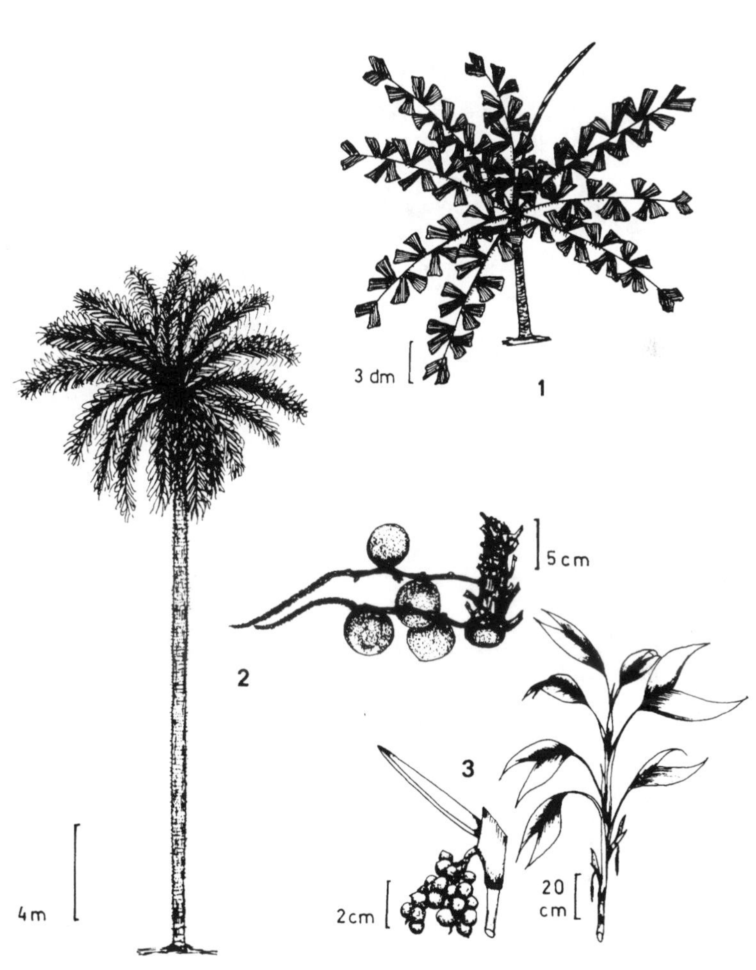

1 - *Aiphanes*

2 - *Acrocomia*

3 - *Bactris*

Palmae
(Spiny - B)

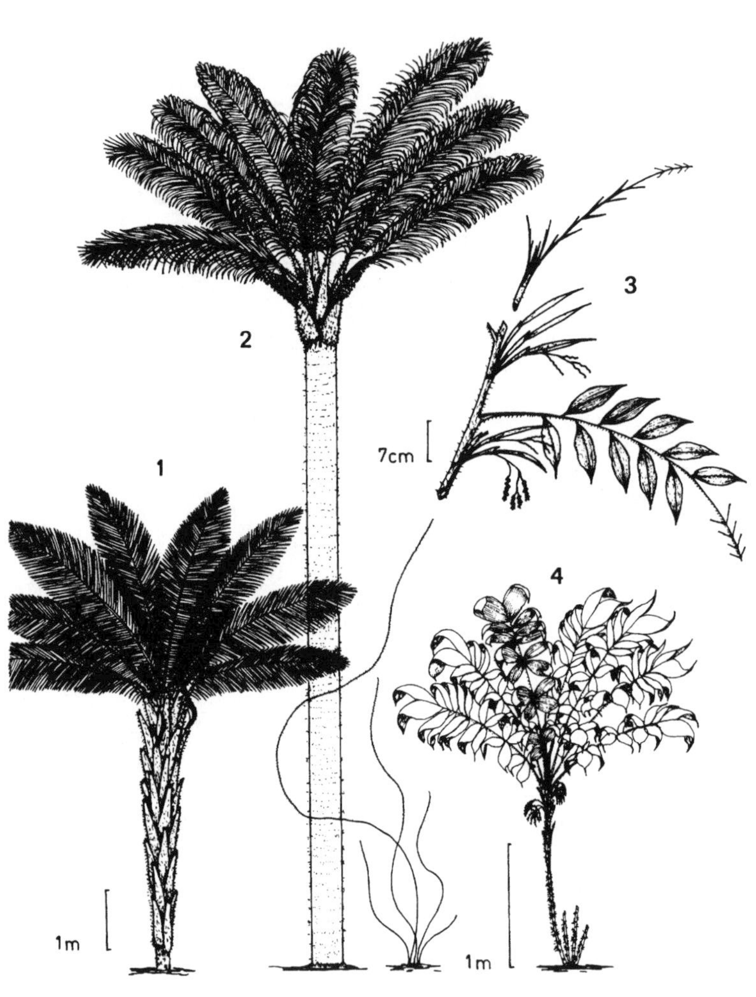

2 - *Astrocaryum*

3 - *Desmoncus*

1 - *Astrocaryum*

4 - *Bactris*

2Bb. Erect plants

(i) Spiny — (All spiny palms in our area are cocosoid except a few species with trunk spines or short prickles on petiole margins.)

Acrocomia (ca. 10 spp.) — Large solitary palms of dry areas, usually with very frayed leaflet tips, the leaflets held irregularly at many different angles. Technical character is that female perianth splits to base while *Bactris, Aiphanes, Astrocaryum* have cupular flowers with fused sepals and petals.

C: tamaco

Aiphanes (38 spp.) — Small, sometimes stemless, similar to *Bactris* but the leaflet tips irregularly truncately chopped off. Fruit small, fleshy.

Bactris (240 spp.) — Small, often colonial spiny palms with acute-tipped leaflets. (A very few species virtually lack spines [always present at least on apical margin of pinnae]: see below). Fruit usually more or less fleshy and berrylike. Differs from *Astrocaryum* in male flowers scattered among pistillate flowers and not immersed and in the less developed inflorescence rachis.

C: corozo de lata, macana, chontadurillo; E: chonta; P: ñejilla

Astrocaryum (50 spp.) —Large usually solitary spiny palms with acute-tipped leaflets. Fruits large, hard. Essentially a larger version of *Bactris* except inflorescence rachis thick and well developed with the male flowers densely clustered at ends of inflorescence branches and somewhat immersed.

C: palma malibú, chambira, yavarí (*A. jauari*); E: mocora; P: hiuri-rima, huicungo, chambira

(ii) Nonspiny palms — (All cocosoid palms without spines have the typical thick "bony" 3-pored endocarp. Male flowers are generally needed to separate many of the (too?) closely related genera of this group. (*Attalea* is the oldest name in the group which includes the first four genera below and a good case can be made for lumping at least *Orbignya* and *Scheelea,* perhaps also *Maximiliana,* with it).

Scheelea (10 spp.) — Mostly tall palms but some species stemless. Technical character is the thick elongate more or less cylindric petals and 6 stamens.

C: palma real; P: shapaja

Palmae
(Attalea Relatives)

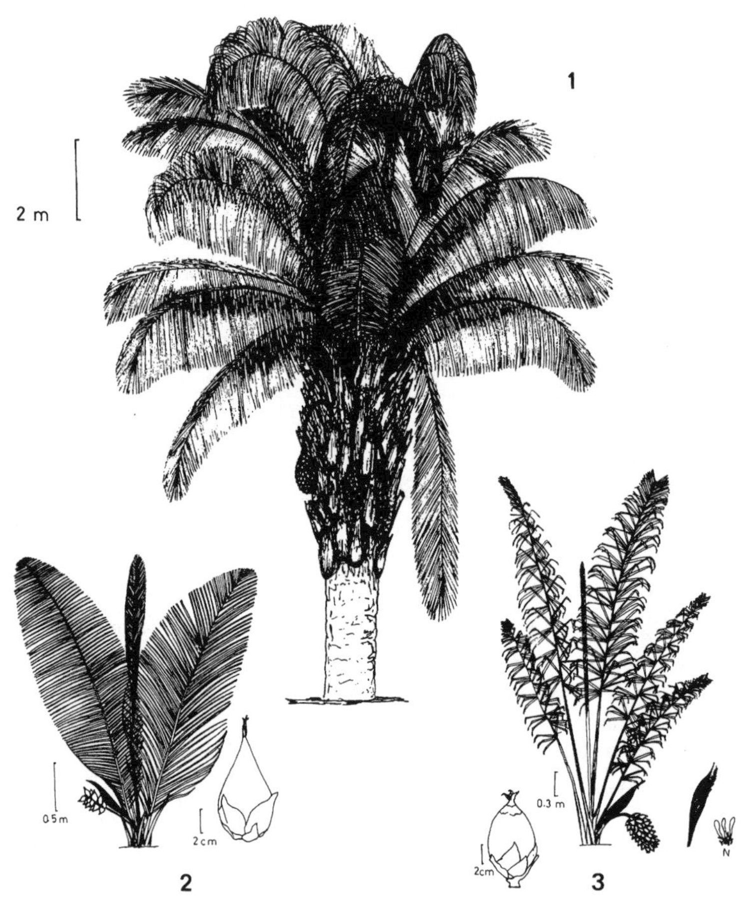

1 - *Scheelea*

2 - *Attalea*

3 - *Scheelea*

Orbignya (12 spp.) — Sometimes tall, often stemless. Technical character is male flowers with broad flat petals and many stamens having elongate contorted anthers. The Brazilian "babassu" is an *Orbignya.*

P: catarina

Attalea (22 spp.) — Frequently short-trunked or stemless but sometimes large massive palms. Technically characterized by the narrow lanceolate flat petals with many stamens (rarely only 6) having nontwisted anthers.

E: palma real; P: chonta

Maximiliana (1 spp.) — Medium-tall palms with trunk thick and somewhat jagged from persistent leaf bases, always with very long leaves and the leaflets irregularly arranged with respect to rachis. Technical character is the male flower with elongate anthers far exceeding the minute thick petals.

P: inayuca

Syagrus (32 spp.) — Mostly tall palms best represented in seasonally dry forests; leaflets nearly always irregularly arranged with respect to the horizontal rachis, but in one peculiar species (or population?), from Colombia completely, undivided. Technical character is that same inflorescences bear both staminate and pistillate flowers (unlike *Attalea* group) and the male flowers have thick elongate petals and 6 stamens. Fruit similar to *Scheelea* but "putamen" rugose inside with 3 smooth bands.

P: inchaui

Elaeis (1 spp., plus 1 in Africa) (incl. *Corozo*)— Inflorescence very dense, almost completely hidden by the leaf bases. Flowers immersed in inflorescence branches, the male and female inflorescences separate; stamen filaments united. The American genus *Corozo* has been merged with that of the formerly monotypic African oil palm.

C: manteca negrita; P: puma yarina, peloponte

Cocos (1 sp.) — Restricted to seashores except where cultivated. Leaves typically with a yellowish cast and with stiffer and narrower leaflets than other taxa; also vegetatively characterized by the typically curved and slightly swollen trunk base. The giant fruit (coconut) is unmistakable. Important technical characters are the very large globose pistillate flowers at base of rachillae and flat petals on male flowers.

C, E, P: coco

Palmae
(Miscellaneous Large Pinnate-Leaved)

3 - *Ceroxylon*

2 - *Welfia*

1 - *Elaeis*

4 - *Cocos*

Palmae
(Stilt Roots: Iriarteae)

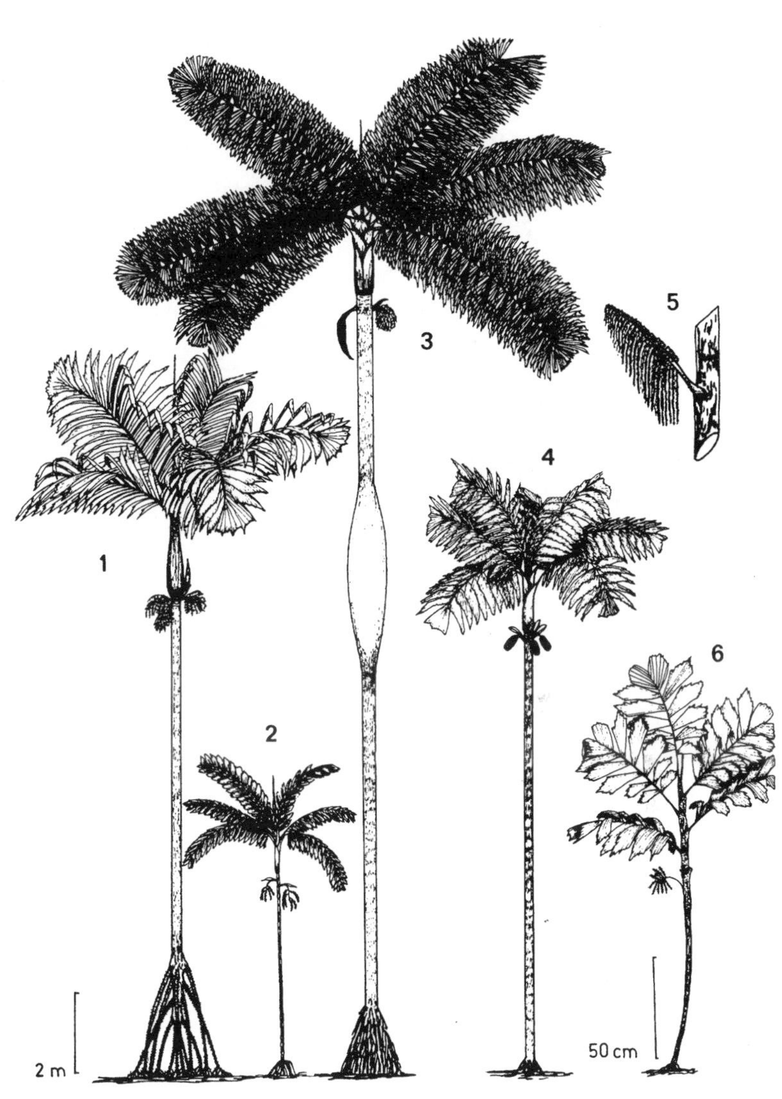

3 - *Iriartea*

5 - *Dictyocaryum*

1 - *Socratea*

4 - *Wettinia*

2 - *Catoblastus*

6 - *Iriartella*

2C. Fruits with thin nonporate endocarp — (All the stilt palms and nearly all the small, nonspiny, understory palms belong in this group as do a few other well-known larger palms — *Welfia, Ceroxylon, Oenocarpus, Euterpe*)

2Ca. Stilt roots present — Leaflets more or less triangular in shape with cuneate bases and more or less truncate apices (but often irregularly splitting into several segments); (Iriarteoid alliance). The first four genera have the inflorescence solitary and male and female flowers in triads on the same inflorescence; the last two have several inflorescences per node and separate male and female inflorescences.

Iriartea (1 sp.) — Large palm with close-together nonspiny stilt roots (can't see through root cone). Trunk often with swollen middle. Leaflets irregularly split and held in different planes, the basal segments widest. Inflorescence terete, curved and pendent in bud, resembling a longhorn's horn; flowers with 12–15 stamens.

C: barrigona, bombonaje, bombona; E: pambil; P: huacrapona

Socratea (5 spp.) (incl. *Metasocratea*) — Usually large palm with open spiny stilt roots (easy to see through root cone). Leaflets irregularly split and held in different planes (except *S. salazarii*), the apical segments wider. Inflorescence dorsoventrally compressed and erect in bud; flowers with >20 stamens.

C: zancona; E: crespa; P: pona

Iriartella (2 spp.) — Small understory palmlet few meters tall, but with distinct stilt roots. Like a very reduced version of *Socratea.*

C: zanconcita; P: casha ponita

Dictyocaryum (3 spp.) — Large mostly high-altitude palms, the stilt roots relatively dense (cf., *Iriartea*). Leaflets split into several more or less equal segments irregularly held in different planes, grayish-waxy below. Flowers with 6 stamens (unlike multistaminate *Iriartea* and *Socratea*).

C: barrigona, bombona paso

Wettinia (10 spp.) (incl. *Wettiniicarpus*) — Large palms with dense stilt roots making sharp angle with trunk, rather short and not always obvious. Leaflets relatively narrow, split and in different planes (*Wettiniicarpus*) or not split and all in one plane (*Wettinia*); inflorescence with very densely crowded flowers and fruits on very thick short axes, few-branched or consisting of a single spike; fruits villous or subspiny, angled from pressure of adjacent fruits.

C: memé; E: gualte

Catoblastus (15 spp.) (incl. *Acrostigma, Catostigma*) — Medium to large palms with dense rather short stilt roots (not always obvious). Leaflets relatively narrow, usually not split and all in same plane. Inflorescence long-branched, the branches slender; fruits ellipsoid, not densely appressed. Probably *Acrostigma* (homogeneous endosperm, pistillate flowers with 3 equal carpels at anthesis) and *Catostigma* (homogeneous endosperm, stigma sessile) are not adequately separated from *Catoblastus* (ruminate endosperm, pistillate flowers with 1 large and 2 small carpels at anthesis, stigmas on prominent styles), which is itself hardly separable from *Wettinia.*

C: crespa

2Cb. Stilt roots completely lacking or very thin, densely appressed and not very obvious

(i) Geonomoid alliance — Inflorescence rachis thick, the flowers in distinct pits in the rachis; two bracts (or bract scars) subtending inflorescence; mostly small understory palms. Technical character: bases of petals of staminate and pistillate flowers fused into tube.

Geonoma (80 spp.) — Small to medium-sized understory palms; flowers with usually 6 (rarely 3 or 7–9) stamens (male) or a 6-lobed tube (female); fruit mostly round; inflorescence <2.5 cm diameter, sometimes spicate, sometimes variously branched. Generic limits in this group have tended to fluctuate and are based mostly on technical floral characters. Of the following genera only *Welfia, Asterogyne* and *Calyptrogyne* were accepted by Wessels Boer in his monograph. Moore also accepted *Pholidostachys* and West Indian *Calyptronoma.* Burret, the previous palm specialist, accepted several additional segregates. I follow Moore's concepts here. *Geonoma* (and its segregates) is one of the two commonest genera of small palms, easily recognized by the flowers in pits in the rachis. Vegetatively *Geonoma* can be told from *Chamaedorea,* the other common understory genus by having the main secondary veins of the leaf relatively poorly differentiated and tending to dry the same gray-green color as rest of leaf; *Chamaedorea* has dark green-drying leaves and two types of "secondary" veins, large ones which dry a contrasting pale tan and smaller ones which usually dry greenish in addition to fine venation similar to that of *Geonoma.* One of the technical characters for distinguishing *Geonoma* (*s.s.*) from its relatives is that the bracts covering the flower buds are elevated, splitting at anthesis with the two lateral parts pushed aside, whereas, in *Pholidostachys* they are laterally overlapping; and in *Asterogyne* and *Calyptrogyne* they are immersed in the rachilla. The pistil of *Geonoma* is 1-celled with a basifixed style and the fruit mesocarp generally nonfibrous; the related genera have the bracts covering the flower pits immersed and lidlike in bud, then rolling back (but not splitting) and a 3-celled pistil with central style giving rise to a fruit with fibrous mesocarp.

C: rabihorcado; P: palmiche, calzón panga

Welfia (1 sp.) — Large palms (by far the largest of the *Geonoma* alliance). Characterized by numerous stamens (ca. 40) in male flowers and many staminodes in pistillate flowers; inflorescence branches very characteristic, long and pendulous, very thick (>2.5 cm diameter). Fruit almond-shaped, partially immersed in the very thick rachis.

C: amargo

Asterogyne (4 spp.) — Small undergrowth palms, nearly always with undivided leaves. Differs from *Calyptrogyne* in the inflorescence bract near base of peduncle and flowers in bud covered with a distinct rounded upper lip. Male flowers with 6–24 stamens and separate anther thecae on a bifurcate connective (rare in *Geonoma*). Fruits apically keeled when dry, with longitudinal apically fused mesocarp fibers.

Calyptrogyne (11 spp.) — Small understory palms, often stemless. Leaves irregularly divided, unlike *Asterogyne,* and the inflorescence bract inserted just below flower-bearing part of inflorescence. Inflorescence usually spicate with a circumscissile bract and with a long peduncle, rarely branched; male flowers with 6 stamens and united anther thecae; female flowers with petals forming caducous cap. Fruit rounded at apex and with anastomosing mesocarpic fibers.

Pholidostachys (4 spp.) — Medium-sized palms with irregularly divided leaves differing from *Calyptrogyne* in having long slender petioles. Inflorescence branched or spicate with a very short peduncle, typically obscured by leaf bases and laterally overlapping floral bracts; male flowers like *Calyptrogyne,* female flowers with valvate noncaducous petals. Fruit like *Calyptrogyne.*

(ii) Inflorescence rachis thin, the flowers not sunken in pits (or in very shallow depressions), the buds evident — (*Euterpe* has thicker inflorescence rachis with the flowers in depressions) The rest of the palms fall into 3 or 4 natural groups which can usually be separated by general aspect as well as by number of bracts (or bract scars) on the inflorescence and arrangements of the flowers. Within these groups several of the genera below are closely related to each other and separable only on technical characters.

The next genus is distinctive in having 5 "spathes" (or bract scars) subtending inflorescence.

Ceroxylon (20 spp.) — The famous Andean wax-palm, often growing above the limits of other arborescent vegetation, very rarely below 1800 meters. Small to very large palms with waxy leaves and conspicuous smooth pale stems with conspicuous sometimes greenish annular rings.

C: palma de cera

Palmae
(Small: Geonoma and Relatives)

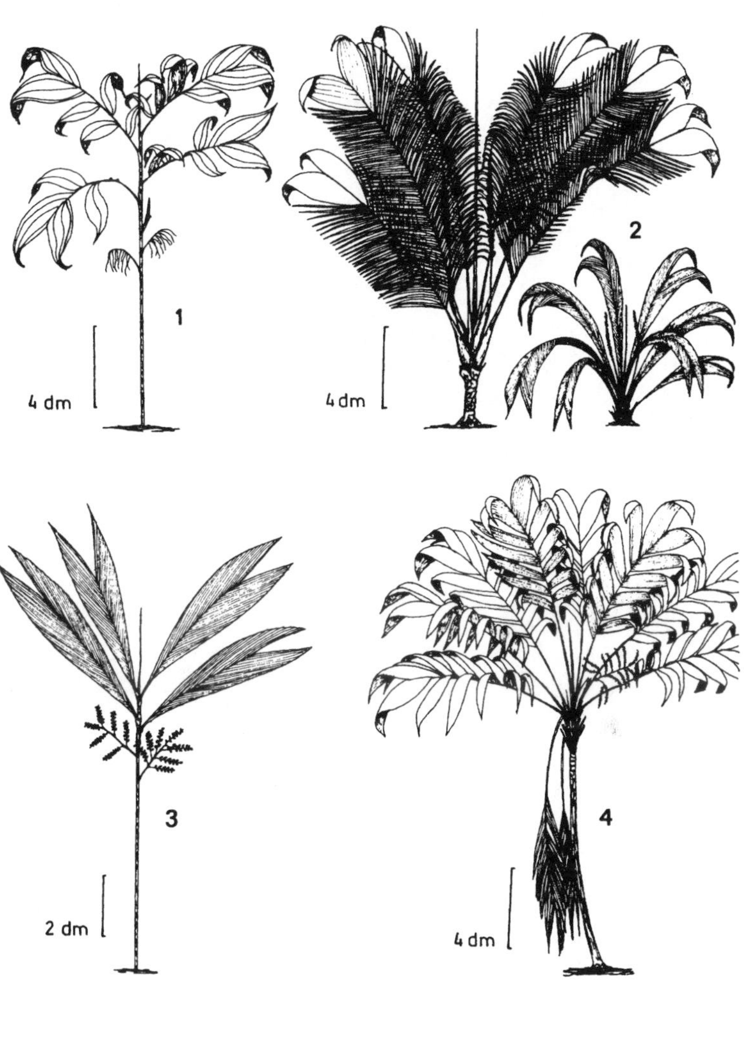

1 - *Wendlandiella* **2** - *Geonoma*

3 - *Hyospathe* **4** - *Calyptrogyne*

The next two genera have (3–)4 bracts (or bract scars) subtending inflorescence, 2(–3) of the scars all the way around peduncle well above base, the other 2 at base.

Chamaedorea (ca. 100 spp.) (incl. *Morenia*) — This is the second, main, understory palm genus, but is best-developed in Central America. It differs from *Geonoma* in having a thin rachis with the flowers and fruits not born in pits (though sometimes in shallow depressions). Usually dioecious (*Morenia* is reputedly monoecious). The fruits are small to medium-sized berries, the trunk usually green and ringed, and the inflorescence usually born from the trunk. A few species are almost stemless and several have undivided leaves and/or inflorescences, just as in *Geonoma.* The leaflets characteristically dry dark green with contrasting tannish main veins unlike *Geonoma.* Best developed in Central America.

P: palmiche

Synechanthus (3 spp.) — Often clustered small to middle-sized palm; very characteristic erect, many-branched, almost broomlike inflorescence typical; flowers in unique "acervulae" (= lines of several flowers, each line with a basal female flower and several male flowers. Vegetatively similar to *Geonoma;* stems green and ringed; main leaflet veins drying tannish. Exclusively trans-Andean.

The next six genera have 2 spathes (or scars) subtending inflorescence — Two groups: The first three genera are small understory palms similar to *Chamaedorea;* the last three are mostly large trees, all evenly pinnate with the leaflets in a single plane (a few species of *Prestoea* are understory palmlets and some may have simple leaves as can *Reinhardtia*).

Wendlandiella (3 spp.) — Tiny (essentially herbaceous) slender-stemmed, dioecious, understory palmlets similar to *Chamaedorea* but the flowers in series similar to *Synechanthus.* Inflorescence supposed to be intrafoliar and subdigitately branched with 2 spathes.

Hyospathe (7 spp.) — Slender-stemmed, monoecious, understory palms with erect, slender-branched inflorescences with well-developed rachis (similar to *Synechanthus* but smaller). Petals of male flower narrow and valvate, in bud characteristically sticking almost straight out from rachis; fruits berries, typically elongate-ellipsoid (stigmatic residue basal).

Reinhardtia (5 spp.) — Tiny (essentially herbaceous) understory palmlets <1 m tall with undivided leaves (or a few irregular basal segments). Inflorescence completely unbranched or with 2–3 short branches. Mostly Central American, barely reaching northwest Colombia.

Palmae
(Miscellaneous Pinnate-Leaved)

3 - *Oenocarpus* (*Jessenia*)

1 - *Euterpe*

5 - *Oenocarpus*

2 - *Chamaedorea*

4 - *Synechanthus*

Oenocarpus (incl. *Jessenia*) (9 spp.) — Large and solitary or smaller and clustered; characterized by a distinctive inflorescence with a cluster of pendulous branches from a short thick base, whereas, *Euterpe* and *Prestoea* have erect inflorescences with a well-developed central axis. *Jessenia* (*O. bataua*) is no more than a larger version of *Oenocarpus* characterized by more (9–20 vs. 6) stamens and the pinnae below with peltate to sickle-shaped instead of simple trichomes.

C: dompedrito, mil pesos (*O. bataua*); E: chapil, mil pesos (*O. bataua*); P: sinamilla, ungurahui (*O. bataua*)

Euterpe (12 spp.) — Medium to large palms with prominent crownshaft; leaves with many narrow leaflets basically in same plane but usually drooping characteristically. Inflorescence with a distinct straight suberect axis, the bracts subequal and inserted near each other; inflorescence branches white- to brown-tomentose, the flowers in distinct pits in the axis. Most commonly in swampy lowland habitats but also occurring in cloud forest to 3000 meters.

C: asaí; P: huasaí, sinamilla

Prestoea (20 spp.) — Small to medium palms differing from *Euterpe* in poorly developed or no crownshaft; bracts unequal with the inner inserted well above outer; flowers superficial on slender usually brownish to reddish glabrous or slightly pubescent inflorescence branches with thickened "pulvini" at bases; mostly montane habitats.

RAPATEACEAE

An exclusively herbaceous family mostly restricted to the Guayana highlands and poor-soil areas, with a single genus epiphytic. The family recognizable by the dense inflorescence heads borne at the end of long peduncles and usually subtended by bracts. Vegetatively characterized by the leaves either linear and graminoid or twisted sideways above basal sheath, and thus, equitant rather than having dorsal and ventral surface.

Only one genus is widespread in our area, *Rapatea* (20 spp.), restricted to swampy habitats and characterized by the long peduncle with a dense terminal head of yellow flowers closely subtended by two large triangular foliaceous bracts. The epiphytic genus, *Epidryos* (3 spp.), occurs in very wet lowland forests of the Chocó and adjacent Panama; it is a tank epiphyte with equitant leaves and few-flowered sedgelike heads but lacking leaflike subtending bracts. Seven of the thirteen genera are known to occur in the Guayana region of Colombia including *Saxofridericia* (9 spp.), *Guacamaya* (1 sp.), and probably red-flowered

Rapateaceae, Smilacaceae, Strelitziaceae, and Velloziaceae

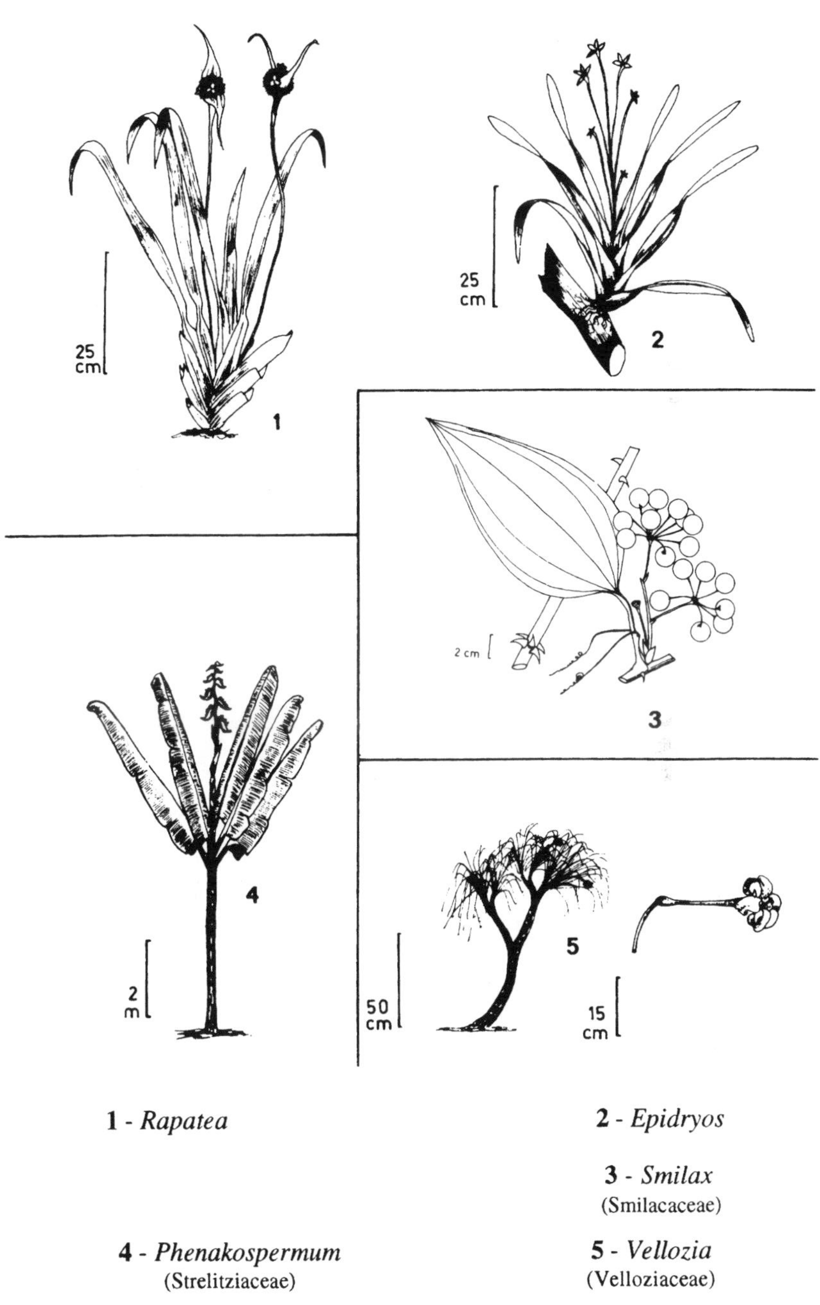

1 - *Rapatea*

2 - *Epidryos*

3 - *Smilax*
(Smilacaceae)

4 - *Phenakospermum*
(Strelitziaceae)

5 - *Vellozia*
(Velloziaceae)

laja-dwelling *Kunhardtia* (1 sp.) all with equitant leaves, and *Schoenocephalium* (5 spp.), *Monotrema* (4 spp.), *Spathanthus* (2 spp.), and *Duckea* (4 spp.) all with linear graminoid leaves.

SMILACACEAE

Mostly rather slender dioecious vines, easily recognized by more or less spiny green stems and broad dicot-looking 3-veined leaves, very distinctive in the paired tendrils arising from partway up the petiole; when the tendrils are not developed, the sheathlike petiole base from which they arose (or should have arisen) is clearly visible. Another vegetative character is the base of the leaf which is usually slightly decurrent onto the petiole apex (unlike *Dioscorea*). Inflorescences umbellate with small flowers or round berries. The only other vine with smooth +/- prickly green stems is *Dioscorea* which occasionally has stem spines but never petiole tendrils and also has a more regular venation network with more or less parallel and well-defined transverse secondary veins.

Smilax (350 spp., incl. n. temperate and Old World)
C, E, P: zarzaparilla, uña de gato

STRELITZIACEAE

In our area a single species of banana-like tree, vegetatively much like an overgrown *Heliconia,* to 10 or more meters tall and with well-developed trunk. *Phenakospermum* differs conspicuously from *Musa* in the strongly 2-ranked leaves held in the same plane (although less rigidly so) as in its close relative *Ravenala,* the widely cultivated Madagascar traveler's "palm". The massive erect terminal inflorescence is *Heliconia*-like with several large rigid greenish spathelike bracts, each enclosing several cream flowers which emerge sequentially as in *Heliconia* and are probably bat-pollinated. The family is often included in Musaceae but differs in the fruit capsular with arillate seeds rather than fleshy and berrylike.

Phenakospermum (1 sp.) — Widespread on relatively infertile lateritic soils in lowland Amazonia, sometimes constituting a conspicuous element of disturbed or secondary forest.

P: abaca

TRIURIDACEAE (Figure 1)

Small achlorophyllous saprophytic herbs found in the leaf litter of the mature forest floor (one species usually on terrestrial termite nests). Often superficially look more like small fungi than flowering plants. A most inconsequential family, of significance mainly as a curiosity on account of its unusual habit and for its phylogenetic position as one of the primitive apocarpous monocots. The only remotely similar flowering plants are saprophytic orchids and Burmanniaceae, which have very different flowers with inferior ovaries. *Thismia* of the latter family has three filiform appendages like *Triuris* but they arise from the margin of a sympetalous cupular flower rather than constituting the apices of free tepals.

Sciaphila (8 spp.) — The spindly usually reddish stem bears a spicate or narrowly racemose inflorescence of small multipistillate flowers, in fruit the flowers becoming small round reddish dense clusters of tiny follicles. The commonest species usually on terrestrial termite nests.

Triuris (3 spp.) — Plants mostly consist of single (occasionally 2–3) mushroomlike whitish flower at end of leafless hyaline stem. The most distinctive feature of the flower is three long tails that hang down and spread out from the "cap", the whole flower tending to resemble an octopus-like miniature ghost with a row of "eyes" formed by the several round sessile anthers.

VELLOZIACEAE

A characteristic mostly shrubby family of the Brazilian and Guayana Shields which barely reaches our area. Most taxa vegetatively distinctive in the thick dichotomously few-branched stems with terminal clusters of linear sclerophyllous bromeliac-like leaves. The shrubby taxa are easy to recognize by the peculiar thick stems, much of the thickness consisting of the persistent frayed fibrous remains of the leaf bases. Some taxa have reduced stems, the leaves then forming a Bromeliaceae-like basal rosette, but the solitary long-pedicellate flowers are very different from the well-developed bracteate inflorescence of Bromeliaceae.

Vellozia (122 spp.) — A few species, characterized by the thick-stemmed habit and white hawkmoth-pollinated flowers, occur in Guayanan Colombia, mostly on rock outcrops.

Barbaceniopsis (3 spp.) — One stemless magenta-flowered species is endemic to upland dry forests in the Apurimac Valley of Peru.

Zingiberaceae

Large-leaved monocots with narrowly elliptic to oblong leaves having more or less sheathing bases (the sheath open in *Renealmia,* closed in Costoideae), distinctive in the fine, uniform, strongly ascending, parallel lateral veins (much more strongly ascending than Marantaceae or Musaceae, less so than Commelinaceae) and in the ligular flap on upper margin of sheath at junction of petiole and stem (this more or less annular and ochrea-like in Costoideae with closed sheath). While none of our Zingiberaceae is woody nor scandent, several species of all three native genera can be very large coarse herbs to several meters tall and several centimeters dbh. There are two distinctive subfamilies (sometimes recognized as separate families), Zingiberoideae with open leaf sheaths and distichous leaves, and Costoideae with closed leaf sheaths and spirally arranged leaves. The former is additionally characterized by a pungent gingerlike vegetative odor (sometimes only in the roots), the latter by the unique spiral-staircase disposition of the leaves. Inflorescence usually with conspicuous bracts, these frequently brightly colored and often agglomerated into a thick dense spike. Flowers conspicuous and zygomorphic with a single stamen and often with a large expanded petal-like staminode (labellum).

We have only three native genera and a naturalized one (plus several cultivated taxa). *Renealmia,* characterized by open leaf-sheaths and relatively small flowers, either in a relatively open inflorescence or the bracts reddish, belongs to Zingiberoideae along with naturalized *Hedychium,* unmistakable in its large white hawkmoth-pollinated flowers. *Costus* and its close relative *Dimerocostus* belong to Costoideae, with spiral-staircase leaves with closed sheaths.

Costus (70 spp., plus 30 Old World) — Unique in spiral-staircase leaf arrangement and (except for Commelinaceae) in the closed leaf sheath. Inflorescence very typical, a congested spike with strongly overlapping imbricate often brightly colored bracts. Flowers usually large and conspicuous, frequently red or mottled-reddish, only rarely white or yellow (and then the labellum <6 cm wide).

C: matandrea; E: caña agria

Dimerocostus (3 spp.) — Vegetatively an unusually large canelike *Costus,* differentiated from that genus by 2-locular (vs. 3-locular) ovary, the calyx longer than the uniformly green bracts (the inflorescence thus less uniformly congested than in *Costus*), and the large white or yellow flower (labellum >7 x 8 cm when expanded). Our two common species, one with white and one with yellow flowers, are sometimes treated as subspecies.

Zingiberaceae

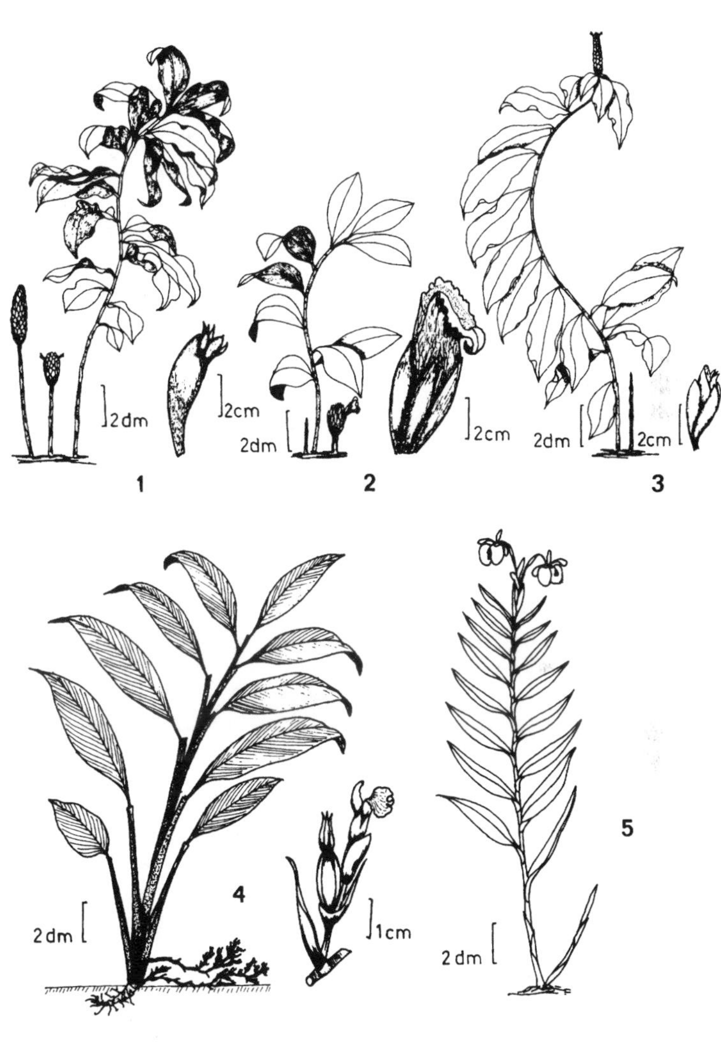

1 - *Costus* **2** - *Costus* **3** - *Costus*

4 - *Renealmia* **5** - *Hedychium*

Renealmia (55 spp., plus 25 in Africa) — Our only native Zingiberoideae genus, thus easily recognized by the open leaf sheath, distichous leaves, and frequent presence of a gingerlike vegetative odor. The inflorescence of many species is branched; spicate inflorescences may be *Costus*-like with red bracts but these are less coriaceous and less closely appressed than in *Costus.*

E: platanillo oloroso; P: mishqui panga

Hedychium (1 introduced sp.) — A very common weedy species of moist ditches and roadsides, similar in vegetative characters to *Renealmia* but very different in the large white very fragrant hawkmoth-pollinated flower. The combination of densely imbricate green bracts in a spicate terminal inflorescence is found in none of our native Zingiberoideae.

C: flor de muerto

Acanthaceae

Mostly herbs or subshrubs with conspicuously bilabiate flowers, usually subtended by conspicuous bracts and/or bracteoles, and very characteristic narrowly obovate 2-valved fruits with elastic dehiscence. Vegetatively characterized by uniformly opposite leaves, either entire (usually) or with spiny dentate margins (uplands); often with cystoliths on upper surface. Nodes tend to be swollen when fresh, contracted when dried. The flowers are often red and hummingbird-pollinated.

One genus (*Mendoncia*) is a liana, easy to recognize by the unique cross section, and three are trees (*Bravaisia, Trichanthera, Suessenguthia*); in addition, a few of the herb genera have some shrubby or subscandent species. The taxonomy of the family is based almost entirely on the stamen number (2 vs. 4) and number of anther cells (1 or 2).

1. Vines or Lianas

Thunbergia (cultivated and escaped) — Large lavender bignonlike flower in commonest species; flowers salmon with deep-maroon throat in the other commonly escaped species.

Mendoncia (66 spp., plus few in Old World)(sometimes segregated as a separate family) — Highly atypical in Acanthaceae both in the single flowers borne individually or several together in the leaf axils and partially enclosed by pair of enlarged bracteole-like foliaceous calyx lobes and in the fleshy drupaceous fruit, enclosed by the pair of persistent calyx lobes. The genus is absolutely unmistakable in flower or fruit. Some species can become high-climbing lianas with very anomalous stem vascularization (many distinct bundles of vascular tissue in cross section).

(A few other genera are scrambling or subscandent, especially several species of *Dicliptera,* with the typical narrow acanth calyx lobes.)

2. Trees — Two genera of acanth in our area are middle-sized trees, both with the petiole bases distinctly interconnected across the node; a third, monotypic *Suessenguthia,* is usually a small tree.

Bravaisia (5 spp.) — Mostly Central American; only northern Colombia in our area. Usually multiple-trunked, the individual trunks to 20 cm or more diameter. Leaves +/- puberulous below, tapering to long thin pubescent petioles. Very characteristic white Bignoniaceae-like flowers.

Trichanthera (2 spp.) — Mostly rich-soil areas, especially at middle elevations; found mostly along rivers or cultivated as living fence posts.

Weak-wooded and often multitrunked but definitely a tree. Flowers dull dark red and openly campanulate with exserted stamens, presumably bat-pollinated. Leaves larger and less pubescent than *Bravaisia* but, at least, the long petiole tannish-puberulous.

C: nacedero; E: palo de agua

Suessenguthia (1 sp.) — Large *Ruellia*-like shrub or small tree 2–5 meters tall, endemic to southwestern Amazonia. Corolla pubescent and rather gesnerlike, magenta with ca. 4 cm-long narrowly tubular-campanulate tube and 5 very similar round or slightly bifid lobes. Inflorescence rather capitate with pair of large basal bracts (cf., *Sanchezia*). The 4 stamens in two pairs, one long-exserted, the other subexserted.

(*Aphelandra*)— A few species are more or less arborescent but easy to recognize on account of the characteristic *Aphelandra* inflorescence, spicate with conspicuous overlapping bracts.

(*Sanchezia*) — Some species may be weak-stemmed shrubby treelets.

3. Herbs and Subshrubs — There are too many herbaceous acanth genera, mostly separated on technical characters related to the stamens and bracts, to treat completely here. All have the typical acanth fruit. Cystoliths are present on the upper leaf surface in most genera (not *Aphelandra* and a few of the 2-staminate genera). A few of the larger or more characteristic herbaceous genera are:

3A. Four stamens

Aphelandra (ca. 200 spp.)— Easily recognized by the very characteristic spicate inflorescence with conspicuous tightly overlapping, often brightly colored bracts. There are 4 fertile stamens with 1-celled anthers; the strongly bilabiate flowers are mostly red and hummingbird-pollinated.

Encephalosphaera (2 spp.) — Upper Amazonian subshrub, very like spiny serrate-leaved type of *Aphelandra,* but occurring at lower altitudes and with broader, rounded (though with sharply spiny terminal apicule) bracts.

Ruellia (over 50 spp., depending on generic concepts) — The main 4-staminate genus beside *Aphelandra;* besides lacking the characteristic *Aphelandra* inflorescence *Ruellia* differs from *Aphelandra* in having 2-celled anthers and generally less bilabiate flowers.

Blechum (6 spp.) — Common weedy herbs, the inflorescence similar to *Aphelandra* but with loosely appressed, thin, foliaceous, green bracts and small, pale, purplish flowers.

Acanthaceae
(Woody Taxa: Trees, Shrubs, Vines)

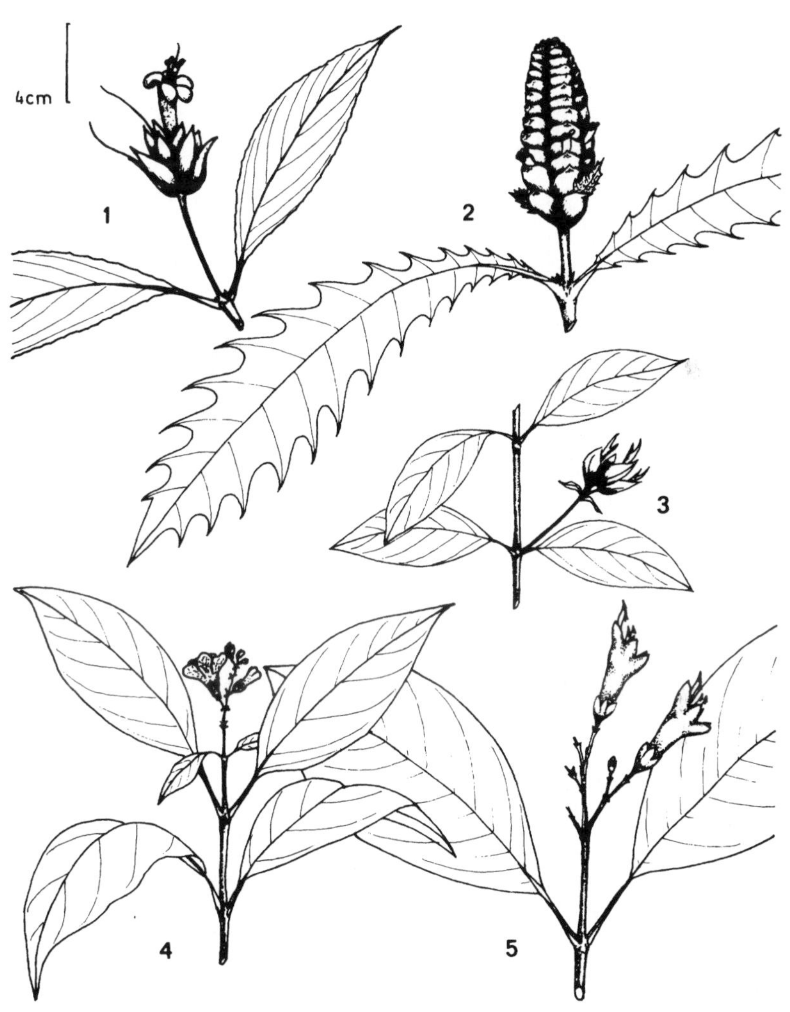

1 - *Suessenguthia*

2 - *Encephalosphaera*

3 - *Dicliptera*

4 - *Bravaisia*

5 - *Trichanthera*

Acanthaceae
(Herbs and Climbers)

1 - *Aphelandra*

2 - *Justicia* (*Beloperone*)

3 - *Mendoncia*

4 - *Razisea*

5 - *Thunbergia*

6 - *Justicia* (*Jacobinia*)

Teliostachya (9 spp.) — Somewhat similar to *Blechum* but unequal calyx segments and more rigid and pointed bracts.

Hygrophila (24 spp., plus ca. 50 in Old World) — Common along streams; characterized by the sessile axillary fascicles of small white flowers.

Dyschoriste (45 spp., plus ca. 50 in Old World) — Flowers in axillary fascicles as in *Hygrophila* but pale magenta or lavender rather than white.

Barleria (4 spp., plus ca. 120 in Old World) — Flowers yellow and with 5 more or less equal corolla lobes and very unequl calyx lobes (some species have only 2 stamens). Inflorescence bracts large, overlapping, serrate, and green.

Herpetacanthus (10 spp.) — Longer stamen pair with 2-celled anthers, shorter pair with 1-celled anthers.

3B. Two stamens

Razisea (ca. 3 spp.) — Distinctive in the narrowly tubular shortly bilabiate red flowers with 2 acuminate lips, the individual lobes fused, these apically curved in bud and arranged in narrow spikelike raceme; anther 1-celled.

Other cloud-forest genera with 2 stamens having 1-celled anthers (and usually long-exserted) include *Kalbreyeriella* (1 sp.,) (very like *Razisea* but short bracts and corolla longer-beaked in bud); *Habracanthus* (differing from *Razisea* in openly paniculate inflorescence, often magenta flowers, and the corolla either urceolately swollen with reduced lobes or narrowly tubular and strongly bent apically); *Hansteinia* (4 spp., differing from *Razisea* in the openly paniculate inflorescence and the broadly urceolate red corolla narrowed at mouth and with 5 reduced lobes); *Stenostephanus* (6 species, shrub 1–2 m tall, glandular-viscid on inflorescence and to some extent the stem; flowers tubular-campanulate above narrow basal tube, strongly curved but more obtuse in bud than *Razisea* or *Habracanthus,* pink to magenta in color).

Justicia (about 500 spp., plus few in Old World and Temperate Zone) — The main genus of 2-staminate acanth, characterized by small bracts, strongly bilabiate corollas, often with colored chevron-shaped markings on lower lip, and the absence of staminodes.

Tetramerium (28 spp.) — Dry-area herbs characterized by basically rather dense spikes with leaflike bracts (similar to 4-staminate *Blechum*). Most *Justicia* species have narrower less-leaflike bracts. Technically *Tetramerium* is differentiated from *Justicia* by the septum (with the seed-ejecting retinacula) separating from the wall of the dehisced capsule.

Acanthaceae
(Herbs - A)

1 - *Teliostachya* **2** - *Cylindroselenium* **3** - *Blechum* **4** - *Hygrophila*

5 - *Justicia* **6** - *Ruellia* **7** - *Hansteinia*

Acanthaceae
(Herbs - B)

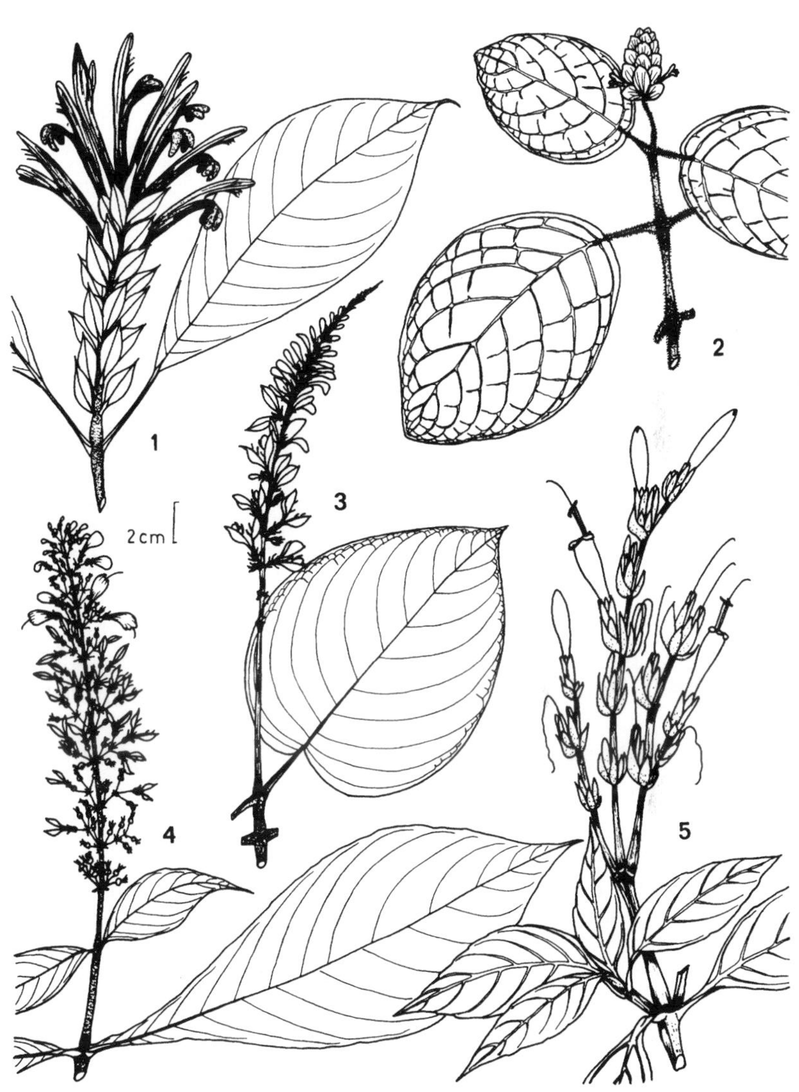

1 - *Pachystachys* **2** - *Fittonia*

3 - *Pseuderanthemum*

4 - *Stenostephanus* **5** - *Sanchezia*

Sanchezia (30 spp.) — Rather large, usually red, *Aphelandra*-like bracts but the inflorescence branched and the bracts larger, blunter, and more loosely appressed than *Aphelandra.*

Odontonema (40 spp.)—Similar to *Justicia* but with staminodes, the flowers typically red and hummingbird-pollinated (rare in *Justicia*).

Pseuderanthemum (65 spp., also in Old World) — Similar to *Justicia* but staminodes present and corolla lobes more or less equal.

Dicliptera (150 spp., plus few in Old World)—Similar to *Justicia* but stems +/- 6 angled (rather than round or tetragonal) and flower subtended by two partly united bracts. Fruit similar to *Tetramerium* in the septum (plus attached retinacula) separating from capsule wall but the stem of *Tetramerium* is never hexagonal.

(*Barleria*) — Very unequal calyx lobes; yellow flowers; may be 4-staminate also.

Other *Justicia* relatives with 2 stamens and 2-celled anthers include *Chaetothylax* (8 spp.) and *Chaetochlamys* (7 spp.) (small herbs with dense spikes and conspicuously bristly linear bracts, with 4 or 5 calyx segments, respectively); *Elytraria* (7 spp.) and *Nelsonia*); *Cylindroselenium* (1 Peruvian sp.) (differing from *Justicia* in lacking bracts and the less zygomorphic deeply 5-lobed small corolla); *Pachystachys* (12 spp.) resembling *Aphelandra* in long red flowers and conspicuously overlapping bracts but only 2 stamens as in *Justicia*); and *Fittonia* (2 spp.) with variegated leaves having contrastingly pale veins).

Actinidiaceae

Trees, mostly of middle-elevation cloud forests where they are often common and tend to be rather weedy. Traditionally treated as part of Dilleniaceae with which it shares such vegetative characters as leaves with numerous straight parallel, often strongly ascending, secondary veins, serrate margins, and frequently rough-pubescent surface. The multistaminate white flowers are also similar to Dilleniaceae except in the phylogenetically critical character of being syncarpic rather than apocarpic. Three genera in Asia, only one reaching the Neotropics.

Saurauia (80 neotropical spp., plus ca. 160 in tropical Asia) — The only neotropical genus of Actinidiaceae, and almost exclusively middle-elevation. Leaves characteristically serrate, usually rough-pubescent (or at least strigose) with many close-together veins, often with petioles of

Actinidiaceae, Aizoaceae, Alzateaceae and Amaranthaceae (Trees and Opposite-Leaved Vines)

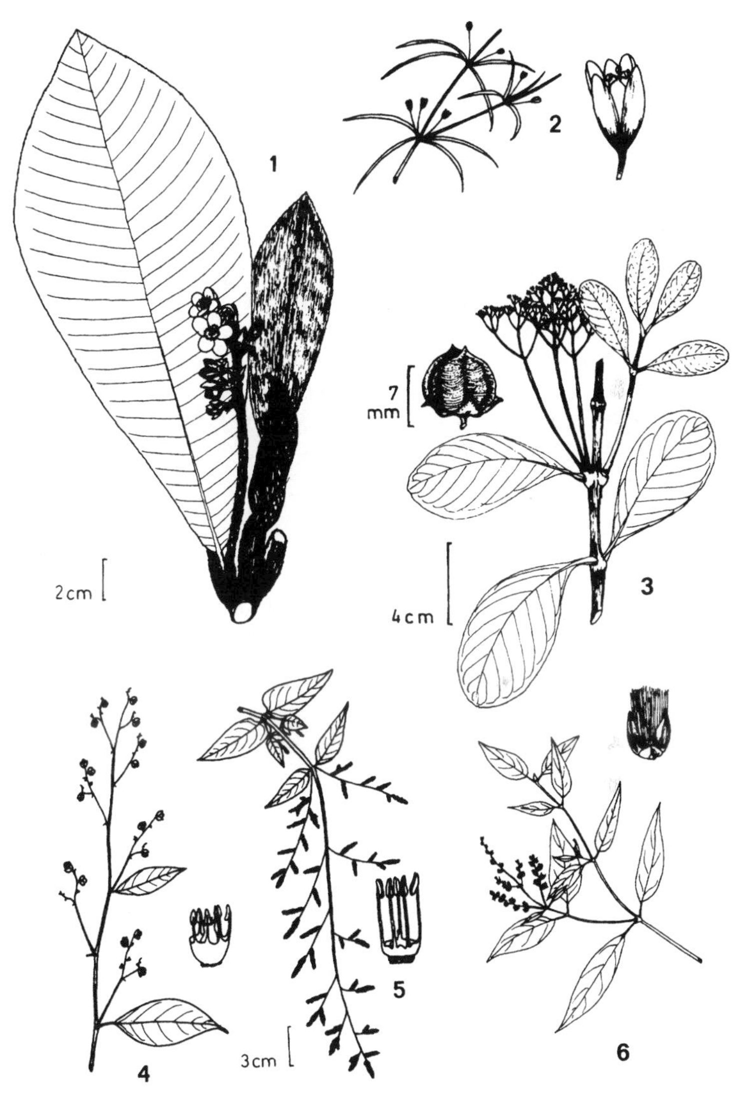

1 - *Saurauia* (Actinidiaceae)

2 - *Mollugo* (Aizoaceae)

3 - *Alzatea* (Alzateaceae)

4 - *Pleuropetalum* **5** - *Iresine* **6** - *Pfaffia*

somewhat variable length. The fruit is an irregular rather inconspicuous berry with numerous minute seeds. At least one Ecuadorian species (*S. magnifica*) is essentially glabrous and looks very much like *Meliosma* on vegetative characters, but petiole base not enlarged and woody.

AIZOACEAE

Leaf-succulent herbs, mostly along beaches, or dry-area weeds; very like Portulacaceae but lacking petals, the brightly colored tepal-like perianth parts of some genera being actually composed of staminodes. Large-flowered taxa (e.g., some cultivated African *Mesembryanthemum*) have an uncanny superficial floral resemblance to *Aster*. Molluginaceae (*Glinus, Mollugo*), frequently segregated on account of axile placentation and anthocyanins rather than b-cyanins. Usually recognizable by opposite very succulent leaves with more or less connected petiole bases; leaves of less succulent-leaved genera usually strongly anisophyllous and/or whorled. Alternate-leaved *Tetragonia* similar to Portulacaceae but distinguished by mostly whitish-appearing leaves from the silica-containing trichomes.

1. Weeds with Whorled Nonsucculent Leaves (= Molluginaceae)

Mollugo (20 spp., incl. Old World) — Narrow almost linear leaves and long-pedicellate flowers; widespread weed.

Glinus (12 spp., incl. Old World) — Prostrate weed, especially in sandy places; leaves elliptic, very strongly anisophyllous, stellate-pubescent; flowers inconspicuous, in sessile axillary clusters.

2. Alternate Leaves

Tetragonia (50-60 spp., incl. Old World) — Dry areas, mostly restricted to coastal Peru and dry western slopes of Peruvian Andes. Leaves rather rhombic and more or less succulent, mostly conspicuously whitish from the silica-containing trichomes (except cultivated *T. expansa* or New Zealand spinach).

3. Opposite Succulent Leaves

Trianthema (1 sp., plus 190 Old World) — Often more or less prostrate herbs, typically on sea cliffs. Leaves opposite but very anisophyllous, the opposing leaf bases connected by membranelike tissue (cf., Chloranthaceae).

Aptenia (1 sp.) — A monotypic genus native to South Africa. Flowers magenta and rather attractive from the enlarged staminodes; established as weed in dry inter-Andean valleys of Colombia.

Sesuvium (8 spp., incl. Old World) — Mostly prostrate beach plants. The most strongly succulent neotropical aizoac genus. The leaves are thick-linear in shape and almost round in cross section. Flowers magenta and often relatively showy from the petal-like staminodes.

ALZATEACEAE

A single locally common but rarely collected species restricted to wet montane cloud forest between (1200–)1800–2200 meters. Formerly placed in Lythraceae, but looks more like *Clusia* (Guttiferae) with very coriaceous, opposite, obovate to elliptic, usually sessile or subsessile leaves with faint secondary veins not differentiated from the parallel intersecondary veins; trees with tendency to multitrunked growth form, stilt roots, and sometimes strangler habit. Different from Guttiferae in the strongly tetragonal young branches and inflorescence and in the jointed more or less swollen nodes, the latter suggesting *Mouriri* but that lowland genus not tetragonal-angled. Inflorescence a terminal panicle with small whitish apetalous flowers with a deeply 5-dentate calyx. Fruit a small round laterally compressed 2-valved capsule, subtended by persistent calyx.

Alzatea (1 sp., 2 subsp.)
P: raja fuerte

AMARANTHACEAE

Mostly opposite-leaved herbs (often prostrate or clambering with two genera (*Iresine, Pfaffia*) becoming lianas), but a few genera have alternate leaves, and two of these are woody, one a liana (*Chamissoa*), the other a shrub or small tree (*Pleuropetalum*). The leaves are always entire and (in our area) membranaceous. The opposite-leaved lianas can be distinguished from other families by the tendency to longitudinally striate branchlets (especially *Iresine*), the more or less swollen and jointed node (sometimes contracted when dry) often crossed by an incomplete sometimes, +/- V-shaped line and usually with a row or band of appressed whitish trichomes along it (and/or with white appressed trichomes covering the axillary bud). Alternate-leaved *Chamissoa* is very similar to glabrescent climbing species of *Tournefortia,* but can be consistently separated by the tendency for the petioles to be longer and have flexed bases. *Pleuropetalum* is vegetatively nondescript, but the black-drying leaves are more membranaceous than in most rain-forest taxa and are uniformly narrowly elliptic, acuminate, and cuneate at base to a long petiole. All native

Amaranthaceae, in our area, have small usually greenish or tannish flowers, except for *Pleuropetalum* with dry scarious perianth. The one-seeded fruit is normally small and dry (a utricle), but opens to expose a small arillate seed in *Chamissoa* and is a fleshy berry in *Pleuropetalum.*

1. Distinctly Woody Small Trees or Lianas

1A. The first two genera have alternate leaves.

Pleuropetalum (5 spp.) — An understory shrub or small tree 1–3 m tall occurring in rain forest on rich soil; the only really arborescent amaranth genus in our area. Leaves narrowly elliptic, membranaceous, long tapering to base and apex. Unique in the family in the fleshy yellowish to blackberry-like fruit, this subtended by the wine-red calyx lobes. Inflorescence sparsely paniculate. More likely to be confused with Phytolaccaceae than with other amaranths.

Chamissoa (7 spp.) — A large canopy liana, mostly in montane cloud forest but also occurring in lowland forest on good soil, especially along forest edge and in second growth. Leaves alternate; branchlets more or less longitudinally striate. Differs from similar species of *Tournefortia* most consistently in the long slender petiole with flexion at base.

1B. The next three genera have opposite leaves.

Iresine (80 spp., incl. Old World) — Varying from subwoody subscandent second-growth plants to true lianas, the latter differing from *Chamissoa* in opposite leaves, and from very similar *Pfaffia* in more strongly longitudinally striate-ridged branchlets. The flowers smaller and inflorescence more diffusely branched than in other taxa.

E: camar6n

Pfaffia (50 spp.) — Ours mostly vines or lianas, very similar to *Iresine* in opposite leaves and paniculate inflorescence but much more pilose around the base of perianth. The flowers usually dispersed along inflorescence branches but sometimes floriferous portion contracted and inflorescence *Alternanthera*-like. The technical distinguishing character is the staminal tube with 5 variously toothed or laciniate lobes and no staminodia.

(*Alternanthera*) — Some *Alternanthera* species are more or less scandent but never true lianas.

Amaranthaceae
(Herbs and Alternate-Leaved Vine)

1 - *Chamissoa* 2 - *Alternanthera* 3 - *Gomphrena*

4 - *Celosia* 5 - *Achyranthes*

6 - *Amaranthus* 7 - *Cyathula*

2. Herbs

2A. The first two genera are erect herbs with alternate leaves.

Amaranthus (60 spp., incl. Old World) — Weedy herbs, the leaves long petiolate, secondary veins close-together and ascending, sometimes with spines at nodes (*A. spinosus*). Panicles with densely spicate branches (denser than *Celosia* except pink-flowered +/- cultivated species).

E: bledo

Celosia (60 spp., incl. Old World) — Three very different species in our area. One, mostly cultivated, has a large, dense pinkish-flowered inflorescence; another (*C. grandifolia*) looks vegetatively like a herbaceous *Pleuropetalum* but has an interrupted-spike inflorescence; the third (*C. virgata*) like an erect *Chamissoa* but the perianth segments longer and narrower.

2B. The rest of the herb genera have opposite leaves — Some species are erect, others prostrate or clambering.

Cyathula (25 spp., incl. Old World) — Weedy herbs with spicate inflorescence, differing from *Achyranthes* in the fruits with uncinate spines.

E: cadillo piche de gato

Achyranthes (100 spp., incl. Old World) — Weedy herb with spicate inflorescence, differing from *Cyathula* in the fruits lacking hooked spines.

E: cadillo, rabo de chancho

Alternanthera (200 spp., incl. Old World) — Highly variable, at least in macroscopic inflorescence and vegetative characters: when in doubt try *Alternanthera;* leaves larger and smaller, narrower and broader than species of other genera, some erect, others prostrate, a few climbing (but not as true lianas). Nearly always with dense +/- capitate inflorescence, but these may be sessile or pedunculate, axillary (usually) or in a terminal panicle. The technical character is a staminal tube with entire lobes, these usually alternating with staminodia.

E: escances

Guilleminea (5 spp.) — Our only species a prostrate villous-stemmed herb of dry inter-Andean valleys. Characterized by the small (sometimes very small) leaves (sometimes obscured by the stem pubescence), more densely arranged than in similar *Alternanthera* species. Inflorescence a sessile axillary glomerule of flowers. Technically differing form *Alternanthera* in perigynous rather than hypogynous stamens.

Froelichia (20 spp.) — Erect spindly weed of dry lowland areas of coastal Peru (and Ecuador?). Looks like *Achyranthes* but with spike reduced to a capitate cluster at end of long peduncle. Also distinctive in the spiny fruit.

Gomphrena (100 spp.) — In the cerrado and adjacent areas, this is a diversified genus that makes little sense vegetatively, and includes some species with very large conspicuous flower heads. The technical character to separate the genus from *Alternanthera* is a narrowly 2–3-lobed stigma. Our species all look more or less like *Alternanthera* but most can be individually separated macroscopically: Common dry-area *G. serrata* has pink perianth parts (differing from the only similarly pink-flowered *Alternanthera* in these sericeous and broad rather than small, narrow, and puberulous). Another species (*G. elegans*) has more appressed +/- sericeous pubescence than do otherwise similar *Alternanthera* species. Several species of *Gomphrena* (but none of *Alternanthera*) are reduced sessile puna plants.

Irenella (1 sp.) — A tenuous herbaceous version of *Iresine diffusa.*

Blutaparon (incl. *Philoxerus*) — Prostrate succulent seacoast herb; essentially an *Alternanthera* adapted to the seashore by nearly linear thick succulent leaves.

(***Iresine and Pfaffia***) —The same species that are normally viny may be erect herbs, especially when young.

ANACARDIACEAE

One of the numerous difficult-to-distinguish Rosidae families with alternate pinnately compound leaves. In addition several genera of anacards are either entirely or in part simple-leaved. In either case the family is usually distinguishable (except from Burseraceae) by its strongly turpentiny vegetative odor. Some simple-leaved species cannot be reliably differentiated vegetatively from Lauraceae although they generally have a more turpentiny odor and more numerous closer-together secondary veins. Compound-leaved species can be differentiated from similar Rutaceae by the different vegetative odor and lack of gland dots and from Simaroubaceae by lack of the characteristic bitter taste (but be careful: many anacards are notoriously poisonous). Both of these families as well as the tree Sapindaceae that might be confused with anacards lack any kind of milky latex. Most anacards are characterized by reddish or vertically red- and white-striped inner bark. Compound-leaved Anacardiaceae species are exceedingly difficult to differentiate from Burseraceae. Anacards always lack the apically swollen or flexed petiolules of most species of the commonest burserac genus (*Protium*); they tend to have thinner, cloudy or submilky latex that dries black, while Burseraceae latex is more resinous and mostly dries whitish (but some members of both families lack apparent latex or resin). As a rule burseracs are more aromatic but some species of both families lack obvious odors (in anacards, especially *Tapirira* with dark-drying leaf upper surface; in burseracs most species with milky latex). Even in flower the two families can only be definitively separated by technical characters of ovule position (which account for their traditional placement in different orders) but burseracs usually have the stamens arising from inside a disk while in anacards they arise from outside it. All neotropical anacards have 1-seeded drupaceous fruits while Burseraceae often have more than one seed/fruit. The only way to macroscopically distinguish many burseracs and anacards is by recognizing the genera; in the case of *Thyrsodium* and *Trattinnickia,* genera of the two families look more like each other than they do like their confamilials (and tend to be rather randomly intermixed in herbaria: one wonders if they could be confamilial, or even congeneric, despite the reputed difference in ovule placement).

Anacardiaceae genera are mostly readily distinguishable vegetatively (once their family has been established). The few simple-leaved genera (or simple-leaved members of compound-leaved genera) are quite distinct from each other vegetatively and each has a very different fruit. The fruits of the compound-leaved genera are all quite different except *Mosquitoxylon* and *Mauria* which can easily be distinguished by the more numerous leaflets and lowland habitat of the former. The leaflets of most genera (but not *Mosquitoxylon* or *Tapirira* which are uniformly entire) vary from entire- to serrate-margined within the same species.

1. Leaves Simple

Haplorhus (1 sp.) — Endemic tree of high-altitude inter-Andean valleys of central and southern Peru; unique in very narrow, almost linear, essentially sessile, willowlike leaves, and in apetalous flowers in small few-flowered inflorescences.

P: sauce cimmarón

Anacardium (15 spp.) — Trees, often very large, aromatic. Leaf broad, obovate with rounded apex, and usually well-differentiated petiole; trunk base more or less columnar, the inner bark dark reddish; inflorescence openly paniculate; fruit very characteristic with an enlarged, fleshy, edible aril topped by the smaller kidney-shaped fruit-proper (resembling a naked seed).

P: marañón, casho, sacha casho

Campnosperma (2 spp., also in Paleotropics) — Not obviously aromatic, with slight watery latex. Trees of freshwater swamps of Chocó coast, often forming pure stands; a second species in Rio Negro area swamp forests. Base of trunk with narrow stilt roots. Leaves obovate with rounded apex and long-cuneate base decurrent almost to base of petiole (*Anacardium* usually has unwinged petiole). Not noticeably aromatic (unlike *Anacardium*). Flowers sessile or subsessile on few-branched inflorescence. Fruit red or brownish at maturity, less than 1 cm long, asymmetrically ovoid with an acutish apex.

C: sajo

(***Mangifera***) — An Asian genus with one species, the mango, widely cultivated and more or less naturalized in Neotropics. The leaves are aromatic and narrower and more acuminate than in simple-leaved native species; unlike all native anacards in combining simple leaves and five stamens. The large fleshy single-seeded fruit is the well-known mango. Only produces well in areas with pronounced dry season.

C, E, P: mango

(***Mauria***) — A few *Mauria* species have simple leaves.

(***Schinus***) — *S. microphyllus* of the high Andes has small simple leaves on slender spine-tipped branches as do several extralimital species of *Schinus*).

2. Leaves Pinnately Compound — (Sometimes simple or 1-foliolate in *Mauria* and *Schinus*); the first four genera below are wind-dispersed, the other seven have vertebrate-dispersed drupes.

2A. Wind-dispersed fruits with wings or fringe of long hairs

Schinopsis (7 spp.) — Ecologically important dry-forest trees of the chaco region with one disjunct species reaching our area in the patch of dry forest around Tarapoto. Leaves multifoliolate with small asymmetrically oblong, obtuse, sessile leaflets. Fruits ca. 3 cm long, with a single elongate wing.

Loxopterygium (5 spp.) — Dry-area tree, in our area only in southwestern Ecuador and adjacent Peru. Differs from *Schinopsis* in much larger acute leaflets with short petiolules; the leaflets have deeply crenate margins and are distinctly pilose below. The fruits are similar to those of *Schinopsis* but smaller and less than 2 cm long.

Astronium (15 spp.) — Strongly aromatic trees, mostly of dry areas. The very characteristic fruit is 5-winged from the expanded calyx lobes. Five stamens, pedicellate flowers and three separate styles are important technical characters. The commonest species in our area has smooth bark with contrasting whitish and reddish patches, but some extralimital species have almost the opposite, with thick deeply ridged bark. Tends to flower and fruit while deciduous in the dry season.

C: quebracho

Anacardiaceae - A

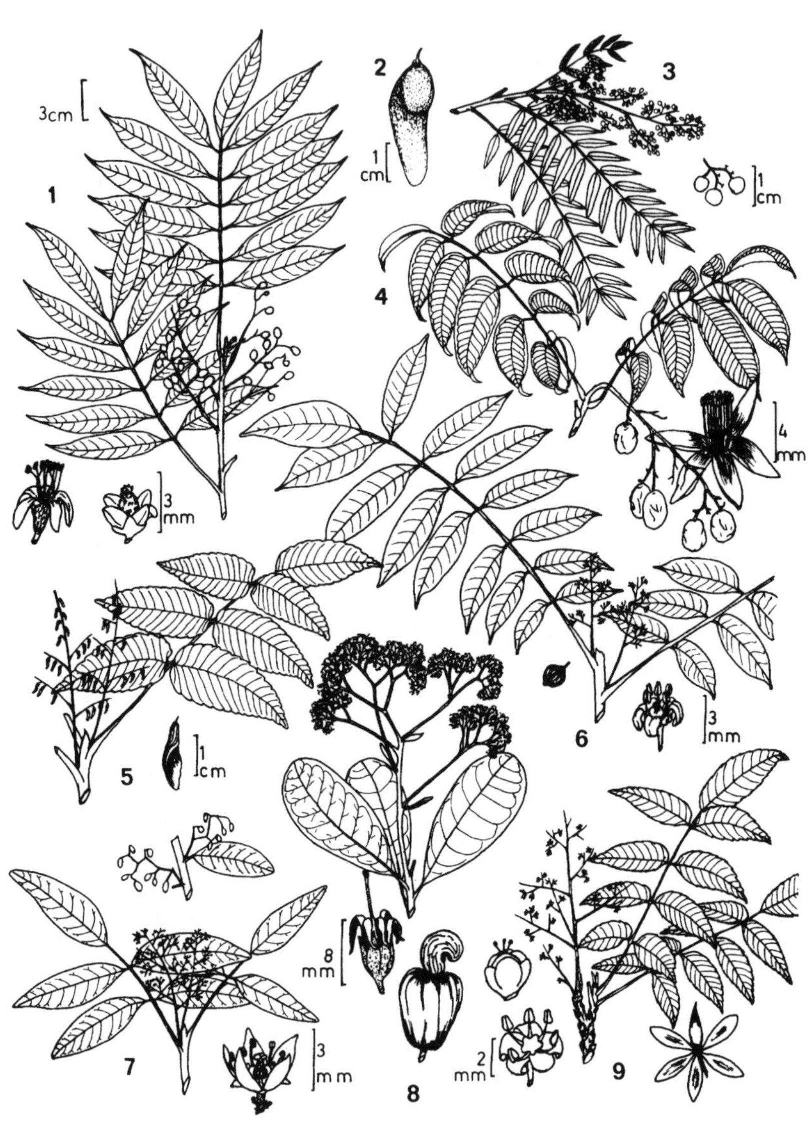

1 - *Tapirira* 2 - *Schinopsis* 3 - *Schinus*

4 - *Spondias*

5 - *Loxopterygium* 6 - *Toxicodendron*

7 - *Mauria* 8 - *Anacardium* 9 - *Astronium*

Anacardiaceae - B

2 - *Haplorhus*

1 - *Campnosperma*

3 - *Mosquitoxylon*

4 - *Ochoterenaea*

6 - *Thyrsodium*

5 - *Mangifera*

Ochoterenaea (1 sp.) — A very characteristic second-growth tree of the Colombian Andes with a characteristic flat, spreading crown. The inflorescence is also conspicuously flat-topped, unlike any other anacard, but superficially quite reminiscent of *Brunellia* which grows in similar habitats. The unique fruit is small, flattened, and dry with a thick marginal fringe of very long hairs.

2B. Fruits bird- or mammal-dispersed — (The first two genera have five stamens [as does wind-dispersed *Astronium*]; the others have ten.)

Mosquitoxylon (1 sp.) — A monotypic large tree, mostly of Central American wet forests (also Jamaica); although not yet collected in northern Colombia it has been found just across the border in Darien. It has clear milky sap and red inner bark. Flowers sessile (unique in 5-stamened genera) in a large open panicle. Drupes small (6–9 mm long), red, laterally compressed, similar to *Mauria.* Leaves 11(–29)-foliolate with very characteristic smallish, asymmetrically oblong, obtuse-tipped leaflets, usually rather densely pilose below.

Toxicodendron (15 spp., mostly n. temperate) — A single species in South America. Small trees common in disturbed middle-elevation forests. The plant is aromatic and produces a rather watery white latex that eventually turns purplish or black and produces a very strong allergic reaction (much worse than poison ivy). Leaflets characteristic in extremely asymmetric bases and the petiolules characteristically red when fresh. Fruits small and round, white with the caducous exocarp leaving a pale mesocarp crossed by conspicuous dark streaks.

C: manzanillo; E: alubillo, compadre, caspi; P: itil, incati, maico

Schinus (30 spp.) — Shrubs or small trees of dry inter-Andean valleys; leaves simple (and borne on spine-tipped branchlets) or 5(–27)-foliolate with sessile leaflets, in the common species these narrower than in any other area anacard and usually rather serrate on a subwinged rachis. Fruits small, red, globose with a hard crustaceous endocarp.

Mauria (20 spp.) — Trees or shrubs mostly of Andean cloud forests. Leaves simple to 7-foliolate; simple-leaved species have larger leaves and lack the spine-tipped branchlets of *Schinus microphyllus;* compound-leaved species have petiolules unlike *Schinus* and the leaflets are coriaceous and prominulous-reticulate above and below, and sometimes conspicuously pilose; there are fewer leaflets per leaf than in other compound-leaved anacards. Fruit a compressed drupe, red or orangish, less than 1 cm long, with a chartaceous endocarp.

Thyrsodium (ca. 6 spp., also in W. Africa) — Multifoliolate leaf with extremely asymmetric leaflet bases, usually drying blackish or intricately prominulous reticulate both above and below (or both). Inflorescence a large openly branching terminal panicle with conspicuous brownish pubescence; fruit usually puberulous and rather asymmetrically elliptic, >1 cm long.

Tapirira (15 spp.) — Forest trees, not, or very slightly, aromatic, lacking latex. Leaves 5(–15)-foliolate. The only common species *T. guianensis* is extremely widespread and prevalent in most types of lowland forest. It is almost invariably mistaken in the field for an unusually multifoliolate species of *Trichilia* when sterile, but the characteristic blackish-drying leaf upper surface (and sometimes reddish undersurface) of the dried leaflets is unmistakable; the leaflet bases are unusually strongly asymmetric. Fruit oblong, smaller than *Spondias* and *Thyrsodium* but larger than *Mauria* or other genera, turning black at maturity (unique in family in Neotropics except possibly some *Thyrsodium*).

C: palo de gusano, caimito; P: huira caspi, purma caspi

Spondias (4 spp., plus 7 in Old World) — Strongly aromatic trees. Leaves 11(–25)-foliolate, the leaflets with very characteristic translucent secondary veins connected by an equally translucent marginal vein. Drupes oblong. Three–five centimeters long, with a large stone surrounded by abundant edible pulp, usually turning yellow or red at maturity.

E: jobo, ciruelo (*S. purpurea*); P: uvos, taperiba (*S. purpurea*)

Several more genera occur in Central America and the Caribbean including *Comocladia, Cyrtocarpa, Metopium, Pachycormus, Pistacia,* and *Pseudosmodingium;* the additional genus *Lithraea* occurs in dry parts of southern South America.

ANNONACEAE

Easy to recognize to family, even when sterile, by the combination of Ranalean odor, frequently strong bark (= "carahuasca"), 2-ranked leaves (except *Tetrameranthus*), and typically myristicaceous branching. Almost all species are trees, but a few *Annona* species and one extralimital *Guatteria* are lianas. The inner bark sometimes forms a black ring around the edge of the trunk slash as in Ebenaceae. At least one *Unonopsis* (plus an extralimital *Oxandra*) species can have a distinct trace of red latex. Flowers and fruits extremely characteristic. Fruits either a cluster of monocarps from the apocarpous ovaries of a flower or these fused into an at least externally segmented fruit as typified by the well-known cultivated species of *Annona*); in either case the seeds with ruminate endosperm. Beetle-pollination is apparently typical of almost all Annonaceae, the pollinator often being trapped for a day inside the flower.

Often difficult to recognize to genus, especially when sterile. Easy genera to identify when sterile include *Ruizodendron* (leaves elliptic, whitish below, fairly small and with a pronounced tendency to be borne on short shoots from the main branches), *Tetrameranthus* (leaves spirally arranged (unique in family) and with stellate trichomes and typically unusually long petioles), *Fusaea* (leaf with strong marginal vein), some species of *Annona* (a few of the species have stellate leaf trichomes, unique except for spiral-leaved *Tetrameranthus*, some *Duguetia*, and a very few *Rollinia*); and some have distinct axillary domatia (also in few *Rollinia*) *Xylopia* (typically strongly pendent horizontal branches and frequently sericeous pubescence on twigs, leaves, or floral parts; leaves frequently narrow and coriaceous with reduced secondary venation and/or prominent intersecondaries giving a characteristic smooth undersurface), *Trigynaea* (leaves usually sub-3-veined at base). Some species of *Oxandra* and *Pseudoxandra* have *Xylopia*-like leaves with the secondary venation not very prominent; the commonest Peruvian *Oxandra* species has conspicuous stilt roots and narrowly cordate leaf bases. *Duguetia,* with prominent lepidote scales or sometimes stellate hairs on the leaf undersurface (visible to naked eye) is the easiest genus to recognize vegetatively. The very large and variable genus *Guatteria* tends to have unusually straight and often close-together secondary veins and the midvein immersed above. *Unonopsis,* also large and commonly encountered, has the midvein raised above and the secondary veins more curved towards the leaf margins and generally further apart; the tertiary veins are often parallel and/or unusually inconspicuous giving the undersurface of most species a characteristically smoothish appearance.

Only twelve genera have over a dozen neotropical species — *Guatteria, Annona, Duguetia, Rollinia, Xylopia, Unonopsis, Oxandra, Anaxagorea, Cremastosperma, Desmopsis, Malmea, Cymbopetalum.*

1. Dehiscent Monocarps (*Xylopia* Alliance) — Three genera of neotropical annonacs are characterized by dehiscent monocarps. These are large pantropical *Xylopia* and *Anaxagorea* plus *Cymbopetalum.* Of these *Anaxagorea* is unique in lacking an aril. Both *Anaxagorea* and *Xylopia* have narrow valvate thickish petals (the combination of all three of these petal characters is restricted to these two genera). *Cymbopetalum* has unmistakable large thick-petalled pendent flowers suspended from very long ebracteate pedicels. All three genera are vegetatively recognizable individually as indicated below.

Xylopia (ca. 40 spp., 100 in Old World) —Several seeds in each monocarp, the monocarps often rather elongate and usually contracted between the arillate seeds. The ultimate extreme in myristicaceous branching

with many species having the side branches strikingly pendent. Often characterized vegetatively by sericeous pubescence; the leaves often rather narrow and coriaceous, usually with reduced secondary venation and/or prominent intersecondaries giving a characteristic smooth undersurface; calyx lobes more strongly connate than in other genera.

C: guanábana, rayado; P: espintana (*X. densiflora, peruviana, micans*), espintana de varillal (*X. parviflora*), pinsha callo (*X. benthamii*), yaurache caspi (*X. cuspidata*)

Anaxagorea (21 spp., plus 4 Old World) — Unique two-seeded explosively dehiscent monocarps, with the ventral margin expanded and the dorsal one straight, and shiny black seeds. Vegetatively characterized by the coriaceous leaves (typically drying a characteristic olive color) which have minute, brownish, microscopically stellate trichomes.

C: rayado; P: bara

Cymbopetalum (28 spp.) — Very distinctive when fertile by the long ebracteate pedicels; the pendent nonaxillary flowers have thick broad large boat-shaped petals and are unmistakable. The fruits typically have only a few monocarps which are larger and thicker-walled than in *Xylopia* and *Anaxagorea;* they are pendent on the long pedicel, also. Vegetatively, the rather membranaceous leaf with midvein raised above, an acute base, and a very short petiole is characteristic.

2. Syncarpous Fruits (*Annona* Alliance) — Another natural group consists of the four syncarpous mammal-dispersed genera: *Annona, Rollinia, Raimondia,* and less-related *Fusaea.* The fruits of these genera are very difficult to distinguish from each other although vegetative and floral characters are distinctive. The flowers of this group are nonaxillary, unlike *Guatteria* and its relatives (although both can be ramiflorous: look for the leaf scars). Vegetatively differ from most members of the nonsyncarpous genera (but not from *Guatteria*) by the midvein impressed above (except *Fusaea*), and from most *Duguetia* in lacking lepidote scales (although stellate hairs are found in both groups).

Annona (ca. 70 spp., plus few in Africa) — Typical syncarpous fruit and flowers with usually broad thickish imbricate petals; some species distinguished vegetatively by stellate trichomes. A few species of *Annona* (e.g., *A. acuminata*) have small fruits with the syncarp opening irregularly to expose arillate seeds. Two savannah species have spiny branches (unique); several species are lianas, along with extralimital *Guatteria scandens,* the only neotropical annonacs to adopt this common Old World annonaceous growth-form.

C, E: guanábana (*A. muricata*); P: anona, sacha anona, carahuasca

Annonaceae
(Dehiscent Monocarps Plus *Raimondia*)

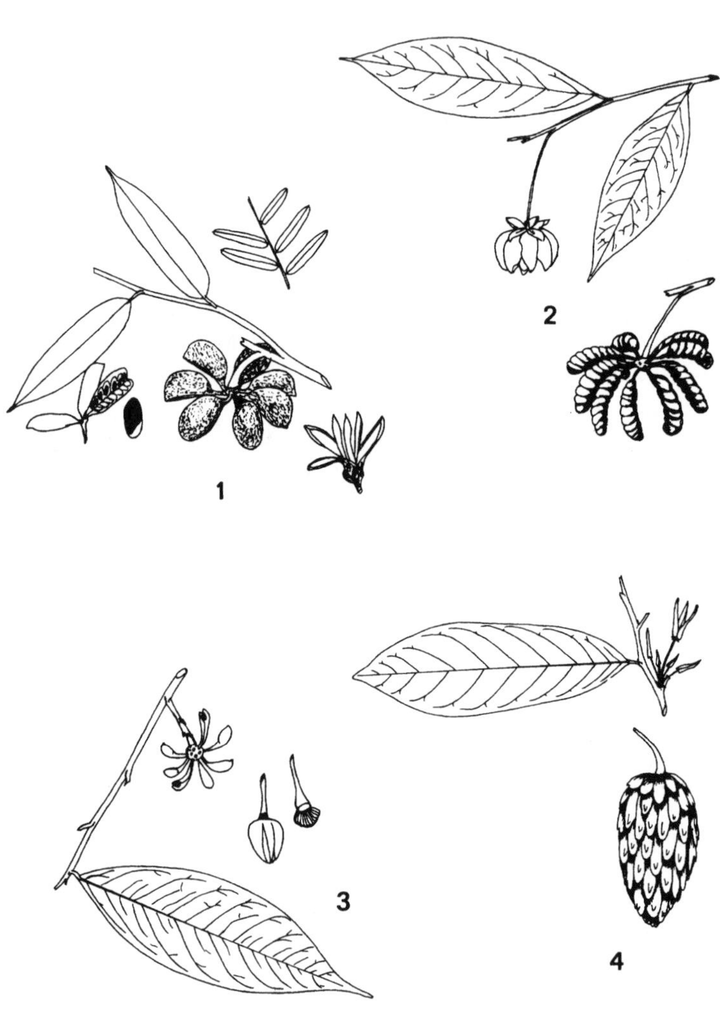

1 - *Xylopia*

2 - *Cymbopetalum*

3 - *Anaxagorea*

4 - *Raimondia*

Annonaceae
(Syncarpous or Appressed Monocarps [See Also *Raimondia*])

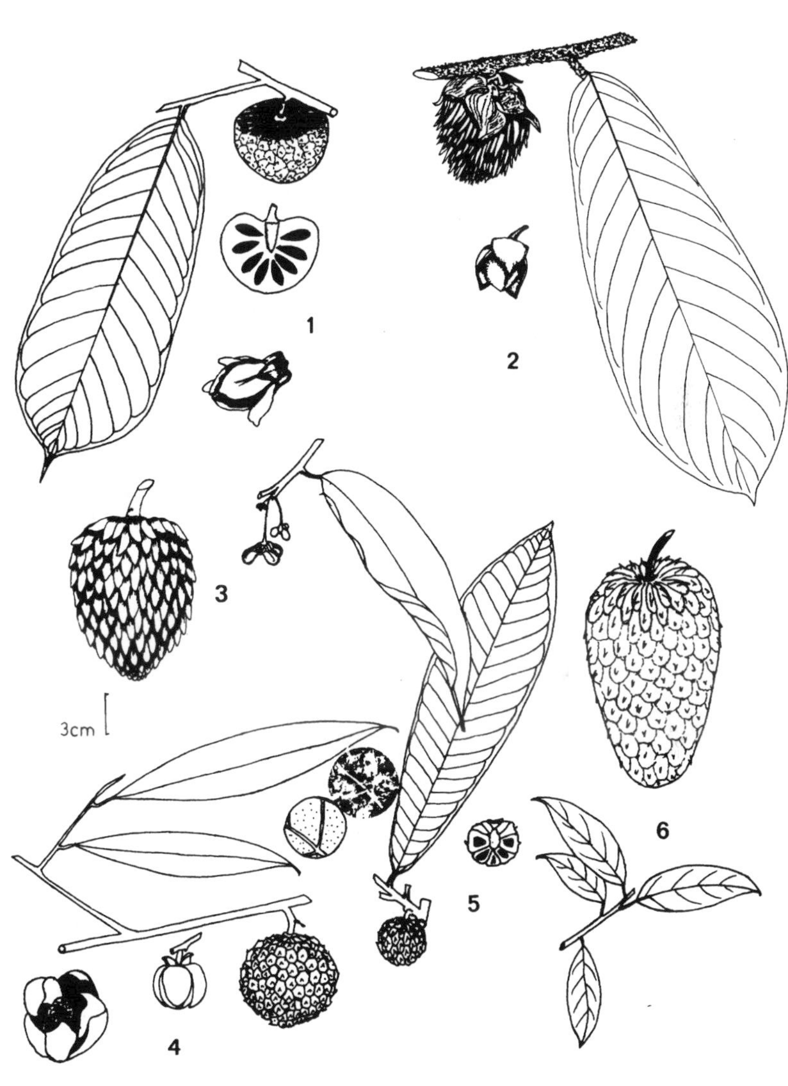

1 - *Fusaea*

2 - *Annona*

3 - *Rollinia*

4 - *Duguetia* (*D. odorata*)

5 - *Duguetia* (*D. macrophylla*)

6 - *Annona*

Rollinia (incl. *Rolliniopsis*)(45 spp.) — A derivative of (and perhaps congeneric with) *Annona,* differentiated by the unique striking laterally winged petals. In fruit *Rollinia* and *Annona* cannot always be distinguished although a distinctive tannish leaf undersurface is found only in many *Rollinia* species and stellate leaf pubescence is more common in *Annona. Rolliniopsis* differs in the monocarps appressed rather than fused (cf., *Duguetia*) but there are intermediates.

E: chirimoya; P: anona, sacha anona

Raimondia (2 spp.) — Mostly in middle-elevation cloud forests. Differs from Annona in being monoecious with twisted petals longer and narrower than in any species of *Annona* (though approached by *A. cherimola*). Its fruit is also more elongate than in any *Annona.*

Fusaea (3 spp.) — Differs from *Annona* in calyx lobes fused into patelliform collar (fused petals and/or sepals are unique to *Annona* and *Fusaea,* but the sepals are always less completely fused in *Annona*); in fruit *Fusaea* can be distinguished from *Annona* by the rim around its base formed from the calyx. The distinctive marginally veined leaves (with numerous straight secondary veins) of *Fusaea* are quite different from those of *Annona,* and the midrib is usually plane or slightly raised above.

P: sacha anona

3. Fruits with Appressed (But Not Fused) Monocarps (*Duguetia* Alliance) — These genera have fruits intermediate between the syncarpous ones of the *Annona* alliance and the separate monocarps characteristic of most genera. The only one of these of any significance is *Duguetia;* the other two genera may not be closely related to each other or to *Duguetia* despite the similar fruits. Most species of *Duguetia* easy to recognize vegetatively by the lepidote scales.

Duguetia (75 spp.) — Easily recognized by the lepidote scales or stellate trichomes on the leaves (easily visible with the naked eye); another vegetative character is the midvein impressed above. Flowers usually tan (from the lepidote scales), often with rather broad thick blunt petals. Fruit has basal collar +/- as *Fusaea* but unlike *Annona* or *Rollinia.* A few species strikingly flagelliflorous with the flowers and fruits borne at ground level.

E: piñuelo; P: tortuga caspi

(*Rollinia*) — A few species (*Rolliniopsis*) have fruits like *Duguetia* with appressed separate monocarps.

Duckeanthus (1 sp.) — Flowers like *Guatteria* but fruit similar to *Duguetia* except for lacking the basal collar formed by persistent calyx base. Leaves with distinct marginal vein like *Fusaea* and with the midvein raised above. Only known from the Rio Negro area and perhaps not entering our region.

4. Nonappressed Indehiscent Monocarps — The majority of neotropical Annonaceae have separate nonappressed monocarps. In fruit many of these genera can be distinguished by whether the monocarps are large or small, sessile or stipitate, thick- or thin-walled, few or numerous, single-seeded or several-seeded, and by their seed structure. The majority of taxa have leaf midveins raised above (main exception: *Guatteria*); *Tetrameranthus* is unique in spiral leaf arrangement.

4A. Genera with large subsessile monocarps (all several-seeded); typically only 1(–2) monocarps per fruit

Diclinanona (3 spp.) — Monocarp very large and thick-walled, usually single with few (3–8) seeds; inflorescence very branched, with unisexual largish, long-petalled flowers; vegetatively distinguished from most other annonacs by its rather large glabrous coriaceous oblong leaf with the midrib impressed above.

P: tortuga blanca

Porcelia (6 spp.) — Fruit large and very like *Diclinanona* but leaf thinner and more or less membranaceous, slightly asymmetric at base, and the midvein raised above; flowers nonaxillary, ebracteate, and open in bud (like *Guatteria*).

(***Sapranthus***) (7 spp.) — Mostly Central American but reaches Panama and likely northwestern Colombia. Cauliflorous, typically with large blackish maroon flowers, often blooming while almost leafless in the dry season. Fruits as large as those of *Diclinanona* from which *Sapranthus* is easily distinguished vegetatively by its always pubescent leaves.

Tetrameranthus (6 spp.) — Distinctive in the non-2-ranked spirally arranged leaves usually with strikingly long petioles (= not very annonaceous in appearance), and the leaf trichomes stellate (otherwise only in *Duguetia, Annona* and few *Rollinia*). The fruits mostly have 2–3 monocarps and are smaller than in *Diclinanona* and *Porcelia;* the monocarps are always sessile and often rather oblique. Four-merous flowers are another unusual character, as is the branched inflorescence.

4B. Genera with (usually few) small subglobose 1–few-seeded subsessile monocarps (*Oxandra* alliance) — The fruits of these genera are usually few-carpellate, often reduced to single monocarps, and may not look at all annonaceous; they have relatively delicate small white flowers. The leaves are typically rather small and either notably coriaceous and glossy or sub-3-veined.

Annonaceae
(Few-Carpelled Fruits with Subsessile Monocarps)

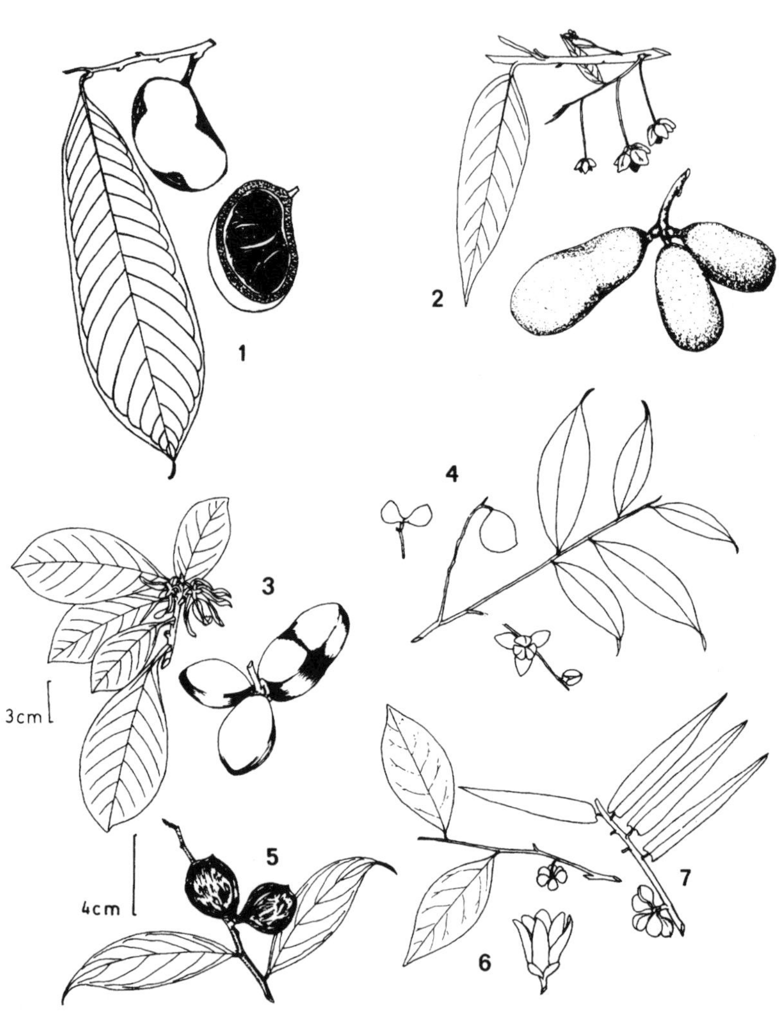

1 - *Diclinanona*

2 - *Porcelia*

3 - *Tetrameranthus*

4 - *Bocageopsis*

5 - *Trigynea*

6 - *Oxandra* (*O. espintana*)

7 - *Oxandra* (*O. xylopioides*)

Bocageopsis (4 spp.) — Vegetatively characterized by small coriaceous leaves mostly with distinctly asymmetric bases (the most asymmetric bases of any annonac) and the midrib raised above. Conspicuous black inner bark ring, thus easy to confuse with *Diospyros.* Flowers minute, usually in fascicles below the leaves; fruits often in fascicles, the several monocarps tiny, short stipitate, usually thin-walled and 1–2(–4)-seeded.

P: espintana

Trigynaea (5 spp.) — Monocarps few (often only one developed) subglobose, with woody wall, typically with more seeds (4–8 or more) than relatives. Leaf rather small and obovate with indistinctly sub-3-veined slightly asymmetric base. Flowers differ from other genera of this group in being ebracteate and distinctly supra-axillary.

P: espintana

Onychopetalum (4 spp.) — Similar to *Bocageopsis,* but characterized by narrow petals with hooked apices. Fruit usually a single barely substipitate globose monocarp with thicker walls than *Bocageopsis* and *Oxandra.* Leaves like *Caraipa* with parallel tertiary venation, the midvein raised above.

Oxandra (22 spp.) — Most species have small, typically rather glossy, leaves with inconspicuous secondary venation (cf., *Xylopia*). The stipes, pedicels and flowers are all small and the flowers usually are subtended by many little bracts. The flowers are white, unlike *Guatteria.* Typically there are several thin-walled monocarps per fruit, sometimes only one. One common species has a *Xylopia*-like leaf with a unique abruptly acute-cordate base and stilt roots (espintana negra).

P: espintana (*O. xylopioides*)

4C. Genera with (typically many) small, ellipsoid to globose, mostly one-seeded, stipitate monocarps (*Guatteria* alliance) — This group includes *Guatteria,* by far the largest genus of the family, plus two closely related good-sized genera, *Cremastosperma* and *Malmea* plus several minor segregates. It also includes the large genera *Unonopsis* and (mostly extralimital) *Desmopsis.* Most members of this group are clearly bird-dispersed. Except for *Guatteria* and *Ephedranthus,* the leaf midvein is usually raised above.

Guatteria (150 spp.) — By far the largest genus of neotropical Annonaceae. Flowers solitary (rarely in branched inflorescence) axillary (usually ramiflorous), almost always green to greenish-yellow, unusual in the "open" flower-buds and pedicel jointed below middle (rather than at base as in most other genera). Monocarps uniformly single-seeded, usually narrowly ellipsoid, usually turning black in contrast to the red or magenta stipes. Secondary veins of leaves typically rather straight and parallel, sometimes not very inconspicuous; midvein nearly always impressed above.

C: cargadero; P: carahuasca, bara, zorro caspi (*G. pteropus*)

Ephedranthus (5 spp.) — Monocarps of our species larger and longer stipitate than in *Guatteria* and orangish-yellow at maturity (rare in *Guatteria*); flowers smaller than *Guatteria,* but similarly open in bud. The leaf is intermediate between *Guatteria* and *Unonopsis* with the midvein impressed above (like *Guatteria*) and a smoothish undersurface resembling *Unonopsis;* the secondary veins are unusually straight and close-together, especially for a barely chartaceous-leaved annonac.

Malmea (13 spp.) — Superficially very like *Guatteria,* differing chiefly in the leaf-opposed or terminal flowers with basal articulation and often ciliate petals. The midrib is usually prominently raised above (very rare in *Guatteria*), and there are usually fewer secondary veins.

Cremastosperma (17 spp.) — Very close to *Malmea,* and possibly could be lumped with it; thus, also similar to *Guatteria.* The leaf of *Cremastosperma* is usually very distinctive in the midvein raised above, longitudinally ribbed, and notably broad (at least at base). The official technical difference is that the ovules of *Cremastosperma* are mostly apical (sometimes lateral) while those of *Guatteria* are basally attached. *Cremastosperma* tends to have larger monocarps and longer pedicels than *Guatteria* and the fruits may be yellow at maturity unlike most *Guatteria.* One cauliflorous species has a branched inflorescence.

P: carahuasca

Guatteriella (1 sp.) and ***Guatteriopsis*** (4 spp.) — Segregates of *Guatteria* based on valvate rather than imbricate petals (*Guatteriopsis*) and flattened monocarps with resinous walls (*Guatteriella);* likely better retained in *Guatteria.*

Froesiodendron (2 spp.) — Unusual in supra-axillary flowers. The few monocarps are several-seeded and distinctly swollen ventrally, contrasting with the relatively straight dorsal margin. Leaves are rather membranaceous with the midrib raised above.

Ruizodendron (1 sp.) — Very characteristic transversely ellipsoid asymmetrically attached monocarps (unique). Vegetatively easy to distinguish by the elliptic leaves which are whitish below (smaller than in similarly glaucous-leaved *Annona* species) and mostly borne on few-leaved, short-shoot, lateral branchlets; the midvein is raised above unlike *Annona* and *Rollinia.* Flowers small, white, with narrow petals.

Annonaceae
(Multicarpelled Fruits with Stipitate Monocarps; Broad Petals)

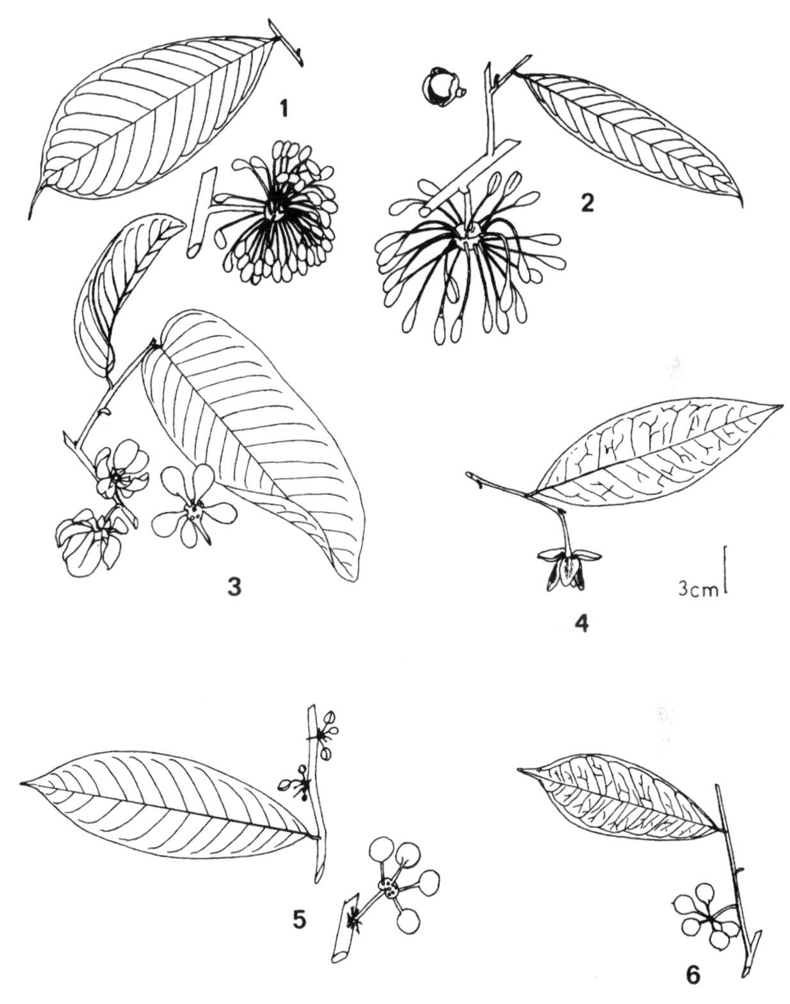

1 - *Guatteria* **2** - *Cremastosperma*

3 - *Guatteria* **4** - *Malmea*

5 - *Unonopsis* **6** - *Pseudoxandra*

Unonopsis (27 spp.) — Monocarps typically vary in seed number (1–2[–3]) even from the same flower; the majority of *Unonopsis* monocarps are completely globose and the majority of neotropical Annonaceae with globose monocarps belong to *Unonopsis* but there are exceptions. Typically there are fewer monocarps than in *Guatteria.* The seed ("hamburger"-shaped fide G. Schatz) is very distinctive in the strong median rib and pitted surface. Most species of *Unonopsis* have branched inflorescences, a character found elsewhere only in very different *Tetrameranthus* and *Diclinanona* (plus a very few species of *Guatteria*). The flowers of *Unonopsis* are much smaller than in *Guatteria* and are usually white or cream rather than green or greenish. Vegetatively there is a strong tendency for the secondary veins to be farther apart and more strongly curving upward at the margins than in *Guatteria;* the undersurface is characteristically smooth between the prominent secondary veins, a feature especially noticeable when dried. The midvein is raised above, unlike *Guatteria* or the *Annona* alliance and the tertiary veins are often parallel.

E: candelo (*U. magnifica*); P: carahuasca, icoja (*U. floribunda*)

Pseudoxandra (6 spp.) — Secondary veins inconspicuous and intersecondaries often present, the leaf thus with a typical smoothish undersurface, the midvein raised above; also distinct in a marginal vein close to leaf margin. Usually only 3–4 globose monocarps per fruit with rather short stipes (but longer than in *Oxandra*). The only common species occurs mostly in seasonally inundated forests.

Desmopsis (18 spp.) — Mostly Central American (also in Cuba) but reaches northern Colombia. Monocarps (bird dispersed) usually shortly cylindrical with 2–several seeds; when dry there is a prominent transverse central constriction between the seeds. The yellow flowers are almost always subtended by unusual leaflike bracts. The leaves tend to be rather membranaceous and puberulous.

Stenanona (incl. *Reedrollinsia*) (10 spp.) — Mostly Central American but recently discovered in western Colombia. Related to *Desmopsis* by the short cylindrical to globose dark red-purple few-seeded, bird-dispersed monocarps, but distinct in fly-pollinated flowers with long-caudate petal apices. Vegetatively distinctive in golden-ferruginous hirsute pubescence, subcordate leaf base, and short thickened petiole.

In addition to the genera included here, there are a number of entirely extralimital neotropical genera including *Tridimeris* (3 spp., Mexico), *Heteropetalum* (2 spp.) and *Pseudephedranthus* (1 sp.), both of the upper Rio Negro, and *Bocagea* (2 spp.), *Cardiopetalum* (1 sp.), and *Hornschuchia* (6 spp.) of eastern and southern Brazil.

Annonaceae
(Multicarpelled Fruits with Stipitate Monocarps; Narrow or Very Small Petals)

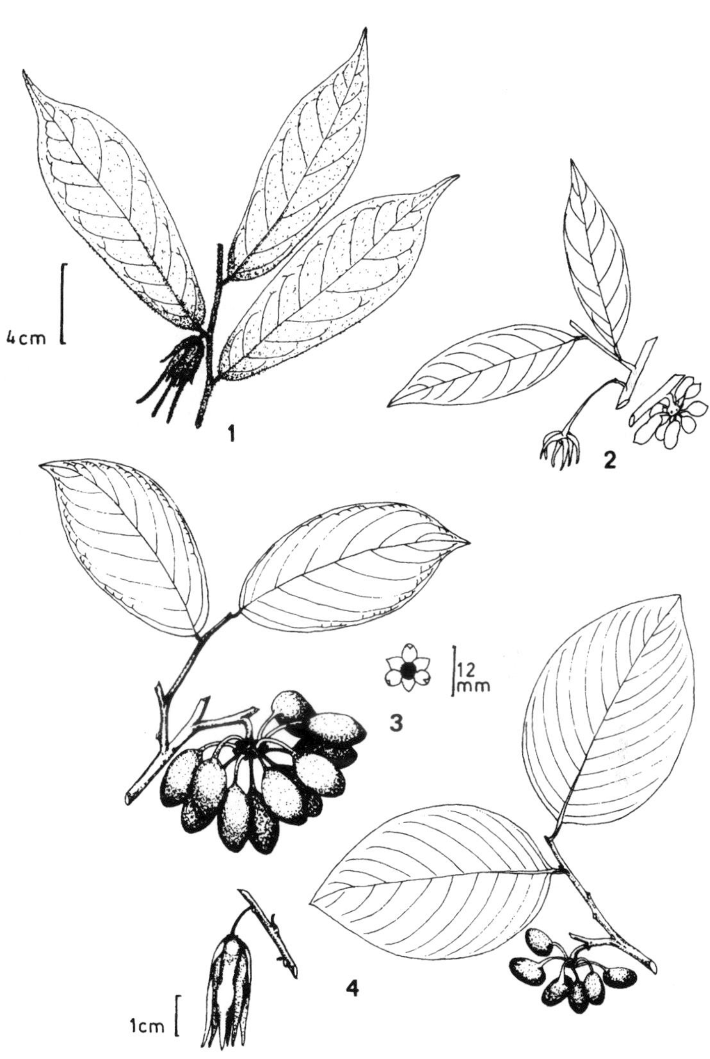

1 - *Stenanona*

2 - *Desmopsis*

3 - *Ephedranthus*

4 - *Ruizodendron*

Apocynaceae

Mostly trees and lianas characterized by usually profuse latex, simple entire leaves, and flowers with distinctive conically appressed anthers. Fruit a berry or follicle, often in pairs; if dehiscent, the seeds usually arillate or winged or plumed. The climbers are vegetatively undifferentiable from Asclepiadaceae except for their generally woodier stems (and perhaps in lacking a greenish layer under the bark). Differs from Guttiferae, the only other family having trees with opposite leaves and latex, in more freely flowing, exclusively (in opposite-leaved taxa) white latex and in usually lacking (except in a few Amazonian taxa with very profuse latex) the characteristic Guttiferae terminal bud arising from between the hollowed petiole-bases of the terminal leaf pair. Alternate-leaved taxa are vegetatively difficult to distinguish from Sapotaceae, differing in the nonswollen petiole base and typically more free-flowing latex. *Aspidosperma* sometimes has little latex or lacks latex in the trunk, but many species differ from any sapotac either in having red or orangish latex or in having a strongly fenestrated trunk; the species most similar vegetatively to Sapotaceae have leaves with *Clusia*-like venation but with secondary veins more differentiated (slightly thicker than "tertiaries" though not raised) than in *Micropholis* and relatives. Alternate-leaved apocs differ from lactiferous Euphorbiaceae in uniformly entire leaves, lack of petiolar glands, and/or uniform-lengthed petioles.

This family can be conveniently divided into four main groups: 1) trees with opposite or whorled leaves; 2) trees with alternate leaves; 3) vines with opposite or whorled leaves and lacking glands on midvein above; 4) vines with opposite leaves and glands on midvein above.

1. Trees with Opposite (or Whorled) Leaves — (Flowers always white to cream except *Stemmadenia* and few *Bonafousia* spp.)

1A. Fruits round to obovoid, fleshy, indehiscent, large (> 1 cm diameter); mostly large trees

Couma (8 spp.) — Large tree with large, broad, whorled leaves having straight conspicuous close-together secondary veins forming nearly a right angle with the midvein. The only other tree apoc genus with whorled leaves is *Rauvolfia* with very different small berry-fruits and smaller leaves with few curved secondary veins. Both fruits and latex edible.

C: popa, chicle; P: leche caspi, leche huayo

Ambelania (3 spp.) — Amazonian lowland trees, vegetatively distinct in the narrowly oblong leaves with acutish apices and uniquely faint secondary veins and absence of visible tertiary veins; in addition sclereids in

Apocynaceae
(Trees with Opposite Leaves and Large Indehiscent Fruits)

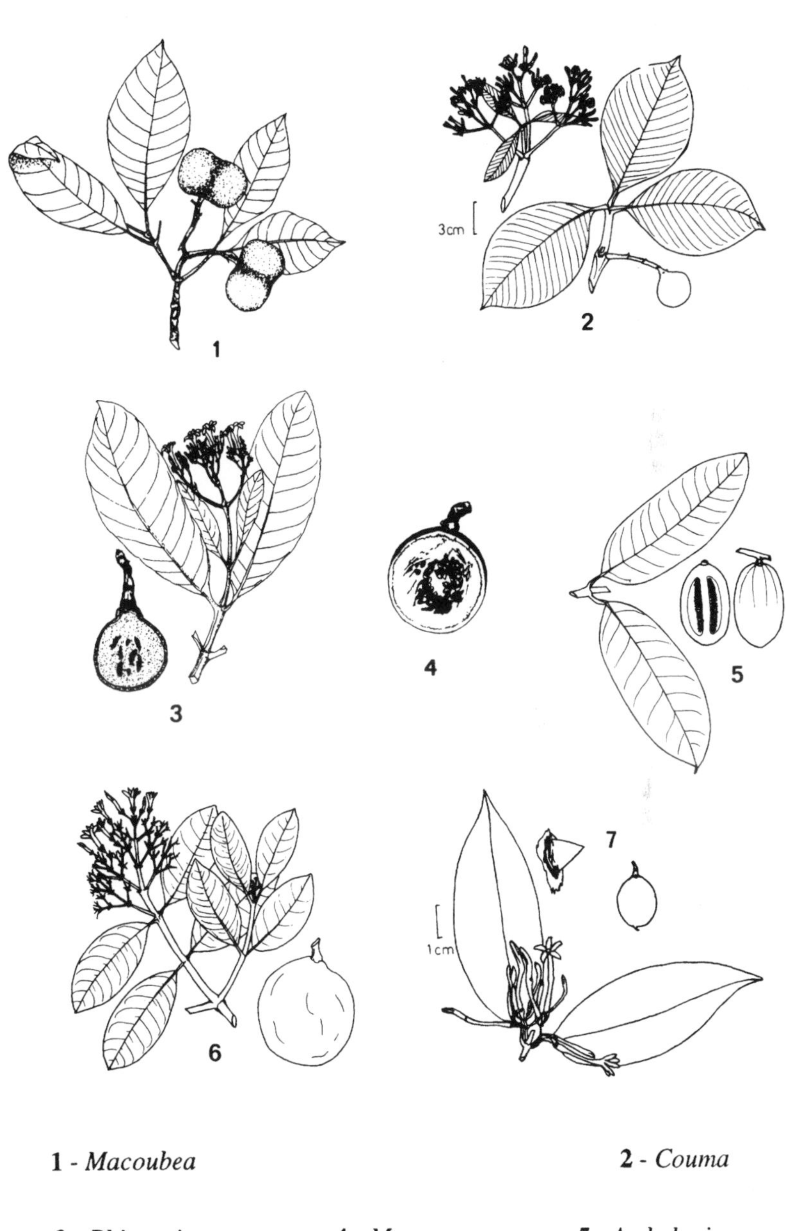

1 - *Macoubea* **2** - *Couma*

3 - *Rhigospira* **4** - *Mucoa* **5** - *Ambelania*

6 - *Parahancornia* **7** - *Lacmellea*

the leaves give a more rigid texture than in other genera. Inflorescence axillary and fruit large and two-celled.

P: cuchara caspi

Lacmellea (20 spp.) — Medium to large moist- and wet-forest trees, occasionally with spiny trunks. Distinctive among indehiscent apocs in the relatively small leaves with acute to acuminate tips, the secondary veins rather close-together, and the usual presence of a few parallel but usually faint intersecondaries. Guttifer-type bud lacking and nodes with only faint interpetiolar lines. Inflorescence axillary; fruit smaller than in related taxa and with fewer (sometimes one) seeds. Latex sweet unlike somewhat similar *Parahancornia.*

P: chicle caspi

Mucoa (2 spp.) — Large Amazonian trees with axillary inflorescences like *Lacmellea* and *Ambelania* but the twigs compressed and the leaves much larger and broader with obtuse apex and the faint, closely parallel tertiary veins more or less perpendicular to the secondaries. Fruit 1-celled unlike *Ambelania* and *Neocouma.*

Rhigospira (1 sp.) — Tree with very distinctive sharply quadrangular branchlets and large thick leaves with finely parallel tertiary venation more or less perpendicular to the secondaries. Inflorescence terminal. Pulp of fruit dark red to maroon (unique).

Neocouma (2 spp.) — Restricted to poorly drained forest on sand, mostly in the upper Rio Negro region. The large very thick leaves lack intersecondaries and have relatively few (7–13) strongly brochidodromous secondary nerves; texture hard from sclereids. Inflorescence terminal with small flowers (tube < 1 cm) and 2-celled pulpy fruit.

Molongum (3 spp.) — Barely reaching our area in the seasonally inundated black-water igapos of the upper Rio Negro. Leaves characteristic in being small with conspicuous intersecondary veins parallel to and indistinguishable from the secondaries (as in extralimital *Hancornia*). Inflorescence terminal, with thin-walled few-seeded fruits.

Spongiosperma (6 spp.) — A Guayana Shield relative of *Molongum* and *Ambelania,* restricted to riparian habitats and presumably water-dispersed. Differs from *Neocouma* in the 1-celled nonpulpy fruit, larger corolla (tube >1 cm) and more numerous (>15) secondary veins with parallel but very indistinct intersecondaries.

Parahancornia (8 spp.) — Large trees, mostly of poor-soil areas with characteristic small obovate-oblong round-tipped leaves on flattened

Apocynaceae
(Trees with Opposite Leaves and Berries or Dehiscent Fruits)

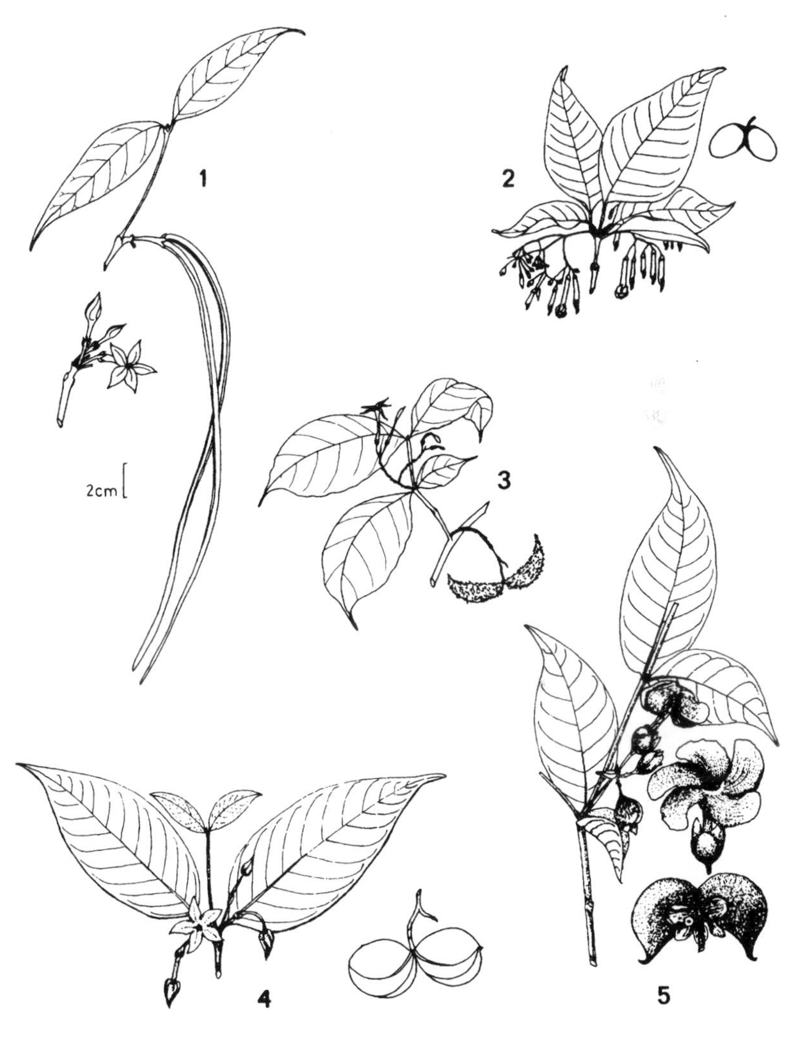

1 - *Malouetia*

2 - *Rauwolfia*

3 - *Stenosolen*

4 - *Tabernaemontana* (*Bonafousia*)

5 - *Stemmadenia*

twigs; intersecondary veins present but poorly developed. Terminal bud not guttifer-like and interpetiolar lines inconspicuous. Latex bitter unlike vegetatively somewhat similar *Lacmellea*.

P: naranjo podrido

Macoubea (2 spp.) — Large trees, quite unrelated to other apocynacs with indehiscent fruit and closer to *Tabernaemontana* and allies. Leaves large, broad, not very coriaceous, the tip obtuse to acutish (cf., *Couma* but opposite), distinctive in lacking intersecondaries and the tertiary venation more or less reticulate. Terminal bud strongly guttifer-like giving rise to noticeably "jointed" twig. Fruit very different from *Ambelania* relatives in being strongly lenticellate (often asymmetric and/or broader than long).

1B. Fruits dehiscent — (Not always in *Bonafousia*?), although usually thick and fleshy; seeds with arils (usually bright orange); mostly shrubs and small trees

Tabernaemontana (incl. *Peschiera*) (ca. 33 spp., plus many in Old World, depending on generic taxonomy) — Mostly shrubs or small trees, a few species reaching canopy. Inflorescence usually many-flowered; flowers white, often with exserted blue-green anthers, or anthers at base of tube (*Peschiera*).

E: huevo de berraco, cojón; P: sanango

Stemmadenia (15 spp.) — Differs from *Tabernaemontana* by calyx lobes and corolla relatively large and the flowers yellow or cream; fruits always short and thick. Mostly Central American, absent from Amazonia.

Stenosolen (5 spp.) — A *Tabernaemontana* segregate differing by flowers few per inflorescence and with very narrow rather elongate corolla tubes and calyx lobes, and by fruits with soft spines outside.

Bonafousia (incl. *Anartia*) (26 spp.) — Shrubs or small trees with globose to ovoid fruits. A *Tabernaemontana* segregate (not always recognized), differing in flowers usually yellow with anthers near middle of tube and the tube interior hairy below anthers, also in small axillary inflorescences. Leaves often rather large.

Woytkowskia (2 spp) — Shrub very like *Tabernaemontana* except for the paired linear follicles up to 20 cm long.

1C. Fruit a small or smallish berry; shrubs or usually small trees

Rauvolfia (33 spp., plus 65 in Old World) — Shrubs or small (rarely large) trees with whorled leaves (as in *Couma* but much smaller and often of different sizes). Berries borne on more or less umbellate inflorescences, sometimes apically bifid.

P: misho runto

1D. Follicles dry and much longer than wide; paired; seeds covered by hairs; mostly small trees

Malouetia (ca. 20 spp., plus few in Africa) — Shrubs or smallish trees with paired follicles. Seeds with hairs all around body (not in apical tuft as in apoc lianas). Flowers small, white. (This genus is related to Apocynoideae [= vines] because of sterile anther bases.)

P: chicle

2. Trees with Leaves Alternate or Spiral

Aspidosperma (80 spp.) — Large trees; latex sometimes red or orange; trunk often fenestrated. Flowers mostly small and whitish. Fruit very distinctive; a woody, usually round (sometimes narrowly oblong) and compressed, follicle; the seeds thin and round or oval with membranaceous wing surrounding the body which has long threadlike stalk from its center.

C: costillo redondo, carretto (*A. polyneuron*); E: naranjillo, naranjillo de mono, naranjo, naranjo de mono; P: remo caspi (series *Nitida*), quillobordón, pumaquiro

Geissospermum (5 spp.) — Large Amazonian trees, similar to *Aspidosperma* but fruits ellipsoid and not compressed, the seeds unwinged. Vegetative and floral pubescence more sericeous and the characteristic fine venation more raised-reticulate.

Thevetia (9 spp.) — Mostly shrubs, distinctive in more or less round, rather large, indehiscent fleshy fruit. The only Peruvian species, *T. peruviana* of inter-Andean valleys, has distinctive very narrow leaves.

E: chilca (*T. peruviana*)

Vallesia (10 spp.) — Shrubs of very dry areas with flowers small and in reduced axillary inflorescences; fruits very distinctive, a small curved (banana-shaped) white berry. Leaves characteristically narrow, along zigzag olive twigs.

Plumeria (7 spp.) — Thick-branched, dry-forest trees with large infundibuliform flowers and tiny inflorescence bracts. Fruit woody, shaped like banana prior to opening, flat when dehisced; seeds with a thickish body and brownish asymmetric wing on one end. Commonly cultivated, but not native south of Colombia.

E: frangipani

Himatanthus (7 spp.) — Middle-sized to large moist- and wet-forest trees. Related to *Plumeria* but bracts of inflorescence larger and calyx small and reduced. Fruit like *Plumeria* but the seeds with wing concentrically surrounding the thin body (which lacks the stalk of *Aspidosperma*).

C: caimito platano; P: bellaco caspi

Apocynaceae
(Trees and Shrubs with Alternate Leaves)

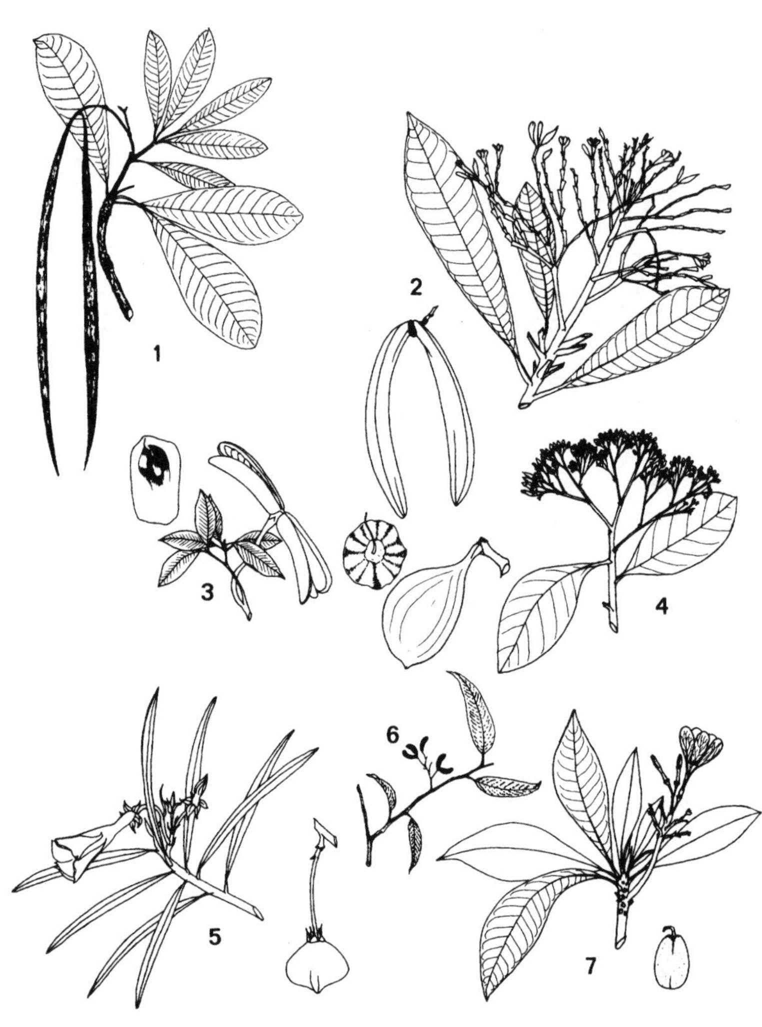

1 - *Laxoplumeria* **2** - *Himatanthus*

3 - *Aspidosperma* **4** - *Aspidosperma*

5 - *Thevetia* **6** - *Vallesia* **7** - *Plumeria*

Laxoplumeria (3 spp.) — Large wet-forest trees with leaves larger, thinner and more obovate than in most relatives. Flowers small and in much-branched inflorescences. Fruits very slender, the seeds with narrow elongate body covered by long brownish trichomes.

3. Lianas with Opposite Leaves with Glands on Midrib Above, at Least at Base — (These three genera have completely fertile anthers suggesting primitiveness within subfamily Apocynoideae.)

Mandevilla (over 100 spp.) — Mostly weedy vines (sometimes canopy lianas, especially in cloud forest). Leaf bases cordate in most species (in apoc lianas virtually unique). Leaves always with nectaries on midrib, sometimes also on upper petiole.

Allomarkgrafia (4 sp.) — Similar to noncordate *Mandevilla* species, but the inflorescence usually much branched; bracteoles often conspicuous unlike *Mesechites.*

Mesechites (10 spp.) — Slender weedy vines closely related to *Mandevilla,* differing in the usually branched inflorescence, the upper part of corolla tube narrower and the flower color (usually whitish with green center). Leaves not cordate and with nectaries only at base of blade (as in asclepiads, but unlike most *Mandevilla*).

(*Forsteronia*) has few species with leaf glands. — They have very small white flowers and are always woody canopy lianas.

4. Lianas with Whorled or Opposite Leaves Lacking Glands on Midrib Above — (Mostly members of Apocynoideae which usually have the anthers only partly fertile, with conspicuous sterile basal auricles.) The first three genera are very distinctive, the others are mostly difficult to tell apart vegetatively.

4A. Miscellaneous distinctive genera

Allamanda (15 spp.) — Our only climbing apoc with whorled leaves. Also very distinctive in the large campanulate yellow flowers and spiny syncarpous fruit. Commonly cultivated and naturalized; mostly Brazilian, one rare arborescent native species in inter-Andean Marañon Valley.

E: bejuco de San Jose, copa de oro; P: campanilla de oro

Anechites (1 sp.) — Slender vine with very characteristic hooked hairs on stem and inflorescence causing plant to stick like bedstraw. Fruit reduced to single seed, sticky and exozoochorous (unique). Mostly Caribbean region, south to coastal Ecuador.

E: tuba

Peltastes (7 sp.) — Unique in peltate leaf.

Apocynaceae
(Lianas with Small Flowers or Leaves with Midrib Glands)

1 - *Condylocarpon* **2** - *Mesechites*

3 - *Mandevilla*

4 - *Allomarkgrafia*

5 - *Forsteronia* **6** - *Anechites*

Apocynaceae
(Lianas with Large Flowers and Nonglandular Leaves)

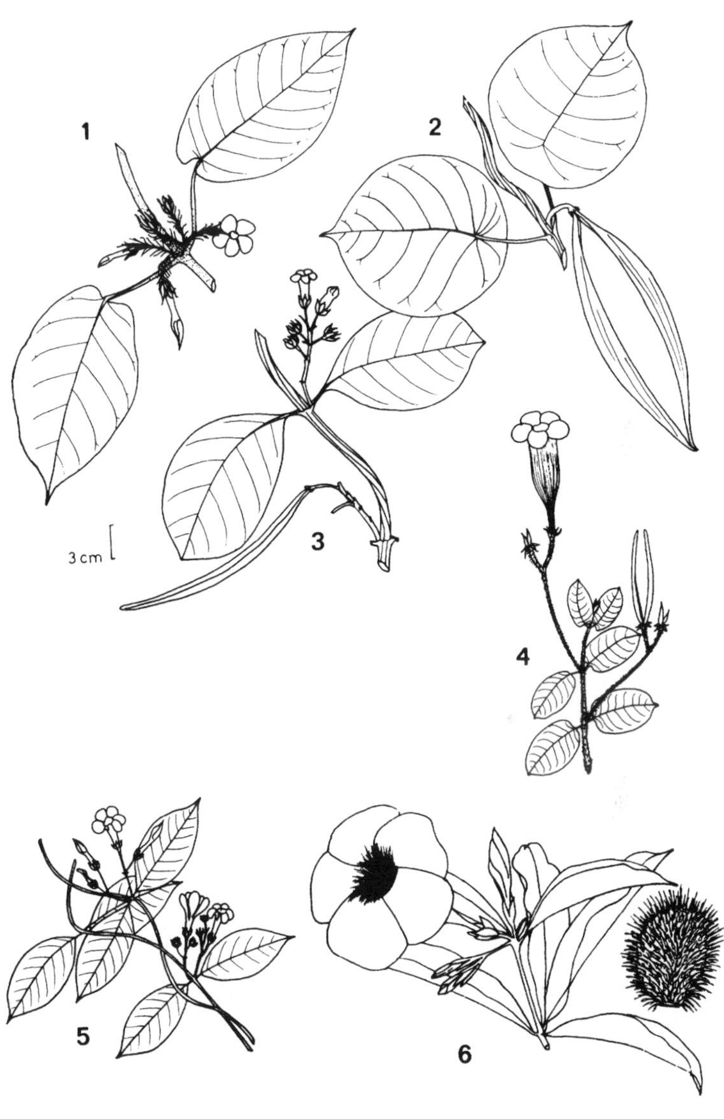

1 - *Macropharynx* **2** - *Peltastes*

3 - *Prestonia*

4 - *Rhabdadenia*

5 - *Odontadenia* **6** - *Allamanda*

4B. The next three genera, mostly canopy lianas, have small salverform corollas and the fruits either very narrow or constricted between seeds.

Condylocarpon (15 spp.) — Swampy inundated forests. Flowers <5 mm long. Fruit breaking into 1-seeded segments, the seeds otherwise unwinged.

Forsteronia (46 spp.) — Terra firme forests. Flowers <5 mm long, the anther tips usually exserted. Fruit not breaking into segments.

Secondatia (7 spp.) — Similar to *Forsteronia* but the flowers 1 cm long and anthers included.

4C. The next six genera have larger flowers and a usually broader fruit.

Odontadenia (30 spp.) — Mostly canopy lianas. Corolla campanulate, usually yellow, occasionally red, without annular ring around mouth of tube. Leaves usually with characteristic closely parallel tertiary venation.

P: sapo huasco

Prestonia (65 spp.) — Vines or lianas, the latter sometimes with conspicuously corky bark. Flowers (mostly light yellow) with a thickened annulus around corolla mouth and appendages inside corolla tube (can be thought of as halfway to Asclepiadaceae). Vegetatively characterized by always having toothed nectaries in the leaf axils. (Some ascleps, e.g., *Marsdenia rubrofusca,* look very much like *Prestonia* vegetatively.)

Echites (6 spp.) — Vines with much-twisted salverform corolla and very narrow acuminate calyx lobes. As now defined, mostly West Indian and Central American, one species to northern Colombia.

Macropharynx (4 spp.) — Characterized by large leaves (like *Peltastes* but not peltate) and conspicuously laciniate calyx lobes.

Rhabdadenia (4 spp.) — One common mangrove species with white flowers; one freshwater swamp species with large campanulate magenta flowers.

Laubertia (6 spp.) — Like *Prestonia* (i.e., with thickened annulus around corolla mouth) but flowers magenta and leaves (in South America) subcordate.

AQUIFOLIACEAE

Rather nondescript trees and shrubs with simple, usually conspicuously coriaceous leaves, often with characteristic but not very obvious faint blackish tracing of tertiary venation below; the margins often serrate (sometimes spiny-serrate), usually at least inconspicuously serrulate; undersurface usually light green with minute blackish punctation-like dots (might be confused with Myrsinaceae). Bark slash always white and nondescript except for a very characteristic thin greenish outer layer. Inflorescence axillary or ramiflorous, the flowers very characteristic, always in fascicles, with a broad disk and almost always four (rarely five) small spreading white petals.

Ilex (150 spp., plus 250 in Old World and Temperate Zone) — Trees and shrubs (one western Colombian species is apparently a liana); mostly of high altitudes. Several species reach low altitudes in the Chocó and there is one common lowland Amazonian species (*I. inundata,* with entire characteristically black-drying leaves) of seasonally inundated tahuampa forests. The genus is rich in caffeine and the source of the famous "mate" tea of Argentina and adjacent countries as well as other similar beverages.

P: guayusa

ARALIACEAE

Mostly trees, sometimes shrubs or treelets; often hemiepiphytic (especially *Schefflera*). Characterized by more or less 3-veined (or palmately lobed or compound or pinnately bicompound) leaves, of different sizes and with different-length petioles, these often with sheathing bases and a persistent stipulelike appendage (ligule) that approximates the conical terminal stipule of Moraceae. Usually with a rather rank vegetative odor especially from the trunk or branches. Inflorescences very characteristic, the ultimate clusters of small inconspicuous flowers or fruits arranged in umbels, these usually compounded into often large and complex terminal racemes or panicles. Fruit usually small blackish and berrylike, sometimes drier and more or less longitudinally 5-furrowed (flattened in *Didymopanax*).

Neotropical genera are generally easy to distinguish on vegetative grounds. *Dendropanax* and *Oreopanax* have simple leaves, those of the former, mostly from lowland forest, are mostly glabrous and entire while *Oreopanax,* mostly in middle-elevation cloud forests, has mostly conspicuously pubescent and usually more or less lobed leaves. *Schefflera* and *Didymopanax* have palmately compound leaves; the former is mostly in middle-elevation cloud forests and typically epiphytic or hemiepiphytic; the latter has few area species, but includes a common secondary-growth tree of lowland forests. One neotropical genus has bipinnately compound leaves, *Sciadodendron,* mostly of dry regions and reaching our area in northern Colombia and the inter-Andean valleys of southern Peru.

1. Simple Leaves

Oreopanax (120 spp.) — Trees (sometimes epiphytic or hemiepiphytic) mostly of middle-elevation cloud forests (only epiphytic climbers sometimes to lower elevations). Leaves usually palmately lobed, at least in part, often deeply lobed and nearly compound (a few extralimital taxa actually compound), usually conspicuously puberulous below, often with tannish trichomes. Umbels usually reduced to capitate heads and racemosely arranged in terminal panicles.

Dendropanax (75 spp., incl. Old World) — Mostly lowland forest trees (some species shrubby treelets). Leaves entire, unlobed (sometimes 3-lobed in juveniles), glabrous, usually 3-veined at base, clustered near branch apices and of conspicuously different sizes, the larger ones with much longer petioles than the smaller. Stipular ligule much reduced. Flowering and fruiting umbels open and with long pedicels, rather few umbels per inflorescence.

2. Palmately Compound Leaves

Schefflera (800 spp., mostly Old World) — Mostly epiphytic or hemiepiphytic trees or climbers of middle-elevation cloud forests, sometimes becoming fair-sized terrestrial trees. Leaves with well-developed ligule and almost always palmately compound (in a very few extralimital species simple and in Ecuadorian *S. diplodactyla* the leaflets subdivided and the leaf thus bicompound). Inflorescence umbels typically clustered into large complex terminal inflorescence with racemosely or paniculately arranged lateral branches.

C: yuco

Didymopanax (ca. 40 spp.) — The common species a tree of lowland forest second growth and easy to recognize from a distance by the typical branching pattern and reddish-tan leaf undersurface that can be seen from a distance when the wind blows; juvenile leaves have long marginal hairs and look very different from those of adults. The two other species in our area are restricted to poor sandy soil. The ligule is smaller than in *Schefflera*. The main technical difference is the flattened dry 2-parted fruit (rather than fleshy and round or 5(–9) ridged, but *S. megacarpa* with a dry 5-ridged fruit is intermediate and the genus may not be adequately differentiated from *Schefflera.*

E: puma maqui; P: sacha uvilla

Aquifoliaceae and Araliaceae

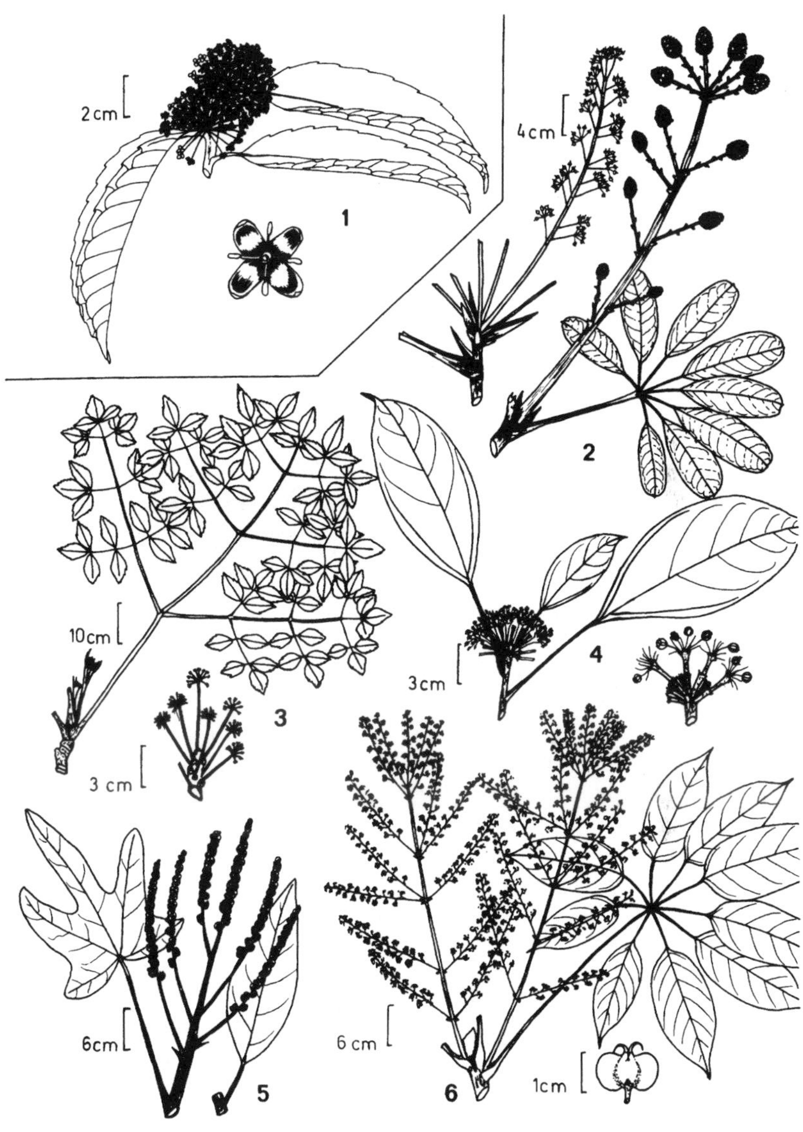

1 - *Ilex* (Aquifoliaceae) **2** - *Schefflera*

3 - *Sciadodendron* **4** - *Dendropanax*

5 - *Oreopanax* **6** - *Didymopanax*

3. Bipinnately Compound Leaves with Small Serrate-Margined Leaflets

Sciadodendron (tentatively incl. American spp. of *Pentapanax* which are intermediate between it and *Aralia*) (ca. 4 spp.) — Mostly dry-area trees, especially in second growth. Reaching our area only along the seasonally dry north coast of Colombia and in the dry inter-Andean valleys of southern Peru.

Aristolochiaceae

Lianas (a very few erect herbs) with characteristic medicinal or Ranalean odor, strikingly anomalous, sectioned, stem anatomy and unique flowers and fruits. The strangely shaped flowers are maroon or mottled maroon and cream or yellow, have unpleasant odors, and function as fly traps to gain pollination. The fruits are capsules, dehiscing incompletely to form a hanging basket with vertical slits in the sides through which the thin, winged, wind-dispersed seeds are shaken out a few at a time. The leaves are variable (as in *Passiflora* and probably related to the same kind of coevolution— in *Aristolochia* with the *Aristolochia* swallowtails of genera *Parides* and *Battus*) but generally palmately nerved or 3-veined and usually cordate with an often truncate-topped basal sinus. The basal venation of the *Aristolochia* leaf is usually quite distinctive, typically with three main veins ascending towards the leaf apex and two lateral veins, one on either side, forming the margin of the sinus and then branching to send a main vein into the cordate basal lobe; all the lateral veins tend to look rather irregular and poorly defined. The petiole is often angled and sometimes borne from a raised projection on the stem; the petiole base is almost always extended as a thickened nodal ridge or decurs downward as an unusually raised striation on the far side of the vertically striate branchlet. A number of *Aristolochia* species are cauliflorous and many are notorious for having giant flowers.

Aristolochia (ca. 180 spp., plus almost the same number of paleotropical and n. temperate spp.) — The only genus of the family in our area, *Aristolochia* species may be canopy lianas but are especially well represented along the edges of wet, lowland forests.

E: zaragoza; P: zapatito de difunto (= deadman's shoe)

There are two additional monotypic genera, *Euglypha* and *Holostylus*, in Paraguay and southern Brazil.

Aristolochiaceae and Asclepiadaceae

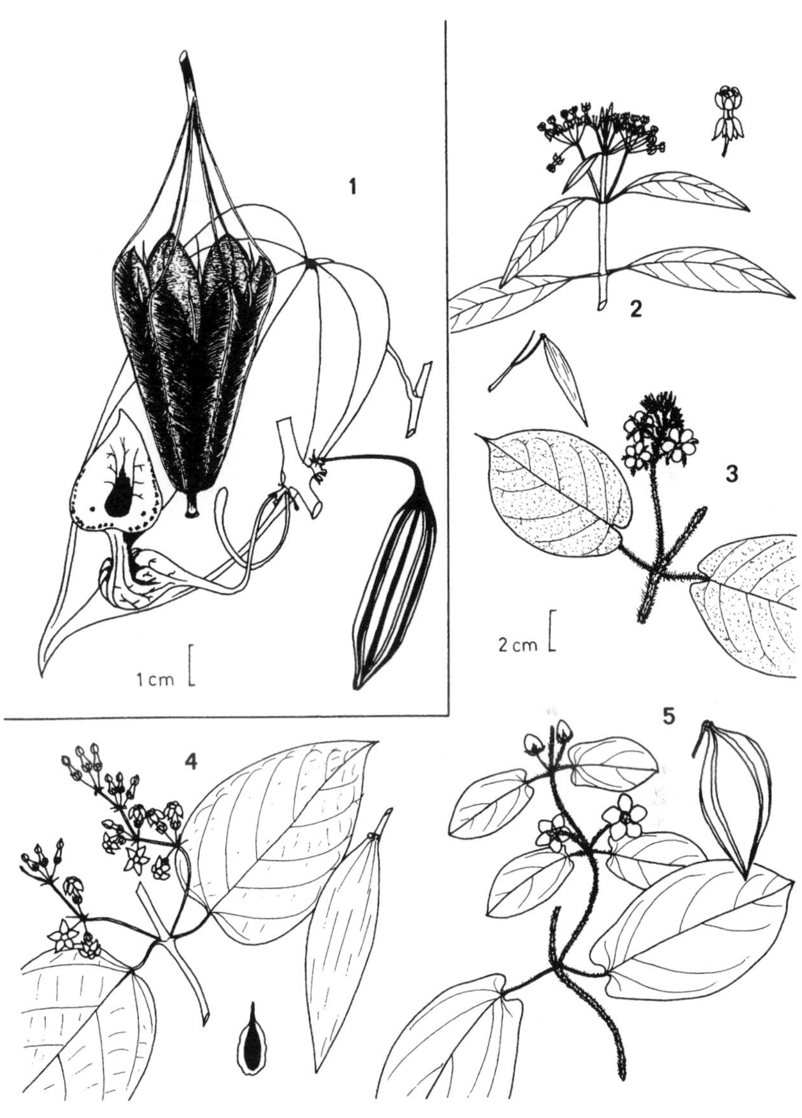

1 - *Aristolochia* (Aristolochiaceae)

2 - *Asclepias*

3 - *Fischeria*

4 - *Marsdenia*

5 - *Matelea*

Asclepiadaceae

Mostly herbaceous or subwoody vines, very closely related to Apocynaceae and similarly characterized by the combination of opposite (always simple and entire) leaves and milky latex. The leaves of most ascleps, but relatively few apocynacs, have conspicuous glands at base of midvein above (and *Prestonia,* the genus most likely to be confounded with ascleps, lacks these). The main vegetative difference is in habit with Apocynaceae climbers nearly always woody lianas while most Asclepiadaceae are thin vines occurring in disturbed areas or tree falls. Ascleps mostly have cordate leaf bases but only *Mandevilla* is cordate among apocynacs. Another difference is the inflorescence (the scar of which is often visible in sterile condition) which is extra-axillary and cymose in neotropical ascleps (though sometimes appearing axillary when large and branching) but axillary in apocs. The relatively few ascleps that become woody lianas tend to have thick corky bark more frequently than do apocs, and it has been suggested (B. Ursem, pers. comm.) that they have a green layer just inside the bark that is lacking in apocynacs. At least *Marsdenia, Matelea,* and the *Metalepis* group of *Cynanchum* become true woody lianas, though mostly in dry forest.

The main differentiating feature between the two families, which are combined by some authors, is the flower. Asclepiadaceae have a complicated flower with the pollen in pollinia, a usually broad open corolla, and a flat peltately expanded pentagonal stigma; a few (mostly extralimital) ascleps have larger more tubular flowers and look like apocynacs (some *Macroscepis* and closely related *Schubertia,* both very like hairy *Prestonia;* also cultivated *Cryptostegia* and *Stephanotis*). The fruit is a follicle with plumose seeds, similar to most apocynac climbers but frequently complexly ridged or tuberculate (unlike any climbing apocynacs) and usually only one follicle develops (except a few *Cynanchum* species), whereas, both carpels usually develop into follicles in apocynacs. Generic taxonomy is in a state of flux in Asclepiadaceae and some authors recognize many segregate genera, especially among taxa related to *Cynanchum.* Worse, generic differentiation of Asclepiadaceae is based to a larger extent on technical characters of the flower than in most other families. Nevertheless, many ascleps can be confidently assigned to genus even when sterile.

1. Erect Herbs or Subshrubs

Asclepias (120 spp., incl. USA) — Distinguished by the typical milkweed flower with rotate corolla lobed to base and with reflexed lobes. In our area represented only by the well-known pantropical weed, orange- and red-flowered *A. curassavica.*

E: viborana, venenillo

Calotropis (2 naturalized spp.) — Dry-area weedy shrub with large subsessile cordate obovate leaves (the bases subclasping). Corolla campanulate and lobed to middle, pinkish to purplish.

Nephradenia (10 spp.) — A *Blepharodon* segregate, small erect herbs of sandy savannas with long linear grasslike leaves and few inconspicuous flowers.

(Ditassa and Cynanchum) — Some species are erect savanna herbs with ericoid leaves.

(Matelea) — One species, *M. rivularis,* is an erect rheophytic upper Amazonian herb with narrow leaves.

2. Vines

2A. There are four main vine genera: *Cynanchum* (smallish nondescript whitish flowers); ***Matelea*** (corolla mostly large, rotate, green to brownish, the buds pointed from the strongly spiralled lobes); ***Marsdenia*** (corolla with a tube, small, brownish to reddish or greenish); and ***Gonolobus*** (flowers like *Matelea* but shorter less spiralled lobes forming flat-topped bud).

Cynanchum (150 spp., incl. Old World) — *Sensu stricto,* mostly restricted to Andean uplands but rather poorly defined and variously lumped and split. Typically with small leaves, often racemose (panicle in *Metalepis;* umbel in *Metastelma*) inflorescence and small whitish flowers. Technical floral character is the corona scales laminar, separate and not elaborated (= basal group of ascleps and not very well characterized). Flowers of some of the taxa here are tiny (<5 mm across); others have larger flowers, the petals narrower than in *Gonolobus* or *Matelea.*

E: angoyuyo

Marsdenia (100–150 spp., depending on taxonomy) — Vegetatively the most apoclike of our genera. Often woody lianas, especially in dry-forest. Leaves usually +/- coriaceous, cordate or not (some look very like *Prestonia*); dry-area taxa usually strongly pubescent. Main technical character is that the flower has a short corolla tube. The flowers are usually small, brownish-red to green or variously spotted or splotched, typically with the throat closed by an annulus or dense pubescence. Fruit usually densely pubescent, not ridged, unique among our taxa in being distinctly woody.

Matelea (130 spp.) — Several species are true woody lianas (but only in dry-forest?). Leaves and stem usually pubescent with unique combination of long brownish hairs and short bulbous ones; bark always conspicuously thick-corky when lianescent. Corolla large and green to brownish, the lobes shorter and more symmetrical than in *Gonolobus* and barely spiralled in the flat-topped bud. Fruit of many species conspicuously tuberculate. *Gonolobus* and *Marsdenia* can always be separated vegetatively by the leaf characters but *Matelea* overlaps both (especially *Marsdenia*); most pubescent *Marsdenia* and *Matelea* leaves are more broadly ovate than *Gonolobus.*

E: condurango

Macroscepis (8 spp.) — Usually pilose vines vegetatively very distinctive in a strong rank odor somewhat like that of tomato leaves; latex usually rather watery. Usually placed in *Matelea* but differs in a +/- tubular corolla (in some species the tube elongate and resembling Apocynaceae); differs from *Marsdenia,* also with corolla tube, in the much larger flowers. Winged fruit more like *Gonolobus* but with 7 rather than 5 wings.

Gonolobus (200 spp.) — Leaves always membranaceous and cordate (or at least with basal angles), usually somewhat oblong in outline; leaves and stem glabrous or variously pubescent but without combination of long and bulbous trichomes that characterizes most *Matelea*). Flower green and completely open (rotate), the lobes longer and more asymmetrical than in *Matelea* and more strongly spiralled, forming a distinctly beaked-acuminate bud; the technical characters to separate it from *Matelea* are anthers with fleshy dorsal appendages, simple cupulate corona (often complex in *Matelea*), and strongly pentagonal (rather than round) style apex. Fruit usually conspicuously longitudinally 5-winged.

2B. The rest of the genera are smaller (at least in our area), all with white to cream or greenish-white flowers.

Oxypetalum (150 spp.) — Mostly Brazilian and south temperate (where often erect rather than scandent); only a single common species in our area (and two rare ones). Ours characterized by the remarkable elongate linear corolla lobes 15–20 mm long. Vegetatively recognizable by the membranaceous, rather narrowly ovate cordate, long-acuminate leaves.

Blepharodon (15 spp.) — A *Cynanchum* relative differing in corona of 5 separate saclike segments. Leaves glabrous, usually rather small, coriaceous, the base rounded, apex apiculate. Inflorescence usually a raceme, corolla lobes sometimes conspicuously ciliate.

Sarcostemma (30 spp., incl. Old World) — A *Cynanchum* relative with largish usually white flowers in an umbellate inflorescence, mostly occurring in open swampy areas (and the leaves glaucous and often narrow), nonswamp species have flowers with very broad, shallowly lobed petals (cf., Convolvulaceae); one species becomes prostrate puna herb. The technical character is a corona of 5 basally connate sacs. Leaves usually rather small, variable in having the base acute to cordate.

Asclepiadaceae

1 - *Blepharodon* **2** - *Sarcostemma*

3 - *Oxypetalum*

4 - *Cynanchum* **5** - *Metalepis* **6** - *Gonolobus*

Fischeria (6 spp.) — Distinctive in the conspicuously erose-margined corolla lobes. Plant usually conspicuously pilose with long brownish multicellular trichomes; leaf rather large and membranaceous, the base cordate. Closest to *Matelea* from which it differs technically in vesciculate corona lobes.

Metastelma (60 spp.)— Essentially a small-leaved *Cynanchum* with small umbellate inflorescence and the flowers +/- tubular. Leaves always mucronate in our area. Technical characters include free corona lobes and valvate corolla lobes. Follicles unusual in both usually developing, each with 5 seeds.

Metalepis (4 spp.) — Large-leaved *Cynanchum* relative with a paniculate inflorescence. Usually treated as subgenus of *Cynanchum* but differs in the large ovate usually cordate leaves. Definitely a canopy liana, at least in seasonal forest. Flowers small and white, the lobes pubescent inside.

Tassadia (20 spp.) — A *Cynanchum* segregate characterized by the small glabrous leaves and large, complexly branched inflorescence with tiny yellowish-green flowers. In our area mostly in seasonally inundated riverside forest.

Ditassa (ca. 75 spp.) — Poorly represented in our area, mostly in dry savanna-type vegetations. Similar to *Cynanchum* in small flowers and small leaves; differs in the leaves typically linear or sublinear (frequently +/- ericoid) and often erect stature. Inflorescence usually with very few small inconspicuous flowers. Very close to *Blepharodon;* main technical character is a corona with 2 opposite ranks.

Vailia (1 sp.) — Fairly common at higher elevations in Colombia and Ecuador. According to D. Stevens this a *Blepharodon* look-alike but actually related to *Marsdenia.*

BALANOPHORACEAE

Root parasites that look far more like fungi than like higher plants. The only identification problem is to figure out that they are angiosperms. Completely lacking chlorophyll and the whole plant whitish or pinkish or red or brown. The vegetative thallus grows underground and only the inflorescences emerge above ground. As treated below, they form a putative evolutionary series from heavily bracteate not very mushroomlike *Scybalium* and *Langsdorffia* to completely mushroomlike *Corynaea* and *Helosis.*

Balanophoraceae and Basellaceae

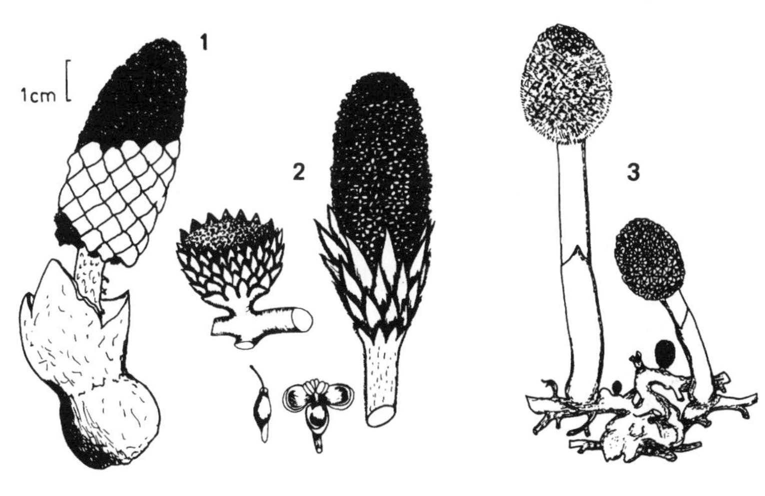

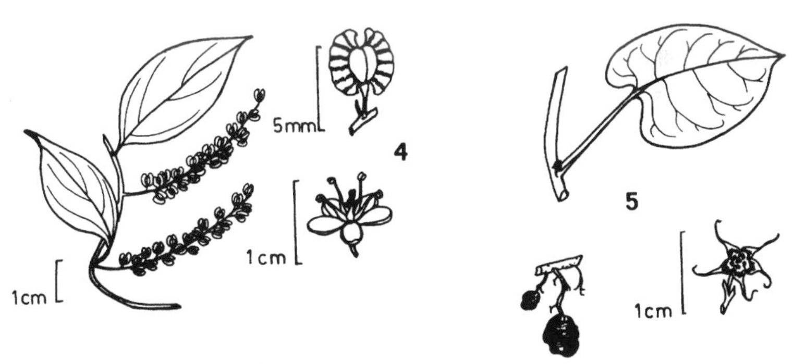

1 - *Ombrophytum* **2** - *Langsdorffia* **3** - *Helosis*

4 - *Anredera* (Basellaceae) **5** - *Ullucus* (Basellaceae)

Scybalium (4 spp.) — Montane cloud forests. Strongly scaled or bracteate inflorescence stalks with a well-defined discoid or ellipsoid fertile apex like *Langsdorffia* but the apical triangular bracts not enlarged to form apparent involucre.

Langsdorffia (1 sp.) — Montane cloud forests. Strongly scaled or bracteate inflorescence stalk with swollen apex and half-round or broadly conical fertile apical portion that appears involucre-like from the elongate apical triangular bracts that surround it.

Lophophytum (3 spp.) — Wet lowland forests below 1000 meters. The basal sterile portion of the inflorescence stalk bracteate as in *Scybalium* but much shorter; fertile portion of inflorescence elongate and tapering gradually toward apex, with flowers borne on the numerous short lateral branches.

Ombrophytum (4 spp.) — In our area only in wet lowland forests but in Bolivia and Argentina one species reaches the puna. Characterized especially by the ruptured basal cupule surrounding inflorescence base. Rather intermediate between *Helosis* and *Corynaea*, on one hand, and *Lophophytum* on the other. Fertile portion of the inflorescence as in *Lophophytum* (but with peltate apex to female branches) but inflorescence stalk naked as in *Helosis*.

P: aya ullo (= deadman's penis)

Corynaea (1 sp.) — Montane cloud forests. Similar to *Helosis* but several inflorescences borne directly from a large woody potato-like tuber. (Figure 1).

Helosis (1 sp.) — Common lowland species with naked mushroom-like stem and ellipsoid inflorescence covered by hexagonal scales; inflorescences borne singly from the underground thallus.

P: aguajillo

Basellaceae

More or less succulent herbaceous vines or prostrate herbs with succulent alternate leaves with usually rounded or obtuse apices (sometimes with inconspicuous apicule), mostly broadly ovate; the venation usually more or less pinnate but mostly more or less plinerved at cordate base; the petiole more or less winged at least at apex, the base decurrent onto angled stem. Inflorescence usually axillary, a spike or narrow raceme (or the leaves suppressed and several racemes compounded into terminal panicle), or rarely a dichotomous cyme (*Tournonia*); flowers small, 5-parted, apetalous, white or translucently greenish or orangish.

Basella (1 sp., plus 5 in Old World) — Our species a cultivated and naturalized vine with small sessile flowers cupular from partly fused tepals and scattered rather far apart along thickish inflorescence. Fruits sessile black, single-seeded, pea-sized.

Anredera (10 spp.) — Herbaceous to woody vines with leaves obovate to (usually) broadly ovate, the apex often acute to acuminate (unique in family). Inflorescence slender and tenuous, racemose or +/- spicate, the flowers (with spreading tepals fused only at base) and fruits close together along it. Fruits smaller than *Basella* and sometimes (*Anredera sensu stricto*) minutely winged, sometimes fleshy (*Boussingaultia*). Rather similar to *Trichostigma* (Phytolaccaceae) but the leaves broader and more succulent.

Ullucus (1 sp.) — A prostrate high-Andean herb, often cultivated for the edible tuber. Flowers translucent orangish, very distinctive in the caudate tails on petals.

P: olluco

Tournonia (1 sp.) — High-altitude cloud-forest vine of Colombian Andes. Unique in the family in having the broadly ovate leaves definitely palmately veined and with slightly and irregularly serrate margin. The axillary inflorescence a dichotomously branched cyme with minute greenish flowers.

BATACEAE

The entire family consists of a single genus with two species, only one of which occurs in the New World. *Batis maritima* is a sprawling coastal shrub or subshrub with opposite succulent linear leaves and succulent swollen catkinlike inflorescences, the reduced flowers hidden by the bracts, the whole catkin forming a succulent green ellipsoid compound fruit. Typically forming large colonies in salt marshes or around the edges of mangroves.

Batis (1 sp., plus 1 in New Guinea)

BEGONIACEAE

A mostly herbaceous family but with a few soft-wooded treelets (occasionally to 6–8 m tall) and lianas; a number of species epiphytic or lithophytic. Stems usually succulent, conspicuously enlarged and jointed at nodes (with conspicuous scars from the fallen stipules). Leaves mostly asymmetrically ovate, usually 3-veined at base (on only one side

Bataceae, Begoniaceae, Berberidaceae, and Betulaceae

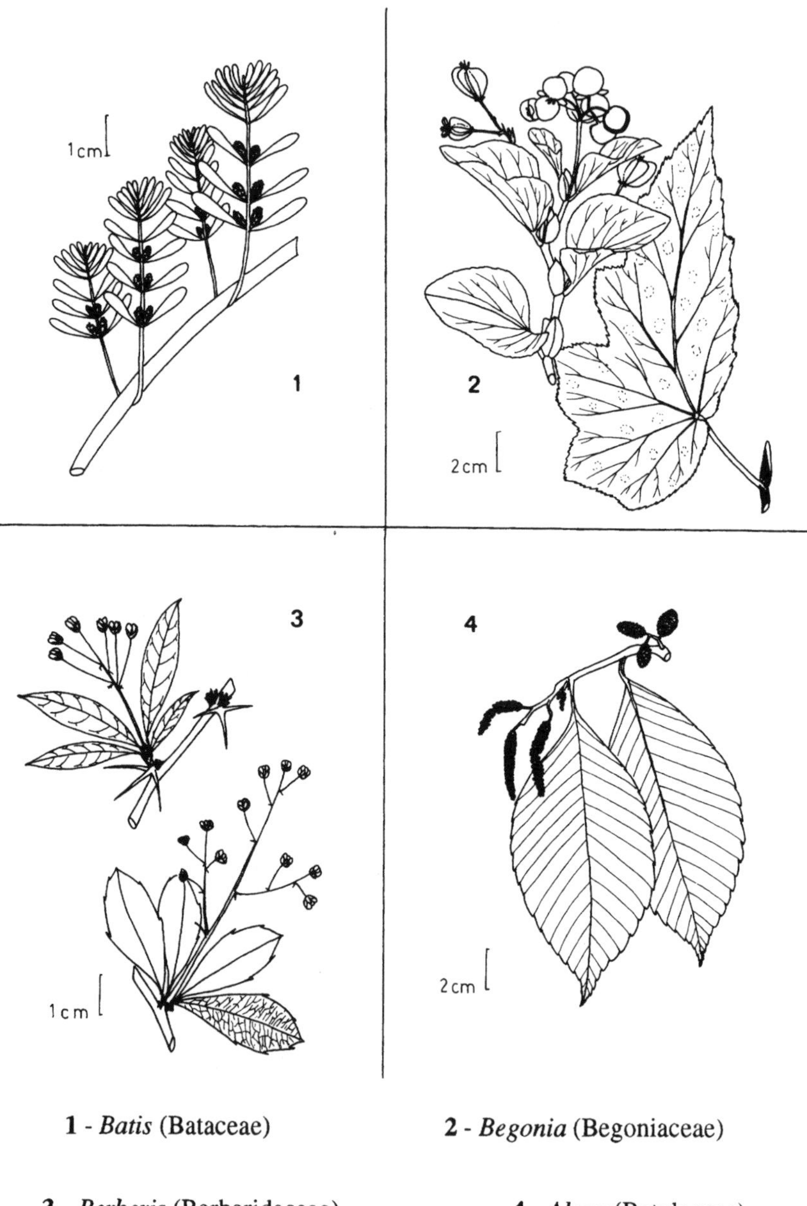

1 - *Batis* (Bataceae)

2 - *Begonia* (Begoniaceae)

3 - *Berberis* (Berberidaceae)

4 - *Alnus* (Betulaceae)

in most strongly asymmetric species), mostly serrate (typically with broad shallow teeth or lobes in addition to fine serration), the relatively few entire-margined species always conspicuously succulent and usually very strongly asymmetric or deeply cordate or peltate. When climbing, with adventitious rootlets at nodes. Male and female flowers separate but in same inflorescence. Male flowers with 4 valvate petal-like segments (two larger = calyx and two smaller = petals) and usually numerous stamens (only 4 = *Begoniella*), the female with 2–5 imbricate petaloid segments surmounting a usually 3-winged inferior ovary. Fruit a dry usually 3-winged capsule or only one of the wings developing.

Begonia (incl. *Begoniella, Semibegoniella*)(580 spp., plus 620 in Old World) — One common cloud-forest species (*B. parviflora*) is a treelet to 4–8 m tall and 3 cm dbh, with large very asymmetric leaves shallowly jaggedly serrate and finely ciliate-serrate, asperous above; inflorescence dichotomously branched, flat-topped, with small white flowers. *Begonia microphylla* and allies are subwoody or climbers with small (<2 cm long), obovate leaves usually with 2–3 subapical lobes and conspicuous stipules; similar to some *Pilea* but leaves strictly alternate.

BERBERIDACEAE

High-altitude shrubs and small trees, mostly at and above timberline. Characterized by spine-tipped or spinose-margined leaves clustered on bracteate short shoots. Most species have trifurcate spines (unique to family). Flowers always yellow. Fruits glaucous-blue, round to spindle-shaped, capped by the remains of the truncate patelliform stigma above a short narrow neck.

Berberis (450 spp., mostly in Old World)

BETULACEAE

Common trees of montane second-growth forests, vegetatively unique in our area by the doubly toothed leaf margin (with teeth over secondary vein endings slightly larger). Also characterized by the close-together straight ascending secondary veins with rigidly parallel tertiary veins perpendicular to the secondaries. Both male and female flowers in catkins, the former much longer than latter.

Alnus (3 spp., plus 30 in USA and Old World) — Perhaps only the single species *A. acuminata* is in South America.

C, E, P: aliso

Bignoniaceae

A very distinctive family in its opposite compound leaves; indeed, the great majority of lowland forest taxa with opposite compound leaves are Bignoniaceae. This is the most important neotropical liana family and the lianas are especially distinctive in having the terminal leaflets of their compound leaves converted into a tendril (unique among opposite-leaved climbers); another unique feature is anomalous secondary growth forming a 4–32-armed crosslike structure in stem cross section. Tree taxa mostly have palmately compound leaves; the only other local genus with palmately compound opposite leaves is *Vitex* which differs from area Bignoniaceae (except *Godmania*) in having the cuneate leaflet bases less sharply differentiated from the generally indistinct petiolules. (A number of taxa of other families have opposite 3-foliolate leaves, but these are rare in erect area Bignoniaceae). Our few genera of pinnately leaved Bignoniaceae (*Tecoma, Jacaranda,* one species of *Digomphia*) are individually distinctive. The few taxa with simple leaves are difficult to recognize to family in sterile condition. All have opposite leaves (usually +/- 3-veined at base) except two indehiscent-fruited bat-pollinated genera — *Amphitecna* with alternate leaves and *Crescentia* with alternate fascicles of leaves.

Bignoniaceae have very characteristic flowers and fruits. The tubular flowers are large and showy, most having 2 pairs of stamens inserted near the base of corolla tube with the divergent anther thecae of each pair held together inside the roof of the corolla tube. Most Bignoniaceae are bee-pollinated, but a few genera have corollas adapted for hummingbird-pollination, hawkmoth-pollination, butterfly-/small-bee-pollination, and even bat-pollination. Bignoniaceae fruits are mostly dehiscent by two valves to release winged wind-dispersed seeds. One anomalous but mostly extralimital group (tribe Crescentieae) has large indehiscent fruits with wingless seeds embedded in their pulp; another (*Schlegelia* and *Gibsoniothamnus*) is intermediate with Scrophulariaceae and composed of epiphytic and hemi-epiphytic plants with coriaceous simple leaves and mostly small berrylike fruits subtended by the cupular calyx.

The notorious reputation of the liana genera as difficult to recognize is undeserved since most of them can be readily identified even in sterile condition. Four liana genera (*Memora, Pleonotoma, Tourrettia, Eccremocarpus* plus two atypical *Arrabidaea* species) are unique in mostly bipinnate or biternate leaves and another group of four liana genera are unique in hexagonal twigs (*Pithecoctenium, Amphilophium, Distictis, Haplolophium,* see also irregularly hexagonal *Pyrostegia*). The numerous liana genera with 2–3-foliolate leaves and nonhexagonal twigs can be vegetatively distinguished by presence or absence of foliaceous pseudostipules, petiolar and interpetiolar glandular fields, distinctive leaf odors, and whether the tendrils are simple or trifid or the branchlets terete or tetragonal, although a few genera are polymorphic for each of these characters. Here these taxa are arrranged in four groups. One has simple tendrils, more or less tetragonal twigs and lacks interpetioloar gland fields (*Mussatia, Cydista, Clytostoma*). Another has round twigs, yellow flowers, mostly simple (or minutely trifid) tendrils and mostly lacks interpetiolar glandular fields (*Adenocalymna, Anemopaegma, Callichlamys, Periarrabidaea, Spathicalyx*). A third (fourteen genera) has nonyellow (white to magenta or orange) flowers and round twigs, usually with petiolar or interpetiolar glands. The final "group" is a miscellany characterized by hollow stem and punctate leaves (*Stizophyllum*), trifid "cat's-claw" tendrils (*Macfadyena, Melloa*), hawkmoth-pollinated flowers and either a rank or almond vegetative odor (*Tanaecium*) or minute strongly bilabiate white flowers and usually a clove odor (*Tynanthus*).

1. Leaves Simple, Sometimes Alternate

1A. Fruits hard-shelled and indehiscent with seeds embedded in pulp (tree calabashes); leaves alternate or in short-shoot fascicles; trees

Crescentia (5 spp.) — Leaves in fascicles; seeds small; flowers tannish with reddish markings and triangular-pointed corolla lobes; tahuampa (*C. amazonica*) or cultivated and escaped (*C. cujete*); (A Central American species has 3-foliolate leaves.)

C: totumo; E: mate; P: totumo, huinga, pate

Amphitecna (18 spp.) — Leaves large, coriaceous, alternate; seeds larger and thicker than *Crescentia;* flowers greenish-white with completely fused nonpointed lobes. Mostly Central American, in our area restricted to beaches and mangrove swamps except in extreme northwestern Colombia.

C: totumillo, matecillo; E: calabacillo

1B. Fruits dehiscent and with winged seeds (or the wings reduced for water dispersal); leaves opposite or whorled; trees or lianas

Romeroa (1 sp.) — Small tree with whorled leaves and yellow flowers; endemic to Magdalena Valley.

Delostoma (4 spp.) — Opposite leaves, magenta or red flowers; Andean, both in "ceja de la montaña" and dry inter-Andean Valleys.

Bignoniaceae
(Trees with Indehiscent Fruits and/or Simple Leaves)

1 - *Crescentia* **2** - *Parmentiera*

3 - *Amphitecna* **4** - *Exarata* **5** - *Digomphia*

6 - *Delostoma* **7** - *Romeroa*

(*Arrabidaea*) — One Peruvian *Arrabidaea, A. platyphylla* has consistently simple leaves. In Peru it is an opposite-leaved liana with small narrow-tubed white flowers.

(*Tecoma*) — A coastal Ecuadorean shrub, *T. castanifolia* has simple leaves and yellow flowers like *Romeroa,* but the leaves differ in being serrate and opposite.

(*Tabebuia*) — One Peruvian and one Colombian *Tabebuia, T. insignis* (*s.l.*) and *T. striata,* respectively, have simple opposite coriaceous leaves; both are trees with white flowers.

(*Digomphia*) — One shrubby area *Digomphia* of Guayanan white-sand savannas has simple leaves; it has bluish flowers.

1C. Fruits indehiscent, berrylike, with tiny angulate seeds, subtended by cupular calyx; mostly epiphytic shrubs or hemiepiphytic climbers; opposite leaves

Schlegelia (20 spp.) — Hemiepiphytic lianas climbing appressed against trunks by adventitious roots; opposite (rarely whorled) very coriaceous leaves; often cauliflorous; calyx truncate, not winged. Noticeable pseudostipular, appressed-conical, axillary bud scales are a good vegetative character.

Gibsoniothamnus (9 spp.) — Mostly Central American, barely reaching northern Colombia. Epiphytic shrubs with narrowly tubular magenta (or white with magenta calyx) hummingbird-pollinated flowers; fruit a berry; calyx lobes usually pronounced, often very long or laterally winged. An infamous case of a genus which falls between the cracks of the taxonomic system: various species were described as or referred to Verbenaceae, Bignoniaceae, Gentianaceae, Ericaceae, and Gesneriaceae, while the genus was described (as Scrophulariaceae) only in 1970.

Exarata (1 sp.) — Chocó endemic. Large tree with deeply furrowed trunk. Large, bee-pollinated flowers like *Tabebuia,* but fruits like *Schlegelia.*

2. Leaves compound

2A. Trees (or shrubs) — (*Argylia* of the southern Peruvian coastal lomas is an herb.) Fruits dehiscent (except fleshy-fruited *Parmentiera,* mostly Central American) and leaves compound; fruits dehisce perpendicular to septum (Tecomeae).

2Aa. Leaves bipinnate (to simply pinnate [even simple in one species in Brazil]); flowers blue or bluish

Jacaranda (49 spp.) — Distinctive in blue or blue-purple flowers, elongate staminode with glandular-pubescent tip, and flat oblong-elliptic fruit. *J. copaia,* which has a characteristic unbranched juvenile form topped by a tuft of giant bipinnate leaves is one of our commonest wet-forest, second-growth species.

C: gualanday; E: arabisco; P: huamansamana, ishtapi (*J. glabra*)

(***Memora cladotricha***) — One species of mostly scandent *Memora,* yellow-flowered *M. cladotricha,* is an erect treelet. Its leaflets are larger than those of area *Jacaranda.*

2Ab. Leaves pinnate (to simple)

Digomphia (3 spp.) — A reduced usually shrubby *Jacaranda* with large foliaceous calyx and simple or simply pinnate leaves; mostly Guayana area, west into Amazonian Colombia.

Tecoma (12 spp., plus 2 African) — Serrate leaflets; flowers usually yellow, sometimes orange-red and hummingbird-pollinated; mostly in dry inter-Andean valleys; also along Caribbean coast. Often cultivated.

C: chirlobirlo, flor amarilla, quillotocto; E: cholan, fresno; P: campanilla amarilla

2Ac. Leaves palmately compound (or 3-foliolate)

Argylia (12 spp.) — The only herbaceous neotropical bignon, a single species reaching the southernmost Peruvian lomas.

Tabebuia (100 spp.) — The main genus of palmately compound Bignoniaceae; fruits terete, elongate, with winged seeds; flowers tubular-campanulate, conspicuous (often "big bang" flowerers), usually yellow or magenta.

C, E: guayacán; C: cañaguate (*T. chrysantha, T. ochracea*), polvillo (*T. billbergii*), roble (*T. rosea*) P: tahuarí

Godmania (2 spp.) — Small deciduous-forest trees. Flowers smaller than *Tabebuia,* orangish-yellow below, brown above; fruit differs from *Tabebuia* in being loosely coiled. Leaflets narrowed at base into a shorter more poorly demarcated petiole than in *Tabebuia.*

C: cacho de chivo; E: aceituno

Cybistax (1 sp.) — Tree of dry areas along Amazonian base of Andes. Differs from *Tabebuia* in flowers green with large deeply 5-lobed calyx and in fruits thick, oblong, longitudinally costate.

P: yangua

Sparattosperma (2 spp.) — Tree of seasonally dry areas of southern Peruvian Amazonia. Flowers pale pink or white with pink in throat, broader than in *Tabebuia.* Seed wings of separate trichomes rather than membranaceous as in *Tabebuia.* Vegetatively distinctive in the resinous young growth.

Bignoniaceae
(Trees [and an Herb] with Palmately Compound Leaves and Dehiscent Fruits)

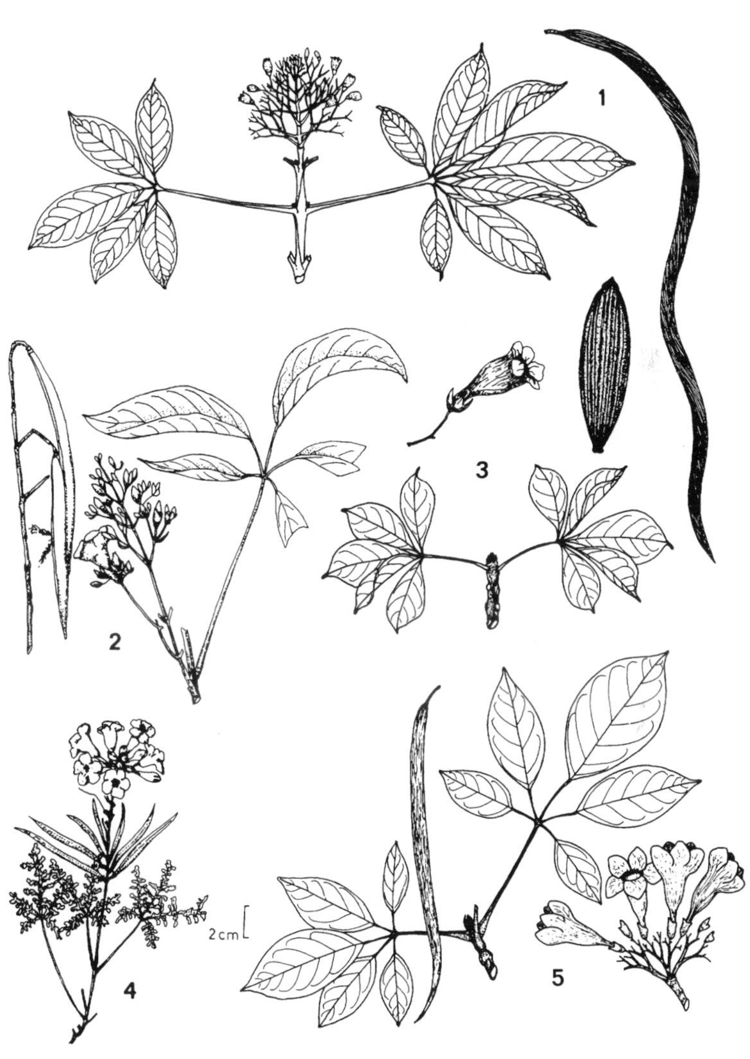

1 - *Godmania*

2 - *Sparattosperma*

3 - *Cybistax*

4 - *Argylia*

5 - *Tabebuia*

Bignoniaceae

(Trees with Pinnate Leaves, Herbaceous Vines with Bicompound Leaves and a Simple-Leaved Hemiepiphyte)

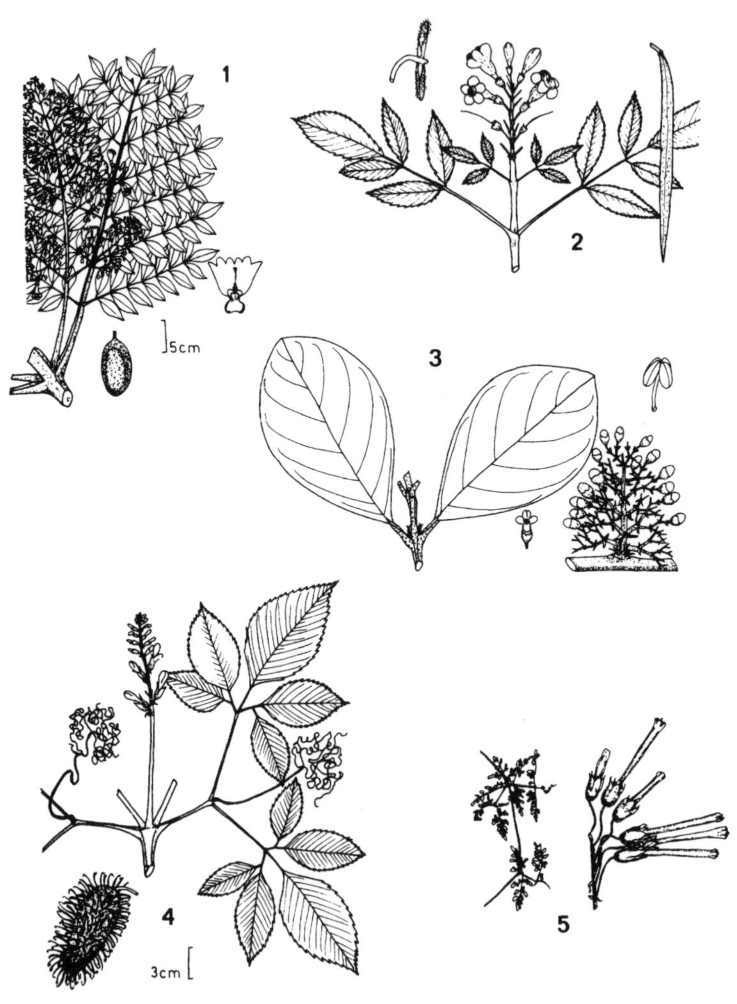

1 - *Jacaranda*

2 - *Tecoma*

3 - *Schlegelia*

4 - *Tourrettia*

5 - *Eccremocarpus*

Parmentiera (10 spp.) — Mostly Central American, in our area only reaching extreme northwestern Colombia. Unlike other compound-leaved South American Bignoniaceae in the indehiscent fleshy fruit (= tribe Crescentieae) and white bat-pollinated flowers with a spathaceous calyx.

C: árbol de vela, palovela

2B. Lianas — Fruits dehiscent; mostly dehiscing parallel to septum (= Bignonieae)

2Ba. Leaves mostly bipinnate or biternate

Memora (33 spp.) — Stem round unlike *Pleonotoma.* Flowers bright yellow, usually with conspicuous bracts or bracteoles (very like *Adenocalymna,* except for bipinnate leaves). Sometimes an erect treelet with peculiar whitish bark and extremely hard wood (*M. cladotricha*).

Pleonotoma (15 spp.) — Unique among bipinnate-leaved taxa in the stem acutely tetragonal. Flowers pale yellow; inflorescence without noticeable bracts or bracteoles.

C: bejuco de puno; P: estrella huasca

Tourrettia (1 sp.) — Ceja de la montaña and coastal lomas; annual vine with burlike exozoochorous fruit. Fertile flowers greenish, sterile flowers at top of inflorescence red.

Eccremocarpus (6 spp.) — Wiry high-Andean vine; flowers long-tubular and hummingbird-pollinated.

(*Arrabidaea*) — One species (*A. inaequalis*) usually has biternate leaves but unlike other biternate taxa has magenta flowers; two other species can have the lower leaves biternate, both with white flowers unlike the above genera.

2Bb. Leaves 2–3-foliolate (-palmately 5-foliolate)

(i) Twigs acutely hexagonal with raised angles — (All with trifid or many-branched tendrils)

Pithecoctenium (4 spp.) — Tendrils many-branched; vegetative trichomes simple; flowers white with a curved tube; fruits large densely echinate.

P: peine de mono

Amphilophium (8 spp.) — Differs from *Pithecoctenium* in the tendrils simply trifid and vegetative trichomes dendroid. Flowers unique in our area in being pseudocleistogamous, the lobes remaining closed at anthesis. Calyx unusual in having a distinctive frilly outer margin.

C: bejuco de oroto

Bignoniaceae
(Lianas: A – Ce)

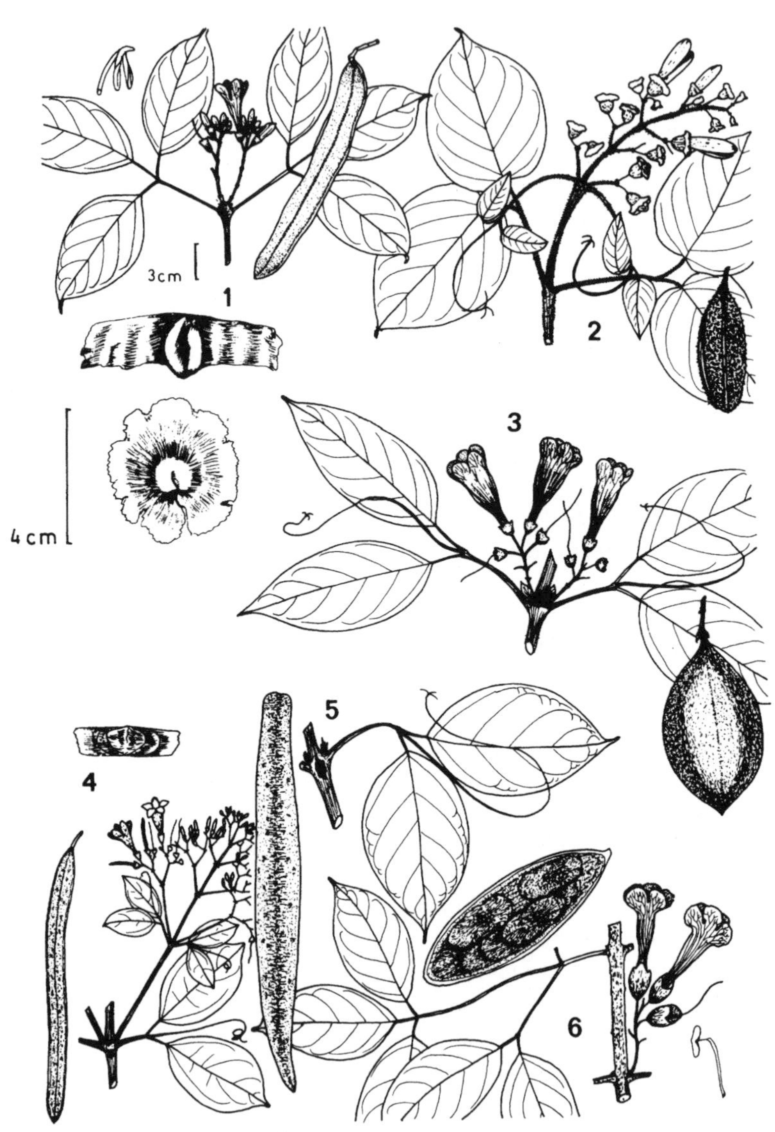

1 - *Adenocalymna* **2** - *Amphilophium*

3 - *Anemopaegma*

5 - *Ceratophytum*

4 - *Arrabidaea* **6** - *Callichlamys*

Bignoniaceae
(Lianas: Cl – H)

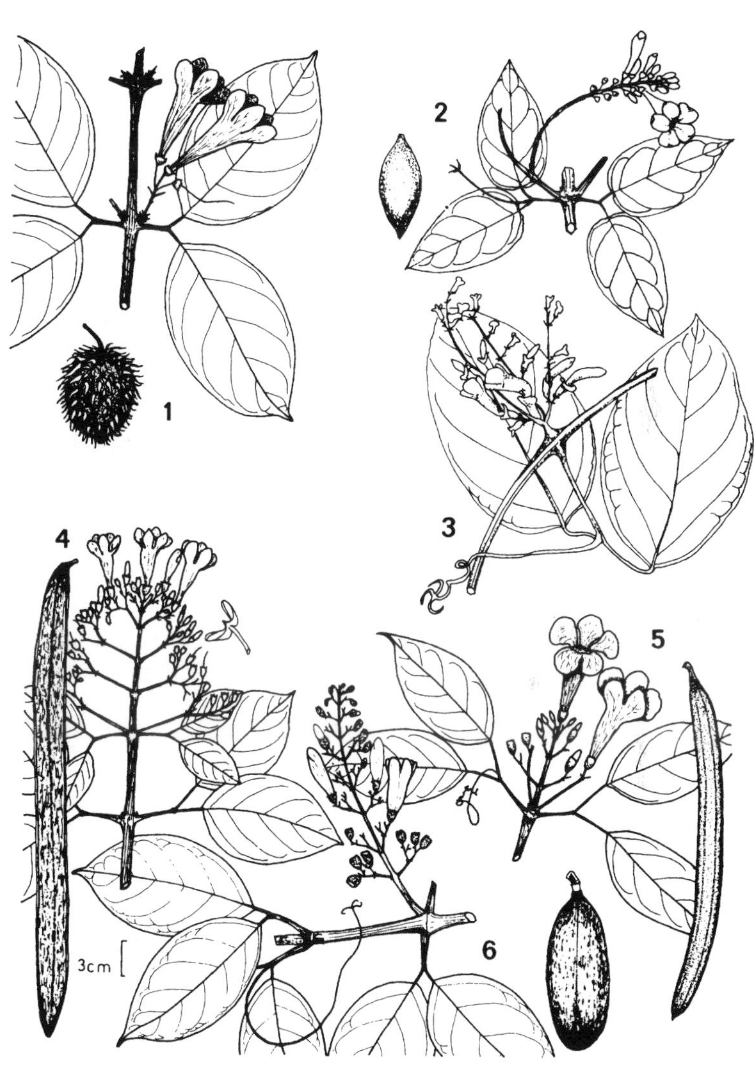

1 - *Clytostoma*

2 - *Distictis*

3 - *Haplolophium*

4 - *Cuspidaria*

5 - *Cydista*

6 - *Distictella*

Distictis (12 spp.) — Stem much less conspicuously hexagonal than *Amphilophium* and *Pithecoctenium;* corolla usually white or magenta, with typical bignon shape.

Haplolophium (4 spp.) — Calyx with frilly outer margin like *Amphilophium* but the corolla lobes reflexed.

(ii) Twigs more or less acutely tetragonal, usually with raised ribbed angles — (All have simple tendrils and lack gland fields between petioles)

Mussatia (2 spp.) — Characterized by prominent leafy pseudostipules; large flattened but woody fruit; flowers mottled yellow and maroon.

P: chamayro

Cydista (6 spp.) — The commonest species is only obscurely tetragonal; tetragonal-stemmed species have leafy pseudostipules (in South America). The main technical character of both this and the following genus is absence of a nectary around base of ovary (associated with mimetic pollination strategy and "multiple bang" phenology); fruits of the two genera are very different, thick and echinate in *Clytostoma,* smooth and flat in *Cydista;* in flower they can only be distinguished with certainty by the smooth ovary of *Cydista* vs. the glandular-pustulate one of *Clytostoma.*

C: campana, bejuco esquinero (*C. diversifolia*)

Clytostoma (10 spp.) — (Close to *Cydista,* see above.) The best vegetative character is the narrow and clustered pseudostipules resembling miniature bromeliads in the leaf axils. The commonest species has only obscurely tetragonal twigs.

(iii) Other genera with obvious unique distinctive features (hollow stems or cat's-claw tendrils or hawkmoth- or butterfly-/small-bee-pollinated flowers) — (All have 2–3-foliolate leaves and terete branchlets; lack interpetiolar gland fields [except a few *Tanaecium* species and one *Macfadyena*]; all have strongly trifid tendrils except *Tanaecium,* one *Tynanthus,* and one *Stizophyllum.*)

Stizophyllum (3 spp.) — Unique in hollow twigs; pellucid-punctate leaves; very narrow long pencil-like fruit. (Tendrils simple to trifid.)

Tanaecium (6 spp.) — Long white hawkmoth-pollinated flowers. Vegetatively with strong almond (cyanide) odor (*T. nocturnum*) or rank-smelling. Also characterized by simple tendrils and terete woody capsules (usually with thick wingless seeds).

C: calabacillo prieto, mata ganado (*T. exitiosum*)

Bignoniaceae
(Lianas: L – Man)

1 - *Mansoa* **2** - *Mansoa*

3 - *Lundia*

4 - *Macranthisiphon*

5 - *Macfadyena* **6** - *Macfadyena* (juvenile)

Bignoniaceae
(Lianas: Mar – Pa)

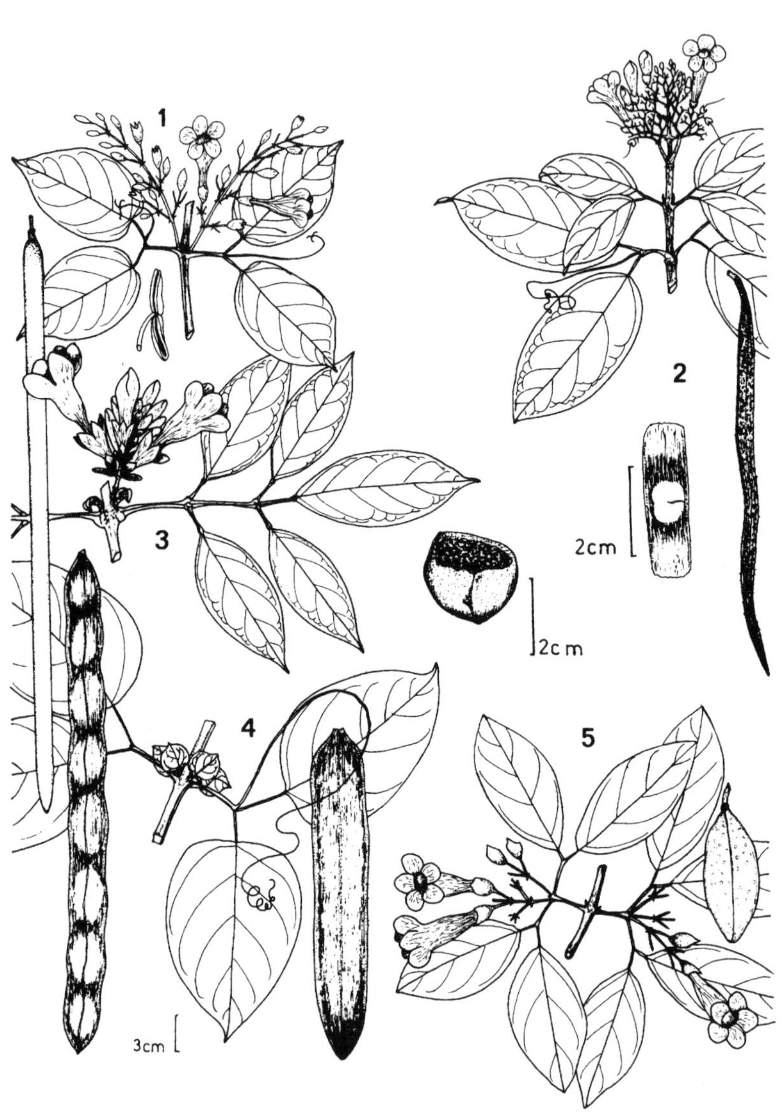

1 - *Martinella* **2** - *Paragonia*

3 - *Memora*

4 - *Mussatia* **5** - *Melloa*

(*Spathicalyx*) — One rare species has *Tanaecium*-like hawkmoth-pollinated flowers, but trifid tendrils and lacks vegetative odor.

Tynanthus (14 spp.) — Very small "butterfly/small bee-pollinated" white flowers (smallest in family). Vegetatively usually with strong clove odor. Fruits elongate, flat with the margins somewhat raised, either narrower than in other genera or with the margins winged.
P: clavo huasca

Macfadyena (4 spp.) — Strongly trifid cat's-claw tendrils and distinctive appressed-climbing small-leaved juvenile form. Also characterized by yellow flowers, rather large green membranaceous calyx, long, linear fruits.
C: bejuco de murciélago, uñita

Melloa (1 sp.) — Vegetatively, and in flower, like *Macfadyena* but the tendrils less like cat's claws and twigs with whitish lenticels. Very different in unique thick woody fruit splitting longitudinally into 4 valves.
C: mata cangrejo

(iv) Other yellow-flowered genera — (All with tendrils simple or minutely trifid and round twigs.) (Fruits mostly elliptic and/or +/- thick or woody (except *Periarrabidaea* and dendroid-pubescent *Spathicalyx,* both with interpetiolar gland fields).

Adenocalymna (39 spp.) — Very close to *Memora* but only 2–3-foliolate leaves. Racemose inflorescence characteristic in the prominent bracts and/or bracteoles. Leaves usually drying dark and often with cartilaginous margin. Seeds with thick bodies. Tendrils simple and interpetiolar gland fields lacking.

Anemopaegma (46 spp.) — Most distinctive in the stipitate elliptic or ellipsoid, usually flattened fruit, the seeds usually very flat with the body completely surrounded by wing (except in few water-dispersed swamp species). Calyx cupular and usually truncate. Tendril usually trifid, simple in a few species; interpetiolar gland fields rare. One species is palmately 4–5- foliolate (*A. orbiculatum*).
C: bejuco cuchareto

Callichlamys (1 sp.) — The very large spongy yellow calyx is unique. Fruit also characteristic, large and woody, similar to *Anemopaegma* but woodier, longer, more elliptical in outline and not stipitate. Tendril simple and interpetiolar glands lacking. Vegetatively characterized by very large leaflets with dendroid trichomes in lateral nerve axils below.
C: botecito

Periarrabidaea (1 sp.) — Distinctive (but subtley so) in the calyx not cupular and gradually narrowed at base and in the distinctly triangular-pointed corolla lobes. Fruit unlike above genera in being thin and linear. Tendrils trifid and interpetiolar gland fields present. Pseudostipules subulate as in *Paragonia.*

Spathicalyx (2 spp.) — The yellow-flowered species unique in the conspicuous yellow leaves just below inflorescence. Also unusual in dendroid trichomes, even on the linear yellowish-tan fruit. Tendrils trifid and inconspicuous interpetiolar gland fields present.

(v) Genera with orange, red, or magenta (sometimes white) flowers and round twig — (Tendrils all simple except as otherwise noted.) (The majority of species of all genera except *Distictella,* monotypic *Macranthisiphon* and *Saritaea,* and the last four genera have gland fields between the petioles and/or at petiole apices.) Fruits mostly linear and +/- flat (except *Ceratophytum, Distictella, Xylophragma,* and a few miscellaneous water-dispersed species.

Arrabidaea (75 spp.) — The commonest genus of bignon vine. Characterized by the combination of magenta flowers (white in few species), simple tendrils, flattened linear fruit, and frequently the presence of gland-fields between the petiole bases.

E: bija (*A. chica*)

Ceratophytum (1 sp.) — Vegetatively characterized by the unique combination of interpetiolar glandular fields and distinctive appressed-conical, subulate, pseudostipules (shared only with *Periarrabidaea*). Also distinctive in flowers white, rather large and thick, trifid tendrils, and the elongate woody, subtetragonal fruit.

Distictella (14 spp.) — Not very distinctive vegetatively, the best character being combination of trifid tendril with nonglandular node lacking pseudostipules. Flowers unique among terete-stemmed taxa in being white with curved base (cf., *Pithecoctenium*). Fruit smooth-surfaced, woody but flattened, oblong-elliptic.

Macranthisiphon (1 sp.) — Vegetatively, and in fruit, like *Arrabidaea,* but the hummingbird-pollinated flowers long, narrow and orange-red. Endemic to the dry-forest of coastal Ecuador and adjacent Peru.

Saritaea (1 sp.) — Like *Arrabidaea* but the flowers are larger, there are conspicuous foliaceous pseudostipules, and the 3-veined leaflets narrow to base. Endemic to northeastern Colombia, but widely cultivated.

C: campanilla

Bignoniaceae
(Lianas: Pe – R)

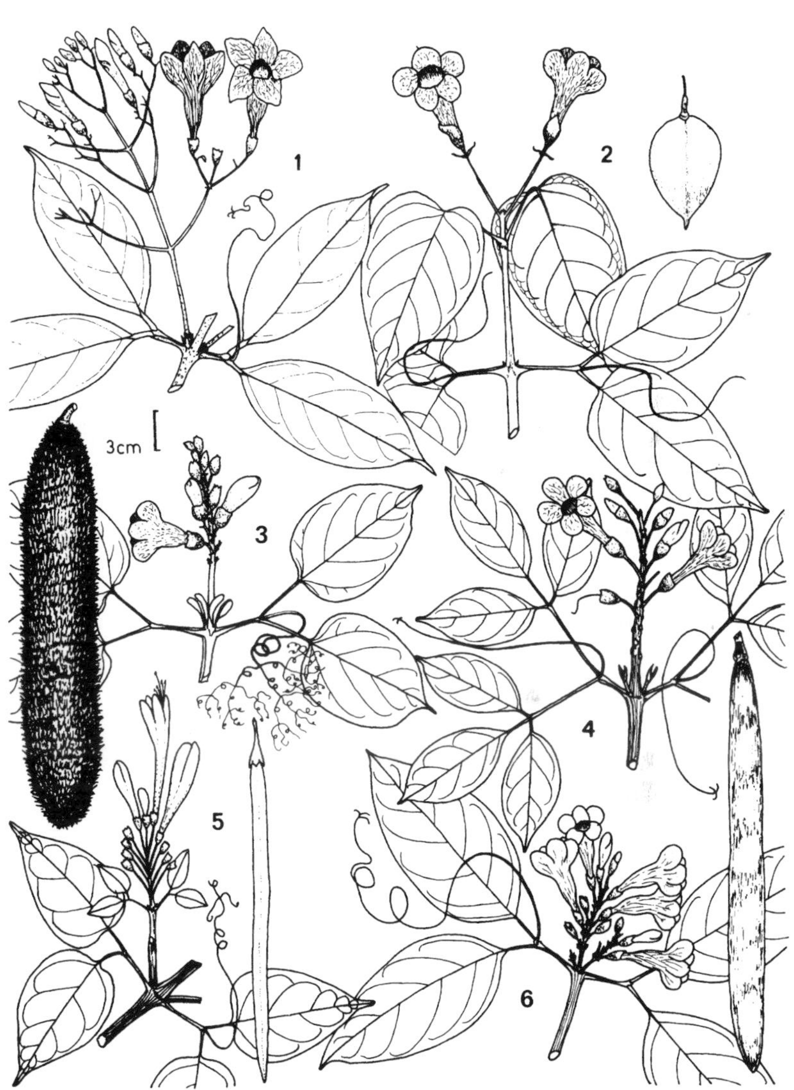

1 - *Periarrabidaea* **2** - *Phyrganocydia*

3 - *Pithecoctenium* **4** - *Pleonotoma*

5 - *Pyrostegia* **6** - *Roentgenia*

Bignoniaceae
(Lianas: S – X)

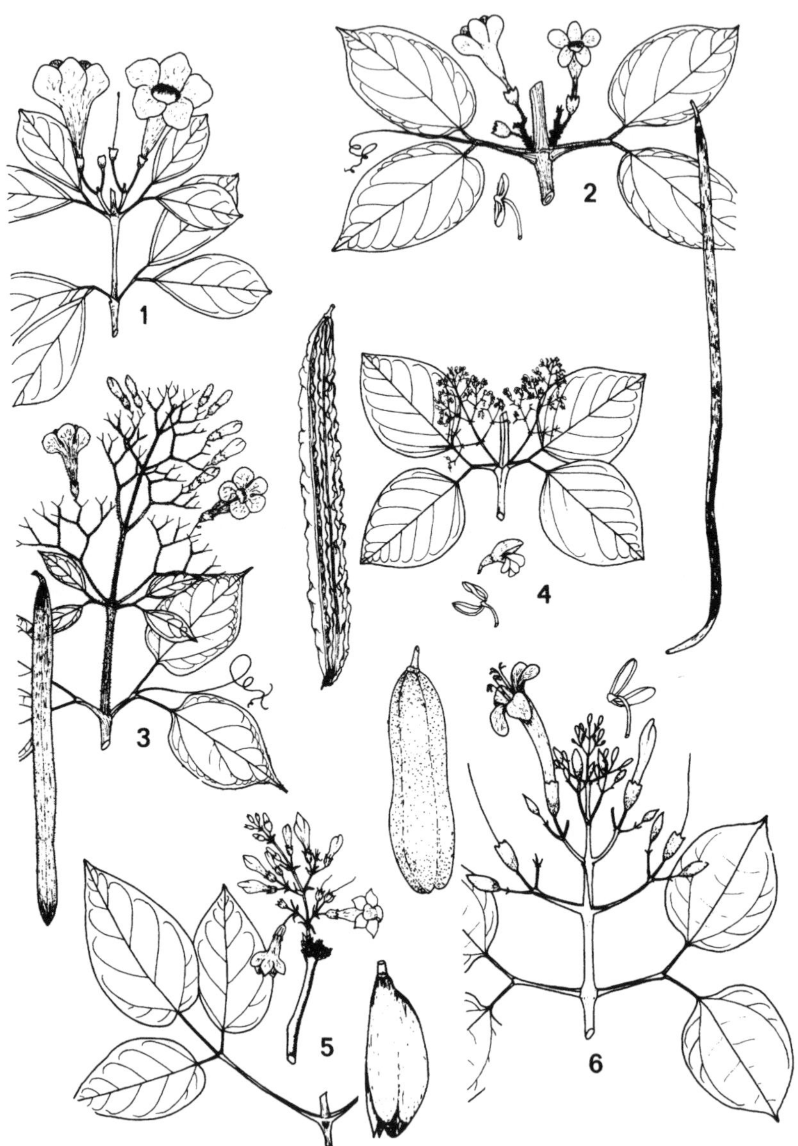

1 - *Saritaea* **2** - *Stizophyllum*

3 - *Spathicalyx* **4** - *Tynanthus*

5 - *Xylophragma* **6** - *Tanaecium*

Paragonia (2 spp.) — Similar to *Arrabidaea* but tendrils usually minutely bifid and with glands on upper side of petiole apex but not at nodes. The appressed-conical subulate pseudostipules are also characteristic (cf., *Ceratophytum* but lacking the interpetiolar gland field). Also distinctive in a distinctly convex sandpaper-surfaced linear fruit and sweetish vegetative odor.

E: huachamoza

Lundia (12 spp.) — Like *Arrabidaea* but villous anther thecae and fruit usually pubescent (rare in *Arrabidaea*). Conspicuous interpetiolar glandular fields as in *Arrabidaea* but tendrils often trifid unlike any *Arrabidaea.*

Cuspidaria (14 spp.) — Like *Arrabidaea* but the anther thecae conspicuously bent and reflexed forward in middle. Fruit usually distinctive in pair of raised submarginal or submedial ridges or wings on each side (similar winged fruits only in some species of *Tynanthus* and one *Adenocalymna,* the latter much woodier).

Xylophragma (4 spp.) — In flower like *Arrabidaea* except for thicker ovary. Vegetative pubescence mostly dendroid (rare in *Arrabidaea*). Fruit very different from *Arrabidaea,* oblong-elliptic, woody but flattened, the surface smooth or with raised lenticels.

Mansoa (15 spp.) — Several species distinctive in unique garliclike vegetative odor. Usually with gland-fields both between petioles and at petiole apices, a unique combination. Tendrils trifid or simple and disk-tipped (only in minute-leaved appressed-climbing *M. parvifolia*).

C: bejuco de ajo; P: ajo sacha

Martinella (2 spp.) — The most distinctive vegetative features are a swollen interpetiolar ridge and the usually gray-green drying twigs and leaves. Tendrils usually trifid and gland fields completely absent. The fruit is characteristic in being very long and thin; calyx irregularly 2–3-labiate.

C: raíz de ojo; P: yuquillo

Phryganocydia (3 spp.) — The characteristic feature is the strongly spathaceous whitish calyx. Interpetiolar gland fields are uniformly lacking and tendrils are simple. A useful vegetative character is the strongly lepidote, +/- resinous young growth.

Pyrostegia (4 spp.) — Our species unique in the narrowly tubular red-orange hummingbird-pollinated flowers with narrow valvate corolla lobes. Tendrils trifid. Twigs slightly irregularly 6-angled.

Roentgenia (2 spp.) — Like *Cydista* but tendrils minutely trifid and the twigs never tetragonal; interpetiolar gland fields lacking. Our species with purple-streaked white flowers and distinctive narrow inflorescence bracts.

There are a number of additional genera in eastern and southern South America, as well as, a few in the Antilles and northern Central America.

BIXACEAE

All five species of the single genus have remarkably similar leaves, evenly ovate or oblong-ovate with entire margins, a palmately 5-veined truncate to broadly subcordate base, and a long slender petiole with distinct but short pulvinus at its apex. The stellate trichomes of most Malvalean families are absent, instead with scattered reddish peltate scales on leaf undersurface, these sometimes becoming stalked (= gland-tipped rufescent trichomes) on twigs or petioles. Another vegetative field character is distinctly yellow or yellow-oxidizing inner bark, sometimes with a thin orangish layer (from the sap?) just under the thin outer bark; the twigs often also contain a trace of reddish-orange latex. The rather large 5-petaled white to pink flowers have numerous yellow stamens with free filaments. The fruit is always a two-valved capsule, usually ovoid and spine-covered but sometimes strongly flattened and nonspiny, the seeds covered by orange-red aril. Might be confused with various entire-leaved Malvales but has shorter pulvinus and lacks stellate trichomes; a few *Bauhinia* (e.g., *B. brachycalyx*) have very similar leaves but with shorter petiole; no similar plant has orangish sap.

Bixa (5 spp.) — Small to large trees, one species widely cultivated, the others restricted to lowland forest where mostly found in second growth or light gaps.

C, E, P: achiote

BOMBACACEAE

Trees, often very large with characteristic columnar trunks and flat-topped spreading crowns. The majority of genera have palmately compound alternate leaves, typically more or less clustered near the tips of thick branchlets. Genera with simple leaves have the typical 3-veined base of Malvales and are only separable from some genera of Tiliaceae and Sterculiaceae by knowing the genera themselves, unless

Bixaceae and Bombacaceae (Three-Lobed Leaves)

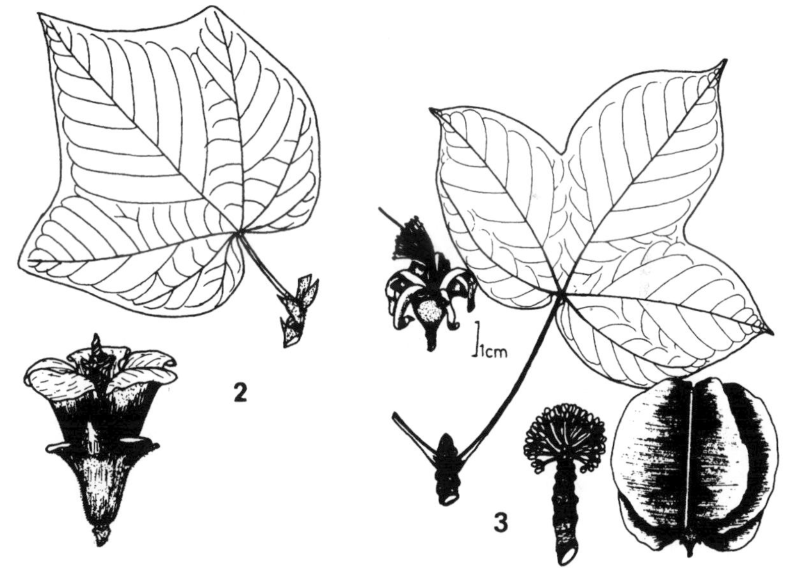

1 - *Bixa* (Bixaceae)

2 - *Ochroma*

3 - *Cavanillesia*

flowers or fruits are present. An unusual characteristic common in several genera of Bombacaceae is spines on the trunk. Bombacaceae are differentiated from most other Malvalean families by the fused filaments, a feature shared with Malvaceae. Familial separation of the mostly herbaceous or shrubby Malvaceae and the almost exclusively arborescent Bombacaceae is rather blurred and one genus, *Hampea* which only barely enters our area, is switched back and forth between the two families; perhaps the only definitive difference is the spinulose pollen of Malvaceae.

Generic taxonomy is largely based on fruit characters and the number and degree of fusion of the stamens. Bombacaceae may be conveniently, if artificially, separated into two groups, one with simple leaves and one with compound leaves. The latter including the great majority of species, contains three natural groups, all with capsular fruits. Those species with many stamens and more or less globose seeds usually surrounded by a woolly fiber derived from the endocarp (kapok) are frequently treated as congeneric under *Bombax sensu lato.* In this group are: *Pochota, Eriotheca, Pseudobombax, Rhodognaphalopsis,* and *Pachira.* (*Pachira* and some species of *Pochota* lack kapok). The second palmately compound-leaved group includes *Spirotheca, Ceiba,* and *Chorisia,* closely related and separable only by degree of fusion of the staminal tube; they differ from the *Bombax* alliance by having only 5(–10) stamens. The third compound-leaved group has winged seeds and includes *Bernoullia* and *Gyranthera.* In addition, one species of generally simple-leaved *Catostemma,* unique in a fruit with a single large ellipsoid seed, has palmately compound leaves.

The simple-leaved genera are more diverse. One (*Cavanillesia*) has a winged fruit. Two others (*Huberodendron* and *Septotheca*) have dehiscent capsules with winged seeds and are closely related to compound-leaved *Bernoullia* and *Gyranthera. Catostemma* has a dry ellipsoid fruit with a single large ellipsoid seed; *Scleronema* is similar but the fruit rounder and more asymmetric. *Ochroma* has narrow dehiscent capsules with wingless seeds embedded in woolly kapok. *Quararibea* and its segregate genera *Patinoa* and *Phragmotheca* have indehiscent more or less fleshy fruits. *Patinoa* differs from *Quararibea* in having the seeds covered with a woolly tomentum, *Phragmotheca* in having plurilocular anthers partitioned by transverse septae.

1. Palmately Compound Leaves — Sometimes with spiny trunks and/or smooth green or green-striped bark

1A. *Bombax* alliance — Stamens numerous; trunk spiny only in *Pochota quinata*

Pseudobombax (19 spp.) — Leaves unique and unmistakable, in the leaflets continuous with swollen petiole apex, and the petiolule bases not

Bombacaceae
(Compound Leaves; Many Stamens)

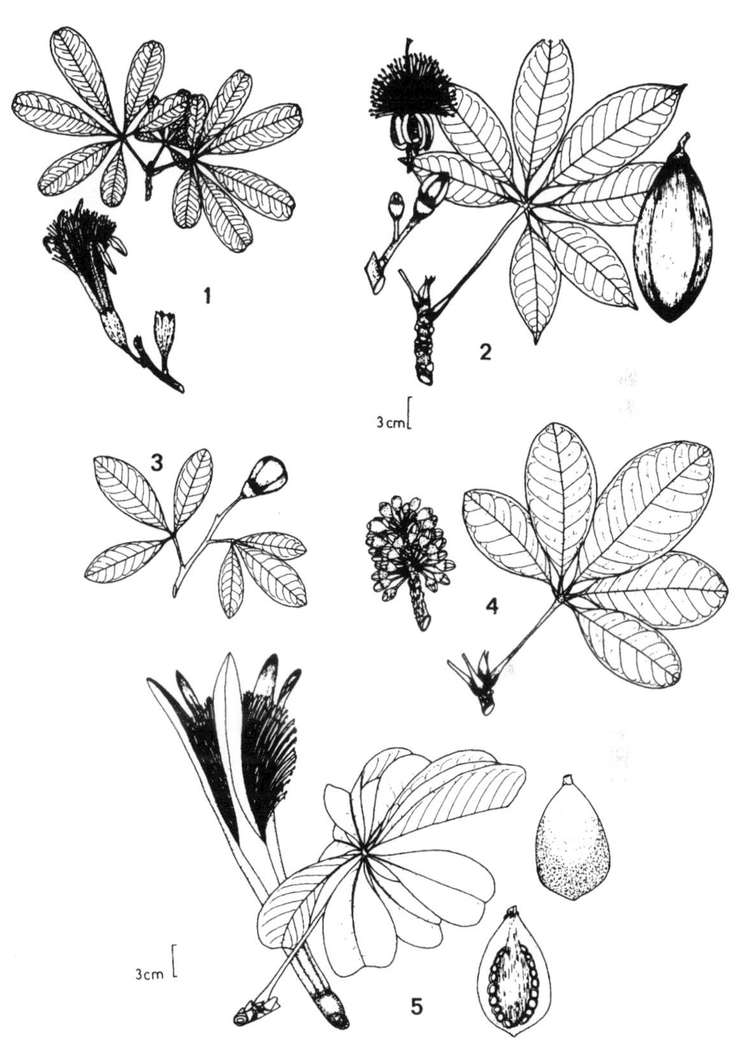

1 - *Pochota*

2 - *Pseudobombax*

3 - *Rhodognaphalopsis*

4 - *Eriotheca*

5 - *Pachira*

jointed. Trunk unarmed, usually with green vertical stripes separated by pale grayish bark; flowers primarily bat-pollinated with denser stamens having thicker filaments than in other genera (powder-puff effect). Fruits woody, 5-valved with abundant kapok.

C: lano, munguba, ceiba barrigón; E: beldaco; P: punga (*P. munguba*)

Eriotheca (19 spp.) — Flowers smaller (<5 cm long) than in other multistaminate compound-leaved genera. Leaves typically membranaceous, pubescent below and often with toothed margins; coriaceous-leaved lowland species have lepidote scales and tannish color below. Fruit smooth and tan, globose (often very small, occasionally to 6 cm long) to ellipsoid, with abundant kapok.

C: lano; P: punga de altura

Pochota (incl. *Bombacopsis*) (20 spp.) — The main neotropical component of *Bombax s.l.* (*Bombax s.s.* is restricted to the Paleotropics). Flowers hawkmoth-pollinated, rather large (7–25 cm long), with very many (100–1000) stamens having long slender reddish filaments. Trunk of most widespread species (*P. quinata*) spiny, the other species without spines. Leaflets entire, usually less coriaceous and with two veins more prominently raised than in *Rhodognaphalopsis*. Fruit woody and 5-valved, more or less ellipsoid; seeds varying from small and embedded in kapok to large and lacking kapok, the latter approaching *Pachira* but smaller.

C: ceiba tuluá; E: ceiba

Rhodognaphalopsis (9 spp.) — A characteristic genus of small twisted trees or shrubs of poor sandy soils, mostly in the upper Rio Negro-Guayana area with several species in Amazonian Colombia but none in Ecuador and only one in the campinarana forests of Amazonian Peru. Close to *Bombax*, from which it differs technically primarily in the pollen; sometimes merged with *Pochota*. Usually easy to recognize vegetatively by the characteristic very coriaceous leaflets with whitish or reddish lower surfaces (from the dense indument of lepidote scales) and mostly more or less plane surface with very indistinct secondary and tertiary venation. Fruits obovoid with a truncate apex, 5-valved, tan-lepidote; seeds embedded in abundant kapok.

P: punga de varillal

Pachira (2 spp.) — Similar to *Bombacopsis* but with mostly larger flowers (usually > 20 cm long) and very large subwoody capsules completely filled by the large angular seeds and lacking kapok. Calyx in flower larger than *Bombacopsis* (1.5–3.5 cm wide). Only two species, one a characteristic element of riverine and swamp forests, the other having the largest flower of any Bombacaceae in our area.

C: lano, ceibo, sapotolongo; P: sacha pandisho, punga de altura (*P. insignis*)

Bombacaceae
(Compound Leaves; Few Stamens; Kapok Seeds)

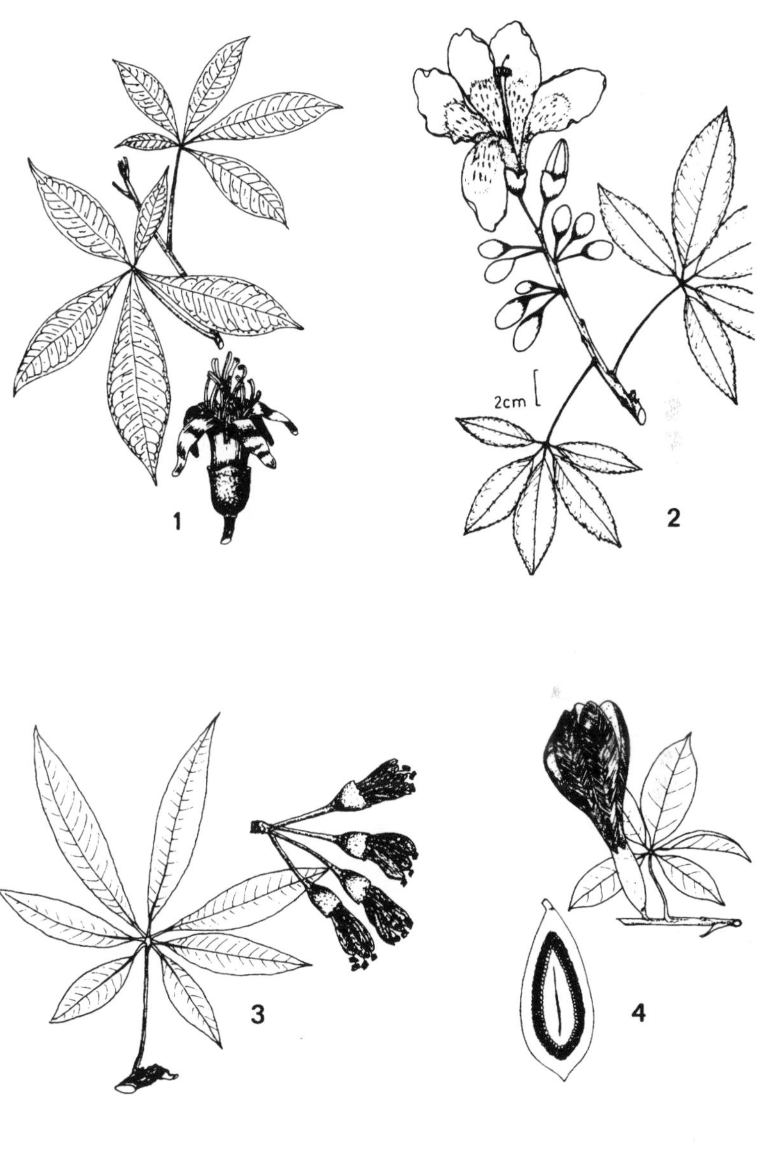

1 - *Spirotheca*

2 - *Chorisia*

3 - *Ceiba* (*C. pentandra*)

4 - *Ceiba* (*C. samauma*)

(*Catostemma*) — One species of the Magdalena Valley has palmately compound leaves (see below).

1B. *Ceiba* alliance — Stamens few (5–10); trunk usually spiny, at least when young; seeds with kapok

Ceiba (10 spp.) — Typically giant emergent trees with characteristically spreading crowns and prominent buttresses. Trunks with spines at least when young, sometimes conspicuously green (*C. trichistandra,* juvenile *C. pentandra*). Leaflets entire. Flowers rather small (ca. 3 cm long) in commonest species (*C. pentandra*) to very large (to 12 or more cm), the 5 stamens connate into a tube only basally. Fruit a 5-valved ellipsoid woody capsule with small seeds embedded in abundant kapok. *C. pentandra* (lupuna) was formerly the main plywood species of the Iquitos area but has now been seriously depleted locally; kapok comes from the woolly capsule lining.

C: bonga; E: ceibo; P: lupuna, huimba (*C. samauma*)

Chorisia (5 spp.) — Very close to *Ceiba* and perhaps not adequately distinct for generic recognition. Differs in having the anthers sessile and clustered together at tip of the unlobed or barely 5-lobed staminal column. Trunk always aculeate. Flowers mostly pink or red, usually flowering spectacularly while leafless. Leaflets usually serrate (entire in *C. integrifolia*). Fruit as in *Ceiba.*

P: lupuna

Spirotheca (7 spp.) — A segregate of *Ceiba* differentiated by elongate spirally twisted several-celled anthers and truncate rather than lobed calyx; mostly in middle-elevation forests. Leaflets with secondary veins more strongly prominulous and closer together than in *Ceiba.*

1C. *Bernoullia* alliance — Stamens few; trunk unarmed; seeds winged

Bernoullia (2 spp.) — Unarmed trunk; large, woody, 5-valved capsule with winged seeds 5–7 cm long. Flowers unmistakable, small, red-orange, bird-pollinated, in multiflowered one-sided scorpioid racemes. One species in nuclear Central America and one at the northern tip of the Cordillera Occidental of Colombia.

Gyranthera (2 spp.) — Unarmed; capsule as in *Bernoullia* with similar winged seeds but the flower very different, 15–20 cm long (with a 7 cm long calyx) and similar to *Ochroma* or moth-pollinated species of *Quararibea.* Differing from *Quararibea* flowers in the anthers elongate, spirally twisted and transversely septate. Only known from northern Venezuela and eastern Panama but collected very near the Colombian border in Panama.

Bombacaceae
(Winged Seeds; Simple or Compound Leaves)

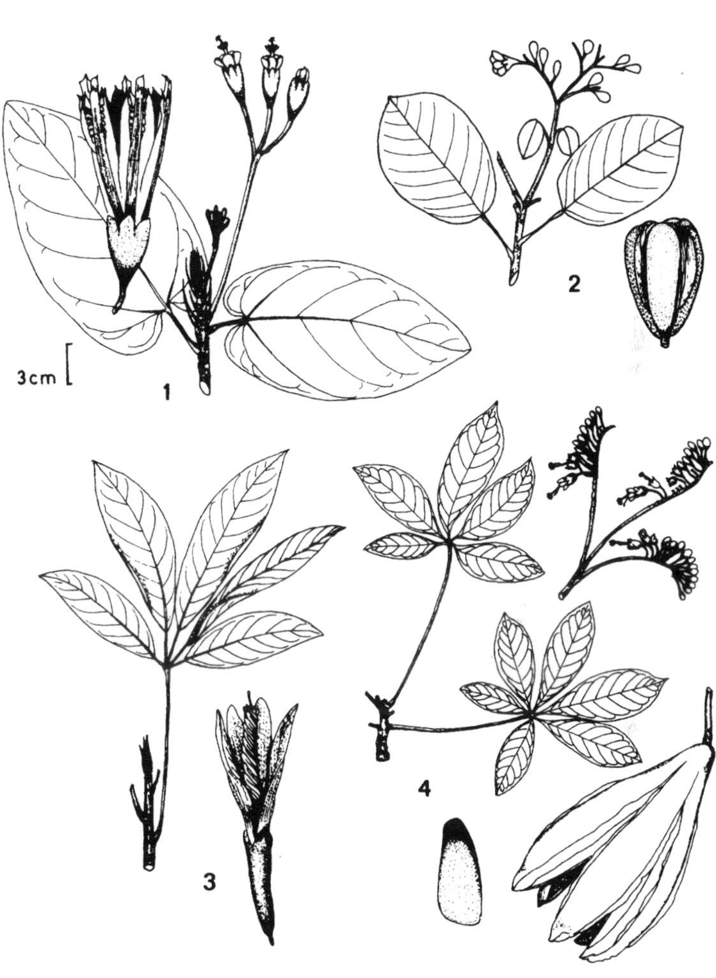

1 - *Septotheca* **2** - *Huberodendron*

3 - *Gyranthera* **4** - *Bernoullia*

2. Simple Leaves; Trunks Never Spiny

Ochroma (1 sp.) — One of the most common and distinctive genera of the Neotropics. Famous for the extremely light wood, useful for making rafts and model airplanes. Although widespread in early second growth, it is not omnipresent and is apparently restricted to areas with at least moderately fertile soils. Easy to recognize by the very large sub-3–5-lobed pubescent leaves, persistent stipules, large (ca. 15 cm long with 8–10 cm long calyx) infundibuliform flower, or the long narrow (ca. 2.5 cm wide) subwoody 5-valved fruit with small seeds embedded in abundant kapok.

C: balsa; E: balsa, boya; P: topa, palo de balsa

Cavanillesia (3 spp.) — Another of the most distinctive of all neotropical tree genera. The trunk is essentially a giant hollow cylinder filled with balsalike pith; the smooth papery reddish bark is extremely characteristic as are the regular trunk rings in the best known species. The unique fruit is large (ca. 10 cm long) and indehiscent with 5 very broad papery longitudinal wings. The flower is small for Bombacaceae, ca. 2 cm long, and with the slender filaments free for most of their lengths. Completely dominating some forests as in the moist forest of Darien, Panama, and some dry forests of the Guayaquil region of coastal Ecuador.

C: macondo; E: pijio; P: lupuna colorado, puca lupuna, lupuna bruja

Catostemma (8 spp.) — Characteristic of the poor-soil Guayana Shield area but with an endemic species in the Magdalena Valley. All species have simple leaves (sometimes 3-foliolate in juveniles) except a palmately compound-leaved one from the Magdalena Valley. The fruit is absolutely distinctive: It is indehiscent and ellipsoid with the single very large seed enclosed in a dry subwoody pericarp. The flowers are small with numerous stamens.

Scleronema (5 spp.) — Large trees. Flowers small, with a short staminal tube and 5-lobed calyx. Leaves elliptic and not very acute. The only species to reach our area is easily recognized vegetatively by the finely tannish-tomentose leaf undersurface with close prominently raised parallel tertiary venation. The asymmetric fruit resembles a single coccus of *Sterculia.*

Huberodendron (5 spp.) — Probably most closely related to compound-leaved *Bernoullia* on account of the large woody 5-valved capsules filled with winged seeds. Large emergent trees characterized by large buttresses. The flowers are small (1.5–2 cm long) white, and borne in panicles. The glabrous leaves are broadly ovate and distinctly articulate to the apex of the long slender petiole.

C: carrá

Bombacaceae
(Simple Leaves; Indehiscent Fruits)

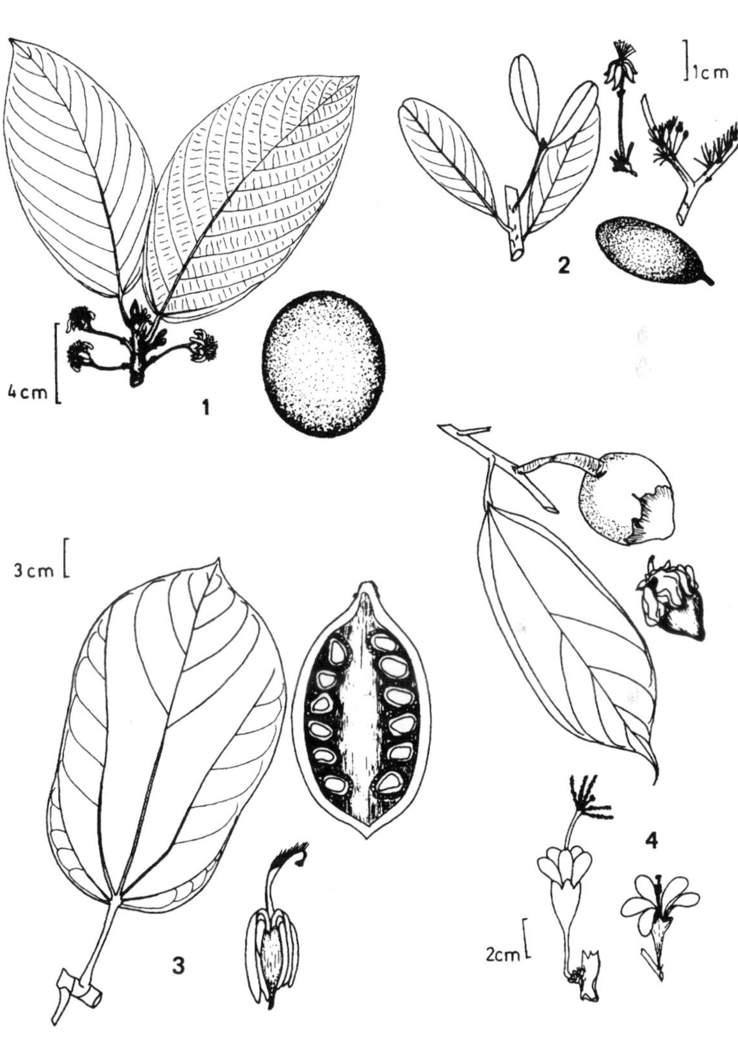

1 - *Scleronema* **2** - *Catostemma*

3 - *Patinoa* **4** - *Quararibea*

Septotheca (1 sp.) — Large upper Amazonian tree of seasonally inundated forest, similar to *Huberodendron* in having a woody capsule with winged seeds. Vegetatively and in flower much like *Quararibea,* with ovate, deeply cordate leaves reminiscent of those of *Q. cordata.* Several flowers are borne together at the end of a long peduncle (unlike the fasciculate flowers of *Quararibea*) and the anthers differ from *Quararibea* in being septate.

Quararibea (incl. *Matisia*) (over 50 spp.) — The largest genus of Bombacaceae. In wet forests on fertile soils (e.g., Rio Palenque, Ecuador) it may be the most prevalent nonpalm tree genus with as many as five strictly sympatric species. Characterized by axillary or cauliflorous flowers borne single or in fascicles and having an elongate staminal column, sometimes apically 5-lobed with numerous sessile anthers at the apex; probably mostly pollinated by nonflying mammals. Vegetatively, most likely to be confused with *Theobroma* (Sterculiaceae) but differing in the characteristic myristicaceous branching pattern and the tendency to have pairs of leaves on poorly developed short-shoot branchlets. The fruits, subtended by the usually persistent and often conspicuously expanded calyx, are indehiscent with from 1–5 seeds embedded in the more or less fleshy mesocarp. Species with a 5-celled ovary and lobed staminal column are sometimes segregated as *Matisia* but some species are intermediate and the striking palynological differences do not correlate with the morphological ones.

C: castaño, bacao, bacaito; E: molinillo, zapotillo, zapote de monte (*Q. coloradorum*), zapote (*Q. cordata*); P: sapotillo, machin sapote, sapote (*Q. cordata*).

Patinoa (3 spp.) — A segregate of *Quararibea* differentiated by the large fruits with numerous woolly lanate seeds and lacking a persistent calyx. In Amazonia mostly in seasonally inundated forest. The pulp of *P. almirajo* of Chocó is edible, but that of *P. ichthyotoxica* is used as a fish poison.

C: almirajo

Phragmotheca (2 spp.) — A segregate of *Quararibea* which differs only in having pluricellular septate anthers. Vegetatively the two *Phragmotheca* species are characterized by a rather dense and shiny ferrugineous-lepidote leaf tomentum.

C: baltran, sapote

Boraginaceae

A predominantly herbaceous family, though mostly woody in our area, where herbs (better represented outside the tropics) mostly occur in dry areas. Vegetatively characterized by usually alternate, often stiff-pubescent asperous leaves and stem and the tightly scorpioid inflorescence (or the inflorescence cymose with often scorpioid branches). The woody genera are also best represented in dry-forest but *Cordia* and *Tournefortia* also have many species in lowland moist- and wet-forest. We have only one important tree genus (*Cordia,* also with many shrubs and a few climbers) and one of lianas (*Tournefortia,* also with some erect shrubs and small trees, especially at higher elevations). In addition the southern tree genus *Saccellium* (with lanceolate leaves and Rhamnaceae-like venation) occurs in the dry Huancabamba depression and a few species of predominantly Central American and Antillean *Bourreria* and (spiny) *Rochefortia* reach northern Colombia (and Venezuela). Tree species of *Cordia* (but not the shrubs) are easy to recognize by the characteristic nodes with a leaf arising from each branch dichotomy and held parallel to the dichotomy; two common species have variously swollen nodes inhabited by ants. Some arborescent upland *Tournefortia* species have opposite serrate leaves; climbing *Tournefortia* species are mostly utterly nondescript when sterile but are always entire and usually blackish-drying.

When fertile, *Tournefortia* (like the herb genera) is easy to recognize by the strongly one-sided scorpioid inflorescence branches (and also by the salverform-tubular probably mostly butterfly-pollinated flowers. *Cordia* and *Borreria* lack the typical inflorescence. Style division is a useful character to separate the woody genera: *Cordia* (and some *Saccellium*) have the style twice-forked with 4 slender stigmas; *Tournefortia* (and its herbaceous segregate *Heliotropium*) have the style entire and with a single conical stigma; *Bourreria* (and extralimital *Ehretia*) are intermediate with once-divided style and 2 stigmas. *Cordia* and *Tournefortia* have fleshy drupes; the herb genera small dry fruits, mostly splitting into 4 nutlets. Again *Bourreria* is intermediate with a large dry fruit splitting into 4 segments.

1. Woody Trees, Shrubs and Lianas

Bourreria (50 spp.) — Small trees or large shrubs, widespread in Antilles and Central America but in our area restricted to northern Colombian dry forests. Characterized by alternate (or in alternate clusters) smoothish, entire leaves. Calyx valvate and leathery, closed and apiculate in bud, irregularly bilabiate at anthesis; flowers rather large for family, white, tubular-infundibuliform with exserted or subexserted anthers. Fruit pyramidally sharply 4-angled, fragmenting along angles into 4 large angular cocci.

Rochefortia (3–5 spp.) — A mostly Central American and Antillean genus, barely reaching our area in the driest part of coastal Colombia (and Venezuela). Our only species is a spiny-branched dioecious tree with

Boraginaceae
(Trees and Shrubs)

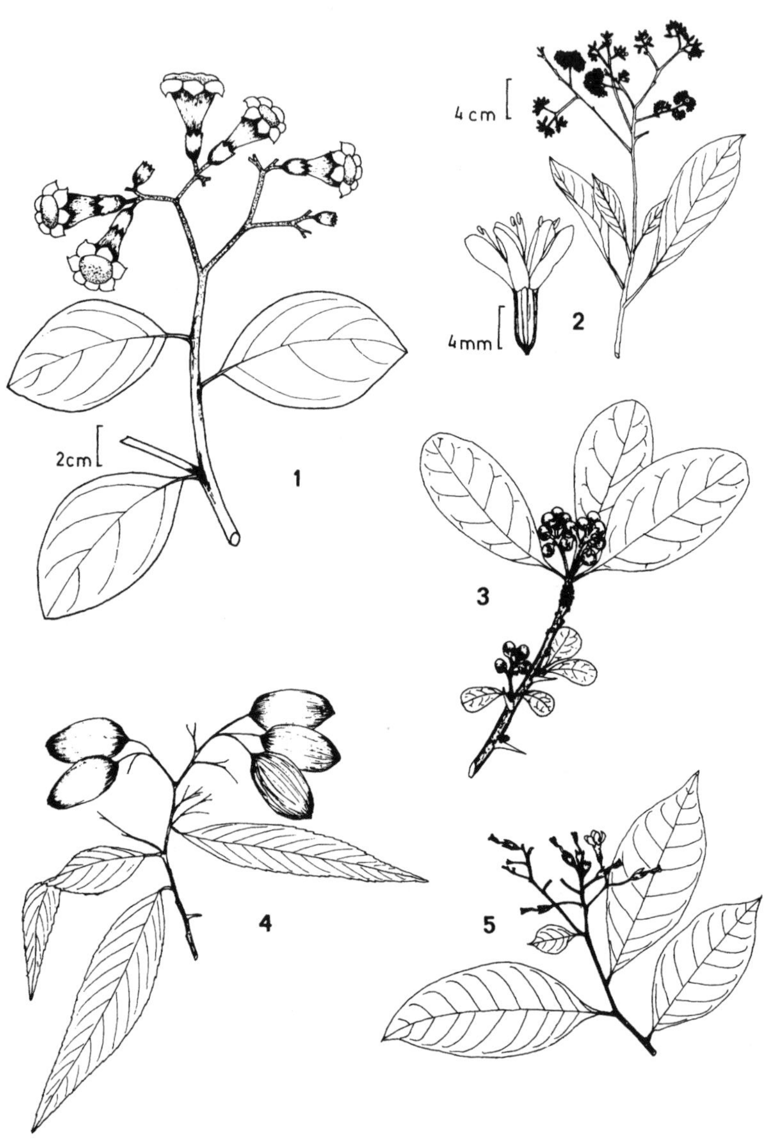

1 - *Cordia* (*C. lutea*)

2 - *Cordia* (*C. alliodora*)

3 - *Rochefortia*

4 - *Saccelium*

5 - *Ehretia*

Boraginaceae
(Lianas and Herbs)

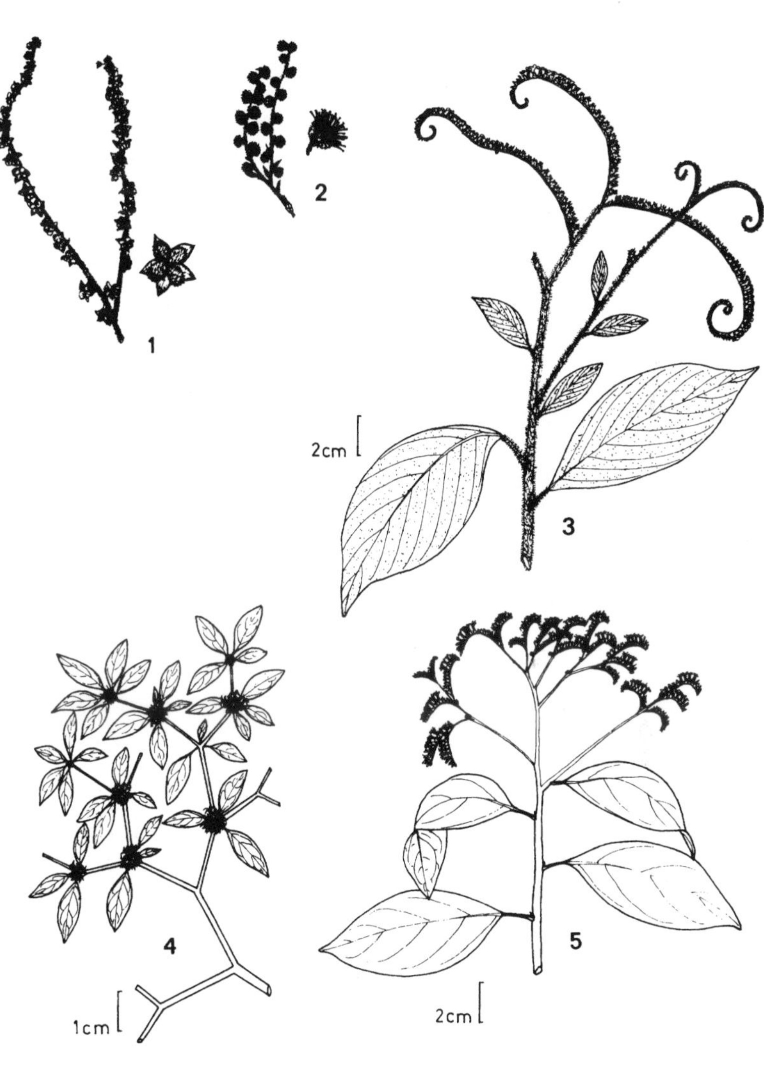

1 - *Cynoglossum* **2** - *Hackelia*

3 - *Heliotropium*

4 - *Tiquilia* (*Coldenia*) **5** - *Tournefortia*

entire obovate leaves mostly clustered on short-shoots in spine axils or near branch apices. Very similar to *Achatocarpus* (Phytolaccaceae) and *Pisonia macracantha* (Nyctaginaceae) and with similarly dark-drying leaves but the secondary veins more conspicuously raised. Inflorescence small and axillary, shorter than adjacent leaves; fruits round, subsessile, more or less clustered at end of peduncle.

Cordia (250 spp., plus 75 in Old World) — Shrubs to large trees, the trees often with tall slender trunk and rather small round crown. Most tree species with very characteristic branching, with a leaf arising from within each branch dichotomy; two common species with variously swollen hollow nodes inhabited by ants. Shrub species lack a leaf in the branch dichotomies and ours all have serrate leaves. Inflorescence of shrubs often spicate, sometimes even globose, that of trees usually more or less openly cymose-paniculate. Flowers usually small white, and short-salverform, but *C. lutea* has yellow openly tubular-infundibuliform corollas. Fruit a more or less fleshy drupe, usually white or black at maturity, except *C. alliodora* (section *Gerascanthus*) group with corolla lobes dry and expanded for wind-dispersal (and vegetatively distinctive in stellate trichomes).

C: uvita mocosa (*C. dentata*), solera or canaleta (*C. alliodora*)

E: alatripe, laurel (*C. alliodora*)

Saccellium (3 spp.) — In our area apparently restricted to the dry part of the Huancabamba depression where it is locally very common. Similar to *Cordia*, but with the dry expanded (ca. 2 cm long) calyx enclosing fruit and with the apex nearly closed. The leaves narrowly oblong-lanceolate, more or less entire or irregularly serrulate, the ascending close-together secondary veins similar to Rhamnaceae. Flowers small, white, inconspicuous.

Tournefortia (150 spp., incl. Old World) — Most (ca. 10) of the common species high-climbing lianas of lowland forest. Very nondescript, with entire alternate, pinnate-veined leaves and normal stem morphology. A few species are cloud-forest trees, sometimes with mostly opposite leaves. Differs from *Cordia* in the strongly scorpioid inflorescence branches with the flowers all along one side of the branches and in the style not split at apex, from *Heliotropium* in usually woody habit and the fleshy (often white or orangish) fruit; flowers white to greenish and always more or less salverform with reflexed lobes.

E: maíz de gallo

2. Herbs — (The first six genera have white flowers, the second two (usually) yellow, the last three blue)

Heliotropium (250 spp., incl. Old World) — Intermediate between *Tournefortia* and the other herb genera. Mostly dry-area weeds, commonly on sandbars, often prostrate, with conspicuously scorpioid inflorescences as in *Tournefortia.* Differs from *Tournefortia* in the habit and in the fruit dry and splitting into 2–4 nutlets.

E: rabo de alacrán

Cryptantha (100 spp., mostly N. Am. and Old World) — Small herbs, mostly in dry sandy places, leaves always linear and scabrous; flowers white, small, inflorescence tightly scorpioid. Almost always with stiff, obnoxious hairs.

Tiquilia **(*Coldenia*)** (25 spp.) — Prostrate pubescent subwoody herbs nearly always growing in sand, the small usually grayish leaves mostly clustered toward branch apices. Flowers tiny, whitish or pinkish. Most species in coastal lomas or on beaches.

Plagiobothrys (100 spp., incl. N. Am.) — Hispid high-Andean herbs with tiny white flowers in reduced inflorescences (or solitary) and narrow sublinear leaves in part in basal rosette. *Allocarya* differs in lacking the rosette of basal leaves but is probably congeneric.

Pectocarya (10 spp., incl. N. Am.) — Small loma and Peruvian western cordillera herbs with linear leaves and minute flowers sessile in the leaf axils. Like *Hackelia,* but the nutlets uncinately pubescent rather than with hooked spines.

Moritzia (5 spp.) — Erect herbs mostly of eastern Brazil, but with one Andean paramo species, characterized by the basal rosette of long narrow scabrous-hispid leaves, dense inflorescence of small white flowers, and the calyx cylindrical, usually only one nutlet developing.

Lithospermum (60 spp., incl. N. Am. and Old World) — Asperous, perennial herbs of Andean valleys, usually with yellow or yellowish flowers (prostrate *L. gayanum* has white flowers). Stigmas bifid.

Amsinckia (50 spp., incl. N. Am. and temperate S. Am.) — Small, erect, yellow-flowered, hispid-asperous herbs, in our area found only in coastal lomas. Differs from *Lithospermum* primarily in being annuals and in having solitary stigmas.

Cynoglossum (50–60 spp., mostly Old World) — High-altitude herbs with small blue flowers and a small 4-parted fruit, the nutlets covered with hooked spines.

Hackelia (40 spp., incl. Old World) — Erect high-altitude herbs, usually with small blue flowers. Fruit like *Cynoglossum* but unlobed and covered with longer hooked spines.

Brunelliaceae and Burseraceae

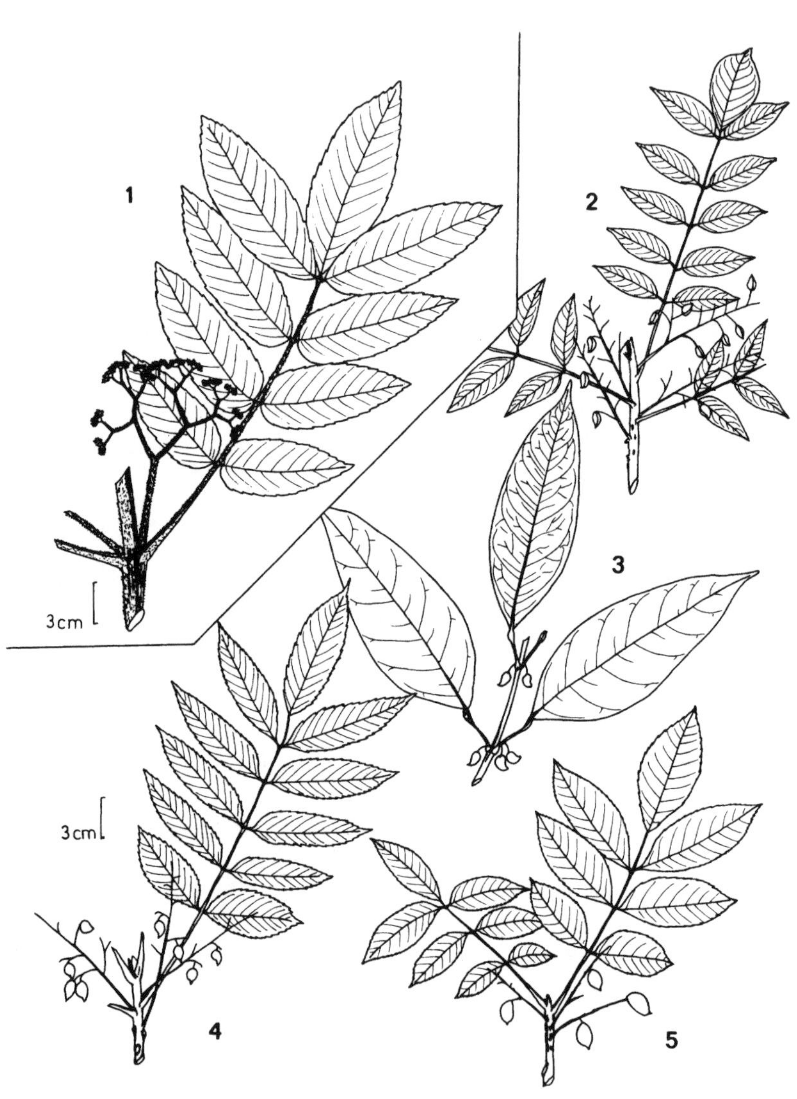

1 - *Brunellia* (Brunelliaceae) **2** - *Bursera*

3 - *Protium*

4 - *Crepidospermum* **5** - *Crepidospermum* (*Hemicrepidospermum*)

(Borago) — Cultivated *Borago* is sometimes escaped or naturalized at high altitudes. The flowers (reminiscent of *Solanum* with anthers in cone around style) are blue but larger than in *Cynoglossum* and relatives. Leaves with conspicuously winged petiole.

Many other genera occur outside our region, especially in the dry areas of Mexico, Chile, and the Brazilian Shield.

Brunelliaceae

Second-growth or light-gap trees, entirely restricted to mesic montane forests (except for two species in lowland Chocó). While this book was in press, a new analysis appeared which shows that this family is best included in Cunoniaceae. The opposite (or ternate) mostly pinnately compound leaves are almost always more or less densely pubescent below, usually somewhat coriaceous and with closely serrate (sometimes merely serrulate and rarely subentire) margins; the venation prominulously reticulate below; and the numerous prominent secondary veins making obtuse angle with midvein, typically ending in marginal teeth. The few simple-leaved species have leaves identical to leaflets of compound-leaved species. The nodes usually have a conspicuous interpetiolar ridge, usually with 2–4 tiny subulate stipulelike projections along it. The characteristic axillary inflorescence is usually flat-topped and repeatedly dichotomously branched. The small apetalous flowers have slightly fused sepals, 10–12 stamens, and several (to six) apocarpous pistils each carpel developing into a small beaked follicle, these radially arranged and pubescent, often with urticating hairs.

Brunellia (51 spp.)

Burseraceae

Trees (rarely shrubs), usually easy to recognize by the pinnately compound leaves and strongly aromatic, often incenselike or turpentine-like vegetative odor. The flowers are small, rather nondescript, usually cream, greenish, or tannish. The fruits may be single-seeded or dehiscent to reveal a typically bright red inner capsule and a seed covered by a rather succulent whitish aril. The families most likely to be confused with Burseraceae are Anacardiaceae (which differs mainly in technical characters of ovule number and orientation) and Sapindaceae (which differs in lacking the vegetative odor and in having the stamens of the similarly small, but often complexly and strikingly pubescent flowers arising from inside rather than from outside the disk). When sterile or in flower *Zanthoxylum* of the Rutaceae might be confused with Burseraceae but differs in usually having spines on the trunk and branches, a yellow slash, and in having a different vegetative odor, either more citruslike or somewhat rank but never pungently incenselike; unlike Rutaceae, Burseraceae are only rarely inconspicuously pellucid punctate. In fruit, *Ophiocaryon* (Sabiaceae) might be confused with *Dacryodes,* but the much more asymmetric orientation of the keeled fruit is distinctive. *Trichilia,* the odd-pinnate genus of Meliaceae, has a sweetish rather than turpentiny odor (and the odor is restricted to the trunk).

Burseraceae almost always have traces of resinous white latex either in the twigs or as widely scattered droplets in the bark slash. Even when not immediately visible, dried white aromatic latex droplets (or a whitish powdery crystalline mass) almost always eventually develop in a day-old slash. Trunk wounds are also characterized by the whitish-drying aromatic resin in contrast to Anacardiaceae where trunk wounds are often characterized by a black-drying resin. Although Sapindaceae lianas have nonaromatic white latex, trees of that family lack latex. The bark, at least of young burserac trees, often is rather smooth; it may be very characteristic, e.g., thin reddish and papery peeling, even in adult trees, in *B. simaruba,* one of the commonest and most widespread species, grayish and horizontally crossed with narrow pale lenticellar marks in many species. Some *Protium* species have characteristic often rather dense and curved stilt roots.

The genera are mostly defined by technical floral characters. The two genera with consistently 3-parted flowers and indehiscent fruits (*Dacryodes* and *Trattinnickia*) form part of a natural group (tribe Canarieae) but do not hang together vegetatively. *Dacryodes, Bursera,* and *Protium* have free or nearly free petals, the former two with 2–3-lobed ovary and stigma and single-seeded indehiscent (very tardily dehiscent in *Bursera*) fruits, the latter with 4–5-lobed ovary and stigma and either dehiscent or several-seeded fruits. *Bursera* often has 4–6-parted flowers while *Dacryodes* flowers are uniformly 3-parted; *Dacryodes* also has a generally larger fruit with a characteristically wrinkled-ridged surface when dry. *Tetragastris* and *Trattinnickia* have the petal bases fused into a tube, the former with 4–5 sepals and petals and a rather large dehiscent fruit usually with 5 carpels and broader than long, the latter with 3 sepals and petals and a small round one-seeded indehiscent fruit. *Crepidospermum* (and its sometime segregate *Hemicrepidospermum*) are close to *Protium* and have 5-parted flowers with free petals; they are easily characterized by sharply and finely serrate leaflets.

Burseraceae

1 - *Protium* **2** - *Tetragastris*

3 - *Dacryodes* **4** - *Trattinnickia*

Bursera (80 spp., mostly in Mexico) — The most distinctive genus. Characterized by always very strongly aromatic odor, thin deciduous leaves, often clustered at ends of thick twigs. Leaflets usually coarsely and bluntly serrate and often with a winged rachis; when entire, the petiolules (even the terminal one) are short, slender, and unflexed (a unique character combination). The fruits are distinctive, small, very tardily dehiscent, one-seeded, usually more or less trigonal (compressed ovoid in subgenus *Bullockia*).

C: indio en cuero (*B. simaruba*), tamajaco (*B. graveolens*)

Protium (pantropical with ca. 90 spp. in world, mostly in Neotropics) — The largest and most heterogeneous genus. Petiolules apically flexed, and "pulvinate" (at least on terminal leaflet), typically long and more slender than in other genera (except *Bursera*); leaflets sometimes rather thin (unique in entire-margined, nonasperous taxa except for short-petioluled *Bursera*. Inflorescence always axillary, usually reduced and more or less fasciculate, when open usually with very slender axis and branches. Fruits usually reddish, of two types, either asymmetric and laterally dehiscent with one or two seeds or symmetrically 5-carpellate with woodier valves (as in *Tetragastris*), the inner surface of valves conspicuously red, the seeds partly covered by whitish succulent aril. Trees sometimes with stilt roots or stilt buttresses, these often rather dense and with a kneelike curve. Usually has 8–10 stamens.

C: anime, caraño, animecillo; P: copal, copalillo (*P. unifoliolatum*)

Crepidospermum (5 spp., including *Hemicrepidospermum*) — Easily recognized when sterile by the rather finely serrate leaflet margins. The only other genera with evenly serrate margins are *Bursera*, where the margins are much more coarsely serrate and the leaflet smaller and blunter if the leaflets are nonentire; and a few *Protium* species, where the serrations are irregular, the latex white, and the terminal leaflet always has apically flexed (pulvinulate) petiolules. Often many-foliolate, the lateral petiolules short, not flexed, often merging with leaflet base. Leaves pubescent, especially on rachis. Fruits usually puberulous, smaller than most *Protium* species. Latex clear and not apparent. *Crepidospermum* has 5 stamens; *Hemicrepidospermum* differs only in having 10.

C: anime; P: copal

Dacryodes (pantropical with ca. 55 spp. in Old World, 22 in Neotropics) — The fruit is very characteristic, indehiscent, single-seeded, ellipsoid (rarely globose), with a characteristic wrinkled-ridged, usually tannish, surface when dry. Flowers always 3-parted. Vegetatively not very distinctive, the leaves always glabrous, entire and coriaceous, often with large leaflets. The secondary veins usually dry, pale, or white-margined below, the margin sometimes is somewhat cartilaginous and the tertiary venation is more or less intricately prominulous below. The petiolules are rather thickish, sometimes short, and may be apically flexed or not.

C: anime blanco, caraño; P: copal

Buxaceae and Cactaceae

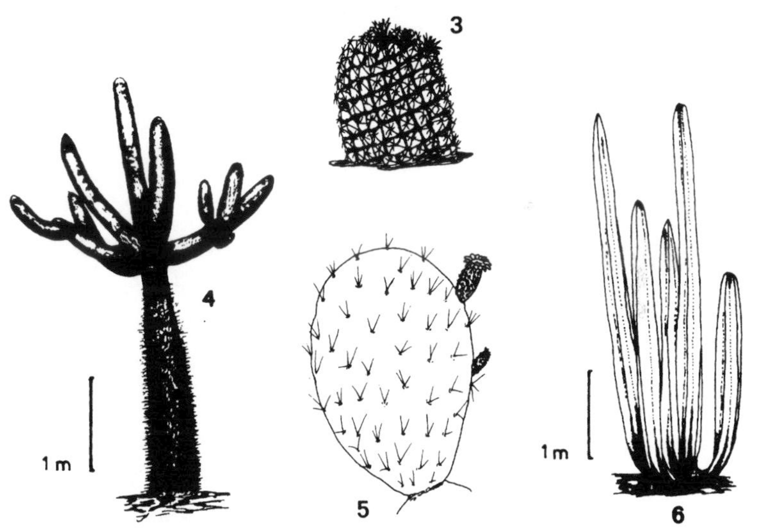

1 - *Styloceras* (Buxaceae) **2** - *Buxus* (Buxaceae)

3 - *Mammillaria*

4 - *Browningia* **5** - *Opuntia* **6** - *Neoraimondia*

Tetragastris (9 spp.) — Close to *Protium* (some species switched back and forth) but with fused petals (and entire cotyledons); a few *Protium* species have petals irregularly fused but not thick, fleshy and forming cap as in *Tetragastris.* A better technical floral character is that the anther bases are continuous with the filament rather than well-defined as in *Protium.* Leaflets like *Protium* but the lateral petiolules short and poorly differentiated from the leaflet base (i.e., not pulvinate); leaf rachis drying reddish-black (rare in most other genera). Fruits usually symmetrically 5-carpelled, wider than long and with a truncate base, typically larger than most *Protium,* but indistinguishable from 5-carpelled type. Bark fissured and shed in large irregular plates, unlike *Protium.*

P: copal

Trattinnickia (11 spp.) — Related to *Dacryodes* by the uniformly 3-parted flowers and indehiscent fruit. Vegetatively usually distinguishable by having petiolules shorter and thicker than in *Protium,* these may be flexed at apex or not. The leaflets are more or less asperous below (unique) and often numerous; when not conspicuously asperous, thick-coriaceous with more or less intricately reticulate pale-drying venation below. Fruits small and almost round, 1-seeded and indehiscent, the surface often finely intricately wrinkled but on a much smaller scale than in *Dacryodes;* the endocarp ("seed") is bony rather than cartilaginous as in *Dacryodes* and also differs in having an apical notch betraying its origin from 2–3 fused carpels. Inflorescence always terminal with a well-developed central rachis. Less asperous species very similar to *Thyrsodium* (Anacardiaceae), where see discussion.

P: copal

The only other neotropical genera are *Commiphora,* formerly exclusively paleotropical, to which two Mexican *Bursera* species have recently been transferred, and recently described Mexican *Beiselia.*

BUXACEAE

As a family, characterized, vegetatively, by coriaceous, sub-3-veined (usually from above base) leaves with smoothish surfaces (tertiary venation slightly or not at all prominulous) that dry the same olive color as many hippocrats and loranths and have the petiole base noticeably decurrent-ridged on twig. One of our two genera (restricted to limestone) has opposite leaves, the other (mostly Andean) alternate. The small unisexual flowers are borne in axillary clusters or small racemes or the male flowers in spikes. In fruit the family is completely unmistakable: the round fruits bear 2 or 3 long slender apical horns.

Buxus (37 spp, also ca. 35 in Old World) — Recently discovered in Colombia on limestone near the Caribbean coast. Easy to recognize vegetatively by the opposite leaves with petiole bases attenuating into strong ridges, the twigs thus rather irregularly 6-angled. Capsule with 3 slender subapical horns, these splitting in half with dehiscence so each valve 2-horned. Male flowers clustered in axils or in small racemes.

Styloceras (4 spp.) — Andean cloud forests plus a recently discovered species in lowland Madre de Dios. Very different from *Buxus* in alternate leaves and the fruit having only 2 horns, these 1–2 cm long. Male flowers usually in spikes; female flowers reduced to nothing but a naked ovary with 2 giant spreading stigmas.

CACTACEAE

With one exception, a very distinctive family, unmistakable in the succulent leafless, usually spiny stem, the spines arising from small felted structures called areoles. Many species become large and treelike, but some are epiphytic, shrubby, or pincushion-like. Best represented in dry areas, but the epiphytic genera mostly occur in rain forests. One genus, *Pereskia,* which includes lianas as well as trees and shrubs, has normal leaves and could easily be mistaken for Phytolaccaceae (especially the liana species for *Seguieria*) except for the longer more numerous spines. The flowers of Cactaceae are also distinctive, mostly single, large and sessile, and having inferior ovaries and numerous perianth parts and stamens. The fruits are always fleshy (usually red or white at maturity) with very small black bird-dispersed seeds.

The family can be easily divided into three natural groups: *Pereskia,* with regular leaves, no glochidia, and stalked flowers; *Opuntia* with sessile, open (rotate) flowers, glochidia (= areoles with clusters of tiny irritating hairs), and tiny terete early-caducous leaves on frequently laterally flattened stem segments; and *Cereus* and relatives (tribe Cereeae or Cacteae) completely lacking leaves and glochidia, the flowers sessile as in *Opuntia* but tubular (except *Rhipsalis*). *Pereskia* and *Opuntia* are uniformly terrestrial; several genera of *Cereus* relatives are epiphytic (*Rhipsalis, Pfeiffera, Epiphyllum,* etc.). True large candelabriform tree-cacti with well-developed trunks occur in our area only in *Cereus, Armatocereus, Browningia, Echinopsis* (*Trichocereus*), *Espostoa,* and *Neoraimondia.* The genera and generic groupings used here follow Bradleya 4: 65–78, 1986. Illustrations, in part, from J. Madsen, *Flora of Ecuador.*

1. Leaves Present and Persistent (Glochidia Absent); Flowers Pedicellate

Pereskia (16 spp.) — Lowland to middle-elevation dry and moist forests, in the latter mostly in disturbed sites such as along streams. Most species trees or shrubs. Liana species very like *Seguieria* except for much longer spines on stem; tree species always with numerous, more or less persistent spines, at least on branches (the main trunk of *P. bleo,* a dry-forest canopy tree of northern Colombian dry forest, becoming nonspiny).

C: guamacho

2. Leaves Terete, Usually Tiny and Early-Caducous (Often Only on Juveniles); Stems with Glochidia (= Clusters of Small Irritating Hairs) at Areoles, Unribbed; Flowers Sessile and Open (= Rotate)

Opuntia (incl. *Cylindropuntia, Tephrocactus*) (200 spp., incl. N. Am.) — Most species (*Opuntia sensu stricto*) have broad laterally flattened stem joints; these are restricted to dry areas and may become rather large trees but are typically shrubs. *Cylindropuntia* has many-jointed branches with terete joints. *Tephrocactus,* a low puna plant, is unbranched or few-branched and has short, clustered terete joints, often conspicuously grayish- or whitish-pubescent. Normally terrestrial.

C, E, P: tuna

3. Leaves and Glochidia Absent; Flowers Sessile and Tubular (Except Rotate in *Rhipsalis*); Stems Mostly Variously Ribbed; Variable in Habit but Frequently Epiphytic (Unique in Family)

The groups below (numbered following the IOS "consensus") are arranged in a logical sequence from nonspiny and highly epiphytic to increasingly spiny and somewhat epiphytic to erect and treelike to reduced and pincushion-like. The first two groups are mostly epiphytic or viny, often with slender or variously flattened or triangular stems; *Rhipsalis* and relatives have small flowers, with the perianth rotate and with few slender segments, *Epiphyllum* and relatives have larger flowers with tubular perianths with more segments. The three groups most closely related to *Cereus* are mostly erect columnar cacti, often becoming trees, but sometimes reduced to globose pincushion-cacti (*Melocactus, Echinopsis-Lobivia*). The sixth group, mostly in Mexico and North America, differs from the rest of the Cereeae by having the flowers and spines at different areoles; represented by only a single species of *Mammillaria* in our area.

3A. Epiphytic with small rotate flowers — "GROUP II"

Rhipsalis (50+ spp., incl. 2 African) — Mostly lowland moist- and wet-forest epiphytes, much-branched with slender, round or flattened, elongate, generally nonspiny (except in juveniles) joints (looks more like *Psilotum* than a cactus). Flowers tiny and nontubular with few slender perianth segments.

Pfeiffera (5 spp.) — Mostly south temperate, reaching upland Peru. Perhaps better lumped with *Rhipsalis* from which it differs in ribbed joints and spiny fruits.

3B. Mostly epiphytic or vinelike (or with single joints if terrestrial); flowers tubular — "GROUP I"

Epiphyllum (15 spp.) — Epiphytic with flat spineless joints. Corolla long and tubular-infundibuliform, night-flowering (mostly sphingid-pollinated).

Disocactus (15 spp.) — Epiphytic, differing from *Epiphyllum* in shorter day-blooming flowers with tube shorter than or same length as lobes. *Pseudorhipsalis* (incl. *Wittia*) (4 spp.), also epiphytic, spineless, with flat joints and short tubular corolla, is essentially the South American equivalent of *Disocactus* and perhaps not generically distinct.

Hylocereus (15–20 spp.) — Mostly Antilles and Central America, reaching only northern part of our area. Epiphytic, especially in drier areas; characterized by the triangular stems and long nocturnal *Epiphyllum*-type flowers with conspicuous bracts on ovary.

Selenicereus (10 spp.) — Mostly more or less creeping, differing from *Hylocereus* in the stems usually not triangular, variously ribbed, fluted or angled; flowers elongate (sphingid-pollinated) as in *Hylocereus* but with inconspicuous bracts.

Acanthocereus (10 spp.) — Mostly Central American, only reaching northern Colombia. Terrestrial and suberect to 2 or 3 m tall (but more slender than most erect *Cereus* segregates), or (usually) arching and more or less clambering; one of our species with triangular stem, the other 4–6-ribbed

C: pitay

3C. Tree-cacti (plus globose lowland *Melocactus*); *Cereus* and its closest relatives (next five genera) are characterized by the nonscaly ovary (naked or spiny or with tufts of short hairs). — "GROUP III"

Cereus (25–30 spp.) — Typically large cylindric-branched dry-area tree-cacti with elongate, infundibuliform, nonspiny, sphingid-pollinated flowers. Sometime segregate *Monvillea* (15 spp.) differs in being more slender and more or less clambering.

C: cardón, jasa

Pilosocereus (incl. *Pilocereus*) (40–50 spp.) — Closely related to the Central American/Mexican genera centered around *Cephalocereus* and characterized by having the flowers borne from a cushionlike pseudocephalium of densely clustered hairs; differs from *Cereus* and *Armatocereus*

Cactaceae
(Erect)

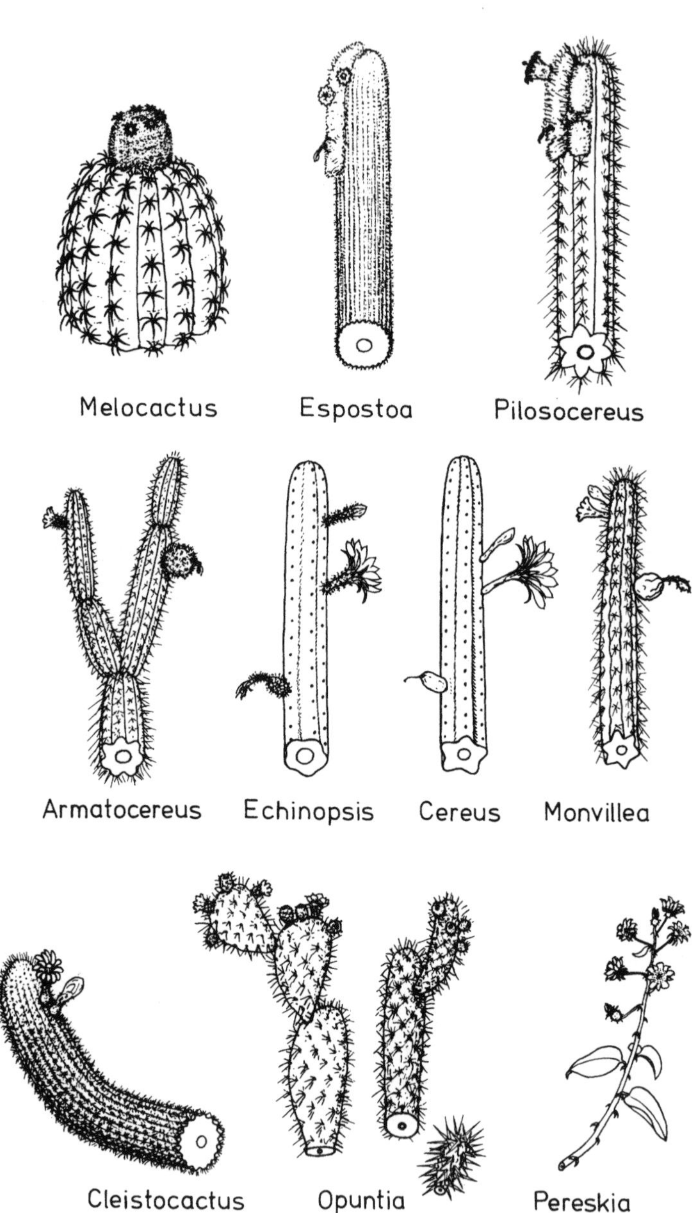

Cactaceae
(Epiphytic and/or Scandent)

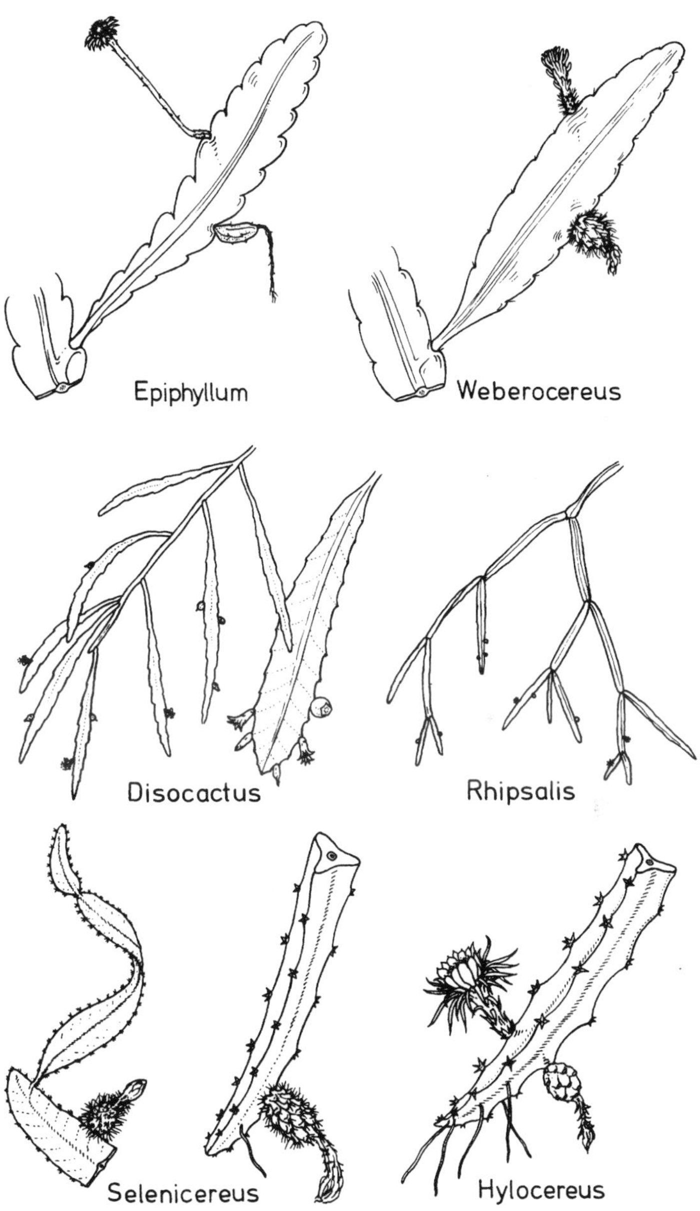

(which lack pseudocephalia), in having woolly hairs at the areoles in addition to the spines; differs from *Espostoa* (which also has pseudocephalia) in fewer and stronger ribs (6–9 vs. > 18 in our area) and spines (<20 per areole vs. > 30). Slender erect branches. Two species in our area: *P. colombianus* in the Dagua Valley and *P. tweedyanus* in Ecuador.

C: cardón

Armatocereus (10 spp.) — A South American segregate from *Lemaireocereus* (= *Stenocereus*). Large much-branched Andean tree-cactus with strongly erect stems; differing from *Cereus* in spiny flowers and fruits. Pachycereeae is no longer separated from the other close *Cereus* relatives.

Jasminocereus (1 sp.) — The Galapagos version of *Armatocereus.*

Calymmanthium (1 sp.) — Northern Peru, 3–4-ribbed and erect, becoming very densely spiny on ribs; characterized by a unique perianth with the perianth invaginated inside the tube apex in bud. Like *Acanthocereus* and *Neoraimondia* in the spiny calyx tube and fruits.

Neoraimondia (2 spp.) — Lower western slopes of Peruvian Andes. Becoming rather large but little-branched tree-cacti with very distinctive thick 4–5-ribbed strongly angled stems, quite smooth between the angles. Unique in the small woolly flowers borne several together on elongate woody areoles along the stem-angles.

Corryocactus (incl. *Erdisia*) (10–20 spp.) — Dry Andes of southern Peru. Tall and columnar but branching only at base, or more slender with red flowers (*Erdisia*).

Melocactus (30 spp.) — Globose pincushion-cactus characterized by the well-developed densely bristly hairy terminal cephalium (like woolly cap on top of the globose stem). In Colombian species (coast, Magdalena and Cauca Valleys) the spines white and curved; single species reaches western Andean slopes of central Peru.

C: cabeza de negra, pichiguey

3D. Tree-cacti with irregular branching — "GROUP IV"

Browningia (7 spp.) — A single species of tree-cactus in our region, occurring on the middle-elevation Pacific slopes of southern Peru. Very distinctive in irregular growth form (its designer should have flunked introductory architecture) with a tall cylindric trunk crowned by cluster of irregular branches of differing sizes and often growing in different directions; long spines on trunks but not the branches. Flowers shorter-infundibuliform than close relatives.

3E. Tree- or shrub-cacti (or puna pincushion-plants); the main group of *Cereus* relatives in our area, including both large tree-cacti and variously reduced forms but always with thick, more or less cylindrical, erect (reduced to pincushion in *Echinopsis/Lobivia* and *Rebutia*) stems. Differs from *Cereus* group in scaly ovary. — "GROUP V"

Echinopsis (incl. *Trichocereus, Lobivia*) (50 spp.) — As currently defined (i.e., including *Trichocereus*), the major genus of Andean tree-cacti, but also including high-Andean puna pincushion-cacti (*Lobivia*). Differs from *Cereus* in the scaly ovary with hairlike axillary bristle-spines.

Espostoa (7 spp.) — Exclusively Andean. Most species are several meters tall and have round, columnar multiribbed trunks and branches with tips usually covered by dense trichomes. Flowers borne on pseudocephalia as in *Pilosocereus* (which has fewer, stronger ribs), but unlike *Echinopsis.*

Weberbauerocereus (5 weak spp.) — Middle elevations of Pacific slope of Peru. Probably close to *Echinopsis* (*Trichocereus*), subtree-cacti but mostly basally branched and 2–4 m tall, with 16 inconspicuous ribs and dense pubescence.

Haageocereus (30 weak spp.) — Dry lower western slopes of Peruvian Andes. Basally branching and only 1–3 m high with numerous (>12) not very conspicuous ribs. Perianth shorter, stouter, and more openly campanulate than in *Trichocereus.*

Cleistocactus (incl. *Borzicactus*) (many species)— Dry inter-Andean Valleys of Ecuador and Peru. Mostly low and bushy, branching only at base, densely spiny with rather slender, +/- erect branches. Flowers slender, +/- red (hummingbird-pollinated), strictly tubular in typical *Cleistocactus,* the limb more or less expanded in *Borzicactus.*

Oreocereus (incl. *Arequipa, Matucana, Oroya*) (15 spp.) — Dry Andean slopes. Plants low and forming large dense clusters with erect, thick, unbranched, ca. 10-ribbed stems. Flowers often in pseudocephalia.

Mila (1 sp.) — Central Peru. Low clustered cylindrical-stemmed cactus closely related to *Echinopsis;* differing from most *Cereus* relatives by having diurnal yellow flowers about as wide as long.

Neoporteria (25 spp.) — Mostly south temperate, reaching southern Peru.

Rebutia (30 spp.) — Mostly southern Andes, in our area a small puna pincushion-cactus, very closely related to *Lobivia* group of*Echinopsis* but not ribbed, having raised tubercles instead. Flowers red or orange (hummingbird-pollinated), borne from side of plant.

3F. Barrel-cacti differing from all the above genera in having flowers and spines at different areoles. — "GROUP VI"

Mammillaria (150 spp.) — Mostly southwestern USA and Mexico, barely reaching northern South America with a single species on the northern coasts of Colombia and Venezuela. A globose pincushion-cactus with tubercles rather than ribs, thus, unlike any other member of the family in our area (except *Rebutia* from puna of southern Peru).

There are many other genera both in Mexico and central America and in sub-Amazonian South America.

Calyceraceae

Mostly south temperate herbs, barely reaching our area in the puna of southern Peru. Characterized by the flowers in a composite-like head, usually subtended by spiny basal bracts. Two of our genera, *Moschopsis* (8 spp.) and *Calycera* (20 spp.), are stemless rosette plants with sessile heads, the former with unarmed achenes, the latter with armed achenes. The third genus, *Acicarpha* (5 spp.) is a herb with narrow, usually dandelion-like toothed leaves with clasping bases and heads subtended by an involucre of thick spines.

Campanulaceae

These are easily recognized by the typical flower with inferior ovary, conspicuous usually strongly bilabiate tubular corolla, and the anthers fused in a column around the stigma. Vegetatively, the combination of alternate serrate leaves and milky latex with usually herbaceous habit is indicative. Only one genus becomes distinctly woody in our area: *Siphocampylos* which is often a thick-branched shrubby treelet near the Andean tree line; a few *Burmeistera* species are soft-wooded climbers or hemiepiphytes.

Several genera are perennial herbs restricted to high altitudes in the Andes, mostly above tree line — *Lobelia* (200 spp., mostly north temperate or Old World), unique in corolla dorsally split to base, the flowers usually blue to lavender; *Wahlenbergia* (12 spp. in South America, many in Old World), characterized by actinomorphic funnel-shaped blue flowers and narrow leaves; *Lysipomia* (21 spp.), tiny caespitose or mosslike plants of highest Andean bogs with

Calyceraceae and Campanulaceae

1 - *Acicarpha* (Calyceraceae) **2** - *Centropogon*

3 - *Sphenoclea*

4 - *Hippobroma* **5** - *Siphocampylus*

6 - *Burmeistera* **7** - *Lobelia*

small flowers, usually white with purplish spots or striations); *Hypsela* (1 sp., plus several in Australia), creeping dwarf puna herb with broad leaves and white tubular slightly bilabiate corolla. There are also three lowland herb genera. *Hippobroma* (1 weedy sp., plus 10 in Australia; *Isotoma* is a synonym) is common as a weed in tropical lowland areas and characterized by the incised-margined dandelion-like leaves and regular long-tubular, hawkmoth-pollinated, white flowers. *Diastatea,* is a tenuous annual weed with tiny blue flowers and a superior ovary which looks more like a scroph than a campanulac. *Sphenoclea* (1 sp., sometimes segregated as Sphenocleaceae) is a succulent semiaquatic herb with minute white flowers congested into a thick, densely spicate, fleshy inflorescence.

The three main genera in western South America, all best represented in cloud forest, are *Burmeistera, Centropogon* and *Siphocampylos. Siphocampylos* has a dry capsular apically dehiscent fruit, the other two more or less fleshy berries. *Siphocampylos* is also distinctive in its leaves which tend to be thicker and strongly raised-reticulate and whitish-pubescent below. *Burmeistera* and *Centropogon* are usually easily distinguished by the greenish to maroon often singly borne flowers of the former contrasted with the more numerous bright red or yellow flowers of the latter. The technical character for separating these two genera is the closed anther column with stiff bristly collecting hairs or triangular appendages at the tips of the shorter anthers in *Centropogon* contrasted with an apically split anther tube with glabrous or softly pilose anthers in *Burmeistera.* The latter genus may be a scrambling herbaceous cloud-forest vine or even epiphytic; the former is more uniformly herbaceous.

Burmeistera (82 spp.) — A few cloud-forest species are subwoody climbers or pendent-branched epiphytes or hemiepiphytes, but most are herbs, all with fleshy fruits and only a few distinctive greenish to maroon flowers.

Centropogon (230 spp.) — Mostly more or less succulent herbs but a few species may be semishrubby. Characterized by the often numerous brightly colored red or yellow flowers and fleshy fruit.

Siphocampylos (215 spp.) — A few high-altitude species are soft-wooded shrubby treelets, mostly with whitish flowers. Differs from *Burmeistera* and *Centropogon* primarily in the dry capsular apically dehiscent fruit; vegetatively, often distinguished by the thicker leaves whitish-pubescent and with strongly raised-reticulate venation below.

CAPPARIDACEAE

A typical and important element of lowland dry and thorn-scrub forests where the dense crowns of dark leathery evergreen leaves of Capparidaceae species often make them one of the most distinctive elements in otherwise deciduous dry-season forests. Vegetatively a sometimes rather nondescript family mostly of shrubs and small trees with a few (in our area) herbaceous genera (and occasional lianas in the diverse genus *Capparis*); usually characterized by simple, entire-margined, often sclerophyllous, always alternate leaves. Many species have conspicuous often tannish-lepidote scales. The leaves are palmately 3 to many foliolate in our herbaceous genera (except a few *Podandrogyne* species) and in a few trees (*Crataeva,* extralimital *Forchhammeria*). Simple-leaved species can usually be recognized by either the very different leaf size and petiole lengths of the terminally clustered leaves, or (when the leaves are uniformly alternate and with similar-sized petioles) by the presence of a small, raised, patelliform, axillary gland on the twig just above each petiole base. When fertile very easy to recognize by having the ovary or fruit borne on a well-developed stipe or gynophore; in fruit the receptacle and perianth scars always form a conspicuous thickening in the middle of what otherwise might appear to be a simple pedicel. The perianth parts are usually in fours and the stamens numerous. The fruits are very long, thin, and dehiscent in most herbs, vs. round subwoody, and indehiscent in most tree genera; however, the largest woody genus *Capparis,* spans the entire range from elongate dehiscent fruits to round indehiscent ones.

The herbaceous genera often have sticky glandular vegetative parts, foetid odors, and palmately compound (or 3-foliolate) leaves; *Podandrogyne* differs from *Cleome* in the orange or red-orange flowers, peculiar fruit dehiscence and sometimes simple leaves. Three-foliolate *Tovaria,* often recognized as its own family, has small greenish 6–8-merous flowers lacking a gynophore, and a small round berrylike fruit. All woody genera except part of *Capparis* have globose indehiscent fruits. In our area the only 3-foliolate tree genus is *Crataeva.* The large diverse genus *Capparis* and three small almost exclusively dry-area genera — *Steriphoma, Morisonia,* and *Belencita* — are all closely related. *Belencita* with subsessile evenly ovate leaves, has large solitary rather *Magnolia*-like white flowers and a subspathaceous calyx. *Morisonia,* usually ramiflorous, differs from *Capparis* in the sepals basally fused and filament bases pilose; the leaves are always rather large, coriaceous, oblong-elliptic, and have petioles of varying lengths (but in part long), whereas, no long-petioled *Capparis* species has large coriaceous truncate-based leaves. *Steriphoma* has bird-pollinated bright orange flowers, rather narrow slender-petioled leaves, usually with truncate bases and a tendency to become subpeltate. *Morisonia, Belencita, Crataeva,* and some *Capparis* species have globose subwoody fruits; *Steriphoma* and most *Capparis* species have elongate, often moniliformly contracted fruits.

1. Trees and Shrubs

Crataeva (2 spp. plus few in Old World) — A characteristic medium-sized tree of poorly drained, seasonally swampy habitats, especially in strongly seasonal climates. The only woody capparid with 3-foliolate leaves in our area. The flowers are rather greenish with reddish filaments and tend to have a garliclike odor. The leaflets are rather smooth and subsucculent and may also have a rather garliclike odor.

E: jagua de lagarto; P: tamara

Capparis (250 spp. incl. Old World) — By far the largest Capparidaceae genus. Although commonest in dry forests, some species also occur in wet and even cloud forests. There are two basic leaf types, one with strikingly different petiole lengths and terminally clustered leaves and the other with typically evenly alternate short-petioled leaves, sometimes characterized by a stalked axillary gland. The multistaminate 4-merous flowers are open a single night and have separate sepals and quickly caducous white petals. The fruits vary from long, narrow, and irregularly dehiscent to globose and indehiscent.

C: olivo, naranjuelo; P: tamara

Morisonia (4 spp.) — Similar to *Capparis* from which it differs technically in the basally fused sepals. Its leaves are always oblong-elliptic, with the base +/- truncate (rarely in part subpeltate) and the apex usually obtuse; they have different-lengthed petioles and are very coriaceous or intricately reticulate, sometimes with erose margins. All of these leaf characters can be found in *Capparis* (except peltate base and erose margins) but never together. Flowers usually more or less ramiflorous, mostly greenish or greenish-cream, sometimes in long racemes. Fruits globose, tending to be larger and woodier than in most *Capparis* species.

Steriphoma (8 spp.) — Calyx and petals bright orange at anthesis (unique); calyx fused into tube that splits irregularly apically at anthesis; stamens relatively few and long-exserted. Fruit elongate (but may be rather large and thick). Vegetatively, can be rather tentatively distinguished from *Capparis* by thinner, less coriaceous leaves having the different-length petioles in part long and slender and the sometimes subpeltate base; more coriaceous-leaved species have more slender petioles than similar *Capparis* species while thinner-leaved species have narrower blades with more truncate bases than similar *Capparis* species.

Capparidaceae

1 - *Belencita* **2** - *Morisonia*

3 - *Steriphoma* **4** - *Crataeva*

5 - *Capparis* **6** - *Cleome* **7** - *Podandrogyne*

Capparidaceae and Caprifoliaceae

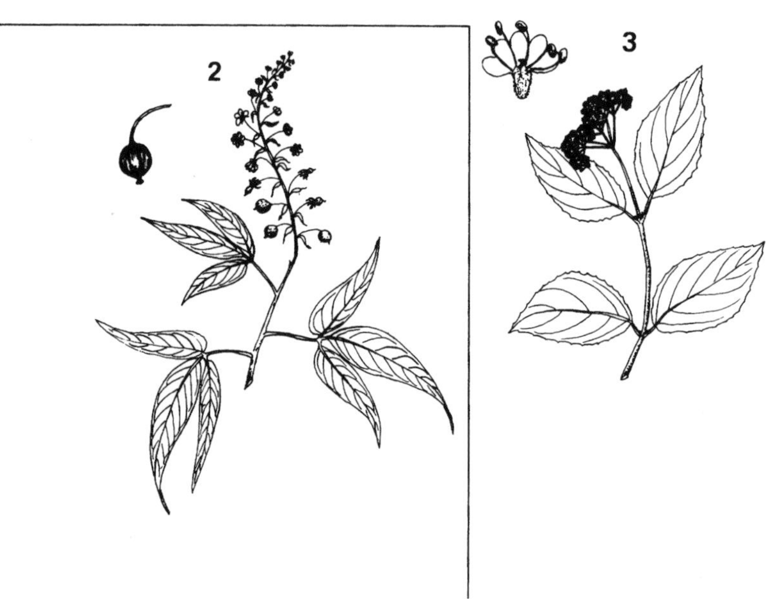

1 - *Sambucus*

2 - *Tovaria* (Capparidaceae) **3** - *Viburnum*

Belencita (1 sp.) — Endemic to dry forests of northern Colombia and Venezuela. Vegetatively, differs from *Capparis* in consistently ovate, mostly subcordate leaves and rather shaggy pale tan twig indumentum; all petioles are short and of same length. Calyx lobes fused, splitting subspathaceously. Fruits are large and globose.

2. Herbs and Subshrubs

Cleome (150 spp. , including Old World) — Leaves 3-foliolate to palmately 7–9-foliolate; plant usually either somewhat spiny or glandular pubescent and/or foetid. Fruit elongate, dry and thin-walled, with a central partition or replum and rather similar to Cruciferae.

Podandrogyne (10 spp.) — Mostly cloud-forest and wet-forest herbs; essentially a *Cleome* with red or orange-red flowers converted to hummingbird-pollination and borne in a flat-topped inflorescence; the fruits, though long and very narrow as in *Cleome,* are also distinctive in lacking a replum and having spirally dehiscing valves and conspicuously white-arillate seeds.

Tovaria (2 spp.) — Our only species an Andean upland weed, the other Jamaican. Traditionally treated as a distinct family, differing from Capparidaceae in 6–8-merous flowers, unstalked ovary, and small globose berrylike fruit. Its 3-foliolate leaves and other aspects of its general habit are very similar to *Cleome* and other herbaceous capparids, although the lax terminal raceme of greenish flowers is more like *Phytolacca.*

Caprifoliaceae

Small montane forest trees, always with opposite leaves and a strong interpetiolar line or ridge. One area genus has pinnately compound leaves, the other simple; the former has finely but rather unevenly serrate leaflets with stipel-like glands between the upper leaflets; the latter usually has at least some leaves inconspicuously toothed near apex, and is also vegetatively characterized by the few strongly ascending secondary veins and tendency to be puberulous at least on the veins below and petiole. Both genera have more or less flat-topped inflorescences, with small white flowers having inferior ovaries. Although the petals are fused at the base, this is not at all obvious and it is, thus, extremely difficult to separate *Viburnum* from *Cornus.* The fruits of both genera are round berries. Perhaps the best distinguishing character to differentiate *Viburnum* from *Cornus* is the conspicuous straight interpetiolar scar contrasting with the decurrent petiole bases of *Cornus.*

Sambucus (40 spp., mostly Old World) — Pinnately compound, (occasionally with 3-foliolate basal leaflets, thus in part, bipinnate); mostly found near settlements in cloud-forest areas.

Viburnum (200 spp., mostly Old World) — Simple-leaved; exclusively in moist, montane forests.

Several other north temperate genera reach Mexico or northern Central America.

CARICACEAE

Typically small succulent trees (*Carica*) but some species of *Carica* are tiny understory treelets less than a meter tall and one is a vine, while most *Jacaratia* are large canopy or subcanopy trees. The leaves are almost always palmately compound or lobed but are merely basally 3-veined in a few subshrub or small treelet *Carica* species. Milky latex is usually present, at least in the leaves and young branches. The combination of milky latex and palmately lobed or palmately compound leaves could only be confused with Euphorbiaceae being similar respectively to some species of genera like *Jatropha* and (extra-areal) *Johannesia.* Palmately compound species are remarkably similar to some Bombacaceae except for the milky latex and generally more succulent leaves. Most species are dioecious (often cauliflorous) with the usually larger female flowers perhaps functioning as mimics of the male flowers; a few species are polygamodioecious or monoecious and sex change is apparently not infrequent.

Carica (22 spp.) — Easily distinguished by the simple, usually deeply lobed, and rather succulent leaves. Frequently tiny treelets 1–2 m tall with red berrylike cauliflorous fruits; one species is a liana and some are small trees of second growth; all have rather succulent weak-wooded trunks. Most commonly seen is the unmistakable cultivated and second-growth species *C. papaya;* an upland species of hybrid origin (*C. x heilbornii,* the babaco) is also frequently cultivated for its fruits.

C, E, P: papaya; E: papaya de mico, babaco (*C.x heilbornii*)

Jacaratia (6 spp.) — Medium to large trees with palmately compound leaves, often with spiny trunks and strongly resembling some Bombacaceae; however, this is the only genus in our area that combines milky latex (though only in young branches and leaves) and palmately compound leaves.

C: papayuelo; P: papaya caspi, papaya de venado

Caricaceae

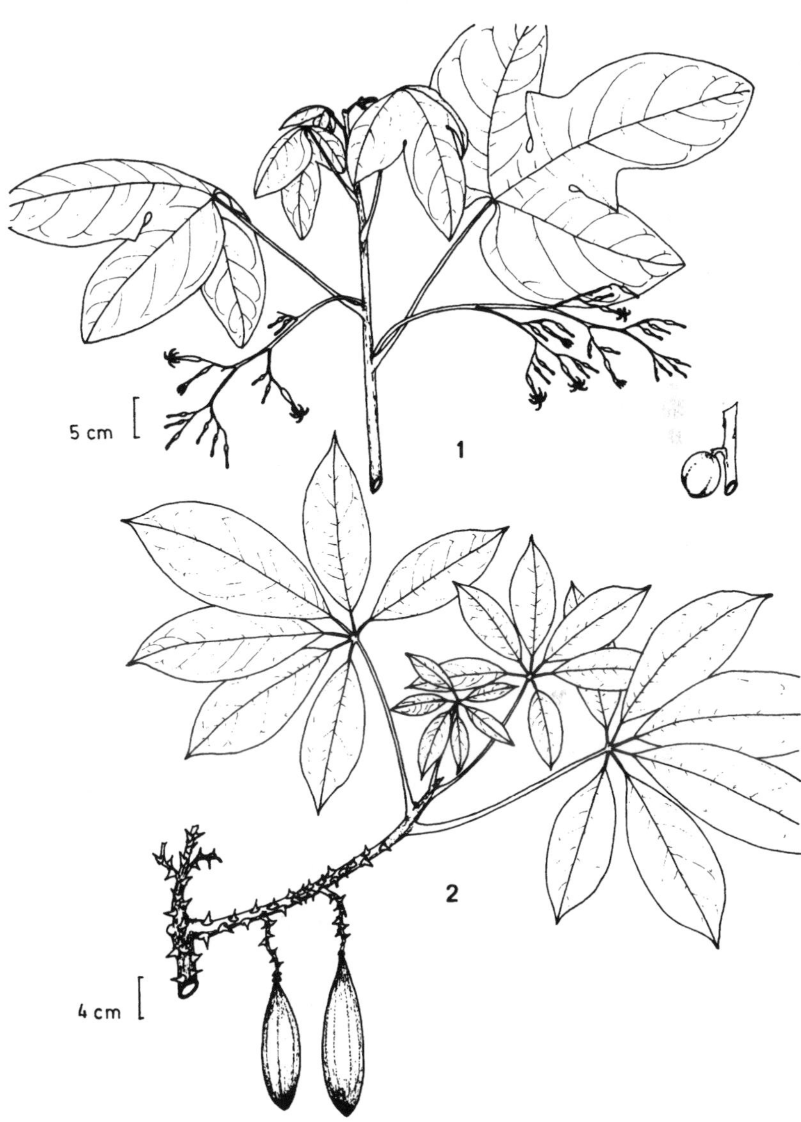

1 - *Carica*

2 - *Jacaratia*

Caryocaraceae

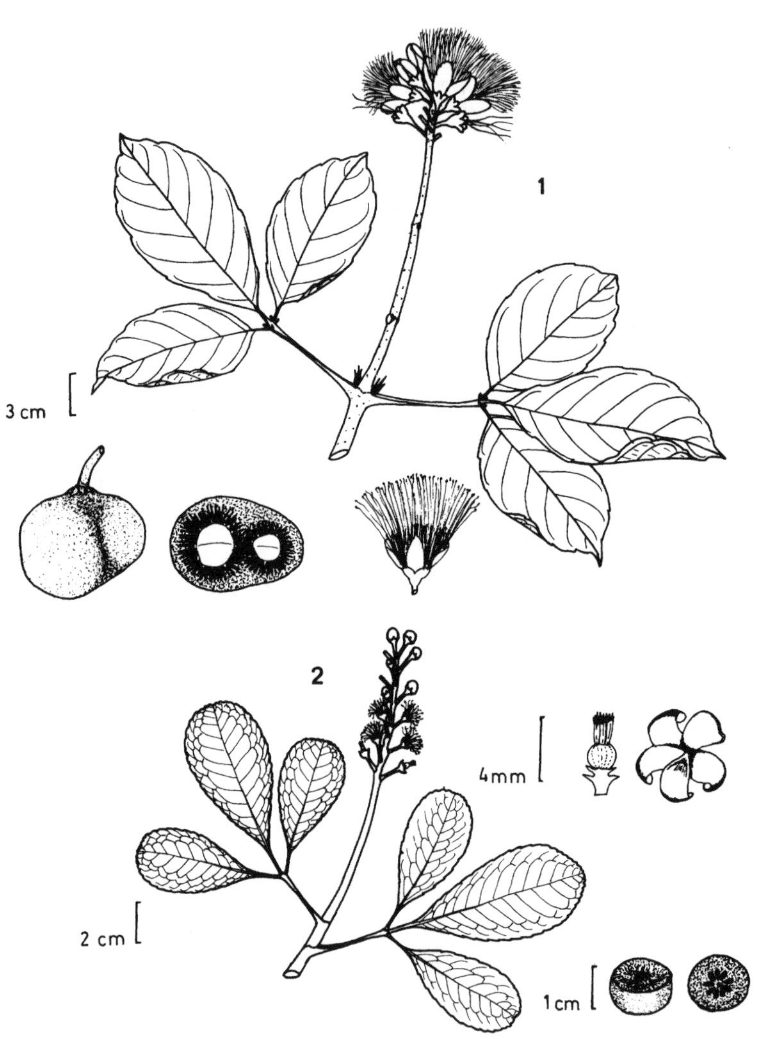

1 - *Caryocar*

2 - *Anthodiscus*

CARYOCARACEAE

Mostly large canopy and emergent trees, the trunks often exceeding 100 cm dbh, never with noticeable buttresses. The leaves always 3-foliolate. Leaves opposite in the largest genus (*Caryocar*) and very distinctive, often having pair of conspicuous, large, round glands at the petiole apex and usually serrate-margined. *Anthodiscus* has alternate leaves characterized by crenate margins and usually obtuse apex. Both genera have more or less conspicuous narrowly triangular, rubiac-like, terminal stipules that fall to leave scars; these interpetiolar in *Caryocar* and sometimes almost completely annular in *Anthodiscus*. In both genera the bark, at least of large trees, is always characteristically deeply ridged. The flowers are multistaminate with the petals falling at or soon after anthesis. Fruit 1–several-seeded and indehiscent, green at maturity, in *Caryocar* with a very characteristic spiny tuberculate endocarp.

Caryocar (15 spp.) — Canopy trees with opposite dark-drying leaves with usually serrate or serrulate margins; a pair of large glands at petiole apex distinguishes several species. Flowers large, bat-pollinated, with 5 large overlapping calyx lobes and five petals that fall shortly after anthesis. Fruit very distinctive in the spinose or tuberculate endocarp, the spines hidden inside the subfleshy mesocarp. Kernel often edible.

C: genené; P: almendro

Anthodiscus (9 spp.) — Canopy and emergent trees with alternate, crenate-margined, usually obtuse leaflets; vegetatively easy to confuse with *Allophylus* (Sapindaceae) but a much larger tree with prominent stipule scars and leaflets usually lacking the acute or acuminate apex of *Allophylus;* the petiolule bases tend to be slightly thickened and roughish unlike *Allophylus*. Flowers yellow, much smaller than in *Caryocar*, very distinctive with the petals apically fused to form a calypterate cap that falls as a unit at anthesis.

P: tahuarí, chontaquiro

CARYOPHYLLACEAE

A large mostly north temperate herbaceous family represented in our area mostly by Eurasian weeds. The simple entire leaves are nearly always opposite and usually have their bases connected across the swollen node. The petals of most taxa are distinctive in having notched or bilobed (sometimes even fringed) apices, and sometimes in being stalked, and the sepals of many genera (but only *Silene* and *Melandrium* in our area) are united into a distinctive tubular calyx. About 20 genera occur naturally or naturalized in

Caryophyllaceae

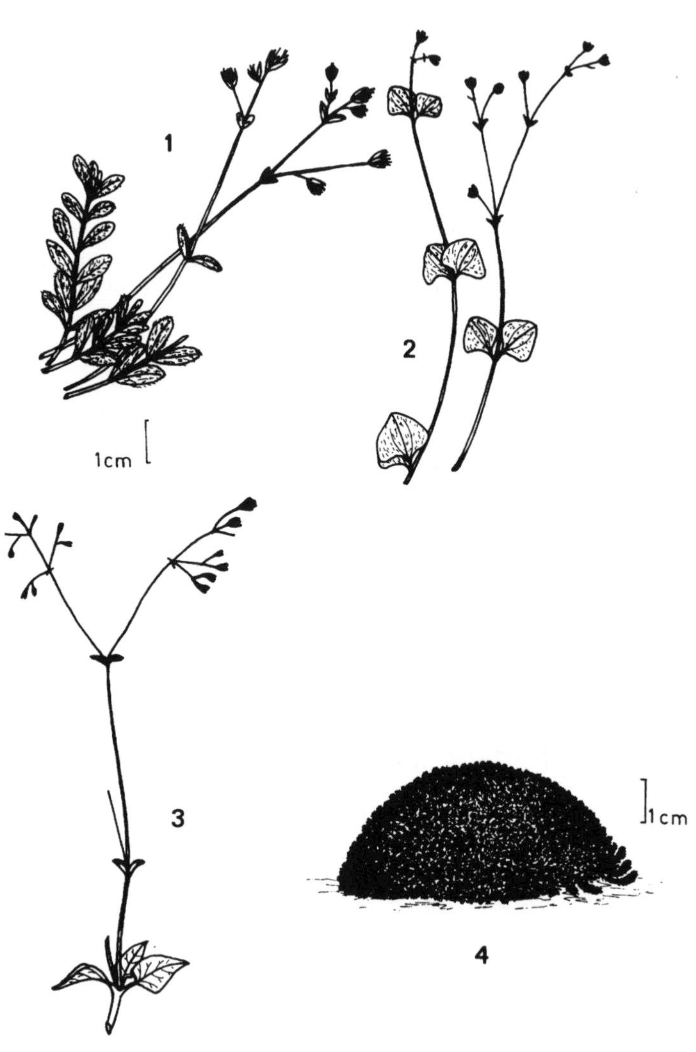

1 - *Cerastium* **2** - *Drymaria*

3 - *Stellaria* **4** - *Pycnophyllum*

Celastraceae

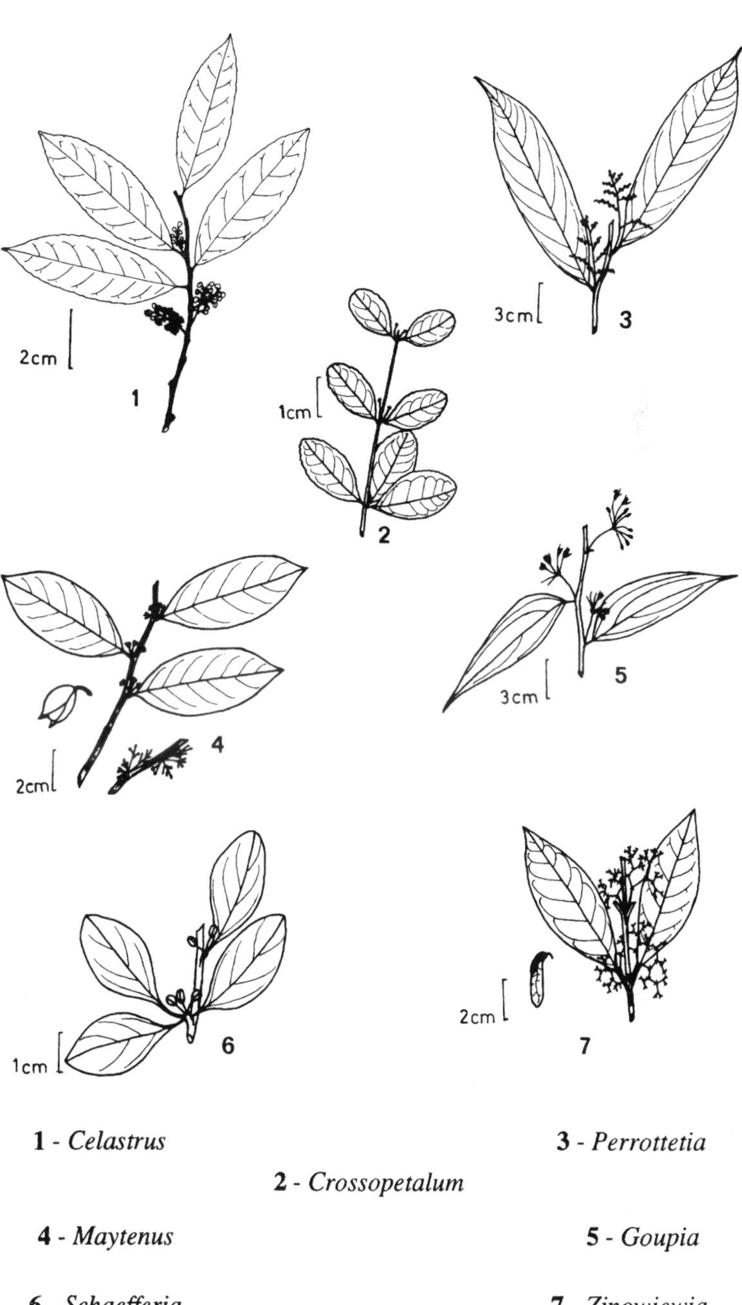

1 - *Celastrus*

2 - *Crossopetalum*

3 - *Perrottetia*

4 - *Maytenus*

5 - *Goupia*

6 - *Schaefferia*

7 - *Zinowiewia*

the Neotropics, with autochthonous *Pycnophyllum* (17 spp.) a distinctive Andean cushion-plant with minute densely bractlike leaves (technically characterized by noncapsular fruits and the absence of stipules) the most important of these, and with mostly north temperate *Cerastium, Arenaria, Paronychia,* and *Spergularia* having significant secondary radiations in the high Andes as does neotropical *Drymaria* (44 spp., plus few in Old World). Only *Stellaria* and *Drymaria* have a few weedy species reaching the tropical lowlands of our area where they are easily recognized by their prostrate habit and the small, thin, ovate leaves.

CELASTRACEAE

A family of trees and shrubs plus a single genus of cloud-forest liana. Much better represented in Central America and the Antilles than in South America, except for the very large genus *Maytenus.* In Central America most Celastraceae have opposite leaves but in South America nearly all are alternate. The only scandent genus, *Celastrus,* is characterized by alternate crenate-serrate leaves and usually strongly white-lenticellate twigs. We have only one opposite-leaved genus, *Zinowiewia,* occurring in upland Andean cloud forests where it is a large, sometimes emergent, tree (another opposite-leaved genus reaches coastal Venezuela and may be in Colombia). This is a distinctly nondescript family whose best recognition character is the twig, which is usually irregularly angled from the decurrent petioles and is often zigzag and/or greenish when fresh. The inflorescence is always axillary (or in part ramiflorous below the leaves), usually dichotomously branched or with the flowers fasciculate (except *Celastrus* [raceme], *Goupia* [irregular umbel], and *Perrottetia* [pyramidal panicle]), and the flowers are small and mostly greenish or greenish-cream. *Goupia,* an important timber tree, is a monotypic genus which is very atypical for Celastraceae but easy to recognize individually by the conspicuous linear stipules on the young twigs and the characteristic asymmetric-based blackish-drying leaves with strongly ascending secondary veins and prominulous parallel tertiary venation. The fruits are usually irregularly 2–3-parted capsules with red- or orange-arillate seeds but some taxa have winged samaras.

1. ALTERNATE LEAVES

1A. Lianas

Celastrus (5 spp., plus 25 in Old World) — Our species are cloud-forest lianas with crenate-serrate, oblong-elliptic, alternate leaves and the twigs usually prominently white-lenticellate. Inflorescence an axillary raceme of small whitish flowers.

1B. Trees

Goupia (1 sp.) — A late second-growth or light-gap tree, becoming an emergent, mostly on poor clay soil. Very characteristic in the coriaceous, asymmetric-based, oblong-ovate, blackish-drying glossy leaves with very strongly ascending lateral veins (often sub-3-veined from above base), also with characteristic finely parallel prominulous tertiary venation perpendicular to midvein. The margin is usually serrulate to distinctly serrate. The young branches have conspicuous linear stipules (these especially evident in juveniles) which leave noticeable scars on older branches. Inflorescence irregularly umbel-like, the tiny greenish to yellowish flowers with narrow valvate petals and borne on conspicuously different-length pedicels.

C: chaquiro

Perrottetia (20 spp., also in Old World) — Mostly upland Andean cloud-forest trees with narrowly elliptic to oblong-elliptic leaves, usually with the petioles red when fresh and more or less grooved above. Leaves of upland species all distinctly serrate or serrulate, but the lowland Chocó species (*P. distichophylla and P. sessiliflora*) entire. Twigs mostly zigzag, at least near apex, usually somewhat angled from the decurrent leaf bases, and often drying dark, the older twigs often with conspicuously pale small raised lenticels. Inflorescence very characteristic and different from other Celastraceae — pyramidal with a well-developed central axis and lateral branches more or less perpendicular to it, the flowers tiny and greenish. Fruits numerous, small (2–3 mm), round, reddish.

Maytenus (225 spp., mostly Old World) — Trees or shrubs with usually coriaceous, alternate leaves, and usually more or less decurrent petiole bases and striate-angled often greenish twigs. Upland species usually have serrate leaves (often finely so), but lowland taxa are mostly entire and very nondescript. Larger-leaved lowland species with entire leaves tend to have the secondary veins inconspicuous. A species common in coastal Ecuador dry-forest, *M. octogona,* is distinctive in roundish coriaceous leaves with sinuous margins. Inflorescence axillary or ramiflorous below the leaves, usually fasciculate, but sometimes slightly branched, the tiny flowers usually greenish, the fruit typically obovoid, splitting in half to reveal red-arillate seed.

P: chuchuhuasi (*M. krukovii*)

Gymnosporia (3 spp., plus ca. 100 Old World) — Understory trees of rich-soil forests at low and middle elevations. Essentially a segregate of *Maytenus* differing in dichotomously branching cymose inflorescence and tendency to irregularly 3-parted fruit. Leaves entire and larger and thinner than most *Maytenus,* olive or grayish-olive with lighter main veins below, the secondary veins strongly arcuately ascending. Tiny flowers greenish to white.

Schaefferia (16 spp.) — Shrubs or small trees of very dry forest, the small, obtuse, elliptic, olive-drying leaves alternate or clustered in alternate short-shoots. Leaves entire and with strongly ascending very inconspicuous secondary venation. The twigs very distinctive, green when fresh and strongly angled from the decurrent petiole bases, slender but usually +/- spiny apically. Flowers green, subsessile or pedicellate in axillary fascicles, the pedicels elongating at least below the small, 2-parted, ellipsoid orangish fruits.

2. Opposite Leaves

Crossopetalum (incl. *Myginda*) (30 spp.) — A predominantly Antillean genus with opposite leaves. One species, characterized by small, crenate, coriaceous leaves reaches coastal Venezuela and may also be in coastal Colombia. The inflorescences are opposite, axillary, and dichotomous or borne below the leaves and distinctly 3-branched.

Zinowiewia (9 spp.) — Opposite-leaved cloud-forest trees with coriaceous entire leaves and the petiole bases tending to be decurrent on the irregularly tetragonal twigs. A distinctive feature often present is the rather V-shaped pair of raised lines descending below some nodes. Twigs dark-drying and somewhat rough from the minutely raised inconspicuous lenticels. Fruit very characteristic, a narrow, curving, one-winged samara with the seed body along one side of the base.

There are many additional Central American and Antillean genera (a number of which reach Panama). In South America the only additional genus is *Plenckia,* a cerrado tree which could reach the Pampas del Heath, which is characterized by broadly ovate, long-petioled, finely serrate leaves and distinctive narrow, ashlike, wind-dispersed fruit.

Chenopodiaceae

Herbs and shrubs, mostly of dry and saline areas, sometimes leafless and with jointed succulent stems (*Salicornia*) or the leaves variously succulent. Leaves (when present) entire to irregularly lobed-serrate, often somewhat triangular and/or irregularly 3-veined, distinctively grayish in desert taxa. Flowers tiny and greenish. Fruit always a small one-seeded utricle.

1. Extremely Succulent Stems or Terete Leaves

Salicornia (5 spp., plus 30 Old World) — Succulent leafless seashore salt-marsh subshrub with jointed stems and the flowers sunk in stem joints.

Suaeda (5 spp., plus 100 Old World) — Seashore herb with succulent cylindric leaves.

Chenopodiaceae, Chloranthaceae, and Chrysobalanaceae (Shrub: *Chrysobalanus*)

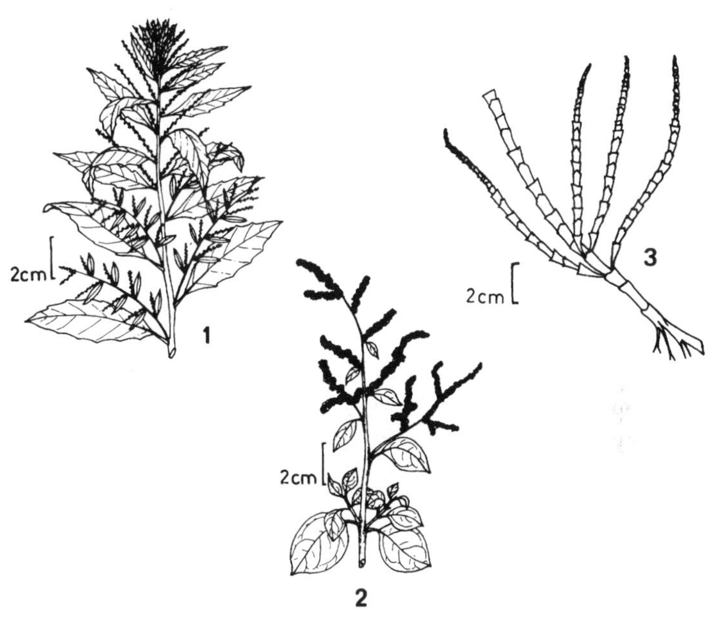

1 - *Chenopodium*

2 - *Atriplex*

3 - *Salicornia*

4 - *Hedyosum* (Chloranthaceae)

5 - *Chrysobalanus* (Chrysobalanaceae)

2. Normal Flat Leaves

Chenopodium (10 spp., plus 100 N. Am. and Old World) — Andean and dry-area herbs. Leaves triangular-ovate to oblanceolate, the margins usually irregularly jagged-toothed, always petiolate, sometimes grayish and sometimes aromatic. Inflorescence typically diffuse with tiny greenish flowers in scatttered tiny clusters.

Atriplex (6 spp., plus 200 n. temperate) — In our area, low loma shrubs with fleshy grayish leaves, the leaves tiny, sessile, ovate, and overlapping each other along branches, or petiolate and larger with irregularly crisped-serrulate margin. Inflorescence typically smaller and denser than in *Chenopodium;* technically differentiated by fruits enclosed by two hardened bracteoles.

Chloranthaceae

One of only two opposite-leaved Ranalean families (the other is Monimiaceae; a few species of Lauraceae are also opposite-leaved). Exceedingly easy to recognize vegetatively by the combination of opposite dentate leaves and primitive odor, and especially by the unique internode with an obvious enlarged sheath between the petioles, this usually further distinguished by the two stipulelike projections on its upper margin between the petiole bases. The plants are usually dioecious with male flowers in dense catkinlike spikes and the smal, naked green female flowers producing fleshy white or blue-black berrylike fruits. This is probably one of the most primitive of all angiosperm families with the ebracteate male inflorescences of some species sometimes interpreted as a strobiloid preflower and with fossil pollen well represented in the mid-Cretaceous.

Hedyosmum (40 spp., plus one in Hainan) — The only American genus, mostly in Andean cloud forests but one species reaches the Pacific coast in Chocó.

C: granizo; E: guayusa

Chrysobalanaceae

Leaves always simple, alternate, and entire. Stipules present at least near branch apices, though often caducous; petioles often with two lateral glands at extreme apex or extreme base of blade above. The inner bark is almost always reddish and distinctive in having a granular texture when rubbed between the fingers. All have single-seeded fruits, usually with hard endocarp, typically the seed having a rather bitter odor and characteristic pinkish or reddish

Chrysobalanaceae

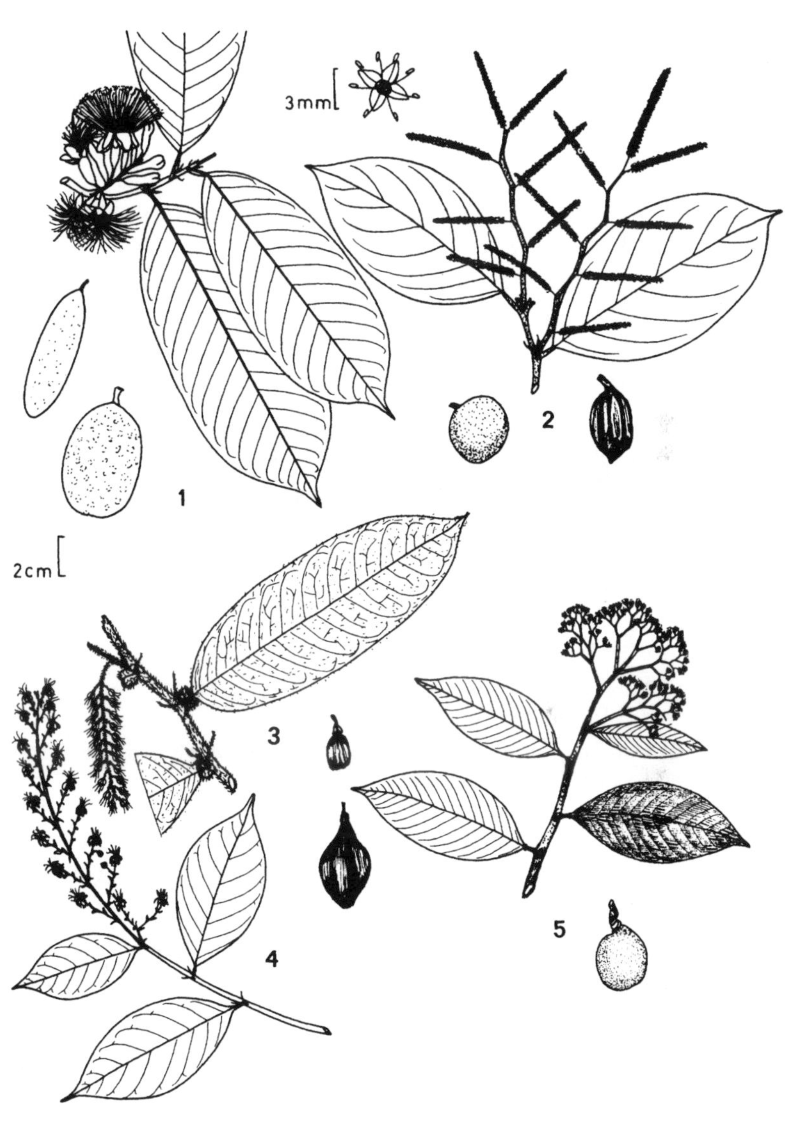

1 - *Couepia* **2** - *Licania*

3 - *Hirtella*

4 - *Hirtella* **5** - *Parinari*

color when cut. Tribal division is based on whether flowers are actinomorphic with ovary at base of the perigynous receptacle (=Chrysobalaneae: *Chrysobalanus* and *Licania*) or bilaterally symmetric with ovary near mouth of receptacle (= Hirtelleae: other genera). This also correlates with flower size — *Chrysobalanus* and *Licania* usually have tiny nondescript flowers, the other genera larger flowers, usually with elongated receptacles and/or conspicuously exserted stamens.

Chrysobalanus (2 spp., 1 also reaching Africa)— In our area a single species of coastal shrub with smallish rather obovate leaves and conspicuous white lenticels on twigs; inflorescence rather small and flat-tipped; outside of fruit fleshy (= cocoa plum), the endocarp longitudinally ridged.

C: icaco

Licania (153 spp., plus 1 in Asia)—Usually canopy trees; the largest genus of neotropical Chrysobalanaceae. Fruit often very large and with a rather dry leathery exocarp and very hard (frequently ridged) endocarp. Leaves often with a white or whitish surface and close-together, rigidly parallel, secondary veins. The white undersurface (when present) of minute scurfy trichomes. Inflorescence usually openly paniculate, with a well-developed central axis, the white sessile flowers tiny and regular. A number of species are characterized by prominent and characteristic leaf galls; the galls of one species were formerly made into capes by the Aguaruna. Vegetatively often difficult to recognize, even to family, especially when the stipules caducous. The leaf bases tend to be acute in *Licania* and truncate in *Couepia* but there are exceptions.

C: carbonero, chano; P: apacharama (white below), parinari (not white below), chullachasi caspi (*L. heteromorpha*)

Couepia (60 spp.) — Difficult to distinguish from *Licania* in sterile condition. Mostly small to medium sized trees. Flowers distinctive, typically large (for family, with very many (14–300) stamens; inflorescence paniculate, often few-flowered, the flowers always pedicellate (unlike *Licania*). Fruits large (typically larger than *Licania*) with thick nonridged endocarp. Leaves difficult to separate from *Licania* (i.e., usually with rigidly parallel secondary veins and whitish surface as in many *Licania*) but often with a loose caducous, appressed cobwebby pubescence when young; most *Couepia* species have truncate or subcordate leaf bases while *Licania* typically has acute leaf bases.

P: parinari

Parinari (16 spp.) — Similar to *Couepia* but ovary with 2 locules and ovules and fruit with a second aborted seed, the endocarp thick and irregularly rough-surfaced. Vegetatively easily recognized by the leaf undersurface with stomatal cavities filled with woolly pubescence (usually visible to naked eye as minute lighter flecks); the secondary veins are rigidly parallel and closer together than in other chrysobalanacs; the tertiary veins, perpendicular to the secondary, are also finely parallel.

Hirtella (94 spp.) — Mostly small to medium-sized trees; inflorescence usually racemose, the flowers typically rather small and inconspicuous except the usually rather few, conspicuously exserted stamens. Fruit usually oblong-elliptic, rather small and berrylike, the exocarp fleshy, the endocarp thin. Leaves nondescript in always lacking the distinctive rigidly parallel venation and whitish undersurface which characterize many other Chrysobalanaceae, often pubescent with rather lax trichomes at least along main veins below; petioles usually short and thick.

C: garrapato; E: macha, quinilla, coquito; P: añallo caspi (species with ants).

There are only 3 other neotropical genera of Chrysobalanaceae: *Exellodendron* (segregate from *Parinari:* smooth fruit surface and thin endocarp), *Maranthes* (Panama and Old World; also related to *Parinari* but many more stamens and a very different and characteristic coriaceous dark-drying glabrous leaf) and *Acioa* (like *Couepia* but fused stamen filaments).

CLETHRACEAE

A single genus of cloud-forest trees or shrubs characterized by alternate simple leaves, conspicuously whitish below from the stellate trichomes, the margin usually +/- remotely serrate or serrulate at least toward apex and the secondary veins usually rather straight and close together. Flowers small and white, always in long narrow spikelike racemes or these terminally clustered into a spicate-branched panicle; similar to Ericaceae except that sepals united only at extreme base and petals not at all; stamens 10 and opening by apical pores as in Ericaceae. Fruit a small round 3-locular pubescent capsule enclosed by the 5 persistent sepals, with numerous tiny, dustlike, usually winged seeds. Vegetatively, most similar to *Styrax* which has a similar indument but differs from most Clethraceae in uniformly entire leaves. Entire-leaved *Clethra* can be distinguished from *Styrax* by the reddish trichomes of the ferruginеous young twigs with longer laxer arms than in more or less appressed-tomentose *Styrax;* in addition the lower leaf surface is more strongly raised-reticulate than in area (but not cerrado) *Styrax* species and the characteristic *Styrax* scattering of rufous trichomes over the dense, white, leaf-undersurface indumentum is absent.

Clethra (38 spp., plus 26 Old World)

Clethraceae, Cochlospermaceae, and Columelliaceae

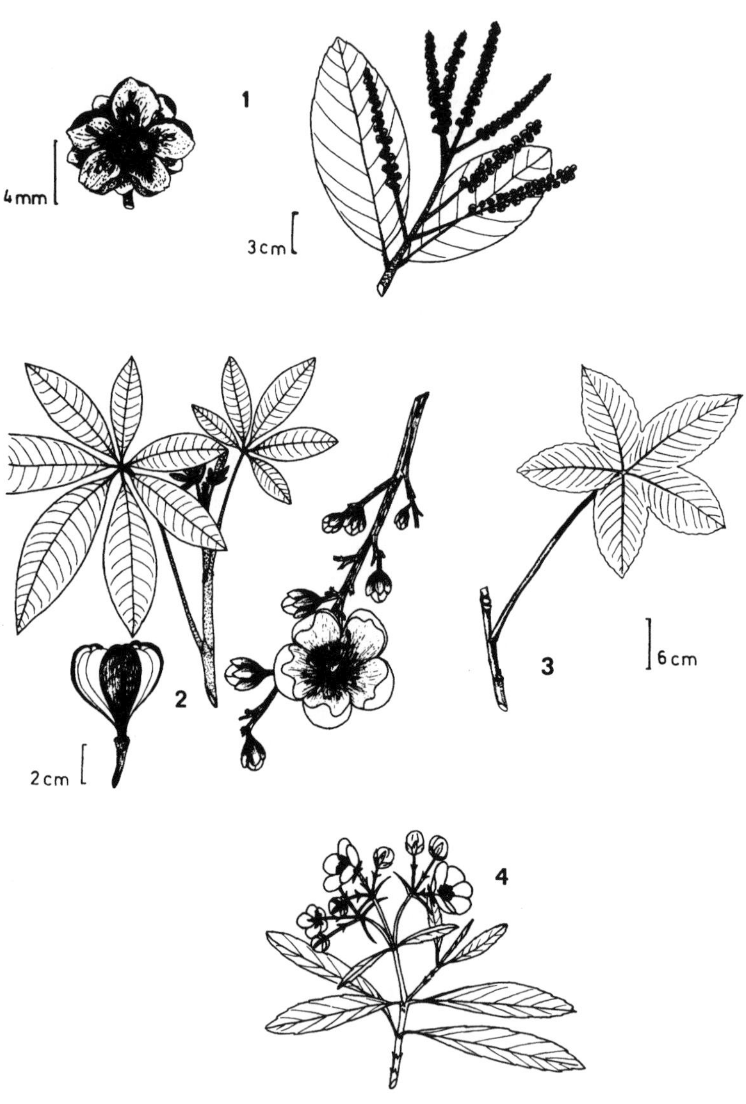

1 - *Clethra* (Clethraceae)

2 - *Cochlospermum* (*C. orinocense*)

3 - *Cochlospermum* (*C. vitifolium*)

4 - *Columellia* (Columelliaceae)

Cochlospermaceae

Small to large soft-wooded trees, characterized by palmately lobed (*C. vitifolium*) or divided (*C. orinocense*) alternate leaves, large yellow multistaminate flowers, and a characteristic obovoid or ellipsoid fruit with valves thinner than in bombacacs but the seeds embedded in kapok just as in that family. Vegetatively, the serrate margins of the palmately lobed species readily distinguish it from similarly lobed tree species of other families (although there are herbaceous and shrubby species with serrate palmately lobed leaves). The compound-leaved species looks very like a bombacac. but has the leaflets generally more nearly sessile (or at least lacking a well-defined petiolule).

Cochlospermum (4 spp., plus 8 in Old World)
E: bototillo, poro poro; P: huimba sacha

Columelliaceae

Andean shrubs with small opposite leaves, opposite petioles strongly connected across node by a line or actual flap of tissue (cf., *Hydrangea*). Leaves distinctly more or less grayish-sericeous, at least below; always apiculate or spine-tipped (sometimes with several thickened subterminal teeth). Flowers yellow, the petals basally fused. Now thought to be closely related to Saxifragaceae alliance despite the fused petal bases.

Columellia (4 spp.)

Combretaceae

Characterized by uniformly simple and entire leaves and tree or liana habit (except the rather shrubby beach/mangrove species *Conocarpus erecta*). The two habit types have little in common vegetatively. Neotropical tree combretacs all have alternate leaves (except the distinctive mangrove genus *Laguncularia*), these typically clustered together at the tips of thick short-shoot branchlets or branch-tips; the leaves frequently have a distinctive pair of glands on the upper petiole surface. Petiolar glands are usually lacking in *Terminalia* but the leaves are then either clearly clustered (and form, with the branches, the well-known *Terminalia* or "pagoda" growth form) or the bark is very smooth and white (cf., Myrtaceae). Rather narrow, sometimes greatly accentuated buttresses are usually present at the trunk base. The inner bark of Combretaceae trees is always yellow, oxidizing tan.

Liana genera have opposite leaves (except a few *Combretum* species) and are mostly defined vegetatively by lack of the characteristic features of similar families. They never have latex or lenticels and both twigs and stems have a hollow mucilage-filled canal(s) at or near their centers; the tertiary leaf venation is often parallel and close together. Woody branchletlike spines are present on the stems of some species (including all of the exceptional alternate-leaved ones). Combretaceae lianas are most easily confused vegetatively with hippocrats and malpighs; most hippocrats differ in serrate or serrulate leaf margins (always entire in combretacs) and combretac leaves never dry the characteristic olive of most entire-leaved hippocrats; combretacs never have the characteristic T-shaped hairs (at least on the petiole) or resultant sericeous aspect of most malpighs, nor do they ever have the raised twig lenticels or petiolar glands or enations that are found in many malpighs. Combretaceae liana bark, even on the twigs, tends to be thick, fibrous, and somewhat exfoliating (unlike Hippocrateaceae and many Malpighiaceae which are relatively smooth-barked), while the stems never fragment into separate cables (unlike many fibrous-barked Malpighiaceae).

The inflorescences of combretacs are usually spicate or narrowly racemose except *Ramatuela* and *Conocarpus* with dense capitate flower clusters; the anthers are almost always exserted and the ovary is inferior; petals are usually absent or reduced with the calyx and stamens providing the main visual attractant for pollinators.

1. Trees

1A. Opposite leaves

Laguncularia (1 sp., also in Africa) — One of the three most fundamental components of neotropical mangroves; unique among area combretac trees in having opposite leaves. The petiolar glands (unique among mangroves) are well-developed and said to excrete salt; pneumatophores, slightly thicker-tipped than those of *Avicennia,* are a unique feature among area combretacs.

white mangrove; C, E: mangle blanco; C: mangle bobo

1B. Alternate (usually strongly clustered) leaves

Terminalia (34 spp., plus 200 Old World) — The main tree genus of Combretaceae, often characterized by the pagoda-like whorls of horizontal lateral branches, each branchlet with the tip turned up and having a terminal cluster of leaves at its apex. The leaves usually lack petiolar glands. Narrow buttresses are usually present and the bark usually has rather fine vertical ridges or is very smooth and white (cf., many myrtacs) The fruit is characteristically 2-winged (unique) but the wings are lost in the water-dispersed

Combretaceae

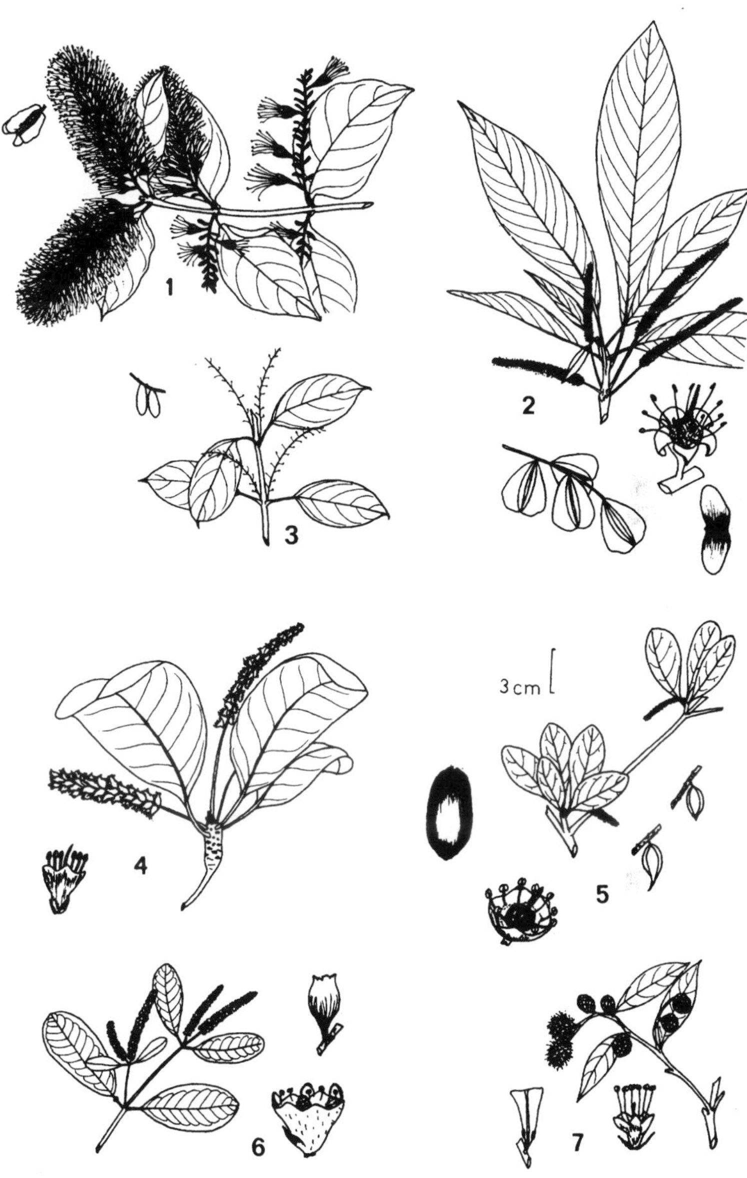

1 - *Combretum* **2** - *Terminalia*

3 - *Thiloa*

4 - *Bucida* **5** - *Buchenavia*

6 - *Laguncularia* **7** - *Conocarpus*

beach tree *T. catappa* (sea almond). The main technical character for distinguishing flowering *Terminalia* from *Buchenavia* is versatile anthers.

C: guayabillo; E: yuyum, almendro (*T. catappa*); P: yacushapana

Buchenavia (24 spp.) — Canopy trees, especially abundant in seasonally inundated forests; branching similar to *Terminalia* but the petioles usually with more or less conspicuous glands and the fruit an oblong-ellipsoid drupe (versus dorsoventrally flattened in *Terminalia catappa* which also lacks wings); the full length of the anthers is attached to the filament unlike *Terminalia.*

P: yacushapana

Bucida (3 spp.) — A Caribbean coastal genus of Central America and the Antilles, possibly not reaching our area. Vegetatively like *Terminalia* and *Buchenavia* but the pagoda branching more extreme with the erect branchlet tips typically longer and more enlarged and the clusters of small leaves widely separated. Fruits distinctive in being small and urn-shaped with an expanded apical rim.

Ramatuela (3 spp.) — Small trees of black-water-inundated forest, entering our area only marginally in upper Rio Negro drainage. The clustered leaves are sclerophyllous, round-tipped, and more or less sericeous (at least on midvein below). Fruits 4–5-winged and *Combretum*-like but in pedunculate capitate clusters; flowers small and white and also in dense capitate clusters.

Conocarpus (1 sp., also in Africa along with second sp.) — Shrub or small tree of mangrove fringes or beaches. Leaves smaller than in other mangrove trees and also characterized by pair of glands at apex of petiole and domatia in axils of secondary veins below. Some forms very characteristic in silvery sericeous leaves. The distinctive inflorescence is a compact globose head ca. 5 mm in diameter.

C: mangle zaragoza

2. Lianas

Combretum (31 spp., plus ca. 200 Old World) — One of the more important neotropical liana genera (sometimes merely shrubby in dry areas). Easy to tell from most other Combretaceae by its climbing habit and the usually opposite leaves. The few alternate-leaved species have spiny stems. Lack of lenticels, fibrous bark, more or less hollow twigs, and stems with central mucilage canals are other useful recognition characters as are the typically parallel tertiary leaf venation and brownish petioles and young twigs. The 4(–5)-winged fruits are very characteristic as are the dense bright orange (or having red calyces with yellow-green stamens) flowering spikes or racemes of the many species pollinated by perching birds (and sometimes primates); some species have small white flowers in more open inflorescences, and in some the fruit wings are thicker and variously reduced for water dispersal.

P: escobilla (*C. llewelynii*)

Thiloa (3 spp.) — In our area a liana of seasonally inundated riversides. Vegetatively and in flower similar to *Combretum* from which it differs in having only 4 stamens (vs. 8 or more); the fruit of the common species in our area is water-dispersed and unwinged (though slightly 4-angled), but other (extralimital) species have fruits like *Combretum*.

COMPOSITAE

A very large family, best represented in upland areas where it is usually the most speciose family above timberline and in dry montane scrub. Most comps are herbs but 23 neotropical genera contain climbers (especially *Mikania*, one of the largest of all scandent genera), many genera contain shrubs or subshrubs, and a few genera (scattered through most of the tribes) contain trees to 4 m or more tall (*Vernonia*, [*Critoniopsis*], *Pollalesta* [Vernonieae]; *Koanophyllon*, *Critonia* [Eupatorieae]; *Llerasia* [Astereae]; *Tessaria* [Inuleae], *Espeletia* [and segregates], *Montanoa*, *Polymnia*, *Verbesina*, *Clibadium* [Heliantheae]; *Ferreyranthus* [Liabeae], *Paragynoxys*, *Gynoxys*, *Senecio* [Senecioneae]; *Barnadesia*, *Dasyphyllum*, *Gochnatia*, *Gongylolepis* [Mutisieae]). (One Chilean *Dasyphyllum* is a giant tree 30 m or more tall.) Most woody members of the family are vegetatively recognizable by either having blackish or dark inner bark (woody members of Vernonieae, some Eupatorieae [and Heliantheae?]) or opposite rather pungently aromatic leaves 3-veined *above* base (most taxa) (these with milky latex in Liabeae), or narrow distinctly grayish leaves with stellate trichomes (*Tessaria*), or spines either in leaf axils or on branches or as spiny leaf apices or teeth (most Mutisieae), or by pinnately compound leaves terminating in tendril (*Mutisia*). The woody Senecioneae (scandent *Paracalia*, *Pseudogynoxys*; arborescent *Paragynoxys*, *Gynoxys*, *Senecio*) have alternate leaves and generally lack distinguishing characters to indicate that they are Compositae.

Because this is such a large complex family with technical floral characters (the peculiar "flower" is actually a head of small close-together tubular flowers surrounded by involucral bracts) so important in its taxonomy, this treatment is divided by tribes. Several of the tribes are in part differentiated on vegetative characters (leaves grayish-pubescent with cobwebby trichomes in Inuleae; leaves opposite in

Eupatorieae, Tageteae, most Heliantheae, Liabeae; latex present in Lactuceae, Liabeae; leaves strongly aromatic and usually gland-dotted and finely dissected in Tageteae and Anthemideae; leaves and stem spiny in Cardueae, some Mutiseae), but many of the alternate-leaved tribes (Vernonieae, Astereae, Senecioneae, Mutisieae, a few Heliantheae) are not very easy to distinguish (or to recognize as comps) in the absence of flowers.

A few tips on some of the more distinctive genera:

Sessile rosette paramo and puna plants — Mostly *Werneria* (fused involucral bracts) and *Hypochoeris* (milky latex).

Paramo and subparamo plants with large leaves and pachycaul growth form — *Paragynoxys, Espeletia.*

Resinous leaves — *Baccharis, Ageratina, Flourensia.*

Leaves opposite and densely white-pubescent below — *Liabum, Munnozia* and relatives; *Gynoxys.*

Epiphytes — *Tuberostylis* (in mangroves), *Neomirandea, Gongrostylus, Pentacalia.*

Compound leaves — *Mutisia* (with terminal tendril), *Bidens, Cosmos.* (2-bristled exozoochorous achene), *Hidalgoa* (vine), *Tagetes, Porophyllum* (glandular pellucid), *Ambrosia* (wind-pollinated), one *Mikania* sp. (vine).

High-Andean shrubs with scalelike leaves — *Loricaria, Oligandra, Hinterhubera, Mniodes.*

Fleshy fruits — *Wulffia* (vine), *Clibadium* (mostly shrubs), *Ichthyothere, Milleria.*

Lianas — *Condylidium, Mikania, Gongrostylus, Tuberostylis, Bartlettina, Critonia* (Eupatorieae); *Vernonia, Piptocarpha* (Vernonieae); *Baccharis trinervis* (Astereae); *Wulffia, Hidalgoa,* few *Ichthyothere, Salmea* (Heliantheae); *Munnozia, Oligactis* (Liabeae); *Paracalia, Pseudogynoxys, Senecio* (Senecioneae); *Lycoseris, Mutisia, Jungia* (Mutisieae).

For those willing to delve into the mysteries of Compositae taxonomy, Mike Dillon's Key to Tribes for the *Flora of Peru* is reproduced here. The esoteric style and anther characters are well illustrated in the *Flora of Panama* (reproduced in Fig. 96).

Compositae
(Stigmas and Head Details)

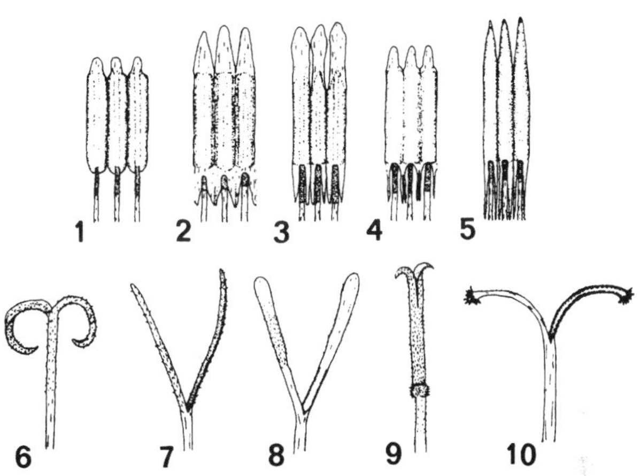

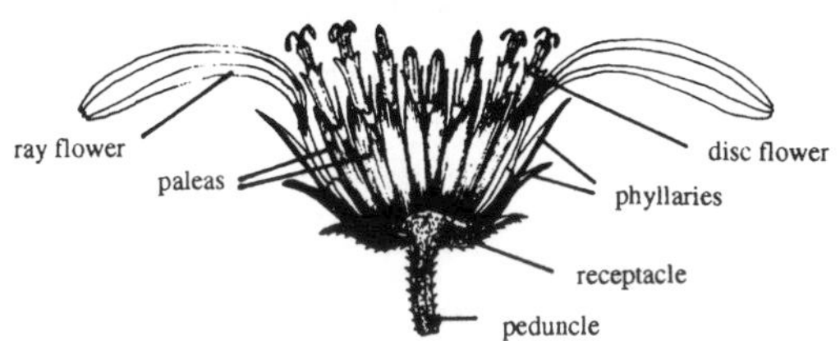

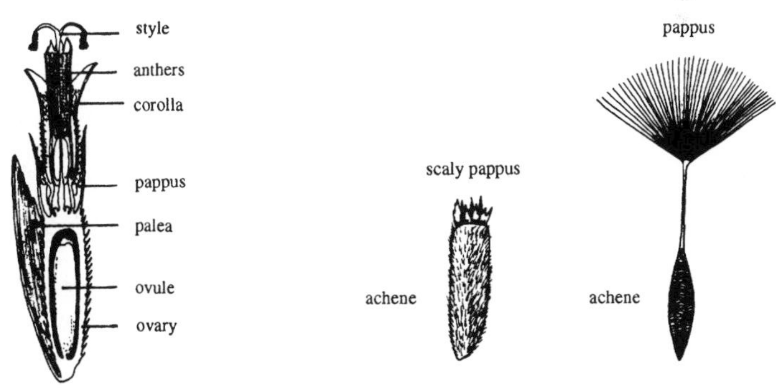

1 - *Eupatorieae* **2** - *Lactuceae* **3** - *Inuleae* **4** - *Inuleae* **5** - *Mutisieae*

6 - *Lactuceae* **8** - *Eupatorieae* **10** - *Senecioneae*
7 - *Vernonieae* **9** - *Cardueae*

KEY TO TRIBES:

1. Heads with staminate or perfect florets towards the middle, the corollas tubular or bilabiate; sometimes with pistillate florets towards the outside; sap usually not milky.
  2. Anther tips with sterile, tonguelike, often hyaline appendages.
    3. Florets all alike, perfect, corollas tubular, not yellow; anthers not tailed, receptacle usually naked.
      4. Leaves alternate; style branches slender, terete, hairy all over, the style shaft apically hairy; anthers auricled (tailed in *Piptocarpha*); hairs often 1-celled..**Vernonieae**
      4. Leaves mostly opposite (except sometimes in the region of inflorescence); style branches gradually expanded near the tips, papillose or short-hairy, the shaft often glabrous; anthers obtuse or rounded; hairs multicellular, often moniliform......................
      ....**Eupatorieae**
    3. Florets often not all alike, corollas often yellow; anthers sometimes tailed; receptacle naked or with paleae.
      5. Leaves mostly not spiny; involucral bracts not spiny; anthers tailed or not; style shaft without an apical ring.
        6. Leaves alternate; style branches flattened-fusiform, sometimes apically appendaged or rounded; anthers tailed or not; receptacle mostly naked; pappus mostly bristles.
          7. Anthers obtuse; style branches often appendaged; achene often compressed; hairs multicellular................ **Astereae**
          7. Anthers tailed; style branches rounded; achene plump; hairs arachnoid.. **Inuleae**
        6. Leaves alternate or opposite; style branches flattened-fusiform, sometimes apically appendaged; anthers not tailed; receptacle with paleae or naked; pappus of bristles, awns or scales.
          8. Pappus of awns, bristles or scales; style branches often appendaged.
            9. Involucre without transparent margins; leaves mostly opposite, often 3-nerved from base or trifoliolate.
              10. Receptacle naked; involucral bracts equal, mostly valvate (biseriate in *Schizotrichia*), with pronounced pellucid glands; leaves glabrous to puberulous, typically bearing conspicuous pellucid secretory cavities or glands and strongly aromatic................**Tageteae**
              10. Receptacle with paleae, squamellae, bristles or merely deeply alveolate (rarely truely naked); involucral bracts unequal, overlapping, 2–many-seriate, lacking pellucid glands; leaves variously pubescent or glabrous, pellucid glands absent.

11. Receptacle with costate paleae enfolding the achenes; achenes usually compressed; pappus of scales, awns or rarely of numerous, strigose bristles; leaves opposite or alternate, mostly eglandular; hairs often verrucose.......... **Heliantheae**
11. Receptacle deeply alveolate, with the margins of the alveolae prolonged into stiff mostly subulate awns, squamellae, or bristles, rarely with true paleae or naked (*Cacosmia, Philoglossa*); achenes usually cylindric to turbinate (2–5–10-angled; pappus generally biseriate, the inner series of bristles and the outer of bristles or squamellae, rarely absent (*Cacosmia*); leaves opposite or whorled in a basal rosette, usually tomentose below...**Liabeae**

9. Involucre with hyaline, transparent margins; leaves alternate, with strong midrib, not 3-veined.
12. Leaves usually dissected, often aromatic; style branches in disc and ray florets truncate, penicillate; pappus paleaceous, coroniform, or absent..**Anthemideae**
12. Leaves entire, not aromatic; style branches of ray florets filiform, glabrous, and of the disc florets undivided; pappus lacking................... **Calenduleae**

8. Pappus of soft, silky, hairlike bristles; style branches not appendaged... **Senecioneae**

5. Leaves and involucral bracts spiny; anthers tailed; style shaft with an apical ring.. **Cardueae**

2. Anther tips sterile, but not differentiated into hyaline, tonguelike appendages; anthers mostly tailed...................................... **Mutisieae**

1. Heads with only perfect florets, the corollas ligulate, 5-denticulate; sap milky..**Lactuceae**

EUPATORIEAE

Characterized by the opposite leaves and the usually flat-topped corymbiform panicle with rather "fuzzy" flowers (from the elongate style branches). The great majority of these genera (except those with 4 flowers/head (*Stevia, Mikania)* and those with reduced pappus (*Ageratum, Piqueria, Adenostemma, Sciadocephala)* have traditionally been included in a *sensu lato Eupatorium.* Especially noteworthy is *Polyanthina* which has alternate leaves and looks more like Astereae.

Compositae - A

1 - *Vernonia* **2** - *Pollalesta*
3 - *Struchium* **4** - *Trichospira*
5 - *Wedelia* **6** - *Bidens* **7** - *Tessaria*
8 - *Wulffia* **9** - *Baccharis* **10** - *Baccharis*

Compositae - B

1 - *Clibadium* **2** - *Ambrosia* **3** - *Elephantopus*

4 - *Eclipta* **5** - *Piptocarpha*

6 - *Erechites* **7** - *Pseudelephantopus*

8 - *Mikania* **9** - *Ageratum*

1. Vines

Mikania (300 spp.) — A huge climbing genus, the largest neotropical liana genus. Characterized by 4 flowers per head and the typical opposite leaves 3-veined above base and more or less aromatic.

E: guaco

Gongrostylus (1 sp.) — Cloud-forest liana, the leaves glabrous, short-petioled, remotely serrate, 3-veined from base; this and the rest of the vine genera differ from *Mikania* in >4 flowers per head; this is the only Eupatorieae vine with persistent involucre, slender style branches, and hirsute style base (= *Ayapana* group).

Condylidium (2 spp.) — Mostly weedy vine (sometimes suberect) with smallish rhombic-ovate leaves distinctive in being glandular below and having winged petiole. Heads very reduced, the flowers small, the inflorescence rather diffuse.

(*Bartlettina*) — A few species are viny; similar to *Critonia* but receptacles hirsute (= *Hebeclinium* group).

(*Critonia*) — A few species viny; closely related to *Koanophyllon* and similarly characterized by strongly imbricate, partly deciduous involucral bracts and flat broadened style tips.

2. Epiphytes

Tuberostylis (2 spp.) — Small-leaved creeping mangrove epiphytes of Pacific coast.

Neomirandea (24 spp.) — Mostly Central American; basically same as *Critonia* (i.e., strongly imbricate, partly deciduous, involucral bracts and flat broadened style tips) but usually epiphytic.

3. Trees and Shrubs

Austroeupatorium (11 spp.) — Mostly subshrubs, but our common species is shrub or small second-growth tree to 4 m tall, characterized by long narrow strongly 3-veined pubescent leaves and small white flowers in rather dense inflorescence. Very close to *Eupatorium, sensu stricto,* and with similarly imbricate involucral bracts and pubescent style bases.

Critonia (33 spp.) — Mostly weedy shrubs or small trees (a few species viny or herbaceous). Mostly glabrous with large serrate leaves and conspicuously jointed, longitudinally striate stems; 8–12 florets per head. (This and the next six genera belong to *Critonia* group, characterized by strongly imbricate, partly deciduous, involucral bracts and flat broadened style tips.)

Koanophyllon (109 spp.) — In our area, mostly shrubs or small trees of seasonally deciduous lowland forest, characterized by glabrous, strongly 3-veined, rather jaggedly serrate, elliptic leaves; related to *Critonia* with flat broadened style tips and strongly imbricate partly deciduous involucral bracts.

Aristeguietia (20 spp.) — Paramo shrubs with oblong perfectly pinnate leaves, usually white-pubescent below and/or bullate and/or finely serrate.

Asplundianthus (9 spp.) — Andean shrubs (rarely viny) with flowers lavender.

Badilloa (9 spp.) — Shrubs similar to *Aristeguietia* but the secondary veins more ascending.

Cronquistianthus (16 spp.) — Nondescript shrubs of dry inter-Andean valleys.

Ophryosporus (38 spp.) — Nondescript Andean shrubs and herbs.

Chromolaena (130 spp.) — Most of ours becoming coarsely woody "herbs", distinctive in long narrow heads and the lavender flower color. (Imbricate completely caducous bracts and reduction of pappus = *Praxelis* group).

Hebeclinium (18 spp.) — Coarse shrubby secondary-growth herbs, the commonest species distinctive in the large very broadly ovate, serrate, irregularly cordate leaves.

Bartlettina (20 spp.) — A mixture of Andean herbs, shrubs and vines; close to *Hebeclinium,* differing in larger heads with >20 florets and the receptacle >2 mm across; differs from *Critonia* in hirsute receptacle.

The next three genera (Alomiinae) are similar to *Gongrostylis* and relatives but have the corolla very constricted apically and broad style branches.

Crossothamnus (1 sp.) — Shrub.

Helogyne (12 spp.) — Shrub of lomas and dry areas, with linear viscous leaves and +/- solitary heads.

Condylopodium (4 spp.) — Shrubs of northern Andean area with distinctively thin, largish leaves.

Stevia (150–200 spp.) — Subshrubs and herbs of dry inter-Andean Valleys; characterized by long narrow head with only 4 flowers; (pappus reduced to awns or bristles = *Piqueria* group).

Ageratina (230 spp., incl. N. Am.) — Only adventive in our region. Mostly subparamo and montane shrubs, typically with resinous crenate leaves (cf., *Baccharis* but opposite). Unusual in involucral bracts not imbricate and notably long slender corolla.

4. Herbs

There are many additional herbaceous genera of Eupatorieae. Most noteworthy are *Adenostemma* (20 spp., incl. Old World) and *Sciadocephala* (4 spp.) on account of their reduced pappus of 3–5 glandular knobs with enlarged apices. Other uniformly herbaceous *Eupatorium* segregates in our area include *Heterocondylus* (12 spp.), *Isocarpha* (5 spp.), and *Ayapanopsis* (12 spp.) (relatives of *Gongrostylus*), *Critoniella* (15 spp.) (herbaceous relative of *Critonia*), and *Fleishmannia* (only adventive). *Polyanthina* (1 sp.) is a common coarse Andean herb to 1.5 m tall with whitish flowers, which is especially noteworthy for the alternate deltoid leaves, highly unusual in the tribe.

Vernonieae

Characterized by alternate leaves, mostly entire or crenate, completely pinnate-veined and +/- oblong to elliptic; when woody the leaves usually white- to gray-pubescent below (except some *Vernonia*, these mostly viney). Woody species always with distinctive blackish layer in inner bark. Two genera contain trees (*Pollalesta, Vernonia*) and two lianas (*Vernonia, Piptocarpha*). This tribe is noteworthy for the tendency to have reduced heads clustered together into secondary heads. The style branches are slender, terete and hairy, the anthers auricled.

1. Trees, Shrubs, and Lianas

Vernonia (1000 spp., incl. Old World) — The large core genus of Vernonieae, including shrubs, trees (especially *V. patens*), and lianas, characterized by the inflorescence tending to have scorpioid branches and the flowers usually magenta (or white).

C: aliso

Piptocarpha (50 spp.) — Our species lianas, mostly of wet lowland forest, the leaves always densely grayish-pubescent with stellate or lepidote trichomes below. Could be confused with *Solanum* vegetatively but has black inner bark.

Pollalesta (24 spp.) — Medium-sized second-growth and savanna trees with corymbose panicles. Leaves densely gray-pubescent below.

P: yanavara, ocuera negra

2. Herbs

The rest of our Vernonieae are herbs, mostly rather weedy and typically with reduced capitula in dense secondary clusters. Most noteworthy are *Elephantopus* (32 spp, incl. Old World; reduced heads in clusters subtended by leaflike bracts), *Pseudelephantopus* (2 spp., inflorescence more evenly spicate and the reduced heads not clustered), *Rolandra* (1 sp., leaves white below and dense axillary inflorescence), *Struchium* (1 sp., seasonally inundated forest, the thin membranaceous leaves crenate-serrate and dense heads sessile and axillary), and *Pacourina* (1 sp., a succulent shrubby swamp herb with spiny-toothed leaves and flowers clustered into very large sessile leaf-opposed heads ca. 3 cm across).

Astereae

This large tribe is characterized by alternate leaves, obtuse non-tailed anthers, a naked receptacle, and usually appendaged style branches. Contains few real trees in our area (only *Llerasia*) but includes a number of shrubs and a few scandent species (mostly *Baccharis*). The flowers are almost never yellow.

1. Trees, shrubs, and Lianas

Llerasia (11 spp.) — The only Astereae genus to become truly arborescent in our area (occasionally to 5 m tall) but usually shrubby (occasionally scandent). Characteristic in the leaves densely whitish-woolly-tomentose below, differing from *Diplostephium* in the longer (usually ca. 2 cm long) narrower few-flowered heads.

Baccharis (400 spp.) — An important Andean genus, mostly of shrubs (*B. brachylaenoides* to 3 m), mostly with small resinous, coriaceous leaves, these usually somewhat 3-nerved, entire or with rather few remote teeth; several species lack leaves entirely and have segmented phyllodial resinous stems. The only lowland species (*B. trinervis)* has resinous entire strongly 3-veined leaves and is usually +/- scandent. Dioecious, the male and female heads usually of different sizes; lacking ray-flowers.

E: alcotan

Archibaccharis (20 spp.) — Dioecious herbs and subshrubs differing from *Baccharis* in leaves less coriaceous, pinnately veined, and serrate.

Diplostephium (90 spp.) — A large genus of high-Andean puna and paramo shrubs with very distinctive leaves, typically ericoid and +/- linear;

stems and leaves below always densely white-woolly-tomentose. Ray-flowers if present narrow, usually inconspicuous, white or purple.

Hinterhubera (8 spp.) — Northern Andes paramo shrub or subshrub with very characteristic reduced, scalelike, densely imbricate leaves.

2. Herbs

At least eight additional exclusively herbaceous genera of Astereae occur in our region, mostly in the high Andes, only *Conyza* and *Egletes* occurring in the tropical lowlands.

Conyza (50 spp., incl. n. temperate and Old World) — Herbs, mostly extratropical but a few species are lowland tropical weeds. Usually with more or less linear leaves; like *Aster* but lacks ray-flowers.

Egletes (12 spp.) — Viscous herbaceous lowland weeds, with characteristic deeply and irregularly dentate (often +/- laciniate) leaves with winged petioles and inconspicuous white flowers.

Oritrophium (15 spp.) — Tussock-forming high-Andean rosette plant with linear leaves, densely white-pubescent below and large solitary aster flower with white or purple rays.

Lagenifera (15 spp., mostly Australia and New Zealand) — Mostly south temperate; ours tiny, high-Andean, subrosette herbs with coarsely serrate leaves.

Noticastrum (12 spp.) — High-Andean aster-segregate with large +/- solitary purple flowers and oblanceolate glandular-sticky leaves in basal rosette.

Grindelia (60 spp.) — Mostly amphitropical, with a single species reaching the inter-Andean valleys of southern Peru, this with narrow sessile resinous leaves very sharply but remotely serrate and solitary large yellow flowers.

Laestadia (6 spp.) — Prostrate small-leaved paramo/puna herb forming mats in boggy places.

Inuleae

Leaves (in our area) alternate and usually with very distinctive conspicuous cobwebby, sometimes +/- sericeous grayish or whitish pubescence. The main technical character is combination of sagittate tailed anthers with unappendaged apically rounded styles. Only two genera (*Tessaria, Pluchea)* becoming arborescent, the former occasionally to 15 m;

Loricaria is a high-Andean shrub occasionally to 1.5 m tall with very characteristic reduced scalelike, densely imbricate, laterally flattened leaves.

1. Trees and Shrubs

Tessaria (1 sp.) — Tree to 15 m tall, common in early riverine succession both along Andean and lowland streams. Easy to recognize by the typical grayish oblanceolate leaves entire or shallowly remotely serrate, the secondary veins strongly ascending and not very evident; flowers pinkish.

P: sauce

Pluchea (40 spp., incl. Old World) — Our species shrubs or small trees (to 3 m tall), mostly along streams in dry areas. Leaves entire or variously serrate, differing from *Tessaria* in the more prominent secondary veins (except resinous *P. zamalloae* of Apurimac Valley). Inflorescence corymbose with lavender flowers. Looks like *Vernonia* except for the scattered cobwebby pubescence (*P. zamolloae* with resinous entire leaves looks like *Baccharis*).

Loricaria (17 spp.) — Highest Andean shrubs or subshrubs, occasionally to 1.5 m tall, very characteristic in the tiny, reduced, scalelike, densely imbricate leaves, these strongly laterally compressed and the branches flat-looking.

Oligandra (3 spp.) — High-Andean shrubs with tiny scalelike leaves like *Loricaria* but plant less densely branched and stem not flattened-appearing.

Chionolaena (8 spp.) — Few-branched paramo subshrub with stiffly coriaceous, sublinear obtuse leaves sericeous above and densely white-tomentose below.

2. Low-Altitude Herbs

Only two Inuleae herb genera occur in the lowlands:

Pterocaulon (25 spp., incl. Old World) — In our area only in disturbed grasslands at middle elevations in northern Colombia. Characterized by strikingly winged stem. Similar to *Gnaphalium* but with densely spicate inflorescence and the narrowly elliptic leaves sharply finely serrate (and densely white-tomentose below).

Blumea (incl. *Pseudoconyza*) (75 spp., mostly Old World) — Erect weed of lowland dry forest with obovate rather sharply serrate leaves. Quite similar to *Egletes* but differs in less deeply serrate leaf margins and erect habit.

3. High-Altitude Herbs

Most montane Inuleae herbs are segregates from *Gnaphalium,* all characterized by more or less linear densely gray-sericeous leaves. These include *Gnaphalium* (150 spp., incl. n. temperate and Old World) (inflorescence tannish-sericeous and +/- corymbose), *Gamochaeta* (inflorescence more congested and spikelike), *Stuckertiella* (2 spp.), *Achyrocline* (20–30 spp., incl. Madagascar) and *Facelis* (4 spp.) (high-puna *Gnaphalium* segregates with narrowly linear leaves); *Antennaria* (mostly n. temperate, with a few high-Andean spp.) looks like reduced *Gnaphalium* with basal rosette of linear leaves. *Chevreulia* (6 spp.) is an even more reduced tenuous high-Andean herb with relatively large solitary flowers. *Belloa* (11 spp.) is a tiny, puna cushion-plant or rosette herb with often small but rather broadly elliptic, gray, sericeous leaves; *Lucilia* (20 spp.) is a similar mat-forming cushion-plant but with narrower leaves. Finally two remarkable high-Andean genera are extreme cushion-plants: *Mniodes* (5 spp.) of the Peruvian jalca looks like a cluster of soft gray rabbit pellets strung together while *Rauliopsis* (2 spp.) of the Colombian superparamos looks quite like a patch of dense, tannish wool growing on the rocks.

Heliantheae

A very large tribe with mostly opposite leaves (except *Ambrosia, Neurolaena, Verbesina, Schkuhria,* most Helianthinae) nearly always 3-veined above base (except some herbs and +/- woody *Zexmenia, Neurolaena*), more or less serrate (often rather scabrous), and aromatic. The fruits always lack the capillary pappus (of thin flexuous trichomes) that is characteristic of most of the family (except *Neurolaena* and *Schistocarpha,* often placed in Senecioneae). The flowers are mostly yellow (especially the ray-flowers, when present), but are white in *Clibadium* and *Ichthyothere, Eclipta, Melanthera,* most *Verbesina,* and some *Bidens,* green in *Garcilassa,* orange in *Hidalgoa,* purplish only in a few *Galinsoga* and an introduced *Cosmos.*

The genera below are arranged in subtribes; although a recent revision subdivides Heliantheae into 35 subtribes based largely on highly technical characters, the traditional broader circumscriptions are followed here. Helianthinae can be thought of as the core group (although unusual in several genera having alternate leaves), characterized by relatively large, usually more or less solitary heads and a paleaceous receptacle, usually with yellow ray-flowers. Melampodiinae is unique in completely lacking pappus (the other groups have a pappus of awns or scales) and a 2-seriate involucre with 5 outer bracts (other groups have 2–multiseriate involucres but rarely with 5 outer bracts); another peculiarity is that only ray-flowers are fertile. Ecliptinae (not always recognized as distinct subtribe) essentially represents a trend from the mostly more or less shrubby Heliantheae core group toward smaller more herbaceous plants (or lianas: *Wulffia*); the main technical character is that the ray-flowers are fertile. Verbesininae (also not always recognized as subtribally distinct) are mostly shrubby and usually have fertile ray-flowers unlike Helianthinae also differing in the generally smaller flowers in larger inflorescences. Coreopsidinae has a characteristic pappus with 2 longer awns (for exozoochoric dispersal) and typical phyllaries having scarious margins and conspicuous longitudinal brownish striations. Galinsoginae is characterized by the conical receptacle; otherwise they are very like the Ecliptinae (i.e., herbaceous,small-flowered Helianthinae). Milleriinae have small heads with a tendency to tight aggregation (e.g., fusion of the bracts), few florets, usually lack differentiated ray-flowers, and tend to have more or less fleshy or oily fruits each representing a single seed enclosed by the variously persistent involucre (fleshy-fruited *Clibadium* and *Ichthyothere* are sometimes placed here and sometimes with Melampodiinae); the leaves are opposite, the receptacle is naked (i.e., lacks bracts) and the small heads have <10 flowers. Ambrosiinae have alternate leaves, greenish, reduced wind-pollinated flowers, and usually deeply lobed or dissected leaves. Finally two subtribes now associated with Heliantheae were traditionally placed in other tribes: Neurolaeninae has setose pappus (unique in Heliantheae) and has traditionally been included in Senecioneae; other distinctive characters are lanceolate scarious paleae (these are stiff and elliptic in Helianthinae) and phyllaries with orangish longitudinal striations (cf., *Bidens* and relatives); one genus (*Neurolaena*) is unusual in alternate leaves. Bahiinae include several genera traditionally placed in the now-dismembered Helenieae and are characterized by a naked receptacle, alternate leaves, and multi(= >10)-flowered heads.

1. Melampodiinae

Defined by complete lack of pappus and the 2-seriate involucre with 5 outer bracts; only the first two genera are woody. *Espeletia* is very unusual in the tribe in alternate leaves, but the other genera are opposite-leaved.

Espeletia (80 spp.) — Very distinctive pachycaul high-Andean paramo plants. Nowadays sometimes split into a number of segregate genera. Unusual in alternate leaves (although the bract-leaves of the inflorescence can be opposite).

Polymnia (incl. *Smalleanthus*) (20 spp.) — Coarse herbs to trees 10 m tall, exclusively montane. Usually with broadly ovate, irregularly conspicuously 3-lobed serrate leaves. Also characterized by large black fruits enclosed by involucral bracts.

Acanthosperma (6 spp.) — Small herbs with echinate burlike fruit.

Unxia (2 spp.) — A segregate from *Melampodium.* Herbs with small flowers, bristly pilose stem, and small subsessile leaves.

Sigesbeckia (9 spp.) — Herbs with characteristic long, narrow, stalked-glandular, outer involucral bracts exceeding the small ray-flowers.

2. Milleriinae

Traditional subtribal character is few-flowered heads and usual lack of ray-flowers. (Only the first two genera, sometimes placed in above group, are woody.)

Clibadium (40 spp.) — Shrubs to small trees, differing (along with *Ichthyothere*) from both other Milleriinae and from Melampodiinae in white flowers and corymbose-paniculate inflorescence. Very unusual in the family in the fleshy fruit.

C, E, P: barbasco

Ichthyothere (18 spp.) — Herbs or subshrubs, sometimes +/- scandent, similar to *Clibadium* and with similar fleshy fruit, but with sparser inflorescence, technically differing in having only 2 ray florets and a glabrous achene.

Milleria (1 sp.) — Weedy herbs, unique in the zygomorphic head with 2 main bracts fused into a cup and a single, large, trifid ray-flower. Also distinctive in the black single-seeded fruit surrounded by the fused bracts.

Delilia (3 spp.) — Small annual herbs with only 2 florets per head, one of these sterile. The curious fruit is round, flattened and surrounded by wing formed from the enclosing bracts.

3. Helianthinae

Characterized by relatively large heads, a paleaceous receptacle, and the usual presence of sterile, yellow, ray-flowers. The core group of the tribe, often with alternate leaves, very unusual in tribe.

Encelia (15 spp.) — Amphitropical, in our area mostly subshrubs of dry coastal and Pacific slope Peru with the thin, opposite, entire, or serrulate leaves characteristically canescent and *Chenopodium*-shaped.

Flourensia (30 spp.) — Amphitropical but with several species of shrubs in Peruvian dry inter-Andean valleys. Close to *Encelia* but the leaves uniformly alternate and resinous rather than canescent. Vegetatively notably resembling *Dodonaea* of the Sapindaceae.

Hymenostephium (4 spp.) — Coarse herbs or shrubs differing from *Viguiera* in the reduced pappus, the achenes epappose or with reduced squamellae.

Pappobolus (38 spp.) — A woody Central Andean offshoot of mostly North American *Helianthus* and also closely related to *Viguiera*, but the leaves usually conspicuously villous or pilose, rather than scabrous.

Syncretocarpus (2 spp.) — Low sclerophyllous shrubs of the western slope of the Peruvian Andes. Sharing the achene type of *Viguiera* but differing in the shrubby habit.

Tithonia (10 spp.) — Shrubby to several meters tall; native only to Central America but widely cultivated for the rather large, showy flowers and +/- naturalized in premontane area roadsides.

Viguiera (160 spp.) — Rather nondescript herbs (in our area), mostly with opposite leaves, rather like small-flowered *Helianthus;* similar to *Wedelia* but ray-flowers sterile. Technically characterized by a pappus of two awns plus some intermediate squamellae. The species that lack pappus are sometimes segregated as *Gymnolomia,* but are probably an artificial assemblage.

4. Ecliptinae

Differs from Helianthineae primarily in fertile ray-flowers; *Wulffia, Salmea,* and a few *Aspilia* are lianas; only *Montanoa* has large shrubs and trees. As here arranged, representing a trend towards reduction in plant and flower size and increasing weediness.

Montanoa (22 spp.) — Montane shrubs or trees to 5 m or more, mostly Central American but also in northern Andes. The very characteristic fruiting head largish and enclosed by numerous conspicuous large thin accrescent involucral bracts.

Wulffia (4 spp.) — A liana with conspicuous largish yellow-rayed flowers and characteristic scabrous rather coriaceous leaves. Vegetatively differs from *Mikania* in more scabrous leaves; several similar Heliantheae have pinnate leaf venation rather than 3-veined from above the base as in *Wulffia.* The fruiting head is very distinctive, consisting of several fleshy blackish berries separated by the exserted paleas.

Aspilia (60 spp., incl. Africa) — Mostly herbs and subshrubs of the cerrado and and subtropical dry areas; our few taxa +/- scandent, with scabrous 3-veined leaves very like *Wulffia.*

Salmea (7 spp.) — Lianas (sometimes shrubs) with few whitish rayless flowers in small heads. Superficially like *Mikania* but with achene with 2 awns instead of pappus. Leaves entire or with minute distant teeth.

Wedelia (70 spp, incl. Old World) — Our species prostrate scrambling herbs with largish solitary yellow flowers; commonest species is most frequent as weed in coastal areas. Leaves 3-veined unlike *Zexmenia.*

Enhydra (10 spp.) — Usually more or less succulent aquatic or marsh plant with narrow leaves and rayless head of whitish flowers.

Melanthera (20 spp.) — Weedy herbs (sometimes scandent?) with broad membranaceous scabrous leaves and globose +/- solitary heads of small inconspicuous rayless white flowers. Looks like *Clibadium* except for reduced few-headed inflorescence and smaller stature.

There are also a number of small, nondescript, herbaceous genera with 1–2 species each which are related to *Wedelia* and *Melanthera,* including *Schizoptera* (1 sp.), endemic to coastal Ecuador, *Thelechitonia* (1 sp.), endemic to Colombia, and a series of widespread weeds. The latter include: *Eclipta* (1 sp.), a common weed with narrow pubescent leaves and single long-pedunculate heads with inconspicuous white flowers; *Baltimora* (2 spp.), weedy dry-area herbs with small yellow flowers in a +/- branching inflorescence with one species reaching coastal Ecuador, the other northern Colombia; *Eleutheranthera* (2 spp.), common sprawling weeds with small inconspicuous short-pedunculate axillary heads of yellow flowers (similar to *Synedrella* but leaves glandular); *Synedrella* (2 spp.), common sprawling weedy herbs with sessile axillary fascicles of ca. 4-flowered, narrow heads of yellow flowers; *Trichospira* (1 sp.), differing from other genera of small weedy herbs in the spatulate leaves conspicuously white below; and *Garcilassa* (1 sp.), an erect wet-forest weed with very reduced (not very complike) heads of few greenish flowers.

5. Verbesininae

Not clearly different from Helianthinae by often fertile ray-flowers; but usually smaller heads and larger inflorescences and tendency to have the achene edges oriented radially; only *Verbesina* and *Zexmenia* (*sensu lato*) are more or less arborescent, the latter also sometimes scandent.

Verbesina (150 spp.) — Mostly montane shrubs or small shrubby trees often to 3–4 m tall. The leaves are *alternate,* often deeply pinnately lobed (species without deeply lobed leaves mostly montane or premontane); usually vegetatively characterized by winged petiole with bases often auriculate or fused to stem to form wing. The rather many heads usually have "loose" not compacted disc-flowers, conspicuous white rays and are arranged in a large flat-topped terminal panicle.

Zexmenia (2 [or 45] spp.) — Shrubs or scrambling vines with the standard yellow-rayed Heliantheae flower; when scandent similar to *Wulffia* but with more pubescent pinnate-veined serrulate leaves. Achenes winged. As recently redefined, reduced to only two species and hardly reaching our area.

Otopappus (9 spp.) — Mostly Central American; very similar to *Zexmenia,* from which it differs in the more conical receptacle and asymmetrically winged achenes with the larger wings extending to apex of pappus awns.

Oyedaea (13 spp.) — Large montane herbs or shrubs with large yellow flowers, characterized by unusual pappus with fused ring of scale-like hairs, two of these much longer and forming awns.

Less significant small herbaceous *Verbesina* relatives that occur in our area include *Steiractinia* (6 spp.), *Leptocarpha* (1 sp.), and *Monactis* (4 spp.), the latter distinctive in alternate leaves like *Verbesina* from which it differs in the reduced heads with only 3–5 disc-flowers and a single ray-flower.

6. Coreopsidinae

Derivatives of Helianthinae with achenes modified for exozoochorous dispersal; none are woody but *Hidalgoa* and a few *Bidens* are scandent.

Hidalgoa (5 spp.) — Vine of wet-forest edges climbing by the hooked petiole bases of the 3-foliolate leaves. Flowers solitary, rather large and conspicuous, the rays bright red-orange.

Bidens (230 spp., incl. N. Am. and Old World) — Weeds with characteristic stiffly awned, zoochorous achenes sticking to fur or clothes by retrorse barbs. Leaves usually 3-foliolate (a few species dissected).

Cosmos (26 spp.) — Weedy herbs very close to *Bidens,* but differing in the uniformly more dissected leaves, in larger ray-flowers and the slender achenes having an apical "neck" below the short retrorsely barbed awns.

Smaller genera of *Bidens* relatives represented in our area include *Cyathomone* (1 sp.), endemic to Ecuador, *Ericentrodea* (3 spp.), *Narvalina* (4 spp.), and *Isostigma* (11 spp.).

7. Galinsoginae

Very close to Ecliptinae except for the conical receptacles; only *Calea* has some shrubby species.

Calea (100 spp, incl. N. Am.) — Mostly subshrubs or sprawling shrubs, the leaves usually gland-dotted. Differs from *Montanoa* in the numerous pappus bristles (and fewer flowers per head?).

Spilanthes (50 spp., incl. Old World) — Small weedy herbs, similar to *Eclipta* but broader leaves and more conical inflorescence with several irregular whitish ray-flowers.

Tridax (26 spp.) — Weedy herbs with or without several ray-flowers, these cream and ca. 3–4 in commonest species. Differing from *Calea* in the leaves usually more deeply toothed or pinnatifid and plant more herbaceous.

Galinsoga (4 spp.) — Small weedy erect herbs with ca. 5 whitish or pinkish ray-flowers. The achene lacks the ciliate angles of *Spilanthes* and head is less conical.

Jaegeria (8 spp.) — Weedy montane herbs, like *Galinsoga* but the involucral bracts expanded into wings at base; rays inconspicuous, yellow in common species.

Sabazia (13 spp.) — Small sprawling upland herbs, mostly Central American but reaching Colombia; like *Galinsoga* but yellow ray-flowers and more disc-flowers (>8).

Less significant small genera of herbs related to *Galinsoga* that occur in our area include *Aphanactis* (4 spp.) and *Geissopappus* (4 spp.), the latter essentially those species of *Calea* that lack paleae.

8. Ambrosiinae

Wind-pollinated derivatives of Helianthinae.

Ambrosia (42 spp., incl. N. Am.) — Only adventive weeds in our area, mostly in dry intermontane valleys, the leaves alternate and usually deeply dissected or lobed.

9. Bahiinae

Traditionally in Helenieae.

This entirely herbaceous group is mostly North American with only a few species of *Schkuhria* (10 spp.) and annual *Villanova* (10 spp.) represented in our area.

10. Neurolaeninae

Traditionally in Senecioneae, due to pappus of fine bristles, but with paleae of Helianthinae — Some area genera of the Galinsoginae like *Calea* and *Geisopappus* may be closer to this group based on floral microcharacters.

Schistocarpha (12 spp.) — Coarse herb or subshrub with broad opposite leaves having petiole bases more or less fused across node. Heads with or (usually) without yellowish rays.

Neurolaena (5 spp.) — Coarse weedy forest-edge herb with alternate pinnately veined leaves, sometimes in part deeply 3-lobed. Flowers yellow, lacking rays.

Tageteae

In our area mostly herbaceous (*Schizotrichia* is shrubby), always very strongly aromatic (formerly part of now-dismembered Helenieae). Characterized by the leaves strongly glandular and either deeply pinnately parted or simple and bristly ciliate at base. Mostly Central American and Antillean.

Schizotrichia (5 spp.) — Shrubs of dry parts of Peruvian Andes characterized by free phyllaries and the pappus of squamellae dissected into 5–10 slender bristles.

Pectis (100 spp.) — Herbs with distinctive opposite, simple, narrow, entire leaves, bristly ciliate at base; small yellow ray-flowers present, unlike *Porophyllum.*

Porophyllum (30 spp.) — Herbs or subshrubs with free phyllaries (unlike *Tagetes*) and pappus of separate bristles; differs from *Pectis* in lacking ray-flowers and the leaves broader and with long slender petioles (almost as long as the small elliptic blade). Flowers greenish and the head distinctively long and narrow.

Tagetes (50 spp.) — Weedy upland herbs, distinctive among the strongly aromatic taxa (Tageteae) in the phyllaries fused one-third to one-half their length. Leaves mostly pinnatifid or pinnately compound.

Liabeae

Formerly included in Senecioneae on account of floral details. Leaves opposite and typically 3-veined above base (unlike Senecioneae *sensu stricto*), often with winged petioles and often conspicuously white-pubescent below. Usually with obvious milky latex. Flowers always yellow (to orangish-yellow), differ from *Senecio* in the involucral bracts in several series. In our area three large genera (*Munnozia, Oligactis,* and *Liabum*) are mostly shrubs or scandent and three small genera are shrubs to small trees (*Cacosmia, Chionopappus,* and *Ferreyranthus*).

Liabum (40 spp.) — The core group of Liabeae, mostly in Andean cloud forests. Mostly coarse herbs or shrubs, the leaves conspicuously white-pubescent below; milky latex obvious.

Munnozia (50 spp.) — Herbs, shrubs or frequently scandent. Characterized by leaves densely white-pubescent below and usually conspicuously triangular (often strongly hastate). (Technical characters include dark-colored anthers and larger heads.)

Oligactis (20 spp.) — Mostly scandent *Liabum* relatives characterized in addition to the habit, by the fewer-branched inflorescence with fewer heads than in *Liabum.* Leaves inconspicuously remotely serrate, more narrowly ovate (or elliptic) than *Munnozia.*

Cacosmia (3 spp.) — Shrubs, distinctive in lacking pappus, unique in Liabeae. Leaves typically narrower than in relatives and often strongly 3-veined nearly to apex (or very narrow and the secondary veins inconspicuous).

Chionopappus (1 spp.) — Mostly streamside shrub of "ceja de la montana" cloud forests, looking much like herbaceous *Erato* but woody and to 3 m tall.

Ferreyranthus (7 spp.) — Shrubs or small trees of Andean forests, but very weakly segregated from *Liabum.* The leaves are densely white below and finely (usually inconspicuously) serrate to barely serrulate.

The other five small genera of this alliance, (*Erato* [4 spp.], *Chrysactinium* [10 spp.], *Philoglossa* [5 spp.], *Paranephelius* [8 spp.], and *Pseudonoseris* [3 spp]) are uniformly herbaceous. Of these at least weedy *Philoglossa* (with strongly 3-veined (or subparallel-veined) leaves with bases connected to form sheath) and the sessile-flowered, high-Andean, rosette herbs of *Paranephelius* (with distinctive densely white-pubescent below, pinnately veined, strongly dentate to pinnatifid leaves) deserve mention.

Senecioneae

Technically characterized by flattened style branches with truncate "penicillate" apex and marginal stigmatic line; the receptacle is naked (nonchaffy), unlike most Heliantheae). The leaves are nearly always pinnately veined and are usually alternate (or in basal rosette). Differs from Liabeae (and nearly all other Compositae) in the bracts of the heads uniform in length and in single series. Includes a number of lianas and trees as well as many herbs and shrubs. The first two genera below are mostly scandent, *Senecio* and *Aetheolaena* are habitally variable, and *Paragynoxys, Scrobicaria,* and *Gynoxys* are shrubs to trees.

1. Lianas

Pentacalia (often included in *Senecio*) — Lianas (and a few herbs) of Peruvian and Bolivian cloud forest, characterized by alternate leaves, either serrate or serrulate or entire and very succulent (then typically with suppressed secondary and tertiary venation). A segregate from *Senecio,* the scandent species differing from *Pseudogynoxys* in the less triangular leaf shape and white to yellow flower color.

Pseudogynoxys (21 spp.) — A mostly scandent segregate of *Senecio,* typically with conspicuous, large-rayed, distinctively orange flowers. Leaves alternate and more triangular in shape than in *Pentacalia* and most *Senecio.* Young stems usually hollow.

2. Trees and Shrubs

Paragynoxys (9 spp.) — Pachycaul rosette tree of northern high-Andean cloud forests, usually with very large leaves that are densely woolly-pubescent below. Vegetatively reminiscent only of some *Espeletia* but the flowers much smaller.

Scrobicaria (2 spp.) — Shrubs with opposite dentate leaves of northern Andean cloud forests. Related to *Gynoxys* but lacks ray-flowers, has palea-like out growths on receptacle, and lacks long subinvolucral bracts outside the phyllaries.

Gynoxys (100 spp.) — Shrubs or trees of upper-montane cloud forest and puna, with mostly opposite leaves strongly white-pubescent below and either entire or very shallowly and remotely subserrate, the base well-marked and +/- symmetrical. Usually yellow-flowered. A very characteristic element of above-timberline Andean vegetation.

Paracalia (2 spp.) — White-flowered shrubs of drier high-Andean slopes, vegetatively very distinctive in broad *Jungia*-like, rather irregularly angle-lobed leaves.

Aequatorium (2 spp.) — Small cloud-forest trees <8 m tall with *Gynoxys*-like, oblong, densely white-below leaves (but more jaggedly serrate and with very asymmetric base). Ray-flowers white and leaves may be alternate or opposite.

3. Diverse in Habit

Senecio (1500 spp., incl. Old World) — One of the largest genera of plants and habitally very diverse. The heads usually have ray-flowers and are characterized by the single row of bracts. The leaves are often opposite unlike most of relatives.

Lasiocephalus (incl. *Aetheolaena*) (20 spp.) — Sometimes included in *Culcitium*. Herbs, shrubs, or lianas of Andean cloud forests, characterized by the usually evenly serrate, alternate leaves either densely white-pubescent below or with clasping bases (typically both); when petiole well-developed, the blade usually +/- hastate.

4. Herbs

Emilia (30 spp., incl. Old World) — Ours naturalized weedy herbs with pinkish to orangish rayless flowers, the few small heads on rather long peduncles.

Culcitium (15 spp.) — Paramo herbs with few large, long-pedunculate flowers (much larger than in relatives) and linear or sublinear densely gray-pubescent leaves in basal rosette.

Erechtites (5 spp.) — Two common weedy lowland herb species with very characteristic pinnately lobed and/or basally clasping alternate leaves. Flowers more or less greenish and obscured by the pinkish pappus; ray flowers absent.

Werneria (40 spp.) — Rosette paramo and puna herbs, typically with sessile heads in center of ring of leaves; distinguished by the connate involucral bracts.

Chersodoma (9 spp.) — Mostly south temperate herbs and subshrubs, in our area limited to the highest Andes of Peru. Characterized by the leaves small and densely white-woolly below.

Dorobaea (1 sp.) — Rosette jalca herb with laciniate leaves and single flower on very long (>30 cm) peduncle.

ANTHEMIDEAE

Mostly extratropical; very poorly represented in our area by a few strongly aromatic high-Andean herbs, ours all with deeply pinnatifid or bipinnatifid leaves and discoid heads lacking ray-flowers.

Cotula (90 spp., incl. s. temperate, and Old World) — Tiny high-Andean herbs with very small roundish solitary pedunculate inflorescence, the disc-flowers whitish, the marginal flowers completely lacking corolla.

Plagiocheilus (5 spp.) — Small high-Andean herbs with round, solitary, pedunculate inflorescences, the flowers yellow, the outer ones with bilabiate corollas.

Soliva (9 spp.) — Acaulescent puna herbs differing from *Cotula* and *Plagiocheilus* in sessile inflorescences and bipinnatifid leaves.

Well-known temperate genera represented only by weeds in our area include *Artemisia, Achillea, Leucanthemum, Matricaria,* and *Tanacetum.*

MUTISIEAE

Now regarded as the least advanced tribe of Compositae, and mostly South American. Technically characterized by the bilabiate corollas (ray- and disc-flowers often not sharply differentiated), sagittate anthers, and characteristic style. The leaves are uniformly alternate (or in basal rosette). Most genera are more or less woody and several are spiny. Subtribes Barnadesiinae (with axillary spines) and Gochnatiinae (without axillary spines) have 5-parted disc-flowers; subtribes Mutisiinae (with style branches rounded at apex) and Nassauviinae (with style branches truncate and with hairs at apex) have the disc-flowers (and ray-flowers) strongly bilabiate or ligulate.

1. LIANAS

Each of the nonspiny subtribes has a predominantly scandent genus: *Lycoseris, Mutisia, Jungia.*

Lycoseris (15 spp.) — Shrubs and (mostly) vines, leaves white-tomentose below, typically strongly parallel-3-veined from above base. Unusual in being dioecious, the heads single, rather large (especially the male) the flowers orange, usually with narrow ray-flowers.

Mutisia (59 spp.) — Vines and shrubs vegetatively very distinctive in the leaves pinnately compound and/or the rachis terminating in a tendril. Heads usually long and narrow, with orange or red flowers.

Jungia (30 spp.) — Mostly lianas, the very characteristic leaves broader than long and usually 3–5-veined from base, with rounded or obtuse lobes. Flowers mostly white (rarely pink) unlike relatives.

2. Barnadesiinae

Andean shrubs or small trees with axillary spines and villous usually more or less actinomorphic disc-florets.

Barnadesia (22 spp.) — Shrubs or trees with both ray- and disc-flowers, unique among the spiny Mutisieae.

Arnaldoa (3 spp.) — Shrubs with unusually large capitula 2–3 cm across and 2.5–6 cm long. Flowers orange to red, more (50–150) per capitulum than in *Chuquiraga* and *Dasyphyllum;* corolla zygomorphic (unique in subtribe).

Chuquiraga (25 spp.) — Shrubs unique in our area in anthers without tails. Also distinctive in all florets the same (unlike *Barnadesia*) and fewer (10–50) flowers per capitulum than *Arnaldoa.* Flowers yellow to orange and leaves one-nerved.

Dasyphyllum (36 spp.) — Shrubs or trees, differing from *Chuquiraga* in 3–5-nerved leaves and white to purple flowers.

Fulcaldea (1 sp.) — Shrub. Flowers reduced to single disc-floret with no rays (unique except for a few Vernonieae very different in plumose pappus).

3. Gochnatiinae

Mostly shrubs or small trees (except scandent *Lycoseris* (see above) and *Onoseris* with 5-parted disc-flowers); differs from above group in lacking spines and in glabrous or slightly puberulous corollas.

Chucoa (1 sp.) — Shrub of northern Peruvian Andes. Distinctive in the oblanceolate leaves with 4–6 pairs of spinulose teeth.

Gochnatia (68 spp.) — Shrubs or trees. Flowers similar to *Plazia* (and unlike *Onoseris* and *Chucoa*) in the corolla split more than one-third of way to base and with linear lobes; unlike *Plazia,* the leaves are evenly distributed along the branches.

Onoseris (29 spp.) — Mostly montane herbs or subshrubs, usually with conspicuously white-pubescent stems and leaf undersides; leaves typically deeply parted or even irregularly compound; heads usually brightly colored, sometimes red, or asterlike, with magenta "rays", the corollas less deeply 5-lobed than in *Gochnatia* and *Plazia.*

Plazia (2 spp.) — Montane dry-area shrubs, very distinctive in the leaves clustered in a dense whorl at tip of branches and subtending the single large (ca. 2 cm long) head.

4. Mutisiinae

The main genus is mostly scandent *Mutisia* (see above). The remainder are mostly herbs; the tribe characterized by bilabiate or ligulate disc- and ray-flowers.

Gongylolepis (12 sp.) — Shrubs or trees, mostly Guayanan.

Trichocline (22 spp.) — Herbs with leaves in basal rosette and whitish-pubescent below; similar to *Chaptalia,* but the heads erect and ray-flowers longer.

Chaptalia (50 spp.) — Herbs with basal leaf rosette and long-pedunculate solitary nodding heads with narrow inconspicuous ray-flowers and conspicuous long pinkish pappus, usually stem and leaf undersides strongly whitish-pubescent.

Related genera of small herbs include *Chaetanthera* (41 spp.), annuals with flowers similar to *Mutisia* but simple narrow etendrillate leaves and *Gerbera* (1 sp., plus many Old World), with basal leaf rosette and erect heads; like *Trichocline* but acute slender trichomes on the achene.

5. Nassauviinae

Mostly herbs except lianescent *Jungia* (see above) and shrubby *Proustia.* Flowers similar to Mutisiinae but the style branches truncate and with hairs at apex.

Proustia (incl. *Lophopappus*) (8 spp.) — Shrubs or subshrubs, rarely spiny, differing from *Trixis* in nonyellow flowers, from *Jungia* in the narrower leaves, and from *Leucheria* and *Perezia* in woody habit.

Trixis (60 spp.) — Herbs or shrubs with rather narrow pinnately veined leaves and yellowish flowers, the outer flowers raylike. Heads characteristic, rather narrow, *Senecio*-like in uniseriate bracts.

Leucheria (46 spp.) — Herbs, usually with leaves in basal rosette; mostly south temperate, north to Peru. Differs from *Perezia* in white flowers and the pappus of plumose bristles.

Perezia (30 spp.) — Herbs, usually with leaves in basal rosette; differing from *Leucheria* in blue to purple flowers and the pappus of scabrous bristles.

Polyachrus (7 spp.) — Loma herbs with pinnatifidly lobed, dandelion-like leaves densely white-pubescent below.

Lactuceae

Very poorly represented in our area, exclusively by herbs with alternate leaves (or in basal rosette), always with milky latex. The leaves are pinnately veined and usually irregularly incised (dandelion-like).

Hypochoeris (100 spp., incl. Old World) — Our only significant Lactuceae species. High-Andean puna and paramo herbs, often forming sessile rosettes (cf., *Werneria*). Very dandelion-like and also like *Taraxacum* in beaked achenes, but differing in overlapping involucral bracts of graded sizes.

Hieracium (1000 spp., incl. Old World) — Only adventive in our area; leaves entire or dentate but never incised, with rather long bristly hairs.

Taraxacum (60 spp., incl. Old World) — Mostly adventive in our area, only in high Andes. Characterized by the typical incised basal rosette of leaves and the large heads with two distinctly different bract sizes.

Sonchus — Adventive in the Andes. Leaves, in part, sessile with clasping leaf bases (some leaves sagittate with winged clasping-based petioles).

Connaraceae

Mostly woody canopy lianas (also including a few species of understory treelets) with uniformly pinnately compound or 3-foliolate alternate leaves. Vegetatively exceedingly similar to Leguminosae and sharing that family's otherwise unique character combination of uniformly cylindrical pulvinuli and pulvinus. All area connaracs have medium-sized to largish leaflets with acutish to acuminate apices, mostly with finely prominulous venation and a characteristic chartaceous texture that is subtley but distinctly different from that of most legume lianas. Unlike many legume lianas, Connaraceae never have red latex, spines, stipules, tiny or round-tipped leaflets, nor anomalous stem structure; tendrils are also absent although woody reflexed branchlets (cf., Hippocrateaceae) are sometimes present. In our area the treelets (all *Connarus*) have pinnately compound leaves and differ from legumes in having (usually inconspicuous) dendroid vegetative trichomes; moreover the great majority of erect legumes are large trees, whereas, erect connaracs are apparently never more than small treelets. The tiny radially symmetrical white flowers, arranged in panicles, are very different from those of any Leguminosae. The distinctive connarac follicle is single-seeded and usually reddish with a conspicuously arillate black seed (occasionally several follicles are produced from a flower).

In our area (though not elsewhere) generic recognition is generally possible even when sterile. The only genus to include normally nonscandent species is *Connarus. Cnestidium* has a characteristic rufous-tomentose pubescence on all its vegetative parts as well as on the inflorescence and fruits. Some species of other genera have a rufous indumentum but of longer trichomes and generally not on the vegetative parts. *Pseudoconnarus,* uniformly 3-foliolate, has more broadly ovate 3-veined leaflets than other genera. In our area climbing species of *Connarus* are usually 3-foliolate, while *Rourea* almost always has pinnately compound leaves, at least in part.

Connarus (51 spp., plus 50 in Old World) — The largest genus and only one in our area with erect treelets as well as climbers. In our area the treelets (but not the lianas, except one Magdalena Valley species) are vegetatively distinguished by (small and inconspicuous) dendroid vegetative trichomes. Our liana species are all 3-foliolate (often with the basal leaflet pair opposite), except *C. wurdackii* (with strikingly villous fruit) and often erect *C. punctata.* In its flower, *Connarus* is unique in a single carpel and in glandular petals; the fruit is distinctive in a more orbicular shape than other genera, with the dorsal margin straight or convex, and in the tiny persistent sepals usually subtending a distinct basal stipe.

Rourea (42 spp., plus many in Old World) — In our area, always scandent. Leaves mostly pinnately 5–7-foliolate at least in part; when more or less uniformly 3-foliolate, the basal leaflet pair usually alternate or subopposite. Very distinctive in fruit with the large accrescent calyx forming a cupule that encloses basal part of an oblong follicle that is either straight or dorsally concave and bent.

Cnestidium (2 spp.) — Widespread in Central America and northern South America, barely reaching on good soils into northernmost Amazonia. A canopy liana, unique in the densely rufous-tomentose pubescence of vegetative parts as well as inflorescence and fruit. Fruit curved and estipitate as in *Rourea* and with similarly large sepals but these narrow, valvate, and not forming a cupule.

Pseudoconnarus (5 spp.) — Woody canopy lianas vegetatively very distinctive in the uniformly 3-foliolate leaves with basal leaflets more broadly (triangular-) ovate than in other genera and distinctly 3-nerved. The leaflets are usually whitish below from the finely papillose undersurface.

Connaraceae

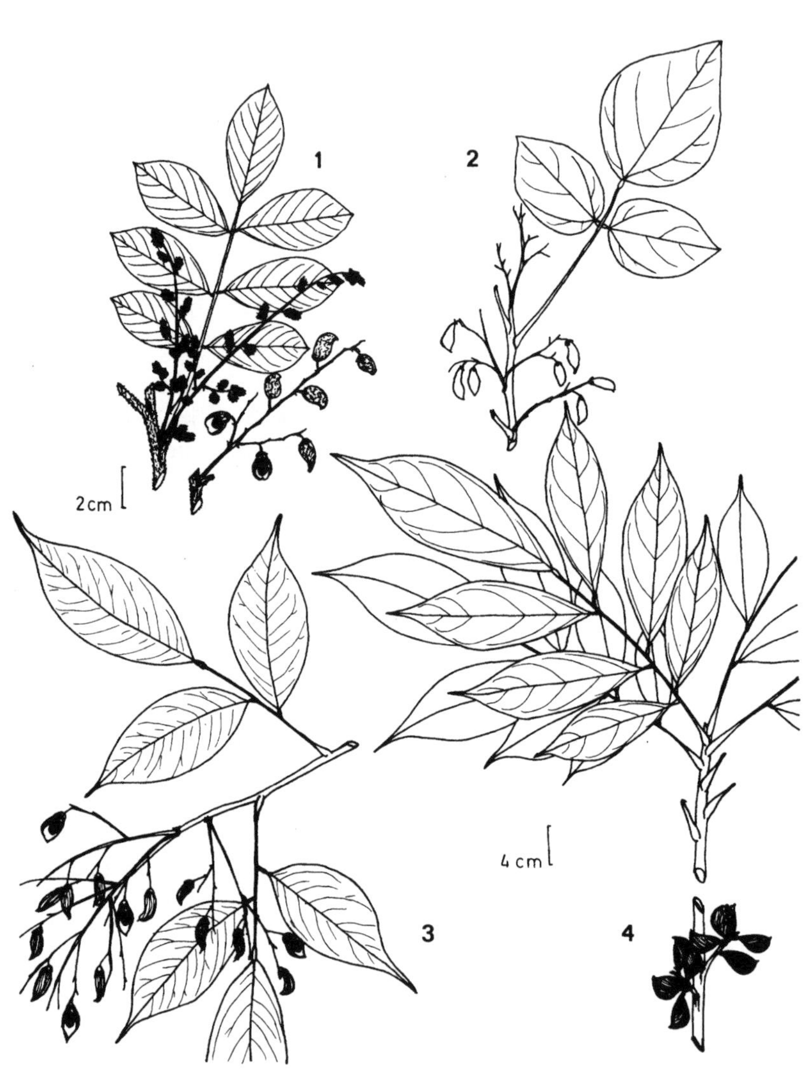

1 - *Cnestidium*

2 - *Pseudoconnarus*

3 - *Rourea*

4 - *Connarus*

5 - *Connarus*

In flower, the stipitate ovary is distinctive; the curved fruit (usually in part with several follicles maturing) is *Rourea*-like but subtended by tiny nonaccrescent calyx lobes as in *Connarus*.

One other genus, monotypic *Bernardinia*, is a treelet restricted to coastal and southern Brazil.

CONVOLVULACEAE

Climbers with alternate simple (rarely palmately compound: *Merremia* or finely pinnatifid: *Ipomoea*) leaves and sometimes milky latex. Mostly slender vines, but a few genera are canopy lianas and some species of *Ipomoea* are shrubs (becoming trees outside our area). A few genera are herbs or prostrate subshrubs and one, *Cuscuta*, is a leafless parasitic achlorophyllous herbaceous vine. The vines are typical morning-glories, usually with broadly funnel-form flowers with the lobes typically completely fused and the stamens each of a different length; vegetatively, these are characterized by their ovate to broadly ovate leaves with cordate bases but essentially pinnate venation, and long petioles. Convolvulaceae may be the only family with essentially pinnately veined vines with conspicuously cordate leaf bases. The lianas are usually characterized by an unusually long, uniform (i.e., lacking glands, swellings, or pulvini) frequently sericeous, petiole and more or less oblong leaf blade, either more or less conspicuously sericeous below or very coriaceous. Stellate trichomes are found in some Convolvulaceae (especially *Jacquemontia*). Conspicuously anomalous stem cross-sections are typical of many of the woody lianas.

Infrafamilial taxonomy traditionally focuses on style and stigma (and fruit) characteristics but the genera are easy to sort on gross morphology as well. Three main habit groups include: (1) the nontwining often prostrate small-flowered herbs and subshrubs of dry areas (*Cressa, Evolvulus, Dichondra*), (2) woody lianas, mostly with campanulate (or tiny and reduced) flowers and oblong noncordate leaves, either with fleshy indehiscent fruits (*Dicranostyles, Maripa,* extralimital *Lysiostyles*) or various dry-fruit types (*Bonamia, Calycobolus,* extralimital *Itzea*), and (3) vines, often herbaceous, with cordate-ovate leaves (sometimes palmately lobed or compound), mostly openly infundibuliform corollas (or red or white tubular and modified, respectively, for hummingbird- or hawkmoth-pollination), and capsular fruits (*Ipomoea, Operculina, Merremia, Jacquemontia, Convolvulus, Tetralocularia*). Intermediate between the latter two groups are two small subwoody genera of swamps and inundated riversides with narrowly oblong leaves and white flowers (*Iseia, Aniseia)* and *Turbina,* a rather slender woody vine with *Ipomoea*-type flowers and leaves but a peculiar helicopter-like indehiscent fruit.

1. Leafless Achlorophyllous Parasite

Cuscuta (ca. 140 spp., worldwide) — Completely unique in its parasitic leafless habit but the tiny flowers are quite morning-glory-like when viewed with a lens. The only plant with which *Cuscuta* could be confused is *Cassytha* of the Lauraceae which differs in having solitary rather than clustered flowers.

2. Nonclimbing Herbs or Subshrubs (Sometimes Prostrate)

Cressa (2 spp., plus 2 in Old World) — Subshrub of high-salt areas. Similar to *Evolvulus* but white flowers and globose stigma. The tiny, narrow, grayish-sericeous leaves are typical.

Dichondra (9 spp., plus few Old World) — Prostrate, dry-area, mat-forming herb with more or less rotund, deeply cordate leaves and tiny white flowers.

Evolvulus (100 spp., plus few n. temperate) — Characterized by the small sky-blue flowers and the narrow, short-petioled, noncordate, often sericeous leaves. Technically differs from *Ipomoea* and relatives by two separate styles and narrow stigmas.

3. Woody Canopy Lianas; Leaf Bases Rounded to Cuneate, Not Conspicuously Cordate (Except Intermediate *Turbina*)— The leaves of all genera are often conspicuously sericeous below and the fruit is usually indehiscent, and sometimes fleshy. The first two genera below have fleshy fruits, the others dry fruits.

Maripa (20 spp.) — The largest liana genus of neotropical Convolvulaceae. The usually magenta flowers are campanulate and with well-developed corolla lobes, unlike most *Ipomoea* relatives. Differs from *Dicranostyles* in much larger flowers and in having conspicuous sepals forming a cup around the basal part of the fleshy ovoid fruit. Leaves are usually very coriaceous and often have a characteristic impressed-venation effect (cf., *Carapa*).

Dicranostyles (15 spp.) — In flower would never be recognized as a Convolvulaceae, the tiny (<5 mm long) greenish or greenish-white flowers with exserted anthers and bifid styles are borne in axillary racemes or narrow panicles. The leaves are nearly always more or less sericeous below. The fruits differ from *Maripa* in lacking subtending sepals.

Convolvulaceae
(Herbs and Vines)

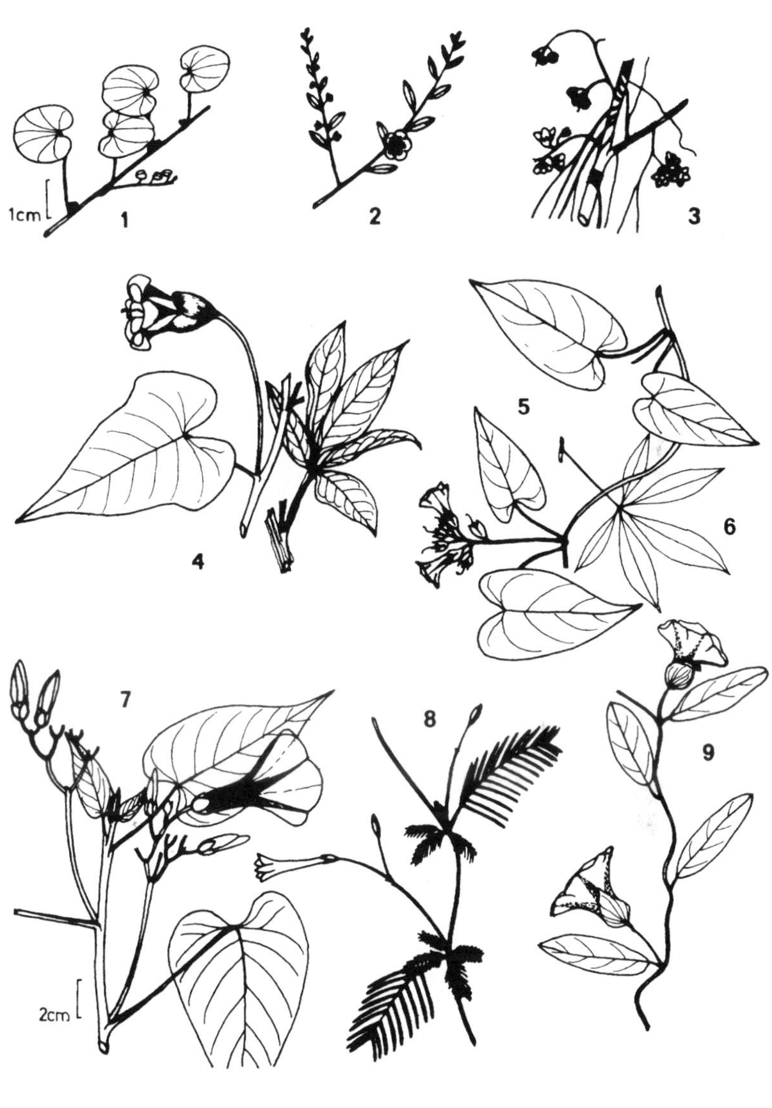

1 - *Dichondra* **2** - *Evolvulus* **3** - *Cuscuta*

4 - *Operculina*

Merremia:
5 - (*M. umbellata*)
6 - (*M. macrocalyx*)

7 - *Ipomoea* **8** - *Ipomoea* (*I. quamoclit*) **9** - *Aniseia*

Bonamia (20 spp., plus 20 Old World) — Differs from *Maripa* and *Dicranostyles* primarily in dehiscent fruit; the seeds usually have numerous long hairs and are wind-dispersed. Leaves always sericeous below. Flowers somewhat intermediate between *Maripa* and *Dicranostyles,* white and smaller than in *Maripa* but of the same shape. Two styles (or a bifid style) with globose stigmas is an important technical character. One tahuampa species has enlarged calyx lobes similar to *Calycobolus,* but these densely sericeous.

Calycobolus (incl. *Prevostoea*) (4 spp., plus 15 in Africa) — Very similar to *Bonamia.* Characterized by having two very large expanded ovate calyx lobes almost covering the campanulate white flowers, these membranaceous and not sericeous unlike *Bonamia* species with similar calyx development. A technical differentiating character is the stigmas less developed and the two styles partly fused rather than completely separate as in *Bonamia.*

(*Jacquemontia*) — A few *Jacquemontia* species can become rather slender woody lianas; they are recognizable by their stellate leaf pubescence and blue infundibuliform unlobed corolla.

Turbina (5 spp., plus few Old World) — Somewhat intermediate between liana group and *Ipomoea.* The leaves cordate as in *Ipomoea* but usually a distinctly woody (though rather slender) liana; the commonest species has strongly sericeous leaf undersides. The main defining character is the fruit, with a rotorlike dispersal unit formed of the dry single-seeded fruit-proper plus the reflexed oblong calyx lobes (this also happens in a few compound-leaved *Merremia* species). The flowers are similar to *Ipomoea* but the large thin, oblong calyx lobes are distinctive.

4. Swamp or Inundated Riverside Plants with Narrowly Oblong Leaves Having Acute to Rounded Bases, Sub-Woody Habit, and White Flowers — These two closely related genera, very easy to recognize by their distinctive narrow leaves, do not fit in either the liana group or with *Ipomoea* and its relatives.

Iseia (1 sp.) — Differs from *Aniseia* in an indehiscent fruit with spongy mesocarp, the corolla ribs pubescent outside, and the outer 2 sepals not enlarged.

Aniseia (4 spp.) — The main feature, aside from the narrow oblong leaves, is the outer 2 sepals very large and foliaceous; fruit a capsule as in *Ipomoea* and relatives.

Convolvulaceae
(Lianas)

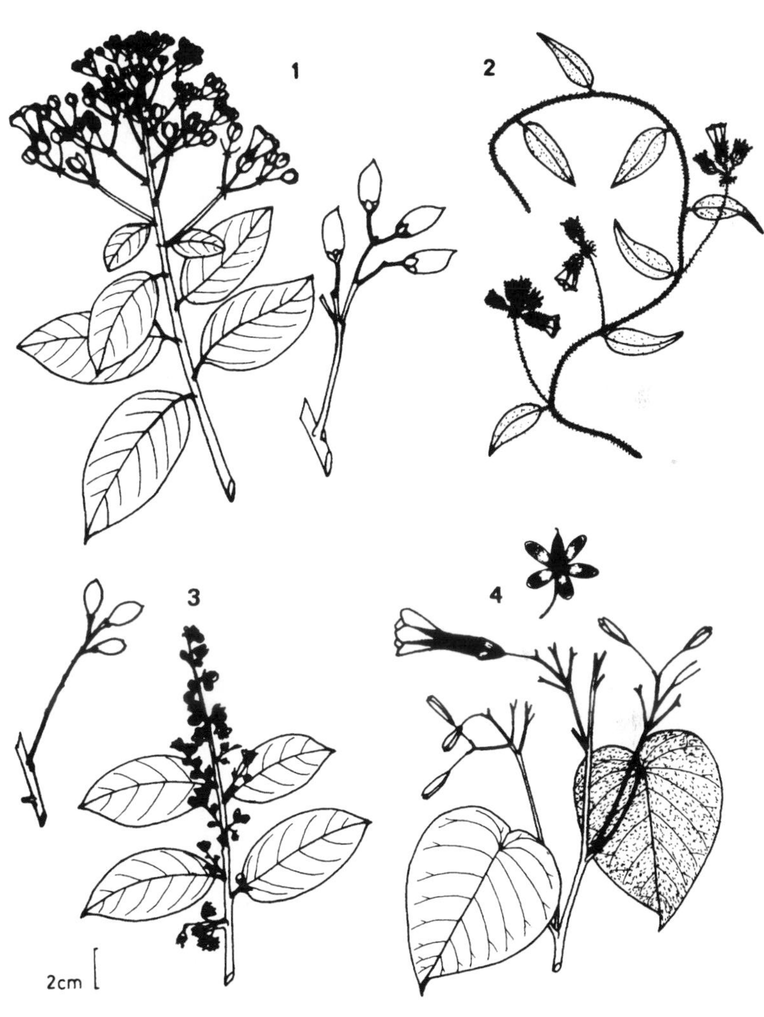

1 - *Maripa* **2** - *Jacquemontia*

3 - *Dicranostyles* **4** - *Turbina*

5. Herbaceous or Subwoody Vines (If Somewhat Woody with Obvious Latex) — A few *Ipomoea* species are large shrubs and some extralimital species are trees, in both cases with obvious latex. Fruits usually a capsule.

Ipomoea (500 spp., about half in Paleotropics) — One of the largest and most important genera of thin-stemmed vines. The great majority of convolv species with ovate cordate-based leaves belong to *Ipomoea,* also defined by the typical morning-glory flower (with lobes usually lacking). A few species have the corolla tubular and red for hummingbird-pollination or elongate-tubular and white for hawkmoth-pollination. Important technical characters are the globose or biglobose stigma, straight stamens, and porate spinulose pollen. No *Ipomoea* has yellow flowers and white flowers are rare except in the tubular-flowered hawkmoth-pollinated group (subgenus *Calonyction*). The fruit is dry, dehiscent, and usually enclosed by the calyx lobes.

Operculina (8 spp., plus ca. 10 in Old World) — Differs from *Ipomoea* in a transversely circumscissile capsule, larger very broad sepals, and the stamens twisting spirally. The common species has large white solitary flowers and is vegetatively easy to recognize by the strongly angled, more or less winged pedicels and branchlets.

Merremia (26 spp., plus ca. 50 in Old World) — Close to both *Ipomoea* and *Operculina.* Most species have palmately compound leaves (unique, but *Operculina* can be deeply lobed); in our area only common *M. umbellata* with yellow flowers, has simple leaves. *Merremia* flowers are yellow (unique) or white (rare in funnel-shaped *Ipomoea*) and have spirally twisting stamens.

Jacquemontia (100 spp.) (incl. *Odonellia*) — Vegetatively usually characterized by stellate-trichomes. Technically differs from *Ipomoea* by ellipsoid stigmas. The flowers are of the typical funnel-form shape but are usually blue (white in a few species). The few blue-flowered *Ipomoea* species usually have much larger flowers than does *Jacquemontia.* Several *Jacquemontia* species (including all the white-flowered ones in our area) have the flowers congested into a unique more or less capitate inflorescence subtended by leaflike bracts. A few species become lianas and might be confused vegetatively with similarly stellate-pubescent Solanaceae from which they differ in the usually cordate leaf bases and longer better defined petioles.

Convolvulus (7 spp., plus ca. 200 in Temperate Zone and Paleotropics) — Mostly a temperate zone and Old World genus; technically differs from *Ipomoea* in the linear stigmas. The few species in our area have smallish (1–2 cm long) white openly funnel-form flowers; a combination found in no *Ipomoea.*

Tetralocularia (1 sp.) — A monotypic endemic of the Magdalena Valley swamps. Leaves like *Ipomoea.* Small white flowers with well-developed lobes (cf., *Bonamia*) arranged in rather many-flowered racemes.

Extralimital Genera:

Lysiostyles — Guayana Shield area; similar to *Dicranostyles* (especially in the tiny flowers); extremely tannish-pubescent leaf undersurfaces.

Itzea — Central America; very like *Bonamia* but the fruit fragmenting into numerous segments.

Coriariaceae

In our area represented by a single species of sprawling shrub or liana, very common in disturbed middle-elevation cloud-forest habitats. Easily recognized by the characteristic branching pattern with the numerous small, sessile, entire, opposite, palmately veined leaves disposed along thin opposite branchlets that look like multifoliolate pinnately compound leaves (cf., *Phyllanthus*). Inflorescence a long narrow pendent raceme with numerous small greenish flowers, followed by small purplish black berries.

Coriaria (1 sp., plus others in Australia and New Zealand) — Although usually seen as a sprawling shrub, it definitely becomes a woody liana inside the forest.

Cornaceae

Only a single dogwood species reaches our area (or South America for that matter). The single Andean *Cornus* has opposite leaves and is exceedingly similar to *Viburnum* but has completely entire margins, more strongly ascending secondary veins, and petiole bases that are notably decurrent onto the somewhat angled tannish-puberulous twig. Unlike *Viburnum* (or any other plant known to me) the main leaf veins of *Cornus* have a mucilaginous tissue so that when a leaf is carefully pulled in half the torn half will hang suspended by the mucilage threads. The difference between the (barely) sympetalous *Viburnum* flower and the separate-petaled *Cornus* one is not at all obvious in practice, and the small round berries (technically drupes) are also similar.

Cornus (1 sp. in S. Am., several others in C. Am., many in N. Temperate Zone)

Coriariaceae, Cornacceae, Crassulaceae, and Cruciferae

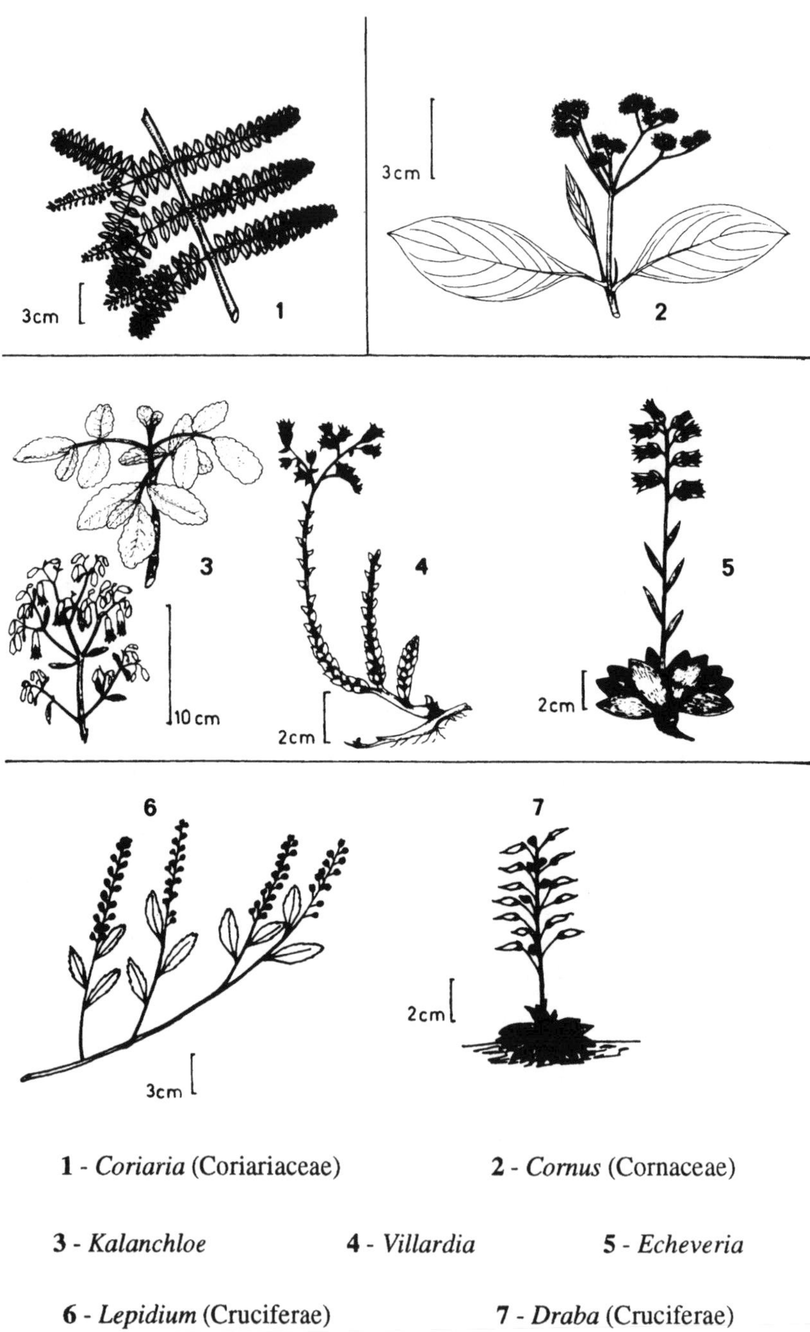

1 - *Coriaria* (Coriariaceae) **2** - *Cornus* (Cornaceae)

3 - *Kalanchloe* **4** - *Villardia* **5** - *Echeveria*

6 - *Lepidium* (Cruciferae) **7** - *Draba* (Cruciferae)

CRASSULACEAE

A family of succulent-leaved herbs very sparsely represented in our area almost exclusively in dry inter-Andean valleys, where *Villardia* (25 mostly Mexican spp.), subshrubs or woody herbs with small leaves clustered at the ends of subwoody branches, and *Echeveria* (150 sp., mostly North American), very distinctive in its succulent leaves in dense rosettes, have undergone secondary radiations. In addition *Crassula,* in our area consisting of tiny (annual) herbs, has a few loma species and a curious aquatic one. Only introduced *Kalanchoe,* characterized by its succulent pinnate leaves, is naturalized in the lowland tropics in our area.

CRUCIFERAE

A large, mostly north temperate, herbaceous family represented in our area almost exclusively in the Andes and mostly by Eurasian weeds. The family, essentially a herbaceous derivative of Capparidaceae, is easy to recognize by the consistent and distinctive floral structure with 4 sepals, 4 petals, 4 long stamens plus 2 short stamens, and a single 2-locular ovary. The bilocular capsule is also distinctive with a central false septum from which the two valves dehisce. The leaves are often coarsely toothed or irregularly incised.While about 50 genera are represented in the Neotropics, most of these are only in the Patagonian region and only one Eurasian *Rorippa* species occurs to any appreciable extent in the tropical lowlands of our area. The most significant high-Andean genera are cosmopolitan *Draba,* with a major Andean radiation and noteworthy for occurring at extremely high altitudes, *Cremolobus* (7 spp.), an autochthonous Andean genus characterized by a Capparidaceae-like gynophore, and primarily north temperate *Lepidium, Cardamine, Sisymbrium,* and *Descurainia,* which have secondary radiations in the Andes.

CUCURBITACEAE

Easily recognized by the uniformly scandent habit and unique placement of the strongly coiling tendrils which are borne from the side of the node at right angles to the leaf axil. The venation is palmate, the leaves often palmately lobed, the margin usually with remote teeth.

The two main groups (= subfamilies) of Cucurbitaceae are easy to distinguish vegetatively by their tendrils. One (Cucurbitoideae) has the tendrils simple or with long spirally coiling branches; the other (Zanonioideae) has tendrils that are forked near the apex and coil both above and below the bifurcation. The former group includes all large-flowered species as well as all species with the irregular, broadly ovate, typically somewhat angulate outline and +/- serrate leaves so characteristic of the family. The latter group has uniformly tiny broadly campanulate flowers and the leaves uniformly entire (or palmately compound with entire leaflets), except for a few species with very few (ca. 4–6 per leaf) thick-triangular toothlike marginal projections.

1. Tendrils Bifid Near Apex and Coiling Both Above and Below Bifurcation — Leaves entire or with a few small, triangular, toothlike, marginal projections; in general the leaves not obviously cucurbitaceous-looking. Flowers small and very broadly capanulate with stamens of the male flowers inserted on the disk and 3 styles; inflorescence (usually even the female one) large and usually openly paniculate (Zanonioideae).

1A. Large many-seeded fruits; flowers with 5 stamens and 3-locular ovary

Fevillea (9 spp.) — Fruits very large, indehiscent (one species tardily dehiscent), globose, completely filled with large lenticular seeds (no pulp). Leaves rather thick and succulent, 3(–5)-foliolate (with entire leaflets) or simple (and entire or with a few small, triangular, toothlike, glandular projections).

P: abiria, habilla

Siolmatra (4 spp.) — Fruits elongate (cf., *Couratari*), dehiscing by an operculate "lid"; seeds winged. Leaves 3–5-foliolate, without the large petiolar glands (cf., Passifloraceae) of compound-leaved *Fevillea.* Except for the dehiscent fruit and winged seeds, very like *Fevillea* and possibly congeneric with it.

1B. Small or winged single-seeded fruits; flowers with 3 stamens and 1-locular ovary

Sicydium (6 spp.) — Leaves strongly cordate, entire. Flowers small; fruit small, black, single-seeded, fleshy, and berrylike; inflorescences (both male and female) paniculate and much-branched (superficially very similar to *Disciphania* of the Menispermaceae).

E: habilla

Pseudosicydium (1 spp.) — Flowers similar to *Sicydium* but fruit a thin, flat-winged samara with the membranaceous wing surrounding seed body (cf., *Pterocarpus*); leaves entire, 3-veined from *below* base, not cordate.

Cucurbitaceae
(Tendrils Bifid Near Apex; Small Flowers)

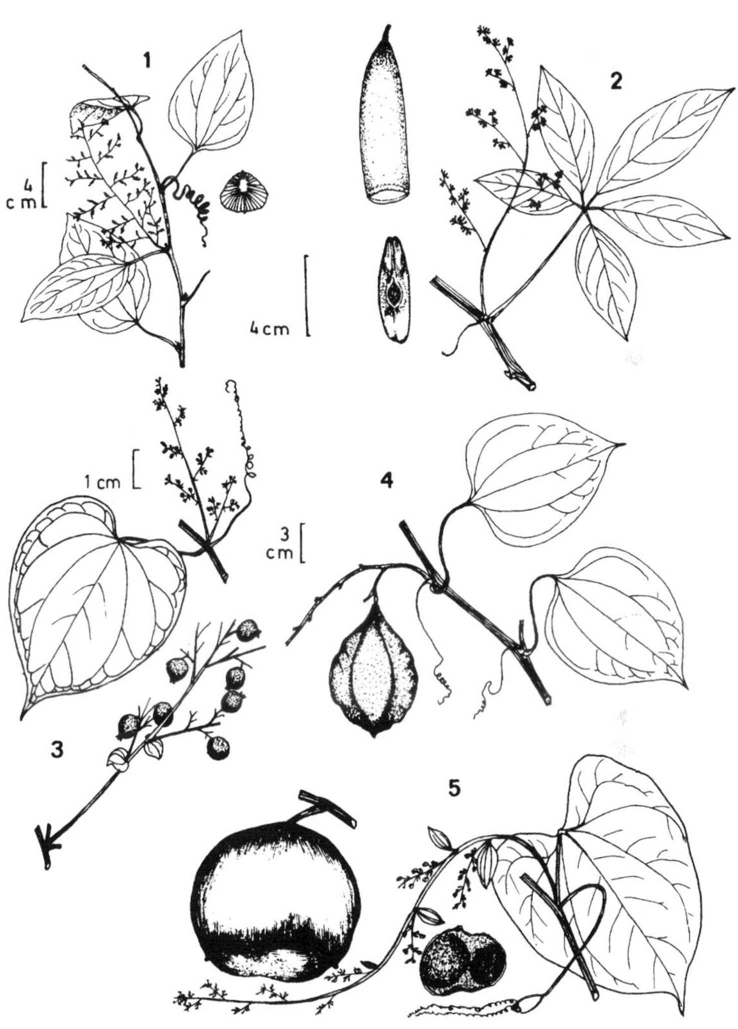

1 - *Pseudosicydium*

2 - *Siolmatra*

3 - *Sicydium*

4 - *Pteropepon*

5 - *Fevillea*

Pteropepon (1 sp.) — Flowers similar to *Sicydium* but fruit a large flattened, thick-winged samara (probably water-dispersed; similar to *Pterocarpus officinalis*). Leaves distinctive in truncate base and tendency to subtriangular shape with basal "corners"

2. Tendrils Simple or Branching Below Middle and Coiling Only Above Bifurcation — Leaves (nearly always) serrate, rank smelling, typically broadly ovate and frequently palmately lobed, usually with somewhat angular outline. Flowers often large (almost always larger than in above group), frequently borne solitarily or the female flowers solitary and the male flowers on a few-flowered, long-pedunculate raceme.

2A. Flowers large, campanulate, white or yellow, never narrowly tubular; fruits large, more or less gourdlike, with many small seeds, the surfaces usually smooth except in one *Luffa* species with operculate apex — Leaves mostly with typical very broadly ovate, broadly angular-lobed cucurbit shape; mostly rather coarse, but nonwoody vines (tribe Cucurbiteae [indehiscent] plus *Luffa* [dehiscent]).

Cucurbita (15 spp., plus a few in N. Am.) — One species native in coastal Ecuador. Otherwise, cultivated and rarely escaped; flowers very large, yellow; fruit = gourd, squash or pumpkin.

C, E, P: sapayo

Sicana (3 spp.) — Cultivated; flowers yellow, fruits cylindric, dark red or purple, fragrant; similar to *Cucumis* but tendril 3–5-branched.

(*Citrullus*) — Leaves pinnately lobed (unique). Native to Africa but widely cultivated and often escaped. Fruit = watermelon.

C, E, P: sandia

Luffa (2 spp., plus 5 in Old World) — Fruit operculate, the surface somewhat spiny in the only native species, smooth in introduced species (these often escaped), the mesocarp fibrous-spongy; flowers large, openly campanulate, yellow. Leaves very broadly ovate, deeply cordate, remotely shallowly toothed, always rough above.

E: estopa, estropajo

(*Cucumis*) — Cultivated and rarely escaped; flowers yellow, tendrils simple; fruit (indehiscent) = melon or cucumber; one species with spiny fruit.

Lagenaria (1 sp., plus 5 in Africa) — Cultivated and rarely escaped; flowers white; tendrils two-branched; petiole apex with 2 conspicuous glands; fruit (indehiscent) = "bottle gourd".

Cucurbitaceae
(Lianas and a Single-Seeded Vine; Tendrils Branching Below Middle or Simple)

1 - *Sechium*

2 - *Cayaponia*

3 - *Psiguria*

4 - *Gurania*

5 - *Calycophysum*

Cucurbitaceae
(Many-Seeded Vines; Tendrils Branching Below Middle or Simple)

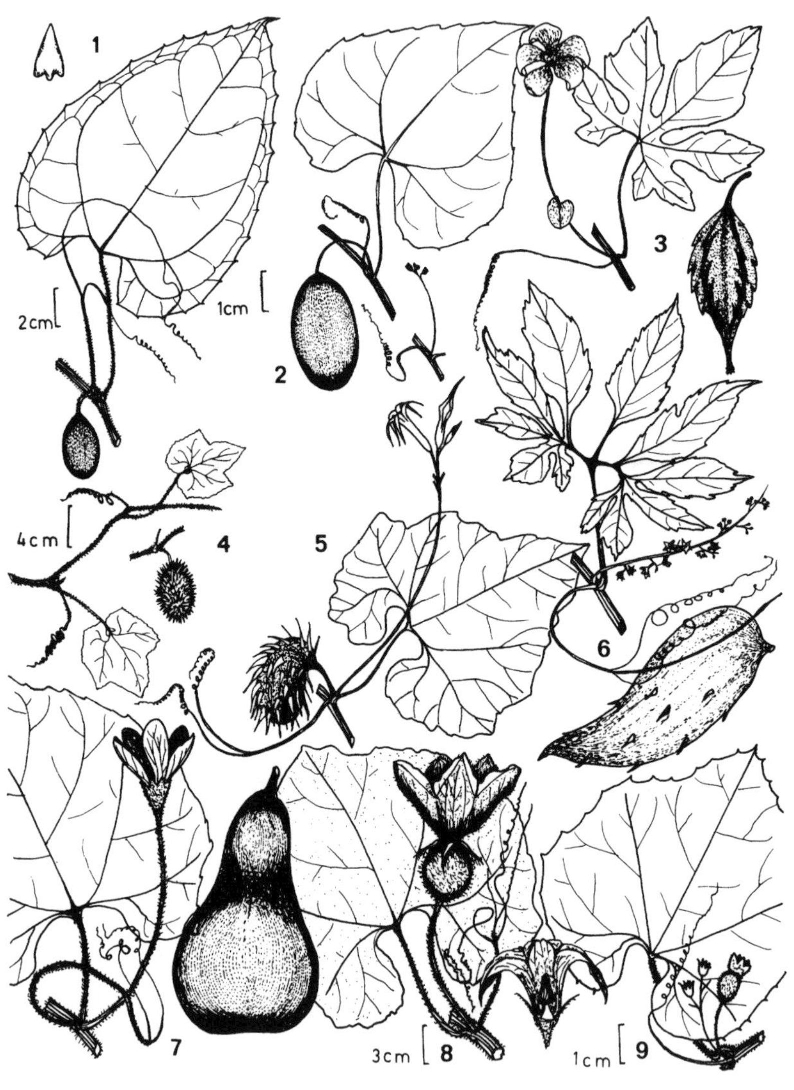

1 - *Selysia* **2** - *Melothria* **3** - *Momordica*

4 - *Echinopepon* **5** - *Rytidostylis* **6** - *Cyclanthera*

7 - *Lagenaria* **8** - *Cucurbita* **9** - *Posadaea*

Calycophysum (5 spp.) — Andean middle elevations, flowers all solitary, calyx inflated (cf., *Physalis*), mostly with distinctly floccose-villous petioles and leaves; the 5 foliaceous lobes often split to near base; fruit ellipsoid; flowers greenish-cream.

(*Posadaea*) — In fruit *Posadaea* (see below) would be confused with this group. It shows the smaller flowers and racemose male inflorescence of next subgroup but a globose to obovoid 8–10 cm diameter fruit.

2B. Flowers small or +/- narrowly tubular; fruits (except Posadaea) not large and gourdlike — Leaves of typical angular-lobed, serrate, broadly ovate Cucurbitaceae form, usually rough-surfaced — inflorescence (at least male) several–many-flowered, racemose or paniculate. Fruits (except *Posadaea*) either: 1) globose to oval and berrylike with a smooth, single-colored, leathery exocarp (e.g., *Cayaponia, Selysia, Melothria*), 2) cylindric, usually striped vertically dark and light green, rather soft and borne on a long usually many-fruited, pendent inflorescence (*Psiguria, Gurania*), or 3) variously spiny or with warty projections and usually dehiscent or single-seeded (or both).

2Ba. Fruit globose to ellipsoid, smooth, baccate, indehiscent (less than 3 cm long [except *Posadaea*]); flowers usually cream to greenish (yellow in *Melothria* and *Apodanthera*) — Leaves with distinctive glands at base of blade.

Cayaponia (60 spp., plus few in Africa and Java) — The largest neotropical cucurbit genus. Leaves entire to deeply lobed and rarely 3-foliolate, usually distinctly rough; inflorescences typically racemose or paniculate, rarely reduced to a single axillary flower (this is the only genus other than the *Fevillea-Sicydium* alliance to have (usually) branched female inflorescences.); fruit a characteristic ellipsoid, smooth, leathery-surfaced, fibrous berry, usually several-seeded.

E: melón de monte

Selysia (3 spp.) — Close to *Cayaponia* but with solitary (female) or fasciculate (male) axillary flowers, unique arrow-head-shaped seeds, and single, relatively large, globose, red fruits on rather long pedicels; leaves cordate, ovate, and subentire to strongly 3-lobed.

Apodanthera (15 spp.) — Mostly coastal and montane dry areas; herbaceous vines; flowers yellow to orange, tubular-infundibuliform with narrow pubescent calyx tube (possibly in part hummingbird-pollinated?); fruits ellipsoid to cylindric, mostly 2–4 cm long.

Melothria (11 spp.) — Very slender weedy vine, vegetatively similar to *Momordica* but leaves usually less deeply lobed and fruits ellipsoid and indehiscent; tendrils uniformly simple (in part forked in *Momordica*).

Posadaea (1 sp.) — Rather like *Lagenaria* in fruit but lacking petiolar glands and with much smaller white male flowers clustered at end of slender peduncle; fruit large (8–10 cm diam) globose to obovoid with a thick smooth exocarp.

2Bb. Flowers orange (very rarely yellow) with calyx tube elongate-cylindric or orange-red; fruits several (–many)-seeded, rather soft and cylindric, vertically light- and dark-green-striped, usually many pendent together on long female inflorescence — Tendrils always simple; leaves usually deeply lobed and frequently 3–5-foliolate.

Gurania (36 spp.) — Calyx usually orange red, its lobes longer than the yellow corolla; leaves usually serrate, rarely compound, often conspicuously pubescent and/or rough-surfaced.

E: zapallito; P: sapaya de monte

Psiguria (11 spp.)(incl. *Anguria*) — Calyx green and cylindric, its lobes shorter than orange-red corolla; leaves glabrous, usually 3–5-foliolate with entire- or serrulate-margined leaflets.

P: sapaya de monte

Dieudonnaea (1 sp.) — Woody liana; flowers red, cauliflorous; leaves and branches densely villous.

2Bc. Fruits mostly dehiscent and/or spiny with rather fleshy projections; sometimes one-seeded; flowers white, greenish or yellow — Slender often rather succulent weedy (or cultivated) vines.

Echinopepon (14 spp.) — Slender, fragile vine +/- densely villous, fruit very spiny, opening by operculum or terminal pore.

Rytidostylis (9 spp.) (incl. *Elaterium*) — Slender weedy vine, very common; flowers yellow with a long narrow tube and narrow acute lobes; fruits green, covered with soft bristles, elastically dehiscent to expose white-arillate seeds; fruits and pistillate flowers solitary.

Momordica (3 spp.) — Slender weedy vine, very common; fruits fleshy, orange, dehiscing to expose small seeds with bright red arils; leaves deeply divided, membranaceous, rather small; flowers yellow, openly 5-lobed, ca. 1 cm across.

C: balsamina; E: soroci, balsamina; P: papailla

Cyclanthera (29 spp.) — Slender rather succulent vines; fruits usually elastically dehiscent, more or less spiny with fleshy tubercles; leaves often palmately compound (one common simple-leaved species, *C. explodens,* is extremely like *Rytidostylis* in fruit but has the tiny, broadly lobed, white flowers of this genus); tendrils simple to branched, fruits and pistillate flowers solitary.

E: cochocho, achogcha de monte

Elateriopsis (5 spp.) — Fruit ovoid, to 6–8 cm long, usually asymmetrically gibbous at base and tapering evenly to apex; usually without spines; male inflorescence long-pedunculate, with solitary female flower at base; flowers greenish-white, openly campanulate, the male with 5 connate stamens; leaves broadly triangular, not lobed, usually entire except for the angles.

Sicyos (25 spp.) — Slender spindly vines similar to *Elateriopsis* but 3 (rather than 5) stamens and clustered female flowers and fruits; fruits usually in pedunculate clusters, ovoid, not dehiscent, dry, usually more or less spiny, usually ca. 1 cm long (= exozoochoric); tendrils 3–5-branched; leaf usually broadly angularly, very shallowly 3–5-lobate, otherwise subentire. The united filaments are an important technical character.

E: cohombro

Sechium (7 spp.) — Fruits ellipsoid, to 20 cm long, *single-seeded;* tendrils 3–5-branched; leaves very shallowly 3-lobed with the lobes broadly acute. Like *Sicyos* in united filaments, but not anthers.

E: achogcha; P: chayote

There are a number of additional extralimital genera. Perhaps in our area, *Franztia,* Central American characterized by fruits with few spines, differing from *Sechium* (and *Elateriopsis* and *Sicyos*) in fused anthers, smaller fruit than *Sechium.*

CUNONIACEAE

Exclusively montane mesic forest trees. Usually with opposite, pinnately compound, leaves with serrate-margined leaflets and distinctive half-round leafy caducous stipules (leaving a prominent interpetiolar line when caducous). The pinnately compound species almost always have a conspicuously winged rachis (unique among serrate-margined taxa in our area). The leaflets, often very small, are blunt-tipped with coarsely serrate to remotely dentate margins, usually thin, glabrate to somewhat puberulous. A few species have simple leaves but these still have the characteristic stipule and have exactly the same form as individual leaflets of the

Cunoniaceae and Cyrillaceae

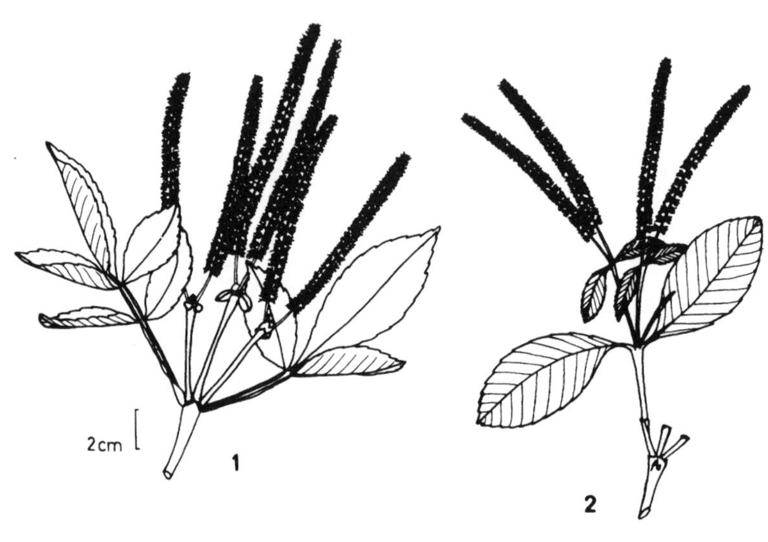

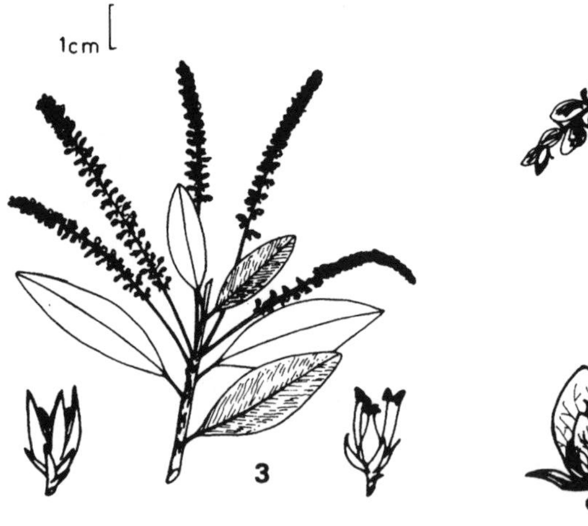

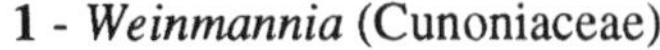

1 - *Weinmannia* (Cunoniaceae) **2** - *Weinmannia*

3 - *Cyrilla* (Cyrillaceae) **4** - *Purdiaea* (Cyrillaceae)

other taxa. Flowers small, white, clustered along the narrow terminal racemes. Fruit small, dry, thin, beaked, splitting incompletely in half to release seed. Brunelliaceae should probably be included in this family.

Weinmannia (70 spp., incl. Old World)
C: encenillo

(***Gumillea***)(1 sp.) is a putative member of this family with alternate leaves known only from the Ruiz and Pavon type from Muna, Peru.

CYRILLACEAE

Small trees of exposed windswept slopes at middle elevations, especially on poor sandy soil. Vegetatively characterized by the narrowly obovate, blunt-tipped, coriaceous, glabrous leaves with poorly developed secondary veins and intricately prominulous tertiary venation above and below. Inflorescence a narrow raceme of small flowers, similar to *Clethra.*

Cyrilla (1 sp.) — Mostly Antilles and Guayana Highlands; barely reaching Guayanan Colombia. Flowers white.

Purdiaea (12 spp.) — Mostly Cuban. Shrubs differing from *Cyrilla* in having lavender flowers, 10 stamens, and especially the 2 outer calyx lobes conspicuously enlarged and serrate-margined.

DIALYPETALANTHACEAE

Large trees of seasonally dry forest characterized by opposite leaves with two pairs of very distinctive completely separate large triangular interpetiolar stipules at each node, the leaves obovate blunt-tipped, coarsely puberulous below, drying dark above. Trunk also very characteristic, gradually expanded and fluted at base and with thick reddish fibrous bark; reminiscent of some sapotacs but no latex. Flowers white, in narrow terminal panicles. Capsule splitting into four parts to liberate very slender winged seeds. Combines Myrtaceae-like flowers with Rubiaceae-like fruits.

Dialypetalanthus (1 sp.) — Barely reaches Madre de Dios.

Dialypetalanthaceae and Dichapetalaceae

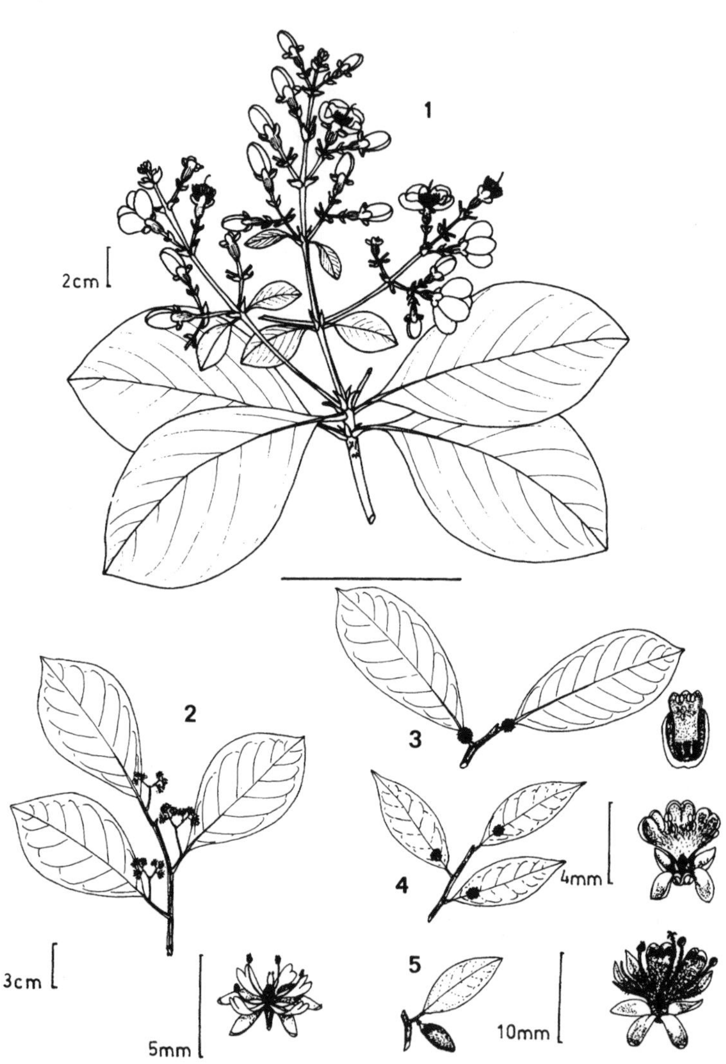

1 - *Dialypetalanthus*
(Dialypetalanthaceae)

2 - *Dichapetalum*

3 - *Stephanopodium*

4 - *Tapura*

5 - *Tapura*

DICHAPETALACEAE

Two genera of trees and one mostly of lianas, all characterized by entire, alternate simple leaves. The lianas (*Dichapetalum*) can usually be recognized by their very characteristic smooth pale bark with scattered darker lenticellate pustules; most tree species of *Dichapetalum* (and some lianas) have conspicuous stipules, sometimes with unusual serrate or fimbriate margins. The tree genera can be recognized by the characteristic uniformly terete (at least below inflorescence scar) tannish-puberulous thickish petiole, often with an apical scar indicating former position of the inflorescence. The family is rescued from taxonomic obscurity by the unusual placement of the inflorescence on the petiole or even leaf blade (rarely axillary in *Dichapetalum*). The inflorescence is dichotomously branched and flat-topped or reduced to a fascicle of often sessile flowers. Flowers always small and nondescript. Fruits rather dry single-seeded drupes, nearly always tannish-puberulous.

Dichapetalum is easily distinguished by the pedunculate inflorescence (usually a pair of peduncles arising from opposite sides of petiole). *Tapura* and *Stephanopodium*, both with sessile or fasciculate flowers (occasionally with a very short peduncle), are very similar and differentiable only on floral characters; both have petals at least in part basally fused unlike *Dichapetalum*.

Dichapetalum (20 spp., plus almost 200 in Old World) — Mostly canopy lianas, usually with smooth pale bark with raised darker pustules. Leaves more or less obovate and often conspicuously bullate or raised-reticulate below, frequently variously pubescent. Easily distinguished from other genera by the pedunculate strongly dichotomous inflorescence. Unlike the other genera, most of the tree species have persistent stipules, often with serrate margins.

Tapura (18 spp., plus 5 in Africa) — Differs from *Stephanopodium* in the flowers minutely zygomorphic with bifid corolla lobes longer than the tube and 3–5 anthers on slender filaments. Vegetatively characterized by the leaves usually either less oblong and/or noncuneate and/or the petiole longer than in *Stephanopodium*.

P: tapurón

Stephanopodium (9 spp.) — Differs from *Tapura* in the flowers radially symmetrical with the short obtuse corolla lobes about equal and 5 sessile anthers. The leaves are shorter and more elliptic than in most *Tapura*, with a shorter petiole not very sharply demarcated from cuneate leaf base.

DILLENIACEAE

Mostly lianas, usually easy to recognize by their papery fibrous reddish bark and the concentric-circle growth rings of the stem cross section. The leaves are very distinctive: often rough and sandpapery, usually with sharp, more or less spine-tipped serrations, at least when young, and with straight, parallel, close-together secondary veins connected by subperpendicular, close-together, strongly parallel tertiary veinlets; the petiole is often narrowly winged (or at least grooved above) and usually more or less continuous with the cuneate leaf base. The multistaminate white flowers of most neotropical species are fairly small, short-lived, and often synchronously blooming. *Curatella,* with asperous stellate-pubescent leaves is the only tree genus in our area. The liana genera are distinguished primarily by their fruits. *Davilla,* always scabrous-leaved, has two enlarged opposing hemispheric yellowish-tan calyx lobes which enclose the fruit (or bud) between them. *Tetracera,* usually scabrous-leaved, has the usually four exposed apocarpous carpels splitting open along the inner face. *Doliocarpus,* vegetatively the least typical Dilleniaceae genus, has usually smooth leaves with mostly more or less entire margins and often lacking the typical straight close-together secondary veins; it has larger rather fleshy reddish fruits, dehiscing into two halves to expose a succulent whitish aril in which the seeds are embedded. Vegetatively, *Pinzona* is similar to *Doliocarpus* in its smooth entire leaves but with close-together secondary veins and the tertiary venation finer than in *Doliocarpus*, the lower leaf surface, thus, plane between the secondary veins; the fruit is very small, medially constricted into two parts, green, and dehisces to show orange-arillate seeds.

Many species are highly regarded as "bejucos de agua" by the campesinos.

Curatella (1 sp.) — Twisted thick-barked tree characteristic of most neotropical savannas. Leaves very scabrous above, stellate-pubescent; flowers and fruits fasciculate ramiflorous.

C: chaparro

Tetracera (15 spp., plus 25 Old World) — Liana, usually with scabrous leaves; the pubescence of lower surface stellate, at least in part; inflorescence paniculate; fruits usually of 4 apocarpous cocci each splitting along upper surface.

P: lija sacha, paujil chaqui

Davilla (17 spp.) — Lianas, mostly in second growth, with more or less scabrous leaves, the pubescence of simple trichomes; inflorescence paniculate; small round fruit enclosed by two opposing yellowish to tannish enlarged calyx lobes.

Dilleniaceae

1 - *Davilla* **2** - *Doliocarpus*

3 - *Tetracera* **4** - *Curatella*

5 - *Pinzona*

Dipterocarpaceae, Droseraceae, and Ebenaceae

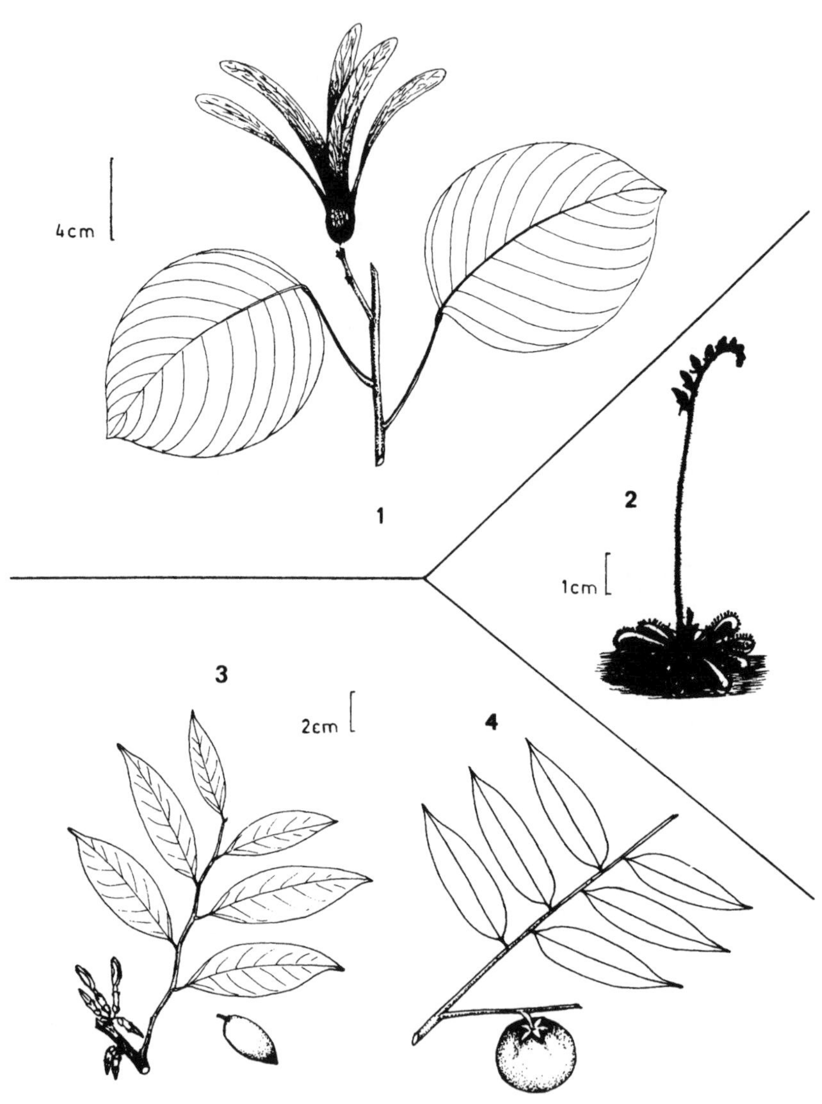

1 - Dipterocarpaceae (undescribed)

2 - *Drosera* (Droseraceae)

3 - *Lissocarpus*

4 - *Diospyros*

Doliocarpus (26 spp.) — Canopy lianas of mature forest; leaves smooth and often entire, secondary veins less consistently close, straight and parallel than in other genera; flowers and fruits fasciculate, usually below leaves; fruits reddish, relatively large, fleshy, splitting in half to reveal watery succulent white aril in which 1–2 seeds are embedded.

Pinzona (1 sp) — Canopy liana; like *Doliocarpus* in smooth entire leaves, but the secondary veins straight and closely parallel; tertiary veins rigidly parallel but very fine so undersurface appears plane; flowers and fruits in fascicles on branches below leaves; fruits tiny, medially constricted, greenish, dehiscing to show orange aril.

Dipterocarpaceae

The dominant tree family in much of tropical Asia, very poorly represented in the Neotropics by two monotypic genera, *Pakaraimea,* in the Guayana Highland region, and a newly discovered genus in Amazonian Colombia. The family is distinctive in the combination of a Malvalean pulvinus with pinnate venation. The Colombian taxon, not yet described, is obviously Dipterocarpaceae on vegetative grounds with very distinctive long-petiolate broadly ovate leaves (reminiscent of those of *Hura crepitans*) with the secondary veins straighter and closer together than in *Sloanea,* the only genus with which it could conceivably be mistaken. The fruit is a samara with 5 very long parallel wings arising from, and loosely enclosing, the base of the globose body.

Droseraceae

Tiny insectivorous herbs of moist poor-soil open areas, unmistakable in the sessile basal rosette of usually reddish leaves covered by glandular secretory trichomes that catch and digest tiny invertebrates; the few tiny flowers are borne near apex of small unbranched leafless raceme.

Drosera (15 spp., plus ca. 80 n. temperate and Old World)

Ebenaceae

A vegetatively rather nondescript family with alternate entire, usually coriaceous leaves, but with two very distinctive vegetative characters. The most diagnostic character (but not present in temperate-zone taxa) is the thin layer of strikingly black inner bark, this forming a black circle around the outside edge of a trunk slash (a similar though less intensely black ring is found in a few annonacs (*Boca-*

geopsis) but these differ in having the primitive odor). Another very useful character is rather large darkish glands on lower leaf surface, typically asymmetrically scattered along (but somewhat away from) midvein (never in axils of secondary veins), occasionally reduced to a single pair near base of midvein (and then very *Prunus*-like). The twigs and main veins below often distinctly rufous- or tannish-pubescent; the twigs are usually smooth and dark-drying when not conspicuously pubescent. The color of dried twigs of many species is blackish and the dried leaves tend to have a blackish or reddish-black tint; most species have leaves reminiscent of Annonaceae but lack the primitive odor and strong bark of that family.

The fruits are very distinctive in the partially fused expanded calyx lobes forming a rather flat, somewhat lobed, subtending cupule under the globose to ovoid, otherwise Sapotaceae-like, fruit; several species have the fruit rufous-hispid. Inflorescences always small and axillary, the small flowers more or less subfasciculate and with narrowly cupular calyces.

Diospyros (80 spp., also 400 in Old World) — Mostly small to midcanopy trees. Essentially as in the familial description above; fruits always globose, unlike *Lissocarpa* and with more expanded calyx. A number of species are conspicuously tannish-pubescent on twigs and main veins below.

P: tomatillo

Lissocarpa (2 spp.) — Trees of sandy-soil forest. Fruit more ovoid and fruiting calyx not conspicuously expanded. Glabrous; leaves drying olive or yellowish-olive, at least below and with tendency to have tertiary veins parallel to the inconspicuous secondaries, the venation (in Peru) prominulous above and below (rather *Heisteria*-like). Flowers with longer petals than *Diospyros* and glands on leaf underside tend to be fewer (apparently lacking on some individual leaves) and closer to midvein.

ELAEOCARPACEAE

In our area consisting of four genera of small to large trees with few obvious shared characters to unite them; indeed Elaeocarpaceae is often included in Tiliaceae from which it differs mainly in lacking the typical Malvalean mucilage ducts (and perhaps in nonvalvate petals, but *Sloanea* is apetalous). *Vallea, Dicraspidia,* and *Muntingia* have basally 3-veined leaves and are obviously Malvalean; *Sloanea* does not. The former three genera are among the very few Malvalean tree genera with persistent stipules; pinnately veined *Sloanea* also sometimes has persistent stipules and is distinctive in the strong tendency to have a mixture of opposite and alternate leaves, even on the same branch. None of these genera has the typical Malvalean pulvinus although *Sloanea* is characterized by a flexed (but not swollen as in most Malvales) petiole apex (cf., Euphorbiaceae, with several genera of which it can be vegetatively confused). *Muntingia*, strongly white-tomentose below, has the characteristic stellate trichomes of Malvales (and is also very viscid from longer gland-tipped hairs); the other two genera have glabrous leaves or simple trichomes. *Vallea* has pink or dark-red flowers, *Muntingia* white flowers, *Sloanea* apetalous greenish flowers. The fruit of *Muntingia* is a small red berry, that of *Vallea* an irregularly and incompletely dehiscent globose capsule covered by small warty projections, that of *Sloanea* a very characteristic 5-valved capsule, often spiny, dehiscing to reveal red-arillate seeds.

Muntingia (1 sp.) — A weedy second-growth tree characterized by the oblong leaf densely white-tomentose below with an irregularly shallowly serrate or biserrate margin and an extremely asymmetric base (cordate and subsessile on one side, cuneate and petiolate on the other). The strongly viscid vegetative parts are covered with a mixture of short stellate trichomes and longer, multicellular gland-tipped hairs. Flowers white and borne singly along the branchlets. Fruit a round red berry.

E: capulín, niguito; P: yumanaza atadijo

Dicraspidia (1 sp.) — Shrub or small tree of southern Central America to northern Colombia. Very like *Muntingia* but with amazing large persistent peltate-serrate stipules and large yellow axillary flowers.

Vallea (1–2 spp.) — Perhaps only a single highly variable species, widespread in upland Andean cloud forests. Characteristic in its broadly ovate (sometimes almost rotund) leaves and the usually persistent foliaceous stipules. The inflorescence is a usually few-flowered panicle with dark red or magenta flowers; the small round fruit is covered by warty projections.

Sloanea (70 spp., plus ca. 40 in Asia) — Large canopy trees of both lowland and middle-elevation forests. Vegetatively variable (glabrous or pubescent, stipules present or not, leaves large or small, margins entire or +/- toothed, petioles long or short) but characterized by a flexed (though not swollen) petiole apex and by the usual tendency to have a mixture of opposite and alternate leaves; highly unusual among neotropical Malvales in pinnate venation. Another good habit character is the usual development of unusually thin, frequently large, buttresses; a few species even have stilt buttresses (and an almondlike vegetative odor). The greenish apetalous flowers are small and inconspicuous but the characteristic 5-valved capsular fruits, usually with a spiny surface, are often obvious.

Elaeocarpaceae

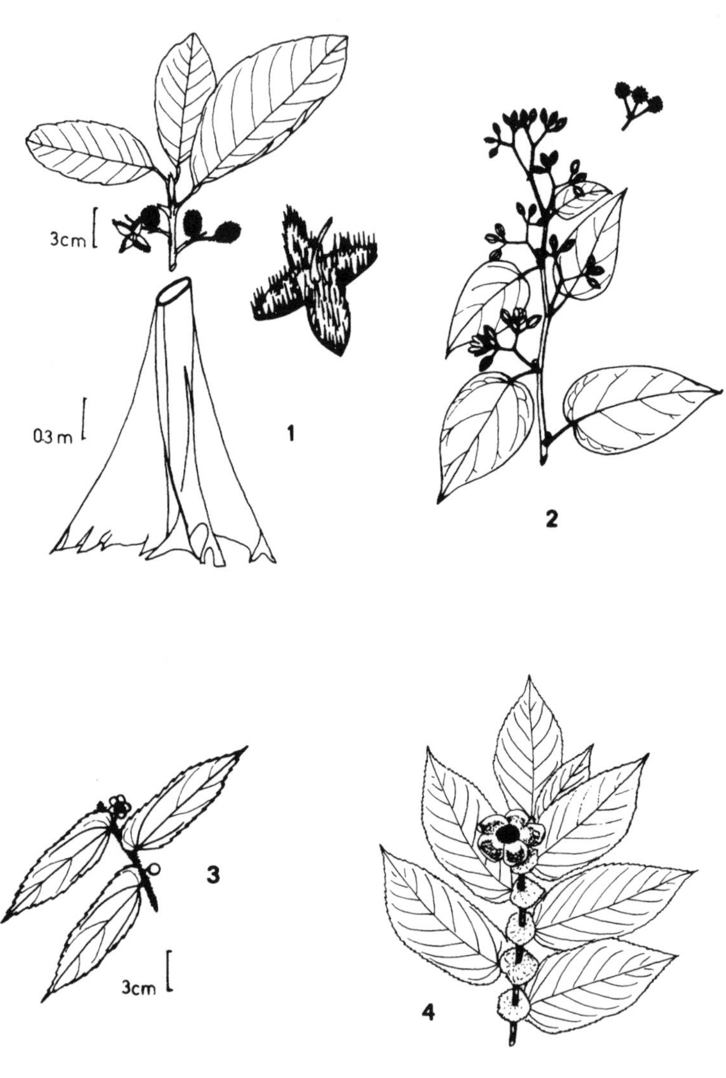

1 - *Sloanea* **2** - *Vallea*

3 - *Muntingia* **4** - *Dicraspidia*

C: embagatado; P: cepanchina, huangana casha, achotillo, pusanga caspi (= almond odor)

Extralimital neotropical genera are in the South Temperate Zone: *Aristotelia* and *Crinodendron.*

ERICACEAE

Very easy to recognize to family on account of the characteristic flower, the corolla thick, usually more or less urceolate, sometimes tubular with tiny reduced lobes, the anthers usually opening by pores. Most genera and species are epiphytic climbers with characteristic coriaceous alternate plinerved leaves, but many of the genera are very difficult to distinguish without flowers. Generic placement is often dependent on technical characters of the anthers (opening by pores or slits; connectives spurred or not, smooth or granular, etc.). The family is extremely prevalent in Andean cloud forests, poorly represented in the lowlands.

1. MONTANE (TERRESTRIAL) SHRUBS AND SUBSHRUBS WITH SMALL TYPICALLY ERICACEOUS URCEOLATE FLOWERS AND USUALLY SMALL LEAVES — Three main genera: *Pernettya, Gaultheria, Vaccinium,* all superficially similar, the first two unusual in the family in having superior ovaries.

Vaccinium (35 spp., plus 300 in Old World and N. Am.) — Differs from *Pernettya* and *Gaultheria* in having an inferior ovary.

Gaultheria (85 spp., plus 125 in Asia) — Differs from *Pernettya* in the calyx becoming succulent and making up part of the fruit; the flowers are always in racemes or panicles.

Pernettya (15 spp., plus 5 in New Zealand) — Similar to *Gaultheria* but the calyx lobes dry and reflexed in fruit and the flowers sometimes solitary in the upper leaf axils.

Gaylussacia (40 spp., plus 9 N. Am.) — Mostly Brazilian; in our area paramo shrubs, like *Vaccinium* in berry-fruit, but the oblong thick-coriaceous leaves larger than in *Vaccinium* or *Pernettya,* either with inrolled or conspicuously serrate margin; common species differs from *Pernettya* in the young stem grayish-pilose and leaf margin ciliate.

Leucothoe (35 spp., plus ca. 13 Asia and N. America) — Very poorly represented in our area. *Befaria*-like shrubs with similar dry capsular fruit and superior ovary but a typical *Gaultheria*-like urceolate flower. Differs from *Gaultheria* and *Pernettya* in the free sepals and dry fruit.

Ericaceae
(Shrubs and Some Distinctive Hemiepiphytes)

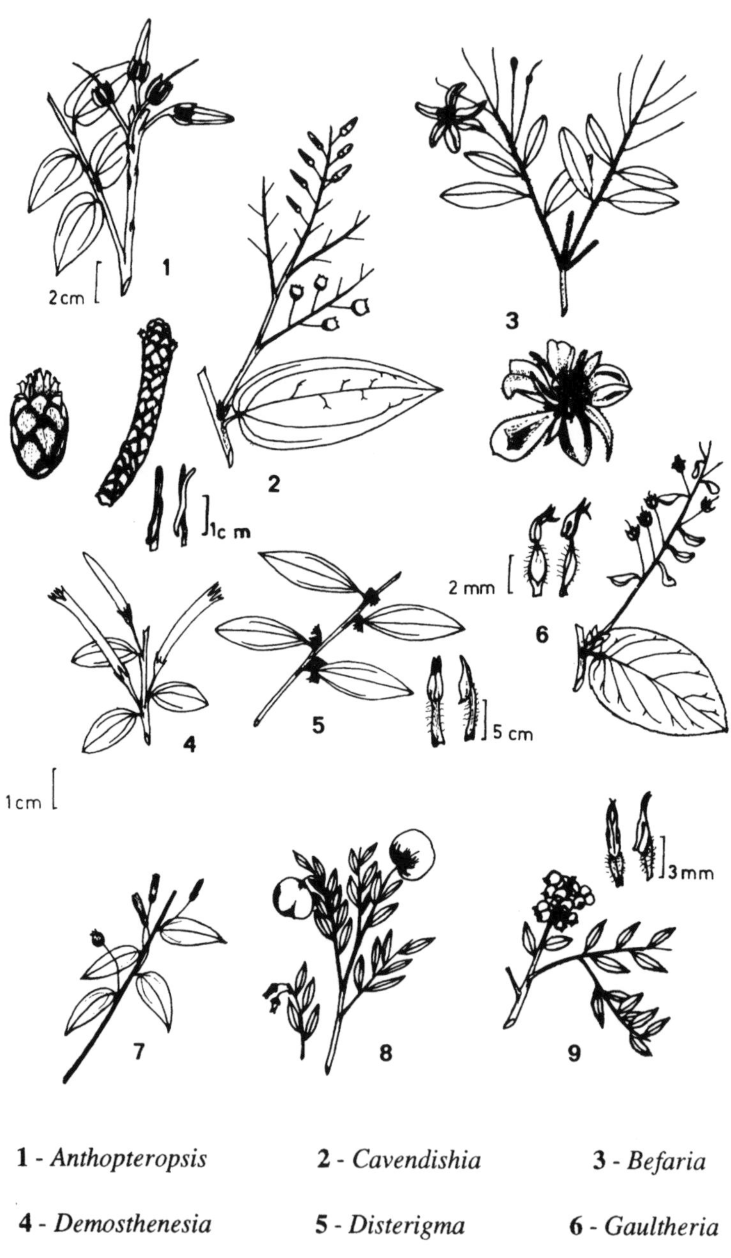

1 - *Anthopteropsis* **2** - *Cavendishia* **3** - *Befaria*

4 - *Demosthenesia* **5** - *Disterigma* **6** - *Gaultheria*

7 - *Sphyrosperma* **8** - *Pernettya* **9** - *Vaccinium*

Ericaceae
(Vines and Hemiepiphytes)

1 - *Ceratostema* **2** - *Psammisia*

3 - *Satyria* **4** - *Thibaudia*

5 - *Macleania* **6** - *Themistoclesia*

Disterigma (35 spp.) — Calyx enclosed by pair of ovate bracts; leaves entire, smooth with the secondary veins not evident.

2. Erect Shrubs with Larger Leaves and Relatively Large Thin-Textured Corolla (Petals Separate in *Befaria*)

Befaria (30 spp.) — Unique in petals separate; free-standing montane shrubs, the flowers usually large and conspicuous. Also distinctive in the fruit a dry capsule and the ovary superior.

Semiramisia (7 spp.) — Leaves medium small, often long-acuminate. Flowers solitary, the calyx entire-margined, the large corolla broadly cylindric-campanulate (to 2 cm wide) with reduced lobes.

3. Small-Leaved, Usually Very Slender, Appressed-Climbing Vines with Small Flowers and Unjointed Pedicels — (i.e., the flower base continuous with the pedicel; unjointed pedicels also present in ***Anthopterus*** and ***Demosthenesia***).

Sphyrospermum (16 spp.) — Small inconspicuous solitary flowers, the corolla usually less than 1 cm long, white or pinkish; flowers borne on rather long slender pedicels (usually longer than flowers) and with non-angled calyx.

Diogenesia (9 spp.) — Similar to *Sphyrospermum* in small leaves and flowers, but an erect shrub and with several flowers together in axillary fascicles.

Themistoclesia (30 spp.) — Similar to *Sphyrospermum* but with 5-angled calyx, flowers red and pedicels shorter than flowers (essentially a hummingbird-pollinated version of *Sphyrospermum*).

4. Robust-Stemmed, Woody, Mostly Hemiepiphytic Plants; Often Pendent-Branched or Shrubby Rather Than Climbing — The flowers usually larger than in *Sphyrospermum* group and with jointed pedicels.

Cavendishia (100 spp.) — One of the easiest ericac genera to recognize on account of the flowers enclosed by conspicuous bracts, at least in bud. Technical character is that alternate anthers and filaments are of different lengths; and the filaments are separate or fused only at extreme base.

The rest of the woody hemiepiphytic climbing genera are distinguishable only by technical characters of the stamens. They include:

Satyria (25 spp.) — Filaments completely connate; anthers of two different lengths (the latter unique except for *Cavendishia*).

Orthaea (15 spp.) — The closest relative of *Cavendishia* (and a few species have similar bracts) but the anthers are equal and dehisce by shorter pores. Corolla or calyx often conspicuously hairy.

Thibaudia (60 spp.) — Defined by combination of connate filaments and smooth flexible anther sacs, these tending to a more elongate dehiscence than in relatives.

Demosthenesia (9 spp.) — Mostly shrubs, the flowers solitary, red, *Fuschsia*-like with narrow lobes, subexserted anthers, and narrow calyx lobes. Differs from *Thibaudia* in pedicel continuous with calyx (a character shared with *Anthopterus* which differs in having calyx wings).

Ceratostema (18 spp.) — Calyx very characteristic, large ([1–]2–4 cm long), cupular and angled, with 5 triangular lobes; corolla swollen at base, with very long narrow acuminate lobes, the elongate exserted anthers held stiffly together.

Macleania (45 spp.) and ***Psammisia*** (50 spp.) — Both have rigid coarsely granular anther sacs, at least in part, and can have separate or connate filaments; *Macleania* typically has smaller, more oblong-based or subcordate leaves than *Psammisia.* The technical differentiating feature is that *Psammisia* has a pair of lateral projections (spurs) on the anther connectives (or at least on half of them) while *Macleania* does not.

Lysiclesia (2 spp.) — Like *Cavendishia* but much smaller bracts (less than 2 mm long) and only 3 calyx lobes, which are distinctively long (3–4 cm), thin, and acuminate, nearly as long as the narrowly tubular white corolla. Very small leaves with no obvious secondary venation.

Plutarchia (12 spp.) — Mostly high-altitude paramo and subparamo erect shrubs. The thick-coriaceous leaves small to very small. Flowers tubular and red, mostly single or in small axillary fascicles. Calyx with conspicuous triangular lobes, usually distinctively membranaceous and deeply split.

Siphonandra (3 spp.) — Our species with nondescript *Macleania*-like greenish-tipped flowers with connate filaments. Vegetatively distinctive in the small, thick, oblong leaves with conspicuous, scattered, large, and rather elongate punctations.

Erythroxylaceae

1 - *Erythroxylon*

2 - *Erythroxylon*

5. Miscellaneous Easy to Recognize Usually Epiphytic Genera

Anthopterus (7 spp.) — Corolla 5-winged, and pedicels not articulated with calyx.

(***Anthopteropsis***) (1 sp., Central America) — Calyx conspicuously 5-winged.

Lateropora (2 spp.) — Anthers opening by slits rather than pores.

Killipiella (2 spp.) — Cloud forests of western Cordillera of Colombia and northwestern Ecuador; scandent-branched epiphyte with very characteristic narrow, parallel-veined, monocot-looking leaves.

Erythroxylaceae

A single genus mostly of shrubs and small trees, only occasionally reaching lower canopy. Leaves always entire and simple, alternate or clustered on short-shoots, usually small, elliptic, and with a rounded apex, nearly always glabrous. Vegetatively distinctive in the usual presence of a pair of faint vernation lines subtending a slightly discolored area paralleling the leaf midvein below, and in the characteristic triangular intrapetiolar stipules, usually brownish or tannish, often longitudinally striate or ribbed, and tending to be clustered together and persistent on short-shoot branches. (When leaves ovate and acuminate in forest species, always with persistent striate stipules.) The small inconspicuous flowers are always borne in axillary or ramiflorous fascicles and are characterized by 10 stamens with the filaments united in tube. Fruit single-seeded, red, narrowly ellipsoid, and subtended by the five narrowly triangular calyx teeth. Another useful character is the conspicuously white-lenticellate young branchlets (cf., *Margaritaria* but glabrous).

Erythroxylon (180 spp., plus 87 in Old World) — In our area concentrated in the Caribbean coastal dry forests. Famous as a source of biologically active alkaloids.

C: maribara; E: mama cuca; C, E, P: coca

Euphorbiaceae

Infamous as one of the most variable of all families vegetatively: If you can't figure out what it is, try Euphorbiaceae (or Flacourtiaceae). The vast majority of species of this family can be rather readily recognized by combinations of milky latex, simple alternate leaves, a pair of glands at the petiole apex or base of leaf blade, and the presence of stipules or stipule scars; however, there are exceptions to every one of these characters. Similarly the fruit of the great majority of Euphorbiaceae is a characteristically 3-lobed, 3-valved capsule that is often crowned by the 3 persistent stigmas and usually fragments at dehiscence, but some genera have very different 2-valved or 4-valved or even drupaceous and indehiscent fruits. The male and female flowers are always separate, borne either monoeciously on the same plant or dioeciously on different plants; the stamens are typically many, the flowers small and apetalous and in advanced genera like *Euphorbia* and its relatives clustered into specialized inflorescences (cyathia) that mimic perfect flowers.

Nearly all of the many herbaceous euphorb species belong to a few large easily recognized and mostly weedy genera—*Euphorbia, Phyllanthus, Chamaesyce, Caperonia.* Three of our genera are woody lianas and two more are mostly thin-stemmed vines; there are occasional scandent species in a few other genera. The recognition problems lie in the trees and shrubs. Many of these genera are recognizable as euphorbs by the presence of a pair of glands near the often flexed apices of frequently different-length petioles (similar glands are found otherwise found only in one group of flacourt genera and a few miscellaneous genera of other families). The combination of these glands and milky latex is definitive for Euphorbiaceae as is the combination of serrate leaf margins with conspicuously different-length petioles having flexed apices. About half our euphorb taxa, including nearly all the climbers, have the leaves basally 3-veined. Several woody taxa (*Pera, Hyeronima, Croton*) are characterized by peltate scales and/or stellate trichomes, a few by urticating hairs (*Cnidoscolus, Tragia,* a few *Dalechampia* species), and at least two (*Pausandra,* some *Croton*) by red latex.

1. Herbs or More or Less Succulent Small Trees or Shrubs, Often with Latex

Euphorbia (incl. *Poinsettia*) (ca. 180 neotropical spp., plus many elsewhere) — Mostly weedy herbs always with milky latex; a few species are small, rather succulent, understory trees with large leaves and abundant latex and a few (mostly cultivated) species are leafless and cactuslike. Leaves membranaceous to succulent, opposite or whorled to alternate; the much-reduced male and female flowers are aggregated together into a very characteristic flowerlike inflorescence (cyathium).

Chamaesyce (175 spp., plus 75 in Old World) — Mostly small weedy herbs; a sometime segregate from *Euphorbia* from which it differs in a different strictly sympodial growth-form, uniformly opposite leaves always with asymmetrical bases, and persistent stipules; *Euphorbia* differs in always symmetrical leaf bases and having the stipules absent or developed into glands. *Chamaesyce* is usually recognized in the Neotropics but typically lumped into *Euphorbia* in North America.

Pedilanthus (14 spp.) — Succulent shrubs or small trees of dry areas; very distinct in the thick chlorophyllous stems and the rather few small, succulent, often rather rhombic leaves; the somewhat shoe-shaped asymmetric red flower is unique.

Phyllanthus (over 200 spp., plus 400 in Old World) — Mostly weeds, lacking latex and with small alternate entire leaves typically each with one to several small greenish apetalous flowers or tiny round 3-parted fruits in its axil. A few species are small trees, with the leaves arranged in a single plane along a slender branchlet, the branchlet looking very much like a pinnately compound leaf unless flowers or fruits are present.

C: chirrinchao, chirrinchao macho; E: culo pesado (*P. juglandifolius*), barbasco (*P. anisolobus*)

Andrachne (2 spp., plus 15 Old World) — Herbs of very dry areas, very like *Phyllanthus* but differing in having petals; vegetatively often distinguishable by having glandular trichomes.

Caperonia (10 spp.) — Aquatic or semiaquatic with hollow stem and lacking latex; most species have the stem coarsely hispid. Leaves usually finely sharp-serrate and with many secondary veins.

Acalypha (285 spp., plus 165 in Old World) — Mostly weedy herbs and shrubs, but also including some small weedy trees and at least one (sometimes?) liana; lacking milky latex. Leaves alternate, membranaceous, serrate, nearly always conspicuously 3-veined from base. Inflorescence a characteristic spike, often with female inflorescences somewhat bottle-brush-like from the long slender reddish stigmas and unique in having flowers subtended by foliaceous bracts; in some species female flowers borne singly at base of spicate male inflorescence.

(*Croton*) — A few species of this mostly woody genus are herbs, characterized by the presence of stellate hairs or lepidote scales.

(*Dalechampia*) — One area species of mostly scandent *Dalechampia* is an erect herb, easy to recognize by the typical large paired inflorescence bracts of the genus.

Euphorbiaceae
(Herbs and Succulent Shrubs)

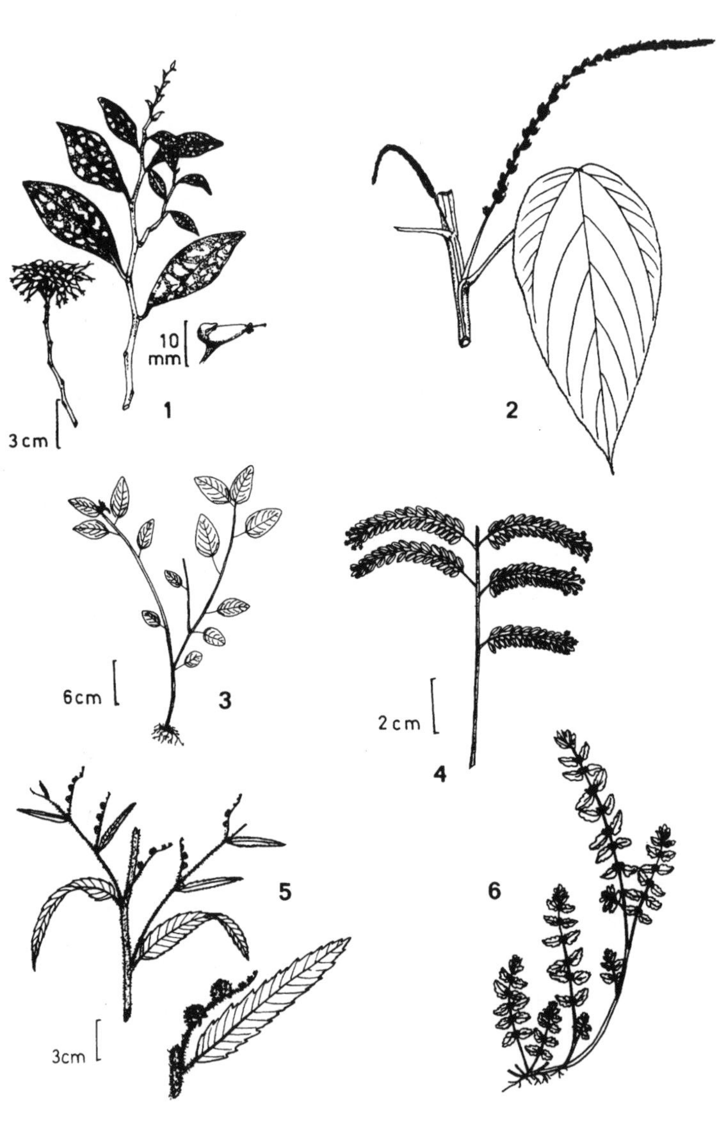

1 - *Pedilanthus* **2** - *Acalypha*

3 - *Euphorbia* **4** - *Phyllanthus*

5 - *Caperonia* **6** - *Chamaesyce*

2. Vines and Lianas, Lacking Milky Latex or (*Omphalea*) with a Peculiar Cloudy-Watery Sap That Sometimes Turns Pinkish or Somewhat Bluish with Oxidation — All have 3-veined (rarely 3-foliolate in *Dalechampia*), often broadly ovate leaves usually with a pair of glands near base of leaf blade or apex of petiole.

Dalechampia (95 spp., plus 15 in Old World) — Mostly vines of disturbed areas, easy to recognize by the pair of large, often bright pink or white bracts that encloses the inflorescence. The 3-parted fruit, often enclosed in the persistent bracts, is usually pubescent and sometimes urticating.

E: ortiguilla; P: manicillo

Tragia (60 spp., plus 40 in Old World) — Urticating vines with small strongly serrate leaves. Male flowers in a very slender spike, the female similar to *Dalechampia* but without the bracts; fruits small, 3-parted, puberulous.

Omphalea (14 spp., plus 6 in Old World) — A very thick liana mostly of swampy places, with characteristic broadly ovate 3-veined leaves with a pair of rather thick glands at petiole apex. The fruit is large, globose, fleshy, and indehiscent. Climbs by a twining tendril-like juvenile branchlet apex. Latex cloudy and spermlike, often turning somewhat pinkish or bluish with oxidation. Outside our area often a tree. Famous as the host plant of the day-flying *Urania* moth.

Plukenetia (incl. *Apodandra* and *Elaeophora*) (14 spp., plus 1 in Madagascar) — A thick-stemmed liana, the common species with palmately 5–7-veined leaves, ovate but with rather squarish basal "corners" and a very characteristic pair of large narrow glands near base of blade above; other species with more elliptic sometimes pinnate-veined leaves, also with prominent pair of glands on more or less auriculate lamina base. Inflorescence a tenuous raceme or racemose panicle. Fruit usually tetragonal (rarely 5–6-parted), strongly longitudinally sulcate, much wider than long. *Apodandra,* with stamens united into a globose receptacle, is best treated as congeneric. Seeds of *P. volubilis* edible.

P: mani de monte, sacha inchi (*P. volubilis*), manicillo

Ophellantha (3 spp.) — A Central American genus (close to Antillean *Acidocroton*), recently discovered in dry part of Magdalena Valley. Ours a very distinctive vine with spines subtending the nodes and with sparsely reddish-hirsute stems and leaf margins; flower subtended by whorl of leaflike bracts.

(*Manihot*) — A few species are scrambling vines, the only vine euphorbs with palmately lobed leaves.

Euphorbiaceae
(Vines and Lianas)

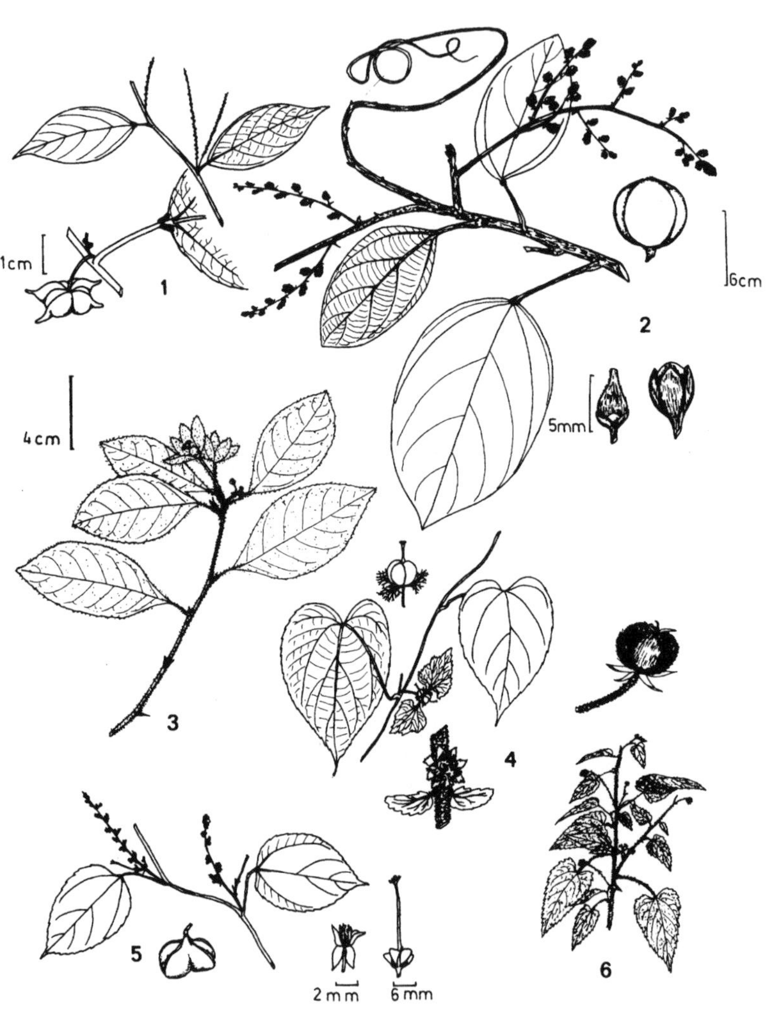

1 - *Plukenetia* (*Apodandra*)

2 - *Omphalea*

3 - *Ophellantha*

4 - *Dalechampia*

5 - *Plukenetia*

6 - *Tragia*

3. Compound or Deeply Palmately Lobed Leaves

Hevea (8 spp.) — Restricted to Amazonia. Unique among our euphorbs (along with *Piranhea*) in 3-foliolate leaves; the small white flowers in a paniculate inflorescence and explosively dehiscent 3-parted fruit are typically euphorbiaceous. Famous as the source of rubber, once the major product of Amazonia.

P: shiringa, jeve debil fino

Piranhea (1–2 spp.) — A small tree of black-water-inundated riversides in the Guayana Shield region. Differs from similarly 3-foliolate *Hevea* in lacking latex and in the much smaller more angular 4-valved fruit (splitting into eight sections at dehiscence). Distinguishable from *Allophylus* by the habitat and more coriaceous more or less crenate-margined leaflets.

Manihot (98 spp.) — Shrubs, scrambling vines, and small trees, mostly in dry-forest or disturbed habitats. Distinctive in the palmately deeply 3–7-parted glabrous leaves, these often with a whitish-waxy coating on the underside. The trees are characterized by thin, very strongly peeling blackish bark. This is the source of "yuca", the staple food of most of Amazonia.

E, P: yuca, manihot

Jatropha (80 spp., plus 90 in Old World) — Mostly shrubs and small trees, often cultivated, especially in dry areas. The flowers have petals, sometimes fused into a tube, and are often brightly colored, sometimes quite ornamental. Our species have glabrous leaves, unlike related *Cnidoscolus*. Can be distinguished from *Manihot* by having the flowers in dichotomous cymes and by the presence of petals.

C: tuatua (*J. gossypifolia*); E: piñón

Cnidoscolus (75 spp.) — Herbs and shrubs of dry areas, characterized by coarse stiff strongly urticating trichomes. The leaves are palmately lobed as in *Jatropha*, the apetalous flowers always white from the petaloid sepals.

C: mala mujer

Ricinus (1 sp.) — An introduced weedy shrub, now widespread in Neotropics. Differentiated from other palmately lobed species by the strongly peltate leaf blade with many gradually tapering lobes, each finely serrate along the margin.

E: higuerilla

(*Dalechampia*) — A few *Dalechampia* vines have 3-foliolate leaves.

Euphorbiaceae
(Leaves 3-Foliolate or Deeply Palmately Lobed)

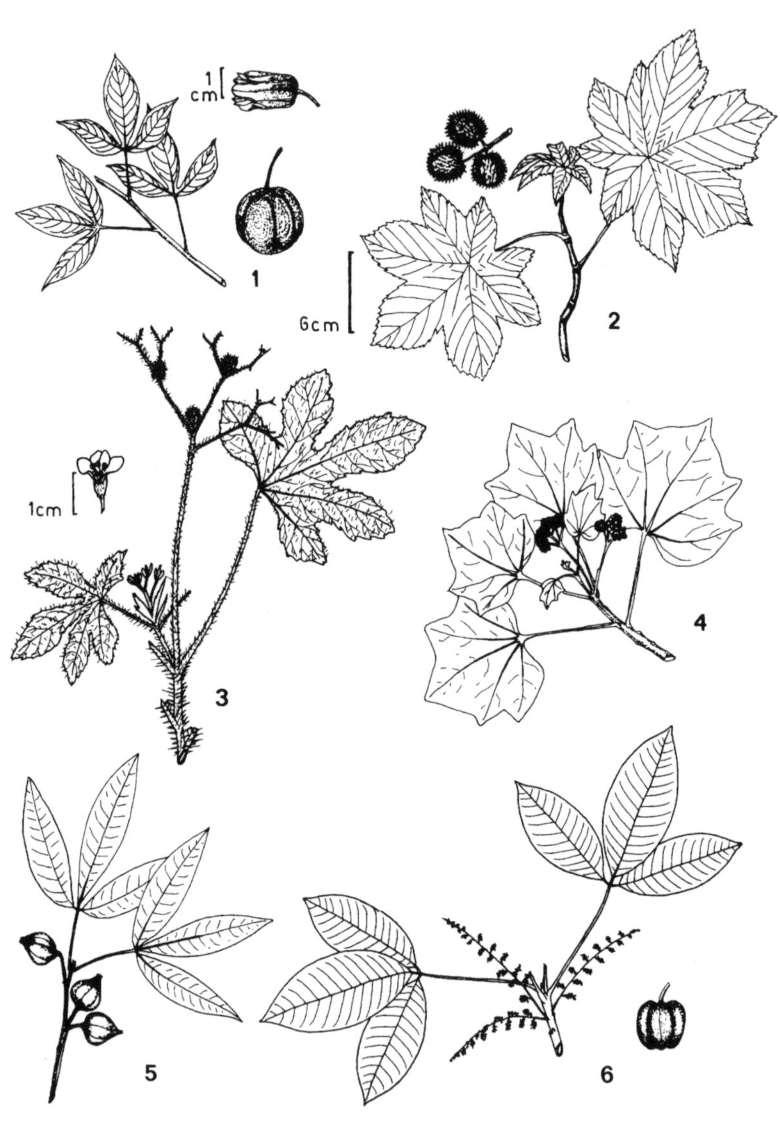

1 - *Manihot* **2** - *Ricinus*

3 - *Cnidoscolus* **4** - *Jatropha*

5 - *Piranhea* **6** - *Hevea*

4. Trees (or Shrubs) with Usually Conspicuous Peltate Scales or Stellate Trichomes on Leaves and Inflorescences (First Four Genera and Few *Chiropetalum*) or Mostly Malpighiaceous Trichomes Mixed with Some Simple Trichomes (*Argythamnia* and Some *Chiropetalum*).

Hyeronima (36 spp.) — Large forest trees especially prevalent in cloud forests, characterized by the usually densely lepidote leaf undersurface, pinnate venation, the paniculate inflorescence with spiciform branches, and the small berrylike drupaceous fruits. Some species have prominent pouchlike stipules when young.

C: pantano, cuacho; P: urucurana

Pera (40 spp.) — Medium to large trees especially prevalent in sandy soils or on exposed wind-swept ridges. Mostly ramiflorous with the multiple short-shoot inflorescences consisting of clusters of few flowers, the apetalous male flowers subtended by pair of small bracts. Fruit 3-parted, more or less obovoid with truncate apex, densely tan-lepidote. Leaves nearly always dark-drying or with whitish undersurface from the scales, lacking glands or with inconspicuous glands at base of midvein above.

Gavarettia (1 sp.) — Trees, mostly Brazilian, recently discovered in Peru. Leaves coriaceous like *Caryodendron* but more obovate and usually with glands below on the cuneate blade base and also on top of petiole apex. Inflorescence a terminal spike. Fruits asymmetrically 2-valved and tan-tomentose.

Croton (incl. *Julocroton*) (400 spp., plus 350 in Old World) — An extremely large and diverse genus; most species are dry-area shrubs or trees of dry or second-growth habitats; a few are weedy herbs. Many species lack latex; when present the latex is conspicuously red or rather orangish. Most species have 3-veined leaves, either serrate or entire but always with stellate or peltate trichomes, and a capsular 3-parted fruit.

C: algodoncillo; P: sangre de grado (*C. lechleri*), shambo quiro (*C. palanostigma*)

Chiropetalum (25 spp.) — Small shrubs or subshrubs, in our area only in dry inter-Andean valleys. Like *Argythamnia* except for deeply and narrowly lobed petals of male flowers. Indument varies from stellate to a mixture of simple and malpighiaceous trichomes. Leaves with pinnate or 3-veined venation.

Argythamnia (17 spp.) — Shrubs of dry areas, poorly represented in our area. Characterized by herbaceous subsessile or short-petiolate leaves with few strongly ascending veins that are usually narrowly elliptic, typically grayish-sericeous below. The inflorescence is axillary, contracted, and subtended by small rather scarious bracts.

Euphorbiaceae
(Trees: Leaves with Peltate Scales or Stellate Trichomes and Lacking Petiole Glands; Latex Red or Absent)

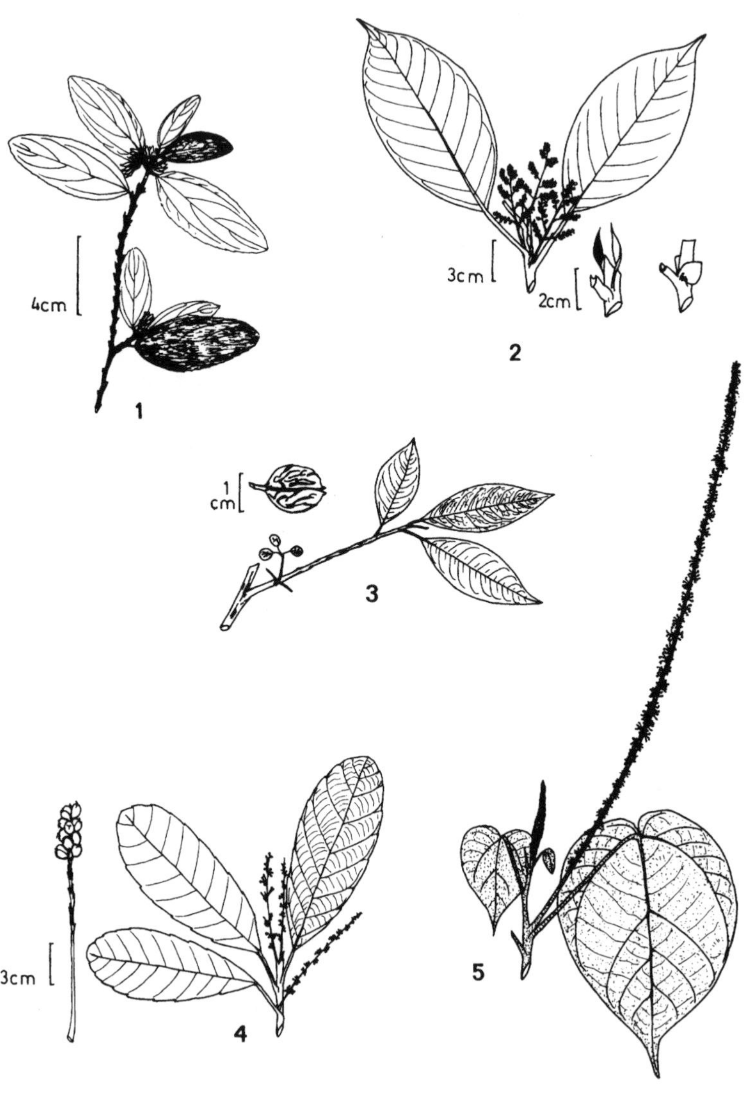

1 - *Argythamnia* **2** - *Hyeronima*

3 - *Pera*

4 - *Gavarretia* **5** - *Croton*

5. **Trees with Conspicuously 3-Veined Leaves** — *Acalypha* and *Bernardia* usually lack petiole or lamina-base glands and have a membranaceous blade with serrate margins; the other genera have glands near petiole apex (or at base of lamina) and a usually more chartaceous or coriaceous texture and entire to weakly or remotely serrate margins.

(***Acalypha***) — Mostly weedy herbs and shrubs (see above) but a few species are trees. All are characterized by strongly serrate leaves and the typical spicate inflorescence.

Alchornea (36 spp., plus 34 Old World) — Mostly large trees and present in most neotropical forests, especially important at middle elevations and on sandy soils. Vegetatively very characteristic in having conspicuous glands in the axils of the basal nerve pair below and also in the crenulate to remotely serrate (but sometimes subentire) margin. The male inflorescence is an open, thin-branched panicle with the flowers in sessile clusters along it; the female inflorescence is usually spicate and often ramiflorous. Fruits unusual in being 2-parted, though topped by the persistent stigmas as in many euphorbs. A useful vegetative character is a rather spongy inner bark that often shows a trace of nonflowing reddish sap when squeezed.

C: escobo

Bernardia (50 spp.) — Essentially a shrubby reduced version of *Alchornea,* occurring in dry or disturbed sites and characterized by rather membranaceous small pubescent serrate leaves, usually short petioles (outside our region the leaves often pinnately veined and may lack obvious axillary glands). The inflorescence is a small inconspicuous axillary spike.

Aparisthmium (1 sp.) — Small to middle-sized tree, common in disturbed areas in Amazonia, especially on poor sandy soils. Closely related to and formerly included in *Alchornea,* from which it differs in the 3-parted rather than 2-parted fruit. Leaves very characteristic in ovate to broadly ovate shape, crenate margins, and especially in having two glandlike stipels at apex of petiole above, as well as, the typical *Alchornea*-type glands in axils of basal nerve pair below.

Alchorneopsis (3 spp.) — Strongly 3-veined with entire to serrulate leaves. Like *Alchornea* but the basal gland pair *outside* basal pair of veins. Inflorescence spicate, axillary, sparsely flowered.

Conceveiba (7 spp., plus 1 recently discovered in Africa) — A common tree in many lowland Amazonian forests. Leaves only sub-3-veined but very characteristic in oblong shape with very remote and shallow serrations, rather parallel tertiary veins perpendicular to the secondary veins, and different-length petioles. Petioles strongly flexed and somewhat thickened

Euphorbiaceae
(Trees with 3-Veined Leaves; Mostly with Glands at Base of Blade)

1 - *Glycidendron* **2** - *Alchornea*
3 - *Conceveiba*
4 - *Alchorneopsis*
5 - *Conceveibastrum* **6** - *Aparisthmium*

at apex and usually with a pair of very inconspicuous apical glands above. Inflorescence terminal, *Alchornea*-like, the male with many spikes from a central rachis, the female a simple spike. Fruits trigonal-globose and characteristically rough-surfaced with 3 conspicuous bifurcate stigmas. Differs from other species with similar but clearly pinnate leaves in the combination of nonentire margins and completely lacking latex.

Conceveibastrum (2 spp.) — Extremely distinctive Amazonian tree with very large broadly ovate finely tomentose leaves deeply cordate at base; petiole apex with conspicuous glands above. Stipules unique, very conspicuous, foliaceous, serrate, acuminate. Inflorescence terminal, the female a thick spike, the male a large panicle.

P: sacha sapote

Glycidendron (1 sp.) — Large Amazonian tree of rich soils. Distinctive in the completely entire, coriaceous, oblong-elliptic leaf 3-veined almost to apex and with a pair of very conspicuous glands at apex of longish petiole above. Inflorescence 1–several rather small axillary racemes; fruits oblong and somewhat stipitate, not dehiscent. Latex present unlike similarly strongly 3-veined taxa.

6. Leaves with Distinctly Pinnate Venation — The first four taxa below are characterized by a pair of glands or glandular enations *below* petiole apex, the next nine by having the flowers and fruits fasciculate or solitary in the leaf axils or along the twigs (and the leaves generally entire, and the petioles short and uniform); the final twenty have variously spicate to paniculate inflorescences (and, except *Gymnanthus*, the leaves either serrate or with slender and/or different-length petioles).

6A. Glandular enations from below petiole apex (rarely absent or vestigial in few *Sapium*)

Sapium (95 spp., plus 25 in Old World) — Common large trees, especially prevalent in succcessional forest. Always with latex and this sometimes caustic. Very distinctive vegetatively in the usually strongly projecting pair of glands borne well *below* petiole apex. The elliptic-oblong leaves also distinctive in having rather close-together not very clearly differentiated secondary and intersecondary veins, and in the usually very finely serrate or serrulate margin. Inflorescence a thickish terminal spike. Fruits 3-parted, fragmenting.

C: cauchillo; E: barbasco; P: guta percha

Tetrorchidium (14 spp., plus few in Africa) — Petiolar glands very similar to those of *Sapium* but the leaf obovate, entire or very remotely and shallowly toothed, intersecondary veins absent and the secondary veins more ascending. Inflorescence usually a few-branched rather tenuous axillary panicle.

C: palo tunda

Euphorbiaceae
(Phyllanthoideae: Trees with No Latex Nor Petiole Glands)

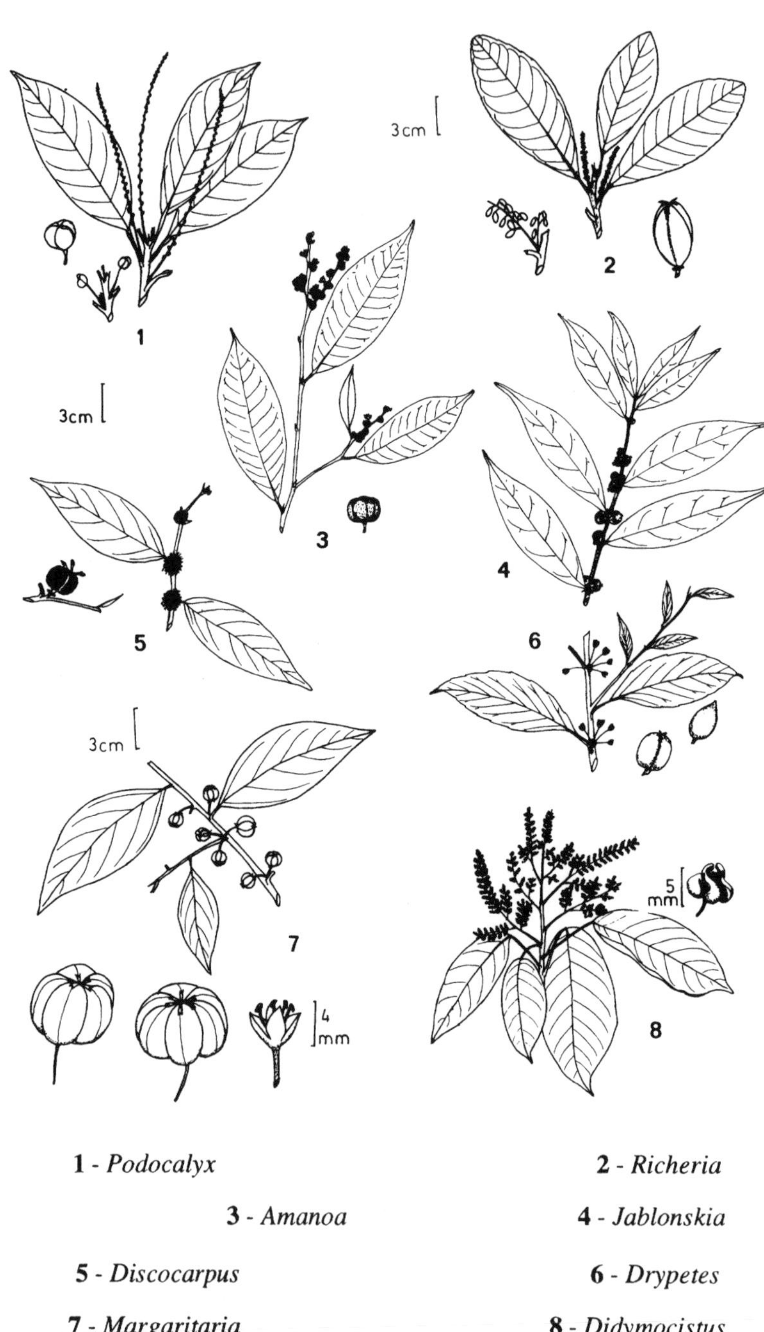

1 - *Podocalyx*
2 - *Richeria*
3 - *Amanoa*
4 - *Jablonskia*
5 - *Discocarpus*
6 - *Drypetes*
7 - *Margaritaria*
8 - *Didymocistus*

Pausandra (12 spp.) — Petiole glands similar to *Sapium* but more clearly representing enations from near the petiole apex. Leaves distinct from *Sapium* in being oblanceolate with a strongly flexed petiole apex and remotely (but usually conspicuously) serrate margin. Usually with red latex in branches (but this not apparent in leaves), unlike the two genera above.

Stillingia (30 spp., plus 3 in Old World) — Essentially a shrubby *Sapium*, mostly in dry areas; leaves finely serrate, small, usually with pair of raised petiole glands.

6B. The next nine genera have pinnate venation and the flowers or fruits fasciculate or solitary in the leaf axils or along the twigs. — All lack latex, have entire leaf margins (in our area), and relatively short uniform-length petioles.

Drypetes (15 spp., plus 185 in Old World) — Small to fairly large trees characterized especially by being very difficult to recognize to family. Relevant characters are coriaceous often rather small shiny leaves with *nearly always asymmetric bases* and the tertiary venation prominulous and subparallel to the secondary veins. The rarely collected flowers are minute and in axillary fascicles. Fruit a small ellipsoid axillary drupe.

Margaritaria (3 spp., plus 8 in Old World) — Small to large trees, both geographically and ecologically widespread but very nondescript. Nevertheless, quite easily recognized by the smallish, thin, entire, rather rhombic-elliptic leaf with a well-developed but short petiole and by the characteristic somewhat zigzag twigs, usually reddish with white lenticels and often somewhat puberulous. The small fruits very distinctive, depressed-globose and 4-parted with blue seeds.

P: mojara caspi

Jablonskia (1 sp., sometimes included in *Securinega* with 24 Old World spp.) — Shrub of seasonally inundated forests. Very *Margaritaria*-like but the leaves subsessile. Flowers in axillary glomerules and the 3-parted fruits in small axillary clusters and completely sessile. Stipule conspicuous and lanceolate, caducous to leave prominent scar.

Savia (14 spp., plus 17 in USA and Africa) — Shrubs on limestone, mostly Antillean and very poorly represented in our area where occurring only in northern Colombia. Very like *Margaritaria* but the sessile fruit 3-parted and the flowers in completely sessile axillary fascicles.

Adelia (5 spp., mostly Central American) — Often spiny or spiny-branched small trees sometimes treated as part of *Bernardia* but differing in entire pinnately veined leaves, these also characteristic in obovate shape and thin texture; another useful vegetative feature is the frequent clustering of tiny pubescent whitish scales in the leaf axils, giving a cushionlike effect.

Euphorbiaceae
(Trees: Pinnate-Veined Acalyphoideae; Latex Red, Watery or Absent; Mostly with Petiole Glands)

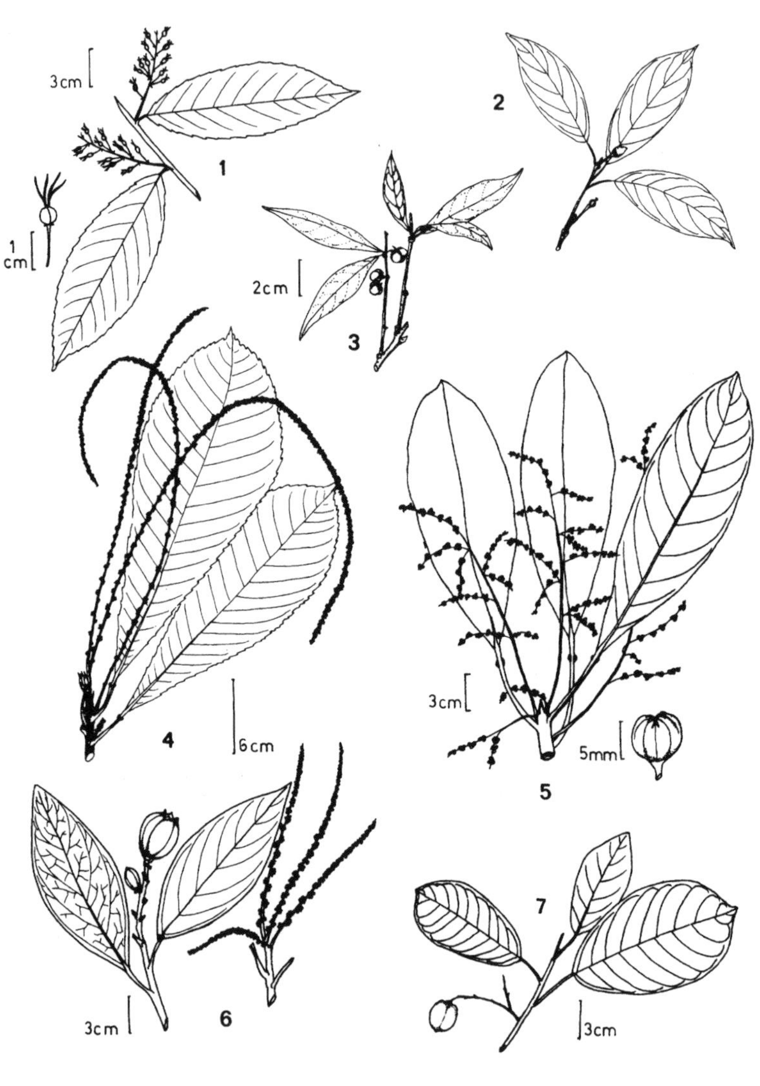

1 - *Cleidion*
2 - *Garcia*
3 - *Adelia*
4 - *Pausandra*
5 - *Tetrorchidium*
6 - *Caryodendron*
7 - *Micrandra*

Chaetocarpus (10 spp., plus few in Old World) — Large tree with oblong-ovate, thick-coriaceous, dark reddish-gray-drying leaves and sometimes a conspicuous small thick-foliaceous stipule. Inflorescence an axillary fascicle, the fruit 3-valved and with a characteristic verrucose surface somewhat reminiscent of *Lindackeria.*

Amanoa (7 spp., plus 3 in Africa) — Vegetatively characterized by elliptic completely entire, coriaceous leaves with cartilaginous margin and short petiole and the intrapetiolar stipules. Flowers fasciculate on axillary short-shoots or these more or less expanded into "spikes" (though still with reduced leaves subtending each flower cluster). Fruit larger and thicker than in other short-petioled genera, 3-valved, very woody, ca. 3 cm long.

Croizatia (3 spp.) — Shrub or very small understory tree, differentiable vegetatively from very similar *Amanoa* in lacking intrapetiolar stipules. Technically differs in being dioecious instead of monoecious, petals pubescent instead of glabrous, and the styles twice (instead of once) bifid.

Tacarcuna (3 spp.) — Midcanopy trees with narrow +/- lanceolate or oblanceolate leaves with short petioles, very like *Amanoa.* Fruits 3-lobed, <1 cm long.

Discocarpus (5 spp.) — *Amanoa*-like coriaceous leaves with a more or less cartilaginous margin and short petiole. Fruits in subsessile or sessile axillary clusters, tan-puberulous, more or less wrinkled-verrucose.

6C. Trees or shrubs with pinnately veined leaves and the flowers (at least male) and usually the fruits (except *Acidoton, Hippomane, Hura*) in spicate to paniculate inflorescences — The rest of the genera are additionally characterized by petioles mostly of different lengths (often in part conspicuously long) and the leaves often serrate or serrulate (always when the petioles short and uniform). The next six genera have short uniform petioles and serrate leaves. *Hippomane* has long slender equal-length petioles, the last nine genera have different-length petioles and mostly entire (except *Nealchornea*) leaves.

6Ca. Short uniform petioles and serrate leaves

Acidoton (6 spp.) — Shrubs with serrate membranaceous leaves, also sometimes vegetatively distinguished by axillary whitish "cushions" similar to *Adelia.* Male flowers in spikes, female in short spike or subsessile fascicle. Fruit 3-valved, frequently solitary.

Adenophaedra (4 spp.) — Shrubs with cuneate oblanceolate to narrowly obovate, shallowly serrate leaves; petioles somewhat intermediate between groups 6Ca and 6Cd, short and of uniform lengths but flexed at apex. Inflorescence as in *Pausandra*, a long sparsely flowered spike; capsule 3-parted.

Sebastiana (95 spp., plus 4 in Old World) — Mostly shrubs and subshrubs (also a few climbers) of Brazilian Shield area; in our area shrubs with serrulate or finely serrate small chartaceous to membranaceous leaves, and short eglandular petioles. Inflorescence a slender axillary spike, usually with flowers sparsely clustered along it. Fruits small and strongly 3-sulcate.

Cleidion (7 spp., plus 18 in Old World) — Essentially a pinnate-veined *Alchornea* with the more or less serrate leaves unusually membranaceous. Both male and female inflorescence an *Alchornea*-like tenuous spike or the male paniculate.

Mabea (50 spp.) — Very distinctive in elliptic leaves nearly always with a serrulate margin and usually noticeably whitish below, the venation brochidodromous or festooned-brochiodromous, usually with some intersecondaries parallel to the secondary veins, sometimes with dendroid (but not stellate) trichomes. Petioles short and of uniform length, sometimes with glands at junction of petiole and blade above. Always with latex and a characteristic whorled branching pattern (Fig. 4). Inflorescence very characteristic, an axillary or leaf-opposed bottle-brush raceme of male flowers with a single female flower at base, the latter much larger and with a long style and 3 coiled stigmas. Capsule 3-parted and fragmenting.

C: chamizo; P: pólvora caspi

Richeria (6 spp.) — Large trees with usually only slightly crenulate-obovate, darkish-reddish drying leaves cuneate to a nonflexed eglandular petiole which is usually slightly enlarged and somewhat woody at base (cf., some Sapotaceae or Annonaceae all of which have completely entire margins). Inflorescence an axillary rather tenuous spike or narrow raceme. Fruit ellipsoid, rather small, partially splitting to extrude the orange-arillate seed.

6Cb. The next two genera are somewhat intermediate between the previous and following groups. One has margins so finely serrate as to appear entire and the other varies in the same species from conspicuously serrate to entire — Both have noticeably long slender petioles, in *Hippomane* of essentially equal lengths, in *Hura*, rather inconspicuously different lengths. Both have caustic latex, a pair of glands at petiole apex, and both are exceedingly distinctive in their secondary venation.

Hippomane (5 spp.) — Beach trees with notoriously poisonous latex and very characteristic leaves with an elliptic to ovate, extremely inconspicuously and finely serrate- or serrulate-margined blade on an

Euphorbiaceae

(Trees: Pinnate-Veined Euphorbioideae; Latex White or Watery-White, Often Poisonous; Leaf Margins Usually Serrate)

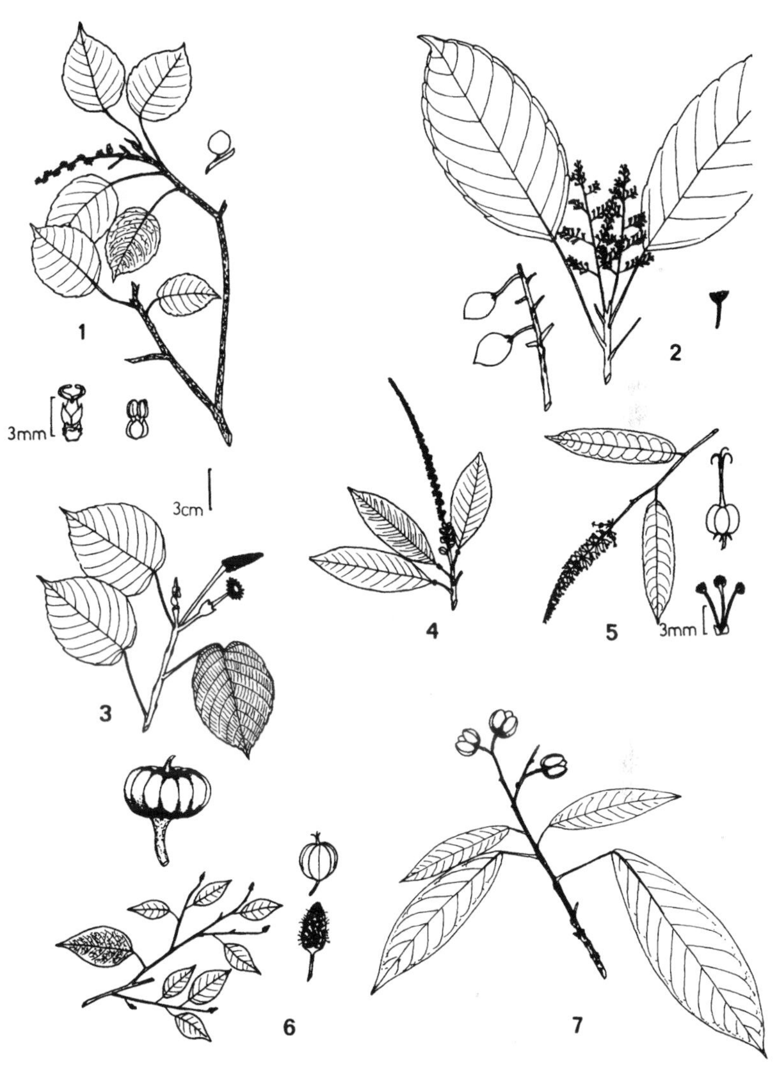

1 - *Hippomane* **2** - *Nealchornea*

3 - *Hura* **4** - *Sapium* **5** - *Mabea*

6 - *Maprounea* **7** - *Senefeldera*

unusually long and slender (but not of conspicuously different lengths) petiole with a small gland pair at its apex above; *secondary veins reduced and hardly evident.* Male inflorescence as in *Sapium;* fruit axillary, sessile, round, indehiscent.

C: manzanillo

Hura (2 spp.) — Large tree, with spiny trunk, mostly found along streams especially in somewhat swampy forests. One of the most distinctive of all neotropical trees on account of the very broadly ovate leaf with rather close-together and parallel prominent secondary veins nearly perpendicular to midvein; margin variable from clearly serrate to apparently entire; petioles long and of somewhat different lengths, with gland pair at apex above. Male inflorescence a thick, dense, almost conical, reddish spike; female flower solitary, the style and stigmas together with a peculiar umbrella-like aspect. Fruit large, depressed globose, explosively dehiscing into numerous segments.

C: ceiba amarilla, ceiba de leche, ceiba de agua; P: catahua

6Cc. The next two genera have entire leaves and equal-length petioles.

Gymnanthes (30 spp., plus few in USA) — Mostly Antillean and Central American shrubs and small trees with elliptic to rhombic oblanceolate, rather coriaceous, entire leaves with short uniform petioles. Commonest species with pair of conspicuous glands in auricle-like basal "lobes" below and thick conspicuous subfoliaceous stipules. Inflorescence racemose; fruit small, 3-parted.

Maprounea (1 sp., plus few in Africa) — Shrubs (in dry areas) to large forest trees with small elliptic or elliptic-ovate, entire, glossy leaves with slender petioles. Inflorescence a small, contracted axillary spike with terminal cluster of sessile male flowers and a single long-pedicellate female arising well below this cluster. Fruits small, round, 3-sulcate.

6Cd. Trees with pinnate venation and conspicuously different-length petioles — All nine of these genera have more or less flexed petiole apex, usually with pair of glands above and all except *Nealchornea* have entire margins.

Nealchornea (1 sp.) — Midcanopy Amazonian tree unusual in combination of conspicuously different-length petioles and nonentire margin. Trunk with characteristic proteinaceous spermlike latex. Leaves oblong, remotely serrate or serrulate, glossy, and with tertiary venation more or less perpendicular to midvein; petioles with pair of glands *below* at extreme base of bluntly cuneate lamina (unique). Male inflorescence a much-branched terminal panicle; fruit large (3 cm diameter), obovate, indehiscent, tannish. Vegetatively very reminiscent of *Conceveiba,* the only other similarly serrate-leaved taxon with conspicuously different-length petioles, but that genus lacks latex and has the leaves somewhat 3-veined at base and with inconspicuous glands at apex of petiole *above.*

P: huira caspi

Senefeldera (10 spp.) — Trees very like *Conceveiba* in oblong-elliptic leaves jointed to flexed petiole but with strictly pinnate veins and entire margin. Unlike *Didymocistus, Nealchornea* and other similar taxa in the leaves very strongly clustered at intervals on the twigs. With conspicuous glands near base of midvein below but usually lacking noticeable glands at petiole apex above. Inflorescence a narrow terminal panicle, the male with bracteate *Phoradendron*-like spicate branches; fruits pedicellate, 3-parted, depressed-globose.

Sagotia (2 spp.) — Nonlactiferous trees mostly of Guayana area poor soils, but reaches Madre de Dios. Leaves rather oblong, with the different-length petioles characterized especially by a double pulvinus at base and apex, the apical swelling very conspicuous (even subwoody), *dorsally grooved,* and somewhat glandular. The terminal stipules resembling Moraceae, falling to leave conspicuous scar; the sometimes in part opposite leaves on flexed petioles are easy to confuse with *Sloanea* but well-developed buttresses are lacking. Inflorescence a rather long-pedicelled, few-flowered, terminal raceme with well-developed petals.

Pogonophora (1–2 spp., also in Africa) — Trees with oblong to elliptic entire coriaceous leaves with tertiary veins more or less parallel and perpendicular to midvein. Petiole strongly grooved and flexed at apex. Inflorescence a small racemelike axillary panicle. The main technical character, in addition to the inflorescence, is the presence of petals.

Dodecastigma (3 spp.) — Trees with large, thick, coriaceous, cartilaginous-margined entire leaves similar to those of *Caryodendron* but larger. Leaves unusual in strong tendency for blade to disarticulate at conspicuously jointed and swollen petiole apex (unifoliolate?). Flowers like *Pogonophora* in having petals but borne in larger subterminal panicles; fruits woody-valved and similar to, but smaller than, *Caryodendron.*

Caryodendron (3 spp.) — Large trees with characteristic smooth and patchily greenish bark and no latex; mostly on rather fertile soils. Leaves elliptic, entire, coriaceous, with cartilaginous margin; glands not on petiole but on upper side of cuneate blade base just above petiole apex. Inflorescence spicate and similar to *Sapium.* Fruits large, globose or subtrigonal, 3-valved, with edible oil-rich seeds.

P: meto huayo, inchi

Didymocistus (1 sp.) — Trees of seasonally inundated forest. Vegetatively reminiscent of *Conceveiba* in oblong leaves with different-length petioles with the apices flexed and with inconspicuous glands above; differs in being completely pinnate-veined and entire. Inflorescence a terminal panicle with densely arranged sessile flowers and fruits. Fruits very distinctive, of two ovate halves, each medially constricted to give the intact fruit a strongly 4-sided aspect.

Garcia (2 spp.) — Small trees entering our area only in moist forest in northern Colombia. Leaves like *Caryodendron* with cartilaginous margin, but less coriaceous and somewhat smaller. Inflorescence reduced to cluster of several long-pedicellate terminal flowers with large maroon tepals. Fruit like reduced *Caryodendron* but tannish-tomentose and with rather finely ridged surface.

Micrandra (13 spp.) — Lactiferous trees of poor sandy soils, especially where poorly drained, characterized by conspicuous stilt roots. Leaves coriaceous, elliptic or oblong, the tertiary veins parallel to each other and perpendicular to midvein; with pair of large glands at petiole apex above. Branchlet apices with *Ficus*-like terminal stipule, this falling to leave conspicuous scar. Fruit rather globose, largish, 3-valved.

P: shiringa masha

(***Gavarretia***) — Some species have only inconspicuous few-branched stellate trichomes.

Fagaceae

Large upland forest trees, in our area restricted to the Colombian Andes, mostly above 1500 meters. Leaves simple, often with an inconspicuously serrate or serrulate margin (lacking the deep lobes of most temperate species), cuneate at base and not strongly differentiated from the short petiole. Characterized by the large size of the trees, by the tendency to have several clustered subterminal buds with bud scales, and by the tendency to round white lenticels on the branchlets. Young growth strongly pubescent (with mostly stellate hairs in *Quercus,* simple or 2-branched in *Trigonobalanus*). As in temperate zone Fagaceae, the flowers are very reduced and apetalous, the male in pendent catkins; the fruit is an acorn or acornlike, subtended by a bracteate cup.

Quercus (150 spp., plus ca. 250 in N. Am. and Eurasia) — Probably only a single variable species in South America. Dominating many Colombian montane forests, but apparently not reaching Ecuador. Vegetative

Fagaceae
and Flacourtiaceae (Odd-Ball Genera: Leaves Opposite or with Reduced Secondary Veins)

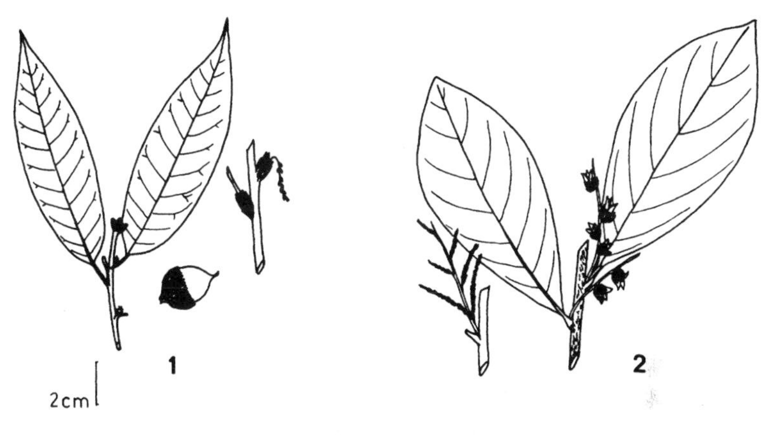

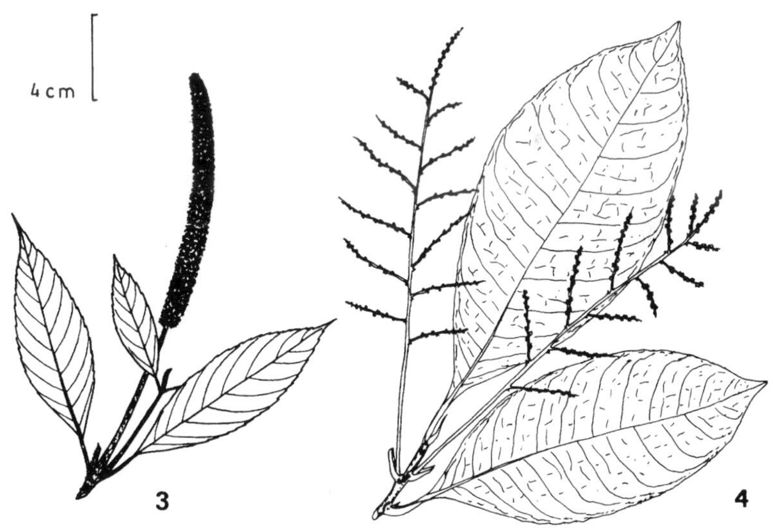

1 - *Quercus* **2** - *Trigonobalanus*

3 - *Abatia* **4** - *Euceraea*

pubescence mostly stellate. Female flowers sessile, axillary, solitary or few together. Fruit a typical oak acorn, subtended by bracteate cupule.

C: roble

Trigonobalanus (1 sp., plus 2 in Southeast Asia) — Only recently discovered in New World, but completely dominates some Andean Colombian forests; occurring mostly at lower altitudes than *Quercus* in forests that have been mostly destroyed. Differs from *Quercus* especially in the paniculately branched female inflorescence and the trigonal 3-parted fruit. Vegetatively distinguishable by mostly simple (or 2-branched) rather than stellate pubescence.

C: roble

FLACOURTIACEAE

One of the two most notoriously heterogenous neotropical families. A good dictum for the beginner is: "If you don't have any idea what family it is, try Flacourtiaceae or Euphorbiaceae." The leaves are always simple, almost always alternate (except montane *Abatia*), usually serrate (or at least serrulate), often with more or less parallel tertiary venation, and frequently minutely pellucid-punctate (look *toward* the light). Inconspicuous stipules are present at branch apices but usually early-caducous. The flowers are typically small, multistaminate, and arranged in reduced axillary inflorescences but may be large (*Ryania*), have only 4 or even one stamen (e.g., *Tetrathylacium, Lacistema*), and be arranged in open and/or terminal inflorescences (many genera). Spines are sometimes present (especially *Xylosma*) and the branchlets tend to be zigzag with 2-ranked leaves.

Taxonomic division of the family is based largely on rather obscure floral characters but the family can be rather easily subdivided into three vegetatively distinct and easy to recognize groups which generally cross the taxonomic lines (plus opposite-leaved montane *Abatia*): 1) the three genera with spiny or subspiny fruit are a natural group, vegetatively characterized by acute to long-cuneate leaf bases and noticeably flexed petiole apices, 2) an artificial group with obtuse-based conspicuously 3-veined leaves which are rarely pellucid-punctate and often have glands at base of blade or apex of petiole (cf., Euphorbiaceae), 3) a group of genera mostly related to *Casearia* which have pinnately nerved leaves, usually pellucid-punctate or pellucid-lineolate, short petioles (or the leaf base long-attenuate onto petiole), and usually serrate margins.

1. Opposite Leaves (Tribe Abatieae)

Abatia (9 spp.) — Montane. Unique in family in opposite leaves, these grayish-pubescsent below, serrate, with a +/- V-shaped interpetiolar ridge. Inflorescence a narrow terminal raceme or spike of small yellow flowers. Vegetatively, looks much like *Buddleja.*

2. Spiny or Subspiny Fruits (or with Longitudinal Papery "Wings")

— Petals more numerous than sepals; leaves with strongly flexed petioles and more or less acute bases, not pellucid-punctate (tribe Oncobeae); the inflorescences form a kind of reduction series from an open panicle in *Lindackeria* to completely cauliflorous in most *Carpotroche.*

Lindackeria (6 spp.) — Leaves entire, with long petioles flexed at apex. Inflorescence a few-flowered usually axillary raceme or panicle of small white flowers. Fruit round, the surface densely covered with small conical projections.

P: huaca pusillo

Mayna (6 spp., plus 1 in Africa) — Small trees or shrubs. Leaves narrowly obovate, serrate, the rather short petioles flexed at apex. Flowers white, in a 1–few-flowered axillary fascicle. Fruit distinctly bristly spiny.

Carpotroche (11 spp.) — Small understory trees, usually cauliflorous. Leaves serrate, usually very long. Flowers white. Fruit with thin longitudinal wings, these often dissected and the fruit (white in these species) with fleshy spinelike projections.

E: caraña; P: champa huayo

3. Leaves Strongly 3-Veined, Not Pellucid-Punctate Nor Pellucid-Lineolate

— (Except rather inconspicuously in *Lunania*), often with glands at base of blade or apex of petiole (cf., Euphorbiaceae), (stamens hypogynous, except *Lunania*).

Prockia (3 spp.) — Shrubby trees, especially in seasonal forest, with characteristic large subleafy semicircular stipules and strongly 3(–5)-veined, heart-shaped, serrate nonpunctate leaves. Inflorescence more or less terminal, 1–few-flowered, the flowers yellow.

Hasseltia (3 spp.) and ***Pleuranthodendron*** (1 sp.)— Very similar vegetatively and extremely difficult to differentiate; both are easily separated from other flacourts by the glands at petiole apex and distinctly 3-veined leaves (but with the lateral veins not ascending into the upper part of leaf, unlike *Lunania* or *Neosprucea*). Vegetatively, they resemble Euphorbiaceae; the only (very subtle) leaf difference between the two genera is that the glands at the petiole apex of *Pleuranthodendron* are lower than those in *Hasseltia,* the latter actually on the blade base. In fruit, the two are very distinctive with *Hasseltia* having a small red berry and *Pleuranthodendron* an ovoid, tan, irregularly dehiscent fruit. *Pleuranthodendron* is a canopy tree with fibrous-ridged bark; *Hasseltia* a common subcanopy species with smoothish bark and often irregular trunk.

Neosprucea (5 spp.) — Wet-forest understory trees. Similar to *Hasseltia* and *Pleuranthodendron* but shorter petioles with less defined apical glands and much more strongly 3-veined, the lateral veins to near apex. Differs from *Lunania* in (bluntly +/- remotely) serrate margins. Flowers and fruits are much larger than in related genera and the axillary inflorescences are few-flowered.

Lunania (14 spp., mostly Antillean) — Subcanopy trees of moist and wet forest. One of the genera that give Flacourtiaceae its bad taxonomic reputation. Very distinctive in its entire, rather oblong, acuminate leaves, more strongly 3-veined than in any other Flacourtiaceae, the lateral veins reaching nearly to apex. Inflorescence long, spicate, and pendent (usually 2–3-branched at base), with small sessile flowers.

(***Macrohasseltia***) (1 sp.) — A Central American wet-forest tree reaching Canal area and likely in northern Colombia. Leaves strongly 3-veined, subserrate with teeth widely separated and glandular, stellate-puberulous, lacking basal glands (except the glandular basal teeth). Flowers apetalous. Fruit yellowish, with 5 thin acute valves, the seeds embedded in cottony wool.

(***Hasseltiopsis***) (1 sp.) — A Central American tree that reaches central Panama and is likely in northern Colombia. Leaves smallish, remotely shallowly toothed, 3-veined, conspicuously pubescent in axils of lateral veins below. Flowers few, largish.

4. Leaves Pinnately Nerved, without Glands, Usually Pellucid-Punctate or Pellucid-Lineolate — Short petioles or leaf base decurrent on petiole; margins usually +/- serrate; stamens often perigynous (= on a disk): tribe Casearieae.

Casearia (75 spp., plus ca.100 in Old World) — By far the largest and most prevalent genus of Flacourtiaceae, and generally rather nondescript. Mostly subcanopy trees distinguished by evenly serrate, minutely pellucid-punctate leaves (A few species are virtually entire and a few nonpunctuate, but never both). The flowers alway small and in axillary fascicles (except one species with the inflorescence branched and cymose, cf., *Laetia*), the fruits 3-valved capsules, usually small but occasionally to 4 cm and +/- woody-valved. The twigs are often rather zigzag and may

Flacourtiaceae
(3-Veined Leaves)

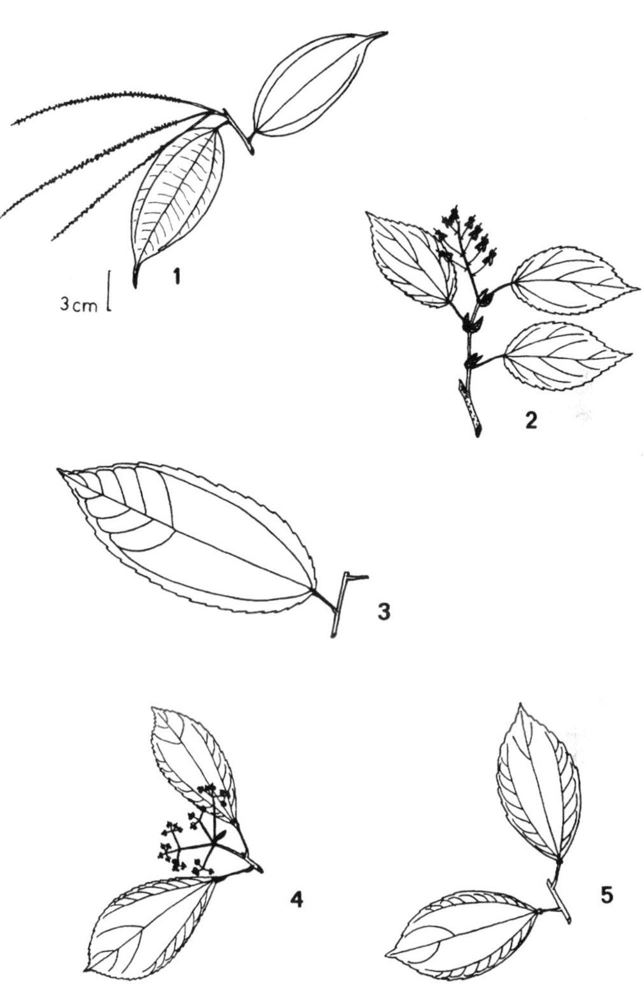

1 - *Lunania*

2 - *Prockia*

3 - *Neosprucea*

4 - *Hasseltia*

5 - *Pleuranthodendron*

have fine white lenticels; inconspicuous caducous stipules are present at the branchlet tips. A few species have spiny short-shoot branches.

C: marcelo; E: llajas, espino del demonio (*C. aculeata*), pinuela de cerro (*C. mariquitensis*), café del diablo (*C. sylvestris*)

Homalium (3 spp., plus 200 in Old World) — Large wet- and moist-forest trees. A largely paleotropical genus with only a few neotropical species. Differs technically from *Casearia* and allies in a semi-inferior ovary (unique in neotropical flacourts). The leaves are pinnately veined and distinctly evenly serrate, the inflorescence a terminal panicle with sessile or short-pedicellate flowers or fruits densely arranged along the rather few branches.

Banara (31 spp., mostly in Chile) — Mostly small second-growth trees with coarsely serrate leaves, usually oblong with a +/- truncate base; the area species have pinnate venation and usually a conspicuous gland pair near petiole apex or at base of lamina (from glandular basal teeth). Inflorescence a rather sparsely branched terminal panicle. Fruit a berry. Aside from the very different inflorescence, it differs from *Casearia* in lacking punctations and the more coarsely serrate margin.

E: guapilte

Xylosma (49 spp., plus 45 in Africa) — Very similar to *Casearia*, even in fruit. It differs technically from *Casearia* in completely hypogenous stamens (and in lacking petals (= tribe Flacourtieae). Vegetatively, *Xylosma* differs from *Casearia* in being much more frequently spiny (sometimes with branched thorns covering trunk) in lacking punctations (also lacking in some *Casearia* species) and in having generally more strongly serrate margins.

P: asta de venado

Zuelania (1 sp.) — A common canopy tree in drier fascies of Central American moist forests, barely reaching northern South America. The trunk is rather smooth with finely flaking bark (= "dandruff tree" fide R. Foster). Leaves oblong, softly pubescent below, the base truncate to subcordate, the margin barely serrulate. Flowers and fruits precociously, while deciduous during the dry season. The very characteristic fruits (not resembling normal flacourts) are large, brownish-puberulous, globose, and fleshy, tardily dehiscent to expose orange arillate interior.

Hecatostemon (1 sp.) — A monotypic segregate from *Casearia* differing only in having more numerous (90–100) stamens. The only species is a small dry-forest tree found mostly in the Venezuelan Llanos where it is very common. It differs vegetatively from most *Casearia* in having a densely tannish-puberulous fruit and somewhat pilose pubescence on the leaf undersurface.

Flacourtiaceae
(Branched Inflorescences or Unusually Large Flowers or Fruits)

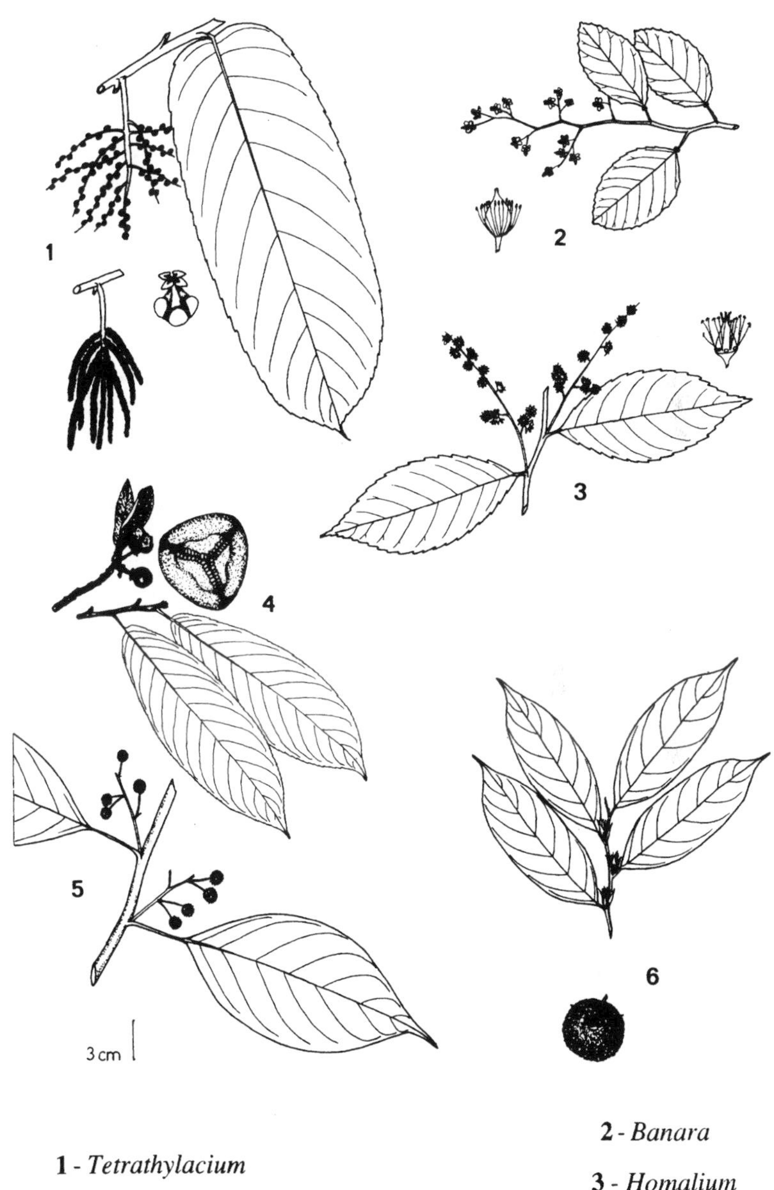

1 - *Tetrathylacium*

2 - *Banara*

3 - *Homalium*

4 - *Zuelania*

5 - *Lindackeria*

6 - *Ryania*

Flacourtiaceae
(Fasciculate Inflorescences; Small Flowers)

1 - *Casearia* **2** - *Mayna*

3 - *Carpotroche* **4** - *Casearia* **5** - *Xylosma*

6 - *Laetia* **7** - *Neoptychocarpus*

Tetrathylacium (2 spp.) — Another of the genera which do not look much like typical Flacourtiaceae. Technically it is characterized by having only 4 stamens and by the characteristic inflorescence which is a contracted panicle with sessile flowers densely arranged along the rather few longish branches. Easily recognized vegetatively by the large oblong leaves with somewhat serrate margins, +/- truncate bases, short petioles, subpersistent conspicuous stipules and noticeably angular twigs. The leaves are puncate as in *Casearia.*

Euceraea (1 sp.) — Tree of poor-soil forests, mostly in the Guayana region, recently discovered in Peru. Vegetatively, very distinctive in the obovate leaf with serrulate margins and barely prominulous close-together secondary veins not well differentiated from intersecondaries, the base long cuneate onto the long petiole. Bark with conspicuous whitish raised-lenticellate pustules, especially near base. Looks more like *Roucheria* (Linaceae) than like Flacourtiaceae. Inflorescence an open pyramidal panicle with minute sessile or slightly immersed flowers scattered along the strongly reflexed lateral branches.

Laetia (10 spp.) — Very similar to *Casearia* but mostly canopy trees. Bark a characteristic smooth whitish color with conspicuous raised blackish lenticels. Differs technically from *Casearia* in lacking expanded receptacular lobes. Inflorescence usually a fascicle, usually above the axils or on branches below the leaves, the pedicels often arising from a kind of cupule; a few swamp forest species with +/- obovate small finely serrate leaves have few-flowered branched inflorescences. The leaves, pellucid punctate as in many *Casearia,* are typically oblong and finely serrate but a few species (*L. cuspidata, L. coriacea*) are essentially entire and look like *Nectandra* (Lauraceae), except for a cartilaginous margin; tertiary venation always strongly parallel and +/- perpendicular to midvein.

P: purma caspi

Neoptychocarpus (2 spp.) — Another anomalous genus which does not obviously belong in Flacourtiaceae, especially when in flower. It is dioecious and has small sessile axillary clusters of flowers which lack petals but have the calyx lobes fused into a tube resembling a sympetalous corolla and subtended by bracteoles which resemble calyx lobes. The leaves are rather longish, conspicuously punctate, and usually slightly serrulate; they dry a characteristic reddish-blackish color.

Ryania (8 spp.) — Shrubs or subcanopy trees with large (for flacourts) solitary flowers and finely serrate, nonpunctate leaves. The fruit is very characteristic, large spongy and reddish surfaced, solitary in the leaf axils, and subtended by the long calyx lobes; it is presumably eventually dehiscent although it persists unopened for long periods. There are more species than recognized in the *Flora Neotropica,* since several of the "varieties" of *R. speciosa* may be strictly sympatric and easily distinguishable, even vegetatively.

(*Lacistema* and *Lozania*) — The two genera traditionally recognized as Lacistemaceae and in older classifications placed with the Amentiferae are included in Flacourtiaceae in *Flora Neotropica* as tribe Lacistemeae. They are characterized by having reduced flowers with single stamens arranged in catkinlike spikes of narrow racemes. See Lacistemataceae.

Gentianaceae

Mostly herbs (sometimes saprophytic) or sometimes weak-wooded shrubby trees (one *Lehmanniella* scandent); except the leafless saprophytes, always with opposite (or whorled) entire leaves. Distinctive in the petiole bases more or less joined across node, the often more or less tetragonal stem thus conspicuously segmented; stipules always absent. When fertile, characterized by the often conspicuous sympetalous flowers with 5 stamens and parietal placentation.

The first four genera below are woody or subwoody, the rest completely herbaceous.

1. Woody and More or Less Shrubby or Treelike

Macrocarpaea (35 spp.) — Essentially an overgrown *Irlbachia;* a spindly, white-flowered, cloud-forest tree to several meters tall. Corolla openly tubular-campanulate.

Symbolanthus (20 spp.) — A hummingbird-pollinated variant of *Macrocarpaea* with larger red flowers with subexserted anthers; restricted to middle-elevation cloud forests.

Lehmanniella (incl. *Lagenanthus*) (3 spp.) — Hummingbird-pollinated shrubs (one species scandent) of northern Andean cloud forests, characterized by the spectacular long-tubular red flowers with more or less constricted neck and reduced corolla lobes.

Tachia (6 spp.) — Understory shrub or treelet unique in the sessile solitary axillary white to yellow flowers. Leaves rather large, elliptic, without obvious secondary veins or 1–2 pairs of very arcuate secondaries.

2. Herbs — (Arranged roughly in sequence from coarse to tiny)

Irlbachia (incl. *Chelonanthus*) (15 spp.) — Often more or less weedy lowland herbs with white to greenish or bluish flowers. Close to Central American/Antillean *Lisianthus* but flower more openly campanulate.

Gentianaceae

1 - *Gentiana* **2** - *Nymphoides* **3** - *Macrocarpaea*

4 - *Symbolanthus* **5** - *Irlbachia*

6 - *Tachia*

Gentianaceae and Geraniaceae

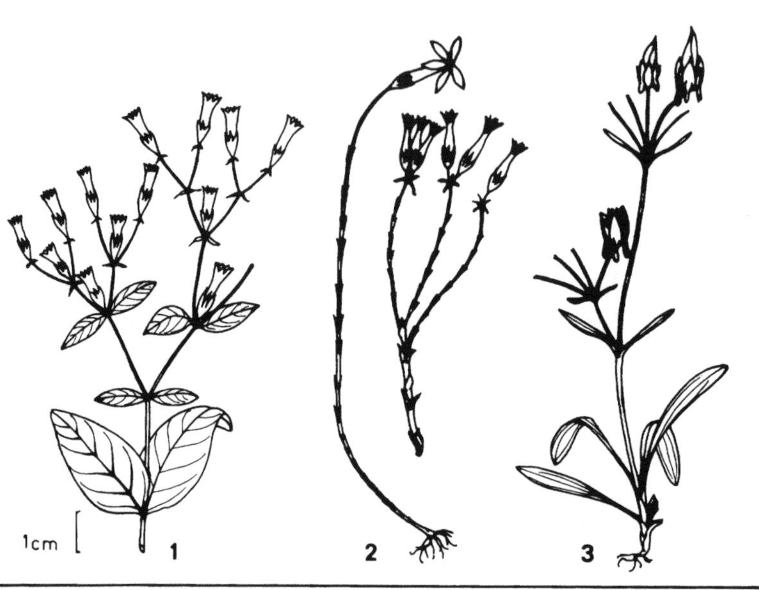

1 - *Schultesia* **2** - *Voyria* **3** - *Hallenia*

4 - *Balbisia* (Geraniaceae)

5 - *Erodium* (Geraniaceae)

6 - *Geranium* (Geraniaceae)

Coutoubea (4 spp.) — Erect lowland savannah herbs, our species with dense spicate terminal inflorescence of small white flowers.

Halenia (ca. 100 spp., plus few in Asia) — Puna and paramo herbs with distinctive 4-parted yellowish flowers, distinctive in basally spurred petals.

Schultesia (19 spp.) — Savanna herbs with the narrowly tubular white to pale yellow or pinkish flowers having medial constriction.

Gentianella (160 spp., plus 90 Old World) — Puna and paramo herbs, separated from *Gentiana* by lacking plaits between the corolla lobes.

Gentiana (20 spp., plus 300 in N. Am. and Old World) — Paramo and puna herbs (sometimes minute) with brightly colored (often blue) variously campanulate flowers, the corolla with plaits between lobes.

Centaurium (incl. *Erythraea*) (19 spp., plus 31 N. Am. and Old World) — Rather tenuous herbs, usually with narrow leaves; flowers pink, narrowly tubular with medial constriction.

Curtia (6 spp.) — Very small tenuous savannah herbs with reduced linear leaves and narrowly tubular white flowers.

Microcala (1 sp., plus 1 European) — Tiny annual loma herb with very small leaves and tiny yellow flowers with cupular calyx.

Nymphoides (1 sp., plus 19 Old World) — Aquatic herb with round deeply cordate water-lily-like floating leaves and fringed white narrow petals. Sometimes placed in Menyanthaceae.

Voyria (18 spp.) — Tiny leafless achlorophyllous saprophytes, common on lowland forest floor.

GERANIACEAE

Two genera are herbs and two Andean shrubs. The herbs are characterized by variously lobed or divided, usually much-incised leaves, pinnately divided in *Erodium*, palmately divided in *Geranium*. The shrubs have more or less sericeous narrow opposite leaves (or these palmately divided to base and sessile so resembling whorl of narrow leaves) and characteristic thin branchlets, with pair of projections at each node, in part persistent and more or less spinescent after leaves fall. All taxa have 5-merous flowers and (except *Balbisia*) very distinctive elongate carpels with the slender styles held together to form long beak; at dehiscence of the fruit this beak splits into the 5 separate individually dehiscent carpels.

1. Shrubs

Balbisia (8 spp.) — Andean shrubs or subshrubs with conspicuous yellow to orange flowers (without elongate "beak") and small, narrow, sericeous, opposite leaves (or the leaves palmately split to base and appearing whorled at each node).

Rhynchotheca (1 sp.) — Spiny shrub vegetatively similar to *Balbisia* but the small opposite leaves more elliptic and less sericeous (sericeous mostly on twigs). Distinctive in apetalous flowers, the 5 caudate carpels as in *Geranium* but conspicuously sericeous.

2. Herbs

Geranium (400 spp., incl. N. Am. and Old World) — Mostly high-Andean puna herbs; also a few loma species and weeds. Leaves palmately lobed or divided, the individual lobes or segments usually entire or with a few deep lobes or incisions near apex. Flowers pink or white; fertile stamens 10; long-beaked fruit pubescent.

Erodium (mostly adventive; 90 spp., mostly Old World) — Weedy herbs introduced from Eurasia, mostly at high altitudes. Differs from *Geranium* in the leaves usually pinnately compound or divided, usually the segments again much divided and/or with finely incised-serrate margins; also in fertile stamens only 5 and fruit beak glabrous.

Gesneriaceae

Mostly herbs or subwoody epiphytic climbers (often pendent) or shrubs of wet- or cloud-forest habitats. Characterized by the opposite (though sometimes strongly anisophyllous) mostly membranaceous to succulent often serrate leaves and the tendency to conspicuous pubescence, at least on stem and upper leaves and usually even on the corolla. The corolla pubescence tends to be pilose or villous with long trichomes, virtually unique in Tubiflorae. Most species of *Drymonia* are +/- woody lianas with glabrous entire leaves but easy to recognize by the cauliflorous flowers with conspicuous broad red sepals and bracts. *Besleria* (and extralimital *Solenophora*) include true small trees, usually identifiable by being obviously pubescent at least on twigs; if not, then the branchlets noticeably angled and often +/- flattened between nodes and sometimes jointed at nodes.

Flowers of all Gesneriaceae are more or less tubular-campanulate, somewhat bilabiate and usually conspicuous; they are often openly campanulate, sometimes narrowly tubular (and hummingbird-pollinated), often with basal spur, stamens (in our area) uniformly 4, usually with the anthers held together in pairs. The calyx differs from bignons in the 4–5 lobes usually free nearly to base (except *Chrysothemis*) and often brightly colored or elaborately laciniate. Fruit a berry (enclosed by calyx lobes), or usually 2-valved capsule (often fleshy and tardily dehiscent) with numerous tiny seeds. Generic taxonomy is in a state of flux; worse, many of the obvious floral characters are misleading.

1. Flowers in Terminal Racemes or Panicles; Terrestrial Herbs with Scaly Rhizomes or Tubers

1A. The first seven genera have relatively large and/or conspicuous red, blue, orange, or lilac flowers.

Kohleria (17 spp.) —Corollas orange to red, the 5 lobes nearly equal; ovary nearly inferior; differs from *Gloxinia* in unridged calyx.

Heppiella (4 spp.) — Andean highlands and Amazonian Ecuador. Looks just like *Kohleria* (except for frequently scandent or epiphytic habit), differing in technical characters: free anthers, capitate stigmas, annular nectaries.

Sinningia (55 spp.) — Like *Kohleria* in red flowers but the two upper corolla lobes much longer than lower three and the ovary superior.

Gloxinia (10 spp.) — Similar to *Kohleria* and *Monopyle* in large campanulate red to lavender or blue corollas but with ridges in floor of corolla tube (unlike more openly campanulate *Monopyle*) and ridged calyx (unlike *Kohleria*). Leaves smooth above and more or less symmetrical at base (unlike *Monopyle*). Commonest species looks like *Kohleria* except for the ridged calyx.

Monopyle (15 spp.) — Herb with strongly anisophyllous leaves, the larger leaves strongly asymmetric at base, always serrate and usually +/- bullate. Inflorescence a diffuse few-flowered panicle; flower very broadly campanulate or somewhat flattened, lavender to white or yellow, lacking ridges in its floor.

Rhynchoglossum (1 sp., plus many in Old World) — Our species (*R. azureum*) vegetatively similar to *Monopyle*. Leaves anisophyllous, exceedingly asymmetric, entire (unlike *Monopyle*); inflorescence a simple raceme, the calyx cupular, flowers lilac and more narrowly tubular than *Monopyle*.

Rhytidophyllum (20 spp.) — An Antillean genus with a single mainland species at Santa Marta, Colombia (and another in coastal Venezuela); our species characterized by elongate narrowly cuneate leaves with petiole winged to base and a diffusely paniculate inflorescence with the red corolla having reduced lobes and long exserted anthers.

1B. The next two genera have very small white flowers.

(*Diastema racemiferum*) — A small herb (<25 cm tall) with tiny white flowers in open raceme.

Koellikeria (1 sp.) — Tiny herb with leaves in basal rosette, always red-violet below, sometimes white-spotted or mottled above. Inflorescence elongate, diffusely racemose with tiny tubular white flowers with red upper lobes.

2. Flowers Axillary, Solitary or in Cymes; Plants Frequently +/- Woody, Often Epiphytic — When terrestrial usually without rhizomes or tubers.

2A. The next five genera are uniformly terrestrial, mostly very small herbs, and have very small white to yellow flowers.

Diastema (18 spp.) — Small herbs with erect stems, the few white to red tubular flowers erect in the calyx and lacking spur and having short lobes; differing from *Achimenes* in being much less bilabiate (and in separate nectar glands). Leaves always rather thin and serrate. More or less intermediate between *Achimenes* and the racemose taxa and one species has a terminal racemose inflorescence.

Cremosperma (23 spp.) — Small herbs with small pubescent leaves, endemic to Chocó region. Flowers tiny, unlike *Achimenes* and *Diastema* in superior ovary, tubular-infundibuliform, usually yellow (sometimes white), often in dense clusters at end of peduncle.

Phinaea (6 spp.) — In our area only in Colombia (mostly Central American). Low terrestrial herbs with erect stem and small leaves more or less apically clustered; corolla white, very small and broadly short-campanulate, nearly actinomorphic.

Napeanthus (12 spp.) — Low acaulescent terrestrial herb with small white very broadly short-campanulate long-pedicellate flowers. Differs from *Phinaea* in superior ovary and in lacking stem.

Reldia (5 spp.) — Small herb with alternate (unique) oblanceolate, pustulate-bullate leaves. The flowers tiny white bilabiate.

Gesneriaceae
(Mostly Scandent or Subwoody)

1 - *Besleria*

2 - *Drymonia*

3 - *Drymonia*

4 - *Columnea*

5 - *Corytoplectus*

2B. The next three genera are low terrestrial herbs but have larger flowers than above genera, with the corollas often red, orange, or lavender.

Episcia (40 spp.) — As currently defined distinctive in having creeping stems (stolons). Low terrestrial herbs, often forming patches in moist places. Flowers solitary, rather large and conspicuous, usually lavender (sometimes whitish) or red, the ovary superior.

Achimenes (23 spp.) — Like an erect *Episcia* with single well-developed stem; leaves rather thin, serrate, and asymmetric at base. Corolla white to red orange or lavender, conspicuous, often strongly bilabiate. Ovary inferior.

Chrysothemis (6 spp.) — A distinctive somewhat weedy succulent herb unusual in the cupular calyx from the almost completely fused calyx lobes. Flowers orangish-yellow, calyx green or red.

2C. The next eight genera are usually either more or less woody or subwoody or are predominantly epiphytic (or both).

Capanea (2 spp.) — Mostly rather sprawling or scandent subwoody epiphytes, sometimes terrestrial; typically found in middle-elevation cloud forests; at least young stems always villous. Corolla large (>3 cm long), campanulate, usually +/- narrowed at mouth, always villous, the tube variously cream or purplish to rose, the very characteristic lobes greenish or greenish-white with purple spots. Calyx lobes lanceolate, split to base.

Besleria (150 spp.) — Always terrestrial, commonly shrubs or small trees, sometimes herbs; often glabrous and often with smallish, nearly always entire leaves. Flowers very characteristic, with relatively small, blunt calyx lobes and a tubular, orange (sometimes red or yellow), glabrous corolla, typically small in size and with short inconspicuous spur. Fruit a berry.

Alloplectus (60 spp.) — Mostly erect terrestrial shrubs or subwoody herbs (occasionally epiphytic), characterized by the densely pubescent red or red-orange corolla tube with reduced lobes and laciniate or serrate, broad red sepals (the latter also found in a few *Drymonia*). Flowers typically on stem below leaves. Differs from *Paradrymonia* and *Nautilocalyx* in ovate to oblong or obovate leaves and from most *Drymonia* in pubescent serrate leaves.

Columnea (incl. *Dalbergaria, Trichantha, Pentadenia*)(200 spp.) — by far the largest neotropical gesner genus (though sometimes split into several genera). Nearly always epiphytic, stems subwoody and pendent or climbing; leaves often very unequal in a pair, typically with conspicuous red margins or apical spots (*Dalbergaria*) or undersurfaces. Corolla usually

Gesneriaceae (Herbs) and Goodeniaceae

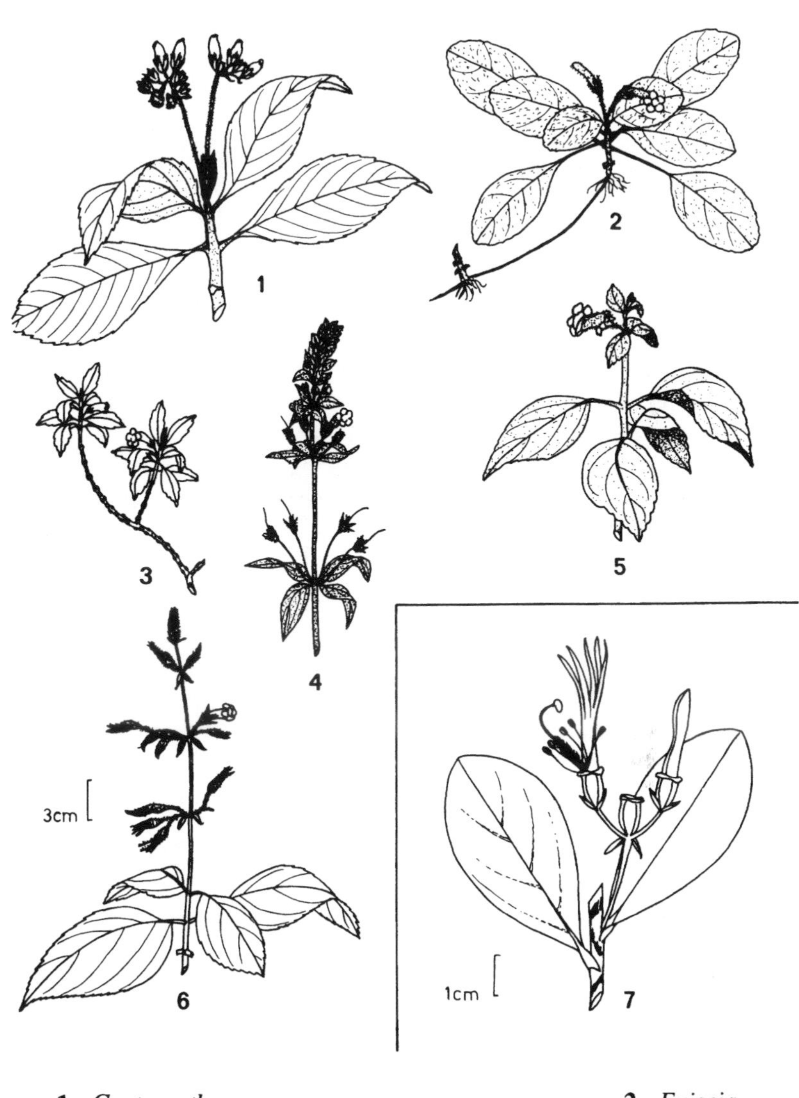

1 - *Gasteranthus* **2** - *Episcia*

3 - *Codonanthe* **4** - *Kohleria* **5** - *Gloxinia*

6 - *Monopyle* **7** - *Scaevola* (Goodeniaceae)

tubular, sometimes bright red and strongly bilabiate (= *Columnea s.s.*) or flowers long-pedicellate (*Pentadenia*), or the 5 corolla lobes more or less equal and with small lobes between them (*Trichantha*). Fruit a berry.

Drymonia (ca. 100 spp.) — A large mostly epiphytic genus of lowland forest shrubs and lianas, one of the few gesner genera to become truly lianescent. Leaves always +/- equal in size; terrestrial species usually have large rather succulent leaves; leaves of the climbers unusual in family in being usually entire and glabrous. Liana species also usually recognizable by the cauliflorous flowers with conspicuous large red calyx and/or bracts, even in bud. Corolla with basal spur; calyx lobes broad, differing from *Alloplectus* in being entire, except *D. serrulata* (where green) and common *D. macrophylla* (looks more like *Alloplectus*). An unusual technical character (shared with *Codonanthe*) is porate anther dehiscence.

Paradrymonia (15 spp.) — Mostly succulent epiphytes (occasionally terrestrial) with short stems and elongate-lanceolate to obovate leaves with long petioles and often appearing to form loose rosette. Flowers sessile or short-pedicelled, usually in dense axillary clusters; calyx lobes very long and narrow or strongly laciniate; corolla white to yellow with reddish or purplish lines. Differs from *Drymonia* in nonliana habit and in lacking anther pores.

Neomortonia (2 spp.) — Very small-leaved root-climbing epiphyte similar to *Columnea* but the corolla broadly infundibuliform and lacking bracts; one species with laciniate lobes, the other with swollen "pregnant" corolla tube (cf., *Gasteranthus*).

Codonanthe (6 spp.) — Ant-garden epiphytic climbers (rarely small subshrubs) with small, extremely succulent, glabrous leaves. The flowers single, with white, rather waxy, narrowly tubular-campanulate corolla having short basal spur.

2D. The next four genera are terrestrial herbs with rather large conspicuous flowers (or, if small, orange).

Gasteranthus (25 spp.) — A segregate from *Besleria* differentiated by being low herbs with fleshy capsules, a nonannular disk, and the larger, usually orange (rarely yellow), openly campanulate corolla with prominent spur and tube often with a peculiar pregnant-looking bulge (the mouth thus on dorsal side). A distinctive vegetative feature is that the leaves have stomates in clusters which are usually obvious without lens.

Nautilocalyx (20 spp.) — Terrestrial herbs with elongate leaves vegetatively similar to *Paradrymonia* but differing in having a well-developed stem, the axillary flowers few in each axil and usually pedicellate, white or cream corolla without dark lines and the flowers less congested. Calyx lobes lanceolate and entire.

Corytoplectus (8 spp.) — Low very pubescent herbs of wet lowland cloud forest, essentially a herbaceous version of *Alloplectus,* and with similar broadly ovate, more or less serrate, red calyx lobes. Besides the habit difference, differs from *Alloplectus* in the terminal inflorescence and the short-tubular, densely villous, yellow corolla.

Resia (1 sp.) — Colombian cloud forests. Terrestrial on rocks. Leaves in terminal cluster, oblanceolate, cuneate to sessile base, the margin +/- crenate-serrate. Inflorescence open, few-flowered, the flowers with long pedicels, corolla orange, looks like typical *Besleria.*

(*Besleria* and *Alloplectus*) — Have a few herbaceous orange-flowered species).

There are many more genera in Central America, the Antilles, and southern and eastern South America.

Goodeniaceae

In our area a single species of beach shrub with succulent obovate leaves with obscure venation and clustered toward ends of thick branches, the base of petiole expanded and subclasping, leaving conspicuous raised scar. Flower white, narrow, tubular, dorsally split, and with long narrow lobes. Fruits ellipsoid, ca. 1 cm long, dark purple.

Scaevola (1 sp., plus 80 in Old World).

Guttiferae

The outstanding vegetative characteristics of Guttiferae are colored (usually yellow or orange) latex and a terminal bud arising from between the hollowed-out bases of the more or less appressed terminal pair of petioles (a configuration with the potential for a distinctly sexual interpretation for those so inclined). Although there are taxa lacking one or the other of these characteristics, all opposite-leaved guttifers with white latex have the distinctive apical bud and all those without the distinctive shoot apex have distinctly yellow or orangish latex. Other useful vegetative characters are the strong tendency to parallel and often close-together secondary and intersecondary veins, the uniformly entire margins (usually with a marginal or submarginal collecting vein), the frequent presence of stilt roots and/or hemiepiphytic habit, and extreme sclerophylly. Most guttifers have numerous stamens and/or thick petals and/or enlarged discoid more or less sessile stigma; nearly all are woody (except *Hypericum*) and several are important timber trees. In our area, nearly all have strictly opposite leaves; the exception is alternate-leaved *Caraipa* (and its close relative *Mahurea*), very difficult to place to family in vegetative state, especially since the latex is often not evident; *Caraipa* is characterized by finely parallel (though sometimes immersed and obscure) tertiary venation more or less perpendicular to the rather straight and close-together secondary veins.

The majority of neotropical Guttiferae are dioecious with more or less fleshy fruits and more than one seed (= Clusioideae); in most of these the fleshy fruit is dehiscent to expose arillate seeds (exceptions: *Garcinia* and *Clusiella* with indehiscent fruits and seeds of the former embedded in pulp and the latter in a clear membrane). Two smaller subfamilies have round indehiscent fruits and perfect flowers, Calophylloideae (*Calophyllum*) with one locule and one seed and lacking the typical shoot apex, and Moronoboideae (*Symphonia, Moronobea, Platonia*) with more than one locule (though often a single seed) and an unusual twisted petal aestivation and characteristic pagoda-style branching; these have distinctly yellow latex and parallel secondary and intersecondary (and also subparallel tertiary) veins but lack the characteristic shoot apex. Subfamily Kielmeyeroideae has capsular fruits with winged (though sometimes minute) seeds and is unique in often having alternate leaves (in *Caraipa, Mahurea,* and extralimital *Kielmeyera*); these taxa all have the tertiary venation reticulate or finely parallel and more or less perpendicular to the secondary veins, perfect flowers with numerous yellow stamens and nonfleshy petals, lack the typical apical bud, and tend to lack obvious latex. Finally, the Hypericoideae have stamens in fascicles, either yellow flowers or woolly-pubescent petals, and the fruit either a berry (*Vismia*) or tiny and capsular (*Hypericum*); woody taxa have orange latex (sometimes faint and only evident in the petioles), lack the typical terminal bud and leaf venation but are often characterized by stellate-rufescent vegetative trichomes and/or punctate leaves.

As here arranged the genera (in part following unpublished concepts of B. Hammel) form a progression from the most typically guttiferous (i.e., *Clusia*) to the least typical.

1. Leaves with Conspicuously Parallel Secondary and Intersecondary Veins, Often Also with Resin Lines — (All except the last four are dioecious and have the typical guttifer twig apex with bud arising from between hollowed-out petiole bases).

Guttiferae
(Hemiepiphytic)

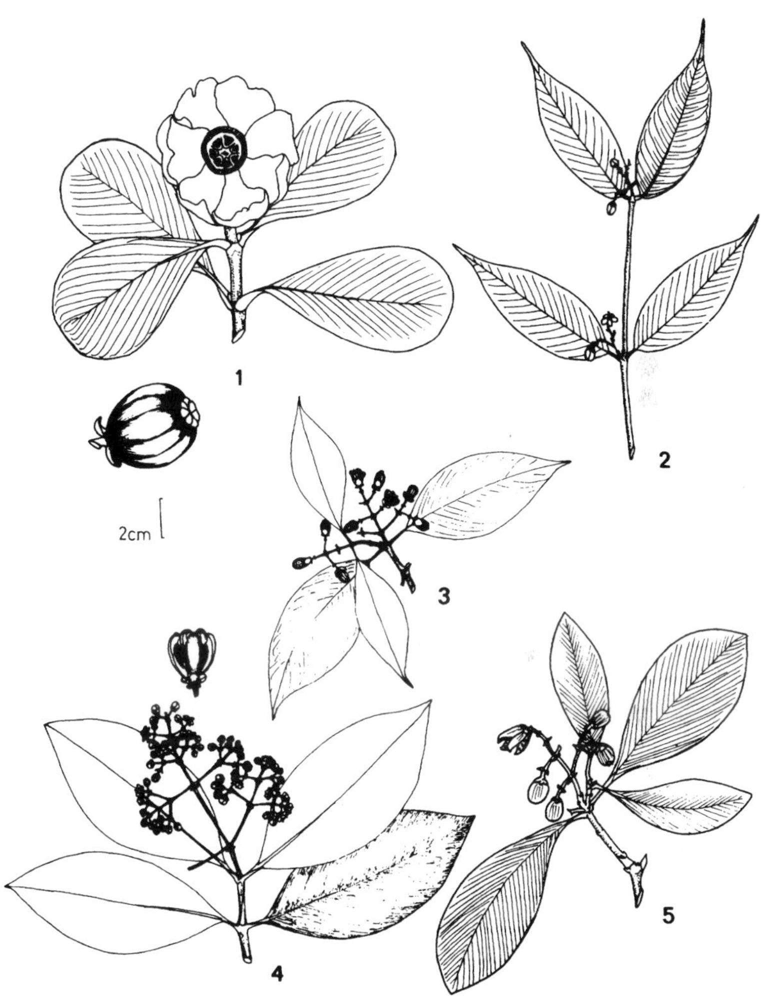

1 - *Clusia*

2 - *Clusiella*

3 - *Pilosperma*

4 - *Quapoya*

5 - *Havetiopsis*

1A. *Clusia* group: Usually epiphytic or hemiepiphytic trees or climbers; seeds small, numerous (except sometimes when fruit very small) — The fleshy but capsular (except *Clusiella*) fruits have characteristic conspicuous apical stigma residues and open into a flat starlike *Clusia* configuration to expose the tiny red-orange-arillate seeds. Leaves often extremely sclerophyllous; latex often white or cream. *Quapoya, Havetiopsis,* and *Oedematopus* are differentiated from *Clusia* by technical floral characters and are all characterized by very thick rather rubber-like usually 4-merous petals (very rare in South American *Clusia: C. amazonica, C. spathulifolia, C. martiana*), and stamenlike staminodia.

Clusia (incl. *Renggeria* and *Decaphalangium*) (300 spp.) — The most characteristic guttifer genus. Usually hemiepiphytic shrubs or trees often with long, straight, woody, free-hanging roots reaching to ground. Inflorescence often reduced to one or a few flowers, these with only slightly succulent petals.

C: chuagulo, mandure, sorquin, gaque; P: mata palo, renaquillo (little leaves), camé (*C. rosea*)

Pilosperma (2 spp.) — Erect hemiepiphytes. Perhaps not adequately separated from *Clusia.* Vegetatively distinguished by branching resin lines clearly visible on upper surface and rather resembling a dendritic stream system (unique). Leaf small and with long drip-tip. Fruits small and rather elongate; the so-called hairs of the seeds are actually a slightly dissected reddish aril not unlike that of some *Clusia* species.

Quapoya (5–6 spp.) — Epiphytic tree very like *Clusia* except for unusual staminal structure with anthers fused into a flat disklike arrangement borne on short stalk. The common species has small narrowly elliptic long-acuminate leaves with faint, strongly ascending, subequal secondary and tertiary veins.

Havetiopsis (2 spp.) — Usually epiphytic climbers. Essentially a *Clusia* with a reduced number of stamens (only 4: unique); the filament is noticeably narrower than the anther, unlike most *Clusia;* fruit also 4-carpeled (but this also found in some *Clusia* species). Vegetatively characterized by reddish exfoliating bark on twigs (but this not unique).

Oedematopus (ca. 10 spp.) — Variable and not clearly differentiated from *Clusia;* usually epiphytic but may be erect or climbing. Differs from *Clusia* in the more succulent petals and from *Quapoya* and *Havetiopsis,* respectively, in lacking a staminal disk and having 8–25 stamens; some *Clusia* species should probably be transferred here.

Clusiella (7 spp.) — Restricted to extremely wet forests. In flower very like *Clusia* but vegetatively distinguishable by being a slender epiphytic vine with short petioles and distinctive narrow leaves usually with long drip-tips. Fruit very different from *Clusia* in being an indehiscent whitish berry with numerous tiny nonarillate seeds; petal aestivation contorted as in Moronoboideae.

1B. Terrestrial trees with fruits capsular like *Clusia* but the seeds relatively large and only one/locule — Latex white or yellow; leaves mostly coriaceous but only rarely unusually sclerophyllous; twig apices of typical guttifer-type; latex white or yellow (rarely not evident).

Tovomita (60 spp.) — Usually with conspicuous stilt roots and a characteristic branching with the leaves clustered on short shoots separated by longer internodes. Latex yellow except in *T. weddelliana* group with distinctive narrow oblanceolate +/- epetiolate close-veined leaves (and strong stilt roots). Fruits warty and brown with the carpels completely reflexed to expose a red to purplish placenta and inner fruit wall. Outer 2 sepals equal and valvate (unique), fused in bud (unique) except in *T. weddelliana* group.

C: zanca de araña; P: chullachasi caspi

Chrysochlamys (incl. *Tovomitopsis* and *Balboa*) (50 spp.) — Lacking stilt roots or with white latex (sometimes latex not evident) or both. Leaves not clustered on short shoots. Fruits smooth and white to reddish outside at maturity, the carpels definitely dehiscent but spreading only slightly to expose whitish placenta and inner fruit wall and usually orange-arillate seeds. Outer sepals unlike *Tovomita* in not being equal nor valvate nor fused over bud; petals not very succulent. *Balboa,* perhaps worth generic recognition, differs in more succulent petals and white arils.

C: zanca de araña, manglillo

Dystovomita (1–2 spp.) — Medium-sized wet-forest tree with stilt roots; differs from *Chrysochlamys* in petiole bases extremely broad and expanded to form an intrapetiolar stipulelike ligule that clasps the stem. Leaves larger and more orbicular than in most *Chrysochlamys* (often > 20 x 15 cm), with lateral veins mostly separated by > 2 cm; fruits only slightly dehiscent.

C: zanca de araña

1C. Trees with indehiscent fruits with 1–few seeds; inflorescence ramiflorous or the flowers terminal on short shoots — Latex almost always bright yellow; stilt roots only in *Symphonia;* (includes genera assigned to three different subfamilies on basis of number of ovules and seeds).

Garcinia (incl. *Rheedia*) (200 spp., mostly Old World) — Dioecious (= Clusioideae) trees with more than one locule and usually two or more seeds/fruit. Leaves characterized by prominulous parallel secondary and intersecondary veins with tertiary venation +/- parallel to intersecondaries;

Guttiferae
(Trees with Fleshy Capsular Fruits; Usually with Stilt Roots)

1 - *Chrysochlamys*

2 - *Tovomita*

3 - *Chrysochlamys* (*Tovomitopsis*)

4 - *Dystovomita*

Guttiferae

(Trees with Indehiscent Fruit, Yellow Latex, and Well-Developed Intersecondary and Tertiary Veins Parallel to Secondaries)

1 - *Garcinia*

2 - *Moronobea*

3 - *Platonia*

4 - *Calophyllum*

5 - *Symphonia*

the veins and margin somewhat wavy and irregular (unique). Strongly *Clusia*-like terminal bud.

C: madroño; P: charichuelo

Calophyllum (4 spp., plus 108 Old World) — Large trees with vertically ridged bark. Leaves with very fine, close-together and undifferentiable secondary and intersecondary veins, but lacking the *Clusia*-type twig apex with hollow petiole bases. Latex usually rather sulphur-yellow, often with a faint greenish tint. Bat-dispersed fruit round, single-seeded, and green at maturity. Unique in single locule (= Calophylloideae).

C: aceite maria; P: lagarto caspi, tornill6n (1 sp.)

Symphonia (2 spp., also in Africa and Madagascar) — Large trees usually with stilt roots and with characteristic pagoda-style branching; latex always bright yellow. Leaves smallish, with secondary and intersecondary veins parallel and similar, also with faint tertiary veins subparallel to these, paired along twigs. Unique bright red flowers with petals twisted together, borne several together at apex of short shoot; flowers perfect and with > 1 locule (= Moronoboideae). Bat-dispersed fruits round and green at maturity.

C: machare; P: navidad caspi

Moronobea (7 spp.) — Large trees, vegetatively very similar to *Symphonia* but without stilt roots; latex bright yellow. Leaves very like *Symphonia* but borne in pairs (sometimes 4) at the ends of short-shoots. Flowers borne singly at apex of short shoot, with convolute white to pink petals. Fruit soft and glaucous-greenish at maturity and with spiral marking.

Platonia (1–2 spp.) — Large tree restricted to Guayana Shield region. Essentially a *Moronobea* with larger more coriaceous leaves and very large pink flowers.

2. Leaves Lacking Strong Intersecondaries, the Tertiary Venation Reticulate or More or Less Perpendicular to Secondary Veins; resin lines lacking but sometimes punctate — (All have perfect flowers and none has a twig apex with hollowed petiole bases). The first four genera have capsular fruits with more or less winged seeds, *Vismia* a berry-fruit, and *Hypericum* a tiny subwoody capsule. The last two genera have the stamens in fascicles and are sometimes segregated as Hypericaceae; the first two have alternate leaves and are intermediate with Theaceae.

2A. Alternate leaves

Caraipa (21 spp.) — Tertiary venation finely parallel (but sometimes not obvious when immersed in sclerophyllous leaves); capsule trigonal-ovoid, usually somewhat asymmetric; flowers small and white.

P: aceite caspi, brea caspi (*C. densifolia*)

Guttiferae
(Shrubs or Trees with Tertiary Venation Perpendicular to Secondaries; Latex Scanty or Orange)

1 - *Caraipa*

2 - *Mahurea*

3 - *Marila*

4 - *Vismia*

5 - *Hypericum*

Mahurea (2 spp.) — Swampy forests. Vegetatively like *Caraipa* but flowers pink and with more narrowly ovoid, woodier capsule with tiny sawdustlike linear seeds.

2B. Opposite leaves

Marila (20 spp.) — Strongly parallel tertiary venation and straight rigid secondary veins. Very characteristic narrowly racemose axillary inflorescence and long narrow capsule with minute seeds.

C: aceitillo; P: pichirina

Haploclathra (5 spp.) — Trees of poorly drained white-sand soil; recently discovered in Amazonian Peru. Our species with cordate leaf bases and rufous-pubescent twigs and petioles. Vegetatively, somewhat intermediate between *Vismia* and *Marila* in the slightly parallel tertiary veins but flowers more like *Caraipa* or *Mahurea.*

Vismia (30 spp., plus 4–5 in Africa) — Mostly second-growth trees with latex orange but not very profuse. Tertiary venation more or less reticulate, not obviously parallel. Fruit a berry and petals woolly (unique).

C: sangre-gallina; P: pichirin

2C. Herbs and shrubs with small ericoid opposite leaves

Hypericum (400 spp, incl. Old World) — One of the most common elements of high-Andean paramos; also tiny weedy herbs. Characterized by yellow flowers and the ericoid growth-form.

HALORAGIDACEAE

Three herbaceous genera in our area, two of them aquatic and the other the distinctive massive herb, *Gunnera.* All are exclusively montane. The aquatic genera are floating *Myriophyllum* with verticillate, finely pinnately divided, immersed leaves; and *Laurembergia,* recently discovered in Colombia, which has sublinear, mostly opposite, sessile leaves frequently with one or two marginal teeth. *Gunnera* (except the reduced paramo species *G. magellanica,* which looks more like *Hydrocotyle*) is an overgrown, essentially stemless rosette herb with large, irregularly lobed, palmately veined, more or less rotund leaves often >1 m across borne on long petioles each with a basal stipule; the cut stem shows small blotches of a bluish color from the nitrogen-fixing blue-green algae that inhabit it. Inflorescence a pedunculate spike or raceme, or pyramidal panicle with well-developed central axis, the maroonish flowers arranged densely arranged along it.

Haloragidaceae and Hamamelidaceae

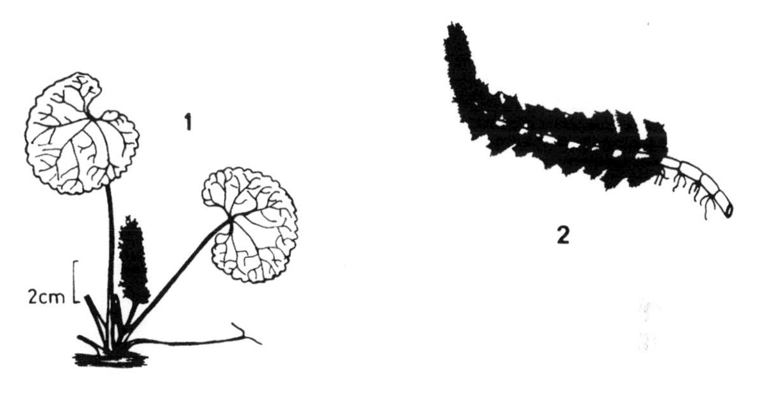

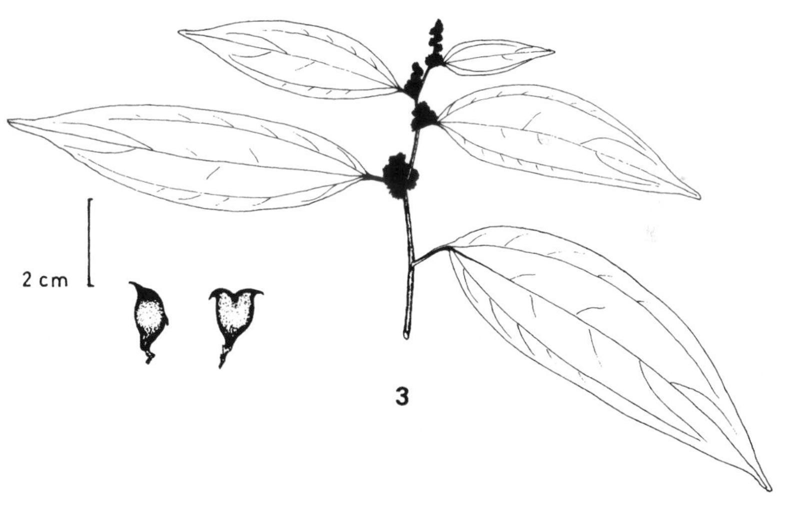

1 - *Gunnera* **2** - *Myriophyllum*

3 - *Matudaea* (Hamamelidaceae)

Gunnera (50 spp., incl. Old World) — Occurs in moist disturbed areas of Andean cloud forests, especially along roadsides.

Myriophyllum (45 spp., incl. Old World) — Aquatic in high-Andean lakes.
C: hoja de pantano

Laurembergia (1sp., plus 3 Old World) — Semiaquatic in moist sand at middle elevations.

HAMAMELIDACEAE

A Laurasian family recently discovered to reach the Colombian Andes. Vegetatively characterized by the typical 3-veined epulvinate Hamamelidaceae leaf. Our genus has oblong ovate entire leaves, strongly 3-pliveined with a slightly asymmetric base, and a short dorsally grooved petiole. There is a striking superficial resemblance to *Mortoniodendron* except for lack of the Malvalean pulvinus. The apetalous flowers are clustered inconspicuously in leaf axils and the small incompletely apically dehiscing capsules are apically elongated as one or two apical projections.

Matudaea (1 sp.) — The dominant species in some middle-elevation Mexican forests but rare and only recently discovered in South America.

HERNANDIACEAE

A small family with two genera of soft-wooded trees (*Gyrocarpus, Hernandia*) and one of canopy liana (*Sparattanthelium*) in our area, all vegetatively recognizable by the strongly and symmetrically 3-veined, mostly long-petioled, alternate, rather rankly aromatic leaves, lacking any hint of a pulvinus. All three genera are completely elenticellate with very finely longitudinally striate, dark-drying twigs. *Gyrocarpus* and *Sparattanthelium* are distinctive in having the 2 main lateral veins arise slightly *below* the base of lamina; at the opposite extreme several *Hernandia* species are subpeltate. Many similarly 3-veined taxa of other families have serrate leaves but all hernandiacs are entire (in part shallowly 3-lobed but otherwise entire in *Gyrocarpus*). All the Malvalean families differ in having a pulvinar thickening at petiole apex and Menispermaceae, similar to *Sparattanthelium,* have a pulvinar flexion. Three-veined Ulmaceae have shorter petioles and tend to be asymmetric at base; Flacourtiaceae and Euphorbiaceae lack odor and often have latex and/or paired glands at tip of petiole. Perhaps most vegetatively similar (especially to nonpeltate

Hernandia species) are the 3-veined species of Araliaceae (mostly *Dendropanax*) which are also rather rankly aromatic but differ in having +/- wrinkled mostly tannish-drying twig bark, in the main lateral vein pair either straight or curving slightly *outward* whereas in hernandiacs the basal lateral vein pair always curves upward; another difference is that araliacs nearly always have a mixture of smaller leaves with very short petioles interspersed with the long-petioled ones. Hernandiaceae lianas have a unique climbing mechanism with axillary hooks formed from spinelike reflexed branches that are apparently the bases of old inflorescences. Inflorescences of all three genera are more or less flat-topped, in *Sparattanthelium* and *Gyrocarpus* from the strongly dichotomous branching, in *Hernandia* with a central rachis but the branches evenly ascending and lower branches progressively longer. Each genus has a very distinctive fruit but these have little in common with each other.

Gyrocarpus (1 sp., plus 2 in Old World) — Large dry-forest canopy tree with rather smooth reddish bark (reminiscent of *Cavanillesia* but lacks the trunk rings), deciduous for much of year. Leaf broadly cordate, sometimes shallowly 3-lobed, otherwise entire, 3-veined from slightly below base of lamina, with very long petiole. Flowers very tiny, in clusters in strongly dichotomous corymbose panicle. Fruit a distinctive samara with nearly parallel pair of long, basally tapering, oblanceolate, subwoody, wings extended above the ellipsoid body.

Hernandia (8 spp., plus 17 Old World) — Canopy trees of extra-Amazonian lowland moist and wet forest (elsewhere especially common along coast). Leaves with somewhat variable petiole length (cf., araliacs but less strongly clustered), always +/- pliveined *above* the base (often even subpeltate). Inflorescence flat-topped, with distinct central rachis and ascending progressively longer lower branches, conspicuously bracteate with pairs of spatulate bracts at nodes and verticil of 4 similar sepal-like bracts subtending each group of ca. 3–4 subsessile buds (or the greenish-white flowers). Fruit looks more like an earthstar fungus than a fruit, the single large seed loosely enclosed in a fleshy, coriaceous, usually whitish cupule with a distinctly jellylike consistency.

E: pechuga

Sparattanthelium (13 spp.) — Canopy lianas usually climbing by reflexed sharp-tipped axillary "hooks" (= bases of old inflorescences). Bark blackish and finely ridged, primitive odor usually obvious. Leaves 3-pliveined or 3-veined *below* base, completely lacking any pulvinar flexion (unlike similar menisperms). Inflorescence perfectly dichotomous, sometimes with thickened nodes (especially in fruit where it sometimes becomes conspicuously whitish). Fruit ellipsoid-cylindric, vertically 5-angled.

Hernandiaceae and Hippocastanaceae

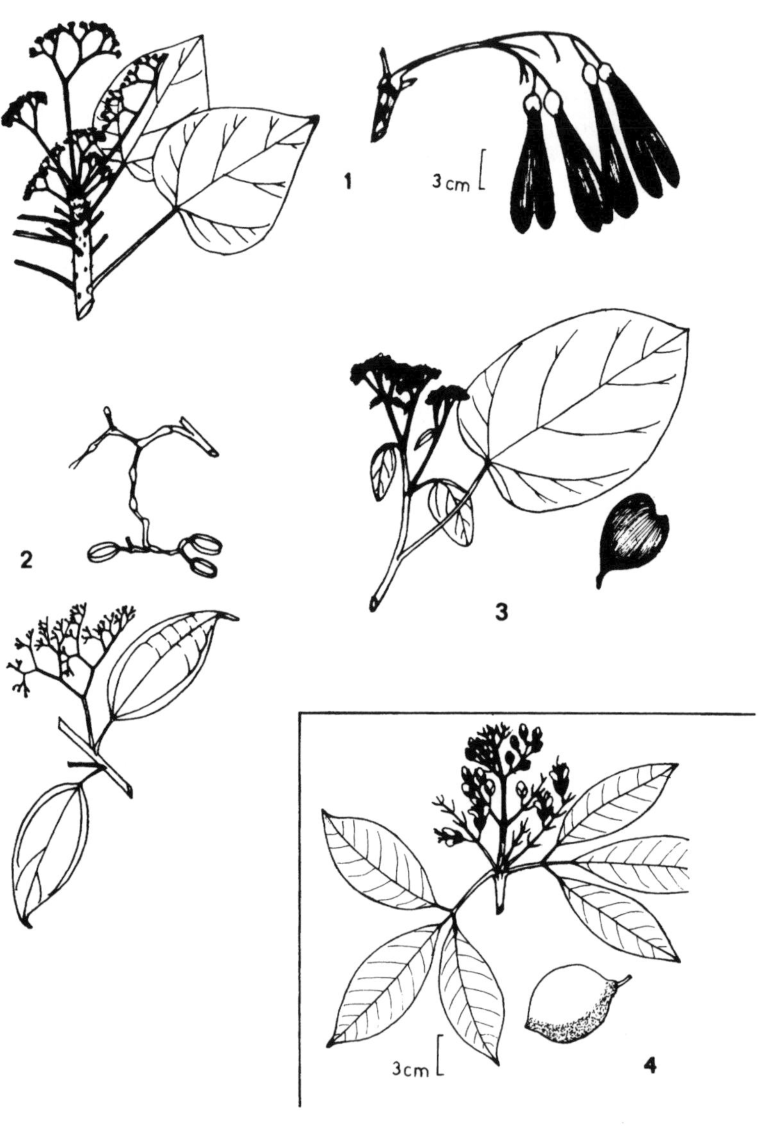

1 - *Gyranthera*

2 - *Sparattanthelium*

3 - *Hernandia*

4 - *Billia* (Hippocastanaceae)

Hippocastanaceae

In our area a single species of large tree with opposite 3-foliolate leaves, restricted to mesic montane forests in the northern Andes where it may be locally dominant. Vegetatively very similar to *Caryocar* but has uniformly entire leaflets and lacks both the gland pair at the petiole apex and the long rubiac-like terminal stipule that characterize that genus. Flowers in a terminal panicle, rather showy, open, the narrow white petals tapering to long cuneate base, with yellow basal spots that turn reddish with age. Fruit asymmetrically ovoid or subglobose, 5–6 cm long, with a conspicuously brownish-lenticellate surface.

Billia (2 spp.)
C: tres hojas

Hippocrateaceae

An almost entirely scandent family, vegetatively usually characterized by simple more or less serrate or serrulate opposite leaves (unique among area lianas) and by the tendency to form peculiar hooked or recurved branchlets that substitute for tendrils. Most species of *Salacia,* several of *Tontelea,* and a very few of other genera have entire leaves but these are nevertheless recognizable by the combination of coriaceous texture, a smooth nonreticulate leaf undersurface (in *Salacia* usually with immersed venation), and especially a characteristic olive color when dry (serrate-leaved species also may dry olive but usually have prominulous tertiary venation and are often membranaceous). Another unusual feature is that twigs often dry the same olive color as the leaves. The stems of some hippocrat lianas have interesting anomalous development, sometimes with a rather chain-like surface of differently oriented alternating internodes and in cross section often with a series of two or more irregular concentric reddish rings connected by widely separated "spokes". All but one (*Hemiangium paniculatum* of the coastal Colombian dry forest which has more or less viny branches) nonclimbing species of Hippocrateaceae in our area belong to *Cheiloclinium* (with conspicuously parallel tertiary venation) or to *Salacia* (with coriaceous leaves, usually large, smooth, and olive-drying).

The characteristic flowers of Hippocrateaceae are small and flat with a well-developed disk and only 3 stamens (one *Cheiloclinium* species has 5), and the axillary (or ramiflorous) inflorescence is dichotomously branched and often small (in *Salacia* frequently reduced to a single flower or fascicle of flowers). The fruits are of two very different types; half the genera have strongly dorsoventrally compressed, 3-valved capsules that release winged seeds while the other half have a more or less fleshy indehiscent fruit, often large and usually with three or six large seeds. The flat 3-parted capsular fruit type is completely unlike that of any other family; the fleshy fruit type is usually distinguishable by the trigonous seed arrangement and often by a tendency to a bluish-glaucous gloss on the fruit. The family is closely related to Celastraceae from which it is distinguished by stamens arising from inside the disk and only 3 (very rarely 6) in number. Although it is possible that Hippocrateaceae are polyphyletic, with capsular-fruited and indehiscent-fruited lineages separately derived from Celastraceae stock, there are so many striking similarities between the two groups that it seems preferable to retain the traditional familial concept.

Despite their small size, the flowers have a wealth of differentiating characters that define most of the genera. Most of the dehiscent-fruited (but not the indehiscent-fruited) genera can also be differentiated by the capsules and seeds. However, only a few genera are easily distinguishable when sterile — *Prionostemma,* with asperous leaves, and *Cheiloclinium,* with leaves drying green with parallel tertiary venation are the most distinctive. Many *Salacia* species are characterized by the largest, most-coriaceous leaves in the family, (also unusual in being typically entire) but other species have smaller, serrulate-margined leaves; nearly all *Salacia* species are also characterized by the smooth rather dull undersurface with immersed tertiary venation and the typical grayish-olive color of the dried leaves. *Tontelea* approaches *Salacia* but the leaves tend to be smaller, more glossy with somewhat prominulous venation, and are very rarely completely entire; the twigs are usually dark-drying with a minutely tuberculate surface unlike the usually smooth olive-drying twigs of *Salacia.* The undersides of *Peritassa* leaves usually dry a characteristic reddish-green to reddish-brown with contrasting light main veins and frequently a cartilaginous margin. *Cuervea* and *Hylenaea,* with similar largish pale-olive-drying leaves with intricately prominulous venation tending to be whitish below can also be recognized, but *Hippocratea* and the rest of the genera are vegetatively separable only by subtle characters of the individual species.

1. Three-Parted Capsular Fruits with Winged Seeds; Leaves Almost Always Serrate or Serrulate, Often Thin in Texture — The first two genera below are unusual in having the 3 capsule valves united by their margins, the last two in having the seed body thickened and the wing much reduced.

Anthodon (1 sp.) — Canopy liana of mature moist and wet forest. Leaves brochidodromous well inside the evenly crenulate-serrulate margin, often somewhat glossy above, the tertiary venation tending to form poorly defined intersecondary veins perpendicular to the midvein. In fruit very distinctive, with the most completely united capsules in the family, the shallow sinus between adjacent capsules rounded. In flower unique in having the relatively long narrow petals finely and regularly serrate (entire to irregularly erose-serrate in other genera), greenish-yellow in color.

Hemiangium (2 spp.) — Mostly erect or subscandent dry-area trees, restricted to northern Colombia in our area. The fruit like *Anthodon* in partially fused capsules but the sinus deeper and acute. In flower characterized by the conspicuous enlarged pulvinate disk that is flattened and glabrous (unlike *Hippocratea*) and surrounds a deeply sulcate ovary (unlike *Salacia*). Leaves smallish, obovate, sometimes rather asperous.

Hippocratea (1 sp.) — Canopy liana with an extremely wide ecological amplitude, but especially prevalent in dry forest. In fruit characterized by a large seed wing much longer than seed body and thickened toward base only along one margin. In flower unique in a conspicuous pubescent flange that crosses upper side of each petal below middle, and also by the conspicuous truncate-conical puberulous disk. The characteristic inflorescence is rather sparsely openly dichotomous with a somewhat one-sided flower arrangement. Leaves usually less membranaceous than in close relatives and often drying rather brownish with not very prominent tertiary venation; young twigs often 4-angled.

P: yaco-yaje

Prionostemma (1 sp.) — The easiest hippocrat to recognize when sterile on account of its asperous leaves (*Hemiangium* may be somewhat asperous but only when an erect tree); could be confused with *Petrea* (Verbenaceae) but that genus has the leaves more cuneate-based or the petiole obviously puberulous. Flowers yellowish-green with expanded disk, similar to *Hemiangium* or *Salacia* but puberulous. The fruits are unique in their asperous surface and also have thicker valves than other genera (except *Hylenaea*). There is a tendency to have red latex.

Pristimera (9 spp.) — In our area mostly lianas of dry or seasonally inundated forest. Closely related to and somewhat intermediate between *Hippocratea* and *Elachyptera*. Vegetatively easy to tell from *Hippocratea* by the greenish- rather than brownish-drying leaves with more obviously reticulate venation prominulous on both surfaces. The seed is like *Hippocratea* except that the wing is usually thickened along both margins. The inconspicuous disk is smaller than in *Hippocratea* and less cupular than in *Elachyptera*. The inflorescence is repeatedly dichotomous and the tiny flowers yellowish or greenish. The stem is at least sometimes distinctively chainlike.

Hippocrateaceae
(Dehiscent Fruit)

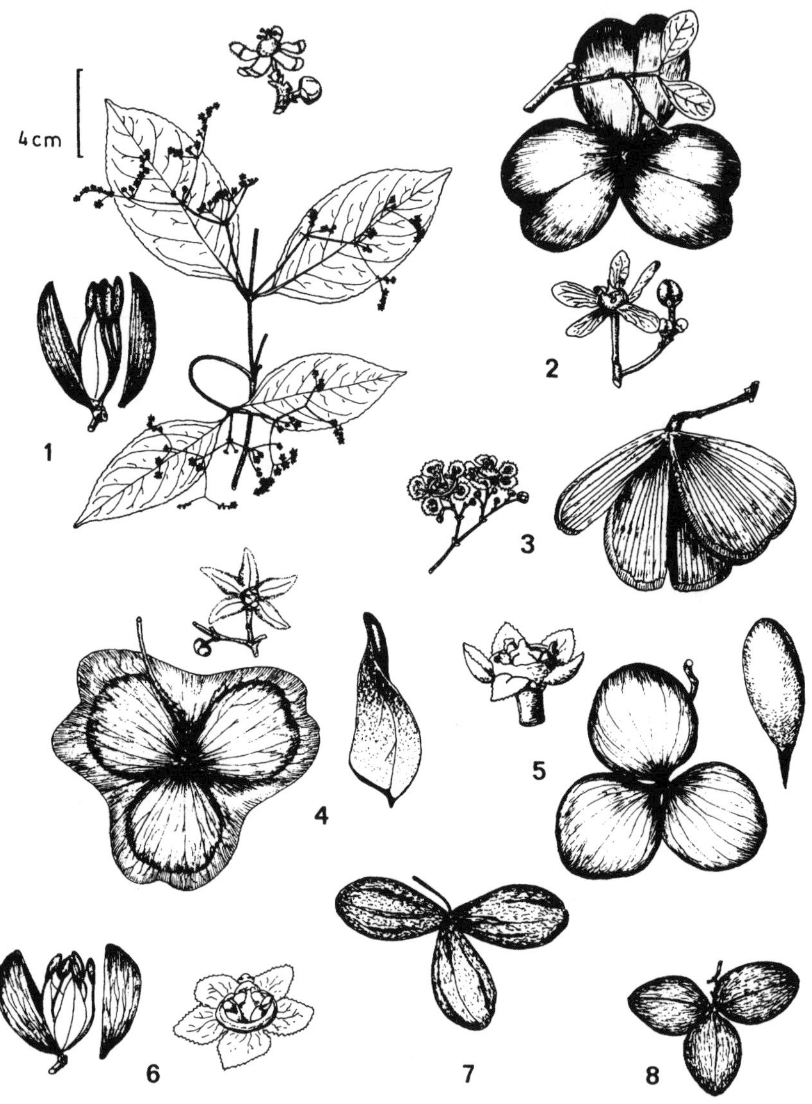

2 - *Hemiangium*

1 - *Hippocratea*

3 - *Prionostemma*

4 - *Anthodon* **5** - *Cuervea*

6 - *Pristimera* **7** - *Hylenaea* **8** - *Elachyptera*

Elachyptera (3 spp.) — Similar to *Hippocratea* from which it differs mainly in the apical seed-wing-proper shorter than the seed-body and instead greatly expanded laterally. The inflorescence is a more densely branched panicle than in *Hippocratea* and has strongly tetragonal branches. The flowers, white unlike most hippocrats, are minute (< 2 mm) with a thinly cupular disk and the petals more or less erect at anthesis. Venation below is more intricately prominulous than *Hippocratea* but leaves drying a similar brownish color (at least below).

Hylenaea (2 spp.) — Canopy liana of rich-soil and swampy areas. In fruit one of the most distinctive genera on account of the distinctly woody capsule valves; the seed-wing is reduced and coriaceous. Inflorescences openly paniculate, the flowers tiny, often on extremely slender and elongate pedicels, unique in the narrow acute or subacute sepals. Leaves much like *Cuervea* with intricately prominulous venation (though with more tendency to finely parallel tertiary venation) and the main veins below drying pale; different in the tendency to 4-angled brown-drying young twigs.

Cuervea (3 spp.) — A thick-stemmed liana, mostly occurring along sea coasts in wet-forest areas, also along rivers. The most distinctive feature is the thick corky seed, adapted for water dispersal and with only a reduced vestigial wing; the capsule, though thin-walled, is less strongly flattened than in most other genera. Inflorescence open and rather few-flowered with the relatively large (> 8 mm across) white flowers characterized by conspicuously erose-serrate sepals and by the broadly cupular disk. Vegetatively distinguished by the intricately prominulous tertiary venation both above and below, the venation below tending to dry pale as compared to the light olive surface; unlike *Hylenaea,* the young twigs dry olive and lack tetragonal angles.

2. Fruits Indehiscent; Leaves Often Coriaceous or Entire, When Membranaceous and Serrate with Closely Parallel Tertiary Venation Perpendicular to Midvein. — In flower, the fruit difference is already apparent and this group distinguishable by the ovary subterete to trigonal but never sharply sulcate.

Salacia (30 spp., plus 90 in Africa) — The largest genus of Hippocrateaceae and one of only two genera to include trees. Most species have characteristic unusually large and/or unusually thick-coriaceous leaves with smooth surfaces and more or less immersed venation; the dried leaves and branchlets tend to be a very characteristic dull olive color and the leaf margins are often entire. The inflorescence may be openly paniculate but in many species is unusual in the family in being much reduced, often to an axillary or ramiflorous fascicle of flowers; the disk, in which the ovary is usually immersed, is thick and fleshy, always much expanded when the

Hippocrateaceae
(Indehiscent Fruit)

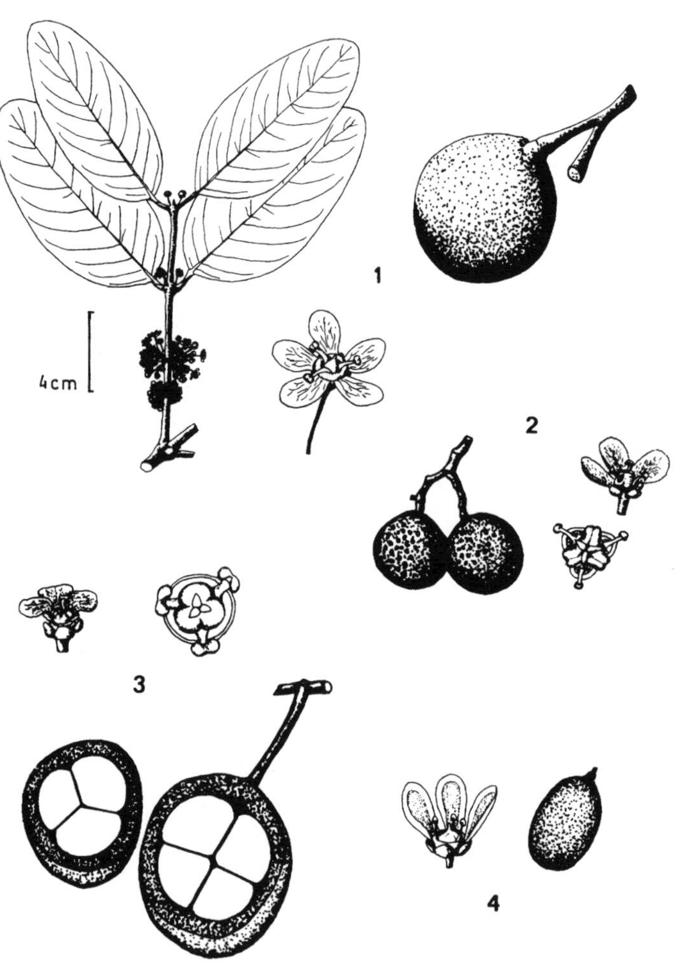

1 - *Salacia*

2 - *Cheiloclinium*

3 - *Tontalea*

4 - *Peritassa*

inflorescence is paniculate, occasionally thinner and bowl-like in ramiflorous species. Fruit usually subglobose, sometimes ellipsoid.

Tontelea (30 spp.) — Canopy lianas. Fruits indehiscent, often large, indistinguishable from *Salacia.* Vegetatively poorly differentiated, the leaves usually larger and/or more coriaceous than in *Hippocratea* alliance; sometimes with parallel tertiary venation as in *Cheiloclinium* but more coriaceous and the margin often more or less entire. Some species have leaves indistinguishable from *Salacia* but most have more prominulous venation and a shinier surface when dry. The inflorescence is always dichotomously branching and the flowers differ from *Salacia* in a cylindrical disk free from the ovary, from *Peritassa* in the reflexed entire petals.

Peritassa (14 spp.) — Moist- and wet-forest lianas (sometimes erect outside area). Vegetatively usually distinguishable by the coriaceous texture, smooth undersurface with immersed venation, and especially by the color of the dry leaves, usually dark (reddish-olive to brownish) with contrastingly pale main veins and often with a cartilaginous margin, the secondary veins weakly or not at all brochidodromous near margin (the few *Salacia* and *Tontelea* with similar leaf color are more brochidodromous). Very close to *Tontelea* which has a very similar yellow flower but with suberect usually conspicuously erose-margined petals and an unusual (for the family) anther dehiscence, the two thecae well-differentiated and opening separately and the connective sometimes even extended.

Cheiloclinium (20 spp.) — Although most species are lianas, the commonest species are small subcanopy trees. Vegetatively characterized by the pronounced tendency to finely parallel tertiary veins perpendicular to the midvein, finely crenate-serrate leaf margins are typical of several common species and one of the erect Amazonian species usually has noticeably 4-angled twigs; the leaves generally dry green, the tertiary veins sometimes darker. The dichotomously branching inflorescence may be large and open or much reduced, often with thickish branches. The flowers are unique in the family in having the disk broken into 3 (5 in *C. anomala*) small pockets, each enclosing the base of a stamen, and in completely lacking a style.

Humiriaceae

A vegetatively rather nondescript family of mostly large trees, usually with red or dark red inner bark and relatively dense dark green crowns. Perhaps the best sterile character is the tendency to crenate leaf margins and festooned-brochidodromous venation in the alternate simple leaves, but *Vantanea* has neither character; when entire (*Vantanea,* some *Humiria*), the leaves always coriaceous and more or less obovate (and in our area drying with characteristic dark reddish color). Young leaves at apex of shoot of most species rolled and mature leaves sometimes with vernation lines. The ellipsoid or ovoid fruits with very hard woody endocarps are unmistakable; the endocarp may be filled with resinous cavities (*Sacoglottis, Schistostemon*) and opens by longitudinal valves (these not obvious in *Sacoglottis* and reduced to part of a starlike arrangement of 5 subapical foramina in *Humiriastrum*). The inflorescence is usually a more or less flat-topped cymose-panicle, terminal (most genera) or reduced to a small dichotomously branching cymose-panicle or a few axillary flowers; the regular 5-parted flowers are nearly always small (except some *Vantanea*) and greenish to whitish (occasionally red in *Vantanea*), the stamens 10–many with filaments connate at base.

The genera in our area can be most easily distinguished by their fruits — the endocarp lacking foramina and obvious valves and with resin-filled cavities in *Sacoglottis* and *Schistostemon,* lacking foramina and with 5 obvious valves extending most of length in *Vantanea,* with 5 conspicuous apical foramina in *Humiria* and *Humiriastrum,* the latter with short valves alternating with the foramina, the former with smaller fruit and valves extending most of length of fruit. In flower multistaminate (with bilocular thecae) sometimes larger-flowered *Vantanea* is distinctive while the other genera have unilocular thecae and only 10 stamens (*Sacoglottis*) or 20 (*Humiria, Humiriastrum, Schistostemon, Duckesia*).

Vantanea (14 spp.) — Large trees not very Humiriaceae-like vegetatively; even the inner bark only slightly reddish. When fertile, easy to recognize within the family by the numerous stamens, the ovoid fruits with endocarp lacking apical foramina and with 5 valves extending most of their length. The only entire-leaved genus of Humiriaceae, the leaves always coriaceous and more or less obovate, in our area drying a characteristically dark reddish color.

P: manchari caspi

Humiria (3 spp.) — Shrubs to large trees of the Guianas and Guayana-influenced Amazonia, mostly on white sand. In our area charac-

Humiriaceae

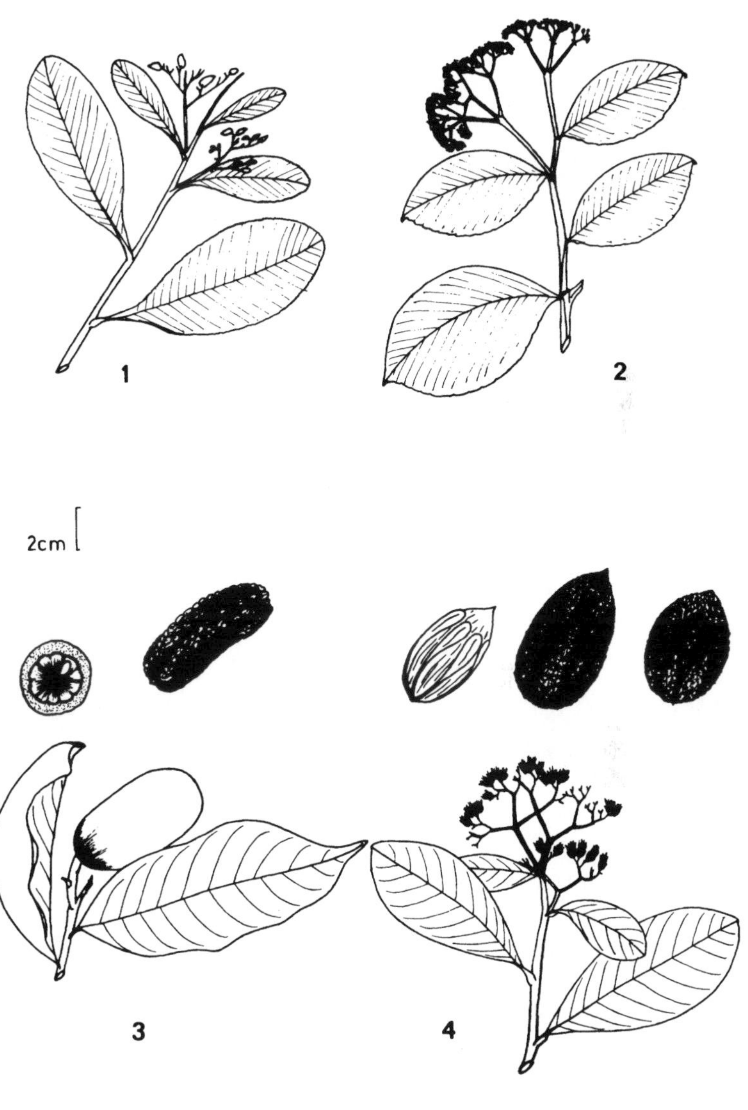

1 - *Humiria* **2** - *Humiriastrum*

3 - *Sacoglottis* **4** - *Vantanea*

terized by the sessile, usually more or less cordate-based, obovate leaves and usually winged young branches. The fruit is smaller than in other humiriacs (about 1 cm long) and has 5 apical foramina. Unique in pilose anther thecae.

P: puma caspi

Humiriastrum (12 spp.) — Large trees with dark-red inner bark and crenate-margined, often reddish-black drying leaves. Differs from *Sacoglottis* in the endocarp with 5 apical foramina, alternating with the shorter valves and lacking resin cavities. Fruits mostly 1.5–2.5 cm long (smaller than most *Sacoglottis,* larger than *Humiria*). Stamens 20 with 2 basal thecae and long connective as in *Humiria* (but with glabrous thecae) but twice as many as in *Sacoglottis.* Flowers always small and inflorescence always terminal.

C: chanul; P: puma caspi

Sacoglottis (7 spp., plus 1 in W. Africa) — Medium-sized to large trees characterized by the woody endocarp with resin-filled cavities (shared only with *Schistostemon*) and by having only 10 stamens (unique). Inflorescence axillary, cymose-paniculate with dichotomous branching or reduced and subfasciculate, always much shorter than subtending leaf. Leaves usually more or less serrulate and with festooned-brochidodromous venation.

Schistostemon (ca. 4 spp.) — Small to medium-sized trees, mostly of the Guianas or Guiana Shield Amazonia, only one exclusively white-sand species in our area. A segregate from *Sacoglottis* that differs primarily in having 20 stamens, five of which are much longer and apically trifurcate. Fruit and inflorescence exactly as in *Sacoglottis.* Leaves generally more prominulously reticulate than *Sacoglottis* and somewhat reminiscent of some *Eschweilera.*

Duckesia (1 sp.) — A widespread Amazonian canopy tree, recently discovered in Peru. Related to *Sacoglottis* with resin-filled cavities but a rather spongy endocarp (unique) and 5 rather obvious valves, these more like *Vantanea,* but the surface more strongly rugose-tuberculate; stamens 20–25 with 4 unilocular thecae. Inflorescence rather open and few-flowered and leaves smallish, thin, and strongly crenate.

There are also two monotypic extralimital genera of Humiriaceae, restricted to Amazonia. *Hylocarpa,* of the upper Rio Negro, is similar to *Sacoglottis* but lacks the resinous cavities and has 30 stamens. *Endopleura* of the Central and lower Amazon, is another *Sacoglottis* segregate which differs in more stamens (20), anthers with 4 unilocular thecae, and a deeply sulcate endocarp that lacks resin-cavities.

Hydrophyllaceae

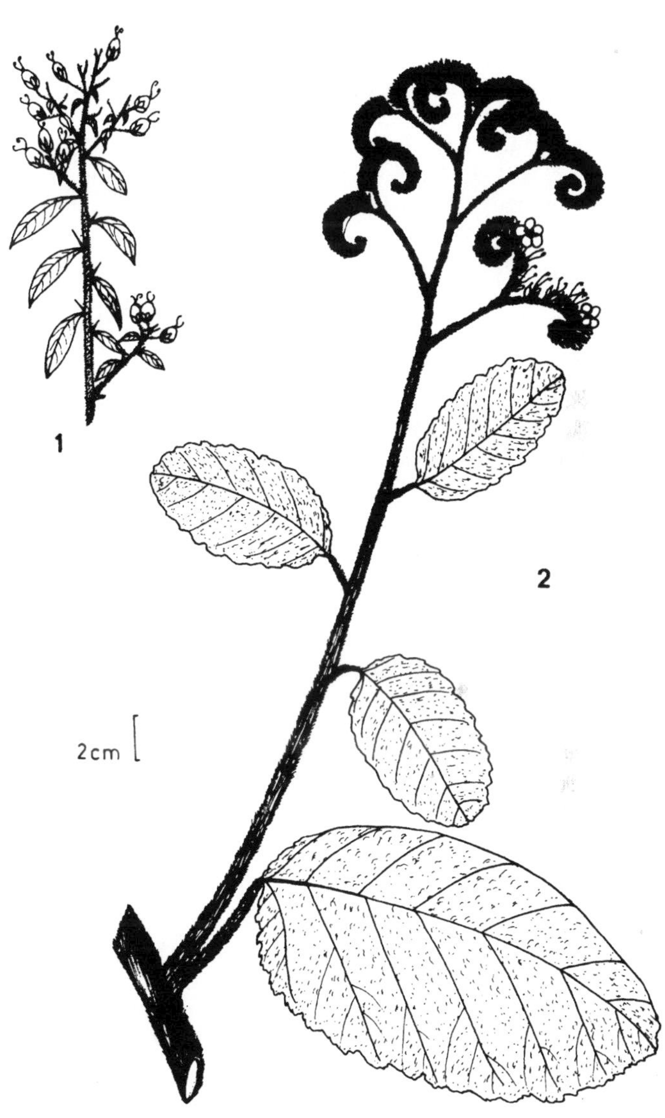

1 - *Hydrolea* **2** - *Wigandia*

Hydrophyllaceae

Herbs and shrubs, mostly characterized by viscid gland-tipped trichomes (also longer stiff urticating trichomes in one *Wigandia* species). The inflorescence is usually scorpioid (i.e., one-sided and curled like a fern fiddle-head) and the leaves are alternate. Technically differs from borages primarily in the numerous ovules on parietal placentae but few borages have the viscid pubescence of this family. Of our four genera, *Hydrolea* has spines, *Wigandia* is usually urticating and has doubly serrate leaf margins, *Nama* has single sessile flowers, and *Phacelia* has a strongly scorpioid inflorescence and usually narrow pinnatifid leaves.

Hydrolea (20 spp., mostly Old World) — A viscid-pubescent spiny shrub of swampy areas. Leaves membranaceous with entire margins. Flowers blue.

E: hierba de la potra

Wigandia (6 spp.) — A subwoody weedy shrub common along premontane and inter-Andean valley roadsides. Characterized by broadly ovate, cordate, doubly toothed leaves, usually densely pubescent, sometimes stinging. Inflorescence densely scorpioid with whitish or greenish flowers 2–4 cm long.

Phacelia (200 spp., mostly N. Am.) — A herb mostly of rocky places above timberline. Leaves ovate and serrulate to narrow and pinnatifid, viscid-pubescent. Inflorescence strongly scorpioid, the flowers purple.

Nama (40–50 spp., mostly in Chile and southwest USA) — Tiny dichotomously branching weedy herb with a sessile white flower in each dichotomy. Leaves oblanceolate.

Icacinaceae

A vegetatively very nondescript family of trees and lianas with simple alternate entire leaves. Almost all species have a groove on the top of the petiole which is often somewhat twisted. Most species either have the leaves drying a distinctive blackish color or rather tannish-green (cf., Olacaceae). In most arborescent genera (and scandent *Pleurisanthes*) the tertiary venation is conspicuously parallel; when the tertiary venation is not perpendicular to the midvein and parallel, conspicuous (sub)axillary domatia may be present (*Citronella*). A rather medicinal odor is typical of several genera.

The fruits of most genera are distinctive: always one-seeded (except in the primitive *Emmotum*) and always with hard endocarp. *Calatola* has large fruits with elaborately sculpted longitudinal ridges on the endocarp which often persist many months on the forest floor (when fresh the ridges are hidden by the mesocarp, however). *Discophora* has an asymmetrically curved rather flattened almost jellybean-shaped, half-cream and half-green to black fruit; and *Citronella* has an oblong fruit with a radially extending partition of the endocarp around which the seed is bent, and thus, horseshoe-shaped in cross section. Except *Metteniusa* (sometimes placed in Alangiaceae), the flowers are very small and inconspicuous. The technical definitive character of Icacinaceae is the ovary with 2 ovules pendent from apex of the single locule.

1. Trees

1A. Genera with conspicuously black-drying leaves; usually with finely serrulate margin or stellate trichomes.

Calatola (7 spp.) — Leaves usually more or less serrate or serrulate, drying blackish. Fruit subglobose to broadly ellipsoid, with very distinctive, elaborately sculpted, longitudinal ridges on surface of the endocarp; male inflorescence pendent and spicate (cf., many Moraceae).

E: erepe

Dendrobangia (3 spp.) — Leaves membranaceous, drying black, entire, with stellate trichomes; fruit compressed-ellipsoid, 2 cm long, the endocarp smooth. Distinct medicinal odor.

P: yodoformo caspi

1B. Genera with greenish or greenish-tan-drying leaves; entire or coarsely spiny-toothed, without stellate trichomes, sometimes sericeous below

Discophora (2 spp.) — Leaves below with a characteristic flat sheen, the tertiary venation noticeably smooth. Inflorescence axillary, paniculate, with minute flowers; fruit unmistakable, small (for the family), rather flattened, arcuate (jellybean-shaped), bicolored, the concave side cream-colored with a median ridge, the convex side green to black with several conspicuous longitudinal ridges.

P: repollito

Citronella (7 spp.) — Leaves thinly coriaceous with intricately prominulous parallel tertiary venation perpendicular to midvein; usually with a noticeable hyaline margin; vegetatively very similar to *Cordia* with leaves from between branch dichotomies. Inflorescence not axillary, usually long and racemiform-paniculate, the short lateral branches subscorpioid with the flowers only along one side; fruit an oblong-ellipsoid drupe with a very thin flesh, the seed folded around a partition extending into center of locule, and thus, horseshoe-shaped in cross section.

Icacinaceae
(Trees: Small Fruits)

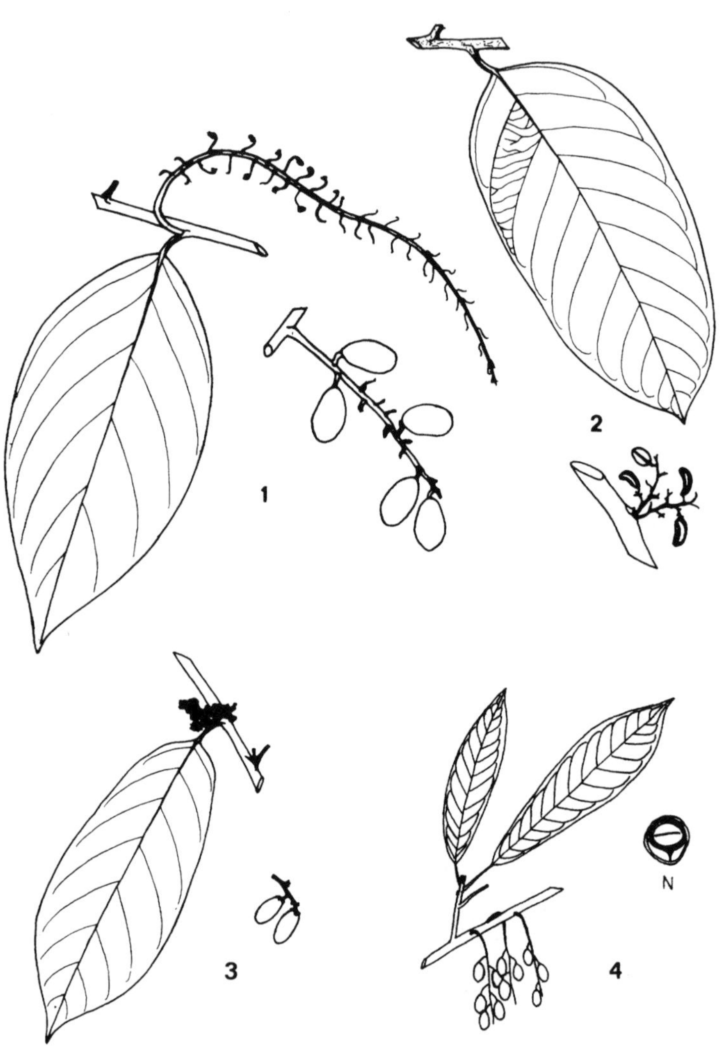

1 - *Citronella*

2 - *Discophora*

3 - *Dendrobangia*
(*D. boliviensis*)

4 - *Dendrobangia*
(*D. multinervia*)

Icacinaceae
(Lianas)

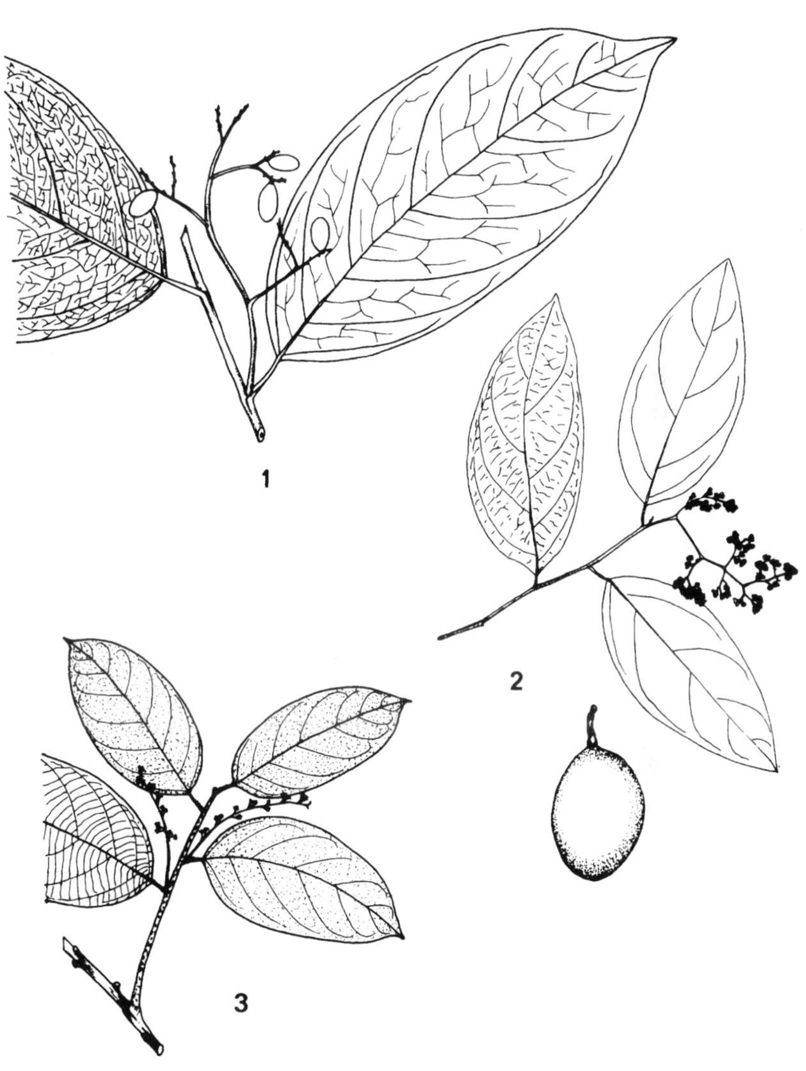

1 - *Lehretia*

2 - *Casimirella*

3 - *Pleurisanthes*

Poraqueiba (3 spp.) — Leaves sericeous-pubescent below, at least when young (unique except for *Emmotum*), the parallel tertiary veins prominulous, unlike *Emmotum*. Inflorescence axillary, paniculate, with minute flowers. Fruit flattened-oblong-ellipsoid, with a thin buttery edible pulp, to ca. 4 cm long. *Poraqueiba sericea*, the umiri, is frequently cultivated in upper Amazonia for the edible fruit pulp.

P: umiri

Emmotum (12 spp.) — Trees of poor-sôil forest. Looks very like *Poraqueiba* except for smoother (with nonprominulous tertiary veins) more strongly sericeous and discolorous leaf undersurface. Fruit smaller (ca. 1 cm across) and rounder than *Pouraqueiba*, technically differing in 2–3 locules rather than one.

Metteniusa (3 spp.) — Leaves with prominently parallel tertiary venation (cf., *Minquartia*), more coriaceous than *Citronella*, less so than *Poraqueiba*. Flowers much larger than in other area genera and more similar to *Ryania* (Flacourtiaceae), especially in the numerous stamens with purple anthers.

C: cantyi

2. Lianas

Leretia (1 sp.) — The single species, *L. cordata*, has a characteristic, openly paniculate, tiny-flowered axillary inflorescence. Leaves short-petioled, with conspicuously reticulate ultimate venation and at least a few T-shaped trichomes on the lower surface, drying grayish. Fruits oblong-ellipsoid, to 4.5 cm long, hirsute when young.

Pleurisanthes (5 spp.) — Canopy liana usually with flattened (+/- 2-parted) stem; leaves below intricately raised-reticulate, pubescent with stiff subappressed trichomes and asperous. Fruits tannish-pubescent and scabrous, ca. 2 cm long. Inflorescence axillary and spicate or narrowly racemose, much narrower than *Leretia* or *Humirianthera*.

Casimirella (7 spp., incl. *Humirianthera*) — Differs from *Leretia* in terminal rather than axillary inflorescence and rufous-pubescent leaves and twigs. Vegetatively more like *Pleurisanthes*, but the trichomes in part +/- stellate or subdendroid and the leaf below with secondary veins less obviously scalariform and the surface not asperous.

Also three extralimital genera, *Mappia* and *Oecopetalum* in Central America and *Ottoschulzia* in the Antilles.

Icacinaceae
(Trees: Large Fruits)

1 - *Emmotum* **2** - *Metteniusa*

3 - *Calatola* **4** - *Pouraqueiba*

JUGLANDACEAE

Large montane forest trees, vegetatively characterized by pinnately compound leaves with mostly serrate leaflets and a distinct rank walnut odor. The chambered pith of the twigs typical of temperate zone Juglandaceae is not evident in ours. The leaves may be either alternate (*Juglans*) or opposite (*Alfaroa*), and the leaflets form a conspicuous progression from larger terminal ones to smaller basal ones, the lowermost sometimes becoming stipulelike; usually somewhat viscid-pubescent or hispid and often with few-rayed stellate trichomes intermixed with simple ones. The amentiferous flowers are highly unusual for South America, the male inflorescence a pendent spike with numerous apetalous flowers; female inflorescence similar or reduced to 1–few naked flowers each with two styles. The round or ellipsoid indehiscent fruits very characteristic in their hard endocarp with the seed partitioned by projections from endocarp inner wall (walnut).

Alfaroa (6 spp.) — Mostly Central American, one species recently discovered in the eastern Cordillera of Colombia. Leaves opposite; fruits (and female flowers) on a spike.

Juglans (7 spp.) — Montane forests. Leaves alternate; fruits (and female flowers) solitary or in cluster of two or three.

C, P: nogal

JULIANACEAE

Thick-branched trees with white latex and pinnately compound leaves with serrate leaflets clustered at branch apices. Probably a florally reduced (wind-pollinated?) derivative of Anacardiaceae, characterized by the minute apetalous flowers lacking a nectar-disk and with a single locule and ovule. Flowers borne while leafless, the male in a small panicle, tiny, reddish, and nondescript. Fruit very distinctive with thick-bodied seed borne at end of a long straight wing formed from the pedicel (resembles an inverted ash samara). In our area occurs only in very dry middle elevations (ca. 1500–1800 m) on the Pacific side of Peru where it is usually the only tree and may form open single-species stands. Leaves deciduous for much of year and the characteristic open rachitic growth-form with few thick naked branches is distinctive.

Orthopterygium (1 sp.)

The only other genus, *Amphipterygium*, occurs in the driest forests of Mexico and Central America.

Julianaceae, Juglandaceae, and Krameriaceae

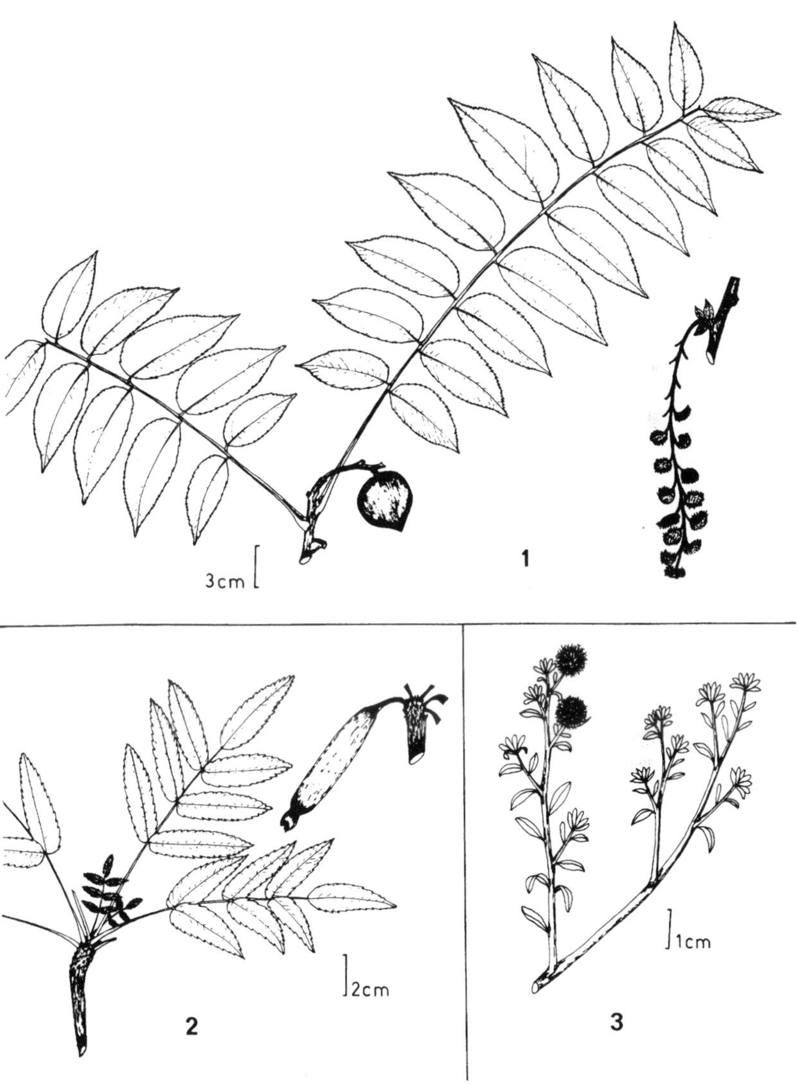

1 - *Juglans* (Juglandaceae)

2 - *Orthopterygium* (Julianaceae) 3 - *Krameria* (Krameriaceae)

Krameriaceae

Largely amphitropical and not very common in our area. Hemiparasitic shrubs of dry areas, characterized by small coriaceous, silvery grayish, alternate, simple leaves with sharp-pointed apices (one extralimital species is 3-foliolate). The legumelike flowers are usually orange or reddish with showy sepals and reduced petals, the two lower petals modified into lipid-secreting glands. The fruit is a small, round, exozoochoric, bristly bur (cf., *Triumfetta*). Although traditionally placed under caesalpinioid legumes, its true relationships are with Polygalaceae.

Krameria (15 spp.)

Labiatae

A mostly herbaceous family, easy to recognize by the mint odor, opposite, nearly always serrate leaves on square stems or branchlets, and nearly always strongly bilabiate flowers. Similar only to Boraginaceae (which are nonaromatic and differ in alternate leaves and/or scorpioid inflorescences) and Verbenaceae (which tend to be less zygomorphic but differ fundamentally (and somewhat tenuously) only in lacking the gynobasic Labiatae style which arises from between 4 nutlets). Other families sometimes confused with Labiatae are Scrophulariaceae and Acanthaceae, both with a capsular fruit, very different from the four nutlets of the Labiatae, and normally have the style persistent after corolla has fallen while corolla and style normally fall together in Labiatae. The characteristic tubular, usually longitudinally ridged, calyx of the Labiatae is very different from these other families. Except for some Verbenaceae, the tree mints are distinctive in their opposite, aromatic, nearly always serrate leaves on square stems; *Lepechinia* is especially similar to *Lippia* (Verbenaceae) but can be vegetatively distinguished from the arborescent species of that genus by its more perpendicular secondary veins. Our few opposite-leaved borages have entire leaves and acanths are entire or serrulate (or spiny-serrulate); the very few entire-leaved *Salvia* species are densely canescent below unlike acanths or opposite-leaved *Tournefortia* (Boraginaceae).

Eleven mint genera which are wholly or predominantly woody or subwoody occur in our area, along with a number of additional herbaceous genera. Most species of *Minthostachys* (tending to be scandent), *Satureja, Gardoquia, Lepichinia, Hypenia,* and *Hyptidendron* are more or less woody; some species of predominately herbaceous *Salvia* and *Hyptis* are also shrubs or small trees and a few *Scutellaria* are subshrubs. *Salvia* and introduced *Rosmarinus* are the only woody genera with 2 stamens. One group of 4-staminate genera (*Hyptis* and relatives) has the stamens held near or within the lower corolla lip; the other 4-staminate genera (*Satureja, Lepechinia* and relatives) have the stamens held near upper corolla lip. The numerous herbaceous genera are distinguished mostly by characters of the stamens (4 vs 2, anthers 1- or 2-celled) and calyx (number of teeth, longitudinal nerves, shape, degree of zygomorphy).

1. More or Less Woody Shrubs or Small Trees (*Minthostachys* +/- Scandent) — The first two genera have two stamens, the others four.

1A. Shrubs or small trees with two stamens

Salvia (700 spp., incl. Old World) — Mostly upland cloud forests; the majority of species herbaceous but many are shrubby and some are soft-wooded trees several meters tall. Arborescent species often have discolorous leaves white below and are almost always serrate (entire only when strongly canescent: *S. acutifolia*). Unusual in family in having only 2 fertile stamens; other native mints with only 2 stamens are small herbs. Inflorescence a terminal spike or raceme; the flowers usually large (>(1.5–)2 cm long) and brightly colored (red, blue, purple), strongly bilabiate, the anthers often exserted; calyx rather large and distinctly bilabiate.

Rosmarinus (introduced) — A strongly aromatic shrub with narrowly linear leaves having recurved margins. The blue flowers differ from *Salvia* in lacking a hooded upper corolla lip and bilabiate calyx.

1B. Shrubs or small trees with four stamens held against lower corolla lobe — (Several with only weakly tetragonal branchlets)

Hyptidendron — Shrubs to fair-sized trees, related to *Hyptis* but more arborescent (ours is *H. arborea,* a common Andean second-growth tree to 15 m tall) and the conspicuous corymbose axillary panicles with larger, tubular, lilac flowers; purple-tinged bracts are another distinctive character. Branchlets not very tetragonal; leaves finely serrate, narrowly elliptic, usually densely pale-pubescent below and with close strongly ascending secondary veins (rather like *Rhamnidium* but the veins more curved).

Hypenia — A recent segregate from *Hyptis,* in our area only in the Colombian Llanos. Slender shrub or xylopodial subshrub with raceme of widely spaced sky-blue flowers on long slender pedicels; stem weakly tetragonal, with petiole bases decurrent onto the conspicuously swollen (jointed-looking when dry) internode, in the common species sparsely hirsute with spreading hairs.

Eriope (28 spp.) — Lowland savanna shrubs or subshrubs, mostly with swollen xylopodia; mostly in the cerrado and barely reaching our area. Inflorescence a slender raceme of violet-blue flowers, the calyces of which have conspicuous tufts of white hairs in the throat.

(*Hyptis*) — Most *Hyptis* species are herbs but a few are somewhat shrubby, all characterized by small flowers with regularly 5-toothed calyces, usually arranged in compact heads subtended by greenish bracts.

1C. Shrubs or small trees with four stamens held against or near upper corolla lobe

Minthostachys (12 spp.) — Mostly upland dry areas; our only (more or less) viny mint. Differentiated from *Satureja* by the 12-veined calyx with equal narrowly triangular teeth and spreading hairs. Leaves small, ovate, at least slightly serrate. Flowers small and white, in dense sessile pilose axillary inflorescences.

Satureja (30 spp., incl. Old World) — Andean shrubs (at highest altitudes reduced to subshrubs or herbs) characterized by the small leaves densely clustered along branches, usually entire, narrowly elliptic to narrowly obovate or even linear and ericoid, sessile or subsessile. Flowers axillary in very short racemes or these reduced to fascicles or solitary flowers; corolla small and white to purple or longer and red or orange; calyx narrow and not very bilabiate, 10-nerved; the expanded anther connective is an important technical character.

Gardoquia (ca. 12 spp.) — Essentially a hummingbird-pollinated segregate of *Satureja* with much larger tubular, usually red or orange (rarely yellow or purple) flowers. Subshrubs of Andean steppes, with leaves densely white-pubescent below or grayish all over; either serrate or sublinear, usually smaller than in *Salvia* (which also differs in only 2 stamens).

Lepechinia (50 spp., incl. *Sphacele* and Old World) — Andean upland shrubs or small trees to 4 m tall (often course herbs in *Sphacele*). Leaves mostly rather large, always puberulous and serrate, typically more or less bullate and rough above. Flowers not very large, essentially a small-flowered 4-stamened version of *Salvia*, but neither calyx nor corolla very bilabiate. Inflorescence a terminal raceme or spike or often paniculately branched with the flowers rather dense along each branch. Very similar to *Lippia* (Verbenaceae), but usually differing vegetatively in the secondary veins more nearly perpendicular to midvein.

Lacistemataceae and Labiatae

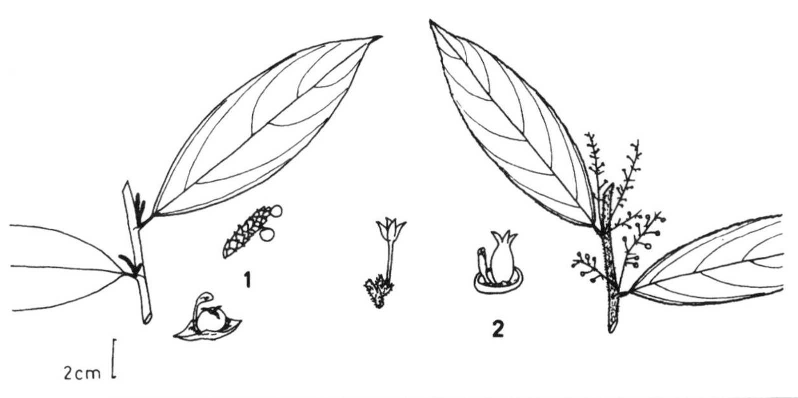

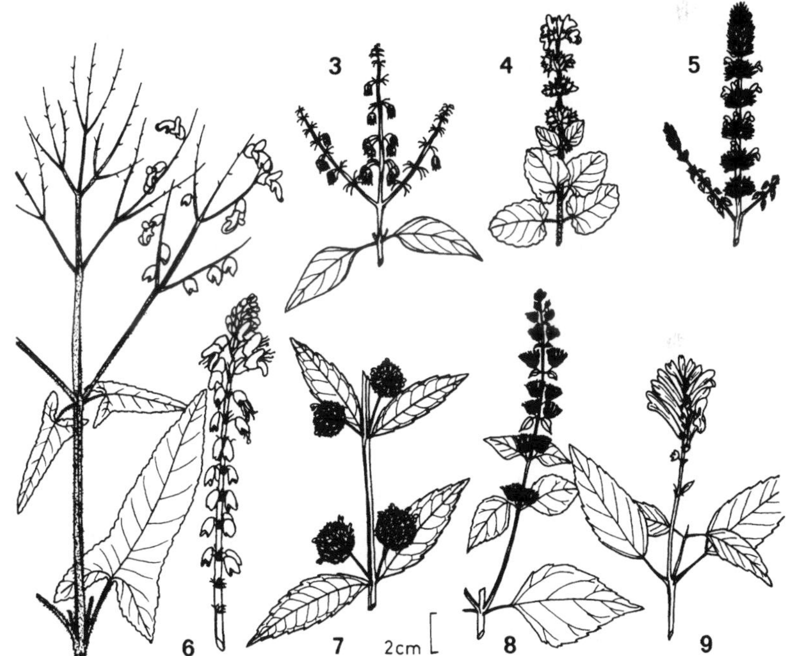

1 - *Lacistema* (Lacistemataceae) **2** - *Lozania* (Lacistemataceae)

3 - *Ocimum* **4** - *Stachys* **5** - *Satureja*

6 - *Salvia* **7** - *Hyptis* **8** - *Hyptis* **9** - *Scutellaria*

2. Herbs, Mostly Weedy and/or Introduced

2A. Herbs with two stamens

Hedeoma (1 sp., plus ca. 30 in N. Am.) — Prostrate weedy high-altitude herb with round leaves.

(***Salvia***) — Many species are erect herbs, usually with brightly colored flowers and strongly bilabiate calyces.

2B. Herbs with four stamens

2Ba. The next eight genera have anthers held in floor of tube or exserted but not held in pairs under the hood formed by upper corolla lobes.

Ocimum (150 spp., mostly Old World) — Our species small weeds characterized by sweetly aromatic thin leaves tapering to long thin petiole, small pinkish flowers verticillate in narrow racemes, and strongly bilabiate calyx with upper lip decurrent to form wing down calyx tube and 3-setose upcurved lower lip, +/- inflated in fruit.

Solenostemon (incl. *Coleus*) (introduced) — Easily recognized by the variegated purplish or pinkish and pale green leaves. Inflorescence a narrow racemose panicle, the flowers bluish, very small (calyx only 2 mm long in flower).

Hyptis (320 spp.) — Large genus mostly in lowlands; a few species are large coarse herbs or somewhat shrubby. Most species with the small flowers in very characteristic pedunculate axillary capitate heads, these usually subtended by involucre of greenish bracts; species with more open cymes having these arranged in large panicles; paniculate species with more diffuse inflorescence than any other mint. Calyces tubular and regularly 5-toothed, mostly 10-veined unlike most other mints.

Catoferia (3 spp.) — Our only species a rare herb of upland Colombia and Ecuador with a compact spike of white tubular flowers with purplish bracts and calyces and the stamens and style conspicuously long-exserted. Looks more like Verbenaceae (e.g., *Clerodendron*) than Labiatae with its weakly tetragonal stem and large thin, serrate, ovate-elliptic leaves with base of blade decurrent onto petiole; the nodes are prominently jointed, at least when dry.

Marsypianthes (5 spp.) — Weedy viscid herb with heads of violet-blue flowers and a regularly 5-toothed calyx that becomes inflated in fruit. Very close to *Hyptis* (especially similar to *H. atrorubens*), distinguished technically by the margins of the concave nutlets thin and fimbriate, and macroscopically by the bracts narrower and more sharp-pointed than in *H. atrorubens.* (MacBride supported generic segregation from *Hyptis* on basis that the name is a pleasant one!)

Teucrium (300 spp., mostly Old World) — Weedy herb of lowland dry areas with thin membranaceous leaves. Distinct from other mints in the spicate inflorescence and in having an unridged calyx, inflated in fruit, from which the small pinkish corolla, which entirely lacks an upper lip, barely emerges. Looks more like *Priva* (Verbenaceae) than like other Labiatae; differing from *Priva* in the more bilabiate flower and more puberulous leaves and young stems.

Mentha (introduced) — Strongly aromatic upland Andean weeds (plus aquatic *M. aquatica*) characterized by terminal spikes or heads of small almost regular flowers usually with exserted stamens and nonbilabiate calyx.

Origanum (introduced) — Erect aromatic weedy herb with small flowers having exserted anthers and arranged in small cymose clusters partly enclosed by small conspicuous bracts which more or less take the place of the usually reduced calyx.

2Bb. The next seven genera have four stamens with the anthers held in pairs under "hood" formed by upper corolla lobes.

Scutellaria (77 spp., also 200 N. Am. and Old World) — In our area Andean herbs and subshrubs. Characterized by the often showy tubular corollas and short obscurely lobed calyx with 2 rounded untoothed lips which close together in fruit and usually have a distinctive humplike dorsal projection (= "skullcap").

Marrubium (introduced) — Weedy gray-tomentose herb with white flowers, the calyx teeth 10, rather than 5 as in other genera, the spreading teeth recurved in fruit for exozoochorous dispersal.

Lamium (introduced) — Eurasian weed, in our area at high altitudes. Leaves sessile with clasping base, suborbicular, broader than long, the apex irregularly deeply toothed; flowers magenta, densely clustered in leaf axils, the calyces subequally 5-toothed.

Stachys (300 spp, incl. Old World) — Plant always viscid-pubescent, the flowers in sessile verticils in interrupted terminal spike with rather few flowers in each cluster; corolla magenta, calyx evenly 5-toothed, not bilabiate, the teeth often spine-tipped.

Leonurus (introduced) — Eurasian weed, unique in area taxa in leaves pinnatifidly deeply divided; otherwise like *Stachys* except the nutlet truncate. Inflorescence with sessile magenta flowers in dense interrupted verticils.

Leonotis (introduced) — Coarse weedy herb of middle elevations; flowers orange, rather large, the 15-toothed calyx with conspicuously spine-tipped teeth, arranged in dense globose many-flowered verticils.

Prunella (introduced) — A creeping weedy herb with short erect spikes of purple bilabiate flowers, the bilabiate calyx closed in fruit.

LACISTEMATACEAE

Subcanopy trees characterized by the alternate membranaceous leaves with parallel tertiary venation perpendicular to midvein and by a tendency for the margin to be finely serrate or somewhat remotely serrulate; when completely entire with conspicuous stipule scars. Essentially a Flacourtiaceae with much reduced flowers and now often placed in that family; the very typical inflorescences are axillary and spicate or spicate-racemose, the numerous tiny yellowish-green flowers apetalous and with a single stamen. Fruit a small incompletely dehiscent berrylike capsule, usually with a single seed.

Lacistema (11 spp.) — Flowers aggregated into small dense narrow spikes, these clustered at nodes and sticking out in different directions. Vegetatively characterized by the stipule caducous to leave conspicuous scar (on young growth the twigs and stipules tend to dry blackish, the stipule scar whitish (reminiscent of *Rinorea*). Most species are essentially entire-leaved. The trunk slash is unusual in being completely white and textureless.

Lozania (4 spp.) — The leaves usually finely serrate and often puberulous; differs from *Lacistema* in the smaller and usually persistent stipules and the inflorescence less congested, spicate or racemose, or rarely several narrow racemes forming a kind of panicle.

LAURACEAE

Except for *Cassytha* (a leafless herbaceous parasitic climber completely lacking chlorophyll), Lauraceae are easy to recognize to family, even when sterile, by the combination of Ranalean odor, non-2-ranked leaves that are often irregularly spaced along the branches and often apically clustered, and lack of strong bark or myristicaceous branching. The leaves are archetypical of the predominant rain-forest leaf morphology, this "lauraceous look" being characterized by elliptic shape with acuminate apex and cuneate base, medium size, glossy glabrous surface and entire margins. Typically there are fewer, more ascending secondary veins than in other Ranalean families. A useful herbarium character is that the secondary veins of all Lauraceae are decurrent on the midvein. Another useful character is the frequent presence of an inrolled or involute leaf base, sometimes developed into a basal auricle (especially in *Nectandra*). When pubescent, Lauraceae often have appressed trichomes and a sericeous aspect; the trichomes are always simple. Lauraceae only very rarely have lenticels on the twigs which are usually green or greenish. Stilt-root buttresses are relatively frequent. Bark is variable, frequently smooth with raised lenticels, but sometimes thick and fibrous or even vertically ridged. The trunk slash usually has a distinctive Ranalean odor; in some species the leaves lack an obvious Ranalean odor and in others the bark does, but the odor seems to be always apparent in at least one or the other. The vegetative odor of Lauraceae can be sweet and spicy, pungently aromatic and kerosene-like, or unpleasant and foetid, depending on the species. Lauraceae flowers are usually small, whitish, and nondescript at first sight, but are absolutely unmistakable with closer examination on account of the 2 or 4 flaplike valves on each anther; though small, these valves can almost always be seen with the naked eye. Lauraceae fruits are also easy to recognize. All are single-seeded and the vast majority have a more or less enlarged, usually red, receptacular cupule surrounding the basal part of the shiny blackish-maturing fruit. Only two (or three, if intermediate *Cinnamomum*, with well-developed but often weakly fused fruiting tepals, is included) significant genera, *Persea* and *Beilschmiedia,* have fruits not subtended by the characteristic lauraceous cupule (in the case of *Cryptocarya* and opposite-leaved *Chlorocardium* surrounded by it).

Lauraceae are exceedingly and notoriously difficult to recognize to genus, even when fertile. Generic placement is primarily based on technical characters of the stamens (number of whorls, presence and size of staminodes, anthers 2-valved vs. 4-valved) and to a lesser extent the fruit. The two largest and commonest genera are theoretically separated by whether the 4 anther valves are arranged in an arc (*Nectandra*) or in two rows (*Ocotea*) but some species are intermediate. Different collections of the same species have been described in different genera with disconcerting frequency, even by the same specialist. Despite these problems most species are morphologically distinct and easy to recognize as taxonomic entities, even when sterile, and many natural groups can be recognized, even on vegetative characters. Given the taxonomic chaos that prevails in Lauraceae, the suggestions for generic recognition given below (based largely on the original observations of H. van der Werff) are no more than useful approximations. In addition to *Cassytha,* sixteen genera of Lauraceae are known to occur in our area. The difference between the three genera lacking cupules in fruit and the numerous cupulate-fruited genera is probably a taxonomically fundamental one, but lowland *Cinnamomum* is usually intermediate, having the tepal bases only weakly fused in fruit. Several of the cupulate genera have characteristic fruits (e.g., *Cinnamomum* with swollen pedicel merging with tapered base of reduced cupule, *Licaria* with double-rimmed cupule, *Endlicheria* with subsessile clustered fruits, *Crytocarya* and *Chlorocardium* with cupule completely enclosing fruit, *Aniba* and *Pleurothyrium* with large brownish warty cupules) but in each case a few species of other genera share the characters. In flower several genera (*Aniba, Licaria, Mezilaurus, Ajouea*) are characterized by unusually small tepals and a tendency to fusion into a floral tube; in contrast *Ocotea, Nectandra, Cinnamomum, Persea* and other genera have relatively large distinct tepals. The only genera with 4-celled anthers are *Ocotea, Nectandra, Persea* (most species), *Cinnamomum* (most species), *Williamodendron* and one species of *Caryodaphnopsis.*

Vegetatively useful characters include: opposite leaves (*Caryodaphnopsis, Anaueria, Chlorocardium*), conspicuously parallel tertiary venation (most *Nectandra*), clustered leaf arrangement in *Endlicheria, Mezilaurus, Williamodendron,* some *Aniba* species and a few *Pleurothyrium* species (plus a few Central American *Cinnamomum*); unusually few (4–6) leaf veins (usually *Endlicheria, Licaria,* or a few *Nectandra* species); unusually long petioles (lowland *Persea, Williamodendron,* some *Endlicheria* species); 3-veined leaves (*Cinnamomum* plus miscellaneous species); unusually intricately reticulate tertiary venation (most *Beilschmiedia,* few *Endlicheria*), and a yellowish-green "matte" sheen on leaf undersurface (mostly *Aniba*). One of the most obvious prospective vegetative characters, a strongly sericeous leaf undersurface, occurs in one or two species of nearly all genera and is useful only at specific level.

1. Leafless Achlorophyllous Parasitic Vine

Cassytha (1 sp., plus 19 in Old World) — A leafless parasitic herbaceous vine without chlorophyll that looks much more like dodder (*Cuscuta*) than like Lauraceae; the anthers have the typical lauraceous anther valves, however.

2. Opposite-Leaved Trees

Caryodaphnopsis (8 spp., plus 7 in Asia) — Segregated from *Persea* on account of opposite leaves and more strongly unequal tepals with a triangular shape and unique outcurved tip. Most species have strongly 3-veined leaves. The fruits are usually large and lack persistent tepals as in

Lauraceae
(Cassytha and Fruits without Cupules)

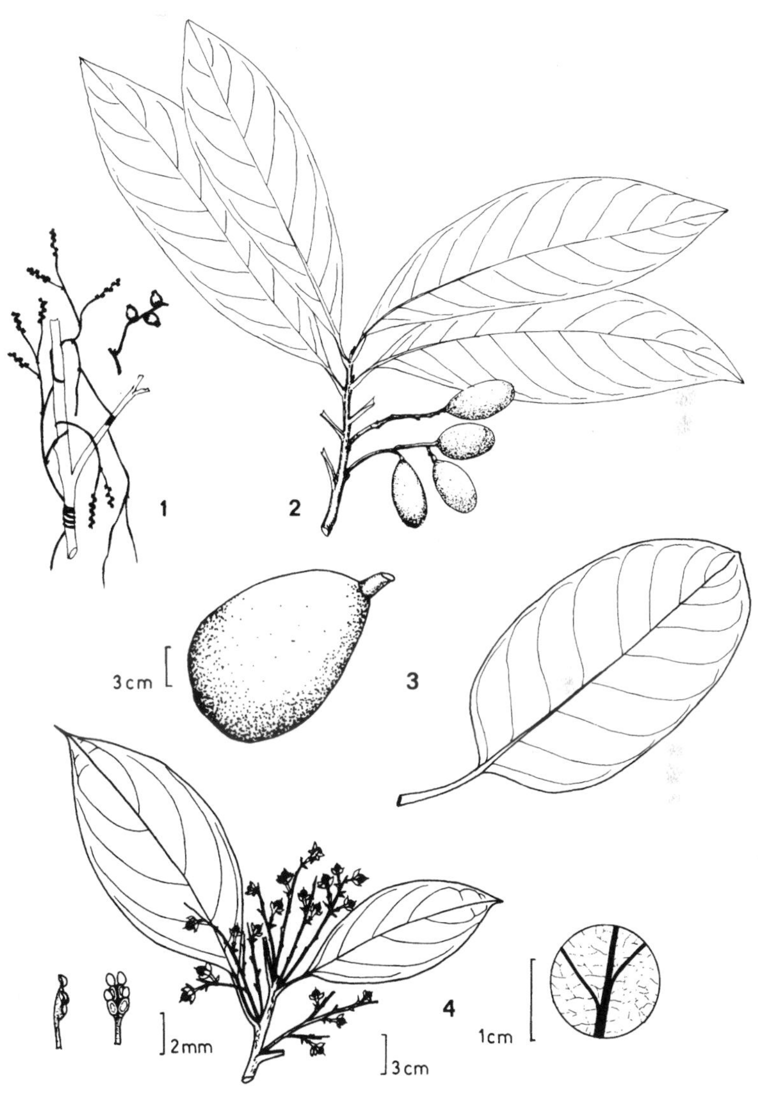

1 - *Cassytha* **2** - *Beilschmiedia*

3 - *Persea* (*P. americana*)

4 - *Persea*

Lauraceae
(Opposite Leaves)

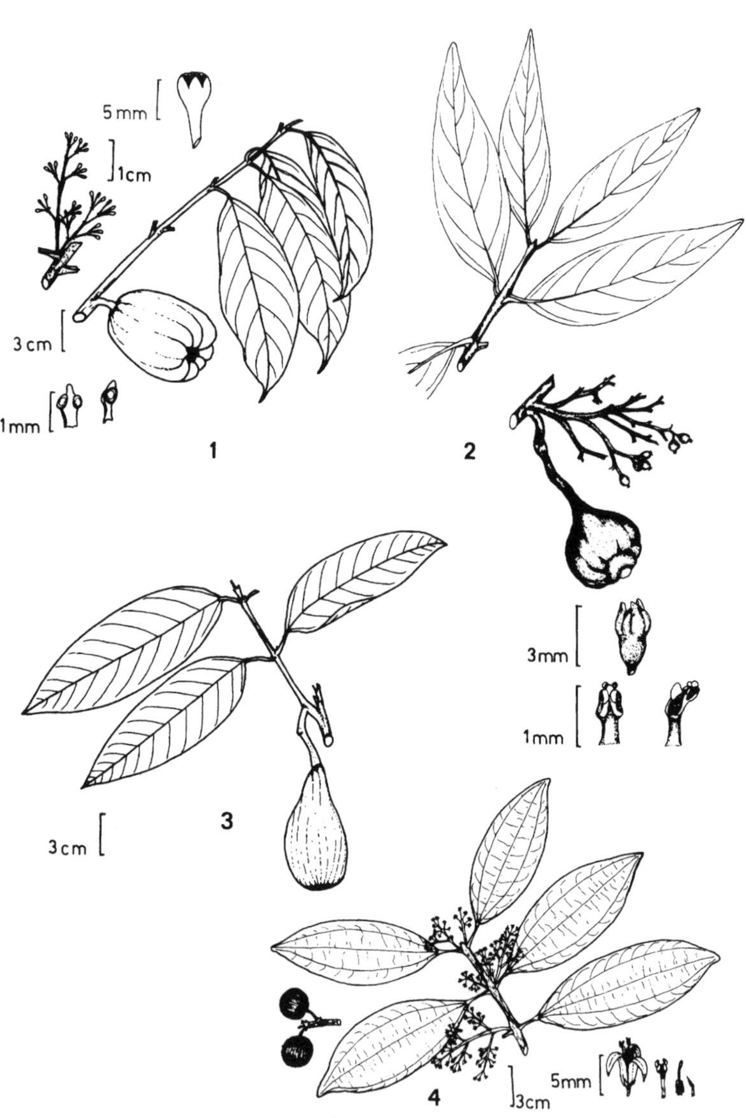

1 - *Anaueria*

2 - *Chlorocardium*

3 - *Caryodaphnopsis*

4 - *Caryodaphnopsis*

Central American *Persea.* Of the two neotropical *Persea* species recently transferred to this Asian segregate, the trans-Andean one has 3-veined leaves and 4-celled anthers (as do several additional new species), the Amazonian one pinnately veined leaves and 2-celled anthers; the newly described species are all 2-celled.

Anaueria (1 spp.) — Opposite, shiny, coriaceous, not at all 3-veined leaves that dry dark (cf., many Nyctaginaceae) are distinctive. There are 3 fertile 2-celled anthers and each inflorescence branch is subtended by 3 bracts. The ecupulate fruit is large and similar to *Persea americana.* The green flowers with dark red stamens are unique in the family.

Chlorcardium (2 spp.) — A recent segregate from *Ocotea,* our species very distinctive in the very large (ca. 5 cm diam) fruit completely enclosed in the cupule. Leaves glabrous and pinnately veined.

(*Nectandra*) — One *Nectandra* species (*N. oppositifolia*) has consistently opposite leaves, distinctive in pinnate venation and the strongly tannish-tomentose undersurface.

3. Alternate-Leaved Trees With Fruit Not Subtended by Cupule (= Fleshy Expanded Fruiting Receptacle)

Persea (82 spp., plus 68 in Old World) — Mostly montane with very coriaceous leaves, frequently somewhat glaucous below; lowland species mostly have relatively long petioles. In fruit all South American species (except cultivated *P. americana*) have persistent tepals; the South American species (but not *P. americana*) all have different tepal lengths, a feature shared only with opposite-leaved *Caryodaphnopsis* and a few species of *Licaria* (with much smaller flowers, so the character not obvious).

C, E, P: aguacate, palta

Beilschmiedia (15 spp., plus 200 paleotropical) — Uncommon in Amazonia; mostly in Central America and southern Brazil. Technically distinguished from *Persea* by 2-celled (vs. 4-celled) anthers. In fruit can be distinguished from South American *Persea* (but not *P. americana*) by the lack of persistent tepals. The fruits are usually ca. 2.5 cm long and elongate; although larger-fruited *Persea* may also have elongate fruits, those with fruits as small as *Beilschmiedia* have round fruits. Vegetatively *Beilschmiedia* is mostly characterized by finely intricate prominulous venation and fairly frequently opposite leaves (always alternate in *Persea*).

P: "añushi rumo" (opposite leaves)

4. Alternate-Leaved (Rarely Somewhat Scandent) with Fruit Subtended by Cupule

4A. Four-celled anthers; leaves never whorled

Ocotea (ca 350 spp., plus few in Africa and ca. 20 in Madagascar) — The main genus of neotropical Lauraceae, especially in Amazonia. Diverse and vegetatively rather variable: In our region if a Lauraceae cannot be recognized by any distinctive characteristic, the chances are high that it will turn out to be *Ocotea*. *Ocotea* sometimes has unique, conspicuously coriaceous obovate leaves with unusually well-developed basal auricles; unlike some relatives, the leaves are never conspicuously tomentose (though sometimes very sericeous as in miscellaneous species of several genera). In flower *Ocotea* can be distinguished from its closest relative *Nectandra* by lacking the (usually) granular inner surface of the tepals. In young fruit, the two genera apparently can be consistently distinguished: The tepals of *Ocotea* fall individually leaving the dried anther remnants subtending the fruit while the tepals of *Nectandra* are circumscissile and fall as a unit taking the anthers with them. One species (*O. gracilis*) is a clambering liana.

C: laurel; E: jigua; P: moena, moena negra (*O. marmiensis*), yacu moena (near rivers), moena amarilla (*O. aciphylla* and *O. costulata*)

Nectandra (ca. 120 spp.) — The second largest and most prevalent neotropical laurac genus and much confounded with *Ocotea*. In lowland Amazonia, *Nectandra* is poorly represented, but in Central America and the Andes it is common and speciose. The majority of species of *Nectandra* can be recognized by the strongly parallel tertiary venation (very rare elsewhere); conspicuously tomentose leaves are also relatively common. One segment of the genus (*N. globosa* group) is vegetatively characterized by fewer lateral veins (4–6 per side) than most other Lauraceae and an almost always slightly inrolled leaf base; this group has relatively large white flowers and a unique sterile anther apex. Another segment of *Nectandra* (*N. cuspidata* group), vegetatively undistinguished, has a characteristic extremely diverticate inflorescence and similar, but smaller, flowers lacking the sterile apical part of the anthers. Another group of species (*N. reticulata* group) has very tomentose leaves and conspicuously inrolled basal auricles. The usual floral feature to separate *Nectandra* from *Ocotea* is having its 4 anther valves in an arc rather than in two rows but this feature is inconsistent and difficult to use; for other differentiating characters from *Ocotea*, see above.

P: moena

Rhodostemonodaphne (12 spp.) — A poorly defined and perhaps unjustified segregate from the *Ocotea/Nectandra* complex, characterized by unisexual flowers with the anthers not clearly differentiated from the broad filaments. Additionally differs from *Ocotea* in the 4 anther cells in an arch rather than in two rows, just as in *Nectandra*, but that genus has

Lauraceae
(4-Valved Anthers, Nonclustered Leaves; Fruiting Cupule Cup-Shaped)

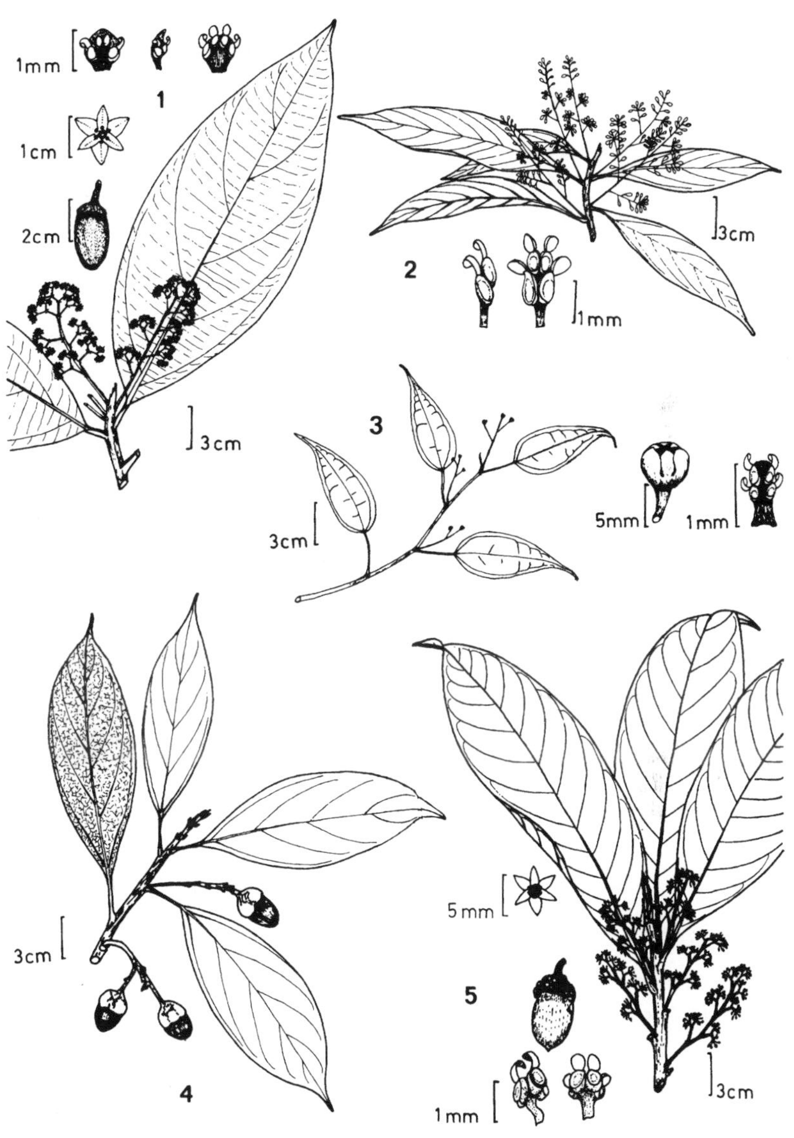

1 - *Nectandra* 2 - *Ocotea*

3 - *Cinnamomum*

4 - *Rhodostemonodaphne* 5 - *Pleurothyrium*

perfect flowers. Vegetatively distinguished by the strongly coriaceous leaves with conspicuously prominulous fine venation above and below and a usually sericeous or otherwise puberulous undersurface.

Cinnamomum (ca. 50 spp., plus ca. 250 in Old World) — Neotropical species traditionally treated in *Phoebe* probably belong in *Cinnamomum.* Close to *Ocotea,* supposedly differing in larger and more conspicuous staminodes and lacking a well-developed cupule in fruit, thus, also approaching *Persea.* In our area (but not in southern and central Brazil) the characteristic leaves are almost always strongly 3-veined with axillary hair tufts. The fruit is distinctive from *Ocotea* and *Nectandra* in having a swollen pedicel that merges into the swollen base of a poorly developed cupule that consists essentially of the separate calyx lobes.

Pleurothyrium (40 spp.) — Typically small shrubs and treelets (unlike most lauracs but similar to some *Endlicheria*). Can usually be recognized vegetatively by the unusually large leaves with relatively numerous straight, and close-together lateral veins (i.e., for length of leaf and compared to other lauracs); frequently with marginal veins, otherwise, very rare in family. One common species with relatively indistinct leaves is characterized by hollow twigs with stinging ants (*P. parviflorum*). Flowers often rather large and green or yellowish-red, unique in the inner anthers having large basal glands that force apart the outer six anthers (each with 2 of its 4 cells lateral) and the tepals, the latter having the same granular surface as *Nectandra,* in several species distinctive in curling outward longitudinally after anthesis (unique in family). The fruit (similar to *Aniba*) is commonly larger than in most cupulate genera (usually 2–3 cm long) and the cupule is usually warty lenticellate. The genus tends to combine floral characters resembling *Ocotea* with fruit (and in some species leaves) resembling *Aniba.*

4B. Two-celled anthers (except *Williamodendron*); leaves often whorled or apically clustered (if not, often with few secondary veins or a "matte" undersurface

Licaria (45 spp.) — Characterized by a nearly always double-rimmed calyx, especially obvious in fruit (otherwise found rarely in only a few species of *Nectandra* and *Ocotea*). Vegetatively usually characterized by leaves with unusually few (4–6) secondary veins; one common species (*L. canella*) has very strongly reticulate leaves. Technical recognition is by the narrowly tubular flowers with three 2-celled anthers.

P: moena

Endlicheria (40 spp.) — Always dioecious, the main technical character; dioecy otherwise is known only in some species of *Ocotea* from which *Endlicheria* is distinguished by uniformly having 9 fertile 2-celled

Lauraceae
(Clustered Leaves, Mostly 2-Valved Anthers; Fruiting Cupule Often Short)

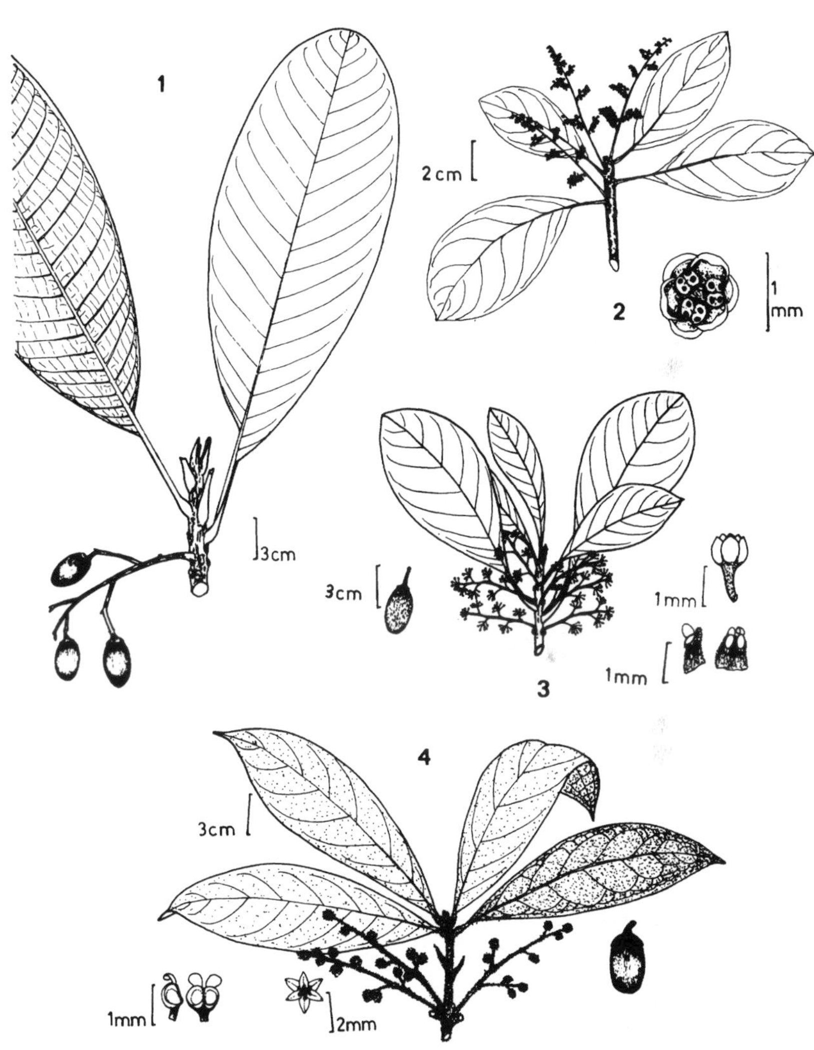

2 - *Williamodendron*

1 - *Mezilaurus*

3 - *Mezilaurus*

4 - *Endlicheria*

anthers. The fruits are often unusual in being subsessile and usually clustered together (although the inflorescence itself may be large and branched); other species are distinctive in the fruiting pedicel enlarged (cf., *Cinnamomum* but with a well-developed cupule). Most species of *Endlicheria* can be recognized vegetatively. This is the most consistently strongly vegetatively pubescent laurac genus, although most genera have a few pubescent species. In our area the great majority of species have obviously pubescent leaves, sometimes sericeous but usually with erect trichomes. A number of *Endlicheria* species are small understory trees with very characteristic clustered-verticillate obovate to oblanceolate subsessile usually conspicuously brown-tomentose leaves (similar verticillate leaves occur only in in *Aniba* and a few understory *Pleurothyrium* species that differ from *Endlicheria* in having a strong submarginal connecting vein (and 4-celled anthers). Several *Endlicheria* species have unusually long petioles, up to 5 cm long and the longest petioles in the family. Several otherwise nondescript species (e.g., *E. sprucei*) have unusually few (4–6) secondary veins, others have entirely glabrous leaves with very pronounced reticulations (cf., *Beilschmeidia*), *E. tessmannii* has secondary veins making a right angle with the midvein (unique in Lauraceae), and *E. dysodantha* dries black with axillary hair tufts (unique except for few *Ocotea* species).

E: jigua; P: moena

Mezilaurus (16 spp.) — Typically has hard wood and occurs in dry forest. Characterized by usually distinctly petiolate leaves, always all clustered at the branch tips (other clustered-leaved species [except *Williamodendron*] have subsessile leaves and only extralimital *Ocotea rubra* has the strictly short-shoot with clustered apical leaves growth-form of *Mezilaurus*); the leaves may be conspicuously pubescent or glabrous. One highly unusual character that is usually present is prominent lenticels on the young twigs. The flowers are like those of *Licaria* (three 2-celled anthers) a genus that never has clustered leaves; moreover the fruit is very different, that of *Mezilaurus* subtended by a small, platelike cupule and never double-rimmed.

P: moena, itahuba

Williamodendron (2 spp.) — A segregate from *Mezilaurus*, from which it differs only in 4 anther cells. Distinctive in large long-petiolate leaves clustered at branch apices, open few-flowered inflorescence, and fruit cupule reduced and platelike, but these characters all found in some *Mezilaurus*.

Aniba (40 spp.) — Both leaves and slash characterized by a pleasantly sweet spicy odor; the wood distinctively yellow. About one-third of species are easy to recognize vegetatively by the distinctive sheen of the conspicously pale yellowish-green "matte" leaf undersurface resulting

Lauraceae
(2-Valved Anthers, Nonclustered Leaves; Fruiting Cupule with Distinctive Rim or Raised Lenticels or Poorly Developed or Completely Enclosing Fruit)

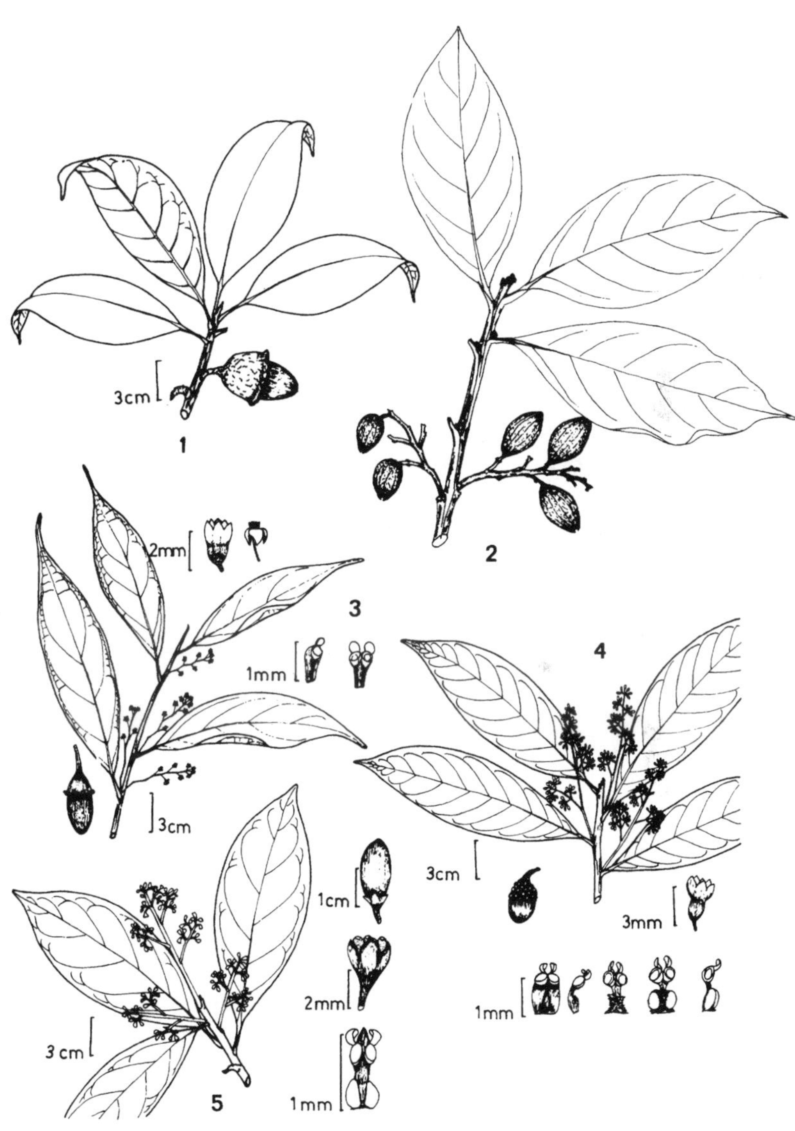

1 - *Licaria* (*L. canella*)

2 - *Cryptocarya*

3 - *Licaria*

4 - *Aniba*

5 - *Aiouea*

from a minutely tuberculate epidermis (otherwise found only in two species of *Licaria* and a few of *Cinnamomum*). Another one-third of the species are vegetatively characterized by verticillate leaves (cf., *Endlicheria*). The flowers are unusually small with a long tube and reduced tepals; they are recogizable when dry by the closely sericeous indument and longitudinally wrinkled floral tube. The fruiting cupule is unusually large compared to the fruit size (at least one-third as long as fruit and usually warty lenticellate (unique except *Pleurothyrium*). The flowers have nine 2-celled anthers.

C: comino; P: canela moena (*A. puchury-minor*), palo de rosa (*A. roseodora*)

Aiouea (20 spp.) — Characterized by a name composed of all the vowels, but by little else. The flowers are small and supposed to be characterized by conspicuous staminodes, but these are lacking in at least one-third of the species. The leaves show no obvious generic characters—many are 3-veined (and tend to dry yellowish below), some large, some conspicuously reticulate, *et cetera*. The leaves of some species dry yellowish and have a cartilaginous margin; the commonest upper Amazonian species has conspicuously reddish-drying young stems and leaf midveins. The inflorescence is often larger and more open than typical in most other genera. The flowers have three to nine 2-celled anthers, and most species have fruits with a shallow cupule on a thickened pedicel (cf. *Cinnamomum*).

Cryptocarya (10 spp., plus ca. 200 Old World) — The neotropical species mostly southern, but a few in Andes. The very distinctive fruit is completely enclosed in the cupule except for a small apical pore. Vegetatively nondescript (extralimital species can have opposite leaves) except that the rather narrow leaves have a distinctly prominulous network of fine venation. The flowers have 2-celled anthers but a deeper floral tube than in vegetatively similar *Beilschmiedia*.

Extralimital genera (all with cupular fruiting receptacles) include: 1) *Dicypellium* (formerly much exploited and now rare; lower Brazilian Amazonia) with 4-celled anthers, the outer three converted into tepaloid staminodes, and double-margined cupule with tepals persistent in fruit, 2) *Litsea* (Central America) with a sub-umbellate inflorescence having bud clusters subtended by large bracts, 3) *Systemodaphne* (or *Kubitzkia*) (Guianas) with a double-margined cupule and fused filaments, 4) *Urbanodendron* (Rio de Janeiro region), and 5) *Phyllostemonodaphne* (Rio de Janeiro region), and three recently described monotypic genera — *Gamanthera* and *Povedaphne* from Costa Rica, the first with a single stamen, the second with the anther cells on the flat tip of columnar stamens, and *Paraia* from Amazonian Brazil with roundish hyaline tepals and peculiar bract scars below the whorled leaves.

LECYTHIDACEAE

Mostly readily recognized when sterile by the characteristic "huasca" odor, the strong fibers in the twig (and sometimes trunk) bark, the pronounced tendency to serrate or serrulate leaf margins (most "entire"-margined species have serrulate margins if you look carefully), and secondary (and usually intersecondary) veins that turn up and then fade out marginally in a characteristic manner. Several brochidodromous genera have the tertiary veins very closely parallel and perpendicular to the midvein. Mostly canopy and emergent trees but two genera are typically unbranched or few-branched pachycaul understory trees with tufts of large terminal leaflets (cf., *Clavija,* Fig. 4). The bark of the large trees is usually rather ridged, sometimes conspicuously so, and typically fibrous; in some species, even trunk bark is so strong that strips are pulled off to use for tying; differs from other strong-barked taxa (e.g., Annonaceae, Malvales) in that the bark pulls off in separate plates rather than as single unit.

The flowers, often large and showy, have many stamens, the filaments often fused into a characteristic androphore that is then variously recurved as a hood over the broad flat stigma. The fruit of most genera is a pixidium, or "monkey pot", with an operculate lid that comes off to release the seeds. The fruits and flowers of most genera are well differentiated and highly characteristic but the leaves of many genera are quite similar. Two genera, *Grias* (with 4 petals) and *Gustavia,* (with 6–8 or more petals) have radially symmetric flowers (lacking an androphore) and indehiscent fruits; most species of these two genera are pachycaul. Two other genera have indehiscent fruits but well-developed complex androphores. *Couroupita* has large globose cauliflorous fruits with small seeds embedded in the pulp while *Bertholettia,* the Brazil nut, has a vestigial operculum and large angular hard-covered seeds completely filling the shell. Two genera with elongate conical or horn-shaped dehiscent fruits have very small flowers in which the androphore is poorly developed (*Cariniana*) or lacking (*Allantoma*); *Cariniana* has seeds with a long broad lateral wing on one side; *Allantoma* has angular narrow seeds with a pointed vestigial wing. *Couratari* also has similarly elongate capsules but a broad wing completely encircles its seed; its larger magenta flowers are often produced while the tree is leafless, and have a complexly folded androphore. The final two genera in our area, *Lecythis* and *Eschweilera,* both with well-developed androphores, have operculate capsules broader than long and large wingless seeds, pendent on a funicle in the former but not the latter.

Lecythidaceae
(Actinomorphic Flowers; Indehiscent Fruits)

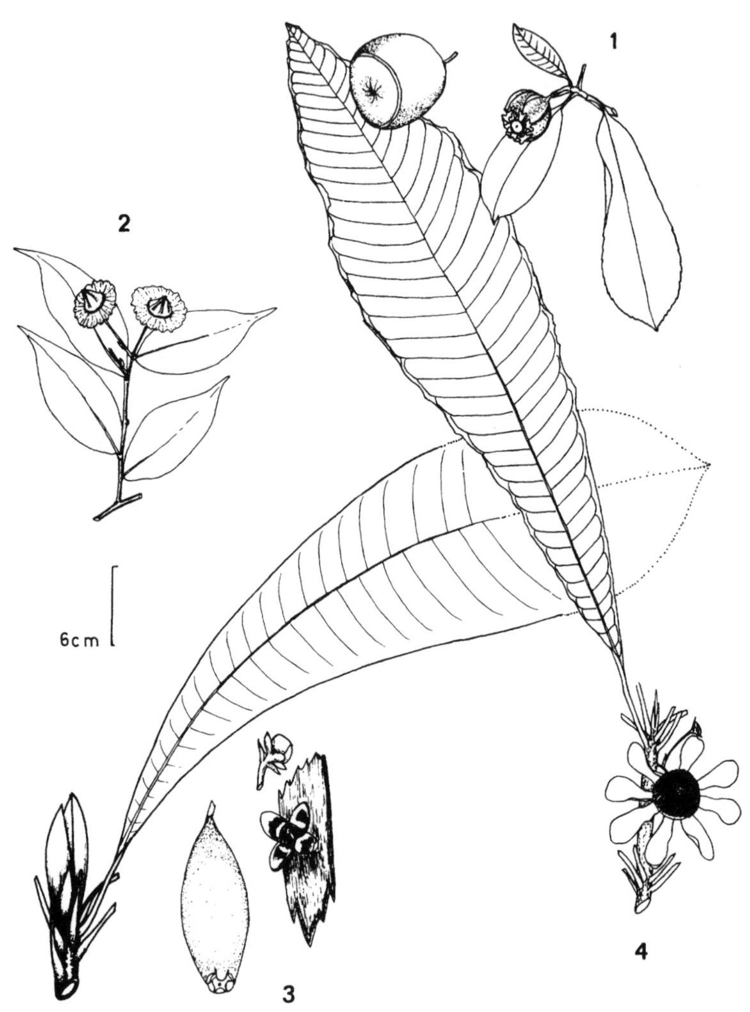

1 - *Gustavia*

2 - *Asteranthos*

3 - *Grias*

4 - *Gustavia*

1. Broad Radially Symmetric Flowers; Indehiscent Fruits with Fleshy Exocarp and Pericarp; Frequently Pachycaul Trees with Very Large Leaves

Gustavia (40 spp.) — Flowers with 6-8 or more petals; fruits usually as broad as long and with 2 or more seeds. Frequently small pachycaul trees with very large terminal leaf tufts; leaf margins usually serrate.

C: manteco, membrillo; E: membrillo; P: chopé, sacha chopé

Grias (6 spp.) — Differs from *Gustavia* in flowers with only 4 petals and fruits with a single seed and much longer than broad. Always a pachycaul tree or treelet (at least when young) with large leaves in terminal tufts, the leaves usually entire (unlike most *Gustavia*).

E: jagua lechosa; P: sachamango

Asteranthos (1 sp.) — Restricted to sandy seasonally inundated river beaches along black-water rivers of the upper Rio Negro. Totally unlike other neotropical lecythids in single axillary apetalous flowers with a broad circular corona formed by fused staminodes; the single-seeded obconical fruit surrounded by a broad circular rim formed by expanded calyx.

2. Flowers Either Small or Strongly Zygomorphic with Androphore (or Both); Never Pachycaul with Terminal Cluster of Large Leaves; Fruits with Woody Exocarp, Usually Dehiscent

2A. Flowers very small, either long, narrow and radially symmetric or with inconspicuously developed androphore hood; fruits elongate, conical or more or less horn-shaped, with a caducous operculum — Leaf tertiary venation very conspicuously finely and closely parallel and perpendicular to midvein, the secondary veins frequently close together and straight.

Allantoma (1 sp.) — A tree of seasonally inundated river margins, probably reaching extreme southeastern Colombia. The leaf longer and more oblong than in any *Cariniana* species, with very straight parallel close-together secondary veins and closely parallel tertiary veins. The very evenly cylindrical capsule with a truncately rounded base and the elongate corky seeds with a vestigial wing on one end are distinctive; flowers longer than wide and with the anthers at many different heights along the inner side of the staminal tube.

Cariniana (15 spp.) — Emergent lowland rain-forest trees. The tiny flowers with filaments fused into a poorly developed androphore hood; fruit elongate, usually more or less cone-shaped with a tapering base; seeds with a single broad elongate lateral wing. Leaves of some species small, membranaceous and closely serrate; tertiary veins always conspicuously parallel and perpendicular to main veins, sometimes as intersecondaries,

Lecythidaceae
(Fruits Either Elongate with Winged Seeds or Indehiscent with Minute Seeds in Pulp; Flowers Zygomorphic to Subactinomorphic)

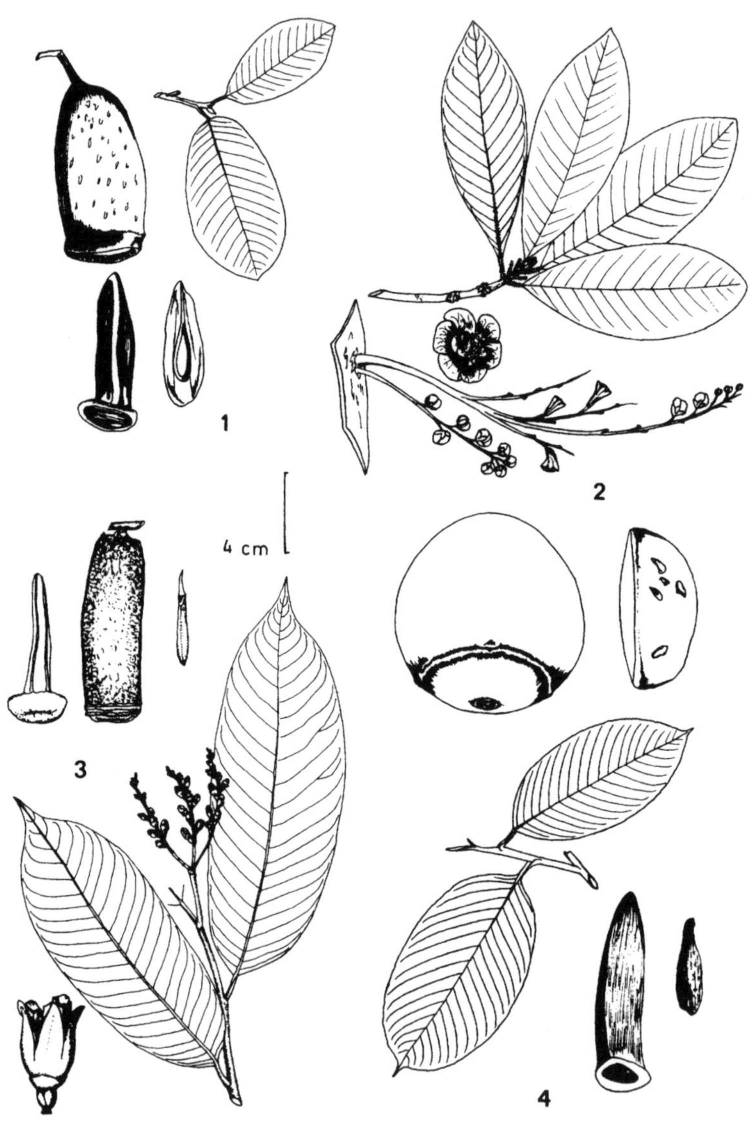

1 - *Couratari* **2** - *Couroupita*

3 - *Allantoma* **4** - *Cariniana*

more frequently very fine; not evident and the undersurface below thus completely smooth in commonest Amazonian species.

C: abarco; P: papelillo, tornillo

2B. Flowers relatively large and with a complex conspicuously developed androphore; fruits indehiscent with a hard woody exocarp — Leaves either unusually large and coriaceous or thin with close-together secondary veins and cuneately tapering to base.

Couroupita (3 spp.) — Emergent trees; cauliflorous, the flowers and fruits borne on short woody branches all along trunk (these branches usually obvious even in sterile condition. Leaves distinctive in family in being thin, narrow and cuneately tapering to base, also unusual in apical clustering, +/- entire margins and close-together secondary veins. Fruit very distinctive, large and globose with a hard thin shell filled with soft pulp, in which the numerous tiny flat seeds are embedded. Flowers unique in the staminodial projections of androphore on outer side of hood apex as well as inside hood.

cannon ball tree; E: bala de canon; P: aya uma

Bertholettia (1 sp.) — Emergent tree, in our area only in Madre de Dios, Peru. Vegetatively characterized by unusually large coriaceous leaves with conspicuous intersecondaries. Flowers and fruits borne on thick branches usually projecting above the foliage; fruit very thick and woody, filled with large hard-coated angular seeds, a vestigial opercular suture present. Calyx unique in being 2-lobed, the flowers yellow, in bud enclosed by calyx (unique), the androphore appendages thick and fused into a thick, nonfimbriate hood.

P: castaña

2C. Flowers relatively large and with a complex conspicuously developed androphore — Fruits woody and dehiscent by caducous operculum. Leaves never with finely parallel tertiary venation, usually with an intricately reticulate network of prominulous fine veins.

Couratari (19 spp.) — Emergent trees; typically with conspicuous magenta flowers borne while leafless. Leaves easily distinguishable by the festooned-brochidodromous venation, the often conspicuous vernation lines and the unique presence of stellate trichomes in some species. Androphore the most complex of the entire family, coiled back on itself three times; fruit elongate, cylindric or campanulate, similar to *Cariniana* but the seed with a thin broad wing around entire circumference.

C: guasconato, fono, carguero; P: cachimbo, machimango colorado, zorrocaspi or machimango blanco (*C. oligantha*)

Eschweilera (83 spp.) — By far the largest and most taxonomically difficult genus; small understory to large emergent trees; ovary 2-locular;

Lecythidaceae
(Fruits Broad and Woody without Pulp; Flowers Zygomorphic)

1 - *Eschweilera* **2** - *Lecythis*

3 - *Bertholettia*

androphore hood coiling inward with a double fold; blunter appendages at tip of hood differentiated from more acute ones farther back. Fruit as wide or wider than long; seeds unwinged nonarillate or with a lateral or enclosing whitish aril, but never pendent from capsule; capsule often rather small and may be much broader than long. Leaves typically with secondary veins curving upward and fading out toward the almost always more or less serrulate margins (rarely distinctly brochidodromous), almost always with well-developed intersecondaries parallel to secondary veins, often the fine venation intricately reticulate-prominulous.

C: guasco, carguero; P: machimango, machimango blanca, machimango negra

Lecythis (26 spp.) — Very similar to and often vegetatively indistinguishable from *Eschweilera;* differentiated by 4-locular ovary, an androphore that does not coil inward and has all hood appendages more or less uniform, and by the seeds hanging on a funicle pendent below the capsule at maturity. Fruits always about as broad as long. One of the commonest species has fruits larger and leaves smaller and more membranaceous than any *Eschweilera* species.

C: guasco salero, salero; E: sabroso, quiebra hacha; P: machimango blanco, machimango colorado

One other neotropical genus, *Corythophora* (4 spp.), which has the two-locular ovary of *Eschweilera* but the androecium structure of *Lecythis,* is known from Brazil.

LEGUMINOSAE

The most important neotropical tree family; also the predominant family of slender vines and including some lianas and herbs as well. Legume trees play essentially the same role in neotropical (and African) forests as the dipterocarps do in SE Asia. Easy to recognize when sterile by the pinnately compound (sometimes bipinnate or 3-foliolate) leaves with leaflets having the *entire petiolule evenly cylindrical* and pulvinate; the leaf itself also has a cylindrically pulvinate petiole base. Exceptions to this character are vanishingly rare (e.g., *Crudia*) and the only nonlegumes that share it are Connaraceae and *Picramnia* (Simaroubaceae), the latter differentiated by its typical rather rhombic-shaped completely alternate leaflets or the bitter taste of its bark or twigs. Many legumes have bipinnate leaves or bifoliolate leaves, both very rare characteristics in other families. All compound-leaved legumes in our region, except one *Vatairea,* have entire margins. The very few simple-leaved legumes are difficult to recognize; in addition to the typical cylindrically pulvinar petiole they tend to have asymmetric bases (cf., *Drypetes*), serrulate margins, and dry a characteristic light olive-green. Unifoliolate legume species are more common but easy to recognize by the 2-parted "petiole" with the typical cylindrical pulvinus of the single leaflet forming its apical part.

The three major legume groups are sometimes recognized as families: these are generally distinguished by the complex bilaterally symmetrical pealike flower of Faboideae vs. the tubular actinomorphic flowers of Mimosoideae (these usually small, with numerous exserted anthers, and densely arranged into a conspicuous spicate, capitate or umbellate inflorescence) vs. the slightly bilaterally symmetric flower with 5 separate and more or less equal petals of Caesalpinioideae. The Mimosoideae and Faboideae usually have a conspicuous rather rank "green-bean" vegetative odor; the Caesalpinioideae often lack this odor. Vegetatively, the Mimosoideae are usually characterized by bipinnate leaves (except *Inga*) with large, usually cupular petiolar and/or rachis glands (except *Pentaclethra, Mimosa, Entada, Calliandra*); the Caesalpinioideae by mostly paripinnate (or bifoliolate) leaves. Most caesalpinioid genera with imparipinnate leaves have the leaflets alternating on the rachis; bipinnate caesalpinioids lack the large petiolar or rachis gland(s) that are found in most bipinnate mimosoids; the Papilionoideae by 3-foliolate or imparipinnate leaves (the latter often with strictly opposite leaflets). Habitwise, mimosoids are usually trees, often with spreading crowns; the vine and liana genera of mimosoids usually have numerous small spines on both stem and leaf rachis (the only nonspiny liana mimosoid is *Entada*). Most papilionates are 3-foliolate vines (unique); papilionate lianas and some tree genera typically have red latex. Caesalpinioids are almost always trees (except a few spiny *Caesalpinia* species and a few species of the very diverse genera *Cassia* and *Bauhinia*); they lack red latex.

Caesalpinioideae

Contrary to the traditional classification, caesalpinioids are now thought to be the most primitive legume subfamily; multiple stamens in mimosoids secondarily derived as are the complex bilaterally symmetrical flowers of Papilionoideae. Tribe Swartzieae is intermediate and is treated here as papilionate. The sepals of caesalpinioids (except *Poeppigia* and *Dimorphandra*) are free (though fused in bud in *Bauhinia* and the*Swartzia* relatives); those of papilionates are united. Habitwise, caesalpinioids, are all trees except morphologically diverse *Cassia,* a few spiny climbing species of *Caesalpinia,* and part of *Bauhinia;* they are the least diverse of the legume subfamilies, but in terms of fruit and floral morphology they are perhaps the most diversified; all of the largest showiest legume flowers as well as the most inconspicuous apetalous ones are caesalpinioid. Caesalpinioids are especially prevalent on the poorest white-sand soils, perhaps related to their symbiosis with ectotrophic mycorrhizae rather than N-fixing bacteria. Excluding the *Swartzia* relatives (to Papilionoideae), the simply compound caesalpinioids belong to five different tribes; all the bipinnate genera, however, are in the same tribe, Caesalpinieae. The bipinnate genera include three usually spiny genera found almost entirely in dry or arid habitats plus several nonspiny genera of lowland rain forest. Unmistakable *Bauhinia* with its incompletely divided 2-foliolate (sometimes simple with palmate venation) leaves is not closely related to other neotropical genera. As now recognized, the other tribes are difficult to define and largely based on trends: toward enlarged bracteoles in Detarieae (where imbricate) and Amherstieae (where valvate), toward anthers opening by small slits or terminal pores (and primitive lack of a hypanthium) in Cassieae, toward bicompound leaves and/or reduced flowers in dense inflorescences and/or spines or swollen ant-inhabited petioles in Caesalpinieae. All Amherstieae and Detarieae have intrapetiolar stipules (at least a connecting line between bud and top of petiole) and all have paripinnate leaves with opposite leaflets except *Crudia* and *Copaifera* with alternate leaflets on odd-pinnate leaves but with the naked rachis apex extended beyond upper leaflet; Cassieae and Caesalpinieae have the stipules lateral or lacking. The pinnate-leaved caesalpinioids are most conveniently subdivided artificially into paripinnate and odd-pinnate genera, the latter including three genera of Caesalpinieae with opposite or subopposite leaflets (*Batesia, Recordoxylon, Campsiandra*), two genera of Detarieae with alternate leaflets and the naked rachis apex extended (*Crudia, Copaifera*), and several genera of Cassieae with alternate leaflets, cymose-paniculate inflorescences, and indehiscent (winged or drupaceous) fruits (*Dialium, Apuleia, Martiodendron,* extralimital[?] *Dicorynia*, and *Androcalymma*).

1. Bipinnate Caesalpinioideae — (Always lacking petiolar and rachis glands)

Caesalpinia (ca. 50 spp., plus 50 in Old World) — Dry-area trees, occasionally spiny (also spiny lianas); leaflets always oblong-elliptic, typically small and numerous; flowers yellow (usually orange-red in cultivated and naturalized *C. pulcherrima*); fruits narrowly oblong, irregularly flattened, indehiscent or tardily dehiscent. Sometimes with strikingly smooth bark.

C: ébano, dividivi (*C. coriacea*); E: clavellina

Parkinsonia (12 spp.) — Spiny arid-area tree, easily recognized by smooth green bark and unique leaves with the 2 long rachi photosynthetic, the numerous tiny leaflets more or less vestigial (as the petiole); flowers yellow, in open raceme; fruits long and narrow, acuminate at both ends,

Leguminosae/Caesalpinioideae
(Bipinnate)

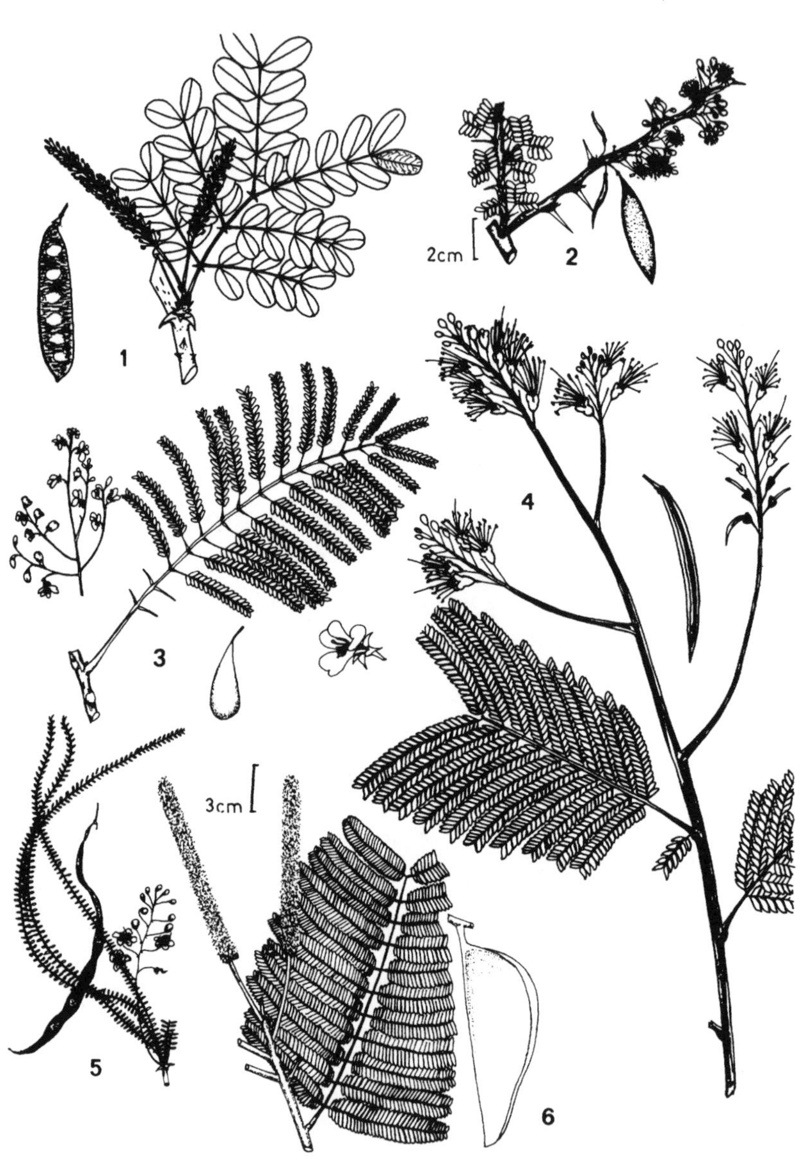

1 - *Caesalpinia* **2** - *Cercidium*

3 - *Schizolobium*

4 - *Jacqueshubera*

5 - *Parkinsonia* **6** - *Dimorphandra*

somewhat constricted between seeds. Sometimes included in *Cercidium*.
C: sauce; C, E, P: palo verde

Cercidium (4 spp., plus 6 in N. Am.) — Spiny arid-area trees, easily recognized by smooth green bark; leaves on short shoots reduced to two pinnae with small leaflets; flowers yellow, very conspicuous, borne while leaves deciduous; fruits indehiscent, flat, cf., *Lonchocarpus*.
C: cuica, yabo; C, E, P: palo verde

Schizolobium (5 spp.) — Our only representative a common fast-growing late-successional species (similar to *Jacaranda copaia* when young but with greenish bark and alternate leaves). Very large leaves with woody rachis and slender side rachises, little narrowly oblong leaflets; flowers rather large, yellow, borne while leaves caducous, and thus, very conspicuous (may be confused with *Tabebuia* from a distance); fruit flat, evenly obovate, apex rounded, base long-tapering, dehiscent to release a single seed with a large wing on one side.
P: pashaco

(*Peltophorum*) (3 spp., plus 5 in Old World) — Native from Antilles to southern Brazil-Paraguay but apparently not in our area where it is commonly cultivated as a street tree; leaves and flowers like *Schizolobium* but an indehiscent wind-dispersed fruit very like that of *Tachigali* (though sometimes 2-seeded).

(*Delonix*) — Native to Madagascar but very frequently planted throughout dry areas of tropics; its bright red-orange, large flowers are unmistakable; vegetatively the pinnate (or even bipinnate) stipule is very characteristic.
flamboyán

Jaqueshubera (3 spp.) — Only on white sand. The neotropical equivalent of *Delonix* with large, red to purple, probably bird-pollinated flowers with long-exserted filaments and borne in a long terminal raceme; the leaves are large with very small leaflets (cf., *Parkia pendula*) and easily recognized by pinnate stipules.
P: pashaquilla

Dimorphandra (25 spp.) — A large and vegetatively rather variable genus. The leaves of some species are strikingly like *Parkia* but the leaflets not curved; the inflorescence very like that of *Pentaclethra*, a dense terminal spike or spikelike raceme with tiny whitish to orange-red flowers; fruit also like that of *Pentaclethra*, stipitate, rather straight on dorsal side, broadly curved on ventral side, with conspicuous slanting fibers and strongly dehiscing.

Leguminosae/Caesalpinioideae
(2-Foliolate or Bifid)

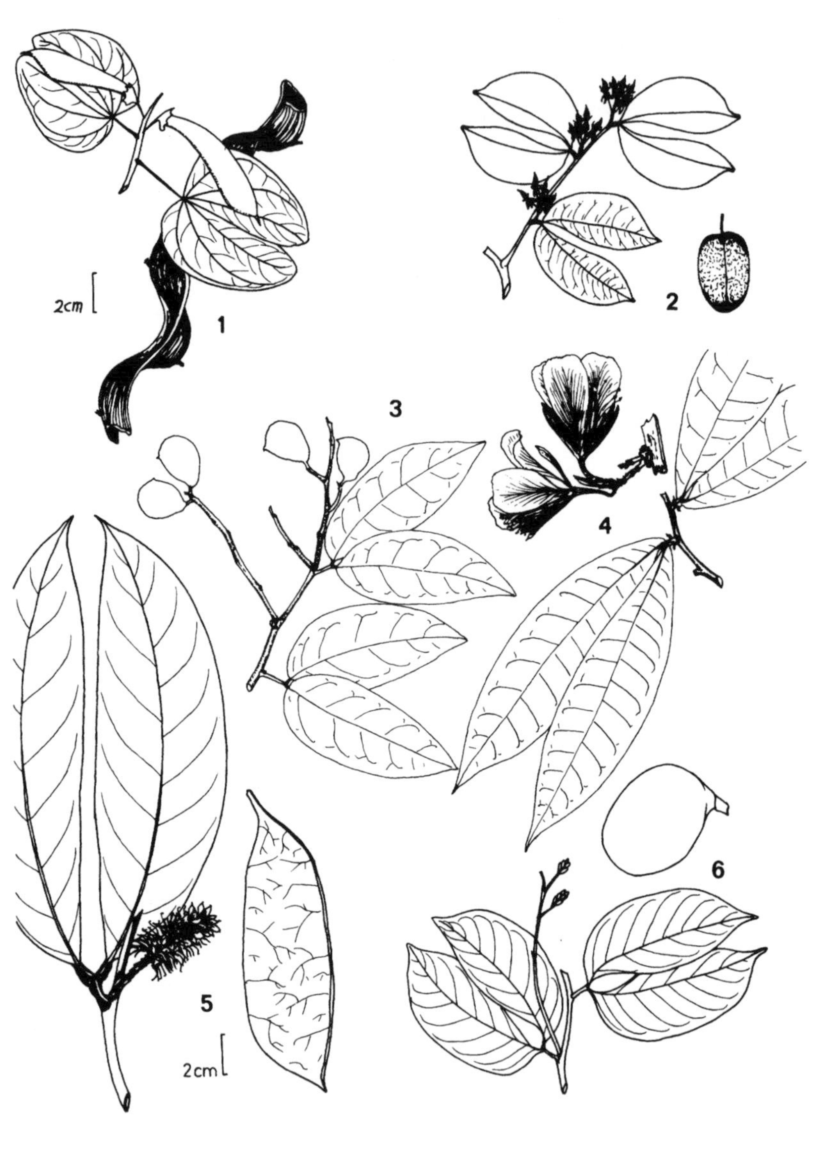

1 - *Bauhinia* **2** - *Cynometra*

3 - *Peltogyne* **4** - *Heterostemon*

5 - *Macrolobium* **6** - *Hymenaea*

2. Leaves 2-Foliolate or Bifid

2A. Leaves incompletely split in half — (Sometimes the two halves completely fused but the venation palmate; the two halves sharing a common pulvinus even if split to near base).

Bauhinia (150 spp., plus 200 in Old World) — The unique leaves are unmistakable; some species are tendrillate lianas, often with conspicuously flattened ''monkey's ladder" stems; others are trees or shrubs, sometimes with spines in dry areas.

C: escalera de mono; E: escalera de mono, uña de gato; P: pata de vaca

2B. 2-Foliolate — The two-foliolate genera of Caesalpinieae are all Detarieae and Amherstieae, characterized by intrapetiolar stipules.

Cynometra (25 spp., plus 45 in Old World) — Usually restricted to tahuampa or streamsides, also in dry forest. Leaflets rather obtuse (minutely emarginate) in our area; the fruit is ellipsoid to subglobose with a slight longitudinal medial furrow bisecting it.

Hymenaea (13 spp., plus 1 in Africa) — Usually large trees. The leaflets coriaceous, smooth or puberulous, less reticulate and the flowers larger than in *Peltogyne;* fruit flattened-ellipsoid, indehiscent. Fruit pulp (aril) edible.

C: algarrobo; P: azúcar huayo

Peltogyne (23 spp.) — In our area, differing from other 2-foliolate genera in having a more strongly intricately reticulate texture and often drying blackish; flowers small, white; fruit smallish (ca. 3 cm or less), flat, single-seeded, often almost triangular with a relatively straight dorsal margin and much expanded "pregnant" ventral margin.

C: tananceo, nazareno

Heterostemon (7 spp.) — Mostly on white-sand substrates in upper Rio Negro and Guayana Shield area. Most species are 2-foliolate but a few (extralimital?) have paripinnate leaves with numerous very trapezoidal leaflets becoming conspicuously smaller toward tip; the often cauliflorous flowers are conspicuous, lavender, with lower petals reduced; fruits flattened and woody, similar to *Macrolobium.*

(*Macrolobium*) — See under paripinnate, below; the 2-foliolate species differ from *Hymenaea* in more numerous and less ascending secondary (and intersecondary) veins and from *Peltogyne* in lacking the intricately reticulate texture.

3. Leaves Paripinnate — The first five genera are Caesalpinieae and *Cassia* is Cassieae, both characterized by stipules lateral or absent; the other genera are Detarieae or Amherstieae, characterized by intrapetiolar stipules (or their scars).

3A. Large water-dispersed single-seeded fruit; flowers minute (< 6 mm long) and radially symmetric; leaves 4-foliolate

Mora (6 spp.) — Our single species is a large tree of mangrove swamps, forming pure stands near the interface between fresh and salt-water. It has the largest seed of any dicot (ca. 10–15 cm long) and is completely unmistakable in fruit; the leaves are 4-foliolate with narrowly oblong-ovate leaflets. Closely related to bipinnate-leaved *Dimorphandra* and sometimes included in it; the flowers are similarly tiny and radially symmetric.

C: nato

Prioria (1 sp.) — A large tree sometimes forming nearly pure stands, mostly in freshwater swamps or riversides; in our area only in northwestern Colombia. The leaves are 4-foliolate with coriaceous elliptic leaflets (broader than in *Mora*), pellucid-punctate. Flowers very distinctive: tiny, apetalous (but the sepals resembling petals and subtended by calyx-like bracts), with ten free stamens. Fruit obovate to suborbicular, to 10 cm long, flattened on one side and convex on the other, with a single large seed.

C: cativo

3B. Wind-dispersed fruits or seeds; leaves 6–many-foliolate

3Ba. Whole fruit wind-dispersed

Haematoxylum (2 spp.) — Spiny arid-area trees, especially in poorly drained places. Leaves few-foliolate, the leaflets obovate, usually broadly retuse with close-together secondary veins, the uppermost pair larger than lowermost; flowers yellow; fruit wind-dispersed, like *Lonchocarpus* but at length splitting open down middle of valve. Closely related to bipinnate *Cercidium* and *Caesalpinia* and an extralimital species may be bipinnate.

C: palo brasil, hala

Poeppigia (1 sp.) — Large trees with many little, narrowly oblong, membranaceous leaflets (although reported in the literature to be imparipinnate it seems to be uniformly even-pinnate). Flowers yellow, in cymose axillary inflorecences shorter than the leaves; fruits wind-dispersed, similar to *Lonchocarpus* but narrow (ca. 1 cm), rather acute at base and apex and with slight ridge along one margin.

Leguminosae/Caesalpinioideae
(Paripinnate; Wind-Dispersed)

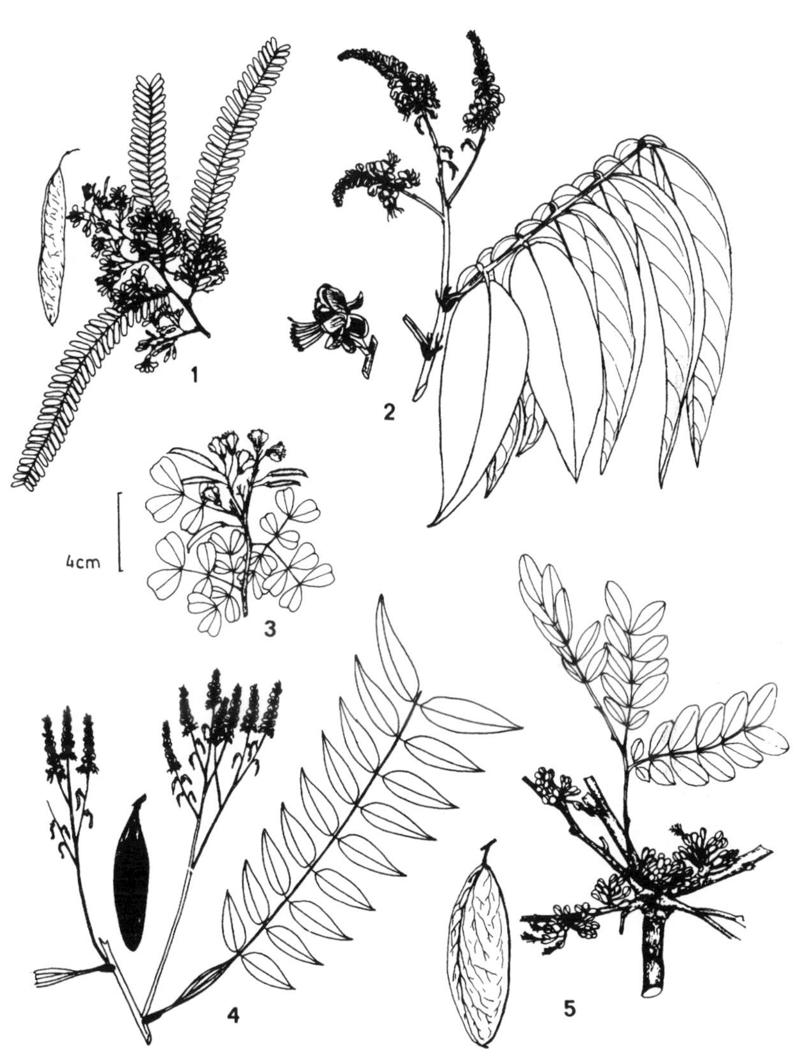

1 - *Poeppigia* **2** - *Sclerolobium*

3 - *Haematoxylum*

4 - *Tachigali* **5** - *Phyllocarpus*

Sclerolobium (30 spp.) — Large trees, mostly on poor soils; probably not adequately distinct from *Tachigali;* although essentially even-pinnate with opposite leaflets at least at base, the terminal leaflet pair often irregular, sometimes with a more or less reduced terminal leaflet; rachis often extended beyond terminal leaflet pair (cf., *Guarea*); petioles sometimes swollen, hollow, and inhabited by ants; sometimes with conspicuous leaflike stipules. Flowers small, usually yellow, in terminal panicles with dense spicate lateral branches; fruit wind-dispersed, very flat with no raised ribs, somewhat narrower at base and apex, the exocarp flaking off prior to dispersal.

P: tangarana, ucshaquiro

Tachigali (24 spp.) — Large trees, very like *Sclerolobium* from which it differs in somewhat zygomorphic rather than regular flowers; the flowers are always yellow or yellowish and usually larger than in *Sclerolobium;* to a greater degree than in *Sclerolobium,* the leaf rachis is usually flat or subwinged, usually swollen and with stinging ants, and always terminates in an aborted free tip. Many (perhaps most) species are monocarpic, the tree dying after producing its single crop of fruits; this phenomenon is not known to occur in *Sclerolobium.*

P: tangarana

Phyllocarpus (2 spp.) — Mostly on rich soils. The leaves are essentially like those of a ca. 6-foliolate *Macrolobium* and the fruit like *Tachigali* but larger with a slight submarginal rib along the slightly more curved dorsal margin; inflorescence reduced, few-flowered, the flowers red, only 3 small petals, the stamens conspicuously longer, flowering while leaves caducous. Bark smoothish except for numerous small orangish pustules.

3Bb. Seeds wind-dispersed

Diptychandra (3 spp.) — Mostly restricted to the Paraguay-southern Brazil area with an undescribed species from Magdalena Valley. Related to above genera, but differs in a dehiscent fruit that releases winged seeds.

3C. Anthers opening by pores —*Cassia*

Cassia (382 spp., plus 140 Old World) (incl. *Senna, Chamaecrista*) — Large and morphologically diverse but forming a clearly natural group characterized by the anthers opening by terminal slits or pores; rachis glands are often present, usually between the leaflets. Although segregated into three genera in a recent monograph, there seems no real reason for splitting *Cassia* more than subgenerically. *Cassia sensu stricto,* all trees, is characterized by the 3 enlarged and curved abaxial filaments having much smaller anthers, and by the straight, elongate, typically more or less cylindric, indehiscent pod with a pulpy or "pithy" interior. *Senna,* mostly shrubs, small trees, and a few climbers, has all filaments straight and shorter than (or not much longer than) their anthers; the fruit is usually smaller, often dry and compressed, either indehiscent or breaking gradually open along the sutures; the leaves are of two fundamental types — large and 4-foliolate, usually with a conical gland between lower pair of leaflets or more numerous and oblong (but larger than in *Cassia sensu stricto*); *Chamaecrista,* mostly herbs with tiny leaflets, has flowers as in *Senna* but subtended by 2 bracteoles; its main feature is a narrow elastically dehiscent fruit with coiling valves.

C: galvis or bajagua (*C. reticulata*); E: aya porotillo (*C. occidentalis*), retama (*C. reticulata*); P: retama (*C. reticulata* and *alata*), mataro (*C. bacillaris*).

3D. Flat, often large, woody-valved fruits with raised margins; flowers large and zygomorphic with longitudinally dehiscent anthers; leaves 4–many-foliolate

Macrolobium (60 spp., incl. Africa) — Trees, often rather small. Three rather distinct leaf types: 2-foliolate (see above), oblong and sessile, and small, narrowly oblong, and multifoliolate; the inflorescence is axillary, spicate, and characterized by the valvate subtending bracteoles and the white 1-petaled flowers with 3 fertile stamens and several maroonish staminodes; fruit always smooth, flat, and rather woody-valved, dehiscing to release one or few seeds, variable in shape from round to very large and irregularly oblong, the edge always as thick as or slightly thicker than face but not ridged.

C: marimbo; P: mashaco colorado, santo caspi (*M. acaciifolia*)

Eperua (14 spp.) — Large trees, often dominating on white sand in the Guayana Shield area. Leaflets always coriaceous, asymmetrically curving, and acuminate, usually very minutely and characteristically prominulously intricate-reticulate below (glaucous and smooth below in *E. purpurea*), always drying blackish or olive, the leaf usually 4–6-foliolate. Fruit very flat, broad, woody-valved, elastically dehiscing; flowers often very conspicuous, white or magenta.

Elizabetha (10 spp.) — Guayana Shield area. Leaves of the multifoliolate *Macrolobium*-type with leaflets all the way to rachis base (i.e., no defined petiole above basal pulvinus); differs from *Macrolobium* in pincushion-type terminal inflorescence of very large, usually red flowers, bombaclike calyx; fruit differs from *Macrolobium* in usually having a rather wrinkled surface.

(*Heterostemon*) — See under bifoliolate leaves, above.

Brownea (12 spp.) — Medium-sized to large trees, often cauliflorous, the young twigs characteristically cross-shaped in cross section. Leaflets usually long, subopposite, acuminate to caudate, at least the basal

Leguminosae/Caesalpinioideae
(Paripinnate; Non-Wind-Dispersed [A])

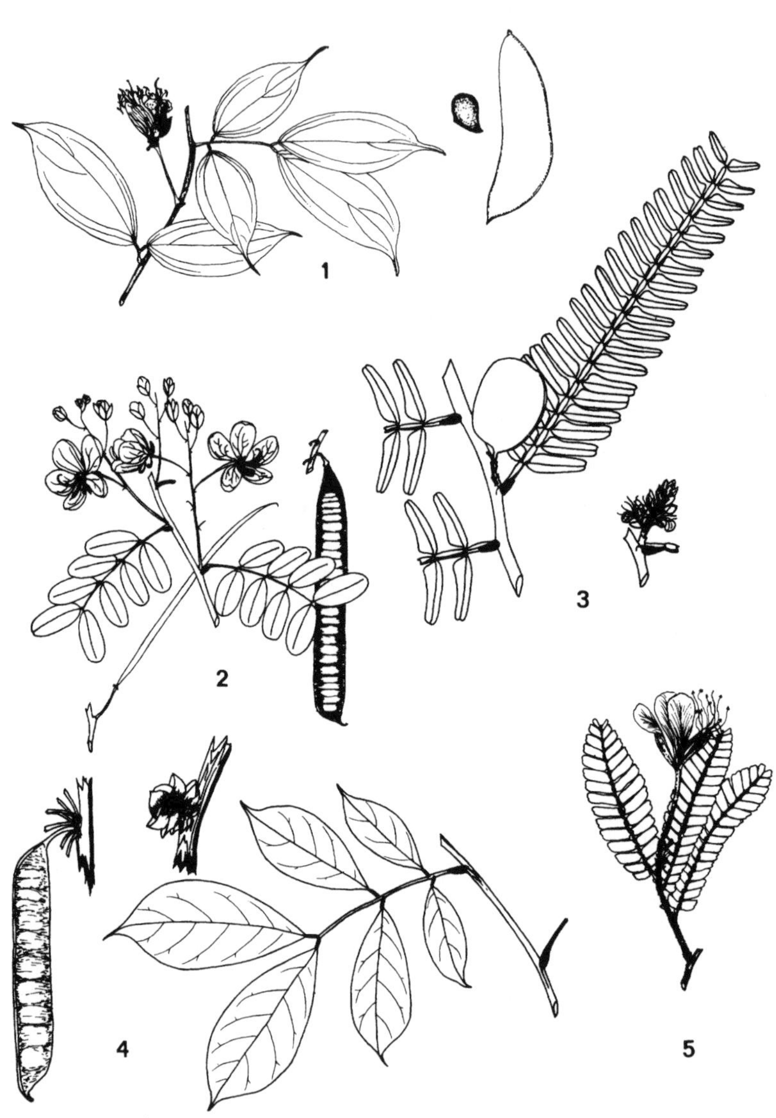

1 - *Browneopsis*

2 - *Cassia*

3 - *Macrolobium*

4 - *Brownea*

5 - *Heterostemon*

Leguminosae/Caesalpinioideae
(Paripinnate; Non-Wind-Dispersed [B])

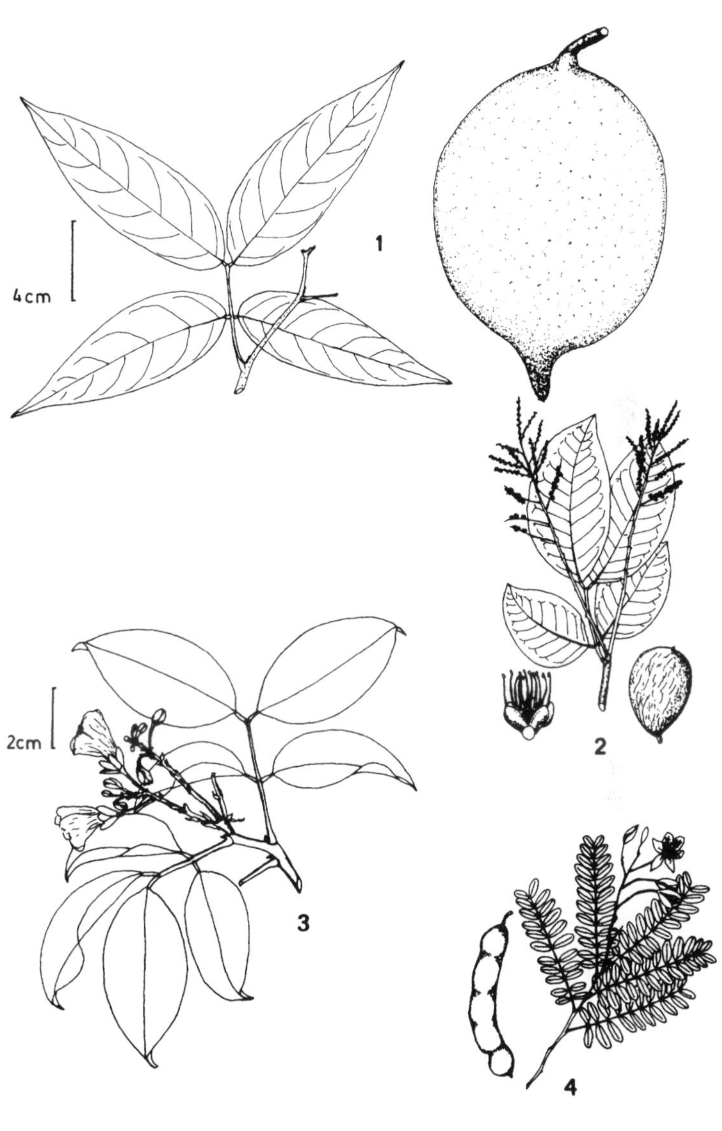

1 - *Mora*

2 - *Prioria*

3 - *Eperua*

4 - *Tamarindus*

ones distinctly caudate; often with a reduced stipulelike basal pair near rachis base. Flowers showy, red-orange, with exserted stamens with the filament bases joined, clustered, sessile and subtended by conspicuous imbricate bracts and bracteoles in bud, these forming an ellipsoid or ovoid inflorescence bud; fruit linear-oblong, similar to *Macrolobium* and *Eperua*.

C: ariza, flor de rosa, clavellino; E: clavellin; P: machete vaina

Browneopsis (6 spp.) — Large trees known mostly from Chocó area, but also in Amazonia. Similar to *Brownea* but bracts and bracteoles lacking, young twigs terrete, basal leaflets not cordate, and inflorescence bud globose; flowers often white.

(*Dicymbe*) (13 spp.) — White-sand specialist trees of Guayana Shield area. Leaflets strictly opposite, 4–6-foliolate, symmetrical, coriaceous, often puberulous; flowers large, whitish (one species bat-pollinated), the calyx densely pubescent.

(*Tamarindus*) — Cultivated and more or less naturalized in dry disturbed areas. Leaves multifoliolate with membranaceous narrowly oblong leaflets; petals greenish-cream with reddish spots and veins, the flowers few and inconspicuous on inflorescences shorter than leaves; fruit subterete, subglobose to elongate and often irregularly constricted.

4. Leaves Odd-Pinnate

4A. Opposite to subopposite leaflets — Usually largish and acute to acuminate, typically drying darkish or grayish (Caesalpinieae)

Batesia (1 sp.) — Amazonian trees; leaflets strictly opposite, with large *Inga*-like gland between at least basal pair of leaflets and sometimes all pairs; rachis flattened and shallowly grooved above (cf., *Dipteryx*), the leaflets oblong-elliptic, with rather close, straight secondary veins minutely tannish-puberulous below and on petiolules and rachis. Flowers light yellow, in terminal panicle; fruit oblong, subglobose, 2–3 cm long, conspicuously longitudinally ca. 8–costate, dehiscing by 4 "valves".

Recordoxylon (2 spp.)— Large Amazonian trees; 5–7-foliolate, the leaflets largish, acute, dark-drying, with truncate base (often very asymmetrically truncate), the margin more or less cartilaginous; large yellow flowers (cf., *Cassia*).

Campsiandra (3 spp.) — Restricted to riverside tahuampa where usually common; leaflets smoothish, strictly opposite to subalternate, rather long and dark-drying, the secondaries and intersecondaries not well differentiated, the rachis always flattened or square-edged above; flowers white to pinkish with red stamens, in corymbose terminal inflorescence. Fruit flat, oblong, like overgrown *Macrolobium* but terminal.

P: huacapurana

4B. Alternate leaflets — Usually rather small and drying greenish (blackish in *Copaifera*) (Cassieae plus *Crudia* and *Copaifera* of Detarieae, the latter two with intrapetiolar stipule scars and the leaf rachis extended beyond "terminal" leaflet.)

Dialium (1 sp., also 40 in Old World) — Large trees. Leaflets rather few (usually ca. 5–7) medium smallish, alternate, ovate, acuminate. Flowers small, apetalous; fruits small, ellipsoid, with thin smooth exocarp over single seed covered with soft, rather dry, whitish, edible aril.

C: abrojo, mari; P: palo sangre

Apuleia (1 sp.) — Mostly in seasonally dry areas (including Tarapoto); leaflets elliptic-ovate to smallish and oblong (2.5 cm), strongly alternate, apex more or less obtuse or retuse. Flowers white, borne in rather umbellate cymes while leaves caducous; fruits like *Lonchocarpus* but conspicuously long-stipitate. Bark gray, smoothish but conspicuously insculpted.

Martiodendron (4 spp.) — Large trees of Amazonia. Leaflets alternate, bases always cordate or more or less truncate, strikingly resembling *Eschweilera* leaves in texture and color; very large conspicuous conical dark bud in axils. Flowers yellow to orange, from long narrow buds; fruits flat, thin, single-seeded, like overgrown *Lonchocarpus* with outlines of the single large 2-winged seed visible as surface ridge on young fruit.

Dicorynia (2 spp.) — Mostly Guayana area, but one species more widespread in Amazonia. Leaflets opposite to subalternate, like *Martiodendron* in cordate to truncate *Eschweilera*-like leaflets and large axillary bud, but flowers white and buds not long and narrow. Fruit similar to *Lonchocarpus* but with submarginal dorsal costa, tardily dehiscent.

(*Androcalymma*) (1 sp.) — São Paulo de Olivença very near Peru border; few-foliolate with yellowish-green-drying, alternate leaflets.

Crudia (10 spp.) — Mostly in tahuampa. Very distinctive vegetatively, the very alternate leaflets oblong-elliptic, short-acuminate, with short flat, twisted nonlegume-looking petiolules; rachis sometimes ending in free tip; flowers apetalous, green to whitish, in long raceme, the sepals falling to leave tan-pubescent ovaries; fruit woody, flat, densely tan-tomentose and with irregularly ridged surface.

P: tushmo

Copaifera (25 spp.) — Large trees. Leaflets (in our area) similar in texture and shape to *Eperua* but alternate, smaller and always dry dark; prominulously intricate-reticulate above and below. Flowers white, in terminal panicle with unbranched spicate lateral branches, apetalous; fruits

Leguminosae/Caesalpinioideae
(Odd-Pinnate)

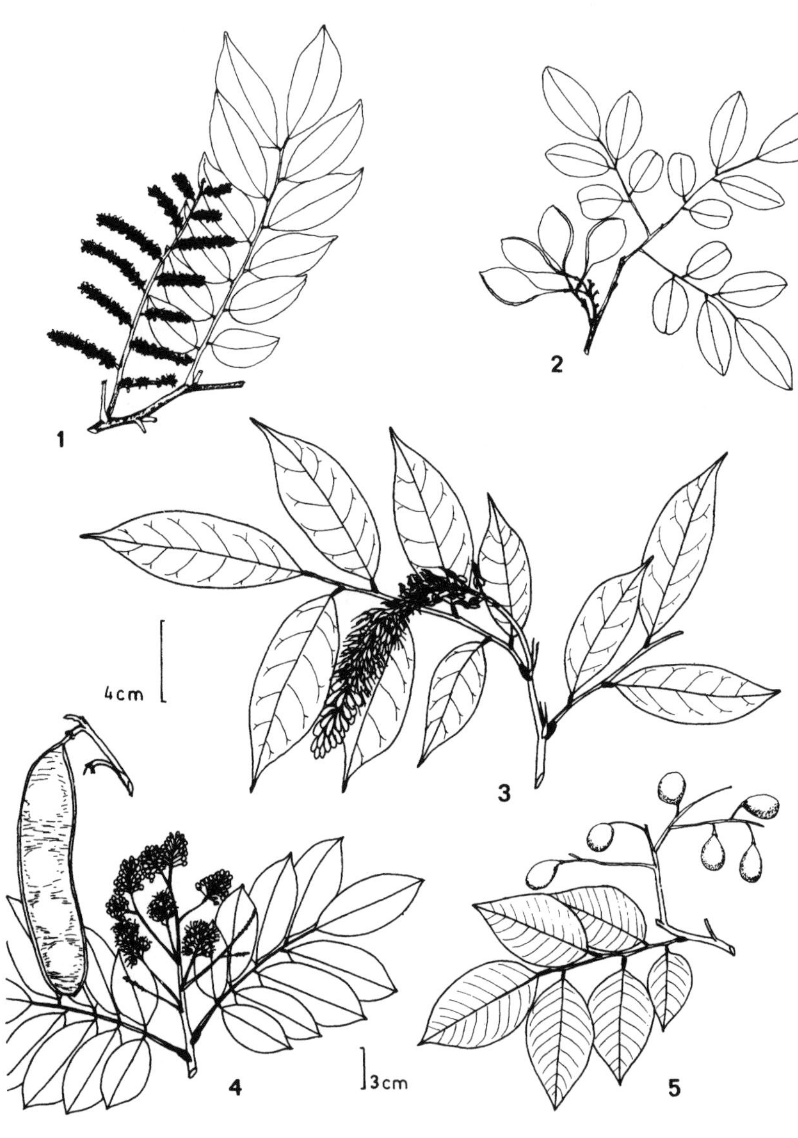

1 - *Copaifera* **2** - *Apuleia*

3 - *Crudia*

4 - *Campsiandra* **5** - *Dialium*

2-valved, small, roundish (flat when young), not strongly compressed at maturity, the single seeds black with usually orange aril.

P: copaíba

Mimosoideae

Easy to recognize by the bipinnate leaves usually with petiolar or rachis glands. *Inga* has simply pinnate leaves but is unique in conspicuous cup-shaped glands between each leaflet pair; *Pentaclethra* lacks the petiolar glands and is very like caesalpinioid *Dimorphandra; Entada, Calliandra,* and *Mimosa* (in our area) also lack the glands, the latter possibly having substituted spines for ants as protective measures. Intrageneric variation in leaflet size and shape is extreme and many mimosoid genera have both little-leafleted and large-leafleted species. Most genera are trees but a few are entirely or predominantly herbaceous or suffrutescent and several spiny genera include some scandent species; only *Entada* is strictly a liana. Spines are quite common in mimosoids and are of two kinds, stipule-derived, and small enations along the branchlets, the latter very characteristic type occuring in all the climbing genera except unarmed *Entada*. Tribal division is based largely on stamen number and fusion: *Mimosa* and its relatives (including all the herbaceous genera) have 10 or fewer free stamens; *Acacia* has more than 10 free stamens; *Pithecellobium* and *Albizia* and their relatives have more than 10 stamens with the filaments fused into a tube.

1. Leaves Simply Pinnate

Inga (350 spp.) — One of the largest and most characteristic neotropical genera. The only simple-leaved mimosoid (along with closely related Brazilian *Affonsea*), the typical cup-shaped glands between each leaflet pair of the paripinnate leaves, frequently with a winged rachis, are completely distinctive. Flowers always of the typical mimosoid form but the inflorescence varies from umbellate to capitate to spicate. Fruits never dehiscent, always elongate, otherwise quite variable from large and almost subwoody to slender and soft, frequently edible either as a vegetable or for the sweet pulp surrounding individual seeds. There is one area species intermediate between *Inga* and *Pithecellobium* that has simple-pinnate leaves but a dehiscent fruit with red lining of the typical *Pithecellobium*-type; that species is currently referred to *Pithecellobium* as *P. rufescens.*

C: guamo, guabo, guaba pichindé; E: guaba, guaba vaina de machete (*I. spectabilis*), guaba mansa (*I. edulis*); P: shimbillo, guaba (cultivated)

2. Leaves Bipinnate

2A. Herbs, subshrubs, or prostrate vines

Neptunia (11 spp., also Old World) — Shrub or subshrub of swamps or dry areas; flowers (at least the basal ones) yellow, in capitate head; fruits very thin-valved, linear-oblong, dehiscent, with tiny winged seeds, stipitate, forming a cluster like a much-reduced *Parkia.*

Desmanthus (25 spp.) — Like *Neptunia* but more herbaceous and usually in weedier nonaquatic habitats, and with sessile fruits and pinkish or whitish flowers.

Schrankia (19 spp., incl. temperate N. Am.) — Weedy, spiny, prostrate, viny herb, very like *Mimosa* and sometimes included therein, but the long, narrow, spiny fruits split longitudinally into 4 valves.

Mimosa (450 spp., plus few Old World) — Variable in habit from herbs to small trees to lianas and in inflorescence from capitate and pink-flowered (commonest) to an elongate white-flowered spike (in a few lianas). Mostly held together by the very characteristic fruit that breaks into transverse segments that detach from the persistent margins. The fruit is often spiny, especially along margins; vegetatively characterized by the small recurved (never stipular) spines, the frequently hairy leaf rachis and branchlets (*Acacia* is always glabrous in our area) and the sometimes digitately arranged pinnae; the petiolar gland is absent.

E: pica-pica

2B. Lianas

Entada (incl. *Adenopodia*) (30 spp., but mostly African) — Our only nonspiny bipinnate legume lianas; leaves lacking petiolar glands; flowers tiny, white, in elongate spikes or spikelike racemes, these sometimes densely clustered together; fruit very large, breaking into transverse segments that usually separate from the persistent margin as in *Lysiloma* or most *Mimosa;* one species has the tips of some leaves modified into tendrils.

E: habilla (*E. gigas*); P: machete vaina

(*Piptadenia*) — Some species are spiny lianas, stems usually angled and with the spines more or less arranged in lines along the angles; liana species of *Acacia* have the spines irregularly arranged on the branchlets.

(*Acacia*) — A number of *Acacia* species are spiny lianas, differentiated in flower by having more than 10 free stamens, and vegetatively by the irregularly placed spinules of the branches.

(*Mimosa*) — *Mimosa* lianas, characterized by the typical fruit that breaks into segments that separate from the persistent margin, are similar

Leguminosae/Mimosoideae
(Lianas and Large Moist Forest Trees)

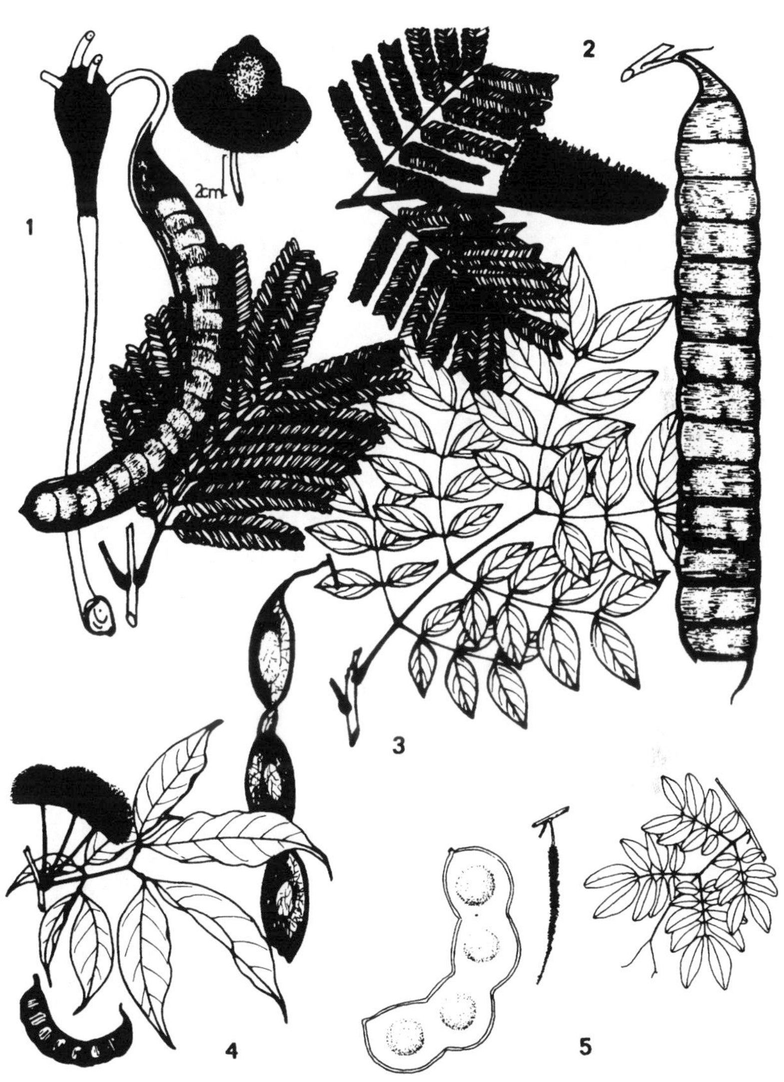

1 - *Parkia*

2 - *Entada* (*Adenopodia*)

3 - *Cedrelinga*

4 - *Pithecellobium*

5 - *Entada* (*E. gigas*)

Leguminosae/Mimosoideae
(Inga and Bipinnate Taxa with Tiny Leaflets)

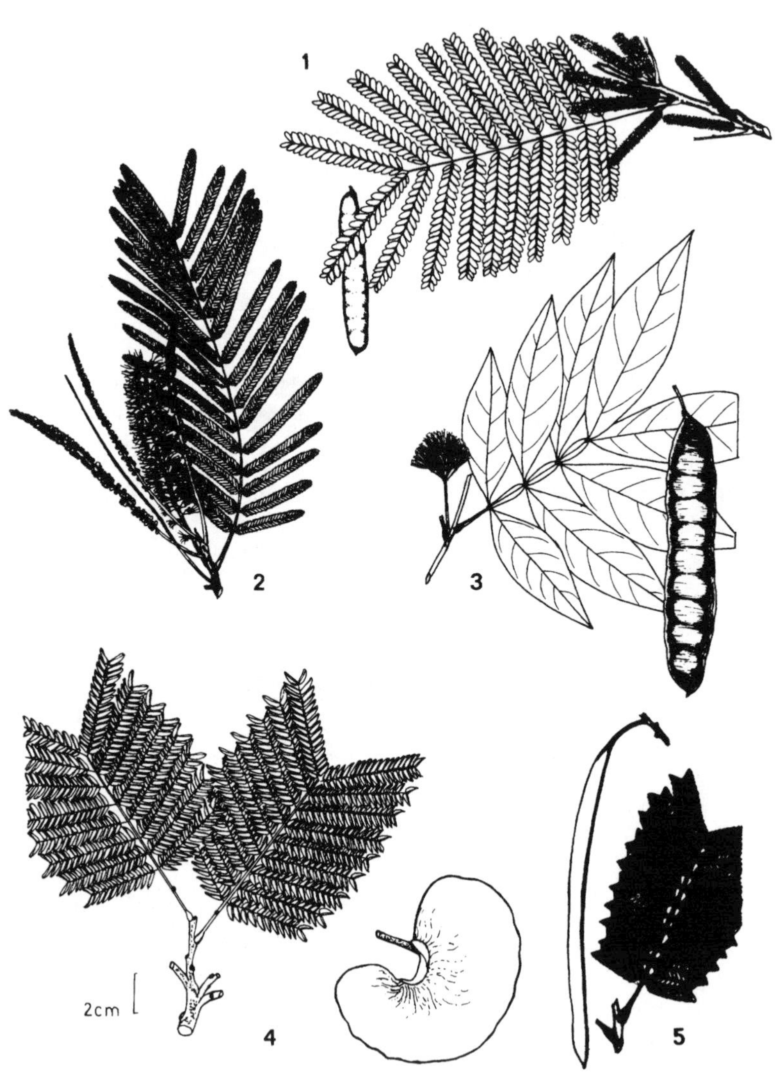

1 - *Stryphnodendron*

2 - *Pentaclethra*

3 - *Inga*

4 - *Enterolobium*

5 - *Piptadenia*

vegetatively to *Piptadenia* lianas in having the spines in lines along branchlet angles, but generally lack petiolar glands, unlike both *Acacia* and *Piptadenia.*

2C. Trees and large shrubs

2Ca. Petiolar and rachis glands lacking

Pentaclethra (1 sp., plus 2 African) — Flower with an imbricate calyx and only 5 stamens is the technical distinguishing feature. Inflorescence consisting of one-several long dense terminal spikes of white flowers; fruit elongate, fusiform-oblong, woody, explosively dehiscing: the only plant capable of breaking a plant press by the strength of its dehiscence! Locally very common and dominant, especially in swampy forests, in Chocó (and Central America); less common in Amazonia.

C: dormilón

Calliandra (100 spp., incl. Old World) — Usually flat-topped smallish trees or rather tortuous shrubs occurring along fast-moving streams. In flower, essentially *Pithecellobium* (and similarly with both few-foliolate and multifoliolate leaf forms), but lacking petiolar and rachis glands (in our area) and with a distinctive erect subwoody fruit, usually with strongly raised margins, that dehisces by elastically recurving. Flowers in umbellate powder-puff inflorescence, usually conspicuous and reddish from the numerous stamens.

P: bubinsana

Zapoteca — A segregate of *Calliandra* from which it differs in a thinner, less woody fruit and a terminal-branched inflorescence with leafy "stipule"-like bracts. Morphologically better retained in *Calliandra* but the two apparently represent convergence from different *Pithecellobium*-like ancestors.

(*Mimosa*) — A few *Mimosa* species are shrubby trees.

2Cb. Petiolar gland present but poorly developed, rather large but flat and more a glandular area than a typical mimosoid gland

Parkia (40 spp., incl. Old World) — Usually emergent spreading-crowned trees; vegetatively distinctive in the flat glandular area of the petiole; large-leafleted species tend to have very characteristic narrowly oblong, slightly curved, coriaceous leaflets and may have opposite leaves. The technical characters include calyx imbricate in bud and the very characteristic, capitate, bat-pollinated, usually long pendent inflorescence. Fruits linear and woody, long-stipitate, borne hanging in groups from the pendent inflorescence.

C: guabo-vaina, dormilón; P: pashaco, pashaco curtidor (*P. velutina*), goma huayo (*P. igneiflora*)

Leguminosae/Mimosoideae
(Shrubs and Trees of Dry Areas)

1 - *Acacia*

2 - *Mimosa*

3 - *Leucaena*

4 - *Albizia*

5 - *Prosopis*

6 - *Calliandra*

2Cc. Petiolar glands and/or rachis glands present, well-developed, usually more or less cupular in form

Acacia (750 spp., incl. many Old World) — Arid- and dry-area trees and moist-forest lianas. Most tree species of *Acacia* have paired stipule-derived spines and more or less fleshy fruits; all lianas have small nonstipular recurved spines and wind-dispersed fruits (see above); flowers sessile in small mostly globose heads, white in lianas, yellow or white in trees. The technical character is the the numerous stamens with nonfused filaments (or rarely very shortly fused).

C: aromo, murrai, cuji

Prosopis (40 spp., plus 3 in Old World) — Trees of arid and dry areas; always with stipule-derived spines, these sometimes very large; leaves like *Acacia* from which it differs in the elongate spicate inflorescence of white or cream flowers and 10 or fewer stamens; fruit indehiscent or breaking up irregularly.

C: trupillo, trupio, cuji

Piptadenia (15 spp.) — Trees and lianas, almost always with small irregular spines on branchlets; characterized by elongate spicate inflorescence branches (or inflorescences) and long very flat, usually narrow fruits, sometimes irregularly moniliform-constricted between seeds; fruits dehiscing along both sutures to release nonwinged seeds.

P: pashaco machete vaina (*P. suaveolens*), siuca pashaco (*P. flava*)

Parapiptadenia (3 spp.) — Large trees with tiny leaflets; very like *Piptadenia* but the seeds winged and flowers red.

Stryphnodendron (20 spp.) — Usually large trees; vegetatively and in flower similar to *Piptadenia* but the pinnae alternate and the branchlets lacking spines; fruit linear and green-bean-like, differing from *Piptadenia* (and *Albizia*) in being fleshier, less compressed, and septate.

Anadenanthera (2 spp.) — Close to *Piptadenia*, differing in being unarmed, in the globose inflorescence, and in the flat dry fruits dehiscing along one suture to release the seeds. The leaves have tiny leaflets lacking obvious venation, looking more like *Enterolobium* than *Piptadenia*. Often dominant in (mostly extralimital) dry forest.

Adenanthera (1 sp., possibly a second in Asia) — Amazonian tree, not yet collected in our area but may occur; fruits like *Pithecellobium* (though with red and black mimetic seeds) but minutely spiny stems like *Piptadenia* and climbing *Acacia;* inflorescence racemose with the flowers pedicellate, unlike sessile-flowered *Stryphnodendron* and (most) *Piptadenia.*

Leucaena (40 spp., plus 2 in Pacific) — Mostly in dry areas and mostly shrubby; looks very like *Albizia,* differing in having only 10 stamens and nonpolyad pollen. Flowers always in capitate inflorescence with the apex swollen *Parkia*-like in fruit; fruit flat, thin, but dehiscing.

Pithecellobium (200 spp., incl. Old World) — One of the largest and most diverse legume genera. The petiole often is very short and often has the gland at the bifurcation; the gland position is quite variable and may be near the base of the petiole, then often extremely large, between petiolules *Inga*-fashion, or not obviously present; leaflets vary from minute to very large and from many to ca. 4 but the leaves are always bipinnate (except for *P. rufescens* which is sometimes placed in *Inga*). Flowers are typically in umbels but may also be in spikes; stipular spines sometimes present but only in species with relatively few large leaflets (spines are never present on rachis and leaf as in most large-leafleted *Piptadenia, Acacia,* or *Mimosa*); fruits also variable, but never thin and wind-dispersed; typically linear and twisted to expose round mimetic seeds set against the dry reddish inner fruit wall; flattened, green-bean-like, perhaps sometimes water-dispersed in the common riverside species. Most of the species with more or less fleshy, indehiscent animal-dispersed fruits or water-dispersed swamp species with flat, dry fruits that break into transverse segments, have been recently transferred to *Albizia.*

C: guabo querre; E: guabilla (*P. longifolium*), bantano (*P. macradenium*), guaba del rio (*P. latifolium*); P: pashaco

Albizia (ca. 100 spp., incl. Old World) — In flower not obviously different from some *Pithecellobium* species but the petiolar gland usually not strongly raised, often elongate; flowers in umbellate inflorescence; fruits flat, straight, stipitate, dehiscent but wind-dispersed with seeds attached to raised margin, or fragmenting (the latter sometimes thicker and water-dispersed). Essentially a wind-dispersed *Pithecellobium.* A few species with indehiscent mammal-dispersed fruits (*A. saman* and allies) moved back and forth.

C: campano (*A. saman*)

Lysiloma (35 spp.) — Mostly Central American and circum-Caribbean; very like *Albizia* from which it differs in the wind-dispersed central part of the fruit separating from the thicker persistent margins. Perhaps better included in *Albizia.*

Enterolobium (5 spp.) — Usually large spreading flat-topped trees, differing from *Pithecellobium* mainly in the very characteristic strongly curved, rather ear-shaped, reniform, indehiscent fruit; leaflets always narrowly oblong, tiny, membranaceous with very acute apex; petiolar gland dark-drying, round, at or above middle of petiole (rather than large and cupular and near base of petiole as in the most similar species of *Pithecellobium*). Bark distinctive, smoothish and gray with fine vertical striations, the furrows somewhat orangish.

C: carito, orejero; P: pashaco oreja de negro, platanilla pashaco

Cedrelinga (1 sp.) — Giant emergent with coarsely ridged reddish bark. At maturity with only four pinnae (each with only a few large acute leaflets), petiolar gland lacking, the glands between petiolules *Inga*-fashion; the inconspicuous tiny white flowers are sessile in tiny poorly defined few-flowered heads. Fruit unique, thin and wind-dispersed, very elongate with marginal rib and prominulous tracing of venation on surface, contracted and spirally twisted at irregular intervals. One of the most important timber trees of poor-soil parts of Amazonia.

P: tornillo, tornillo huayracaspi

Papilionoideae

The Papilionoideae are easily and naturally divided into two major groups, trees with imparipinnate (sometimes reduced to unifoliolate or simple) leaves and vines or lianas with usually 3-foliolate leaflets. Only one tree genus is 3-foliolate, usually spiny-trunked *Erythrina*. A few species of 3-foliolate but usually climbing *Clitoria* are trees (and a very few species [all arborescent] of *Lonchocarpus*, *Machaerium*, and *Pterocarpus* are largely 3-foliolate). The majority of tree genera have indehiscent wind-dispersed fruits with a wide variety of wing types; all yellow-flowered genera have winged fruits, but so do some purple-flowered ones. Several genera have more or less woody, often elastic capsule valves that dehisce to release the hard-coated nonarillate seed; a few genera (all with purple flowers except *Geoffroea*) have very nonlegume-like fleshy indehiscent globose or ellipsoid, usually single-seeded mammal-dispersed fruits, one (*Dussia*, with purple flowers), has fleshy fruits that dehisce to release the large, rather soft, red-arillate seed, and one (*Erythrina*, with characteristic orange or red-orange flowers), a wide variety of fruit types ranging from fleshy and indehiscent to moniliform with mimetic red seeds to dry and apparently wind-dispersed. Tribe Swartzieae, traditionally placed in Caesalpinioideae, fits here much better (presence of stipels, odd-pinnate leaves with opposite leaflets, frequently with a trace of red latex) despite the rather caesalpinioid-like flowers.

The tree papilionates belong predominantly to six tribes. In addition to Swartzieae (characterized by free stamens, calyx entire in bud and usual reduction to a single petal), the large tribe Sophoreae also has free stamens but 5 petals (except *Amburana* and *Ateleia* with one). The small tribe Dipteryxeae is very distinctive in having the upper two calyx lobes greatly enlarged and petaloid. The other three tribes, very closely related, are Dalbergieae, characterized by few (usually one) ovules and indehiscent fruits and Tephrosieae and Robineae, both with generally dehiscent fruits and several ovules, the former with branched, mostly terminal, inflorescences (as in Dalbergieae), the latter with few-flowered axillary racemes. *Lonchocarpus* and *Deguelia* (= *Derris*), the latter lianas, have the indehiscent fruit of Dalbergieae but the wood anatomy and chemistry (most genera used for fish poison) of Tephrosieae. Three-foliolate *Erythrina* and a few species of *Clitoria* of tribe Phaseoleae and perhaps *Apoplanesia* of tribe Amorpheae are the only other tree papilionates in our area.

The vine, herb, and subshrub genera belong to a dozen different tribes, but by far the largest is Phaseoleae, with the great majority of species scandent, which is easily recognized by the almost always 3-foliolate leaves with strongly asymmetrical regularly stipellate lateral leaflets; other climbers are *Abrus*, pinnately multifoliolate with typical red and black mimetic seeds, some *Desmodium* species (tribe Desmodieae with its typical segmented stick-tight fruits) and several *Aeschynomene* with nonsticky segmented fruits. Among the herbs, both *Desmodium* and *Aeschynomene* and relatives (tribe Aeschynomeneae) have segmented lomentiferous fruits, the former with 3-foliolate leaves, the latter with pinnate (or variously reduced) leaves. Other herb and subshrub genera may have simple (*Poissonia, Alysicarpus,* some *Crotalaria*), 2-foliolate (*Zornia*), 3-foliolate (*Orbexilum, Eriosema, Collaea, Stylosanthes,* some *Crotalaria*), palmately compound (*Lupinus*), or pinnately compound (ten genera) leaves.

A few genera are extremely distinctive vegetatively. Some unusual features which characterize certain genera are:

Leaf rachis noticeably winged — *Dipteryx,* some *Swartzia* (also typical of many mimosoids).

Pinnate leaves with stipels between the petiolules — *Andira,* some *Swartzia, Coursetia* (tree species), some *Hymenolobium.*

Leaflets punctate — *Apoplanesia, Myrocarpus, Centrolobium, Myroxylon,* some *Lonchocarpus*.

Leaflets strictly opposite on rachis — *Lonchocarpus, Muellera, Hymenolobium, Gliricidia, Fissicalyx, Ormosia, Acosmium, Swartzia* (some).

Fruits fleshy, indehiscent, subglobose, mostly bat-dispersed —*Andira, Dipteryx, Geoffroea, Muellera* (in part moniliform), *Lecointea.*

Fruits more or less fleshy but dehiscent to reveal arillate seed —*Dussia, Swartzia, Aldina, Bocoa, Zollernia*.

Fruits flat, more or less woody, dehiscent —*Alexa, Clathrotropis, Diplotropis, Monopteryx, Ormosia, Poecilanthe, Gliricidia, Coursetia.*

Leaves 3-foliolate — *Erythrina, Clitoria* (few species are trees).

Leaves simple (or unifoliolate) — *Etaballia, Cyclolobium, Bocoa, Swartzia, Zollernia, Lecointea.*

Woody lianas (see 4A below) (often with stipule-derived spines and/or red latex) — *Machaerium, Dioclea, Mucuna, Deguelia, Abrus, Dalbergia, Clitoria, Canavalia* (few species), *Vigna* (few species). (Many other genera are slender vines.)

1. Trees with Imparipinnate Leaves

1A. Trees with opposite leaves — (A highly distinctive feature that occurs in only two of our genera)

Platymiscium (20 spp.) — Easy to differentiate from other legumes but likely to be confused with Bignoniaceae from which it is vegetatively differentiated by the legume odor, typical pulvinuli and pulvinus, and a conspicuous straight interpetiolar line connecting the tops of opposite petioles Flowers yellow; fruit similar to *Lonchocarpus,* indehiscent, flat, winged, single-seeded, the wing unusual in lacking any kind of costa or rib.

C: trébol

Taralea (5 spp.) — Only one large tree species in our area which may be locally dominant on white sand; the few other Guayana area species are not all opposite-leaved. Differs from *Platymiscium* vegetatively by lacking the conspicuous line connecting opposite petioles; flowers as in *Dipteryx* with two enlarged colored calyx lobes as long as the petals. Fruit broadly elliptic in outline, flat, dehiscent to release a single large flat brown seed. The distinctive bark is smooth and whitish as in many Myrtaceae.

P: tornillo caspi

1B. Trees with alternate leaves

1Ba. Calyx fused, (buds round and closed); stamens free; petals usually lacking or only one (five and subequal in *Aldina* and some simple-leaved taxa) — See also 1-petaled *Amburana* and *Ateleia;* leaves often simple or unifoliolate; if pinnate usually with opposite leaflets and stipels (tribe Swartzieae).

Swartzia (133 spp., plus 2 in Africa) — A large variable genus, always odd-pinnate (or unifoliolate or simple), leaflets usually opposite but sometimes only subopposite, stipels often present, rachis sometimes winged, most simple-leaved species easy to recognize by conspicuously parallel secondary and intersecondary venation. Flowers in raceme, petals one (or none), usually white or yellow; fruits more or less thick and fleshy or subwoody, mostly dehiscing to expose conspicuously arillate seeds. Sometimes with scant red latex.

P: cumaceba, acero shimbillo (*S. benthamii*), frijolillo (*S. obscura*)

Leguminosae/Papilionoideae
(Trees: Opposite or Simple or 3-Foliolate Leaves)

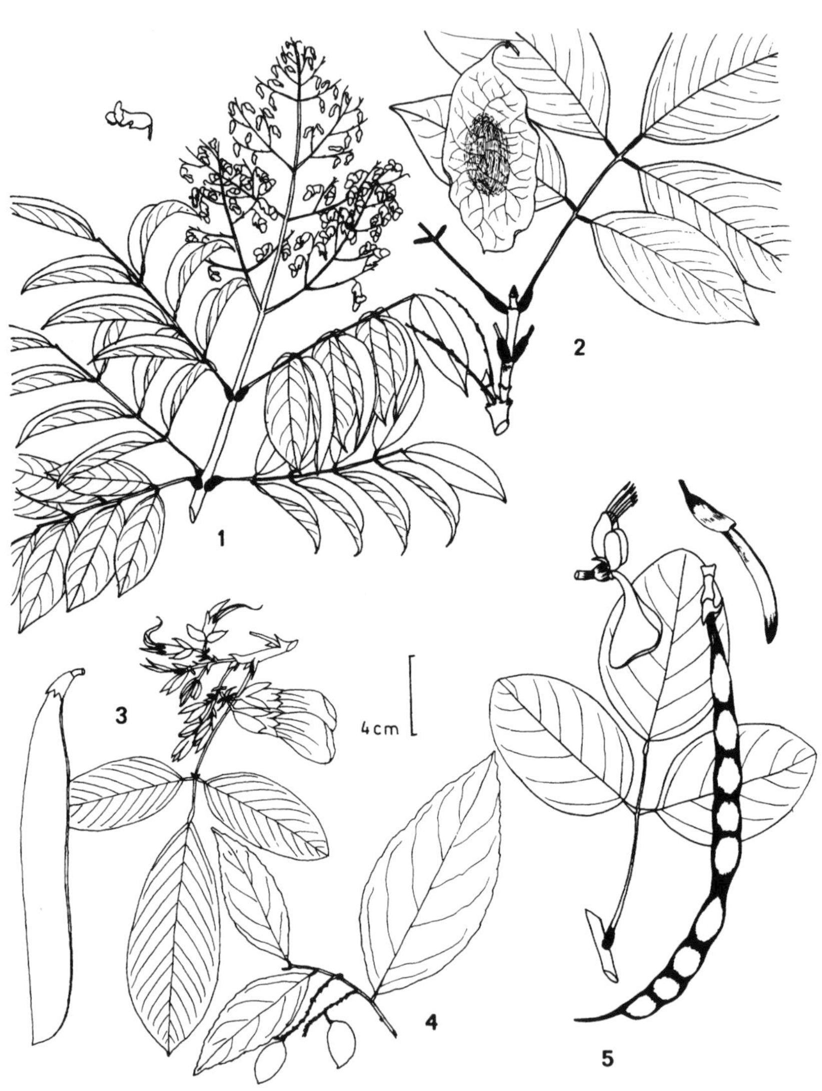

1 - *Taralea* **2** - *Platymiscium*

3 - *Clitoria* **4** - *Lecointea* **5** - *Erythrina*

Aldina (15 spp.) — Large trees of white-sand soils, mostly in Guayana area. Very like *Swartzia* but 5 petals; fruit large, more or less soft and globose, single-seeded.

Bocoa (7 spp.) — Unifoliolate or with very strongly alternate cuneate-based leaflets; unifoliolate species differ from *Lecointea* in having petiolule clearly differentiated from the petiole; also the leaves (leaflets) entire and the mature fruit dehiscent and neither white nor fleshy. Several-foliolate leaves differ from *Swartzia* in the more strongly alternate leaflet arrangement; our unifoliolate species distinctive in the asymmetric leaf blade (cf., *Drypetes*) and in having petiole and petiolule both short. Differs technically from *Swartzia* in basifixed anthers.

1Bb. Indehiscent, winged, wind-dispersed fruits (or their water-dispersed derivatives with variously reduced, thickened or vestigial wings) — All have pinnate leaves except one or two 3-foliolate species of *Lonchocarpus* and *Pterocarpus*. All yellow-flowered papilionate trees (except Swartzieae and a few Robinieae with reduced racemose axillary inflorescences) belong to this group (including *Platymiscium*, see above).

(i) Fruit tadpole-shaped with single relatively thick seed and single asymmetrically elongate (rarely more or less vestigial) wing

1) Wing basal, forming stipe below apical seed

Myroxylon (2 spp.) — Large tree; leaflets with pellucid-punctate dots and lines (but these not always conspicuous), alternate, acute to acutish (sometimes obtuse but rounded oval, not at all oblong), drying yellowish-green, tertiary venation more or less parallel to secondary venation; flowers white to yellow. Fruits with wing base strongly asymmetric and rounded on one side, the middle of face with thickened ridge connecting seed body to wing tip. The resin is the "balsam of Peru" of commerce.
E: balsamo; P: estoraque

Myrospermum (2 spp.) — Probably not adequately distinct from *Myroxylon* by anthers shorter than filaments and the fruit wing reticulately veined; leaflets with conspicuously linear punctations, alternate, membranaceous, strictly oval-oblong and usually retuse. Flowers white; fruit like *Myroxylon* but wing with faintly raised venation and lacking medial thickening. Caribbean area and Central America.

Platypodium (2 spp.) — Large forest tree with characteristically fenestrated trunk; very typical leaves with usually retuse oblong leaflets having close prominulous not well-differentiated secondary and tertiary venation; leaflets very similar to *Geoffroea* but usually larger and the

Leguminosae/Papilionoideae
(Trees: Wind-Dispersed with One-Winged Tadpole-Shaped Fruits)

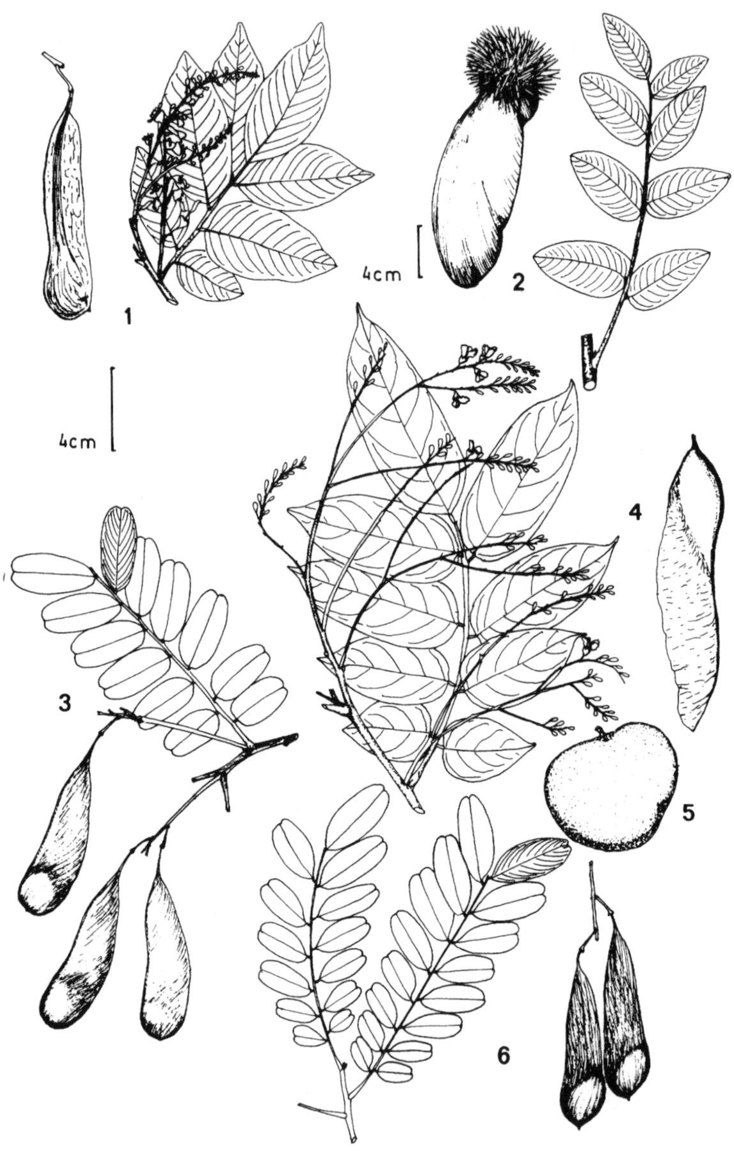

1 - *Myroxylon*

2 - *Centrolobium*

3 - *Platypodium*

Vatairea: **4** -*V. erthrocarpa*
5 -*V. guianensis*

6 - *Myrospermum*

secondary venation more ascending and more prominulous above. Flowers yellow; fruit like *Myroxylon* but gradually attenuate to stipe and lacking lateral thickening.

2) ***Wing apical*** — Sometimes vestigial; sometimes with stipules converted into paired spines (*Machaerium*) or with the leaflets large and strongly yellowish-punctate (*Centrolobium*).

(***Machaerium***) — Vegetatively variable, mostly lianas, often with paired stipule-derived spines (especially the lianas); leaflets usually alternate, may be very small or large; flowers violet or magenta; fruit body more flattened than in other "tadpole-shaped" legumes (not obviously thicker than wing). Usually with red latex.

Paramachaerium (3 spp.) — Very similar to *Machaerium* but the flowers with the wing petals much broader than keel; leaves drying darkish.

Centrolobium (6 spp.) — Very distinctive; mostly in dry forests; leaflets large, more or less puberulous and strongly yellow-punctate; flowers yellow; fruit large with a large *spiny* body (unique) and broad wing.

E: amarillo, amarillo lagarto

Vatairea (7 spp.) — Leaflets alternate, mostly acute to acuminate (oblong and obtuse with rather inconspicuous secondary veins at almost right angles in common tahuampa *V. guianensis*), very unusual in family in tendency to slightly serrate margin. Flowers purple; fruit body much thicker than wing, smooth, the wing narrow and elongate (vestigial in water-dispersed *V. guianensis*).

(ii) Fruit with a small globose body and 5 wings formed by large thin expanded calyx lobes

Apoplanesia (21 spp.) — Not yet recorded but perhaps in dry areas of northern Colombia; leaflets oblong-oval, very conspicuously punctate, the dots orangish when fresh, blackish when dry. Flowers not very legume-like, white with narrow uniform petals, in elongate narrow spikelike racemes; fruit unique for legume, very like that of *Astronium;* red latex.

(iii) Fruit wind-dispersed but with inflated longitudinal pouches rather than wings, eventually dehiscent

Diphysa (15 spp.) — Small dry-forest trees, in our area only in northern Colombia; leaflets membranaceous, alternate, oblong-oval, usually 7–9 per leaf. Flowers yellow, on small axillary raceme; fruit similar to *Crotalaria* but with a dry thin irregularly constricted wall with a definite medial costa on each side.

Leguminosae/Papilionoideae
(Trees: Wind-Dispersed Fruits with Wings Surrounding Seed)

1 - *Diphysa*
2 - *Pterocarpus*
3 - *Hymenolobium*
4 - *Ateleia*
5 - *Lonchocarpus*
6 - *Fissicalyx*
7 - *Piscidia*

(iv) Fruit with four longitudinal membranaceous wings

Piscidia (8 spp.) — Dry-forest trees; leaflets strictly opposite, only 7–9 per leaf, large, puberulous, round-tipped to acutish, more or less obovate, the secondary veins rather close and parallel; twigs with conspicuous whitish lenticels, typically rather elongate, even linear. Flowers light purple; fruit like an enlarged elongate *Combretum* fruit but the wings often irregular-segmented, especially in longer fruits, and breaking apart at maturity.

(v) Fruit planar with the wing more or less surrounding a variously thickened central seed body

Pterocarpus (100 spp., incl. Paleotropics) — One of the most distinctive legume genera on account of the characteristic suborbicular fruit shape with the thin (thicker in water-dispersed species) wing completely surrounding the central seed body, an indistinct apex far to one side is usually apparent. Leaflets alternate, usually finely prominulous-reticulate (except sometimes in characteristically blackish-drying *P. officinalis*); flowers yellow. Usually with red latex, occasionally with hollow twigs or peduncles inhabited by ants.

C: sangre de gallo; P: maria buena, pali sangre

Lonchocarpus (100 spp., incl. Africa) — Common dry-forest trees. (Liana species should be treated as *Deguelia* [= *Derris*]). Leaflets frequently pubescent, variable in size and shape but almost always strictly opposite, and never oblong unless very large. Flowers purple; fruit sometimes with more than one seed; always thin with both margins flat, but often with a slight rib along dorsal margin.

(***Dalbergia***) — Mostly lianas, see below.

Acosmium (16 spp.) — Very like *Lonchocarpus,* but with small white flowers having free filaments; leaflets 5–9 per leaf and strictly opposite, acutish but usually minutely retuse at extreme apex, coriaceous with the secondary veins not very prominent. Fruit flat, similar to *Lonchocarpus* but usually smaller, subwoody, and sometimes more symmetrically subacute at base and apex.

Bowdichia (4 spp.) — A mostly Brazilian genus with one species reaching the Llanos where it is a major savannah component. Leaflets smallish (3–5 cm long), retuse, more oblong than *Lonchocarpus* and more coriaceous than in many relatives; flowers purple, stamen filaments free unlike *Lonchocarpus* and relatives. Fruit thin, 1–several-seeded, rather 1-sided with a distinct rib along dorsal edge, obtuse at apex (with a rib terminating in hooped apicule), attenuate at base.

Ateleia (16 spp.) — Mostly Central American and Antillean, in our area only in northern Colombian dry forest; leaflets numerous, smallish, alternate, acute to acuminate. Flowers distinctive in having a single petal and free filaments; fruit very characteristic, single-seeded, semicircular with a straight-ribbed, short dorsal margin and a much longer round ventral one. (extralimital *Cyathostegia* is very similar but the leaves densely puberulous or retuse).

Fissicalyx (1 sp.) — Restricted to the Caribbean coastal area; leaflets medium to largish, opposite, acuminate, when young subtended by rather long stipules. Flowers yellow; fruit straight, planar except for a median raised rib over the narrow central seed body, with the two broad longitudinal membranaceous wings on either side. Twigs with characteristic roundish raised dark buds subtended by raised leaf and petiole scars.

Hymenolobium (10 spp.) — Mostly very large Amazonian trees, in our area the leaves multifoliolate, pubescent, with *Andira*-like stipels, clustered at branch apices where typically subtended by large bractlike stipules; leaflets numerous, smallish, opposite, elliptic-oblong and retuse; a few extralimital species have regularly placed leaves with acuminate leaflets, these coriaceous and with a fine reticulation. Flowers pink or magenta; fruit flat, membranaceous, round to oblong, with pair of irregular ribs parallel to but rather far from the margins.

P: mari mari

Myrocarpus (2 spp.) — Mostly southern Brazilian, but one species reaches Amazonia; leaflets very distinctive: strongly pellucid-punctate (especially along margin (cf., *Zanthoxylum*) which tends to be finely serrate in Paraguayan-Brazilian species). Flowers small, in bottle-brush-like raceme; fruit elongate with thickened medial area over the narrow central seed (like an elastic *Fissicalyx* stretched vertically); twigs with raised whitish lenticels.

1Bc. Fruits fleshy, indehiscent, subglobose or ellipsoid (moniliformly so in *Muellera*) — (All these genera are predominantly mammal-(especially bat-)dispersed and the fruits are greenish or brownish at maturity.) All have pinnate leaves, often with a tendency to winged rachis or petiolular stipels.

Muellera (2 spp.) — Small tree of mangrove-fringe habitats; unique in having the single-seeded indehiscent dispersal units moniliformly arranged (though only a single segment may develop); leaflets elliptic-ovate, acute, strictly opposite, 5–7 per leaf; flowers purple. The fruits well-known as a fish poison.

Leguminosae/Papilionoideae
(Trees: Fleshy Indehiscent Fruits)

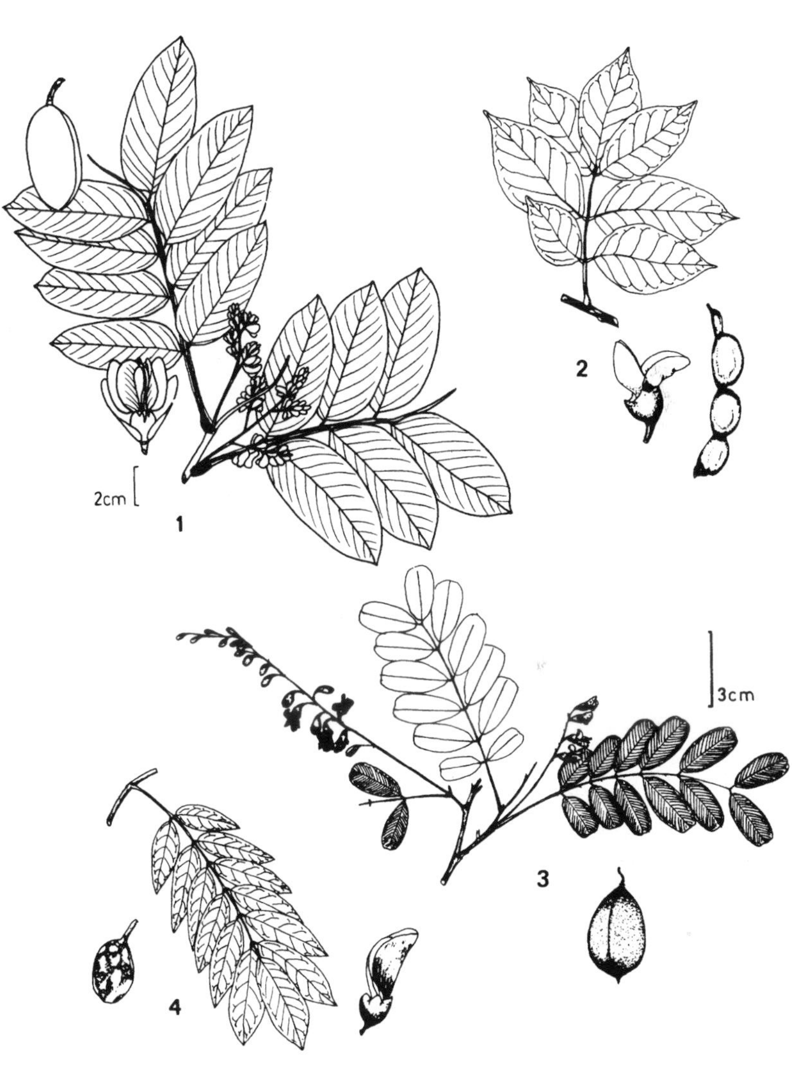

1 - *Dipteryx* **2** - *Muellera*

3 - *Geoffroea*

4 - *Andira*

Leguminosae/Papilionoideae
(Trees: Fleshy Dehiscent Fruits)

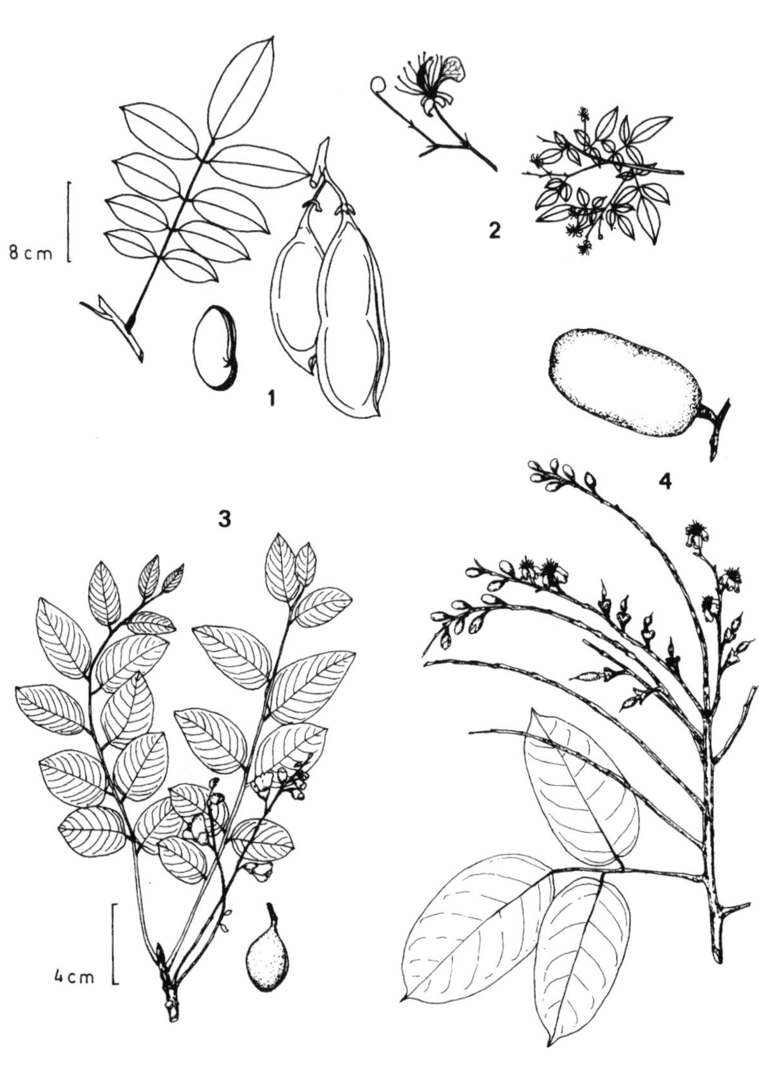

1 - *Swartzia* **2** - *Swartzia*

3 - *Dussia* **4** - *Aldina*

Dipteryx (incl. *Coumarouna*) (10 spp.) — Mostly large emergent trees, typically with smooth, often rather insculpted, salmon-colored bark and a rather contorted trunk and large buttresses; leaves very characteristic in the prominently flattened, usually distinctly winged or subwinged rachis with an apical prolongation along which the usually subopposite leaflets (often with strongly asymmetric midveins) are arranged. Flowers magenta or purple, very conspicuous, characterized by an enlarged petaloid upper pair of calyx lobes longer than the corolla; fruit rather large, single-seeded, flattened-ellipsoid, indehiscent, and with a woody endocarp.

P: shirihuaco, charapillo

Andira (20 spp.) — Trees usually with very characteristic leaves having rather many opposite leaflets with stipels between the petiolules; the commonest species has smallish to medium-sized acuminate leaflets with fairly indistinct close-together secondary and intersecondary veins; the stipels are less prominent in other (mostly extralimital) species and the leaflets may be fewer (even unifoliolate) and obtuse or retuse (but always larger than in *Geoffroea*). Flowers purple with truncate calyx; fruits a single-seeded subglobose to broadly ellipsoid drupe. Twigs lack obvious lenticels, and always have strong legume odor.

Geoffroea (3 spp.) — Dry-area trees with fruits like *Andira* but yellow flowers having dentate calyces and very distinctive leaves with many small alternate elliptic-oblong leaflets with characteristic close-together parallel and finely prominulous secondary veins.

C: silbadero

1Bd. Fruits more or less fleshy or subwoody and dehiscent — Seed arillate; leaves odd-pinnate, usually with subalternate leaflets

Dussia (10 spp.) — Mostly in wet forests. Vegetatively characterized by the largish often puberulous leaflets with rather close-together secondary veins to which the finely parallel tertiary veins are perpendicular. Flowers purple with free filaments; fruit very typical, usually 1-seeded, ellipsoid or ovoid, rather fleshy and appearing indehiscent when young, at maturity dehiscing to expose the large, ellipsoid, red-arillate seed(s). Usually with slight red latex.

E: sangre de gallina; P: huayuro

(*Swartzia* and *Aldina*) — see Swartzieae above.

1Be. Fruits dryish, more or less woody, dehiscent — Leaves pinnate (except 1-foliolate in *Poecilanthe*). As arranged here the genera form a sequence from the largest woodiest fruits to the most slender least woody ones: the first three genera below have strongly woody fruits with broad valves that dehisce by twisting spirally; *Ormosia, Poecilanthe,* and

Leguminosae/Papilionoideae
(Trees: Dry Dehiscent Fruits [A])

1 - *Yucartonia*

2 - *Ormosia*

3 - *Diplotropis*

4 - *Ormosia*

Monopteryx, have subwoody usually single-seeded capsules, and *Gliricidia, Coursetia,* and *Sesbania* have narrow elongate capsules with numerous seeds).

(Platycyamus)(2 spp.) — Not known from our area, but collected in adjacent Brazil on the Rio Acre.

Alexa (8 spp.) — Trees of poor sandy soils; mostly Guayanan. Leaflets alternate, large, coriaceous, often rather smooth and/or glaucous below. Flowers large, probably bat-pollinated, the large pubescent cupular calyx reminiscent of some Bombacaceae; fruits narrowly oblong, flat, rufous-pubescent, the woody valves strongly coiling.

Clathrotropis (6 spp.) — One species common in Magdalena Valley otherwise mostly Brazilian. Leaves ca. 7-foliolate with opposite leaflets large, coriaceous, smooth, and shiny above. Flowers purple, the calyx 5-toothed; fruit broadly oblong, flat, woody valved with one or few large *Dioclea*-like seeds.

Diplotropis (7 spp.) — Amazonian trees; leaflets medium-large, always coriaceous, alternate, either acute to acuminate or elliptic-oblong. Flowers purple to rose, subsessile with a curved calyx; fruits similar to *Macrolobium,* flat, circular-oblong, with a rib along dorsal margin and usually also a fainter rib along ventral margin, dehiscent and not wind-dispersed, one-seeded, subwoody, the curved dorsal margin slightly longer than ventral one with the rib along dorsal margin ending before the nonapiculate apex.

Monopteryx (3 spp.) — Large trees forming single-species stands in poorly drained swampy areas on white sand, the base very large and stilt-rooted, the bark scaling in plates; leaflets alternate or subopposite, coriaceous, in our area rather large, few, acute to acuminate and rather smooth and glaucous below (elsewhere oblong and similar to *Macrolobium* but alternate). Flowers purple, filaments free, with a unique one-sided spathaceous tannish calyx; fruit flat, single-seeded, apparently winged, but elastically dehiscing to release the apical seed. Seeds once used for flour.

Ormosia (50 spp.) — Large canopy trees, characterized by the mimetic seeds which are hard, shiny, and bright red or half red and half black (and widely used for beads). Some species are vegetatively distinctive in the strongly pubescent undersurface of the discolorous leaflets.

P: huayuro

Amburana (3 spp.) — Emergent tree with very characteristic reddish-papery bark (cf., *Bursera simaruba*); leaflets rather many, narrowly oblong-ovate, smallish, membranaceous, alternate. Flower white or yellowish, with single petals; fruit rather elongate (5–7 cm x 1.5 cm) com-

Leguminosae/Papilionoideae
(Trees: Dry Dehiscent Fruits [B])

1 - *Bocoa* **2** - *Bocoa* **3** - *Gliricidia*
4 - *Amburana*

5 - *Clathrotropis* **6** - *Alexa*

pressed, extended as kind of poorly demarcated wing above single basal seed. An important timber tree.

P: ishpinga

Gliricidia (4 spp.) — Frequently cultivated as living fence posts or hedgerows in dry-forest areas; vegetatively rather similar to some species of *Lonchocarpus* in the smallish strictly opposite leaflets, these characterized by being almost rhombic, broadest in the middle and acute to acuminate at both base and apex. Fruits narrow and spirally dehiscing; flowers pink, usually borne during dry season while leaves more or less caducous. Reputed to be poisonous to rodents.

C, E: mata ratón

Coursetia (incl. *Cracca* and *Poissonia*) (38 spp.) — Dry areas; mostly shrubs, especially Andean, but common Caribbean dry-forest species definitely a tree and the species formerly placed in *Cracca* are herbs; leaves variably odd- or even-pinnate, the alternate or subopposite elliptic-oblong leaflets thinly membranaceous, with curious tendency to vary on same leaf from very small at base to much larger at apex. Flowers purple to white and lavender, the leaves of tree species caducous when in flower; fruit very slender, internally segmented, sometimes constricted between the seeds, spirally dehiscing.

(***Sesbania***) — Mostly shrubs or even herbs, the few dry-forest tree species (= *Yucartonia*) characterized by very many narrowly oblong membranaceous leaflets (effect of *Phyllanthus*), white or yellow flowers, and very narrow linear fruits (sometimes 4-angled or 4-winged extralimitally).

E: estaquilla

2. Trees with Three-Foliolate Leaves

2A. Stipels subtending leaflets — Fruits mostly moniliformly constricted, usually dehiscent to expose hard red mimetic seeds.

Erythrina (75 spp., plus 43 in Old World) — Extremely distinctive and well-marked, not only in the 3-foliolate leaves but also in the red or red-orange bird-pollinated flowers (standard folded into tube for hummingbird pollination in most Central American and northern South American species, open for perching-bird-pollination in most Amazonian species. The fruits are also very distinctive, typically constricted between seeds and moniliform, dehiscing to reveal hard bright red mimetic seeds, in a few species thicker, fleshy and irregularly dehiscent to reveal soft whitish seeds; varying to single-seeded, softball-sized and indehiscent in one coastal Ecuadorian species and dry, one-sidedly dehiscing, and possibly wind-dispersed in *E. ulei* and *E. poeppigiana.*

C: chocho, cambulo, búcaro (*E. fusca*); E: porotón, porotillo; P: amasisa, gallito

(Clitoria) — Mostly vines or lianas (see below); one common tahuampa species is a medium-sized tree, vegetatively easily recognized by the close-together, rather straight secondary veins and the presence of stipels subtending the leaflets.

2B. Leaves without stipels

There are occasional 3-foliolate species of normally pinnate-leaved genera, e.g., *Machaerium latialatum* and *Lonchocarpus.*

3. Trees with Simple or Unifoliolate Leaves; Fruits One-Seeded

Etaballia (1 sp.) — Leaves simple, short-petioled, not at all legume-like; inflorescence short and axillary with tiny flowers; fruits small, ellipsoid, single-seeded.

Cyclolobium (4 spp.) — Mostly Brazilian/Paraguayan dry areas; leaves unifoliolate with long petiole and conspicuously darker apical petiolular portion; flowers small, fruits like *Ateleia* but rib on ventral edge.

Poecilanthe (7 spp.) — Characterized by very few leaflets, in the only species in our area unifoliolate with a long petiole and conspicuously finely prominulous-reticulate leaf surface; inflorescences reduced, axillary, calyx 4-dentate; fruit roundish, subwoody, flat, the single seed very thin and almost winged.

Lecointea (5 spp.) — Simple yellowish-olive drying leaves with +/- serrulate to obtusely coarsely serrate margin; 5 petals; round white tasty fruit; not at all obviously a legume although its petiole is very like the petiolules of typical legumes. Belongs to tribe Swartzieae with closed, round buds and free stamens.

P: yutubango

Zollernia (12 spp.) — Perhaps not in our area but in both Amazonia and Central America.Very similar to *Lecointea* in the simple leaves, but these either entire or sharply serrate (cf., holly), not obtusely serrulate. In flower differs from *Lecointea* in the narrowly linear anthers much longer than the filaments.

(Swartzia, Bocoa) — See 1Ba, Swartzieae above.

4. Lianas and Vines

4A. True woody lianas occur in relatively few papilionate genera, several of which are very large and speciose. — Three genera of the mostly arborescent tribe Dalbergieae (incl. *Lonchocarpus*) with imparipinnately compound leaves are predominantly or occasionally lianas:

Deguelia (= *Derris*) with opposite leaflets and *Machaerium* and *Dalbergia*, usually with alternate leaflets. Some species, especially of *Deguelia*, may be very similar to Connaraceae (where see discussion). The other papilionate lianas are *Abrus* (with multifoliolate paripinnate leaves with opposite leaflets), three predominantly woody genera of Phaseoleae (*Dioclea*, *Mucuna*, *Clitoria*), and occasional unusually thick-stemmed species of several other genera of that same tribe. Most papilionate lianas have red sap or latex, sometimes conspicuously so; most *Machaerium* (and a few *Dalbergia*?) have stipule-derived spines.

4Aa. These four liana genera have pinnately compound leaves.

***Deguelia* (*"Derris"*)** — Nonspiny leaflets strictly opposite on rachis, medium to large and acute to acuminate; flowers often white, sometimes purple; fruit thin and indehiscent, of typical wind-dispersed *Lonchocarpus* form (though often with thicker dorsal margin?) or the wings thickened for water dispersal. Roots often used as fish poison. The correct generic name for this concept is still being debated.

P: barbasco

Dalbergia (300, incl. Old World) — Leaves mostly few-foliolate, often in large part unifoliolate, the leaflets always alternate, usually medium to large; inflorescence usually reduced and more or less ramiflorous and/or a somewhat flat-topped panicle. Fruits flat, not subtended by the caducous calyx, typically half-circular, sometimes like thin *Lonchocarpus* (i.e., oblong with central seed body surrounded by thin wing).

E: granadillo; P: palo de la plata (*D. monetaria*)

Machaerium (120 spp.) — Usually with stipule-derived spines, sometimes also with enation-type spines; leaves few to (usually) many-foliolate, the leaflets highly variable in shape and size, alternate on rachis. Fruit with persistent calyx, the seed body basal with an extended apical wing or the wing reduced for water dispersal and very like *Dalbergia* except for the persistent calyx.

Abrus (2 spp., plus 15 in Old World) — Only *A. precatorius* in our area, characterized by the multifoliolate stipellate paripinnate leaves with narrowly oblong opposite leaflets; fruits elongate, compressed, dry, the mimetic seeds bright red and black.

4Ab. The following liana genera have 3-foliolate leaves.

Dioclea (30 spp.) — 3-foliolate, the leaflets usually with relatively (to *Mucuna*) close straight secondary veins; twigs often villous or erect-pubescent. Inflorescence erect, a narrow spikelike raceme with many purple flowers; fruits lacking urticating hairs or very mildly urticating; dehiscent or thick, indehiscent and often rufous-hairy.

Leguminosae/Papilionoideae
(Woody Lianas)

1 - *Dalbergia* **2** - *Machaerium*

3 - *Mucuna* **4** - *Abrus*

5 - *Dioclea*

Mucuna (incl. *Stizolobium*) (120 spp., incl. Old World) — Three-foliolate, the leaflets with relatively few widely spaced secondary veins, the laterals usually with more accentuated basal lobes than *Dioclea;* twigs glabrate or with more or less appressed trichomes. Inflorescence pendent, the large cream, yellow, or orange flowers clustered near tip; fruits often with nastily urticating hairs.

C: pica-pica; E: pasquinaque; P: ojo de vaca

(*Canavalia*) — The few liana species have mostly pendent inflorescences with white or whitish flowers alternating along rachis; fruit narrowly oblong, explosively dehiscing by twisting valves, always with strong raised rib on valve surface.

(*Vigna*) — (See vines).

Clitoria (70 spp., incl. Old World) — (In addition, a few species are trees and several are herbaceous vines) 3-foliolate (rarely odd-pinnate) with leaflets with close-together ascending secondary veins. Flowers pink or purplish, few per inflorescence, unusual in being turned over upside down (resupinate) with the large flat standard forming "landing platform" from the center of which the often more darkly colored sexual parts (enclosed by the other petals) stick up (resembling a mammalian clitoris, hence the name); fruit narrowly oblong.

4B. Slender vines — (Mostly Phaseoleae and characterized by twining stems and 3-foliolate leaves with leaflets subtended by stipels; the lateral leaflets are usually strongly asymmetric and the fruit is almost always a dehiscent pod. The genera with pinnate leaves are mostly Aeschynomeneae with the fruit breaking into transverse segments.)

4Ba. Leaves 3-foliolate

(i) The next eight genera can generally be differentiated by flower color and/or size.

(*Mucuna*) — A few species (= the segregate *Stizolobium*) are more or less herbaceous with narrower fruits, also with urticating hairs.

Rhynchosia (44 spp., plus 150 Old World and 8 in USA) — The only 3-foliolate vine genus with gland-dotted leaves, the leaflets distinctly rhombic. Flowers usually yellow or greenish-yellow, in axillary inflorescence, elongate or reduced to few flowers, the calyx also gland-dotted; fruits dry, few-seeded, dehiscent, seeds small, roundish, shiny, red or black.

Canavalia (29 spp., plus 20 Old World) — Commonest along coasts; unique in tribe in strongly bilabiate calyx; flowers usually pink or lavender; fruit narrowly oblong, subwoody with strongly raised longitudinal rib on each valve.

Leguminosae/Papilionoideae
(Vines)

1 - *Vigna* (*V. vexillata*) **2** - *Pueraria* **3** - *Centrosema*

4 - *Phaseolus*

5 - *Calopogonium* **6** - *Vigna* (*V. caracalla*)

7 - *Cymbosema* **8** - *Clitoria* **9** - *Rhynchosia*

Cymbosema (1 sp.) — Vine of swampy areas; bright red flowers ca. 4 cm long on erect inflorecence; fruit oblong, flattened, like miniature *Dioclea* but with long curved beak, slightly urticating hairs.

Galactia (140 spp., incl. Old World) — In our area usually shrubby vines; mostly grasslands (especially the cerrado); one Andean species is red-flowered shrub; leaves softly puberulous below; small flowers in elongate spikelike raceme, the calyx apparently 4-lobed (unique); fruit softly puberulous, narrow.

Calopogonium (8 spp.) — Common in weedy areas; small blue flowers; stems, rachises and fruits more or less villous; calyx 5-lobed.

Teramnus (15 spp., incl. Old World) — Common weedy vine very like *Calopogonium* but inflorescence lacks conspicuous nodes, the tiny flowers reddish to purplish and fruit narrower and/or more elongate with the apex strongly hooked upward; flowers unusual in alternately aborted stamens.

Cologania (10 spp.) — Related to *Teramnus* but reduced axillary inflorescence with much larger, rather long, narrow pink to purplish flowers (technical character is upper calyx lobes united). Stems and rachises pilose; fruit small, villous, like that of *Calopogonium.*

Pachyrhizus (6 spp.) — Weedy distinctly puberulous large-leafleted vines, similar to *Dioclea* (especially *D. guianensis* and with similar dehiscent, thin-valved, appressed-pubescent fruit) but the leaflets more triangular. Leaflets of some species distinctive in being strongly 3-lobed.

P: oshipa

(***Pueraria***) — Asian, much cultivated as ground cover; rather like a large-flowered *Calopogonium;* stems and rachis villous, the leaflets rather rhombic; flowers bluish or white and lavender; all calyx lobes distinct. Stipules unusual in being extended above and below insertion (also in some *Dioclea*).

(***Desmodium***) — Mostly herbs, see below.

(ii) The next two genera have unique resupinate flowers.

Centrosema (45 spp.) — Slender twining vines; unique (along with *Clitoria*) in upside down (resupinate) flowers; differs from slender twining species of *Clitoria* mainly by standard with conspicuous spur on back; pods narrow, internally septate, with two raised submarginal ribs.

(Clitoria) — Mostly lianas, (see above). Several species are 3-foliolate (a very few species pinnate) herbaceous twiners; flowers always lacking the spurred standard of *Centrosema;* fruit usually larger and subwoody, mostly unribbed or with a single rib.

(iii) The next three genera have the style variously extended or coiled.

Phaseolus (ca. 50 spp.) — Unique in always having uncinate hairs (but ca. 25x magnification needed); flower with style and keel coiled 2–3 revolutions; inflorescence bracts and bracteoles usually persistent in fruit; fruit similar to a green-bean or somewhat more compressed.

E: vaina de manteca (*P. lunatus*)

Vigna (mostly paleotropical, ca. 100 spp. total) — Very close to *Phaseolus* and some species are switched back and forth; stipules differ in being usually peltate or at least with extended basal lobes. In most species the flower differs from *Phaseolus* in straight style and keel, but some species, formerly placed in *Phaseolus* and including the well known woody liana *V. caracalla,* have flowers like *Phaseolus* with 3–5 coils of style and keel.

Macroptilium (20 spp.) — Prostrate vines with few flowers at end of long erect peduncle; differs from *Phaseolus* in the style merely inrolled and becoming sigmoid and the wing petals conspicuously larger than keel petals; fruit very characteristic, very narrow and twisting in a tight woodshaving-like spiral.

4Bb. Pinnate leaves

(i) The next two genera have glandular-punctate leaves.

Poiretia (6 spp.) — Dry inter-Andean valleys; leaves 4-foliolate with stipels; flowers yellow in narrow axillary racemes; fruits segmented like *Aeschynomene.*

Nissolia (12 spp.) — Dry areas; 5-foliolate; flowers yellow, in axillary clusters; fruit apex extended as long flat curved wing.

(ii) The following genera lack glandular punctations.

Chaetocalyx (12 spp.) — Leaves 5-foliolate, with very herbaceous, obovate leaflets; flowers yellow, in axillary clusters; fruit segmented, the segments sometimes rather elongate.

Tephrosia (300 spp., incl. Old World and Temperate Zone) — Odd-pinnate, narrow, rather coriaceous leaflets with very characteristic close-together, strongly ascending secondary veins.

Vicia (150 spp. total, mostly n. temperate) — In our area only in Andes. Sprawling or trailing vines with epulvinate paripinnate usually multifoliolate leaves, the rachis often ending in a tendril; fruit linear, dehiscent.

Barbiera (1 sp.) — Common weedy vine in Amazonia; now often merged with *Clitoria* but has very different long red flowers, pilose stems and rachises and strongly stipellate pinnately compound leaves. Essentially a hummingbird-pollinated *Clitoria.*

(*Clitoria*) — A very few species have pinnate leaves but are easily told by the characteristic upside down flower.

5. Herbs, Shrubs and Subshrubs

5A. Leaves simple or absent

(*Coursetia*) (3 spp.) — The species formerly referred to as *Poissonia* are shrubs of dry inter-Andean valleys; leaves simple, suborbicular, densely white-sericeous; flowers purple.

Alysicarpus (1 sp., plus 30 in Old World) — Prostrate weedy herb; leaves not sericeous, the base rather cordate; flowers little and inconspicuous; fruit segmented but not very compressed.

(*Spartium*) — Widely naturalized in dry inter-Andean valleys; green-stemmed and essentially leafless, usually with a few small simple leaves; flowers yellow.

(*Crotalaria*) — Few species simple-leaved but mostly 3-foliolate, (see below).

5B. Leaves 2-foliolate

Zornia (75 spp., mostly in Old World) — Small weedy herbs with large stipulelike bracts from which the yellow flowers arise; leaves mostly 2-foliolate, sometimes glandular-punctate.

5C. Leaves 3-foliolate

5Ca. The next two genera have glandular-punctate leaves.

Orbexilum (= neotropical *Psoralea*) (50 spp., mostly n. temperate) — Andean upland dry areas; characterized by leaves 3-foliolate and strongly glandular-punctate or pustulate; fruits small, glandular, one-seeded.

Eriosema (140 spp., incl. Old World) — Pubescent shrubs or subshrubs with narrow 3-foliolate leaves with very short petioles; flowers yellow, in reduced axillary racemes; fruit villous, small, narrow.

Leguminosae/Papilionoideae
(Herbs)

1 - *Aeschynomene* **2** - *Desmodium* **3** - *Indigofera*

4 - *Dalea*

5 - *Stylosanthes* **6** - *Zornia*

7 - *Crotalaria* **8** - *Lupinus* **9** - *Lupinus*

5Cb. The next four genera do not have glandular punctations.

Collaea (may = *Galactia*) — Erect upland shrubs or subshrubs; flowers bright red or magenta, fairly large; leaves with narrow sericeous-below leaflets and very short petiole; fruit villous, subwoody, flat, linear-oblong.

Desmodium (50 spp.) — Herbs or subshrubs (a few species are vines), the leaves always 3-foliolate, often with noticeably rhombic leaflets; fruits transversely segmented and breaking into exozoochorous stick-tight segments.

E: pega pega, amor seco

Crotalaria (550 spp., incl. Old World) — Leaves usually 3-foliolate, lacking stipels, rarely simple; flowers yellow, in terminal inflorescence; usually with typical inflated pod.

Stylosanthes (50 spp., incl. Old World) — Grassland herbs; stipules more or less sheathing, united, with the leaf arising from the sinus; leaflets eglandular, narrow, pointed; flowers small, yellow; fruits narrow, jointed.

5D. Leaves pinnately compound

5Da. The following three genera have punctate leaves.

Dalea (250 spp., incl. N. Am.) — Dry Andean uplands; multifoliolate leaves; inflorescence terminal, often dense, the flowers purple or blue; small indehiscent one-seeded fruits.

Weberbauerella (2 spp.) — Andean; leaves multifoliolate, imparipinnate, with glandular excretions obscuring the glandular punctations; flowers brownish-yellow; fruits segmented.

Amicia (7 spp.) — Leaves 4-foliolate, the leaflets obovate with retuse apex; flowers relatively large, red, yellow or magenta, the calyx glandular; fruits segmented.

5Db. The following six genera have nonpunctate leaves.

Adesmia (230 spp.) — In our area, mostly spiny high-Andean shrubs and subshrubs, the spines mostly branched; leaves usually paripinnate, the leaves and stems sometimes glandular-pubescent; flowers yellow to orange.

Aeschynomene (67 spp.) — Weedy herbs with paripinnate leaves with many little close-together leaflets, lacking gland-dots; flowers yellow, few per inflorescence; fruit with (1–)several *Desmodium*-like segments.

Apurimacia (4 spp.) — Shrub 1–2 m, sometimes with trailing branches; leaves odd-pinnate with ca. 7–11 apiculate oblong-elliptic leaflets; inflorescence narrowly racemose, axillary; fruit few-seeded, *Phaseolus*-like but dry and thinner, approaching *Lonchocarpus.*

Astragalus (2000 spp. total, cosmopolitan) — In our area, high-Andean herbs or shrubs, mostly on puna; leaves pinnately multifoliolate, usually grayish-sericeous or very narrow; flowers pink or purplish; fruit inflated, similar to that of *Crotalaria* but with a separate membrane separating 2 seed chambers.

(*Coursetia*) — The dry-area herbs, formerly recognized as, *Cracca* are now lumped with *Coursetia.* They have 5–9-foliolate leaves and a very narrow fruit, more uniformly internally septate than in the shrubby *Coursetia* species.

Indigofera (700 spp., mostly Old World) — Weedy herbs with 5–7-foliolate leaves; flowers reddish, inconspicuous, with caducous petals; fruits small, very narrow, clustered along inflorescence and tending to be subpendent but curving up at tip.

Sesbania (50 spp., incl. Old World) — Also see trees. Mostly weedy herbs or subshrubs with short axillary raceme of few yellow flowers; many narrow leaflets giving a *Phyllanthus*-like effect; fruit very long and narrow, internally septate.

5E. Leaves palmately compound

Lupinus (200 spp., incl. n. temperate) — High-Andean; our only palmately compound legume; flowers blue, in erect many-flowered inflorescence.

LENTIBULARIACEAE

Very small herbs, leafless or with the small leaves reduced to a basal rosette; mostly in open water-logged areas, especially where pH is acidic; sometimes floating aquatics, a few species epiphytic in wet cloud forest. The flowers, though small, are conspicuously large for the tiny tenuous plant; corolla brightly colored, usually yellow, lilac, or purplish, strongly bilabiate, usually with a spur; only 2 stamens. The plants are insectivorous, with viscid leaves (*Pinguicula*) or bladderlike traps on the roots (*Utricularia*) to trap minute invertebrates.

Lentibulariaceae and Lepidobotryaceae

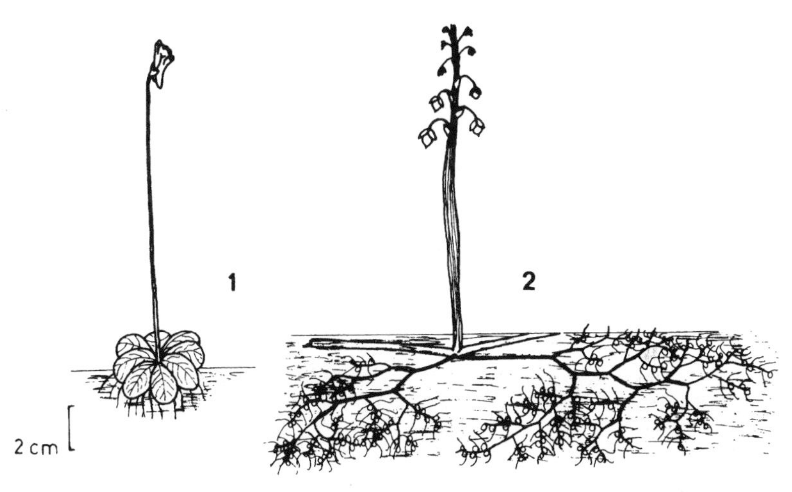

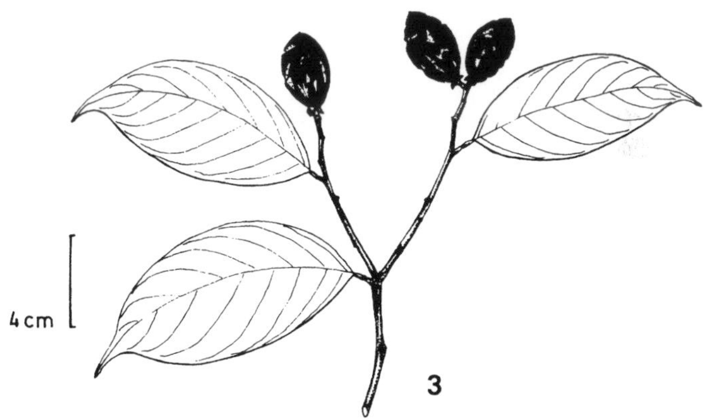

1 - *Pinguicula* **2** - *Utricularia*

3 - *Ruptiliocarpon* (Lepidobotryaceae)

Linaceae

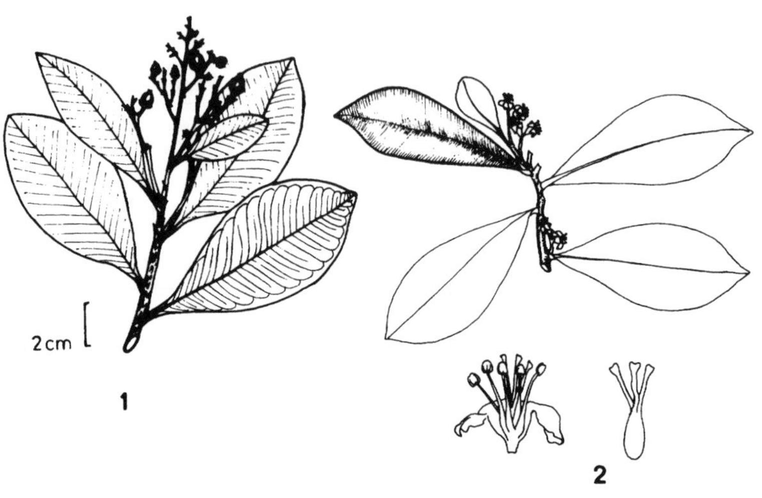

1 - *Roucheria* (*R. humiriifolia*)

2 - *Roucheria* (*R. monsalvei*)

3 - *Linum*

4 - *Ochthocosmus*

Pinguicula (45 spp., incl. N. Am. and Old World) — Uncommon, typically occurring on wet boggy cliff faces or similar habitats. With basal rosette of small, obovate, rather succulent, viscid-pubescent leaves.

Utricularia (180 spp., incl. N. Am. and Old World) — Leafless (although usually with stems variously modified into small narrow (sometimes much dissected in aquatics) photosynthetic organs); roots with macroscopically visible small white nodules which under a lens turn out to be complicated traps for capturing minute organisms.

LEPIDOBOTRYACEAE

A formerly monotypic African segregate of Oxalidaceae recently discovered in tropical America. Ours dioecious, 30 m canopy trees occurring both in Chocó and Amazonia and characterized by long-petiolate unifoliolate leaves with a legumelike pulvinulus. The flowers are tiny and nondescript, but the fruit is distinctive: an ellipsoid dark-drying single-seeded strikingly woody capsule ca. 2.5 cm long and irregularly and incompletely dehiscing to reveal the red-arillate seed.

Ruptiliocarpon (2 spp.)
P: cedro masha

LINACEAE

Two very different genera in our area, one the small herb *Linum* of dry upland areas, the other *Roucheria,* a genus of trees of lowland forests. They have similar flowers with 5 rather narrow petals, distinct styles, and the filaments basally fused into a tube. *Linum* is characterized by yellow flowers, spindly growth-form, and narrow (usually more or less linear) alternate or opposite leaves. *Roucheria* has alternate leaves with strongly parallel secondaries (often very close together and *Clusia*-like) a more or less crenate margin, and a marginal or submarginal collecting vein. When the leaves are not obviously *Clusia*-like (and sometimes when they are) the leaves are of a typical oblong-obovate long-petiolate form. Vernation lines paralleling the midvein (cf., *Erythroxylon*) are usually present. The commonest *Roucheria* (*R. humiriifolia*) is vegetatively similar to some forms of *Humiria* (Humiriaceae) but has leaves larger, thinner, and more gradually tapering to a longer petiole.

Linum (about 7 spp. in our area, over 200 worldwide) — Spindly herbs mostly of upland dry areas, characterized by symmetric 5-petaled yellow flowers and narrow, usually sublinear, opposite or alternate leaves.

Roucheria (incl. *Hebepetalum*) (8 spp.) — Medium to large trees, typically with small rather narrow buttresses. Leaves very characteristic, the secondary and intersecondary veins more or less straight and parallel, frequently very close together and *Clusia*-like, with a submarginal connecting vein, margin usually crenate. Flowers usually yellow (white and with pubescent petals in *R. humiriifolia*, sometimes segregated as *Hebepetalum*). The inflorescence usually has a well-developed central axis (unlike vegetatively similar Humiriaceae) or is reduced and subfasciculate (only when flowers bright yellow and leaves *Clusia*-like).

C: juana se va; P: quillobordón blanco

Two extralimital genera are often included in Linaceae, *Ochthocosmus* of the Guayana Shield area and *Cyrillopsis*, but have a solitary style and are frequently segregated as Ixonanthaceae.

Loasaceae

A mostly scandent (all five genera in part, two exclusively), mostly herbaceous (only a few *Mentzelia* species suffrutescent) family very distinctive in the always serrate and/or deeply toothed and/or lobed leaves (sometimes deeply pinnatifid and even pinnately or palmately compound) which are either strongly urticating (usually with conspicuous stiff trichomes) or +/- asperous (or both). The leaves can be alternate or opposite, sometimes even in same species. The flowers are usually multistaminate and often large and showy, varying in color from white to yellow to red (but not purple nor blue), the two largest genera have outer staminodes fused into 5 very characteristic conspicuous petal-like glands; capsule is obconical, often sticky and exozoochorous from hooked hairs, or spirally twisted and elastically dehiscent.

The first three genera are nonurticating, the first two with small inconspicuous flowers (yellowish in *Gronovia*, white in *Klaprothia*), *Mentzelia* with larger flowers; *Loasa* and *Cajophora* are urticating with large flowers with staminodes fused into 5 distinctive petal-like glands.

Gronovia (2 spp.) — A slender vine of dry lowland forest, characterized by broadly ovate, cordate, deeply palmately lobed alternate leaves, remarkably similar to Cucurbitaceae (but lacking tendrils); whole plant sticky from uncinate trichomes. The flowers small and inconspicuous, greenish to yellowish.

Klaprothia (incl. *Sclerothrix*) (2 spp.) — Two weedy species unique in the family in their opposite, finely doubly serrate, evenly elliptic leaves. The flowers are small (<1 cm across), white, and have only 4 petals

Loasaceae

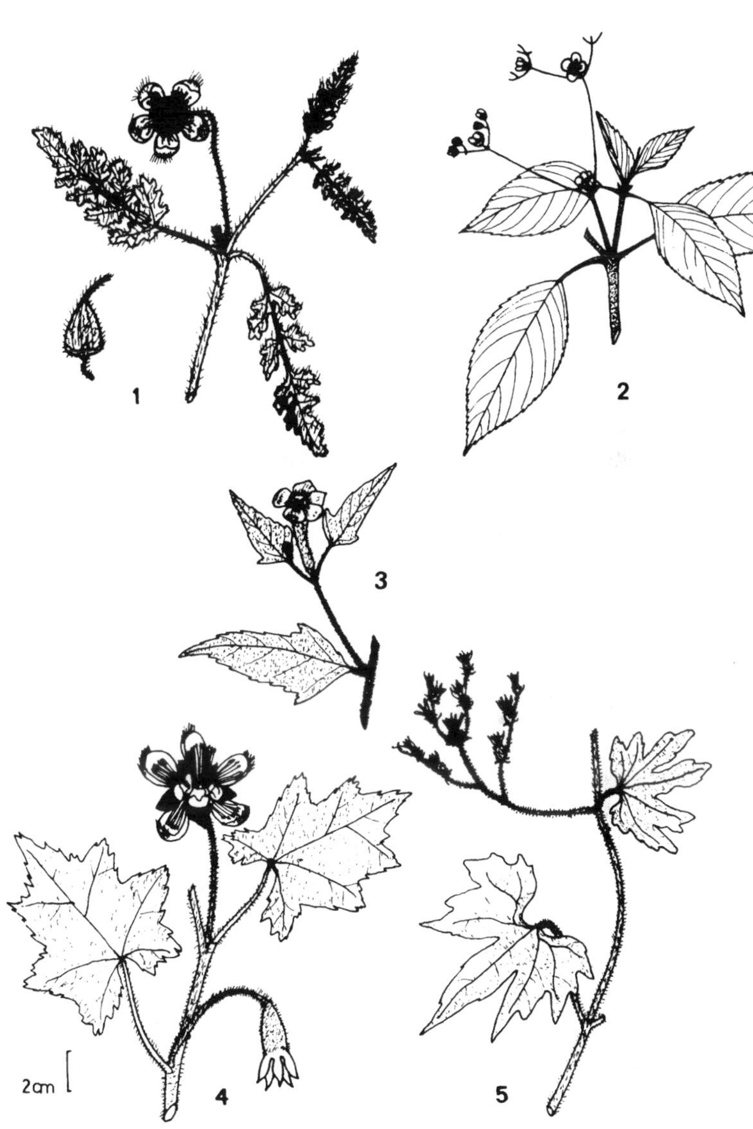

1 - *Cajophora* **2** - *Klaprothia*

3 - *Mentzelia*

4 - *Loasa* **5** - *Gronovia*

(unique). One species is a cloud-forest vine with straight obconical capsule, the other (formerly segregated as *Sclerothrix*) is a herb of low-altitude moist forests with a twisted capsule.

Mentzelia (75 spp., incl. N. Am.) — Plants of dry open or disturbed areas. The leaves vary from opposite to alternate, sometimes even in same species, but are always strongly asperous and more or less triangular with +/- developed basal lobes. Our commonest species (*M. aspera*) is an alternate-leaved vine with yellow to orangish flowers; *M. cordifolia* is usually somewhat shrubby. The capsule is straight and very sticky, longer than in *Klaprothia*. The genus differs from *Gronovia* and *Klaprothia* in larger flowers, and from *Loasa* and *Cajophora* in lacking the distinctive staminodial petal-like glands.

Loasa (100 spp.) — Urticating cloud-forest and wet-forest herbs (a few species scandent), mostly with alternate leaves, the leaves always +/- serrate or jagged toothed, asperous or with long spiny hairs, sometimes palmately or pinnately divided or evenly pinnatifidly compound. The few scandent species have alternate leaves. Flowers white to yellow, orange, or red, with 5 conspicuous petal-like glands formed from staminodia; capsule straight.

E: pringamoza

Cajophora (65 spp.) — Strongly urticating high-altitude (especially puna and paramo) vines with uniformly opposite leaves, more or less triangular in outline and deeply pinnatifidly divided. Essentially the high-altitude version of *Loasa* with very similar flowers (except never white), but easy to separate vegatively since no *Loasa* vine has opposite leaves. Capsule differs from *Loasa* in being spirally twisted.

Several other genera occur amphitropically in Mexico/southwest United States or temperate South America.

LOGANIACEAE

A very heterogeneous family, several of whose elements are often treated as distinct families. In the Neotropics consisting mostly of monotypic genera. All Loganiaceae have opposite simple leaves and 4- or 5-lobed tubular corollas; the fruits are usually 2-valved capsules, often with tiny winged seeds, but the very distinctive genera *Potalia* and *Strychnos* have round indehiscent fruits. The only two significant genera in northwestern South America are *Buddleja* (mostly middle elevations; 4-parted flowers) and *Strychnos* (lowland forest lianas with unmistakable opposite 3-veined leaves). *Buddleja* and its monotypic relatives (*Peltanthera*,

Gomara: Buddlejaceae) almost always have more or less serrate leaves; these are stellate-pubescent (often densely so and tan below) in *Buddleja. Desfontainia* has coriaceous hollylike spinose-dentate leaves; the other genera have entire leaves, ranging from the linear ericoid ones of *Polypremum* to the very large oblanceolate ones of pachycaul *Potalia.*

1. Herbs

Spigelia (45 spp., plus 4 in N. Am.) — Uppermost leaves apparently forming whorl of 4; inflorescence(s) spicate with small round 2-lobed fruits.

E: lombricera

Mitreola (incl. *Cynoctonum*) (1 neotropical species, plus 4 in Old World and 1 in USA) — Upper leaves opposite; inflorescence once or twice dichotomously branched, with 2-horned fruits.

Polypremum (1 sp.) — Beach plant with dense linear ericoid leaves and inconspicuous solitary flowers; mostly circum-Caribbean, especially in dry areas.

2. Shrubs, Trees, and Lianas

Desfontainia (2 spp.) — Andean cloud-forest shrub or small tree with coriaceous, sharply lobed-serrate leaves (cf., *Ilex opaca*) and orange flowers. (Compare *Columellia* with yellow flowers.)

Potalia (1 sp.) — Understory pachycaul treelet of lowland forest, especially on poor soils; long narrow coriaceous leaves in terminal tuft.

P: curarina

Antonia (1 sp.) — Shrub or small tree with thick elliptic, subsessile leaves; flowers look like single-flowered composite with calyx subtended by bracts and epicalyx resembling involucral bracts; inflorescence terminal and more or less umbellate. Cerrado and other dry areas; reaching at least to Bolivia.

Strychnos (70 spp., plus many in Old World) — Easily recognized by opposite conspicuously 3-veined leaves (unique in Neotropics except for very diffrent Melastomataceae and Buxaceae). Fruits globose, usually large. Many species rich in alkaloids (strychnine; some types of curare). Usually canopy lianas, occasionally spindly shrubs.

Buddleja (50 spp., plus 50 in Old World) — Mostly shrubs of middle elevations (a few weedy species in lowlands). Leaves either serrate or

Loganiaceae - A

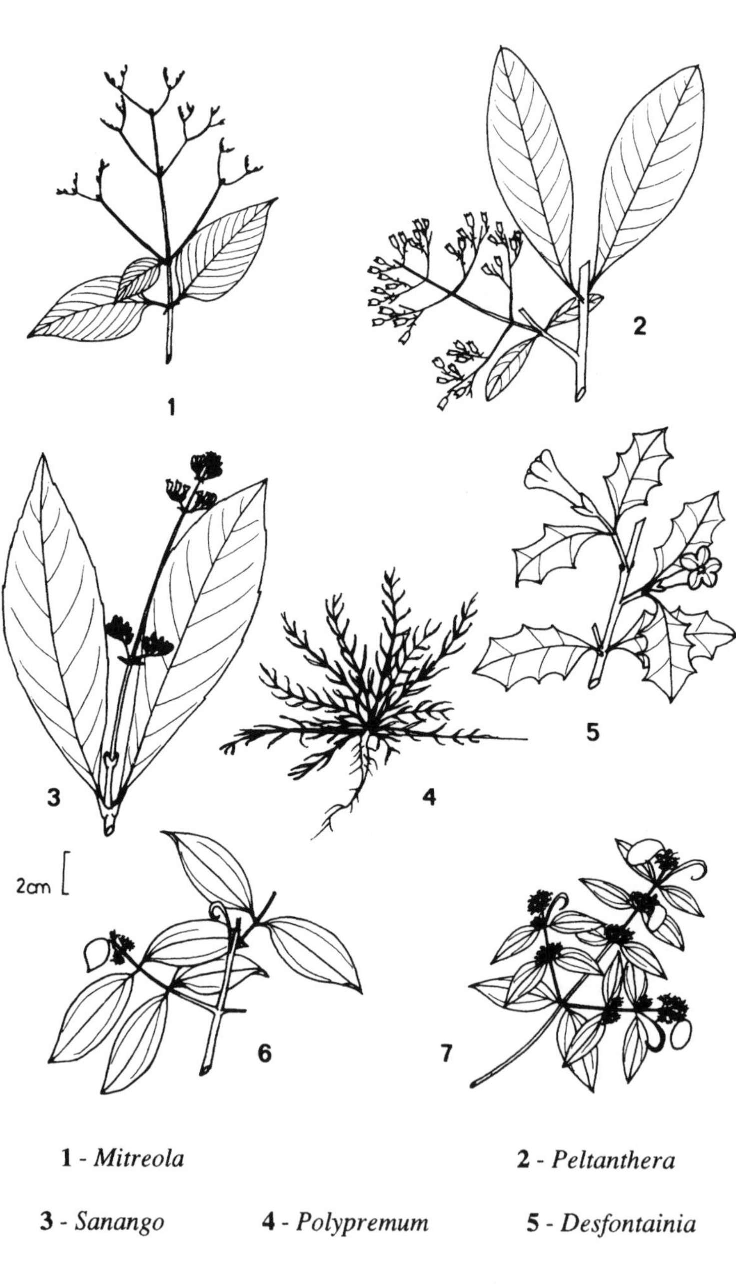

1 - *Mitreola* **2** - *Peltanthera*

3 - *Sanango* **4** - *Polypremum* **5** - *Desfontainia*

6 - *Strychnos* **7** - *Strychnos*

Loganiaceae - B

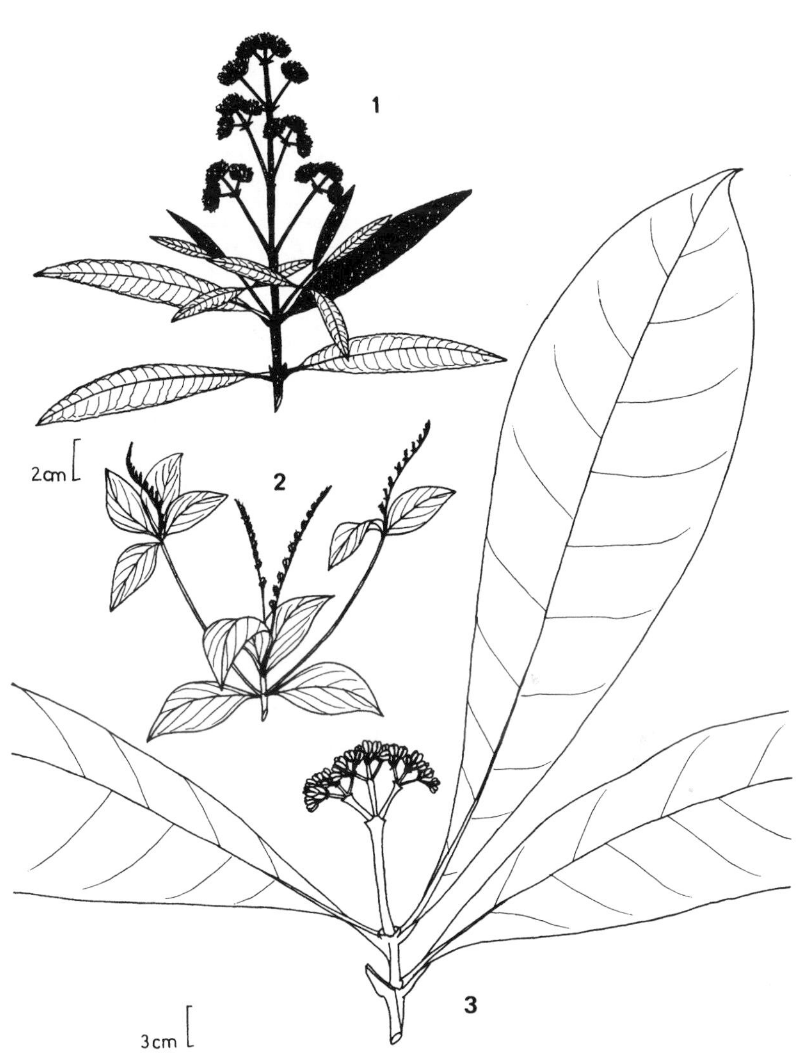

1 - *Buddleja*

2 - *Spigelia*

3 - *Potalia*

densely white or tan stellate-tomentose below (or both) usually narrow or somewhat rhombic, the petiole bases connected or joined by a conspicuous scar; inflorescences narrow, the small flowers usually in dense subsessile clusters along the spiciform branches.

E: lengua de vaca

Peltanthera (1 sp.) — Small tree with much-branched small-flowered terminal inflorescence; looks superficially like many *Psychotria* species but the leaves with serrulate margins and the small fruit a 2-valved capsule; stamens 5, exserted, (cf., thrum flowers of *Psychotria*).

***Sanango* (*Gomara*)** (1 sp.) — Similar to *Peltanthera* but with 4 included stamens (and a staminode) and a circumscissilely caducous corolla; endemic to middle-elevation Peru.

Extralimital Genera:

Gelsemium — Central American vine; other two species in southeast United States and Asia.

Mostuea — Seven African species and one on Guayana Shield.

Plocospermum — Central American; very apocynaceous, especially the fruits.

Emorya — Monotypic; Mexico and southwest United States.

Bonyunia — Shrubs of Guayana Shield with small coriaceous leaves and bivalved capsule with small winged seeds.

Loranthaceae

The most important of the very few actually parasitic neotropical plant families. Usually very easy to recognize by the thick, coriaceous, opposite (but often rather remotely subopposite) leaves that dry a characteristic gray-green (to yellowish-green or blackish) color. The leaves typically have a peculiar rather plinerved type of venation and often a unique rather falcate shape; they are extremely fragile when dried. One genus, *Gaiadendron* (+/- ditypic) is a tree (actually a root parasite) of montane areas; the other genera are supported by the host plant. One group (*Struthanthus* and allies) is typically climbing in habit (technically with adventitious shoots from the stems); the rest of the genera are generally +/- erect "epiphytic" shrubs.

Loranthaceae represent the endpoint of the Olacalean evolutionary lineage characterized by increasingly parasitic habit and an apparently associated loss of some typical angiosperm morphological features. Thus, Loranthaceae mostly lack normal roots and have indistinct ovules, a solitary seed completely devoid of a seed coat, and lack integuments. The family is subdivided into three subfamilies that are nowadays often accorded separate familial rank. The Loranthoideae (= Loranthaceae *s.s.*) have mostly bisexual flowers and a "calyculus" of fused calyxlike bracts subtending the flower; the Viscoideae (Viscaceae) and Eremolepidoideae (Eremolepidaceae) have unisexual flowers and lack a calyculus. The latter are differentiated from Viscoideae by the (usual) presence of external roots (never present in some Central American *Cladocolea*) and lacking the typical mistletoe inflorescence. The mostly Gondwanan Loranthoideae compose a series of floral types ranging from large very conspicuous red hummingbird-pollinated flowers (e.g., *Psittacanthus*) to minute ones embedded in the inflorescence rachis (*Oryctanthus*); all members of the possibly Laurasian Viscoideae have minute reduced flowers embedded in the inflorescences; Eremolepidoideae, largely temperate South American, also have inconspicuous reduced flowers.

The grouping adopted here, emphasizing macroscopically obvious features like flower size and habit, is artificial but conceptually useful.

Common names (same names mostly used for all genera) — C: pajarito, injerto; E, P: suelda con suelda, pisho isma (= bird-shit products!)

1. Terrestrial Tree — (Actually erect root parasite)

Gaiadendron (2 spp.) — Very characteristic element of upper montane neotropical forests. Unique also in bright yellow flower color (white and reminiscent of *Roupala* in an otherwise indistinguishable lower-altitude population.

2. Parasitic Shrubs or Climbers Supported by Host

2A. Large (perfect) flowers, hummingbird-pollinated, (usually) red; shrubby growth-form — Two neotropical groups of Loranthaceae-Loranthoideae have long, usually red hummingbird-pollinated flowers. *Psittacanthus* and its close relative *Aetanthus* lack endosperm; the neotropical genera formerly placed in the now defunct genus *Phrygilanthus* (*Ligaria, Tristerix,* plus white-flowered *Tripodanthus*) have endosperm. The four large-red-flowered genera in our area can also be readily differentiated by flower arrangement in the inflorescence.

Loranthaceae

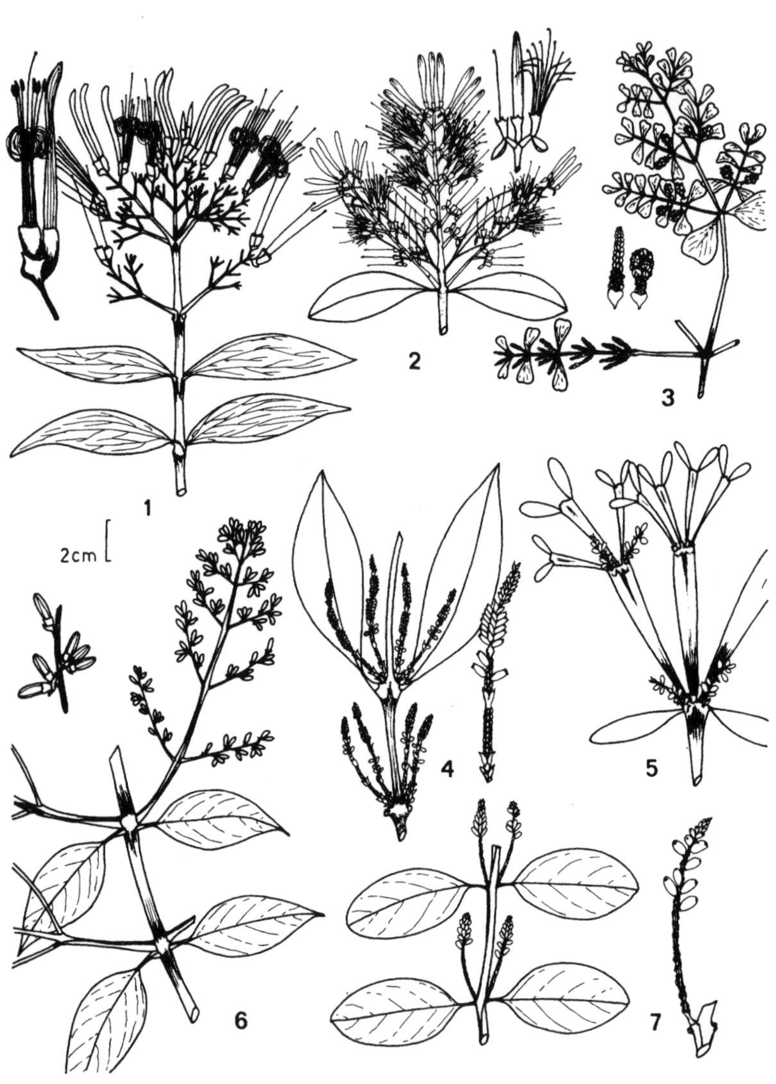

1 - *Psittacanthus* **2** - *Gaiadendron* **3** - *Dendrophthora*

4 - *Phoradendron* **5** - *Phoradendron*

6 - *Phthirusa* **7** - *Oryctanthus*

Psittacanthus (50 spp.) — By far the largest large-flowered genus. Unique (except for closely related *Aetanthus*) among the large red-flowered hummingbird-pollinated loranthacs in having the flowers in diads or triads.

E: hierba de pajarito

Aetanthus (10 spp.) — Essentially a high-altitude version of *Psittacanthus,* technically differentiated by basifixed needle-shaped (rather than versatile) anthers. This genus is restricted to high altitudes while *Psittacanthus* is exclusively lowland.

Tristerix (9–10 spp.) — The major neotropical portion of what used to be called *Phrygilanthus.* The only large-flowered genus with the inflorescence a simple raceme of flowers.

Ligaria (1 sp.) — Similar to *Tristerix* but with single axillary flowers. Mostly Chilean but reaches southern Peru at high altitudes.

2B. Medium-sized (mostly perfect) flowers (few–many mm long) usually whitish to greenish-cream; typically viny climbers rooting from the stem — The flowers always arranged in triads on the rather slender racemose or spicate inflorescence; anthers mostly versatile; (Loranthoideae, *pro parte*).

Struthanthus (75 spp.) — Always climbing in habit; leaves often distinctively pinnate and less obviously loranthlike than in most genera. At least the central flower (and typically all flowers) of each triad sessile; filaments not scalloped. A few *Phthirusa* species (see below) look more like *Struthanthus* except for technical characters.

Tripodanthus (2 spp.) — Flowers white, relatively large (1–1.5 cm) differs from nearly all *Struthanthus* in having all 3 flowers of each triad pedicellate; usually climbing but may be erect and almost treelike. Mostly southern (Argentina, southern Brazil), one species north to Venezuela.

2C. Very small inconspicuous flowers (mostly 1–2 mm long, sometimes longer in *Phthirusa*) — Shrubs (except few *Phthirusa* species); inflorescences often thickened and with the flowers more or less embedded in rachis (*Phoradendron, Dendrophthora, Oryctanthus*).

2Ca. Leaves alternate — (In our area = Eremolepidoideae, often with roots from base of plant).

Antidaphne (1 sp.) — Unique in neotropical Eremolepidoideae in being strictly dioecious; the male flowers are completely naked, consisting only of stamens; flowers are hidden inside bracteate spikes.

Eremolepis (7 spp.) — Differs from *Antidaphne* in being monoecious and the male flowers having petals; mostly south temperate.

(*Eubrachion*) (4 spp.) — Like *Eremolepis* but completely leafless (unique in subfamily but also occurs in some opposite-leaved genera). Known from Jamaica, Venezuela, and south temperate South America but likely also in our area.

2Cb. Leaves opposite (Viscoideae plus few small-flowered loranthoid genera: *Oryctanthus, Cladocolea, Ixocactus*) — Most genera with flowers immersed in rachis; in viscoid genera (*Phoradendron, Dendrophthora*) the inflorescence vertically segmented with the flowers occurring in vertical series separated by bracts.

Phoradendron (190 spp.) — The main viscoid genus; easily distinguished by the typical mistletoe inflorescence vertically interrupted by bracts into distinct segments, each with 2–several vertical series of tiny flowers (or fruits) sunken into pits; exclusively low and middle elevations.

E: suelda con suelda

Dendrophthora (65 spp.) — Essentially the high-altitude replacement of *Phoradendron;* differs technically from *Phoradendron* in 1-celled anthers; mostly occurring in subparamo-level vegetation; often with extremely small reduced leaves.

Oryctanthus (20 spp.) — Superficially like *Phoradendron* but inflorescence confluent (i.e., not broken into discrete sections separated by bracts) with each bract subtending a single flower; inflorescence usually somewhat swollen with the flowers embedded in pits.

Phthirusa (40 spp.) — Rather intermediate between *Struthanthus* and *Oryctanthus.* Differs from former in basifixed anthers, scalloped filaments, and often smaller flowers that are usually reddish. Differs from *Oryctanthus* in lacking the swollen inflorescence rachis. Most species nonclimbing, but commonest species identical in habit to *Struthanthus* and also has *Struthanthus*-like largish white flowers. Leaves almost always drying more or less blackish.

Maracanthus (2–3 spp.) — A segregate from *Oryctanthus,* mostly known from the Maracaibo basin of Venezuela; differs from *Oryctanthus* only in technical characters of pollen and leaf anatomy.

Cladocolea (40 spp.) — Mostly Central American; similar to *Oryctanthus* but differs in the spikes not swollen (vs. usually swollen in *Oryctanthus*), and thus, strongly resembling *Phthirusa* (but the flowers not in triads). Technical differences from *Oryctanthus* include a terminal flower at tip of determinant inflorescence (indeterminate in *Oryctanthus*) and the flowers subtended by bracts (but not by a pair of bracteoles as in *Oryctanthus*).

Ixocactus (1 sp.) — A distinctive monotypic Andean genus, the plants leafless and flat-stemmed (unique in New World Loranthoideae, but also found in some *Phoradendron* and *Dendrophthora*), the single sessile minute 4-parted flowers are also unique.

In addition to the genera treated here, one other viscoid genus, *Arceuthobium* (ca. 15 spp., worldwide) occurs in northern Central America, and several monotypic loranthoid (*Desmaria, Notanthera*) and eremolepidoid (*Lepidoceras*) genera are essentially Chilean.

LYTHRACEAE

Always with opposite simple entire leaves lacking evident stipules; but varying in habit from tenuous aquatic herbs to large canopy trees. Flowers distinctive in having stalked petals arising from rim of a calyx cup that encloses the superior ovary; stamens mostly twice as many as petals and arising from inside calyx cup. Fruit often either a berry subtended by calyx cup or round and capsular with winged seeds. When sterile, the tree and shrub species might be confused with Myrtaceae (although very different in the superior ovary) but the leaves (except *Lafoensia*) do not look very myrtaceous, being usually more membranaceous, having more strongly ascending secondary veins and/or lacking a marginal collecting vein; two genera (*Adenaria* and *Pehria*) have myrtaclike gland-dots on the leaves. Most lythracs share tendencies toward tetragonally angled young twigs and/or longitudinally exfoliating reddish twig and stem bark. Unlike many opposite-leaved families they usually lack interpetiolar lines or ridges (except *Lafoensia*).

The herb and small shrub genera (*Ammannia, Rotala, Cuphea, Lythrum, Heimia*) mostly have the flowers borne singly in the leaf axils (or the leaves reduced to give effect of terminal raceme in a few *Cuphea* species), while the large shrub and tree genera have multiflowered inflorescences, these axillary and more or less contracted in *Adenaria, Pehria, Crenea,* and *Lourtella,* more or less openly paniculate and from the axils of fallen leaves in *Lawsonia* and *Physocalymma,* and in a thick-rachised terminal raceme (or the lower branches sometimes forked) in *Lafoensia.*

There is also a dramatic increase in flower size from the small-flowered herbs and shrubs to large-flowered *Physocalymma* and *Lafoensia.*

1. Herbs — (Sometimes more or less scandent)

Cuphea (250 spp.) — Mostly weeds. By far the largest and commonest genus of the family, and easily recognized by the distinctly asymmetric spurred base of the tubular calyx. The mostly solitary axillary flowers usually have pink to pale magenta petals but there are a few hummingbird-pollinated species with red petals and a more elongate calyx tube. A few species are more or less scandent.

Ammannia (5 spp., plus ca. 25 in Old World and Temperate Zone) — Aquatic. Similar to *Cuphea* but the solitary sessile flowers with shorter calyx, the stem strongly tetragonal, and the linear-oblong leaves with more or less clasping bases.

Rotala (2 spp., plus 3 in Old World) — A reduced and even more strongly aquatic derivative of *Ammannia* with the smaller flowers ca. 2 mm long and very narrow, sometimes linear, leaves with nonclasping bases.

2. Shrubs

Crenea (3 spp.) — A mangrove shrub, often forming large stands. Leaves narrowly obovate or oblong, the twigs strongly tetragonal with winged angles. Flowers white, solitary or in small axillary clusters, white; the round red fruits with lower half enclosed by persistent calyx.

Lythrum (7 spp., plus 45 in Old World) — Riverbeds in dry areas. Like a shrubby *Cuphea* but the calyx without a spur and the branches tetragonal.

Lawsonia (1 introduced sp.) — A straggly dry-area shrub usually with spine-tipped branches and +/- tetragonal twigs. The paniculate inflorescence of small flowers tends to be borne on leafless branches, the small cuneate-based leaves essentially sessile.

Heimia (2 spp.) — A shrub with solitary short-pedicellate yellow flowers rather reminiscent of *Ludwigia* and growing in similar moist weedy habitats. Leaves very narrowly oblong, epetiolate, and the twigs +/- angled.

Lourtella (1 sp.) — Recently discovered in the dry uplands of northwestern Peru. Vegetatively distinctive in the small narrow strongly resinous leaves; the white flowers like a reduced version of *Pehria.*

3. Small Trees

Adenaria (1 sp.) — Shrub or small tree common in weedy areas. The pedicellate white flowers and small round fleshy red fruits in dense axillary fascicles. The narrowly oblong-acute to acuminate leaves unusual in family in being punctate, but otherwise not very myrtaclike in the thin texture and the strongly ascending secondary veins without a marginal collecting vein; twigs slightly 4-angled and usually with bark red and peeling in narrow thin strips.

Lythraceae

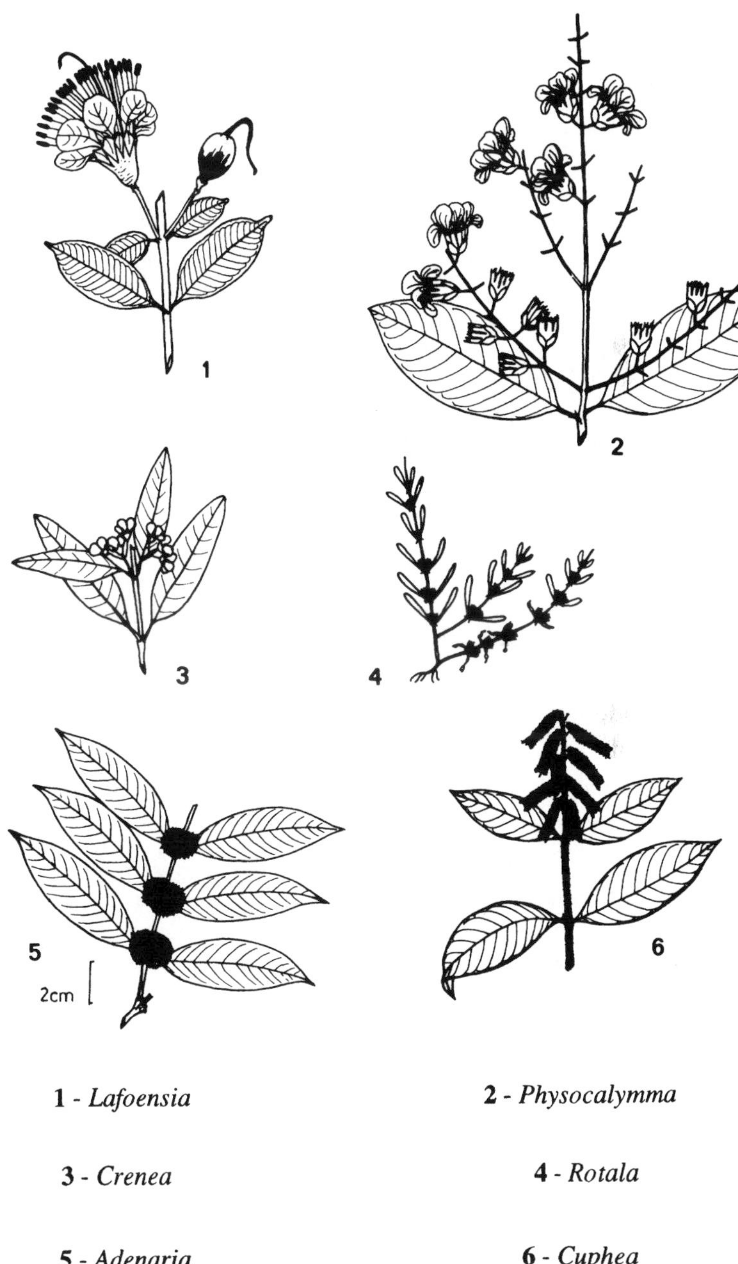

1 - *Lafoensia* **2** - *Physocalymma*

3 - *Crenea* **4** - *Rotala*

5 - *Adenaria* **6** - *Cuphea*

Pehria (1 sp.) — Dry areas of northern Colombia; also Central America. Vegetatively very similar to *Adenaria* and also with punctations, but the axillary inflorescences with longer peduncles and the flowers red. Twigs noticeably puberulous and with flattened angles.

4. Large Trees

Physocalymma (1 sp.) — Canopy tree of dry forests, also persisting in savannahs. Very distinctive in the strongly scabrous leaves (unique among area trees with entire opposite leaves). Flowers magenta and very showy, borne while leaves deciduous.

Lafoensia (12 spp.) — Canopy trees, mostly of moist and montane forests; also in cerrado. The distinctive leaves shiny and glabrous, rather oblong and somewhat smallish, with close-together secondary veins, each pair separated by a parallel intersecondary, the main veins prominulous above and below; interpetiolar lines present unlike other genera of family. The large bat-pollinated flowers are white or tannish and reddish and borne terminally; the roundish capsule dehisces to release flat seeds with a thicker central seed body surrounded by brownish wing.

C: guayacán

Magnoliaceae

Trees with primitive odor, often large and emergent, mostly in middle-elevation cloud forests. Unmistakable in the conspicuous scars surrounding twig at each node, these left from the large caducous Moraceae-like (but spathelike and usually blunter) terminal stipule. Leaves typically rather large and long petiolate, mostly with a conspicuous intricately prominulous network of fine venation; usually (*Talauma*) with the upper side of petiole very conspicuously grooved (unlike any other Ranalean taxon). Flowers usually solitary at branch apices, the pedicel with series of circular scars from fallen bracteoles, always white, usually rather large, with numerous stamens and several to many separate carpels. Fruit woody with a more or less conelike core surrounded by the more or less fused follicles, these dehiscing circumscissilely so that the thick woody outer part of fruit peels off irregularly to expose the red-arillate seeds.

Talauma (66 spp., mostly SE Asian; 16 spp. in Colombia) — Mostly northern Andean cloud forests, also a few species in lowland wet forests. Vegetatively unmistakable in the combination of a primitive odor with a conspicuously grooved petiole.

C: molinillo

Magnoliaceae, Malesherbiaceae, and Malpighiaceae (with 2–3 Cocci)

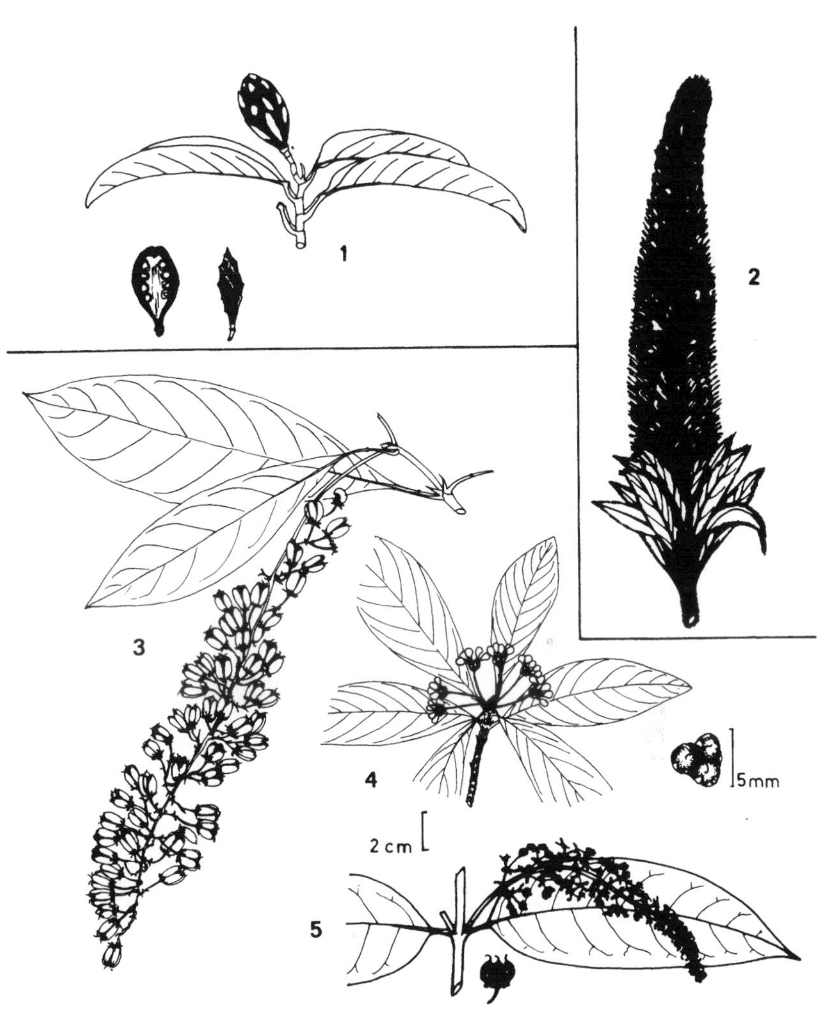

1 - *Talauma* (Magnoliaceae)

2 - *Malesherbia* (Malesherbiaceae)

3 - *Lophanthera*

4 - *Pterandra*

5 - *Spachea*

Dugandiodendron (10 spp.) — Mostly Colombian cloud forests, only a few species reaching Ecuador and Venezuela. Vegetatively distinct from *Talauma* in petioles lacking the conspicuous groove on upper surface.

MALESHERBIACEAE

A small mostly Chilean family consisting of a single genus of erect viscous herbs, some species reaching the dry Andean valleys of Peru. Our species are all easily recognized by the rather dense, narrowly oblong leaves with conspicuous irregular crenate-serrate margins and by narrow large (>3 cm long) tubular, subsessile flowers, with pubescent tubular calyces, and exserted anthers.

Malesherbia (27 spp.)

MALPIGHIACEAE

A family of trees and lianas vegetatively characterized by opposite leaves and the presence of T-shaped "malpighiaceous" trichomes (usually present at least on petioles and young growth of glabrescent taxa). The family is divided about 50/50 between trees and lianas. Another good family character in most of the tree genera is the presence of intrapetiolar stipules (otherwise only found in a very few Rubiaceae). Essentially these represent fusion of the two stipules of a single leaf across that leaf's axil. These are not evident in the shrubby dry-forest genus *Malpighia* (easy to confuse with Myrtaceae but the leaves thinner, epunctate, and with fewer, more separated, less distinct, brochidodromous secondary veins than in comparable myrtacs and also distinctive in a characteristic light olive color and reduced petioles.) In *Bunchosia* intrapetiolar stipules are present but not very obvious (but this genus easily recognizable by the large ocellate glands on underside of lamina near base). Liana genera mostly lack intrapetiolar stipules but *Hiraea* has stipules fused to the petiole and forming a pair of petiolar enations and most genera have inconspicuous interpetiolar stipules (usually caducous or hidden under hairs, sometimes completely lacking in *Heteropterys*). Most malpigh lianas have glands associated with the petiole or leaf base. These may be large and conspicuous and paired at petiole apex (*Stigmaphyllon*), halfway down the petiole (some *Heteropterys*) or at the base of midrib (*Diplopterys,* many *Banisteriopsis*), base of lamina (*Jubelina,* some *Banisteriopsis*), or more scattered over lower surface of lamina (usually ca. 2 pairs near base or in interrupted submarginal row) (*Ectopopterys, Tetrapterys*).

The malpigh flower, designed to attract oil-collecting bees, has one of the most distinctive and constant ground plans of any angiosperm family. Four (or all five) of the calyx lobes each has a pair of large oil-secreting glands abaxially (except *Ectopopterys*), the petals (nearly always yellow or pink) are stalked (sometimes the fifth reduced), the stamens 10 (though some may be infertile) and the ovary 3-locular. The fruits are also distinctive, if for no other reason than that the remnants of the sepals with their unique glands are usually persistent. Tree malpighs have more or less round fruits (often (2–)3-lobed, sometimes fleshy and drupelike, sometimes dry and then typically splitting into 3 cocci) while lianas have winged fruits which typically split into three samaras. Generic level taxonomy of the lianas is largely based on different development and orientation of the samara wings. *Banisteriopsis, Heteropterys,* and *Stigmaphyllon* have a dorsal wing while *Mezia, Tetrapterys, Hiraea,* and *Mascagnia* have lateral wings and genera like *Jubelina* (and a few *Mascagnia*) have the wing expanded in both directions. *Dicella* has a samara-like fruit with wings formed from the expanded sepals instead.

1. Trees and Shrubs with Unwinged Fruits

1A. Fruit splitting into 2–3 cocci

Spachea (6 spp.) — Small to medium trees, in Amazonia mostly in seasonally inundated forest; in well-drained moist forest in northern Colombia and southern Central America. Intrapetiolar stipules present; leaf base below and/or petiole apex with pair of glands (cf., the liana genera). Most species have rather oblong glabrous leaves with well developed intersecondaries (and the tertiary venation also tending to parallel the secondary). Inflorescence an elongate rachis with pedunculate clusters of short-styled pink or white flowers (the technical floral character for separation from *Lophanthera* is slender styles with minute stigmas). The three cocci of the mature fruit have an evenly convex, nonkeeled outer face.

Pterandra (7 spp.) — Small or medium-sized trees, along fast-flowing streams in rocky canyons, in our area known from only a few collections, all from Colombia. Vegetatively very distinctive in pagoda-style branching (cf., *Buchenavia*) with the small leaves clustered at ends of the branches. Intrapetiolar stipules present; leaves eglandular (unlike *Spachea* and *Lophanthera*) and tending to be more or less sericeous below. Flowers whitish-green, in pedicellate clusters at branch apices, borne before leaves fully flushed. Fruits of 3 subglobose cocci.

Lophanthera (5 spp.) — Mostly in black-water-inundated areas of the Rio Negro/Orinoco region. Very close to *Spachea* but the leaves more oblanceolate and long cuneate to base with the gland pair (in our taxa) below middle of petiole. Intrapetiolar stipules are present but narrower than typical in *Spachea.* Also differs from *Spachea* in keeled outer faces of the cocci and (in our area) in yellow flower color.

1B. Fruit indehiscent, often fleshy and drupelike

Byrsonima (150 spp.) — Trees mostly of savannah and cerrado formations especially on the Brazilian Shield, but also becoming rain forest canopy components, especially in areas with poor soils. Vegetatively characterized by always lacking leaf glands and having well-developed intrapetiolar stipules, and in many species by a conspicuous indument of malpighiaceous trichomes The flowers can be either yellow or pink and have a hairy receptacle and 3 distinct subulate styles. The fleshy fruit consists of a single 3-seeded pyrene.

C: guayabillo, mazamoro; P: indano

Burdachia (4 spp.) — Trees of seasonally inundated black-water tahuampa. A water-dispersed relative of *Byrsonima* and sharing the conspicuous intrapetiolar stipules of that genus. Nearly completely glabrous unlike most *Byrsonima* species and also differing in having leaf glands. Vegetatively distinctive in strikingly flattened young branches. The fruit, differing from *Byrsonima* in being nonfleshy, is more or less ovoid (distinctively angled-conical with 3 apical styles in most species) and the flowers are pink.

Bunchosia (60 spp.) — Mostly small understory trees. Intrapetiolar stipules present but reduced and inconspicuous. Lower surface of leaf lamina usually with pair of large ocellate glands, these sometimes near base of midvein and sometimes well out onto lamina; leaves usually drying a characteristic yellowish-olive or the lower surface sericeous and whitish from the dense malpighiaceous hairs. Flowers always yellow and arranged in a more or less elongate racemose inflorescence. Differs from *Byrsonima* in the broader spathulate styles (often more or less fused), the glabrous receptacle, and a drupe composed from 2 or 3 fused but separate pyrenes and capped by the short fused style.

E: cerezo; P: indano

Malpighia (30 spp.) — Small trees or shrubs of dry deciduous forest. In our species the stipules reduced and not evident and ocellar glands usually absent; the leaves nevertheless characteristic in their thin texture, small size, more or less rhombic-elliptic form and/or emarginate apex, few brochidodromous secondary veins, reduced petiole, and characteristic yellowish-olive color. Might be confused with Myrtaceae but too thin in texture and lacking punctations. Flowers usually pink, never yellow, in few-flowered axillary umbel or corymb. As in *Bunchosia* the drupe is composed of (1–)2–3 fused pyrenes topped by spathulate styles (though these unfused).

Malpighiaceae
(Trees and Shrubs with Indehiscent Fruits)

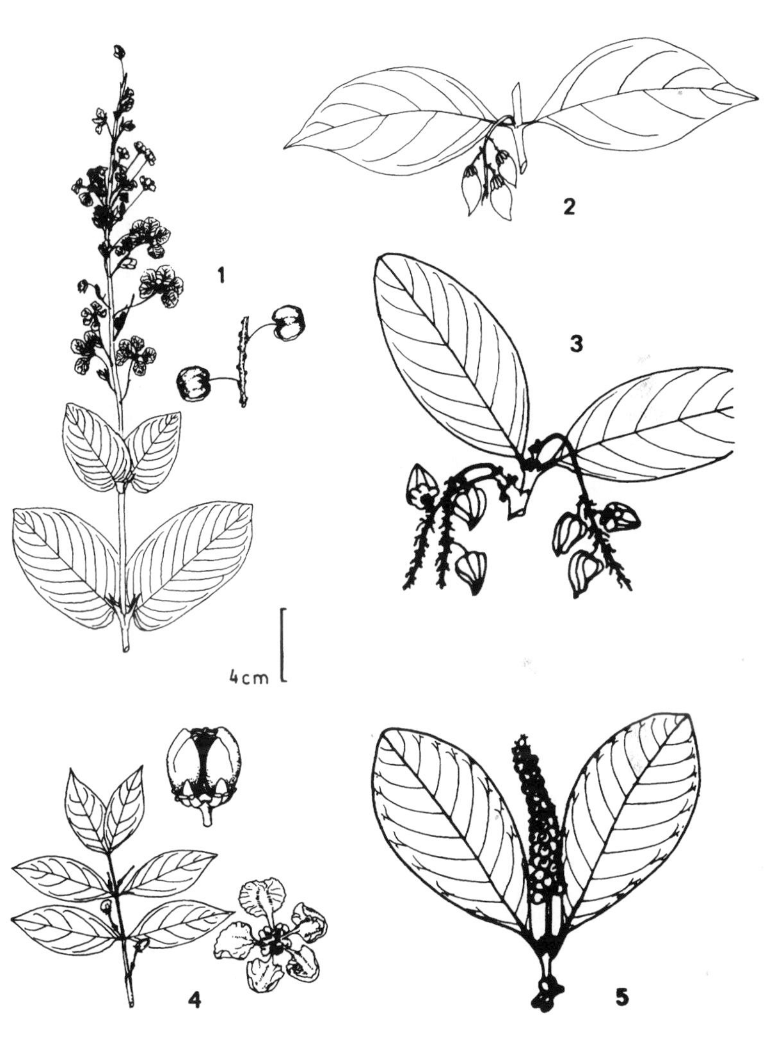

1 - *Diacidia*

2 - *Bunchosia*

3 - *Burdachia*

4 - *Malpighia*

5 - *Byrsonima*

Diacidia (12 spp.) — Small yellow-flowered shrubs of Orinoco area savannas and rock outcrops, especially on "laja". Intrapetiolar stipules present; either the leaves small and sessile or the stipules very large and several centimeters long. Similar to *Byrsonima* but the fruit small and dry and more or less enclosed by the expanded sepals.

2. Lianas with Wind- (or Water-)Dispersed Fruits

2A. Samara wings from expanded calyx

Dicella (5 spp.) — Lowland forest liana vegetatively characterized by rather intricately reticulate fine leaf venation both above and below; very small intrapetiolar stipules usually visible; glands absent or (in our area) sparsely and widely scattered on lamina undersurface. In fruit unmistakable, with the sepals much expanded to form 5 oblanceolate pinnately veined wings, exactly as in *Astronium.*

2B. Only dorsal fruit wing developed

Stigmaphyllon (ca. 100 spp.) — One of the easiest liana genera to recognize vegetatively. *Stigmaphyllon* always has a pair of prominent glands at or near the petiole apex and usually a cordate leaf base, the latter unique among area malpigh lianas (except a few species of very different *Hiraea* and a subcordate *Mascagnia,* both lacking petiole glands). A few species have deeply lobed leaves, also unique. Stems of mature lianas are strikingly broken into separate strands. The relatively few noncordate leaved species have unusually large petiolar glands (larger than in any other malpigh genus). In flower, most *Stigmaphyllon* species are recognizable by having only 4–6 fertile stamens (i.e., with thicker anthers than the others) and by styles with enlarged apex and stigma on the inner angle. In fruit, the samara wing is thickened on the upper margin (like *Banisteriopsis*, but unlike *Heteropsis*).

Banisteriopsis (92 spp.) — Leaf glands usually present either in pair on either side of midrib base below or on lamina base itself (sometimes associated with auricular development of lamina base); in either case well above site of equivalent glands in *Stigmaphyllon.* Another useful feature is sericeous lower surface pubescence, which occurs in many genera but especially frequently in *Banisteriopsis.* The yellow or pink flower is characterized by all 10 stamens fertile and a straight, narrow style with apical stigma. Samara wing thickened on upper margin.

E: napi, nepe; P: ayahuasca

Heteropterys (ca. 120 spp.) — Most species have yellow flowers but a few are pink, the flower very like *Banisteriopsis* but the styles are more or less dilated or hooked with the stigma on the inner angle. The main difference from *Banisteriopsis* is that the samara wing is thickened on *lower* margin. The petiole either lacks glands or has a gland pair near or below its middle.

Malpighiaceae
(Lianas: Fruits with Dorsal Wings)

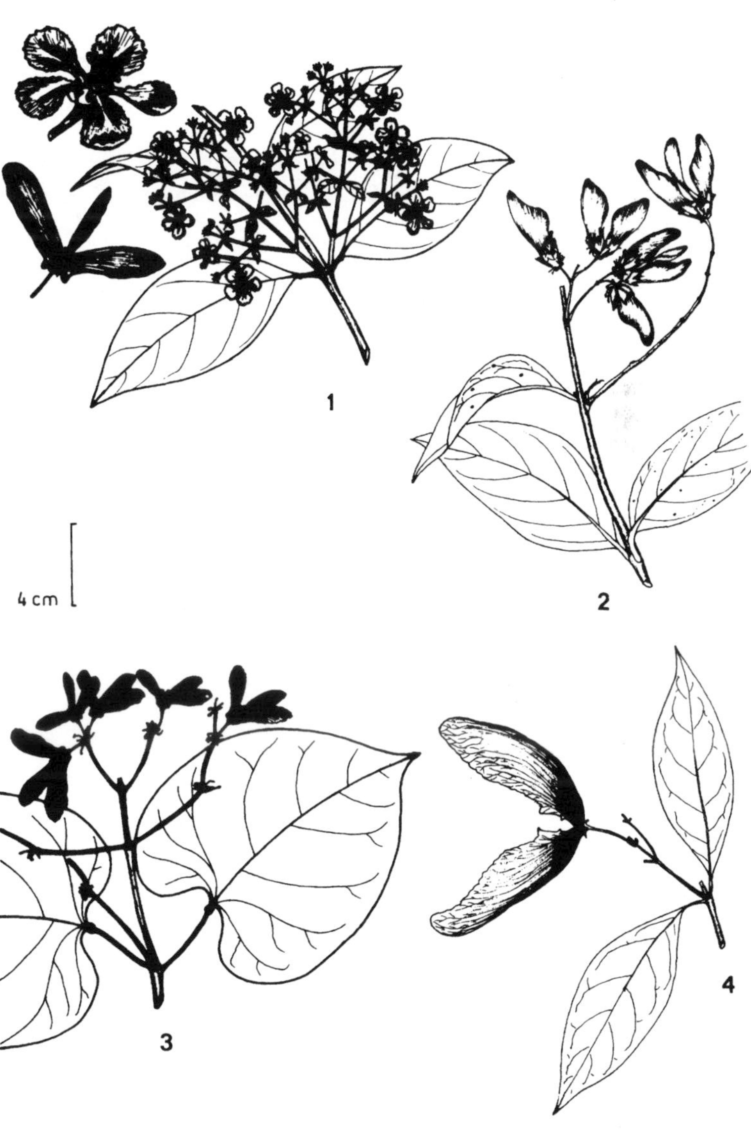

1 - *Banisteriopsis* **2** - *Heteropterys*

3 - *Stigmaphyllon* **4** - *Ectopopterys*

Ectopopterys (1 sp.) — A recently discovered genus, vegetatively characterized by the elliptic leaves distinctly lustrous with the venation somewhat prominulous above and with an irregular series of glands ca. 5 mm in from margin below; glands at petiole apex small or lacking. Flowers remarkable in eglandular sepals and also unusual in some anthers reduced in size (cf., *Stigmaphyllon* but all fertile). Samara superficially like *Heteropterys* but apparently the main wing is actually a dorsally shifted lateral wing (hence the generic name).

2C. Both lateral and dorsal wings developed — (Or water-dispersed with wings forming interconnected series of flanges and orientation not obvious)

Jubelina (6 spp.) — Canopy lianas of lowland forest. Without any definitive vegetative characteristic but has unusually large leaves with gland pair at base of midvein and tends to have twig center pithy (or hollow, at least when dry). The flowers are yellow, the inflorescence conspicuously bracteate and resembling *Tetrapterys.* There are two fruit forms, one rather like *Tetrapterys* with a relatively small dorsal wing and a pair of broader lateral wings (>2 cm wide; another species has shorter wings more like *Hiraea*), the other (presumably water-dispersed) very much resembling *Diplopterys.*

Diplopterys (4 spp.) — Lianas of lowland Amazonian riverside forest (also a Mexican species). Base of midvein with pair of glands. Flowers as in *Banisteriopsis,* of which it is probably a water-dispersed derivative, but differing in the relatively large round fruit which lacks a true dorsal wing and has a series of irregularly interconnected narrow lateral wings.

Mezia (6 spp.) — Canopy vine of noninundated lowland forest, characterized mostly by the pair of rather large ovate bracteoles that cover the buds. Resembling *Hiraea* in fruit, with a large suborbicular principal lateral wing, a suborbicular dorsal wing, and an additional incomplete pair of intermediate lateral wings.

Clonodia (2 spp.) — Lianas of seasonally inundated forests in Orinoco area. Leaves rather small and obtuse and flowers pinkish. The distinctive fruit cocci rather small (<2 cm across), each with ca. 8 small lateral wings and a winglike dorsal crest.

2D. Only lateral wings strongly developed — (Although more or less developed dorsal crest often present)

Tetrapterys (90 spp.) — Usually canopy liana, (but a few species, including the Chocó mangrove shrub, *T. subaptera,* are consistently erect). Vegetatively, often recognizable by the absence of petiole glands, instead

Malpighiaceae
(Lianas: Fruit Wings from Calyx or Both Lateral and Dorsal)

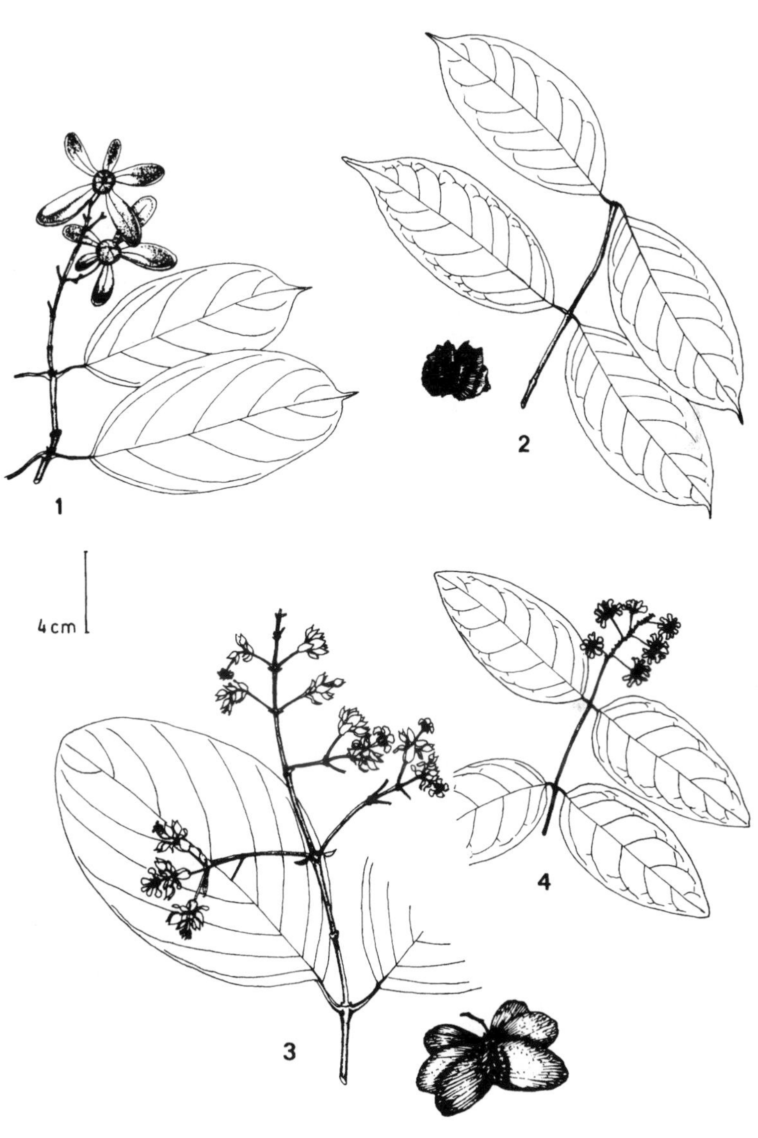

1 - *Dicella* **2** - *Diplopterys*

3 - *Jubelina* **4** - *Clonodia*

Malpighiaceae
(Lianas: Fruits with Lateral Wings)

1 - *Mascagnia*

2 - *Tetrapterys*

3 - *Callaeum*

4 - *Hiraea*

5 - *Tetrapterys*

with usually more than one pair of glands near base of lamina or several glands scattered submarginally. Flowers yellow, in commonest section borne in conspicuously bracteate inflorescences. Samara very distinctive, the wings much longer and narrower than in *Hiraea* or *Mascagnia,* usually with pair of much smaller wings at base in addition to the upper pair of long wings.

Hiraea (45 spp.) — Vegetatively the most easily identified malpigh genus on account of the stipules more or less fused to petiole and forming pair of dorsal enations thereon. Inflorescence a simple axillary or ramiflorous umbel, the pedicels simple and not apparently bracteate (unlike *Mascagnia*); flowers always yellow. Samara with 2 lateral wings and a short dorsal crest, the wings distinctly broader vertically than laterally.

Mascagnia (55 spp.) — Canopy and second-growth lowland lianas with either pink or yellow flowers. Petiole of ours usually eglandular and with pair of minute caducous lateral stipules at its base. Inflorescence larger than in *Hiraea* and more or less racemose or racemose-paniculate, the individual flowers usually on a compound "pedicel", with bracts separating basal portion (which is really a peduncle). Samara very like *Hiraea* but the wings tending to be thinner and more membranaceous.

E: nojarilla

Callaeum (10 spp.) — Our two species vegetatively characterized by rather thin leaves with pronounced glands at base of lamina (distinctly away from petiole apex). The young twigs are conspicuously smooth (but this also occurs in a few *Mascagnia* species). Fruit like *Mascagnia* except for thicker more corky wing bases (in water-dispersed *C. antifebrile* the wings essentially lost) and most species were formerly included in that genus. However, a series of technical floral characters (especially transversely dilated stigmas) indicate that the fruit similarity may represent convergence.

There are a rather large number of additional malpigh genera elsewhere in the Neotropics, especially in the dry areas of central and southern Brazil.

MALVACEAE

Mostly herbs (nearly all Malvalean herbs are Malvaceae), but frequently suffrutescent, a few genera becoming soft-wooded trees. Usually more or less mucilaginous and usually with stellate trichomes (*Thespesia* is lepidote), the typical Malvalean leaves are palmately 3–(7-)veined at base, frequently more or less lobed, usually serrate. Most taxa differ from Sterculiaceae and Tiliaceae in the combination of (almost always) more broadly ovate leaf outline and (usually) serrate or lobed margins (entire in *Thespesia, Hampea, Tetrasida,* some *Wissadula,* two *Sida*). The Malvalean pulvinus is poorly developed (especially in herbs) and usually not apparent (even in intermediate *Hampea*). Malvaceae are closest to Bombacaceae, mostly differing in being shrubs or herbs while all Bombacaceae are trees; however differentiation of arborescent Malvaceae from Bombacaceae can be difficult and one +/- intermediate genus (*Hampea*) has been placed in both families. The most definitive difference is that Malvaceae have spiny pollen while Bombacaceae do not.

In flower most Malvaceae are easily recognizable by the numerous stamens with filaments fused into staminal column surrounding style and/or an epicalyx below the real calyx, the latter often persistent in fruit and never present in related families. The Malvaceae fruit is usually dry, variously separating into cocci (cf., some Sterculiaceae but no Bombacaceae) but capsular in *Hibiscus* and relatives and berrylike in *Malvaviscus.*

The first five genera below include all the real trees (all with leaves either densely pubescent below or serrate); *Malvaviscus* is our only genus with +/- scandent habit (also distinctive in red petals rolled into a tube and berrylike fruit). *Gossypium* and its relatives have capsular fruits and often large flowers, while the taxa related to *Pavonia* (flowers often small, fruits often exozoochorous), to *Abutilon* (fruit with ridged (typically appearing more or less inflated) segments), to *Malvastrum* and *Sphaeralcea* (epicalyx present), and to *Sida* (epicalyx absent) have fruits splitting into variously elaborated cocci or mericarps. Most genera are restricted to dry areas; *Nototriche* and *Acaulimalva* are high-Andean. Excluding the arborescent taxa, only a few genera (e.g., *Abelmoschus, Pavonia, Malachra, Urena,* and *Sida* occur (mostly as weeds) in moist lowland areas.

1. Trees or Large Shrubs — (Fruit diverse but not splitting into cocci except in *Tetrasida* which is unique in 4 calyx lobes)

Hibiscus (75 spp., plus 200 Old World) — Mostly shrubs but a few trees and suffrutescent herbs; leaves always broad and serrate, usually somewhat lobed, distinctive in large showy (variously colored) flowers, evenly 5-dentate calyx, and conspicuous often multisegmented epicalyx. Fruit a largish 5-parted capsule. A convenient base group against which to compare the other woody taxa.

Wercklea (3 spp.) — Small cloud-forest trees differing from *Hibiscus* in spiny trunk and stiff-pubescent (usually more or less spiny) petioles and inflorescence.

Malvaceae
(Trees and Large Shrubs)

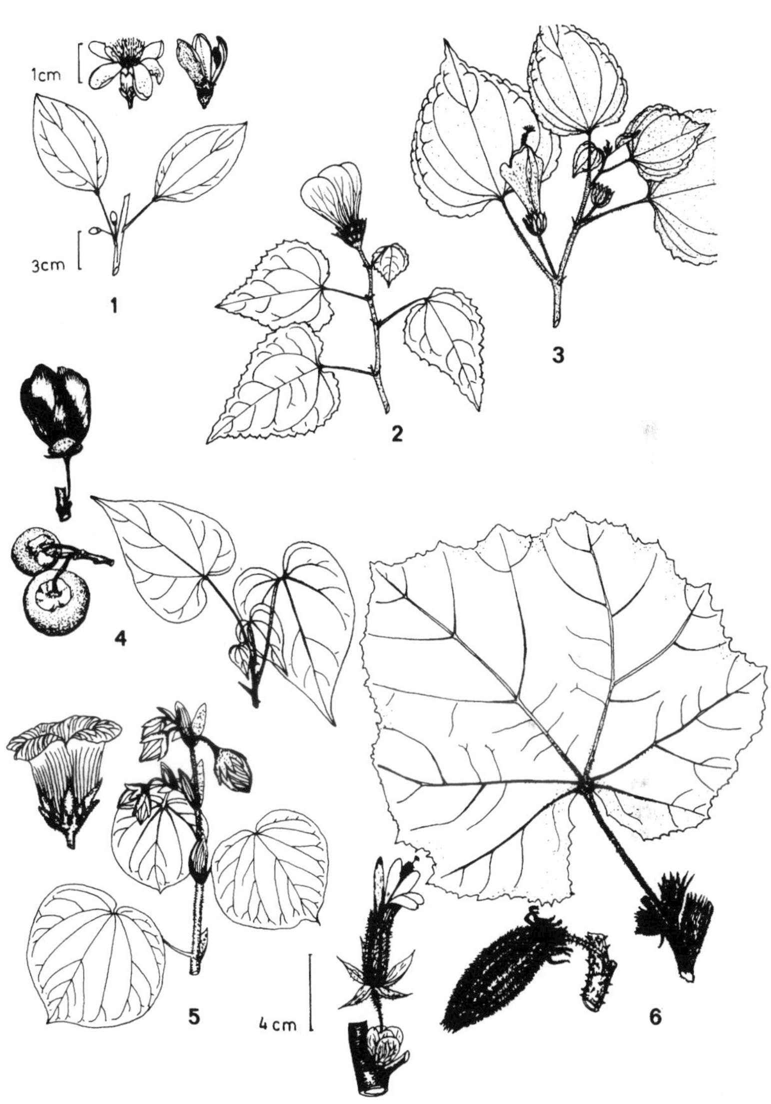

1 - *Hampea* **2** - *Hibiscus* **3** - *Malvaviscus*

4 - *Thespesia*

5 - *Hibiscus* (*H. tiliaceus*) **6** - *Wercklea*

Thespesia (1 sp., plus 13 Old World) — Sea-beach tree with very characteristic evenly cordate-ovate, entire, lepidote leaves (unique) and large light yellow flowers with purple centers. Calyx distinctive, truncate and very broadly campanulate. Fruit not dehiscing, more or less depressed-globose, 2–3 cm across, dispersed by floating. (*Hibiscus tiliaceus* of same habitat has finely serrate leaves more broadly ovate and densely pubescent below.)

C: clemón

Hampea (10 spp.) — Mostly Central American, only reaching Colombia. At least superficially, appears intermediate between Bombacaceae and Malvaceae. Characterized by entire leaves densely tannish-puberulous below (and lacking pulvinus as in Malvaceae) and the Bombacaceae-like ramiflorous flowers in fascicles or solitary and with a very shallowly lobed calyx, in fruit giving rise to a subwoody 3-valved capsule.

Tetrasida (1 sp.) — A small (ca. 5 m) monotypic tree of the Huancabamba area, characterized by entire leaves densely white-tomentose below and a paniculate inflorescence of small *Sida*-like flowers; unique in family in frequently having only 4 sepals.

Malvaviscus (3 spp.) — Mostly large shrubs of lowland or cloud-forest but tending to be scandent, especially in Chocó. Distinctive in the +/- infundibuliform corolla, the red petals auriculate and essentially rolled into tube for hummingbird-pollination. Fruit also unique, fleshy and berry-like.

2. The Next Five Genera Are Coarse Herbs or Subshrubs with Capsular Fruits.

Gossypium (16 spp., plus 23 in Old World) — Dry-area shrub. Unique in epicalyx of large laciniate-margined foliose bracts, the flowers large and cream to yellow. Leaves broadly ovate, deeply lobed (with entire margins) to entire.

Abelmoschus (introduced, 6 spp. in Old World) — Weedy herbs, differing from *Hibiscus* in herbaceous habit and the 2–3-lobed asymmetrically splitting subspathaceous calyx.

Kosteletzkya (1 sp., plus 8 in Africa) — Weedy herb of lowland dry-areas with stiff trichomes (cf., *Malachra*) but flowers white to pinkish with small narrow epicalyx bracts. Leaves serrate and usually divided. Fruit distinctive, small, depressed-globose and strongly 5-angled.

Cienfuegosia (15 spp.) — Herbs or subshrubs of dry areas with entire usually unlobed (sometimes with 3 entire lobes) leaves. Essentially

Malvaceae
(Herbs and Shrubs: Capsular Fruits or Dense Bracteate Inflorescence)

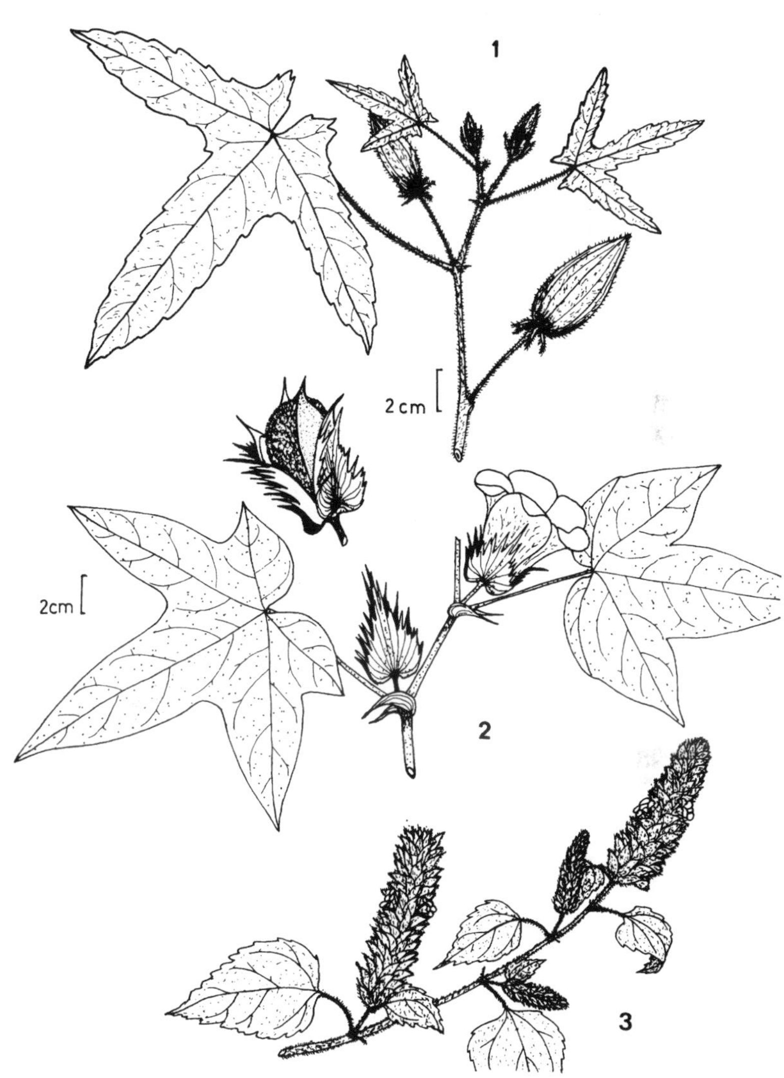

1 - *Abelmoschus*

2 - *Gossypium*

3 - *Malvastrum*

Hibiscus with a gland-dotted calyx, also differing from *Hibiscus* in the smaller, narrower epicalyx bracts and less lobed calyx.

Bastardia (2 spp.) — Viscous dry-area shrubs with unlobed, evenly ovate, serrate leaves, and small yellow flowers. Except for the glandular pubescence and capsular fruit reminiscent of some *Sida* species.

3. The Rest of the Genera Are Herbs or Subshrubs with Fruit Separating into Cocci.

3A. The next three genera (plus *Malvaviscus*) (= tribe Malvavisceae) have 10 stigmas and 5 carpels.

Pavonia (130 spp., plus 50 Old World) — Many species are suffrutescent understory herbs; others are shrubs, mostly in disturbed areas. Vegetatively variable but nearly always with serrate leaves, in most species unlobed or only weakly 3-lobed. Flowers white, pink, red, or yellow, often (especially in herbs) in cluster at tip of peduncles. Fruit of cocci, these frequently with terminal barbs for exozoochorous dispersal (section *Typhalea*).

Malachra (6 spp.) — Weedy lowland herbs, usually with long stiff, sometimes mildly urticating trichomes, the leaves broad and with serrate margins, often deeply lobed. Flowers small, usually yellow (sometimes white or lavender), axillary, enclosed by large foliaceous bracts.

Urena (6 spp.) — Weedy lowland herbs, vegetatively unmistakable in the conspicuous longitudinal gland with a central slit near base of midvein. Leaves shallowly to deeply 3-lobed (even in same species) and densely pale-tomentose below. Flowers small, pink, axillary, short-pedicellate, subtended by fused bractlets. Fruit unique, depressed globose with short spines, differs from *Byttneria* in each of these terminating in whorl of retrorse barbs.

E: amonan

3B. The rest of the Malvaceae are mostly small herbs (*Briquetia, Abutilon, Gaya, Wissadula* are often shrubby) with variable stigma number (frequently 5), the genera mostly differentiated by technical characters of fruit and ovule.

3Ba. The next six genera usually have an epicalyx.

Malva (4 spp. introduced) — Often prostrate herbs with broadly ovate very shallowly lobed or unlobed serrate leaves. The small white or lavender flowers are subtended by an epicalyx and borne in axillary fascicles, the calyx lobes +/- inflated in fruit. Unique technical character is the very thin filiform stigmas (swollen and +/- capitate in other genera).

Malvaceae
(Herbs: Cocci [A – P])

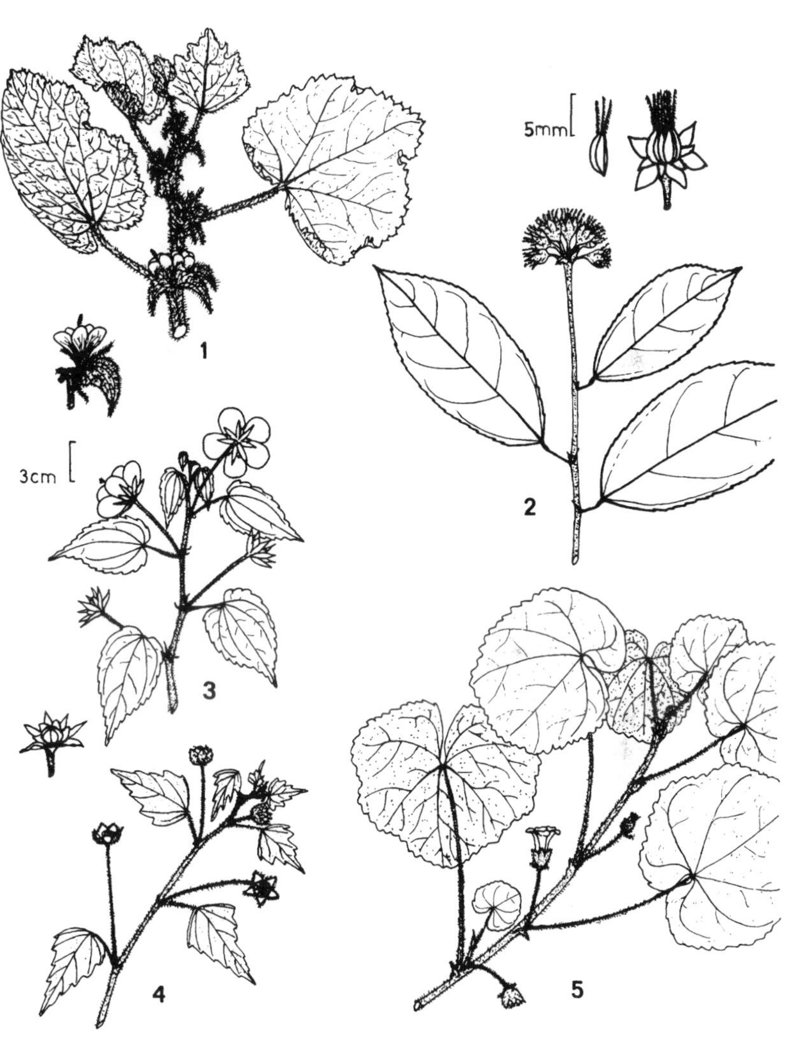

1 - *Malachra*

2 - *Pavonia*

3 - *Pavonia*

4 - *Anoda*

5 - *Malva*

Malvaceae
(Herbs: Cocci [S – W])

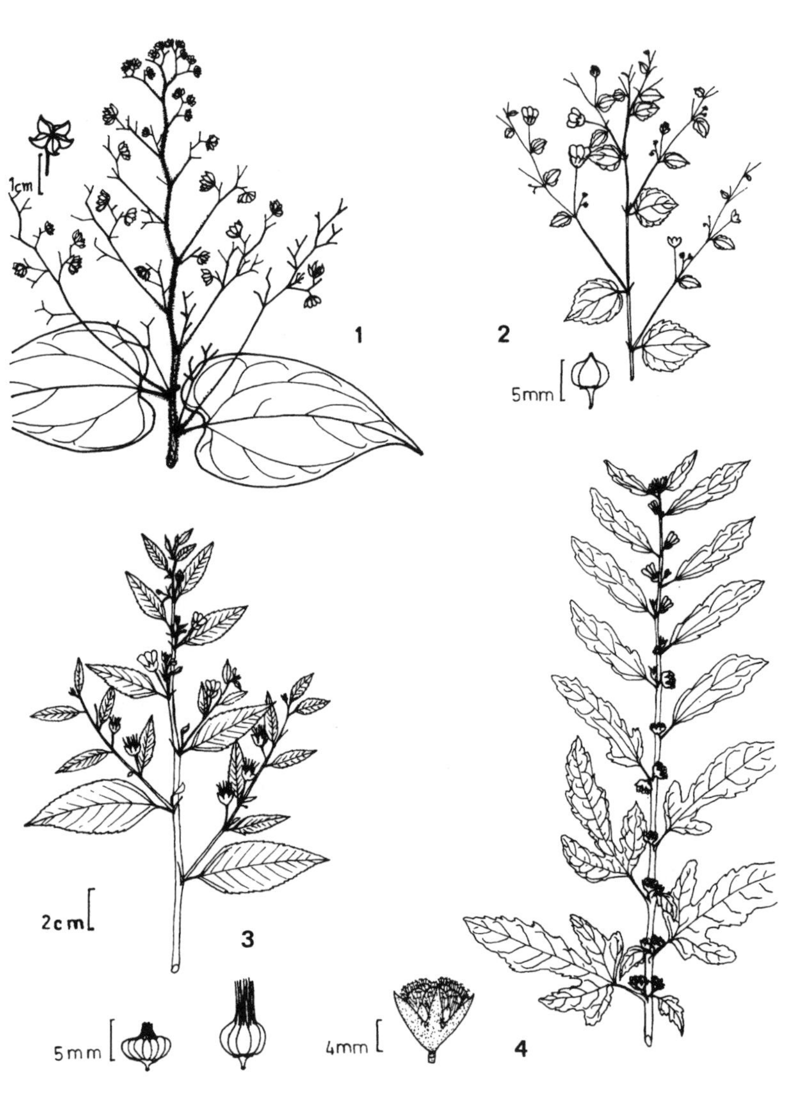

1 - *Wissadula* **2** - *Sidastrum*

3 - *Sida* **4** - *Urena*

Malvastrum (12 spp.) — Herb of dry inter-Andean Valleys with serrate, densely pubescent, divided leaves. Flowers yellow, congested in short axillary clusters or in cylindrical terminal spike. Fruits with 10 (or more) more or less pointed mericarps.

Sphaeralcea (15 spp., plus 20 N. Am.) — An amphitemperate disjunct, barely entering our area in Peruvian inter-Andean valleys. Resembling *Malvastrum* but the carpels 2-parted with an empty upper portion and fertile indehiscent lower part.

Modiola (1 sp.) — One prostrate dry-area weedy species with pinkish-orange flowers and irregularly biserrate leaves. Like *Abutilon* but epicalyx present; differs from *Malvastrum* in solitary flowers.

Tarasa (25 spp.) — Herbs and subshrubs of dry inter-Andean valleys, resembles *Malvastrum* and *Sphaeralcea* but the carpels completely dehiscent. Similar to *Pavonia* but the inflorescence of scorpioid cymes and style branches same number as carpels, also usually with more divided leaves.

Acaulimalva (19 spp.) — High-Andean paramo and puna rosette herbs with thick tap roots and solitary rather large pink to purple flowers. Essentially a high-altitude version of *Malvastrum.* Differs from vegetatively similar *Nototriche* in possession of epicalyx and the petioles, stipules, and pedicels not fused.

3Bb. Epicalyx absent — The first two genera have divided mericarps; the next two have more or less inflated carpels.

Wissadula (25 spp.) — Dry-area herbs or shrubs. Leaves always deeply cordate, entire (in ours) and densely pale-tomentose below. Inflorescence often diffusely paniculate; flowers small and yellow. Fruits 3–6-parted, each mericarp divided by fold of lateral wall.

Briquetia (17 spp., incl. *Pseudabutilon*) — Our species (*B. spicata,* sometimes included in *Wissadula*) a 1–2 m tall weedy dry-area shrub with large broadly ovate slightly serrate leaves, white-tomentose below. Inflorescence spicate with small yellow flowers.

Abutilon (110 spp., plus many Old World) — Dry-area shrubs, mostly in inter-Andean valleys. Leaves serrate and ovate; flowers white or yellow, usually in paniculate inflorescence. Like *Pseudabutilon* but mericarps forming ribbed cuplike flat-topped structure, not split. Carpels characteristically "inflated" in fruit into 5–40 thin-winged or angled segments.

Gaya (20 spp.) — Herb to shrubs of dry inter-Andean valleys with serrate ovate leaves, medium-sized yellow flowers and inflated carpels; resembling *Abutilon* but differs in the seed retained inside the dehiscent carpel.

Sida (100 spp., plus 100 Old World) — Mostly weedy suffrutescent herbs with typically narrow unlobed (or sometimes slightly lobed) not very 3-veined leaves, usually serrate (two species have entire leaves, one of these linear). Flowers almost always yellow; inflorescence axillary, few-flowered, condensed or open.

E: escoba, escoba blanca, escoba verde

Anoda (10 spp.) — Sprawling or viny herb with distinctive +/- triangular, usually hastately lobed leaves. Flowers solitary, frequently lavender (sometimes white or yellow), on long peduncle.

Urocarpidium (11 spp.) — Weedy herb of dry areas with leaves both serrate and lobed and small purple flowers. Segregate of *Malvastrum* from which it differs in annual habit, purple flowers and scorpioid inflorescence.

Palaua (15 spp.) — A loma version of *Nototriche* but the solitary flowers not fused to petiole.

Nototriche (100 spp.) — Low alpine rosette herbs with thick tap roots. Flowers solitary with pedicel fused to petiole and stipules (unique).

Cristaria (40 spp.) — Loma herbs close to *Abutilon* and *Wissadula,* distinctive in the usually deeply pinnatifidly dissected leaves (cf., *Argylia*) and the carpels developing wings.

There are many other neotropical genera, mostly in the drier areas of Central America and subtropical South America.

Marcgraviaceae

A very distinctive, exclusively woody, more or less scandent, mostly hemiepiphytic family, restricted to moist and wet forests. Characterized by the uniformly alternate and entire, frequently dark-punctate, usually succulent-coriaceous leaves; these of two general types, one with secondary venation completely suppressed, the other distinctly myrtaceous-looking with intersecondaries paralleling the secondaries and a more or less prominent marginal or submarginal collecting vein. The petioles are usually short and leaves are frequently subsessile. In most genera the leaves are rolled around twig apex in a manner that suggests the conical terminal stipule of Moraceae; vernation lines are sometimes apparent subsequently. *Marcgravia* has a characteristic juvenile growth-form growing appressed against a supporting trunk against which the overlapping leaves are pressed flat.

The inflorescence is characterized by a unique saccate nectary, and the form and placement of these nectaries largely determines generic placement. The nectaries may immediately subtend each flower (*Souroubea*), each pedicel (*Norantea*), or be fewer in number than the flowers and borne in a circle inside a wheel-like ring of flowers (*Marcgravia*). The segregate genera from *Norantea* (*Schwartzia* and *Marcgraviastrum*, respectively) form a series connecting that genus, which has an elongate narrowly racemose (spicate in *Sarcopera*) inflorescence, and *Marcgravia* with the inflorescence reduced to a terminal whorl; both have nectaries borne on the lower part of the pedicel, *Schwartzia* with the flowers in a short raceme, *Marcgraviastrum* with this contracted and approaching the whorled arrangement of *Marcgravia*. Another distinctive feature is the fruit which consists of a mass of very small red seeds borne in a reddish, spongy, fleshy mass that is revealed when the wall of the round fruit falls away.

The genera below are arranged in a logical sequence from the least modified to most elaborate bracts and inflorescences.

Ruyschia (7 spp.) — Only in middle-elevation cloud forests. Like *Souroubea* in the bracts immediately subtending the flower, but bracts small and nonsaccate. Leaves always with suppressed secondary venation, noticeably asymmetric about the midvein and rather small for the family.

Souroubea (19 spp.) — Inflorescence racemose and with wishbone-shaped nectaries immediately below each flower. Leaves usually broadly obovate, secondary veins suppressed or, if weakly visible, very strongly ascending.

Norantea (2 spp.) — Elongate narrowly racemose inflorescence with straight woody rachis, the short pedicellate flowers subtended by a much larger bright red stalked saccate nectary. Leaves obovate and obtuse, intermediate between suppressed secondary vein type and Myrtaceae-like type.

Sarcopera (10 spp.) — Like *Norantea* but the flowers sessile. Both leaf types represented, those with myrtaclike venation larger than in any *Marcgravia*.

Marcgraviaceae and Martyniaceae

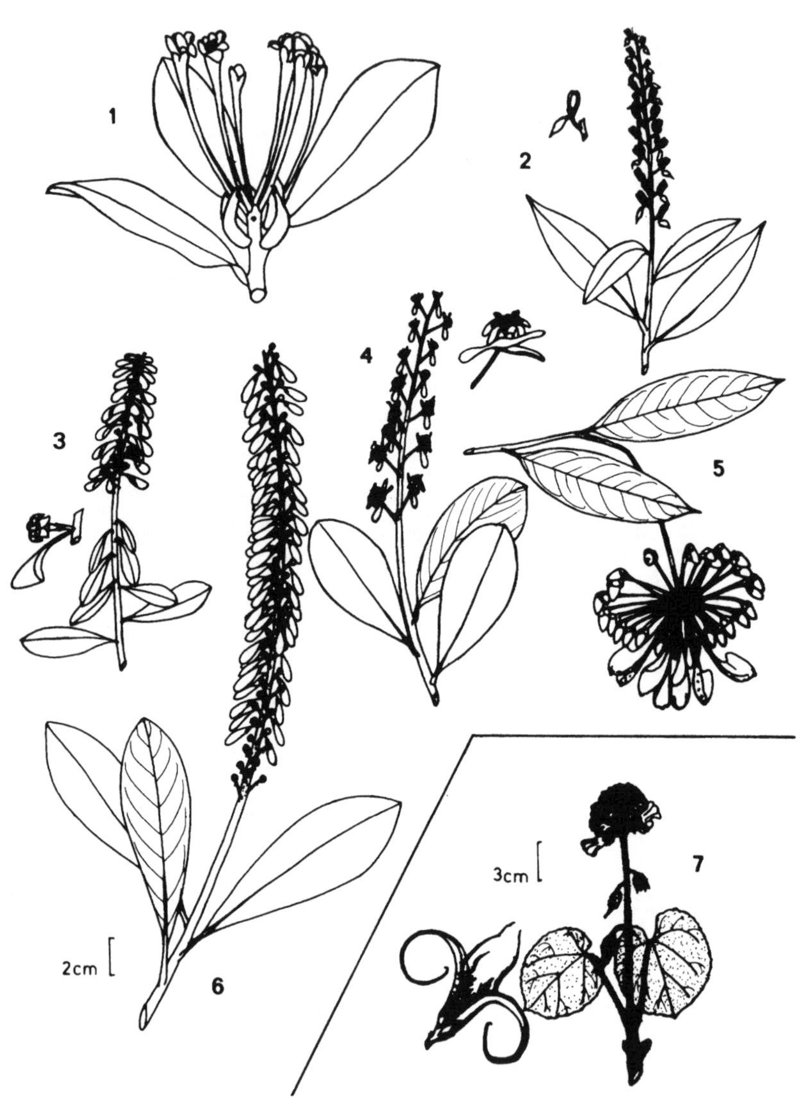

1 - *Marcgraviastrum* 2 - *Ruyschia*

3 - *Sarcopera* 4 - *Souroubea* 5 - *Marcgravia*

6 - *Norantea*

7 - *Proboscidea* (Martyniaceae)

Schwartzia (14 spp.) — Inflorescence short-racemose with each pedicel bearing a saccate greenish or cream nectary in its lower half. All species with suppressed secondary veins.

Marcgraviastrum (15 spp.) — Inflorescence contracted almost into terminal whorl, pedicel with a saccate greenish or cream nectary near its base, these reflexed back from the pedicels. All species with suppressed secondary veins.

Marcgravia (50 spp.) — Inflorescence a well-defined wheel-like whorl of flowers with a smaller whorl of usually greenish long-stalked saccate nectaries in its center. Both leaf types represented; when the secondary veins not completely suppressed, the leaves distinctly Myrtaceae-like with a marginal or submarginal collecting vein and intersecondaries +/- parallel to the secondary veins.

MARTYNIACEAE

Viscid dry-area herbs characterized by very cucurbit-like, broadly ovate, palmately veined, irregularly lobed leaves and curved "devil's claw" like fruit. Our two genera are *Craniolaria* (3 species) in the dry part of northern Colombia, with strongly 3–5-lobed leaf, a white hawkmoth-pollinated flower, and short thick fruit with the apex barely extended, and *Proboscidea* (9 species), in northwestern Peru with a tubular-campanulate orangish or yellow-orange flower and long narrow curving fruit that tapers to the long acuminate apex.

MELASTOMATACEAE

A large family consisting mostly of shrubs but also including herbs, lianas, hemiepiphytes, and large (rarely even emergent) trees. Especially prevalent in second growth and in middle-elevation cloud forests. This is one of the easiest of all plant families to identify in sterile condition thanks to the very characteristic opposite (though sometimes markedly anisophyllous) leaves with one to four pairs of longitudinal veins arcuately subparallel to the midvein and with finer cross veins connecting these perpendicularly. There is also a marked tendency for stellate or dendroid vegetative trichomes. The flowers are also very characteristic, with a cup-shaped hypanthium bearing calyx lobes, petals and stamens on its rim, the latter usually twice as many as the petals and typically with striking, variously appendaged anther connectives, mostly buzz-pollinated with the anthers opening by terminal pores. The fruit of most species is a (usually small) berry but many genera (those with superior ovaries) have capsules in which the hypanthium dries and splits incompletely (only down the sides) longitudinally into (3–)4–5 segments to release the minute seeds.

One melastome group, tribe Memecyleae (*Mouriri* and extralimital *Votomita*), sometimes segregated as a distinct family intermediate between Melastomataceae and Myrtaceae, is highly anomalous in having pinnately veined leaves more like those of Myrtaceae. Memecyleae also often has punctate-looking (from enlarged stomates) leaves similar to Myrtaceae but can usually be distinguished by the more conspicuously thickened and jointed branchlet nodes and (in coriaceous-leaved species) by the *Clusia*-like venation. Memecyleae have flowers with elaborated connectives (but only dorsally) similar to those of other melastomes, but these are uniformly in ramiflorous (or cauliflorous) fascicles (or solitary) unlike most other melastome taxa. In fruit, Memecyleae can usually be distinguished from Myrtaceae by the more oblique position of the flat-topped berry on the pedicel.

Familial subdivision into nine tribes is largely on the basis of fruit (capsular vs. berry), seed shape (straight vs. shell-shaped), and technical characters of the anther connectives. Over half the genera have capsular fruits (mostly from superior ovaries; from inferior ovary in Cyphostyleae) but the great majority of species belong to genera with berry-fruits (and inferior ovaries). The easiest tribe to distinguish is Memecyleae (see above). Four genera of more or less succulent small forest-floor herbs (and epiphytes) (*Triolena, Monolena, Salpinga, Diplarpea*), (our only genera of capsular-fruited Bertolonieae) are also easy to recognize on account of their habit and scorpioid inflorescence, in three of them the fruit distinctively trigonal and opening along the upper valve margins. Two closely related capsular-fruited tribes are characterized by a dorsal spur on the anther connective. One of these, with 4-parted flowers, is represented only by the Andean shrub genus *Monochaetum*, vegetatively characterized by small leaves and often gland-tipped trichomes. The other, centered around *Meriania* and *Adelobotrys* has 5–6-parted flowers, and is mostly vegetatively glabrous; *Meriania*, mostly Andean trees, has large magenta to red or orange flowers, its relatives mostly small white or pinkish flowers. *Adelobotrys* are lowland forest lianas, and *Graffenrieda* mostly lowland forest shrubs or small trees, (technically differing from *Meriania* in the anther pore opposite the basal spur instead of on the same side). *Axinaea* is segregated from *Meriania* on account of technical features of the stamens (and usually white flowers); related genera are mostly small (at least in our area). *Huberia* is a viscous shrub, *Centronia* a middle-elevation cloud-forest tree with anthers like *Graffenrieda* but a very characteristic strongly hispid-setulose calyptrate calyx, *Tessmannianthus* a large tree of lowland Amazonia. Two of the other three capsular-fruited tribes are characterized by ventral anther connective appendages; the third (Cyphostyleae) by a dorsal spur. One of the former is represented in our area only by the very distinctive northern Andean genus *Bucquetia* of paramos and high-altitude (>2800 m) forests with conspicuous 4-merous magenta flowers. The other, *Tibouchina* and relatives, is characterized by mostly 5-parted flowers (4-parted (in our area) in the +/- herbaceous segregates *Pterolepis* and *Pilocosta*). *Tibouchina,* mostly shrubby or subarborescent, mostly has large magenta flowers (except common weedy *T. longifolia* and a few close relatives with small white flowers) while *Brachyotum,* the only other significant genus in this alliance, consists of paramo shrubs with petals rolled into a campanulate tube for hummingbird-pollination. The related genera (*Nepsera, Acisanthera, Aciotis, Arthrostema*) are herbs and subshrubs, mostly small-flowered and mostly found in moist areas.

The berry-fruited taxa include relatively few mostly large (to very large) genera of trees and shrubs divided into two tribes (in addition to myrtaclike Memecyleae). Closely related *Blakea* and *Topobea* have 6-merous flowers and are mostly cloud-forest trees and shrubs, typically epiphytic or hemiepiphytic, with fruits and flowers usually larger than in *Miconia* and relatives; they are further distinguished by two pairs of conspicuous bracts subtending the flowers. *Miconia* and its relatives (tribe Miconieae), mostly shrubs and small trees, make up the overwhelming bulk of the melastome species of our area. *Miconia,* with inflorescence a terminal panicle and mostly regularly 5-dentate calyx, may be taken as the standard against which to compare the rest of these genera. One group of genera (*Ossaea, Leandra*) is characterized especially by narrow acute to acuminate petals. *Chalybaea* and *Huilaea* are small genera of the Colombian Andes with large nectar-producing flowers that look more like *Meriania* and relatives than like *Miconia.* Another anomalous Colombian montane genus is the tiny-leaved prostrate herb *Catacoryne.* Several genera are characterized by conspicuous formicaria on the leaf base or petiole (*Tococa, Maieta, Myrmidone,* a few *Clidemia*). Ramiflorous genera include (in approximate order of decreasing flower and fruit size) *Bellucia* (relatively large (5–)6–8-merous flowers, large glabrous leaves), *Loreya, Henriettea, Henriettella, Killipia,* and some *Clidemia.* This leaves a core group of three large genera, *Miconia, Conostegia* (differing in a calyptrate calyx, truncate at anthesis), and *Clidemia* (differing in lateral inflorescences and usually narrow "setate" calyx teeth).

1. Myrtaceae-Like Leaves (Memecyleae)

Mouriri (81 spp.) — Small to large lowland trees, mostly in Amazonia. Flowers typically melastomaceous with dorsally thickened connectives (though these unique in family in having a distinctive elliptic concave gland), borne in small cymes or fascicles along branches below leaves. Two main leaf types: coriaceous with *Clusia*-like venation of inconspicuous close-together secondary and intersecondary veins (and often conspicuously punctate from the large stomates), or small, thin, and typically long-acuminate. Both vegetatively distinguishable from Myrtaceae mainly by the more strongly jointed, thickened nodes of the branch dichotomies. Fruits usually more obliquely asymmetrical than in Myrtaceae.

P: lanza caspi

2. Capsular Fruits (Mostly from Superior Ovaries) — The first three genera have trigonal fruits opening along dorsal margins, *Salpinga* a longitudinally 10-ridged narrowly campanulate capsule, and most of the other genera at least a few gland-tipped trichomes on calyx and fruit.

2A. Herbs (or small subshrubs)

2Aa. The next four genera have terete stems and scorpioid one-sided inflorescences with 5-merous flowers.

Triolena (incl. *Diolena*) (22 spp.) — Small terrestrial forest-floor herbs with flowers and fruits borne along one side of the characteristic scorpioid inflorescence. Leaves often strikingly anisophyllous.

Diplarpea (1 sp.) — A small rare northern Andean cloud-forest herb that looks like *Triolena* but has conspicuous hairs over the entire plant, including the capsule.

Monolena (10 spp.) — Small succulent cloud-forest herbs, often epiphytic. Like *Triolena* but acaulescent, more succulent, and with larger flowers and leaves.

Salpinga (8 spp.) — Mostly on Brazilian Shield; our species with the flowers sessile along one side of inflorescence and having long narrow hypanthium topped by conspicuous lobes. Technically differing from *Triolena* in the dorsal connective appendage.

2Ab. Tetragonal stems and mostly 4-parted flowers (5-parted in *Desmoscelis, Pterogastra, Acisanthera*).

Desmoscelis (2 spp.) — Weedy herb or subshrub of grassy savannahs on sandy soil. Characterized by long tannish rather stiff stem trichomes, small sessile entire long-sericeous apiculate leaves, and small pink 5-parted flowers, in part borne singly in the leaf axils. A *Tibouchina* segregate differing technically chiefly in elongate connectives with 2 filiform ventral lobes, almost as long as the thecae.

Melastomataceae
(*Mouriri* and Woody Hemiepiphytes)

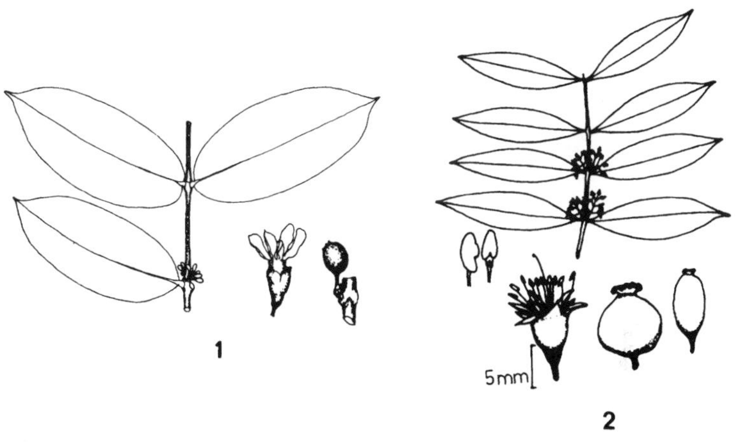

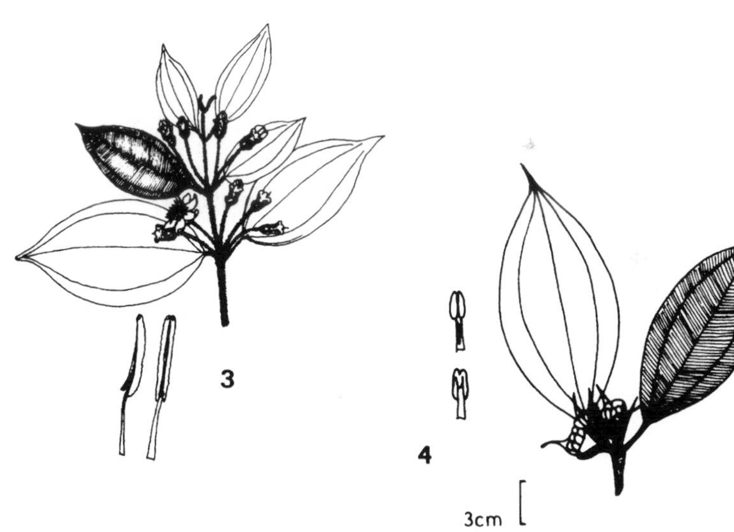

1 - *Mouriri* **2** - *Mouriri*

3 - *Topobaea* **4** - *Blakea*

Melastomataceae
(Herbs with Capsular Fruits)

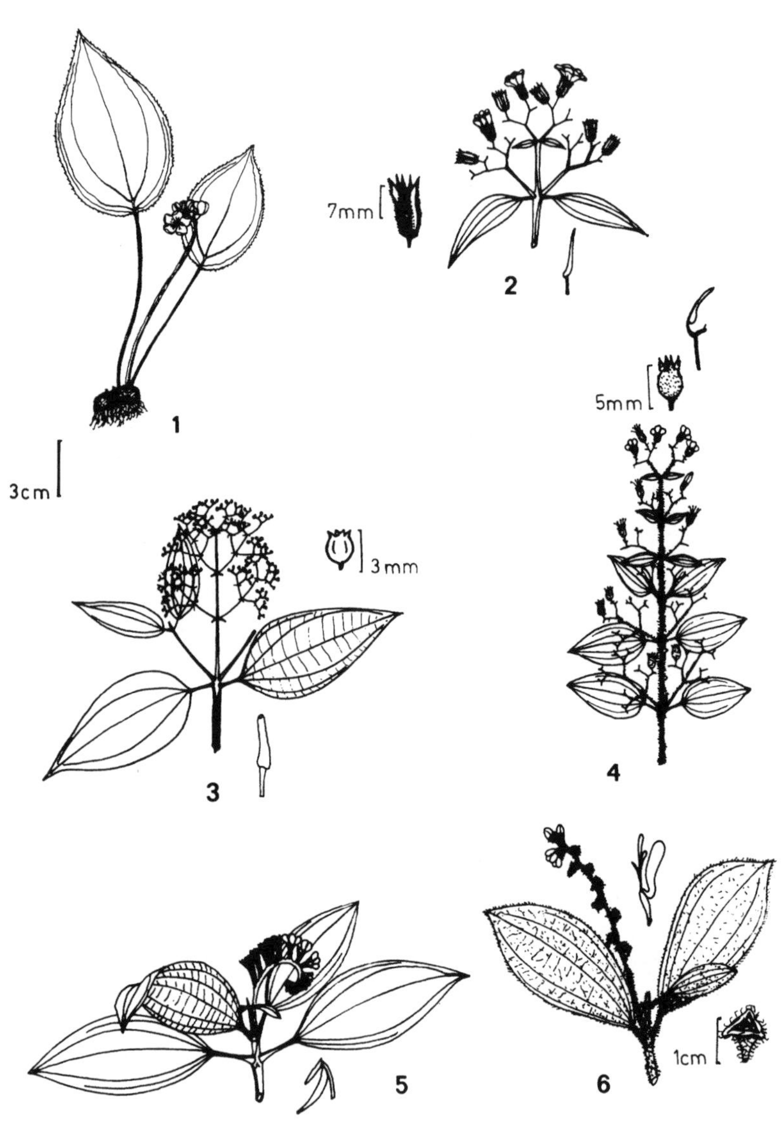

1 - *Monolena* 2 - *Pterogastra*

3 - *Aciotis* 4 - *Ernestia*

5 - *Salpinga* 6 - *Triolena*

Ernestia (16 spp.) — Viscid, glandular-pubescent, small-leaved herbs or subshrubs of open moist sandy places, mostly in the Guayana area with one species reaching northern Amazonian Peru. Generally similar to *Aciotis* but the leaves less membranaceous, and the white to lavender flowers larger, with much longer calyx lobes, and borne in a larger and more pyramidal inflorescence.

Acisanthera (17 spp.) — Our only species a weed with square wing-margined stem similar to some *Aciotis,* the stem also with gland-tipped trichomes. Vegetatively very different from *Aciotis* in smaller leaves and from *Desmoscelis* in much shorter sparser stem pubescence and finely serrate leaf margins. In our area also differs from *Aciotis* in solitary axillary 5-parted flowers with very unequal stamen whorls (or one whorl reduced to staminodes) (stamens equal in *Aciotis*).

Arthrostemma (7 spp.) — Spindly herbs mostly in moist open areas, with strikingly square stem and rather few 4-merous flowers with conspicuous early-caducous magenta petals and sparsely glandular trichomes on young parts.

Pterolepis (14 spp.) — Mostly in the cerrado; in our area represented by a scraggly small-leaved herb of open dry areas characterized by acutely tetragonal stems, small 4-merous (in our area), purplish flowers borne singly or in clusters in apical leaf axils (on either side of a sessile terminal flower); calyx with often glandular trichomes and stiff setae.

Pterogastra (2 spp.) — Herbs of open savannah areas. A *Tibouchina* segregate very similar to *Pterolepis* but with a very distinctive winged calyx, strongly ciliate along the ribs, only. The 4–5-merous magenta flowers are also somewhat larger than in *Pterolepis.*

Nepsera (1 sp.) — Herb or small subshrub of open swampy areas. Easily recognized by the diffuse open capillaceous *hanging* inflorescences and small, very finely serrate, membranaceous leaves. Stems square, sometimes, +/- glandular-pubescent; small white flowers 4-merous with strikingly purple anthers and 3-celled ovary.

Aciotis (30 spp.) — Herbs or subshrubs of low-elevation moist areas, vegetatively usually characterized by strongly tetragonal stems, often with noticeably winged angles. Leaves membranaceous, nearly always with finely ciliate margins, often pale and almost translucent-looking (or suffused with purple) below. Inflorescence characteristic, usually with straight erect terminal rachis and widely spaced pairs of dichotomous cymes of small, whitish, narrow-petaled flowers.

Centradenia (6 spp.) — A mostly Central American genus of herb or spindly shrub with one species reaching the Chocó, where it is found mostly near waterfalls. Our species is white-flowered and rather similar to *Aciotis* but with larger, very thin, strongly asymmetric, +/- oblong leaves with entire nonciliate margins.

Castratella (1 sp.) — A rosette herb of wet Colombian paramos, habitally very distinct from other melastomes in the long-hirsute *Plantago*-like cluster of basal leaves from center of which arises the hirsute inflorescence bearing several rather large bright yellow apical flowers with conspicuously appressed-hirsute hypanthia.

(*Tibouchina longifolia*) — Our only common lowland *Tibouchina* is a white-flowered weedy herb while most other species are magenta-flowered shrubs or trees. This common species can be distinguished by narrow appressed-pilose leaves, somewhat diffuse inflorescence of inconspicuous whitish flowers and characteristic calyx with long narrow teeth (cf., *Clidemia*). There is also at least one similar herbaceous upland *Tibouchina* (*T. kingii*).

2B. Shrubs, trees, or lianas with capsular fruit — These genera are traditionally differentiated based on combinations of anther connective elaboration and number of flower parts; the first eight have dorsally spurred connectives, the next five (*Tibouchina* group) ventrally extended connectives; the last three belong to tribe Microlicieae, differentiated from *Tibouchina* relatives by straight rather than curved seeds.

2Ba. Dorsally spurred anther connectives

Monochaetum (45 spp.) — Viscid-pubescent Andean paramo or subparamo shrubs, vegetatively characterized by small entire leaves often more or less longitudinally "plaited" along the ascending main veins and/or with characteristic pattern of long appressed trichomes on upper surface but not the veins, and usually gland-tipped trichomes. Flowers 4-parted unlike *Meriania* and relatives; hypanthium +/- cylindrical, usually inconspicuously ribbed and slightly contracted at apex below the calyx teeth.

Meriania (74 spp.) — Andean cloud-forest trees, often reaching canopy. An important genus, but lacking a good vegetative gestalt. Leaves often glabrous (or glabrate) and coriaceous; species with large pubescent leaves often have a characteristic interpetiolar stipular ridge or flap of tissue. Best recognized by the large, magenta to red, 5–6(–8)-parted flowers; calyx usually 5-lobed or with 5 submarginal teeth. Technically characterized by anther thecae subulate and with short blunt dorsal appendages.

Adelobotrys (25 spp.) — Lowland forest lianas (occasionally erect in Amazonia), sometimes conspicuously setose with simple or 2-branched

Melastomataceae
(Woody with Capsular Fruits; Dorsal Connective Spurs)

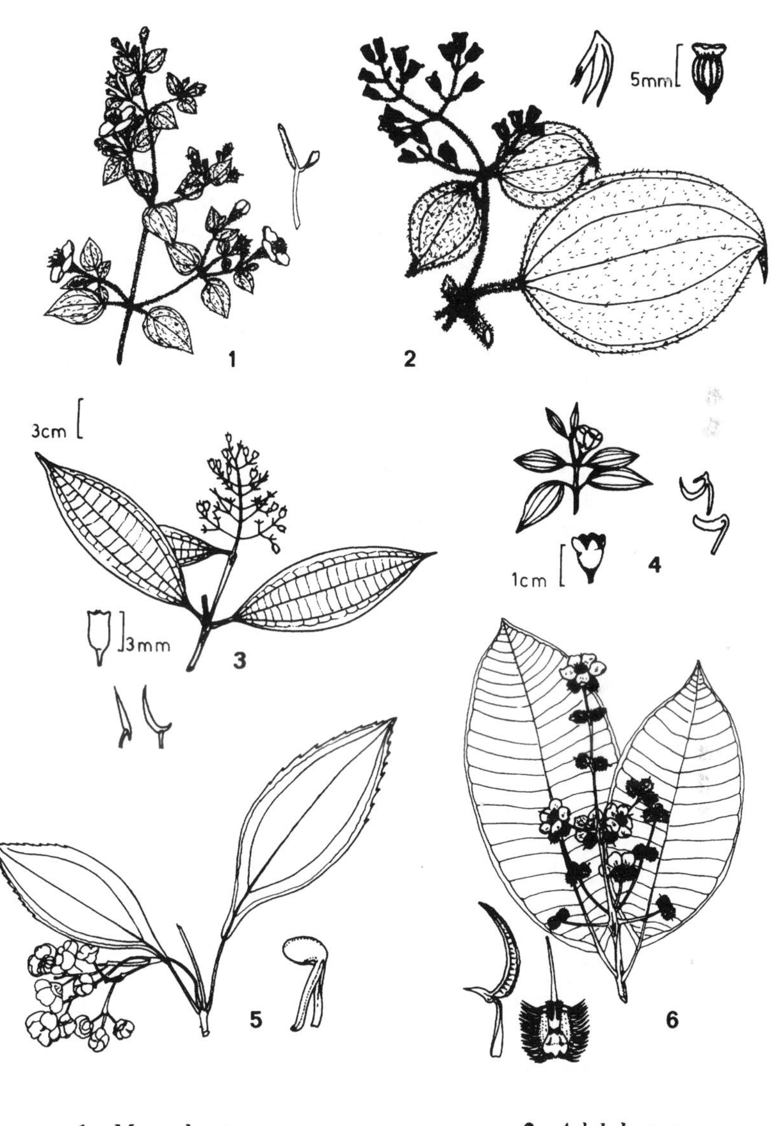

1 - *Monochaeta*

2 - *Adelobotrys*

3 - *Graffenriedia*

4 - *Meriania*

5 - *Axinaea*

6 - *Centronia*

trichomes. Flowers small to medium, usually pinkish (sometimes white), 5-parted, in terminal panicles. Calyx obtusely 5-toothed. Technically characterized by connective with an acute spur bearing a striking appendage held parallel to theca.

Graffenrieda (44 spp.) — Mostly lowland forest shrubs or small trees (commonest Chocó species is a vine with exceedingly caudate leaf apex), almost always with entire leaves which tend to be distinctively large and/or coriaceous, always obtuse to subcordate at base and 3(–7)-veined from extreme base. Flowers small, white or pink, usually with a somewhat irregularly dentate calyx. Technically differing from *Meriania* in the anther pore opposite the basal spur instead of on the same side. In flower rather like *Miconia* (except the irregularly splitting or calyptrate calyx). Can be distinguished from most *Miconia* by the uniformly entire leaves, lack of conspicuous stellate pubescence, and more truncate leaf base; large-leaved species could be confused with *Bellucia* or *Loreya* but differ from both in being 3-veined from extreme base and from *Bellucia* in lacking a papillate lower leaf surface.

Axinaea (30 spp.) — Andean cloud-forest trees and shrubs. Usually vegetatively recognizable by the presence of some kind of auriculate projection at the base of the leaf blade or the very remotely shallowly toothed margins. A relative of *Meriana* with 5–6-parted middle-sized white to magenta flowers in a paniculate inflorescence and having a broad usually truncate calyx. Technically differs from *Meriana* in saccate basal connective appendages. The distinctive thick anther appendages held close together in a ring superficially rather resemble the similarly held thecae of *Blakea.*

Huberia (6 spp.) — Our only species a glabrous-viscous shrub of middle elevations in the Huancabamba region, characterized by small coriaceous 3-veined leaves, 4-parted white flowers and a narrowly campanulate 8-ribbed calyx with triangular teeth. The technical character is a hanging filiform dorsal connective appendage.

Centronia (15 spp.) — Middle-elevation cloud-forest trees. Flowers large and red or magenta like *Meriania,* but anthers like *Graffenrieda* (i.e., with one prominent dorsal connective spur). Distinctive from both in a very characteristic strongly brownish or tannish-pubescent-tomentose or hispid-setulose calyptrate calyx (often also vegetatively hispid) and the usually rather close-together strongly raised tertiary veins below; leaf undersurface typically more or less tannish-tomentose.

Tessmannianthus (6 spp.) — Large trees of lowland Amazonia with glabrous, entire, medium-sized leaves 5-veined from base with the veins running clear into acumen. Flowers small, white, resemble *Miconia* (as do the leaves) but fruit dry, thin-walled, irregularly splitting.

2Bb. Ventrally extended anther connectives and shell-shaped concave-convex seeds

Tibouchina (243 spp.) — Mostly middle- and high-elevation shrubs and small trees, the leaves always more or less asperous on at least one surface. Nearly always pubescent; the great majority of the species can be recognized vegetatively by the 3 main leaf veins reaching all the way to apex, more or less impressed above with the surface between being asperous, this contrasting with the tannish-pubescent lower surface. Usually with relatively large conspicuous magenta flowers (except a few weedy herbs), often setose-pubescent on calyx or vegetative parts. An important technical character is the ventrally bilobed anther connective.

Pilocosta (3 spp.) — A recent upland Andean *Tibouchina* segregate, characterized by 4-merous axillary solitary flowers with a 4-angled hypanthium.

Brachyotum (58 spp.) — Paramo shrubs and small trees with the petals rolled into tube for hummingbird pollination. Petals mostly dark purple (to almost blackish; yellow in *B. ledifolium*) and calyx red. Leaves usually more or less bullate or asperous (from stiff trichomes) above, often with plicate effect from the appressed-pilose indument of upper surface but not along the main veins (cf., *Monochaetum* but plant not viscid).

Chaetolepis (10 spp.) — Paramo shrub with tiny leaves (in one species reduced to scales), small mostly solitary usually yellow axillary flowers. The thick leaves with deeply impressed main veins above are reminiscent of *Brachyotum* but smaller and the flowers are completely different.

Macairea (22 spp.) — Mostly Guayana Highland shrubs, especially on white-sand savannahs, a few species reaching sandy savannahs in Colombia and northern Amazonian Peru. Similar to *Tibouchina* from which it differs in only 4 petals and sepals. Vegetatively characterized by oblong or narrowly oblong-elliptic obtuse leaves with only 3 veins, the lateral pair usually running very close to the margin.

2Bc. Tribe Microlicieae; like *Tibouchina* relatives but the seeds straight and oblong rather than shell-shaped — Leaves (in our area) smallish, usually glabrous and/or resinous, sometimes ericoid, if hirsute with glandular-viscid trichomes.

Bucquetia (3 spp.) — High-Andean paramo or subparamo shrubs and trees, south only to Ecuador. Distinctive in the small glabrous, resinous

Melastomataceae
(Woody with Capsular Fruits; Ventral Connective Spurs or Inferior Ovary)

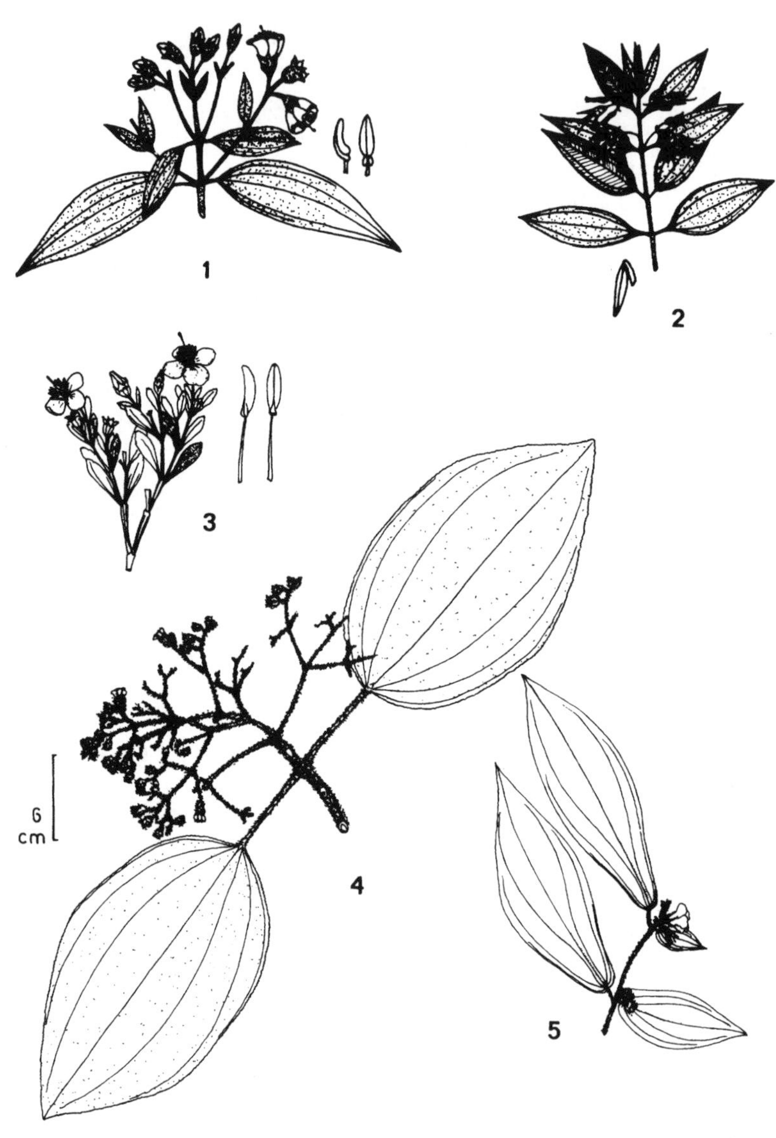

1 - *Brachyotum*

2 - *Tibouchina*

3 - *Bucquetia*

4 - *Alloneuron*

5 - *Allomaieta*

leaves which are nearly entire but subserrate near apex. Tree species vegetatively characteristic in the conspicuous reddish peeling papery bark. Flowers conspicuous, magenta, differing from *Tibouchina* in the oblong pyramid-shaped seeds. Its tribal treatment has been much debated.

Rhynchanthera (15 spp.) — A genus primarily of cerrado subshrubs with conspicuous 5-parted magenta flowers, also occurs sparsely in patches of savannah in Peru as around Tarapoto and in the Pampas del Heath. Differs from *Tibouchina* in the more reddish-purple flower color and in having only 5 fertile stamens (and differently shaped seeds). Ours with narrow glabrous leaves or viscid-hirsute with smallish ovate leaves.

Microlicia (100 spp.) — Mostly extralimital. Small spindly ericoid cerrado and savannah shrubs with small thick stiff leaves held +/- erect along the stems. Barely reaching our area with two species in middle-elevation sphagnum bogs in the Peruvian "ceja de la montana".

2Bd. Inferior ovary (Cyphostyleae) — The next three genera belong to a separate tribe that differs from the other capsular-fruited taxa in the inferior ovary. In addition to the fruit they differ from berry-fruited Miconieae in a single stamen whorl and completely inferior ovary.

Allomaieta (1 sp.) — Cloud-forest shrub of Colombian Andes. Leaves strongly anisophyllous and very similar to *Maieta* though domatia absent or not well developed. Flowers white to pink, much larger than *Maieta* and with only 5 stamens, but in similar reduced axillary inflorescence.

Cyphostyla (1 sp.) — A shrub endemic to middle-elevation Colombian cloud forests. Intermediate between *Allomaieta* and *Alloneuron*. Vegetatively like *Alloneuron* in the asperous (also +/- viscid) finely serrate leaves and with a similar stiffly strigose-hirsute calyx but with 5 rather than 4 flower parts; also differing from *Allomaieta* in the pedunculate terminal inflorescence.

Alloneuron (7 spp.) — Cloud-forest trees of lower Andean slopes. Leaves more or less intricately bullate or reticulate, somewhat asperous either above below. Calyx rather strongly and stiffly strigose-pubescent (cf., *Henriettea* but inflorescence a panicle). Differs from *Allomaieta* in only 4 stamens.

3. Berry-Fruited Shrubs and Trees (or +/- Scandent Epiphytes)

3A. *Miconia* and *Conostegia;* paniculate terminal inflorescence, small regular calyx (5-toothed or calyptrate), and medium to small flowers with broad obtuse petals.

Miconia (1000 spp.) — One of the largest neotropical plant genera, well represented in most moist- and wet-forest types. Most species are shrubs and small trees but a few are large (even emergent!) trees. Stilt roots and strikingly papery fibrous bark are sometimes present and a faint trace of red sap occasionally occurs. Among melastome trees, only *Miconia* has species with such characters as leaves strongly whitish or tan below, large trees with typical membranaceous to chartaceous small to medium leaves, and leaves tapering to a sessile subcordate base. *Miconia* may be thought of as the core of the berry-fruited melastomes, characterized by a paniculate terminal inflorescence, broad obtuse (though usually small) petals, and usually regularly 5-toothed calyx with triangular lobes (occasionally the rim deciduous).

C: mora, tuno, aguanoso; E: huitoto, olutca (*M. impetiolaris*); P: rifari, mullaca

Conostegia (43 spp.) — Differs from *Miconia* in calyptrate calyx, truncate at anthesis. The apex of the pre-anthesis bud is apiculate and also distinctive.

C: mora

(*Clidemia*) — Most species ramiflorous or with axillary inflorescence (see below). Also differs from *Miconia* in the narrrow ''setate'' calyx teeth.

3B. The next two genera are distinguished especially by the inconspicuous flowers with narrow acute petals (in *Ossaea* usually with an exterior tooth).

Ossaea (91 spp.) — Vegetatively not distinguishable from *Miconia* or *Leandra,* although there is a tendency to have thinner, more strongly anisophyllous leaves. Easily separated in flower by the narrow acute petals usually with an exterior tooth and usually by the relatively small few-flowered inflorescences. Fruit tends to be less fleshy than *Miconia* and is longitudinally costate, at least when dry.

Leandra (incl. *Platycentrum*) (175 spp.) — Understory shrubs. Like *Ossaea,* differs from *Miconia* in narrow acute petals. Most species recognizable, even vegetatively, by the dense pubescence of rather straight dark reddish (sometimes tan) trichomes and the tendency to areolate undersurfaces. The characteristic terminal (sometimes pseudolateral) inflorescence has a straight axis with dark spreading trichomes and the flowers are mostly sessile along one side of the lateral branches. Fruits round.

3C. The next two genera, endemic to the Colombian Andes, have relatively large flowers resembling *Meriania* and allies.

Chalybaea (1 sp.) — Tree or shrub of the montane oak forest of the Colombian Andes. Leaves large, ovate, serrate, tan-tomentose and rather

Melastomataceae

(Woody with Berry-Fruits; Mostly with Terminal Inflorescences and/or Ant Domatia)

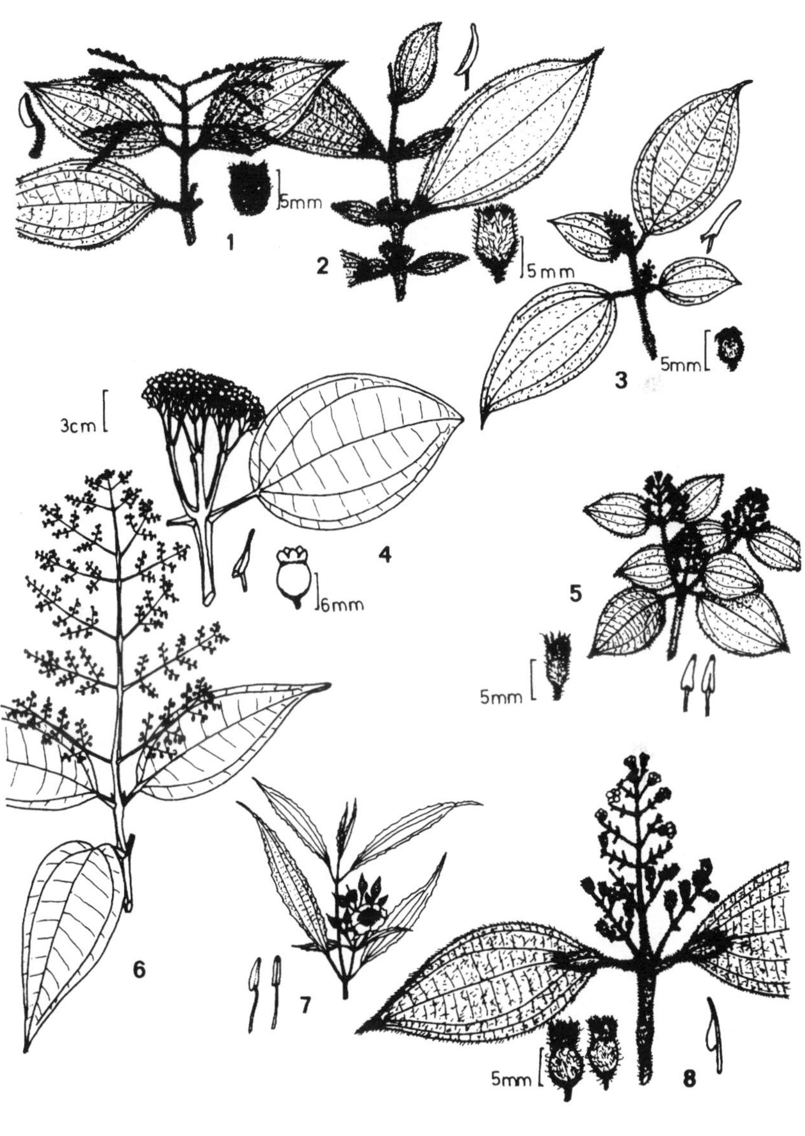

1 - *Leandra* **2** - *Maieta* **3** - *Ossaea*

4 - *Miconia* **5** - *Clidemia*

6 - *Miconia* **7** - *Conostegia* **8** - *Tococa*

viscid below. Inflorescence corymbose-paniculate, long-pedunculate, more or less drooping; petals greenish-white, imbricate to form short tube. Essentially a large-flowered *Miconia* trying to make itself into a *Brachyotum.*

Huilaea (4 spp.) — Colombian Andean cloud-forest trees. Hummingbird-pollinated with the few large (petals >2 cm long) red flowers on long axillary peduncles. Leaves similar to *Chalybaea* in being densely tannish-tomentose below, but smaller. Calyx large (ca. 2 cm wide and long), tannish-tomentose. Fruit large, to 5 cm long, sometimes edible.

3D. The next three genera (and part of *Clidemia*) are characterized by conspicuous ant-domatia.

Tococa (54 spp.) — Shrubs, especially of poor-soil forest understory. Vegetatively nearly all species characterized by having conspicuously swollen hollow ant domatia at top of petiole or base of leaf.

Maieta (3 spp.) — Strongly anisophyllous subshrubs with glandular trichomes and formicaria projecting up from center of broadly auriculate base of short-petioled leaf blade. Inflorescence few-flowered and axillary.

Myrmidone (1 sp.) — An understory herb or subshrub endemic to Amazonian Ecuador, resembling a cross between *Maieta* and *Tococa.*

(*Clidemia*) — Several species have ant domatia at base of blade or on stem (e.g., *C. crenulata, C. allardii*).

3E. The next six genera are mostly ramiflorous or cauliflorous (axillary inflorescences only in many *Clidemia*). The flowers single or in fascicles except *Killipia* and *Clidemia.* — They are arranged in order of increasing flower and leaf size.

Clidemia (165 spp.) — Differs from *Miconia* primarily in lateral (either axillary or ramiflorous) inflorescences and usually narrow "setate" calyx teeth. Tends to have longer stiffer vegetative and inflorescence trichomes than *Miconia* or other genera. As a rule of thumb, a sterile melastome shrub that seems unusually hairy is likely to be *Clidemia.* Sometimes has ant domatia at base of blade or on stem and occasionally climbing.

P: mullaca

Killipia (4 spp.) — Spindly, usually glabrous shrub of Colombian cloud forest, characterized by usually strongly tetragonal stem with winged angles, jointed at nodes (from joined petiole bases). Leaves glabrous, thick, coriaceous, entire, strongly 3–(5-)veined to apex, drying distinctively yellowish below. Inflorescence rather sparsely branching, the small flowers highly unusual in being mostly yellow.

Melastomataceae
(Woody with Berry-Fruits; Mostly Ramiflorous)

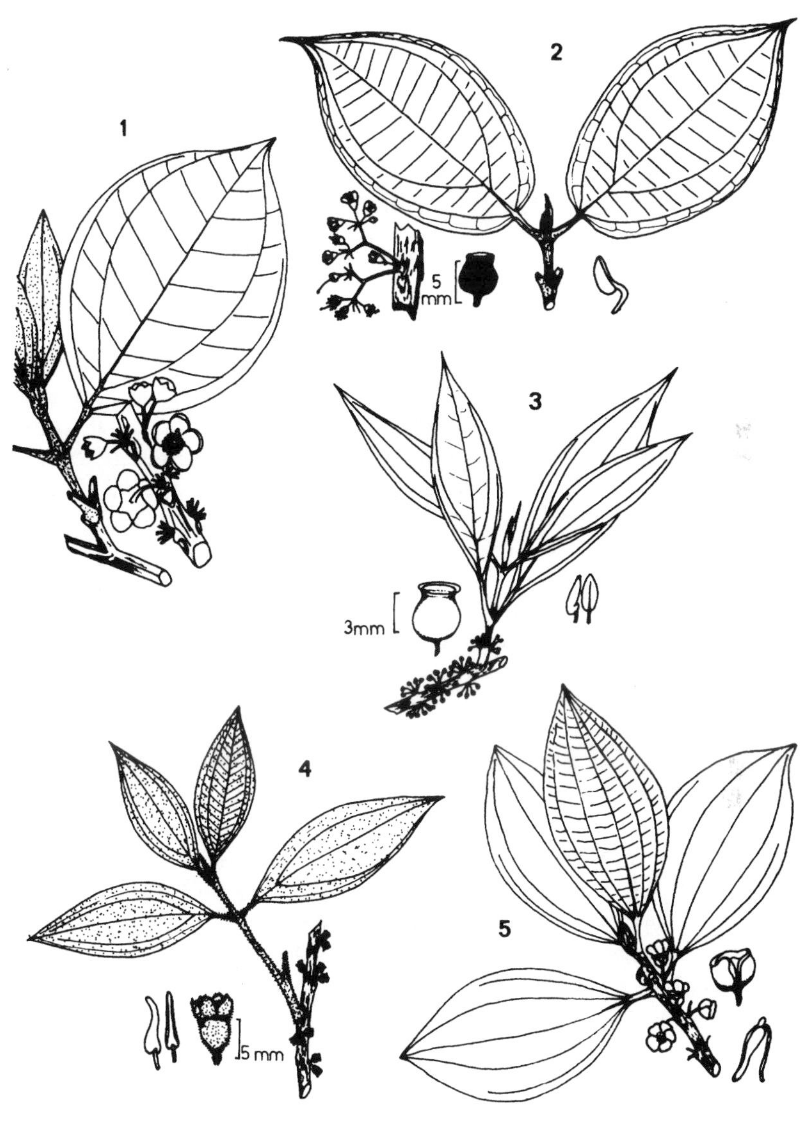

2 - *Loreya*

1 - *Bellucia*

3 - *Henriettella*

4 - *Henriettea*

5 - *Topbaea*

Henriettella (51 spp.) — Mostly Antillean. Understory shrubs or trees mostly in mature forest. Leaves not very distinctive, usually smooth glabrous, and lacking raised cystoliths. Flowers tiny (petals <4 mm long), always borne in fascicles at the nodes below the leaves; calyx smaller than *Henriettea,* usually glabrous or inconspicuously puberulous.

Henriettea (12 spp.) — Lowland Amazonia, mostly in seasonally inundated riverine forest. Leaves usually conspicuously verrucose and/or asperous above from raised cystoliths (= calcium oxalate styloids). Flowers larger than *Henriettella* (petals 10 mm long), and borne in smaller groups (singly or in threes) in leaf axils or from old wood, the calyx conspicuously strigose-pubescent, tapering to base and longer than wide (and than pedicel).

Loreya (13 spp.) — Subcanopy trees of mature lowland forest. Characterized especially by the ramiflorous inflorescence (flowers solitary or in fascicles, the fascicle pedunculate in one species). Leaves 3-nerved from above base and often rather large, thus similar to *Bellucia* but, unlike *Bellucia,* sometimes pubescent, lacking a glaucous lower leaf surface, and occurs in mature forest. Flowers with truncate calyces, smaller than *Bellucia* but generally larger than *Henriettella.*

Bellucia (7 spp.) — Second-growth trees with characteristic large white ramiflorous (sometimes axillary) flowers and relatively large, flat-topped, broadly campanulate, edible yellow fruits. Leaves very distinctive, large and coriaceous, distinctively 3-nerved above base, glabrous (in our area) above and distinctively glaucous below from small papillae.

C: coronillo; E: manzana silvestre, tunguia; P: níspero, sachaníspero

3F. The next two genera (tribe Blakeae) have 6-merous flowers and are mostly epiphytic or hemiepiphytic cloud-forest trees and shrubs, often more or less scandent. — They have fruits and flowers usually larger than in *Miconia* and relatives and are unique in two pairs of conspicuous bracts immediately subtending flower and usually more or less enclosing calyx or calyx base. They differ *only* in anther characters.

Blakea (100 spp.) — Thecae relatively short and thick (>1/2 as wide as long) with two well-separated terminal pores.

Topobea (62 spp.) — Thecae relatively slender (<1/4 as wide as long), linear-oblong with two close-together (or confluent into one) tiny apical pores.

4. Prostrate Herb with Tiny Leaves and Berry-Fruit

Catocoryne (1 sp.) — Colombian cloud forests. A rarely collected and anomalous trailing herb with tiny ovate leaves <5 mm long. Flowers solitary with 4 white petals. Related to *Miconia,* but habitally distinct.

Meliaceae

All of our species are trees and this is the most important neotropical timber family (mahogany, spanish cedar). Vegetatively most genera are characterized by even-pinnate leaves but *Trichilia* (unfortunately the most speciose genus) has terminal leaflets, and a few *Trichilia* species are unifoliolate. Fairly easy to recognize to family by the characteristic flowers with stamen filaments fused into tube at least at base. *Trichilia* often has a poorly developed staminal tube and is the only genus likely to be confused with other families. All neotropical species have distinctive capsular, usually 5-valved fruits. One alliance (subfamily Melioideae: *Trichilia, Guarea, Ruagea, Cabralea*) has the capsules loculidally dehiscent and with arillate, mostly bird-dispersed seeds. The other (subfamily Swietenioideae: *Carapa, Swietenia, Cedrela, Schmardaea*) has the capsules septifragally dehiscent and with nonarillate, usually winged, wind- or water-dispersed seeds. There are two common and very speciose genera, *Trichilia* and *Guarea,* both with arillate seeds. The trunk slash of most genera has a distinctive sweetish odor, whereas *Cedrela* has an unpleasant rank somewhat garliclike odor. *Swietenia* and *Cedrela* are perhaps the most valuable neotropical timber trees; both are characteristic of late second-growth forests and both are notoriously susceptible to the *Hypsipyla* budworm larvae, and thus reputedly not amenable to plantation planting.

1. Arillate-Seeded Taxa

Guarea (35 spp., plus 5 in Africa) — Understory to canopy trees, widespread in the lowland Neotropics. Vegetatively very characteristic in the even-pinnate leaves with a kind of terminal "bud" between the terminal leaflet pair, this sequentially producing additional pairs of leaflets as the leaf gets older; often cauliflorous and often with very large fruits; flowers with anthers inserted *inside* very conspicuous staminal tube.

C: chalde, mancharro; E: caoba (*G. cartaguenya*), chocho (*G. kunthiana*); P: requia

Trichilia (70 spp., plus 16 in Old World) — Understory to canopy trees. Distinctive in family in usually odd-pinnate leaves with ca. 5–7 leaflets (rarely reduced to 1-foliolate and occasionally with more leaflets), the leaflets often alternate or subopposite on the rachis. Vegetatively distinguished from sapindaceous alliance of nondescript pinnate-leaved genera by lacking an aborted rachis apex; flowers differ from other meliac genera (except *Cedrela*) in having staminal tube often poorly formed and anthers inserted at its margin or at apices of the individual filaments. Inflorescence a terminal panicle; fruit mostly smaller than *Guarea,* more likely to be confused with *Matayba* or *Cupania* (Sapindaceae).

E: caigua (*T. hirta*); P: ucho mullaca (hairy)

Meliaceae
(Fruits Capsules with Arillate Seeds)

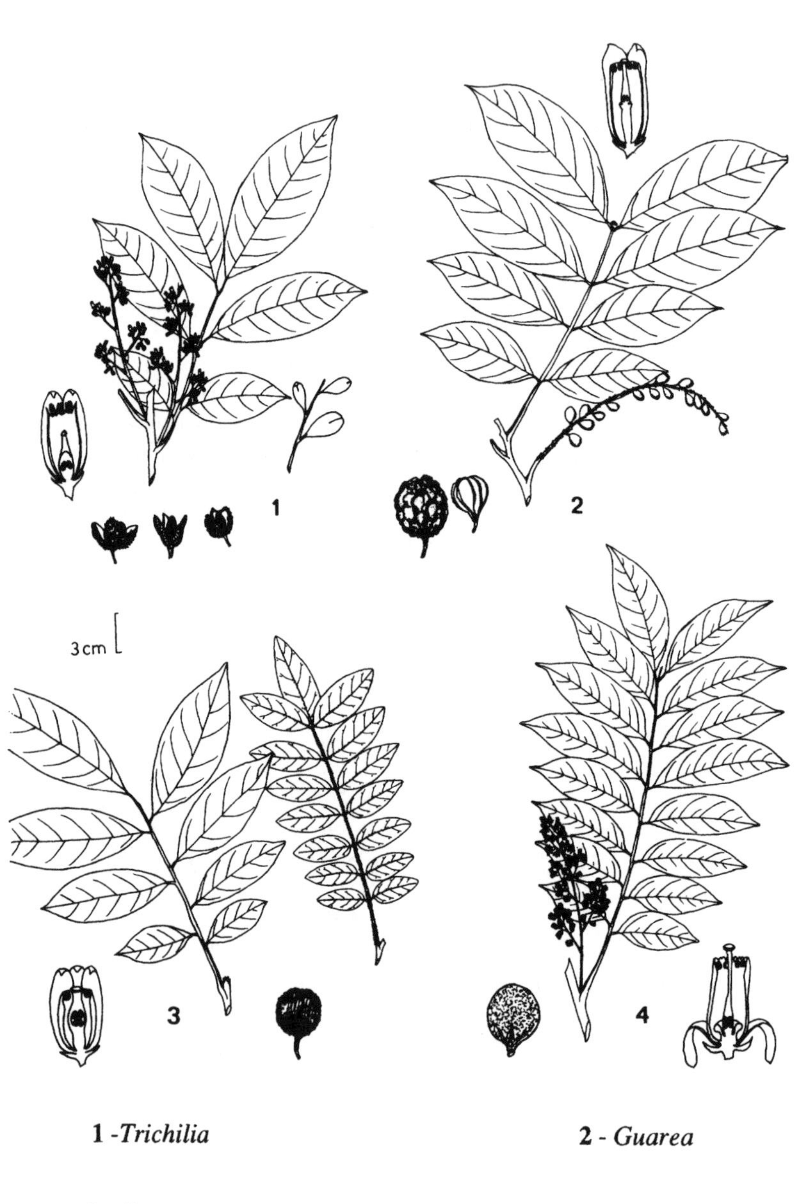

1 -*Trichilia* **2** - *Guarea*

3 - *Ruagea* **4** - *Cabralea*

Meliaceae
(Fruits Large and/or Woody; Indehiscent or Seeds Winged or Leaflets Serrate)

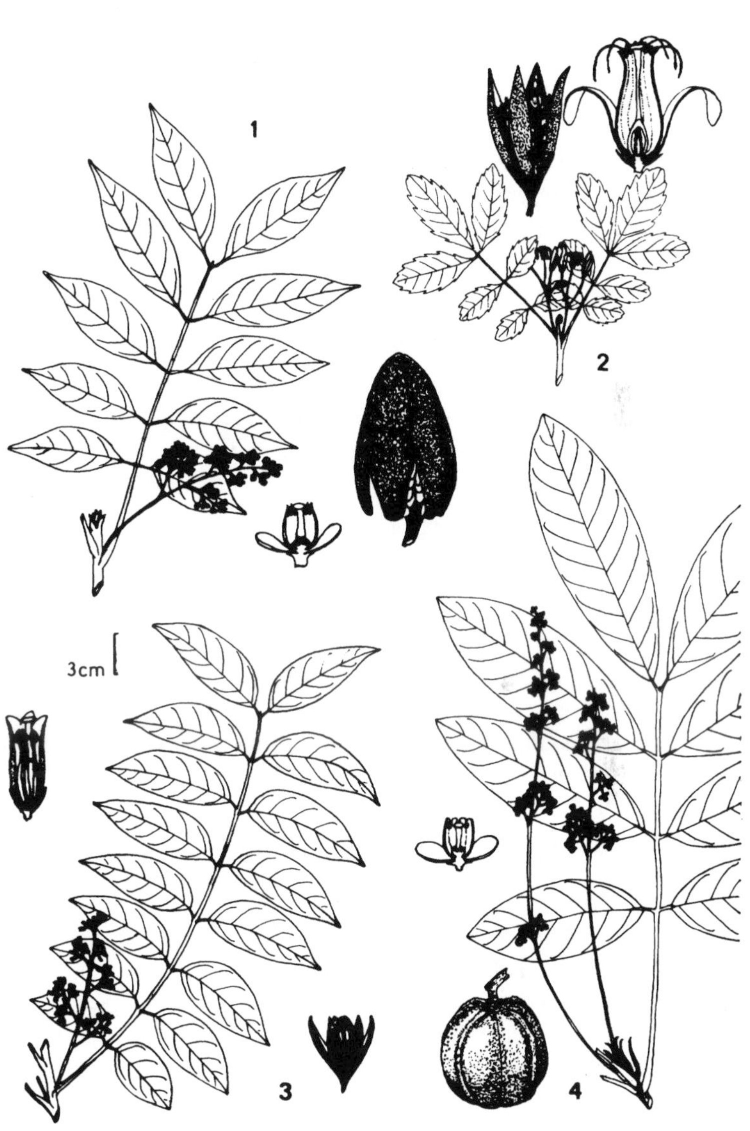

1 - *Swietenia* 2 - *Schmardaea*

3 - *Cedrela* 4 - *Carapa*

Ruagea (7 spp.) — Mostly Andean cloud-forest trees with one species (*R. glabra*) also descending to moist lowland forests on good soil. Flowers and fruits as in *Guarea.* Vegetatively, also looks like *Guarea* but with alternate leaflets, and usually with terminal leaflet. Most species have very pubescent leaves, and any unusually pubescent *Guarea,* especially from higher elevations, is likely to be *Ruagea* instead. The technical differentiating characters are free (rather than partly fused) sepals and glabrous ovary and fruit (rare in *Guarea*).

Schmardaea (1 sp.) — An Andean-area tree, mostly in seasonally dry upland forests, characteristic in the small subsessile obtuse leaflets with distinctly serrate margins (at least apically); lower leaflets progressively smaller. Essentially a 7–9-foliolate *Trichilia* with serrate leaflets puberulous below and much longer (>1 cm) greenish-white flowers (similar to *Melia*).

P: carachugo

Cabralea (1 sp.) — An extremely widespread large tree. Flowers and fruits similar to *Guarea* but the leaves with very many narrow, strongly asymmetric leaflets. Vegetatively looks like *Cedrela* but the secondary veins beneath inconspicuous and lacks a rank vegetative odor.

P: cedro macho

2. Large Seeds, Neither Arillate nor Winged

Carapa (3 spp., also in Africa) — Medium-sized to emergent trees, especially prevalent on rich soils and in swamps. Characterized by a very large globose to slightly 4-angled capsule, the seeds neither winged nor arillate. Vegetatively and in flower, similar to *Guarea;* easy to tell by the large leaves with a distinctive texture: coriaceous with a dull smooth surface and slightly intricately impressed fine venation below. The large fruit with nonarillate seeds is also completely distinctive.

C: tangare; P: andiroba

3. Wing-Seeded Taxa

Cedrela (8 spp.) and ***Swietenia*** (3 spp.) — The two main timber genera of the family (and of the Neotropics), both have woody, five-valved capsules; the capsules of *Swietenia* are much larger and woodier than those of *Cedrela.* The seeds of both are single-winged (and similar except in size); the bark of both is characteristic and prominently vertically ridged. The only neotropical fruits potentially mistakable for *Swietenia* are of a few wing-seeded genera of Bombacaceae (*Huberodendron, Bernoullia*) which have simple or palmately compound leaves. Vegetatively *Swietenia* and *Cedrela* have narrower more asymmetric-based leaflets than other even-pinnate meliacs. *Cedrela* is distinguished vegetatively by the more numerous asymmetric-based leaflets of its even-pinnate leaves and the usual presence of a rather rank, even somewhat garlicy, vegetative odor. *Swietenia* has fewer usually prominently glossy leaflets, also with very asymmetric bases. In flower, *Cedrela* has only 5 stamens, *Swietenia* 8–10.

Cedrela — C, E, P: cedro, cedro colorado, cedro blanco
Swietenia — P: caoba, águano

(***Melia***) — Native to Old World but widely cultivated and escaped, especially in dry areas. The only bi(–3)pinnate-leaved meliac; also unusual in serrate leaflets and vegetatively looks more like an Araliaceae than Meliaceae. The flowers, larger than in any native species except *Schmardaea,* are also distinctive.

P: cinamomo, paraíso

MENISPERMACEAE

In our area an almost exclusively scandent family (a few *Abuta* species are small flexuous treelets) characterized by alternate palmately veined, entire (or 3-lobed but with otherwise entire margins), simple (one species palmately compound) leaves. Although a few other families have climbers with alternate 3-veined leaves none of these has the typical wiry pulvinar flexion at the petiole apex as in most Menispermaceae. Menisperms with the petiole apex lacking the typical wiry thickening (*Disciphania, Odontocarya*) have the petiole base conspicuously flexuous. The lianas of the latter group have relatively soft flexuous stems, frequently with thick corky bark; the former group have stems with conspicuous cambial anomalies with asymmetrically concentric rings of secondary xylem (in many taxa the stem becoming strongly flattened with its "center" near one margin). Menisperms without the wiry pulvinar flexion at the petiole apex might be vegetatively confused with members of several other palmately veined vine families: entire-leaved cucurbits differ in having tendrils; *Aristolochia* and *Dioscorea* have only basal petiole flexions, the former often differing in leafy stipules and/or a more sagittate leaf shape and truncate sinus apex, the latter in the slender wiry branchlets with swollen nodes; *Ampelozizyphus* and *Sparattanthelium* lack any kind of anomalous stem structure, the former also differing in the relatively short stout nonflexed petiole, the latter in having reflexed branchlet bases thickened into "hooks".

All menisperms are dioecious and have rather small flowers (relatively large and green in a few *Disciphania* species); many are cauliflorous. The Annonaceae-like apocarpous fruit, typically consisting of (2–)3 sessile or subsessile monocarps, is distinctive; many of our species have ruminate endosperm and whether this occurs and the form of the variously folded seed are among the most important generic characters. *Cissampelos* is atypical in being a usually weedy vine and has small orange to red berrylike fruits; the other genera are lianas, *Odontocarya* and *Disciphania* with usually black or dark purple fleshy drupes; *Abuta, Anomospermum* and relatives with often red or red orange (or black) hard monocarps; and *Chondrodendron, Curarea,* and relatives with tannish- or grayish-pubescent hard monocarps, these 6 or more per flower in *Chondrodendron* and *Sciadotenia* (unique).

1. Weedy Herbaceous Vines with Solitary Nonapocarpous Fruit

Cissampelos (30 spp., incl. Old World) — Mostly weedy vines with membranaceous, often +/- peltate leaves (unique in area menisperms), usually with apiculate projection of midvein at apex of leaf (unique). Inflorescence with conspicuous leaflike bracts; fruits solitary (i.e., from single ovary per flower), small, round, orange or red-orange, berrylike.

2. Lianas with Pulvinately Flexuous Petiole Base (and Sometimes Also Pulvinate Apex) — Fruit 3 or more per flower (or fewer by abortion), glabrous and with nonruminate endosperm.

Odontocarya (12 spp.) — Leaves broadly cordate, usually membranaceous, typically with a broad shallow basal sinus but the lamina base also shortly extended at point where midvein and adjacent veins join petiole apex. Both male and female inflorescences typically cauliflorous, usually paniculate with a well-developed central rachis and numerous slender side branches more or less at right angles. In fruit the drupes pedicellate with round pedicels.

Disciphania (20 spp.) — Leaf usually broadly ovate, often cordate; the basal lobes of several species more angled than in *Odontocarya,* several other species distinctive in 3-lobed leaves and one species palmately compound. The mostly cauliflorous inflorescences are nearly always spicate and the usually succulent-fleshy black or blackish fruits (in our area) sessile.

Borismene (1 sp.) — Very similar to elliptic-leaved *Disciphania* and *Odontocarya* and with petiole similarly strongly pulvinate at base and apex; leaves distinctive in being rather thin-coriaceous and with finely, laxly prominulous reticulation. Twigs conspicuously finely striate. Fruit a succulent drupe as in *Disciphania* but pedicellate as in *Odontocarya.*

Menispermaceae
(Vines or Soft Lianas; Petiole Flexed at Base; Seeds Nonruminate)

1 - *Disciphania*

2 - *Odontocarya*

3 - *Odontocarya*

4 - *Borismene*

5 - *Cissampelos*

Menispermaceae
(Woody Lianas; Petiole Wiry-Pulvinate at Apex; Seeds Nonruminate)

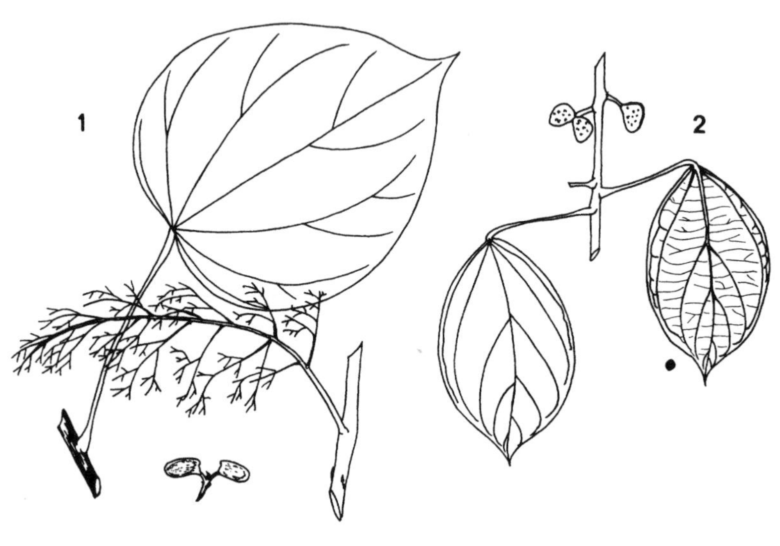

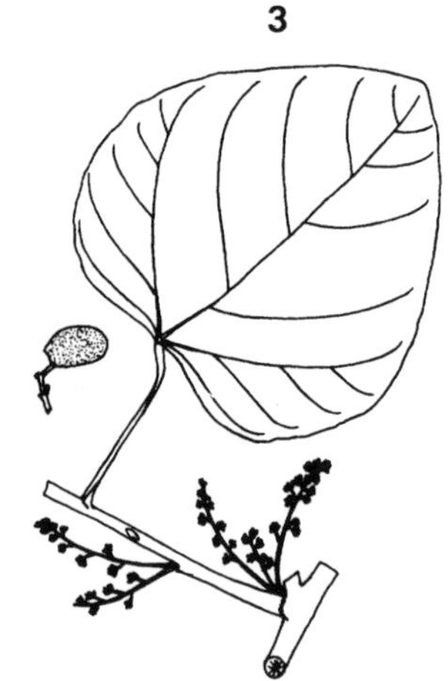

1 - *Curarea* 2 - *Sciadotenia*

3 - *Chondrodendron* 4 - *Hyperbaena*

3. Lianas with Petiole Flexed and Wiry-Pulvinate at Apex, Not Flexed nor Twisted at Base — (Endosperm ruminate or absent and replaced by fleshy cotyledons).

3A. The first five genera have seeds with fleshy cotyledons and lacking endosperm — (Fruits usually tomentose and tannish or grayish), leaves always palmately veined and frequently broadly ovate and densely whitish-tomentose below) (inner tepals valvate in first four, imbricate in *Hyperbaena*).

Chondrodendron (3 spp.) — Leaves densely white-tomentose below, broadly ovate, palmately 3–7-veined at base but otherwise pinnate-veined with several pairs of secondary veins, these beginning below middle of midvein. Drupes smaller and more cylindrical than relatives, usually >3 per pedicel, contracted near base to form short stipe, densely tomentose.

Curarea (4 spp.) — Very similar to *Chondrodendron*. Leaves more completely palmately veined than *Chondrodendron,* with only single pair of secondary veins arising from midvein well above middle; densely (usually white-) tomentose below. Leaves membranaceous to chartaceous, thinner than in similarly tomentose *Abuta* species. Technically differs from *Chondrodendron* in only 3 carpels (vs. 6) and the drupes sessile on the rays of 3-branched torus.

Cionomene (1 sp.) — Leaves coriaceous, +/- bullate, very broadly cordate with obtuse apex and white-tomentose below; vegetatively similar to *Chondrodendron* but venation more nearly pinnate, the more numerous straighter secondary veins strongly ascending and connected by parallel perpendicular tertiary veins below. Distinctive mainly in male flowers with the 5 mm long inner sepals united for most of length into a solid narrow tube at the apex of which are borne the small petals and stamens. Drupe tomentose and on 3-rayed torus like *Curarea* but very large (5 cm long).

Sciadotenia (18 spp.) — Similar to *Curarea* and *Chondrodendron* but the leaves either glabrous or soft-tomentose (but not white) below, usually not as broadly ovate, the secondary venation more strongly scalariform-parallel. Rather intermediate between *Curarea* and *Chondrodendron,* differing from the former in 6 rather than 3 carpels (and potentially drupes) per flower and from the latter in the drupes sessile on a stipelike column.

Hyperbaena (19 spp.) — Leaves small, glabrous, and not strongly 3-veined in the only South American species (*H. domingensis*), which vegetatively resembles *Orthomene schomburgkii,* the commonest species of that genus (though the cylindrical fruit completely different in lacking ruminate endosperm); leaves differ from that species in being more strongly 3-veined with the lateral vein pair farther from margin. The main technical character is the imbricate inner tepals.

3B. The next five genera have seeds with ruminate endosperm folded around embryo. — (The first three have the seed U-shaped; the last two straight or J-shaped).

Telitoxicum (6 spp.) — Our only menisperm genus with strictly pinnately veined leaves. In addition distinctive in the venation prominent or the surface very minutely and intricately beaded-reticulate below. A useful sequence (from not 3-veined to strongly so) in the genera with leaves similar to *Telitoxicum: Telitoxicum, O. schomburgkii, Hyperbaena, Anomospermum.*

Abuta (31 spp.) — Our largest menisperm genus and rather diverse vegetatively. Always with palmately 3–5-veined leaves and more or less parallel scalariform secondary veins. The fruit of 3 rather hard drupes, these sessile (but variously contracted at base) and usually rather cylindrical.

Caryomene (3 spp.) — A relative of *Abuta* from which it differs by the thick, woody drupes and the palmately 3–5-veined leaf blade densely papillate-pruinose below, drying pale green with straw-colored main veins and cartilaginous margin.

Anomospermum (8 spp.) — Leaves glabrous and strongly 3-veined, in most species with intricately reticulate glossy surface (cf., *Orthomene schomburgkii* but stronger, less marginal lateral veins), but common *A. grandifolium* (recognizable by leaf size) has smooth texture. Fruits more broadly ellipsoid (>1.5 cm wide) and attached more to one side than in *Orthomene* to accomodate the J-shaped (rather than straight) embryo.

Orthomene (4 spp.) — The commonest species (*O. schomburgkii*) has glabrous small leaves with very faint (only 3-veined by act of faith) *marginal* nerve pair and smooth lower surface from the reduced secondary venation. Second common species of our area has very different strongly 3-veined +/- bullate leaves and hirsute branchlets, petioles and main veins below. Fruits differ from *Anomospermum* in being straighter, more narrowly ellipsoid, and attached nearer base (the style scar thus terminal).

Two other north temperate genera reach Mexico and there are at least three more genera in central and eastern Amazonia. One of these, *Synandropus*, a monotypic lower Amazonian genus unique in having dentate leaves, was reported in the *Flora of Peru* as likely in that country but has not been collected in our area.

Menispermaceae
(Woody Lianas; Petiole Wiry-Pulvinate at Apex; Seeds Ruminate)

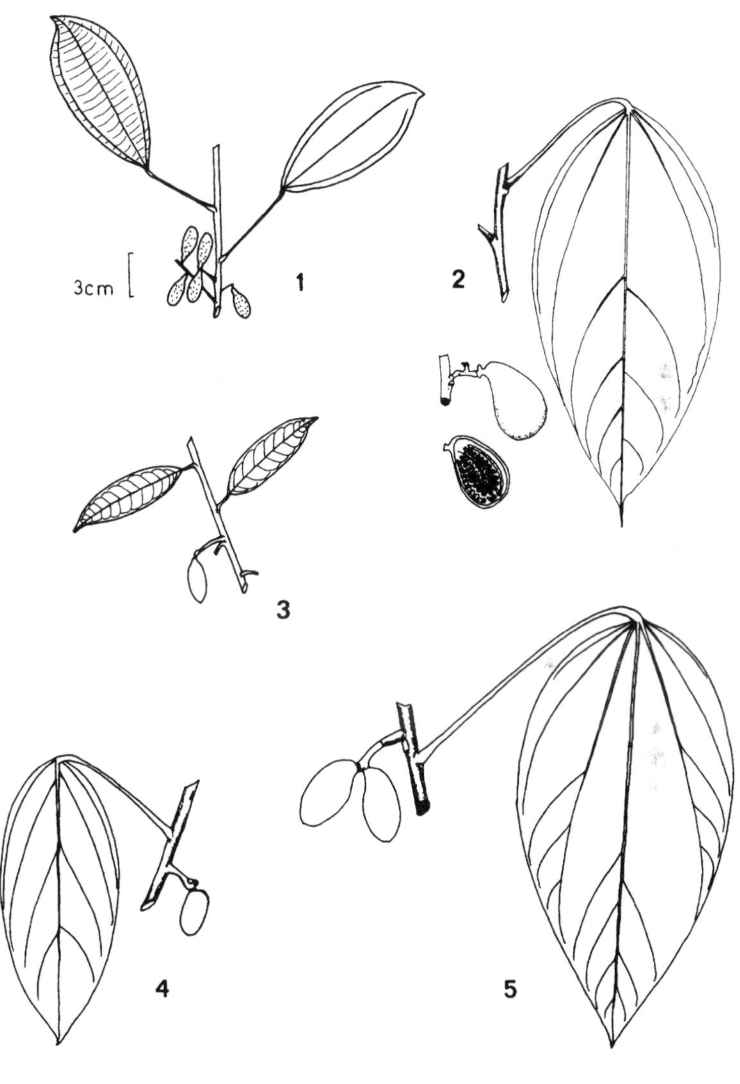

1 - *Abuta* **2** - *Caryomene*

3 - *Orthomene*

4 - *Telitoxicum* **5** - *Anomospermum*

Monimiaceae

Generally characterized by the combination of opposite leaves and primitive Ranalean odor. This is one of only two Ranalean families with opposite leaves (the other is very different Chloranthaceae with an obvious sheath enclosing the node [plus a very few Lauraceae]). Most monimiacs have more or less remotely serrate leaf margins unlike *Hedyosmum* which has close-together teeth or Lauraceae which are uniformly entire. Our two genera are not closely related and are easy to distinguish. *Mollinedia* species, usually with subcoriaceous leaves, often lack an obvious vegetative odor, but they always have a very distinctive leaf that is rather Myrtaceae-like except for a very few widely scattered teeth (and even some completely entire individual leaves). *Siparuna,* which always has a strong vegetative odor and usually more or less membranaceous leaves, has some species with entire leaves but these have stellate or appressed-stellate trichomes unlike any laurac. The tiny greenish or tannish flowers of Monimiaceae are dioecious or monoecious and the floral parts hidden inside the expanded receptacle and always borne on rather small axillary inflorescences (sometimes reduced to single flowers). The fruits are also characteristic with several seeds borne on an expanded receptacle (*Mollinedia,* which is thus quite reminiscent of the *Guatteria* group of Annonaceae except for the opposite leaves) or with fruitlets (which resemble small arillate seeds) enclosed inside a fleshy berrylike receptacle that eventually dehisces irregularly to expose them (*Siparuna*). The characteristic vegetative odor of Monimiaceae ranges from pungently foetid to pleasant and citruslike.

Mollinedia (80 spp.) — Small to medium-sized trees most prevalent in coastal Brazil but also fairly frequent in Andean cloud forests and occasional in lowland Amazonia. The leaves, superficially Myrtaceae-like, are unmistakable in their very few (often only 1–2) sharply accentuated marginal teeth (although some individual leaves of some species lack the teeth altogether) even though the vegetative odor is usually weak or lacking in our area; the rather few secondary veins, brochidodromous noticeably far from margin, are also distinctive. The annonac-like fruits, exposed on the receptacle surface throughout development, are more sharply jointed at their base than in Annonaceae, this especially obvious when the individual fruitlets are stipitate.

Siparuna (120 spp.) — Shrubs or usually small trees (very rarely to 30 m tall), widespread in both lowland and Andean moist and wet forests. The vegetative odor is always pronounced and may be foetid or citruslike and extremely pleasant. The usually membranaceous leaves, usually finely and somewhat irregularly serrate or serrulate along most of the margin, are

Monimiaceae
and Moraceae (*Dorstenia* [Herb] and *Poulsenia* [Prickles on Twigs and Stipule])

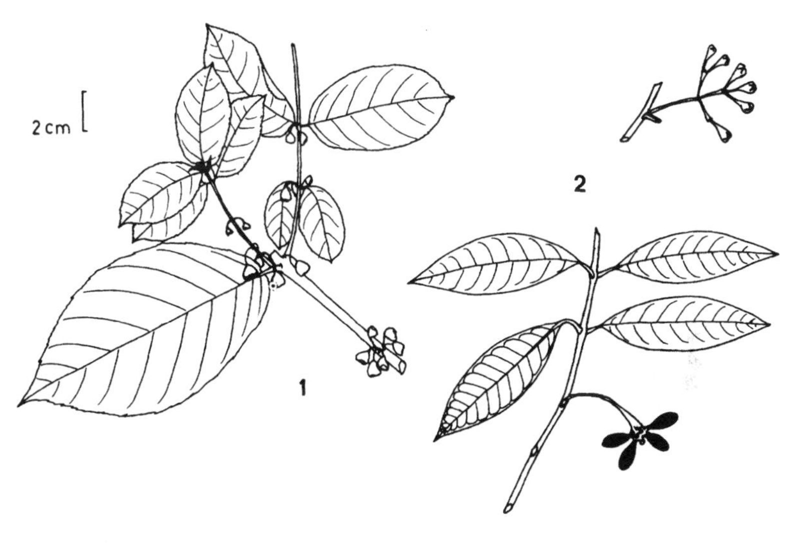

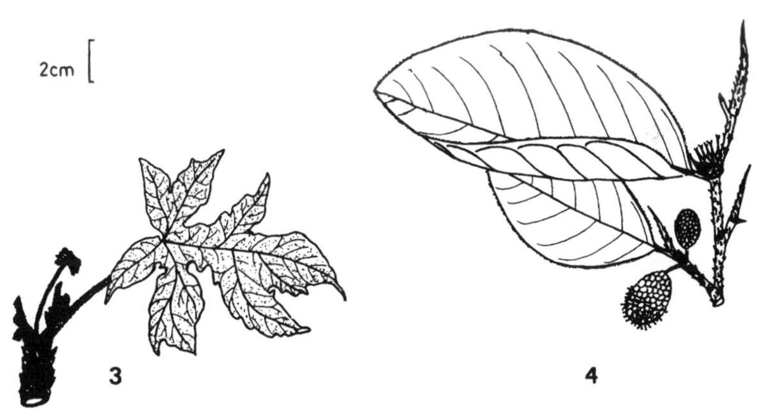

1 - *Siparuna* **2** - *Mollinedia*

3 - *Dorstenia* (Moraceae) **4** - *Poulsenia* (Moraceae)

usually distinctly stellate or appressed-stellate-pubescent; when entire, stellate trichomes or peltate scales are clearly evident below. The characteristic small flowers consist of a flat-topped receptacle with a central hole leading to the chamber in which the tiny stamens or apocarpous pistils are embedded. The red (occasionally yellow) fruits, berrylike from the fleshy enclosing receptacle, are very irregularly and tardily dehiscent at maturity.

E: limoncillo; P: picho huayo, izula huayo

There are several other genera in temperate South America (*Laurelia, Peumus*) and southeastern Brazil (*Hennecartia, Macropeplus, Macrotorus*) plus an additional monotypic genus in lower Amazonia, *Bracteanthus,* a rather large *Siparuna*-like tree but with much larger, indehiscent, yellow, mammal-dispersed fruits.

Moraceae

One of the most important neotropical tree families, especially on fertile soils, and by far the most important strangler family (*Ficus, Coussapoa*). The three main distinguishing features of Moraceae are: 1) a distinctive conical stipule that covers the apical bud (and leaves an obvious circular or semicircular scar on falling), 2) a characteristic leaf venation with the strongly marginally brochidodromous lowermost secondary veins either making a noticeably different angle with the midvein than the others or obviously closer together (or both), and 3) milky latex. The leaf venation character is always present and unmistakable with practice (although a few species of other families are easy to confuse until the Moraceae gestalt is well established). The conical stipule is an exceedingly useful characteristic but is not always apparent in *Sorocea, Trophis,* and allies, and in some other genera may not be discernible when the leaves are flushing. About half of the Moraceae have a uniquely colored pale tannish latex exactly the color of "cafe con leche" (but only in the trunk: twig latex is white); others have dark brown latexlike sap restricted to the terminal branchlets (Cecropieae), strongly tannish-yellow (but somewhat watery) latex (some species of *Naucleopsis*) or plain white latex (*Brosimum, Ficus,* etc.); a few species of *Trophis* and *Sorocea* lack obvious latex, at least in the branchlets. Moraceae tend to have smooth-barked cylindrical often more or less ringed trunks sometimes with thick basal bulges but mostly lacking prominent buttresses (except *Maquira, Poulsenia, Ficus*). A number of species have red pustules on the bark (e.g., *Maclura, Morus, Clarisia racemosa*). *Cecropia, Pourouma,* and *Coussapoa* (when terrestrial) have prominent stilt roots and lack obvious latex in the trunk. Although Moraceae are mostly trees or stranglers, one species of *Maclura* is a free-climbing liana; many of the trees are habitally distinctive in the self-pruning branches which tend to have a penislike shape (with predictable ethnobotanical significance) due to the enlarged hemispherical basal sector.

The flowers of all Moraceae are tiny, apetalous, and clustered into often complicated inflorescences of which the fig is the extreme. In some genera (e.g., *Morus, Cecropia*) both male and female flowers are in elongate inflorescences, spikelike or with spikelike branches; in others (*Sorocea*, most *Trophis, Clarisia*), the male inflorescence remains an elongate spike but the female one becomes a relatively few-flowered raceme (pseudoracemose in *C. racemosa*); in others (e.g., *Maclura, Batocarpus*) only the male inflorescence is spicate while the female flowers are in a globose cluster, while in yet others (e.g., tribes Brosimeae and Castilleae) both male and female flowers are in more or less dense globose inflorescences; finally in the extreme case of *Ficus* the flowers are on the inner surface of the hollow expanded receptacle.

One anomalous genus is herbaceous (*Dorstenia*). *Ficus* is also quite distinct from the rest of the family on account of its peculiar fruit as are *Cecropia* and its allies which are intermediate with Urticaceae, and *Poulsenia*, unique in the prickles on stipules, leaves, and twigs. Below, the rest of the genera are grouped by tribes — Castilleae, basically dioecious (often monoecious in *Castilla, Helicostylis, Perebea*) and with the male flowers grouped into globose clusters; Brosimeae, basically bisexual inflorescences, mostly with male flowers at least in part on surface of same expanded receptacle in which single female flower is immersed; and Moreae, with spicate male inflorescences and spicate to racemose female inflorescence.

1. Herbs

Dorstenia (Dorstenieae) (45 spp., plus 60 in Old World) — Totally distinct in being a herb, often stemless, also vegetatively characterized in our area by the triangular to deeply 3–5-lobed leaves. Only the inflorescence suggests that this is a Moraceae; the inflorescence is a flat receptacle at the end of a long peduncle with the upper surface densely covered by minute greenish or blackish flowers (essentially a pre-fig that has not yet closed to form a syconium).

2. Prickles on Twigs, Stipules, and Leaf Undersurface

Poulsenia (1 sp.) — A canopy or emergent tree restricted to fertile soils and occurring in both lowland and middle-elevation forests. Absolutely unmistakable in the small prickles on its stipule, twigs, and leaf undersurface. The rather broadly and angularly oblong leaves are distinctive, even without the spines, as are the rather abrupt "knees" on the narrow buttresses. The male inflorescence is a small globose cluster and the sessile irregularly angulate female inflorescence includes 3–9 individual flowers subtended by fleshy involucral bracts with free acute tips.

E: majagua, damagua

3. Tribe Cecropieae; Leaves Palmately Lobed or Very Broadly Ovate and Strongly 3-Veined, and with Closely Parallel Tertiary Veins More or Less Perpendicular to the Secondaries; Prominent Stilt Roots; Latex (When Present) Restricted to Shoot Apices and Dark Brown; Inflorescences Branched (Only the Male in Many *Coussapoa* Species) — Rather different from rest of family and sometimes placed in Urticaceae or segregated as a separate family but conveniently retained in Moraceae.

Cecropia (ca. 100 spp.) — One of the predominant genera of early second growth; a few species are also canopy components of mature forest, especially at higher elevations. Very distinctive in the palmately (actually radially) lobed (appearing compound in *C. sciadophylla*), peltate leaves, hollow internodes inhabited by ants, and the characteristic pulvinar area at the base of the petiole with glycogen-containing food bodies.

C: yarumo; E: guarumo; P: cetico, pungara (1 sp.)

Pourouma (25 spp.) — Large stilt-rooted trees characteristic of large light gaps in mature forest. Vegetatively intermediate between *Cecropia* and *Coussapoa.* Usually palmately 3–5-lobed and distinct; when deeply multilobed (e.g., *P. cecropiifolia*) can be told from superficially similar *Cecropia* species by the cordate rather than peltate base; when unlobed can be vegetatively distinguished from *Coussapoa* by the stipule scar perpendicular to the twig rather than oblique. Brown latex is usually present in the young branches. Some species have a strong wintergreen odor.

C: uva; E: uva, uvilla; P: uvilla (palmate), sacha uvilla (not palmate)

Coussapoa (45 spp.) — Mostly woody hemiepiphytes, sometimes becoming large free-standing stranglers. Leaves always unlobed, distinguishable from unlobed *Pourouma* leaves by the oblique stipule scar. Intermediate between this group and the rest of the family, thus providing a good reason not to segregate Cecropiaceae (although there are similar links to Urticaceae in the Paleotropics).

E: mata palo; P: mata palo, huasca topa

Moraceae
(Cecropieae)

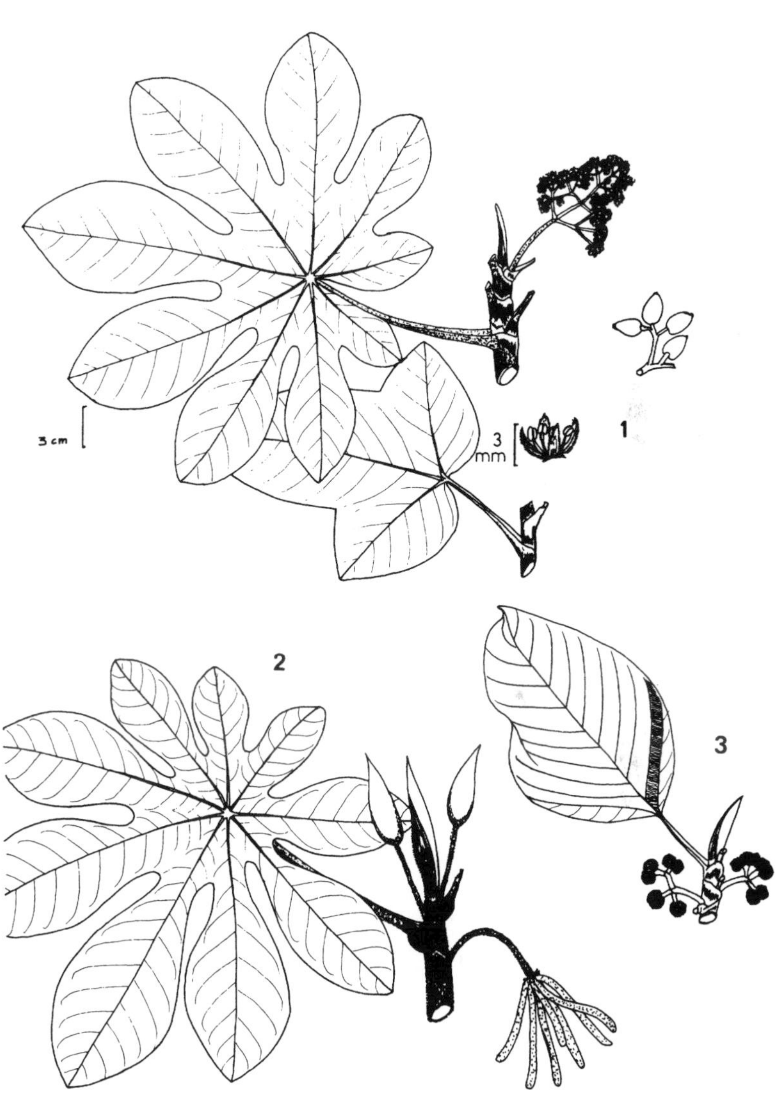

1 - *Pourouma*

2 - *Cecropia*

3 - *Coussapoa*

Moraceae
(Olmedieae)

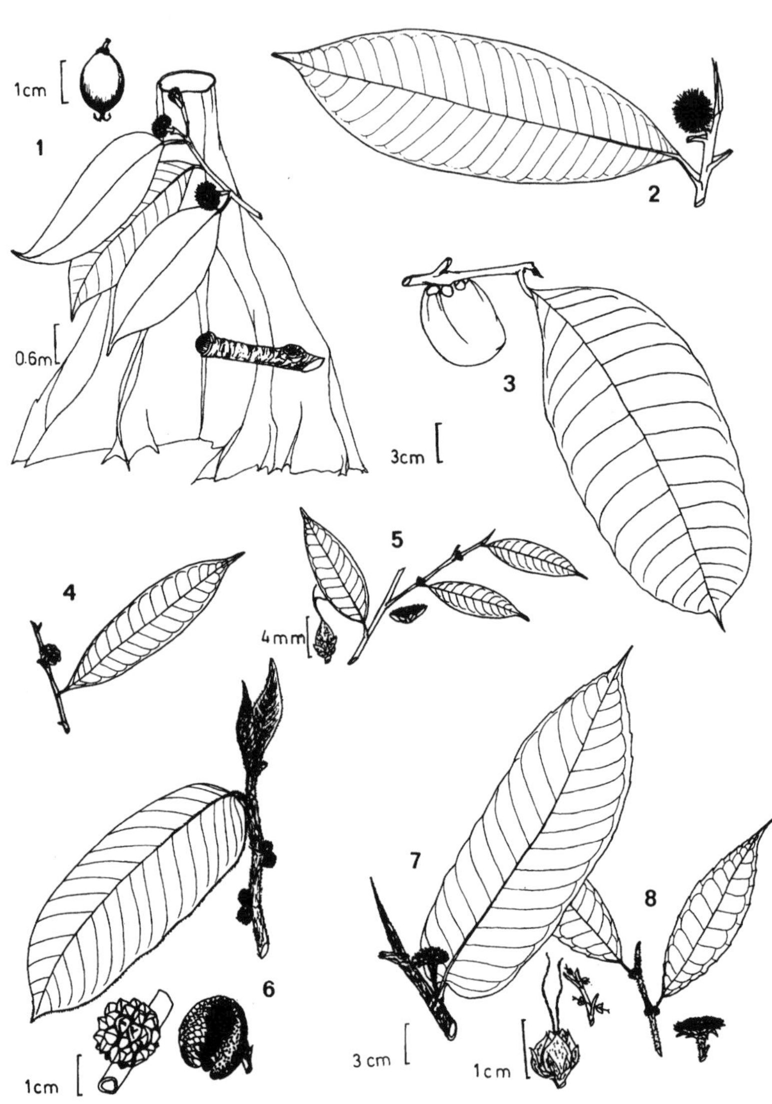

1 - *Maquira*

2 - *Naucleopsis*

3 - *Maquira*

4 - *Helicostylis*

5 - *Pseudolmedia*

6 - Castilla

7 - *Perebea*

8 - ''*Olmedia*''

4. Tribe Castilleae (plus *Olmedia*); the Large Core Group of Moraceae, Characterized by Being Mostly Dioecious, by Both Male and Female Flowers Clustered in More or Less Subglobose Axillary Inflorescences, and by Tannish (Rarely Tannish-Yellow; White in *Castilla*) Latex (Also Found in Some *Sorocea* Species) — The bark is always smoothish and grayish except for the presence of reddish pustules in *Castilla.* This group has the first three genera below with broken stipule scars, the latter four with continuous scars forming a ring around node.

4A. Stipule scars discrete, not forming rings at node

Olmedia (1 sp) — A single common easy to recognize species especially prevalent in somewhat disturbed areas along streams. The leaves are scabrous and dentate to denticulate with an unusual rather falcate shape. The pistillate inflorescences have single discrete flowers not densely clustered into the typical morac inflorescence.

P: llanchama

Maquira (5 spp.) — A rather nondescript morac. The leaves glabrous or sparsely puberulous above and drying characteristically greenish and shiny above; secondary veins forming broad angle with midvein, fewer and spaced farther apart than in somewhat similar *Brosimum.* Pistillate inflorescence typically with several to ten flowers but occasionally one-flowered (*M. coriacea*). The latex is usually "cafe con leche", but cream in *M. coriacea,* a very characteristic large-buttressed emergent tree of Amazonian tahuampas.

P: capinurí (*M. coriacea*)

Helicostylis (7 spp.) — Leaves may be narrow-based or broadly elliptic and *Brosimum*-like but vegetatively characterized by being almost always noticeably hairy below and on twigs, drying dull and brownish above. One common middle-elevation species (*H. tovarensis*) is usually somewhat serrulate apically and is also unusual in single-flowered female inflorescence. Stipules small. Female inflorescences usually several-flowered, discoid or subglobose.

4B. Stipules falling to leave a ring completely surrounding twig at each node.

Castilla (3 spp.) — Large trees characterized by typical branching, a characteristic bark with narrow, horizontal, reddish, raised pustules, and white latex. The leaves are distinctive, hairy, finely serrate, and rather oblong; in the commonest species they have a prominently cordate base. The prominent stipules are hairy and completely connate into a terminal cone and the twigs are also conspicuously hairy.

C, E, P: caucho

Pseudolmedia (9 spp.) — Mostly large canopy and emergent trees with "cafe con leche" latex. The leaves are usually rather small and/or narrow, often with an acute and distinctly asymmetric base; the secondary veins tend to be less prominent below than in most moracs; although mostly macroscopically glabrescent, the lower leaf surface has distinctive small capitate trichomes. Pistillate inflorescences sessile and one-flowered (cf., *Olmedia* but leaves not scabrous). Fruit a single, small, red, ellipsoidal berry with the stigma residues persistent.

E: guión; P: chimicua

Naucleopsis (20 spp.) — Mostly midcanopy or understory species. Leaves entire, usually glabrous (if hairy, the hairs single-celled), at least middle-sized and often distinctively large and oblong; the leaf base is more or less decurrent onto the lateral ridges of the grooved petiole, a character unique to this genus. Several species have a unique tannish-yellow watery latex. The female inflorescence is disk-shaped with the ovaries immersed in the upper surface while the male inflorescence is characterized by having enlarged almost petaloid involucral bracts. The fruit is usually large, green to brown, rather "spiny"-appearing from the projecting bracts.

C: caimo chicle

Perebea (9 spp.) — Mostly small to midcanopy trees of lowland forest. Leaves hairy (always with several-celled capitate hairs), rather unusual in being usually dentate to denticulate, middle-sized to largish, commonly rather asperous. Twigs pubescent but often with shorter trichomes than *Castilla.* The inflorescences are disk-shaped; in the female the ovaries are not immersed (unlike *Naucleopsis*), in bud the male is open and flat-topped unlike the globose closed buds of *Naucleopsis.* Fruits usually pubescent, borne several together on the same receptacle (cf., *Maquira,* but that genus essentially glabrous-leaved).

P: chimicua

5. Brosimeae; Differs from Castilleae in Basically Bisexual Inflorescence and in Female Flowers Immersed in Enlarged More or Less Globose Receptacle — Vegetatively distinct in hook-shaped hairs but these tiny and not macroscopically apparent in most taxa. Latex white.

Brosimum (13 spp.) — Large canopy trees with smooth (strikingly reddish in some species) trunks and white latex; less conspicuously buttressed than the somewhat similar nonstrangling species of *Ficus.* The leaves macroscopically appearing glabrous, tending to be distinguished by broadly elliptic shape with broad base and close-together, evenly brochidodromous secondary veins nearly perpendicular to midvein. The conical stipule always conspicuous and sometimes very long. Often dioecious, the characteristic female inflorescence with a single central female flower immersed in a globose receptacle which is more or less covered with peltate bracts and with the bifid stigma sticking out from its apex; sometimes with male flowers on surface of same receptacle. In fruit, looking rather like a fig but with a single large seed instead of a hollow center. Latex white.

C: palo vaca, sande (*B. utile*), veneno (*B. guianense*); E: sande (*B. utile*), tillo (*B. alicastrum*); P: chimicua, tamamuri

Trymatococcus (3 spp.) — A small tree, mostly in somewhat disturbed areas on poor soil. In our area (*T. amazonicus*), easy to distinguish vegetatively by the characteristic small leaves that are obviously puberulous and more or less scabrous below. When dry these leaves resemble *Sorocea* with prominent whitish-drying venation below but differ in the completely entire margin. The bisexual inflorescence is somewhat like that of *Brosimum* with a single female flower immersed in the receptacle but differs distinctively in having an apical "cap" composed of the clustered male flowers; the fruit is conspicuously scabrous.

Helianthostylis (2 spp.) — Small understory tree. Leaves similar to *Trymatococcus* in being smallish, with pale-drying venation prominent below and having a usually noticeably puberulous (sometimes more or less scabrous) leaf undersurface, but are easily distinguished by the more caudate-acuminate acumen. The globose male inflorescences are very distinctive because of the long thin pistillodes sticking out sea-urchin-like on all sides; bisexual inflorescences and fruits are similar to *Brosimum* but the surface densely tannish-puberulous.

6. Moreae (Incl. Artocarpeae Which Differs in Non-exploding Stamens); Spicate or Racemose Male Inflorescences (Reduced to Discoid-Capitate in Related *Olmedia aspera*, See Above); in the First Three Genera the Female Inflorescence Is Racemose, in *Morus* It Is Spicate, and in *Maclura* and *Batocarpus* It Is Reduced and Globose — Vegetatively, these genera are characterized by the usually rudimentary nature of the conical terminal stipule (which never leaves a ring) and by the marked tendency in most of them (except *Clarisia*) to have serrate or at least serrulate leaf margins (when the margin is nonserrate the undersurface is often somewhat scabrous). These genera (except *Sorocea*) tend to have unusually membranaceous leaves for Moraceae and the tertiary venation below typically is prominent and dries whitish. Our only spiny morac (see also *Poulsenia* with prickles) belongs to this group. *Maclura, Morus,* and *Clarisia* usually have conspicuous reddish raised pustules on the bark.

Moraceae
(Moreae)

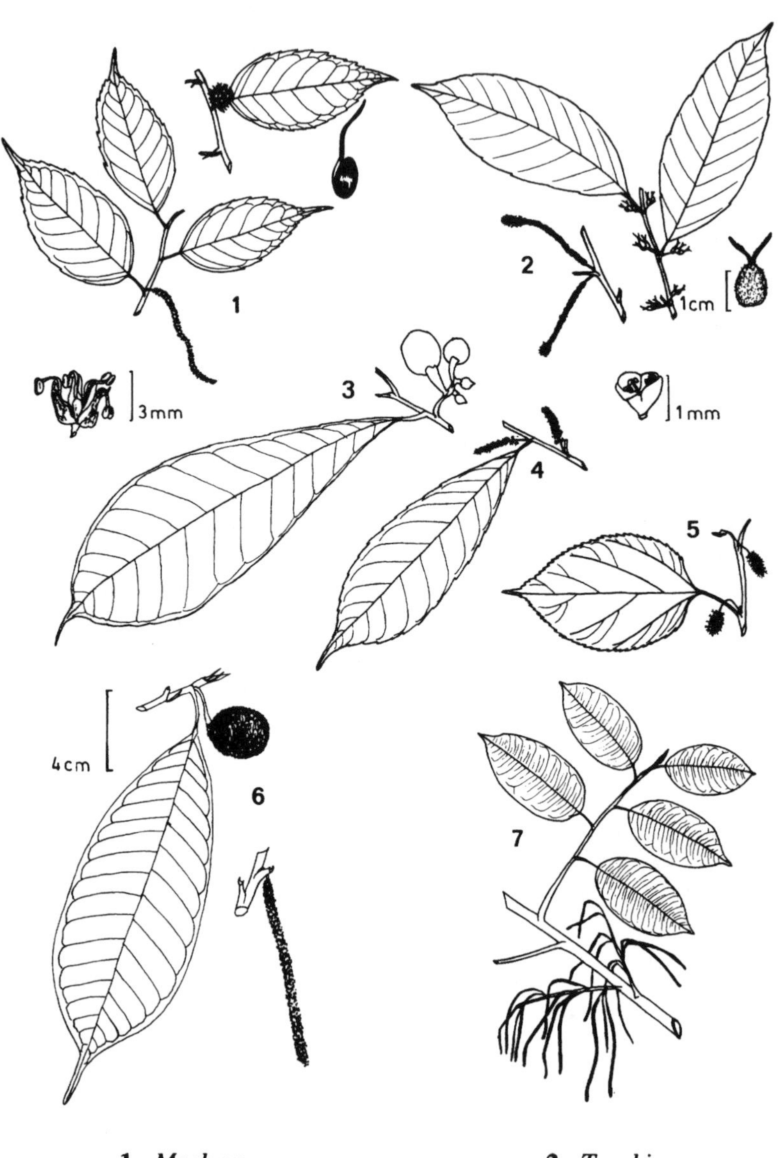

1 - *Maclura* **2** - *Trophis*

3 - *Sorocea* (fruits) **4** - *Sorocea* (male) **5** - *Morus*

6 - *Batocarpus* **7** - *Clarisia*

Moraceae (Brosimeae; Ficeae) and Myricaceae

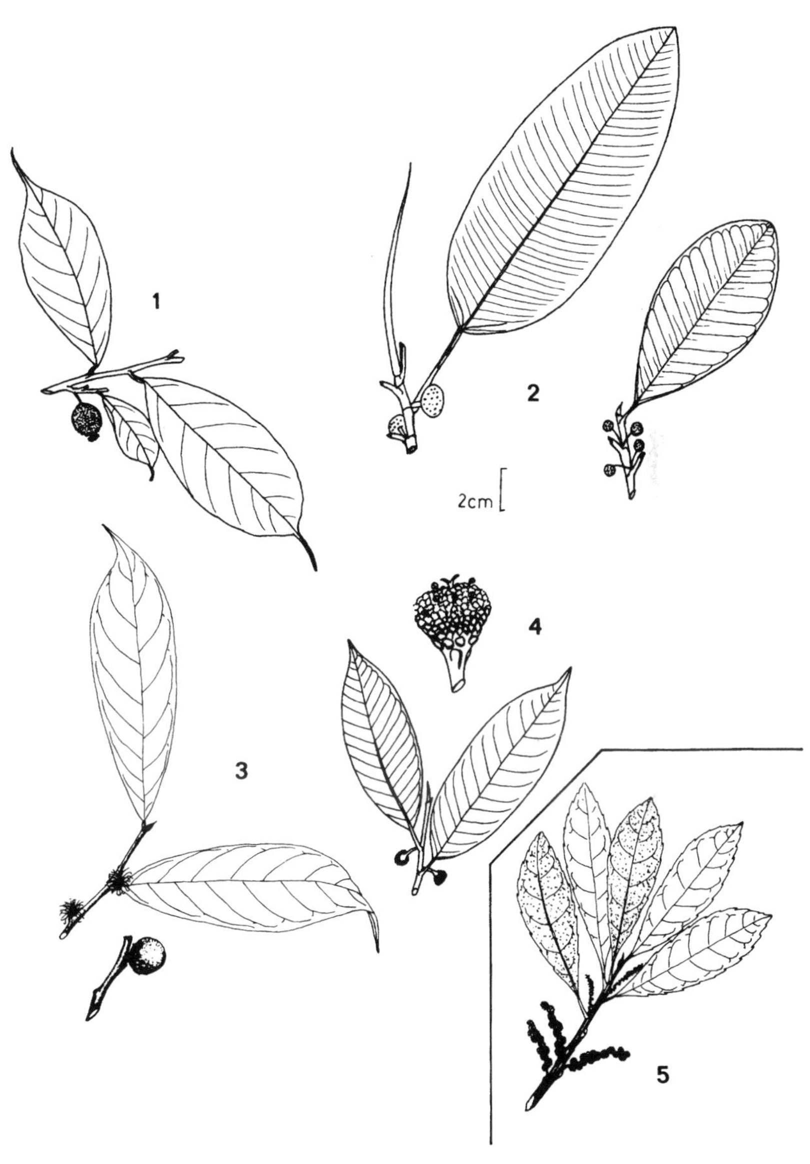

1 - *Trymatococcus* **2** - *Ficus*

3 - *Helianthostylis* **4** - *Brosimum* **5** - *Myrica* (Myricaceae)

Sorocea (17 spp.) — Small to midcanopy trees, usually characterized by more or less serrate or serrulate leaves (sometimes somewhat asperous below when entire) with conspicuously whitish tertiary venation below, even when fresh. Female inflorescence narrowly racemose, the small, fleshy, single-seeded berrylike fruits black and lacking an apical stigma residue.

E: tillo prieto

Trophis (5 spp., plus 4 in Old World) — Small to midcanopy trees of lowland forest. One of the least obviously moraceous of all moracs due to the lack of a well-developed terminal stipule (often present only as an act of faith) and the rather inconspicuous watery latex. As in *Sorocea,* the leaves tend to dry with whitish venation below and the margins are usually somewhat serrate or serrulate. The female inflorescence (except *T. racemosa* where spicate) is racemose as in *Sorocea* but is very different in the conspicuous bifid stigma that also persists on the fruit apex; the red berrylike fruits are less fleshy than in most *Sorocea* species.

E: cuchara

Clarisia (3 spp.) — Large canopy trees of rich-soil lowland forests (except treelet *C. ilicifolia,* with sharply rigid-toothed leaves, which barely reaches Madre de Dios). Completely lacks the moraceous terminal stipule. Also vegetatively characterized by tendencies toward conspicuously asymmetric leaf bases, caudate-acuminate apices, and intersecondary veins more or less paralleling the secondary veins. The contrastingly brownish petiole is also a useful character, especially in *C. biflora* which is otherwise less distinctive. The single-flowered or few-flowered female inflorescence is axillary and usually paired (*C. biflora*), racemosely arranged (*C. racemosa*), or in a capitate ramiflorous cluster (*C. ilicifolia*). The typical trunk has raised reddish pustules (especially *C. racemosa*) differing from *Maclura* in its evenly cylindrical shape. Fruits globose and single-seeded, red (*C. racemosa*) or green (*C. biflora*).

E: tillo serrano (*C. biflora*), moral bobo (*C. racemosa*); P: capinurí (*C. biflora*), tulpay or mashonaste (*C. racemosa*)

Morus (2 spp., plus 1 N. Am. and 4 Old World) — In our area restricted to Andean forests although the European mulberry is cultivated in dry areas. Characterized by the finely and evenly serrate leaflets, sometimes with subpersistent paired stipules; the venation may dry light or dark below, the tertiary network often conspicuously raised-reticulate. Unique in the female inflorescence spicate, like the male. The fruiting inflorescence much narrower and less contracted than in the cultivated mulberry

Maclura (incl. *Chlorophora*) (2 sp. plus 1 in N. Am. and 8 in Old World) — Common species (*M. tinctoria*) a large canopy tree of dry forests and wetter forests on rich soil. Distinctive in the usually spiny branchlets. Leaves characteristically membranaceous, dark-drying, and with an obliquely truncate base, few veins, and a more or less serrate margin. Female flowers in round cluster. The very characteristic trunk has raised reddish pustules, differing from similar *Clarisia* in the usually noncylindrical shape. Second species (*M. brasiliensis*) is a large liana, recently discovered to reach southern Peru.

C: palo mora; E: moral fino; P: insira

Batocarpus (3 spp.) — Canopy trees of rich-soil forests. Distinctive in the largish leaves that are usually more or scabrous below and often irregularly serrate. The venation usually dries whitish below as in *Sorocea* but the female flowers and fruits, borne in a globose cluster, are totally different. The segmented fruit, reminiscent of that of *Duguetia* of the Annonaceae, is tannish-puberulous.

7. Ficeae (Figs); Characterized by the Unique Fruit Consisting of a Round Hollow Invaginated Receptacle with the Inner Surface Lined with the Tiny Flowers or Individual Fruitlets — The great majority of our species are stranglers and nearly all true stranglers in our area are figs.

Ficus (ca. 150 spp., plus 600 in Old World) — By far the largest and perhaps the best known genus of Moraceae. When fertile *Ficus* is absolutely unmistakable because of its unique hollow fruit with the inner surface lined with the tiny flowers or fruitlets; the apical pore through which winged females of the mostly species-specific pollinating wasps enter the fig is closed by a complicated series of bracteoles. The majority of *Ficus* species are stranglers, easily distinguished from *Coussapoa,* the only other genus of Moraceae strangler by the conspicuous white latex and the typical morac leaves lacking the parallel tertiary venation of *Coussapoa.* The conical terminal stipule is always well-developed in *Ficus.* Nonstrangling species of *Ficus* are most similar to *Brosimum* on account of the white latex, well developed terminal stipule and typically close-together secondary veins making a wide angle with the midvein but differ in having longer and/or more slender petioles.

C and E: mata palo (stranglers), higuerón (free-standing); P: mata palo (stranglers), ojé (*F. insipida*)

Myricaceae

Trees (sometimes small and shrubby) of upper elevation cloud forest. Vegetatively unique in alternate simple strongly yellow-gland-dotted leaves; the young growth very densely lepidote-glandular and macroscopically yellowish or tannish. Leaves mostly oblanceolate and more or less coriceous, the margin usually more or less toothed. Twigs strongly ridged from the decurrent petiole bases. Flowers apetalous, in axillary spikes. The round gray-green berrylike fruits with tuberculate surface.

Myrica (ca. 30 spp., mostly in Old World; 2 spp. in the Andes)

Myristicaceae

Easy to recognize to family, when sterile, by the combination of Ranalean odor, regularly spaced 2-ranked leaves, myristicaceous branching, and thin red latex. The leaves of the two commonest genera tend to be more oblong and have more numerous secondary veins than in related families and the petioles are usually short and thick in these genera. The twig bark is not strong, unlike Annonaceae; on the trunk, the bark is often either vertically ridged or peeling in rather thick fibrous plates; stilt roots are typical of some swamp species. The distinctive red latex is often somewhat thin and watery at first; in a few minutes it almost always (except *Osteophloeum*) turns at least distinctly pinkish and usually obviously red. Almost all myristicacs are medium to large trees (*Compsoneura debilis* is a shrub of white-sand catingas with subscandent branches; a few *Iryanthera* species are treelets). The plants are dioecious with tiny tan to greenish 3-parted flowers variously arranged in open multi-flowered panicles (male *Virola*), racemes (most *Iryanthera*) or contracted to few-flowered fascicles (female flowers in many species of several genera); many species of *Iryanthera* are strikingly cauliflorous, others are at least ramiflorous. The fruit of this family is absolutely distinctive, the mesocarp splitting in half to reveal a single large aril-covered seed; the aril is usually bright red (white in *Otoba* and some *Compsoneura*) and frequently conspicuously laciniate (*Virola, Otoba*). The endocarp of some genera (*Virola, Otoba*) is ruminate. Corner says the Myristicaceae seed is the most primitive of any angiosperm.

At least in America, this family is easier to recognize to genus by sterile characters than by fertile ones. *Compsoneura* is easily recognized by the parallel tertiary venation; *Virola* by stellate trichomes, *Iryanthera* and *Otoba* by T-shaped trichomes, the latter with a glaucous or tannish leaf undersurface; *Osteophloeum* has straw-colored latex and a characteristic leaf shape with cuneate base and relatively long petiole.

Compsoneura (9 spp.) — Tertiary venation conspicuously finely parallel (absolutely distinctive in family but also found in a few other families (e.g., Olacaceae, Icacinaceae) all of which lack Ranalean odor and red latex). This genus has more anthers (5–10) than the other neotropical genera (except *Osteophloeum*); a weakly developed thin subentire aril and nonruminate endosperm are other characters suggesting that it may be primitive in the family. The fruit is oblong-ellipsoid, with an entire aril (usually red); a few Chocó area species have white arils and a lepidote-stellate indument. The seeds of some of these are edible.

C: cuangare, sebo, castaña (*C. atopa*)

Virola (40 spp.) — The main myristicac genus. Trichomes of young branchlets and petioles (and usually leaf undersurface) are stellate (unique except for irregularly lepidote-stellate trichomes in monotypic *Osteophloeum* and a few Chocó area *Compsoneura*). Male inflorescence usually openly paniculate (unique). Seeds, and usually fruits, always longer than broad, the latter sometimes conspicuously villous (unique), the seed surrounded by a conspicuously laciniate red aril. The red latex is made into a hallucinogenic snuff by some Amazonian tribes.

C: guangaré, sebillo; E: chispiador, freta dorada (*V. koschnyi*), chalviande (*V. sebifera*); P: cumala, cumala negra (*V. decorticans*), águano cumala (*V. albidiflora*), caupuri (*V. pavonis, V. surinamensis* [stilt roots])

Osteophloeum (1 sp.) — A large and common but poorly collected monotypic tree. The persistently straw-colored tannish latex is a good field character (some species of other genera may have initially watery latex but this soon turns red or pinkish). With experience, easy to recognize by the characterisitc round-tipped, cuneate-based, narrowly oblong-obovate leaves with relatively long petioles; these are somewhat intermediate between *Virola* and *Iryanthera,* with stellate-lepidote trichomes like some species of the former but a coriaceous texture and glabrescent aspect like the latter. The fruits are also somewhat intermediate between *Virola* and *Iryanthera,* being very slightly broader than long at maturity. The rarely collected male flowers have more anthers (12–14) than any other Myristicaceae except one primitive Madagascar endemic. The trans-Andean species is apparently not specifically distinct from the Amazonian one.

C: chucha; P: favorito, cumala blanca

Iryanthera (23 spp.) — The second most important myristicac genus, especially in Amazonia. Vegetatively distinct in the malpighiaceous trichomes (as in *Otoba* but lacking a glaucous leaf undersurface). Typically the leaves are strikingly coriaceous, glabrous-appearing and narrowly oblong with close-together prominent brochidodromous secondary veins. A few species have smaller elliptic, but still glabrous-appearing, leaves. The distinctive fruits (and enclosed seeds) are broader than long (less obvious in the largest-fruited species), also differing from most *Virola* in subentire arils. The flowers are mostly cauliflorous or ramiflorous, the inflorescences usually densely racemose (but more open in the small-leaved species). One

Myristicaceae
(Leaves and Fruits)

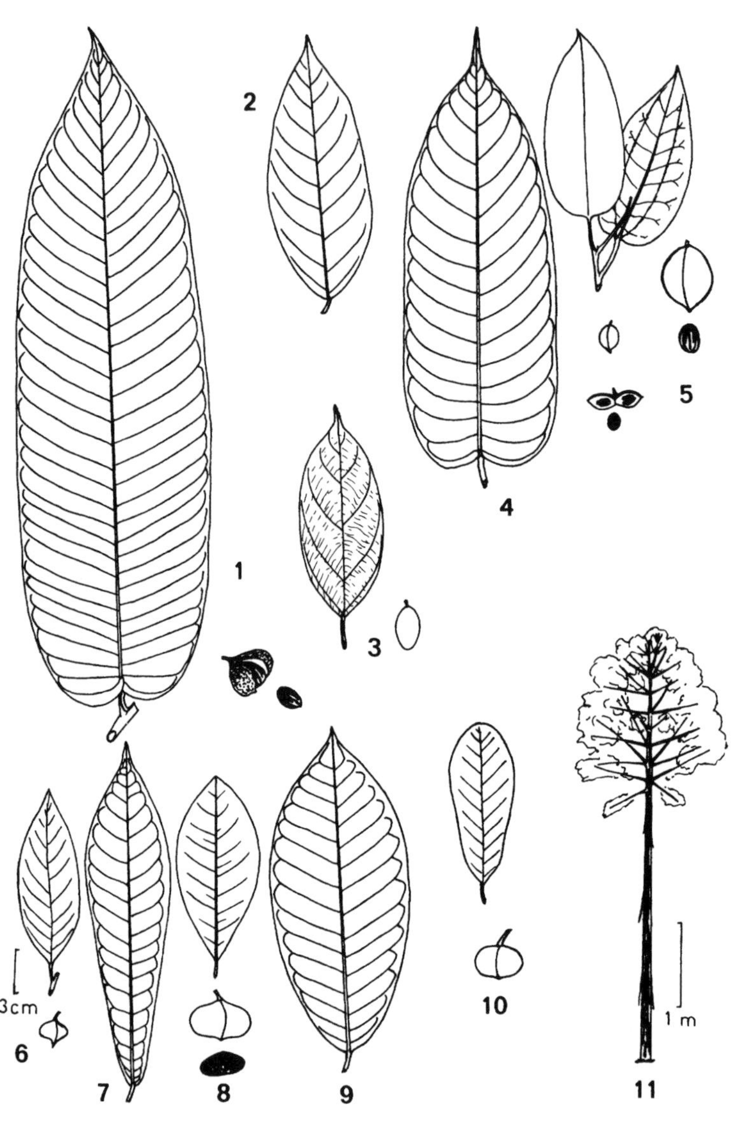

2 - *Virola* **4** - *Virola* **5** - *Otoba*

1 - *Virola*

3 - *Compsoneura*

6–9: *Iryanthera* **10** - *Osteophloeum* **11** - *Iryanthera* (*I. juruensis*)

Myristicaceae
(Inflorescences and Fruits)

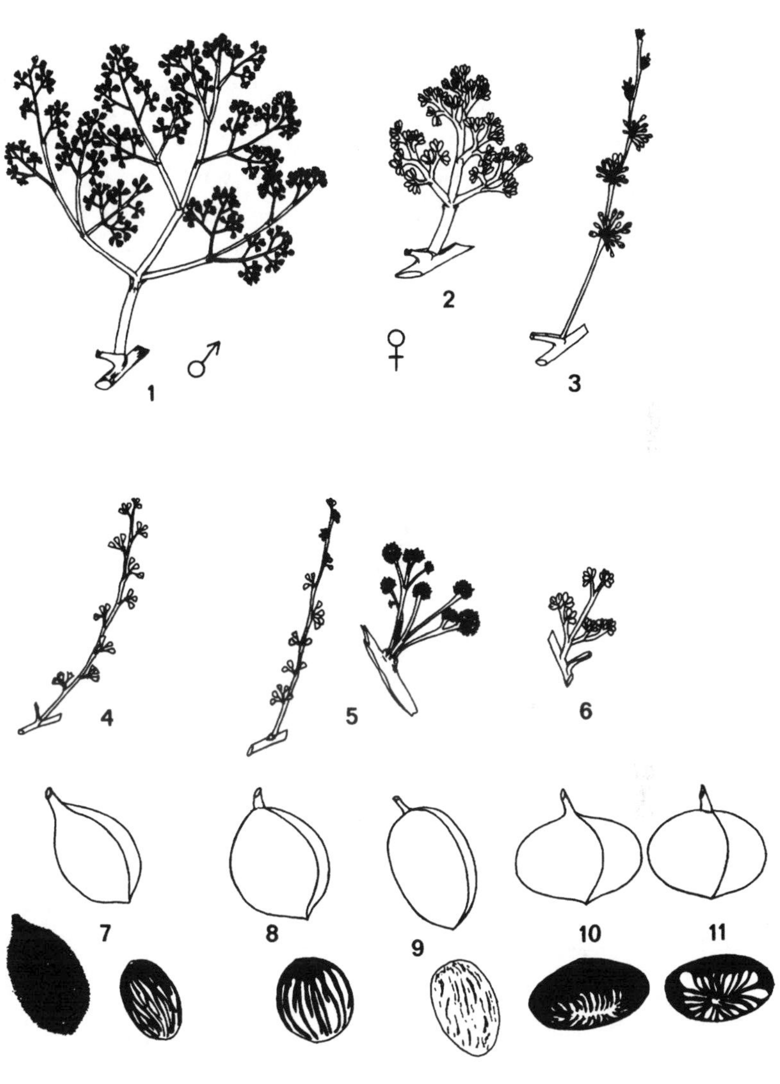

1 - *Virola* (male) **2** - *Virola* (female) **3** - *Otoba*

4 - *Compsoneura* **5** - *Iryanthera* **6** - *Osteophloeum*

7 - *Virola* **9** - *Compsoneura* **11** - *Osteophloeum*

8 - *Otoba* **10** - *Iryanthera*

species is famous in Amazonian Peru as "pucuna caspi", the tree from which blow guns are made.

C: cuangare; P: cumala colorada, pucuna caspi (*I. juruensis*)

Otoba (7 spp.) — Leaves very distinctive in the smooth glaucous or tannish undersurface and the barely prominulous secondary veins which usually fade out well before the margin; vernation lines paralleling the midvein are often conspicuous. Shares the exclusively 2-branched malpighiaceous indument of *Iryanthera* but is otherwise more like *Virola* (with which it uniquely shares ruminate endosperm and a laciniate aril). The fruits are round (sometimes with an elongate apex), green at maturity, and are unique (except for a few *Compsoneura*) in having the aril thin and white; presumably this genus is bat-dispersed as contrasted to the bird-dispersed red-arillate remainder of the family. *Otoba* is the dominant genus in many rich soil upper Amazonian forests.

C: otobo; E: chispiador, sangre de drago; P: aguanillo

MYRSINACEAE

Alternate glandular-punctate estipulate leaves characterize all Myrsinaceae. The distinctive, except in vegetative buds mostly nonpellucid, punctations (storing secondary metabolites rather than secretory fide J. Pipoly) are often elongate, may be blackish or reddish (except in vegetative buds), and are also obvious on petals and calyx. The punctations tend to be most visible *away* from the light and are associated with a characteristic pale green "matte" undersurface in most myrsinacs. In most species the branchlets (sometimes even the pith) have canals in the secondary phloem with the same secondary metabolites that are in the leaf punctations. Typically the leaves are fairly coriaceous and the secondary veins below are often very inconspicuous; a more or less rufous indumentum of usually appressed (usually branched or stellate) scales or trichomes is often present, especially on the young growth. The leaves are usually entire but serrate margins occur occasionally in several genera (*Ardisia, Parathesis, Cybianthus, Geissanthus*). Myrsinac flowers are usually small, rather frequently 4-petaled, and unusual in having stamens opposite the petals. Although the petals are basally fused, this is usually not very obvious. The fruit is always a small one-seeded berry, usually globose and with punctations visible to the naked eye; the essentially separate calyx lobes, also glandular, are persistent at its base. The family is most prevalent at middle elevations, especially in relatively exposed wind-swept situations in cloud-forest environments.

Intrafamilial taxonomy tends to be based almost entirely on technical floral characters and on inflorescence type and position. Perhaps the most distinctive genus is *Myrsine* with sessile or subsessile clusters of axillary and ramiflorous flowers. The other two large core genera are *Ardisia* and *Cybianthus*, the former with paniculate inflorescences having cymose or glomerulose flowers, the latter with the inflorescences usually racemose or with racemose branches, when paniculate usually with large clustered leaves and a more or less pachycaul growth-form; except for the latter group (ex-*Weigeltia*), the two genera are often difficult to distinguish vegetatively although individual species are distinctive.

Myrsine (incl. *Rapanea*) (ca. 50 spp., plus 150 in Old World) — Small trees typical of middle-elevation forests especially in rather exposed situations. Distinctive in the sessile axillary and ramiflorous flowers, typically densely clustered along the twigs below the leaves on suppressed short-shoots. Leaves with obscure or barely prominulous strongly ascending secondary veins, usually rather narrow and subcoriaceous, often slightly asymmetrical, sometimes much smaller and thicker, then with distinctive inrolled margins; young branchlets and petioles usually with more or less appressed-rufous indumentum. The traditional segregation of *Rapanea* with unisexual flowers from mostly paleotropical *Myrsine* is quite untenable.

C: trementino

Ardisia (ca. 100 spp., plus ca. 300 in Old World) — Shrubs and small trees commonest in middle-elevation cloud forests and more diverse in southern Central America (and Asia) than in South America. This is the large morphologically diverse grab-bag core genus of Myrsinaceae and most myrsinacs that seem atypical of the family belong to *Ardisia*. The inflorescence is always paniculate, the lateral branches with the usually 5-parted flowers arranged in glomerules or cymes, typically more or less umbellate. The stamens are always included, the anthers rather long, and the filaments usually short.

Gentlea (6 spp.) — Restricted to cloud and elfin forests. Essentially an *Ardisia* with exserted anthers and long styles; the persistent style also characterizes the fruit. Vegetative parts completely glabrous and leaves smaller and thinner than in most myrsinacs are suggestive of *Gentlea*.

Parathesis (75 spp.) — Similar to (and vegetatively difficult to distinguish from) *Ardisia* except for the valvate perianth and densely pubescent petals and calyx. Two vegetative characters that are strongly suggestive of *Parathesis* are branchlets stellate-tomentose (except *P. glabra*) and a "bizonal" lower leaf surface with stellate pubescence dense near midrib and sparser near margin. The flowers, always 4-merous, have unusually large yellow anthers and are always in panicles. The fruit is distinctive in being ribbed and is also characterized by the pubescent persistent calyx lobes.

Myrsinaceae

1 - *Parathesis*

2 - *Geissanthus*

3 - *Cybianthus*

4 - *Grammadenia*

5 - *Cybianthus* (*Conomorpha*)

6 - *Myrsine*

7 - *Ardisia*

8 - *Stylogyne*

9 - *Stylogyne*

Geissanthus (20 spp.) — Andean shrubs, vegetatively similar to an *Ardisia* or *Cybianthus* with largish leaves and unusually thick twigs but generally with larger leaf scars and more conspicuous denser (black?) punctations. The characteristic large paniculate inflorescence is always terminal unlike *Myrsine* or *Stylogyne*). The main technical character is a calyx that is closed in bud, irregularly rupturing into 2–7 lobes which are also visible in fruit.

Cybianthus (incl. *Conomorpha, Weigeltia,* and perhaps *Grammadenia*) (ca. 150 spp.) — Mostly middle-elevation cloud forests but also in Chocó and lowland Amazonia; rather variable in habit from small subshrubs to subcanopy trees. Leaves usually subcoriaceous, the secondary veins rather close together and more nearly perpendicular to the midvein than in *Myrsine* but often suppressed; leaves and terminal buds often more or less finely rufescent; one section has lepidote scales (macroscopically appearing as granular projections) on the twigs, these unique in the family. One subgenus (the former *Weigeltia* and its allies) is distinctive in having mostly large terminally clustered leaves, a tendency to pachycaul growth-form, and strictly axillary paniculate inflorescence.

Grammadenia (7 spp.) — An epiphytic derivative of *Cybianthus,* vegetatively distinctive in the very narrow, sessile, apically mucronate leaves; sometimes placed in *Cybianthus,* but probably adequately differentiated.

Stylogyne (50 spp.) — Lowland shrubs and small trees. Perhaps the most typically myrsinaceous genus, the unusually large rather thin-coriaceous leaves have very inconspicuous close-together secondary venation and are generally more strongly and densely punctate than in other genera. Even the young stems are punctate (unique except for few *Myrsine*). The small flowers, unusual in having the petals twisted (contorted) in bud, are usually axillary and ramiflorous (occasionally terminal, but mostly in extralimital species), mostly in numerous, usually reduced, inflorescences; some species with much reduced ramiflorous inflorescences resemble *Myrsine* but the leaves are larger and the secondary venation more nearly perpendicular to the midvein. The whitish or pale pinkish flowers and buds contrast with the bright pinkish-red inflorescence. Sexually complicated with androdioecious, andromonoecious, bisexual and dioecious species.

Ctenardisia (3 spp.) — A perhaps artificial genus with one Central American and two South American species, one of which may reach our area along the upper Rio Negro. An *Ardisia* with uniseriate ovules and umbellate clusters of flowers arranged in panicles.

There are also several extralimital genera of neotropical Myrsinaceae including *Synardisia* (one species, Mexico to Nicaragua) notable for long glandular trichomes and campanulate corollas,

Wallenia (Caribbean) with tubular coriaceous corollas with stamens of male flowers exserted, *Solonia* (a monotypic Cuban genus), and *Heberdenia* (one species in Mexico and one in Malaysia), essentially an *Ardisia* with free petals.

MYRTACEAE

As a rule, easy to recognize to family but exceedingly difficult to genus. In the New World they are characterized vegetatively by opposite, simple, pellucid-punctate leaves, an almost unique character combination. They frequently have smooth papery or splotched whitish or reddish bark. The leaves are always entire and typically have close-together strongly parallel secondary and intersecondary veins that end almost perpendicular to a well-developed marginal or submarginal collecting vein. A tendency to more or less sericeous pubescence is also sometimes apparent as are spicy vegetative fragrances. The usually rather nondescript multistaminate white flowers are open for only a single day or night with both stamens and petals falling immediately after anthesis. The 4 or 5 calyx lobes are usually persistent at the apex of the rather flat-topped ovary (except calyptrate *Calyptranthus*) and are often persistent in fruit as well. The fruits of most native species are small, fleshy, and one-seeded but some species like the cultivated guava are multiseeded; cultivated members of the mostly Australian subfamily Leptospermoideae have dry dehiscent fruits and alternate leaves.

Mouriri of the Melastomataceae is vegetatively almost indistinguishable from Myrtaceae. Its typically melastom flowers are very different, however. The fruits are also characteristic in more asymmetric shape, fewer stamen scars, and lines radiating out from the style within the hypanthium. Vegetatively, *Mouriri* can be told by the more jointed nodes and by the leaves usually either coriaceous, subsessile and with suppressed secondary veins, or small and with prominent drip-tips. *Mouriri* leaves also lack pellucid glands but may have similar-looking large stomates.

Although there are many distinctive leaf types, most occur in many different genera. Thus, generic subdivision of the neotropical Myrtaceae is almost completely based on technical characters of stamens, calyx, ovules, and embryo. Worse, many of the characters used taxonomically appear to represent convergence in otherwise distantly related lineages. Thus, the calyptrate flower of the traditional concept of *Eucalyptus* clearly evolved independently in three different lineages and taxonomic splitting of that well-known genus may be unavoidable. (The commonly cultivated Andean species thus becoming *Symphyomyrtus globulus.*)

The three main groups of fleshy-fruited Myrtaceae usually recognized are distinguished by the embryo and easily separated by observing the seed: leafy much folded cotyledons and an elongate radicle in the Myrciinae (three large genera of about 100 species each —*Myrcia, Calyptranthes, Marliera* — plus two extralimital genera with about 40 species each and a monotypic endemic genus of Juan Fernandez); cotyledons thick and fleshy (fused and indistinguishable in *Eugenia*) and the radicle very short in Eugeniinae (the huge genus *Eugenia* plus a few extralimital genera, three segregates with more or less elongate calyx tubes, and *Myrcianthes* with the cotyledons thick but clearly differentiated); spiral or uncinately curved with elongate radicle and very short cotyledons in Myrtinae (the large genus *Psidium,* two tiny-leaved high-altitude genera, and a half-dozen small or largely extralimital genera).

In flower, these three groups can also usually be differentiated although there are a number of exceptions; the flowers are 5-merous and produced in a compound multi-(often 30 or more) flowered panicle in Myrciinae, 4-merous and produced in axillary racemes (or, in part, ramiflorously as single, sometimes sessile flowers) in *Eugenia* and its allies, and either dichasially branched (each branch terminating in a flower subtended by 2 branches or by 2 other flowers) or single-flowered and axillary (and always pedicellate) in most myrtinoid genera. Mostly West Indian *Pimenta* has a multiflowered paniculate inflorescence of 4- or 5-merous flowers, and thus, resembles Myrciinae; *Myrcianthes* is intermediate between Myrtinae and Eugeniinae with axillary dichasial inflorescences (or these reduced to single axillary flowers) as in the former but almost exclusively 4-parted flowers (and thick embryo) more like the latter; *Calycolpus,* usually with relatively large flowers, combines the 5-merous flowers typical of Myrtinae with the short axillary racemes of Eugeniinae. *Psidium* and *Campomanesia* also have larger flowers than the other native genera.

Vegetatively, subtribes and most genera are difficult to distinguish. Exceptions include high-Andean *Ugni* and *Myrteola* with tiny thick leaves, and *Campomanesia,* with a tendency to unusually large leaves with widely spaced secondary veins, with thin membranaceous texture, slender petioles, widely spaced arcuate secondary veins that loop rather than connect to a distinct marginal collecting vein, and parallel tertiary venation in most species. *Pimenta* has especially fragrant spicy smelling leaves (sometimes with [unique] citrus or eucalyptus odors in Antillean species); *Campomanesia* is characterized by a pinelike odor. *Myrcia,* (sect. *Myrcia*), *Marliera,* and *Calyptranthes* are usually characterized by a distinctive rubiac-like terminal "stipule" of young leaves with tiny subtending buds. Most species of

Calyptranthes have malpighiaceous trichomes, whereas, such hairs are found only sporadically in other genera. An evenly dichotomous branching pattern usually indicates *Calyptranthes* but other genera are often only slightly less evenly dichotomous. A characteristic wing on the twig occurs in some species of a number of genera; in *Calyptranthes* this wing is perpendicular to the plane of the leaves; in *Marlieria* it is lateral below the decurrent petioles. *Psidium* typically has relatively prominent secondary veins with irregularly (but not intricately) raised tertiary venation below, a tendency to nonappressed pubescence, and sometimes an almost erose-serrulate margin (unique).

1. Inflorescence a Many-Flowered Panicle; the Flowers Usually 5-Merous — Embryo usually with folded leafy cotyledons (subtribe Myrciinae).

Myrcia (perhaps 300 species) — The core group of Myrciinae, characterized by well-developed central inflorescence axis and open 5-lobed calyx (rarely 4-lobed). Exceedingly variable vegetatively, but often with especially large conspicuous punctations and strong spicy vegetative odors.Vegetatively usually characterized by a rubiac-like terminal "stipule" formed by the young leaves and having tiny subtending buds; malpighiaceous trichomes frequent (rare elsewhere).

Marliera (100 spp.) — Differs from *Myrcia* in the calyx closed in bud or nearly so, at anthesis, the apex thinner walled and flaring, splitting irregularly into 3–5 lobes. Malpighiaceous trichomes are frequent and the twigs are sometimes winged (from decurrent petiole base), the wing running straight down from node. Probably an artificial genus with some species belonging to *Myrcia* and others to *Calyptranthes* (fide B. Holst).

Calyptranthes (ca. 100 spp.) — The extreme in the trend toward apically closed calyces; one of the easiest genera to recognize in flower because of its unique circumscissily caducous calyx apex, leaving a ring in fruit. Vegetatively, usually characterized by sericeous malpighiaceous hairs (these sometimes present but much rarer in other genera); in our area mostly with rather closely parallel, often indistinct, secondary and tertiary venation except a few species (*C. plicata, C. spruceana*) with large subsessile more or less cordate-based leaves. When the twig is ridged it has the ridge perpendicular to the plane of the leaves, unlike *Marliera* which has lateral ridges below the decurrent leaf base.

Pimenta (ca. 18 spp.) — Mostly Antillean, apparently only two species reaching northern Colombia (also one in Brazil), one 4-merous and one 5-merous. Since the embryo is spiral, *Pimenta* is not closely related to the *Myrcia* alliance despite sharing their multiflowered paniculate inflo-

Myrtaceae
(*Myrcia* Relatives with Paniculate Inflorescences and Leafy Folded Cotyledons Plus *Myrciaria* with Glomerulate Flowers and Thick Fleshy Cotyledons)

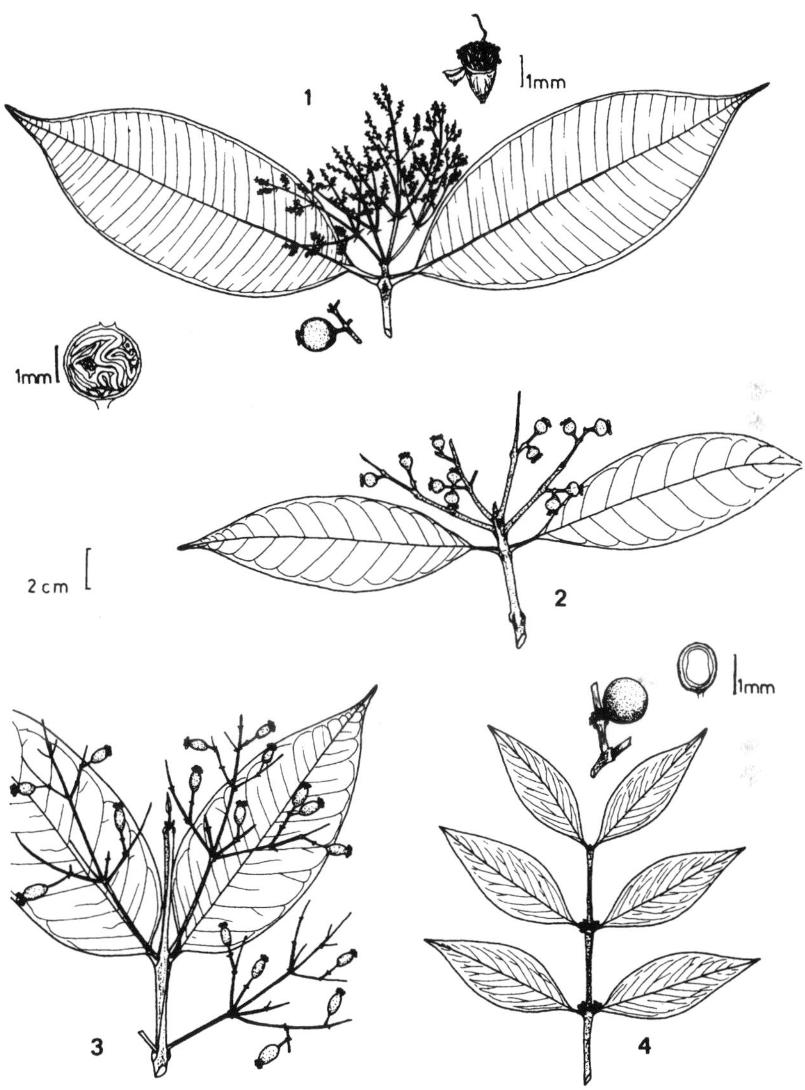

1 - *Calyptranthes*

2 - *Marliera*

3 - *Myrcia*

4 - *Myrciaria*

rescence. The extremely spicy aromatic leaves are coriaceous, medium-sized, usually strongly punctate and typically have the tertiary venation prominulous above as well as below. As in many genera the bark of our species is smooth and white.

"allspice"

(*Gomidesia*) — A mostly Brazilian, and probably inadequately justified, segregate of *Myrcia* from which it differs in the anthers incompletely 4-locular (opening by 4 apical pores in one extreme).

(*Myrceugenia*) (38 spp.) — Chilean and south Brazilian. A strange conglomeration of characters: 4-parted flowers like *Eugenia,* the embryo of *Myrcia,* a 3-locular ovary and mostly solitary axillary flowers).

2. Inflorescence a Raceme or Reduced to Single Flowers; If Not Racemose at Least, in Part, Borne Ramiflorously and the Flowers and Fruits Often (*Siphoneugena, Plinia, Myrciaria*) Sessile (Unique to This Group); Flowers Uniformly 4-Parted (Except Some *Calycolpus*) — Embryo with thick, fleshy cotyledons (or these completely fused) and very short indistinct radicle. Vegetatively often characterized by a sharp-pointed terminal leaf bud (mostly tribe Eugeniinae).

Eugenia (500 neotropical spp., plus many in Old World) — By far the largest genus of neotropical Myrtaceae. The nearly always single-seeded fruit is characterized by an embryo with the thick fleshy cotyledons fused and not obviously delimited (i.e., the seed interior is hard and homogeneous) but otherwise variable from small red and berrylike to several centimeters across. In flower characterized by always having 4 calyx lobes, by the hypanthium not prolonged above the ovary apex, and by the flowers borne in racemes or solitary and ramiflorous (but never solitary and axillary). An important technical character is a bilocular ovary with several to many ovules per locule. The three commonly cultivated Asian species (*E. malaccensis, E. jambos, E. cumini*) should be referred to the exclusively paleotropical genus *Syzygium.* In our area, the flowers and fruits are always pedicellate unlike any species of *Myrciaria* or *Plinia.* Vegetatively, *Eugenia* spans the entire range of variation of the subfamily.

Calycorectes (13 spp.). — Virtually identical to *Eugenia* but the calyx tube prolonged beyond ovary and splitting irregularly into 4–6 lobes. Fruits in our area tend to be pubescent and may be sessile or not; leaves in our area (but not Brazil) are rather large, somewhat discolorous and have the secondary venation raised below. Embryo homogeneous as in *Eugenia.*

Myrtaceae
(*Eugenia* and Relatives with Racemose [or Reduced] Inflorescence, 4-Parted Flowers and Thick Fleshy Cotyledons)

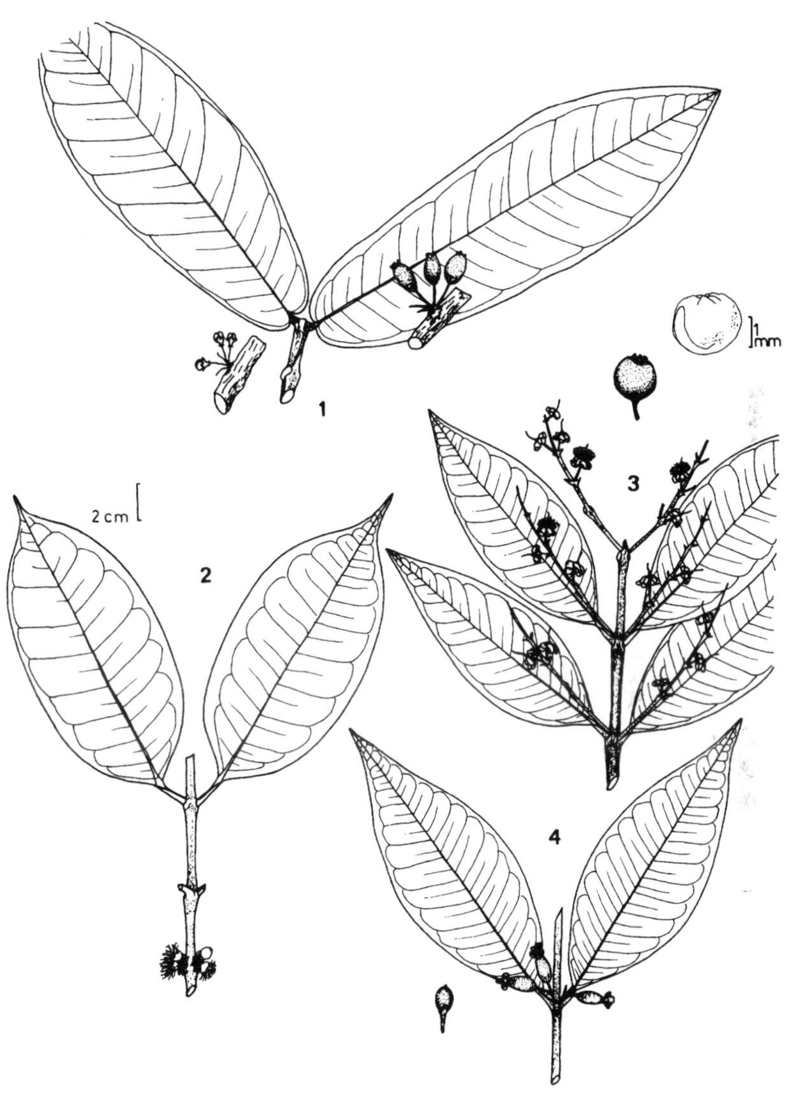

1 - *Eugenia*

2 - *Plinia*

3 - *Eugenia*

4 - *Eugenia* ("*Calycorectes*")

Plinia (10 spp.) — Flowers almost all silky-pubescent and subtended by conspicuous bracts, +/- forming glomerules. Fruits like *Eugenia* except embryo with cotyledons free and plano-convex (as in *Siphoneugena*). Flowers always sessile or subsessile (with one exception) and mostly below the leaves; calyx tube prolonged beyond ovary and irregularly splitting as in *Calycorectes* but only 2 locules/ovule.

Myrciaria (65 spp.) — Lowland trees and shrubs. The end point in the *Eugenia* alliance of the trend to elongate calyx tubes closed in bud (cf., *Calycorectes* and *Plinia*); in *Myrciaria* the calyx apex is circumscissile at base (along with stamens), but, unlike the myrcioid equivalent *Calyptranthes,* this happens subsequent to anthesis. Two ovules/locule are present as in *Plinia.* The flowers are sessile or subsessile in axillary and ramiflorous clusters, and the cotyledons as in *Eugenia.* The leaves are usually small, not very coriaceous, and with accentuated drip-tips; all species have a plane or raised midvein and fine closely parallel secondary veins.

P: camu camu

Siphoneugena (8 spp.) — A single high-altitude species in our area. Related to *Myrciaria* but hypanthium extended above ovary and constricted (and falling off circumscissily as in *Myrciaria*); cotyledons as in *Plinia.*

Myrcianthes (ca. 50 spp.) — Shrubs or small trees of dry, mostly upland, Andean habitats. Although clearly eugenioid in the thick fleshy embryo with a very short radicle, the embryo is distinctive in 2 clearly differentiated free cotyledons. Although there are almost always 4 calyx lobes, *M. quinqueloba,* a Pacific slope Peruvian species, has five. The inflorescences are usually dichasially branched as in myrtinoid group, and both axillary and ramiflorous but may be reduced to solitary axillary flowers. Leaves tend to be more coriaceous and less acuminate than in *Myrciaria.*

Pseudanamomis — A monotypic Antillean genus reaching Venezuela and the Colombian north coast; very close to *Myrcianthes* from which it differs in the cotyledons partly united, fewer ovules (5–6), irregularly branched inflorescence, and 5 deciduous calyx lobes (persistent in *Myrcianthes*). A good vegetative character is lack of a distinct marginal vein.

Calycolpus (10 spp.) — Two species in our area, one in lowland Amazonia and one in the Colombian Andes. Although the flowers are in short axillary racemes (or solitary) as in *Eugenia, Calycolpus* is more closely related to *Psidium* and its allies (Myrtinae) by the uncinately curved embryo with minute cotyledons and an elongate radicle. Unlike most *Psidium,* the sepals are open and reflexed in bud (unique), in the Andean species (*C. moritzianus*) with distinctive appendages. The flowers are relatively large and showy, with a 4–5-locular ovary (unlike *Eugenia*) and usually 5 calyx lobes. The leaves always dry dark and in the Amazonian species (*C. calophyllus*) are medium to largish, and always more or less coriaceous with the secondary veins, though usually faint, clearly stronger than the inconspicuous tertiary venation; our Andean species has distinctively caudate leaves.

(***Campomanesia***) — See below; rarely with ramiflorous mostly 4-flowered inflorescences but uniformly 5-lobed calyx.

3. Inflorescence Dichasially Branched, Usually with 3 to 15 or More Flowers; (the Primary Rachis Terminates in a Flower Which is Subtended by a Lateral Pair of Flowers or Branches) or Reduced to Single Pedicellate Flowers; Flowers Typically Rather Large, Mostly 5-Merous and the Inflorescence (Except Partially Ramiflorous *Campomanesia*) from the Axils of Extant Leaves — Embryo uncinately bent or spiralling, with an elongate radicle (= Myrtinae minus *Pimenta*)

Psidium (100 spp.) — The main pimentoid genus, characterized by the typical curved embryo, the usually 5 calyx lobes nearly always fused in bud, often starting out appressed to petals and then splitting, the 3–4-locular many-ovulate ovary, and the multiseeded fruit with hard seed coats. Unlike *Eugenia,* the flowers are axillary and either solitary or in few-flowered dichasia. Vegetatively, many species can be told by relatively prominent secondary veins with the tertiary venation often raised below but not intricately so, the tendency to nonappressed pubescence, and sometimes an almost erose-serrulate margin.

C, E, P: guayaba, guava

Campomanesia (25 spp.) — Closely related to *Psidium* with a similar 5-lobed calyx but differing in calyx lobes always distinct in bud, the 5–8-locular ovary, larger seeds with softer leathery seed coat, usually longer more spreading calyx lobes (and sometimes a distinctive *Eugenia*-like inflorescence with ramiflorous flowers). Vegetatively, most species are recognizable by having arcuate, rather remote, secondary veins that anastomose with each other without forming a clear marginal vein and often have barbate axils; in our area, most species have rather thin and membranaceous leaves with slender petioles and distinctively parallel tertiary venation perpendicular to midvein. (Extralimital species often have intricately prominulous tertiary veins below.) Unlike *Psidium* the secondary veins are usually closer together near the base than at the center but a few species with strongly ascending, nearly straight secondary veins cannot be vegetatively distinguished from *Psidium.*

Myrtaceae
(*Psidium* and Relatives with Curved or Spiral Embryo and Reduced Cotyledons)

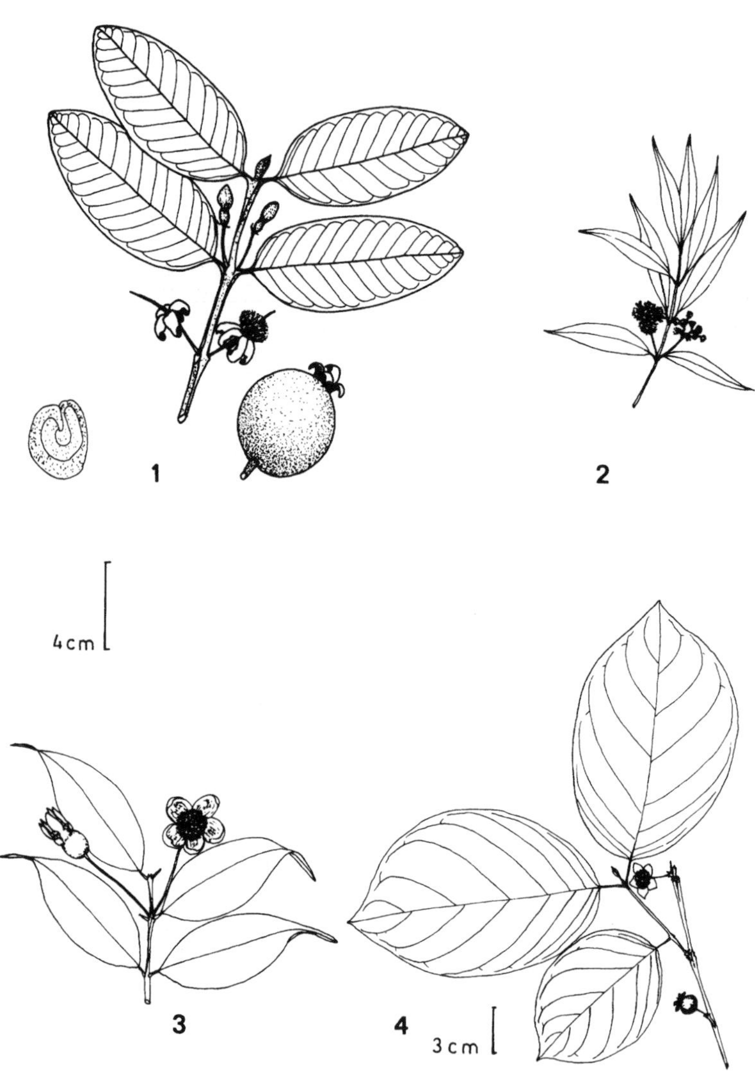

1 - *Psidium* **2** - *Blepharocalyx*

3 - *Calycolpus* **4** - *Campomanesia*

Myrtaceae

(High-Altitude Genera with Small Leaves and/or Few Stamens; Embryo Curved or Spiral with Reduced Cotyledons, Except *Myrcianthes* with Thick, Fleshy Cotyledons)

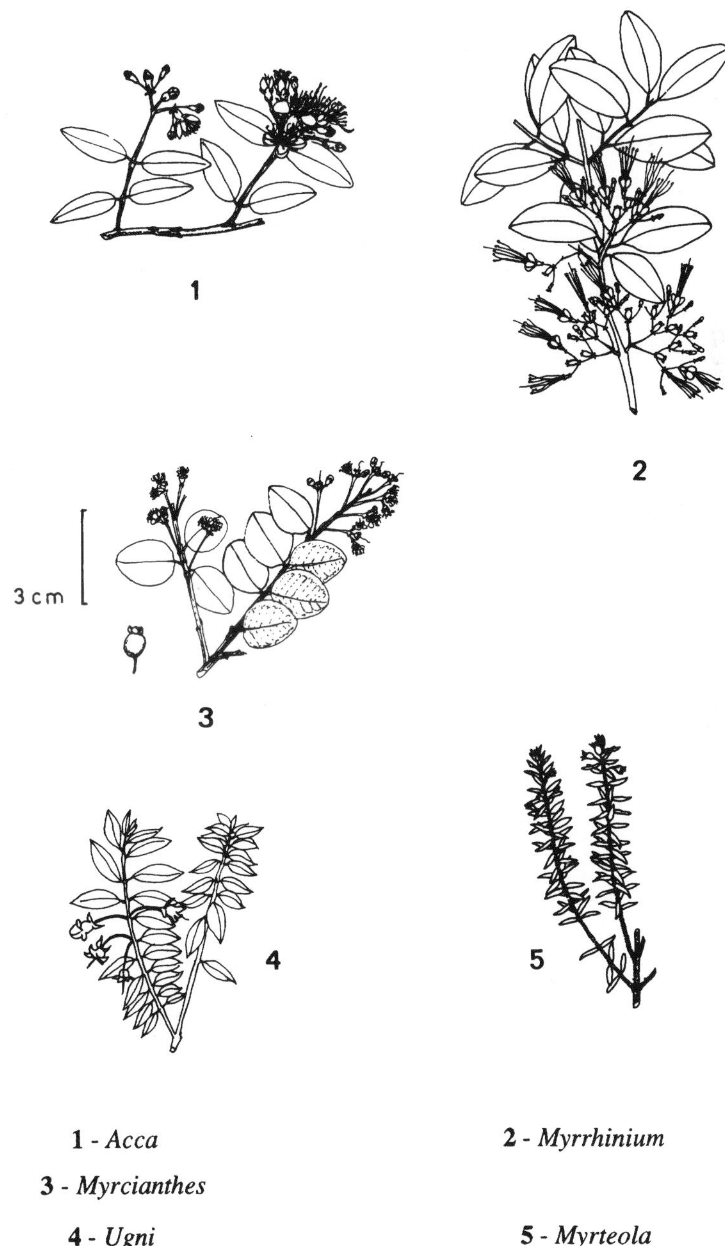

1 - *Acca* **2** - *Myrrhinium*

3 - *Myrcianthes*

4 - *Ugni* **5** - *Myrteola*

Blepharocalyx (3 spp.) — Andean with one southern Brazilian species disjunct in Ecuador and a second species scattered from Peru to the Lesser Antilles. Closely related to *Pimenta* (and *Campomanesia*), and distinguished from the former only by ovule attachment, and lack of strong aroma. Generally characterized by 4-parted flowers in axillary dichasia and unique strongly ciliate calyx lobes that in two species are cleanly deciduous after anthesis. The Ecuador species, vegetatively somewhat distinctive in the small, rather thin, sometimes conspicuously narrow (though less so than *Eugenia biflora*) glabrous leaves.

(*Myrcianthes*) — Inflorescence often dichasially branched, but fruit like *Eugenia.*

4. Upland Cloud-forest Genera with Relatively Few (< 50) Red Stamens — Embryo curved or spiral with reduced cotyledons.

Acca (3 spp.) — Peruvian Andes, disjunct to southeast Brazil and Uruguay (*A. sellowiana,* the pineapple guava, is widely cultivated). Very typical 4-parted flowers with relatively few, ca. 2 cm long, *red* stamens with flattened filaments, these much longer than the petals. Two species have the axillary flowers solitary or in 3-flowered dichasia, the other has them aggregated into pseudoracemes. Vegetatively distinctive in the usually dense villous pubescence.

Myrrhinium (1 sp.) — In our area, restricted to Andean cloud forests above 2000 meters. In flower very distinctive, ramiflorous in dichasially branched inflorescence, with unique red flowers having clawed petals and only 4–8 (usually 6) stamens. Seed with hard coat and the embryo curved. Leaves rather small, coriaceous, narrower than most myrtacs and at least the lowermost on a branch rather obtuse apically.

5. High-Andean (Mostly Above 3000 M) Shrubs or Subshrubs — Tiny very coriaceous dense leaves (< 2 cm long) and small solitary white axillary flowers subtended by persistent foliaceous bracteoles.

Myrteola (ca. 5 spp.) — High-Andean, often prostrate subshrubs, the leaves less than 1 cm long. Differs from *Ugni* in the exerted stamens with subglobuse anthers on filiform filaments and mostly 4-parted, erect flowers, as well as a more extremely ericoid habit.

Ugni (5 spp.) — Andes to Central America and Guayana Highlands; shrubs with rigidly coriaceous leaves up to 2 cm long. Differs from *Myrteola* in the nodding, mostly 5-parted flowers with included stamens and distinctive sagittate anthers.

Nolanaceae

A monogeneric herbaceous (sometimes with suffrutescent base) family of Chile and southern Peru (one species in the Galapagos); in our area almost entirely restricted to the coastal lomas except one species of the Urubamba Valley. Nearly always viscid-pubescent and often with linear leaves. Flowers distinctive, the usually blue (sometimes white or purplish), mostly tubular-infundibuliform, corolla with poorly defined lobes (cf., *Ipomoea*), and the calyx cupular and 5-lobed or dentate. Similar to Solanaceae except for the deeply lobed ovary and a fruit that fragments into (3–)5 nutlets.

Nolana (incl. *Alona*) (only 18 of former 80 spp. accepted in recent monograph but surely many more should have been).

Nyctaginaceae

A generally rather nondescript family, nevertheless usually fairly easy to recognize in vegetative state by the ferrugineous, usually somewhat sericeous, pubescence of the terminal bud. The opposite (frequently in part only subopposite) leaves of the forest genera resemble Rubiaceae when fresh but Nyctaginaceae have no stipule or stipule scar. Another possible source of confusion is Myrtaceae, but most myrtacs are strongly brochidodromous with a strong submarginal collecting vein and those myrtacs that resemble Nyctaginaceae almost always are glandular-punctate. With experience the somewhat succulent texture of the fresh nyctag leaf is distinctive. When dry the great majority of nyctag species turn a characteristic black or blackish color; the rest dry an almost equally characteristic olive. Although the fruit of the tree and shrub species is technically a unique type called an anthocarp (= enclosed by the calyx tube), in practice it is a small nondescript berry remarkably similar to the fruits of *Psychotria* and related Rubiaceae genera. The flowers of the forest genera are tiny, greenish and inconspicuous; the weedy genera have mostly small magenta (sometimes white) flowers.

Generic differentiation of the tropical forest Nyctaginaceae is problematic and based mostly on whether the stamens are exserted (*Guapira, Pisonia*) or included (*Neea, Cephalotomandra*); unfortunately the plants are dioecious and well-developed male flowers poorly collected. The family is more prevalent in extralimital dry areas with only three (of the nine) neotropical genera with alternate leaves (*Bougainvillea, Boldoa, Cryptocarpus*) and five usually weedy herbaceous ones with opposite leaves (*Boerhavia, Commicarpus, Collignonia, Mirabilis, Allionia*) reaching our area in addition to the three or four woody lowland forest genera. At the species level, the lowland tropical Nyctaginaceae comprise one of the most difficult neotropical families: This is the only family for which I cannot sort out meaningful morphospecies from the abundant MO Panamanian collections.

1. Alternate Leaves (Non-Rain-Forest Taxa)

Bougainvillea (18 spp., dry S. Am.) — Spiny lianas or sprawling-branched shrubs or trees, in our area occurring only in coastal Peru and the dry inter-Andean valleys. Very distinctive in the conspicuous magenta floral bracts, and often cultivated as an ornamental.

E: flor de verano; P: papelillo

Boldoa (1 sp.) — Dry-area herbs of the West Indies and Mexico, a single species reaching the Caribbean coast of Colombia.

Cryptocarpus (1 sp.) — A very characteristic shrub (sometimes more or less scandent) of dry sandy deserts, typically one of the first colonists of sand dunes in coastal Peru and Ecuador. Flowers small, greenish, and inconspicuous; leaves rather succulent and somewhat grayish, usually viscid-puberulous.

2. Opposite Leaves

2A. Weedy herbs or montane lianas — All with bisexual flowers and curved embryos, and thus, the typical centrospermous seed; fruits small, dry, either exozoochoric (from small stipitate glands) or with 3–5 small longitudinal wings.

Boerhavia (ca. 50 spp., pantropical) — Common roadside weeds with very small magenta flowers in usually diffusely branched paniculate inflorescences. Differs from *Commicarpa* and *Mirabilis* in lacking involucral bracts.

E: pegajosa

Commicarpus (20 spp., pantropical) — Sprawling herbaceous vine, very like *Boerhavia* but the floral tube (calyx) longer with a medial constriction, thus, more or less campanulate instead of infundibuliform in outline; the fruit is 10 costate vs. 3–5-angled in *Boerhavia.*

Mirabilis (50 spp., mostly N. Am.) — Differs from *Boerhavia* by having a more contracted inflorescence with the flowers subtended by bracts fused into a calyxlike involucre and in a more conspicuous floral (calyx) tube with shallowly lobed (plicate) apex; the most commonly encountered species, *M. jalapa* (4-o'clock) has a much longer floral tube than

Nyctaginaceae
(Alternate Leaves or Weedy Herbs)

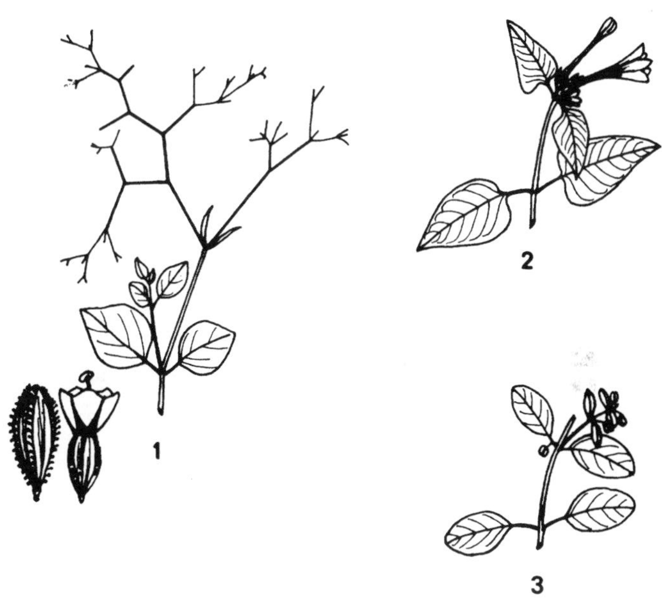

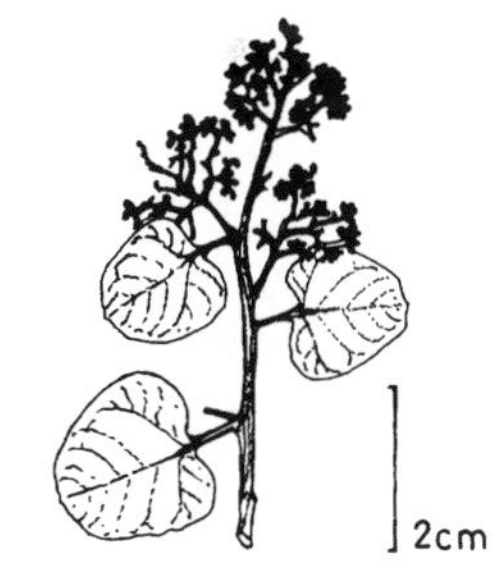

1 - *Boerhavia*

2 - *Mirabilis*

3 - *Allionia*

4 - *Bougainvillea*

5 - *Cryptocarpus*

Nyctaginaceae
(Opposite-Leaved Trees and Lianas)

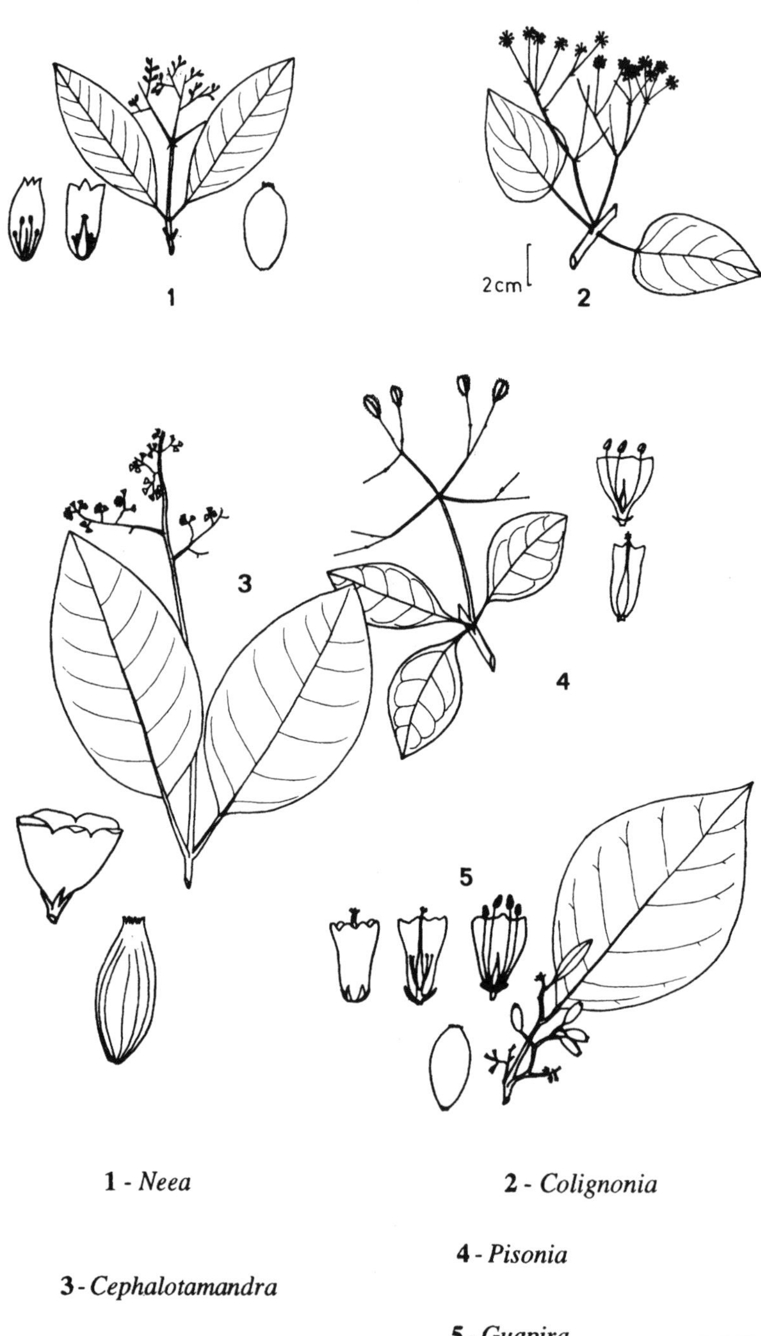

1 - *Neea*

2 - *Colignonia*

3 - *Cephalotamandra*

4 - *Pisonia*

5 - *Guapira*

other nyctags but some species have flowers almost as small as in *Boerhavia.*

E: maravilla (*M. jalapa*)

Colignonia (ca. 10 spp.) — Andean cloud forests and dry inter-Andean valleys. Looks like *Boerhavia* but has white flowers and conspicuous whitish inflorescence bracts. The leaves tend to be verticillate and the plants are usually more or less scandent; at least one cloud-forest species can become a giant liana; with concentric rings of anomalous cambium.

Allionia (possibly only 1 polymorphic sp.) — Restricted to very dry areas, in our area only known from Caribbean coast and Peru. Extremely similar to smaller-flowered weedy species of *Mirabilis* but the 3 involucral bracts free rather than fused. The unique fruit is lenticel-shaped with a thin dentate margin.

2B. Woody taxa of lowland forests

Pisonia (as defined by Willis to include *Guapira* and *Torrubia,* pantropical with ca. 1000 spp.) — Spiny lianas or usually more or less spiny trees of dry forest. Fruit very characteristic, narrowly ellipsoidal and having exozoochorous dispersal via sticky stalked glands, these often arranged in discrete vertical rows on the angles. Male flowers with exserted stamens.

E: uña de tigre

Guapira — Usually small- to middle-story trees; close to *Pisonia* and sometimes included in it but differs in lacking spines and having nonglandular fruits like those of *Neea.* Vegetatively tending to be more arborescent than *Neea* but not distinguishable from the more treelike species of *Neea* without flowers.

Neea (ca. 70 spp.) — Mostly understory shrubs and small trees characterized by included stamens in the relatively (to *Guapira* and *Pisonia*) narrow and elongate male flowers. A few species are larger trees, especially in dry forest. Fruit a berry (apparently with fruitlike galled structures also developing on male plants), typically on a reddish-purple inflorescence, superficially very similar to *Psychotria* and related Rubiaceae.

Cephalotomandra (3 spp.) — Rarely collected and a doubtful segregate, known only from Panama and Colombia. Very similar to *Neea* with which it shares included anthers but with 25–30 (vs. 5–10) stamens and shorter flower more like that of *Guapira* (except for the included stamens). Leaves rather large and olive-drying.

Ochnaceae

Except for one widespread weedy herb (*Sauvagesia*), our taxa all woody shrubs or trees (one possible liana?) with uniformly alternate, glabrous, pinnately veined leaves, always more or less serrate or serrulate. Prominent stipules are always present but early-caducous in many taxa (leaving a prominent annular scar, sometimes rather ochrea like). A few species of *Godoya* have the leaves deeply pinnatifidly divided or almost completely pinnately compound (with coriaceous, conspicuously serrate leaflets, but their bases tending to be incompletely differentiated from rachis). Aside from herbaceous *Sauvagesia* (easily recognized by the persistent stipules with conspicuously ciliate-hairy margins), three very different types of leaf venation occur in our taxa, but each is individually distinctive. The commonest genus (*Ouratea*) has leaves with secondary veins curving strongly near the margin and continuing almost as submarginal veins (typically several of these submarginal extensions parallel to each other at any given point of leaf margin); also with intersecondary veins paralleling basal part of secondaries but ending before the secondaries curve upward. The other common leaf type is obovate to oblanceolate with many straight, rather close-together, rigidly parallel secondary veins and the finely parallel tertiary venation perpendicular between adjacent secondaries; the margins are usually distinctively crenate-serrate. The third leaf type (*Blastomanthus*) has *Clusia*-venation and resembles *Manilkara* but lacks latex and has a retuse apex.

Flowers of woody taxa are always yellow, usually conspicuous. The main floral character is the usually deeply 3-lobed ovary with style arising from its center. Sepals 5, petals (4–)5. Fruit usually a small, narrow, more or less fusiform capsule with very small narrow winged seeds, but in *Ouratea* fleshy with several black "berries" borne on upper surface of fleshy red receptacle.

1. Herbs

Sauvagesia (29 spp.) — Small common weedy herbs with pectinate-ciliate, hairy-margined stipules and mostly solitary, small white axillary flowers. Except for the deeply 3-lobed ovary, one would never suspect that this herb belongs to the same family as the woody taxa.

2. Shrubs or Trees

Ouratea (300 spp., incl. Old World) — Leaves (described above) very distinctive, always more or less oblong-elliptic to oblong-lanceolate with more or less finely serrate margins and secondary veins conspicuously curving upward near margin. Flowers yellow. Very distinctive in fruit with the receptacle swollen and red and the individual ovary segments separately

Ochnaceae

1 - *Blastomanthus*

2 - *Ouratea*

3 - *Sauvegesia*

4 - *Cespedezia*

borne on its upper surface as 3 black berries (resembles Mickey Mouse with the receptacle forming the "head" and the berries the "ears").

P: yacu moena

Cespedezia (6 spp.) — A very distinctive and common late second-growth wet-forest species with very large +/- oblanceolate leaves and a pachycaul juvenile growth-form; stipules very conspicuous and persistent unlike similar taxa (e.g., *Gustavia, Clavija*). Flowers yellow and ca. 2 cm across, with numerous stamens, borne in a large terminal panicle. Capsule fusiform with small narrow winged seeds.

C: paco; P: afasi caspi (= useless pole)

Godoya (5 spp.) — Large Andean cloud-forest trees, very like *Cespedezia* but flowers larger (3–5 cm across), with only 10 stamens, and the stipules caducous. Leaves smaller than *Cespedezia,* narrowly obovate, the margins strongly crenate. Two species have irregularly pinnate leaves with alternate coriaceous serrate leaflets, the bases usually not all differentiated from rachis (= pinnatifidly compound), these distinct from similar Sapindaceae, etc., in the conspicuous stipules at branch apex. Fruits as in *Cespedezia.*

Krukoviella (1 sp.) — Shrub (perhaps, sometimes scandent?) of upper Amazonia. Essentially a small-flowered *Godoya.* Leaves +/- serrulate-denticulate but the margin much less obviously serrate than in *Godoya.* Stipules blunt, subfoliaceous, early-caducous to leave a conspicuous scar, this (along with the petiole base that expands around twig tending to form an ochrea-like ridge (cf., *Coccoloba*).

Blastomanthus (5 spp.) — Small tree of upper Rio Negro area, occurring in black-water-inundated forests and around edges of white-sand savannas. Leaves very different from other members of family, oblanceolate, +/- retuse at apex (and usually also minutely apiculate), very strongly *Clusia*-veined with secondary and tertiary veins completely undifferentiated and the lateral nerves all very fine, close-together, and completely parallel (could only be confused with *Micropholis* group of Sapotaceae but differs in lacking latex and the retuse leaf apex). Flowers like *Ouratea;* fruit an ellipsoid 3-parted capsule.

OLACACEAE

A rather nondescript family in flower or sterile condition but usually easy to recognize by having the calyx (or disc) characteristically expanded in the always single-seeded fruit. The leaves are always simple, alternate, and (in our area) entire. In some genera, including the largest, *Heisteria,* the petioles are characteristically curved (sometimes almost openly U-shaped) and tend to be thicker toward apex. Perhaps more than any other neotropical family Olacaceae are characterized by finely parallel tertiary venation in several genera but this character also occurs in miscellaneous genera of many families. The leaves (usually +/- coriaceous and somewhat olive even when fresh) almost always dry a characteristic olive to blackish color (cf., Loranthaceae). Small amounts of milky latex are present in several genera, at least in the petiole and youngest twigs. One genus (*Ximenia*) has branch spines. The inflorescence is small and axillary, typically consisting of a few fasciculate flowers or sometimes few-branched; in our area only *Minquartia* (easy to recognize by the rather close, straight secondary veins and very strongly parallel tertiary venation) with a spikelike raceme, has the inflorescence even approaching the length of the subtending leaf.

The main characters for distinguishing genera are in the fruits. In fruit *Heisteria* has a +/- flat, usually red (sometimes white or green) expanded calyx, *Chaunochiton* a large papery brown one, *Aptandra* (in the Neotropics) a cupular one enclosing basal half of fruit; *Dulacia* has the calyx almost completely fused to and enclosing the cylindric-ellipsoid fruit which appears derived from an inferior ovary, while *Schoepfia* has a legitimate inferior ovary but the calyxlike collar of bracts at base of the small berrylike fruit obscures this. *Tetrastylidium* (in our area) has the expanded patelliform calyx fused to basal part of fruit while *Cathedra* has a fruit that is at least half-enclosed by a cupule formed from the expanded disc as well as being subtended by two "calyces" (the lower being a collar of fused bracts).

1. GENERA WITH MUCH-ENLARGED CALYCES IN FRUIT; THESE TYPICALLY BRIGHT RED AND VERY CONSPICUOUS

Heisteria (30 spp., plus 3 in Africa.) — Usually small to medium-sized trees; the expanded frilly calyx usually red, occasionally green (the fruit itself then sometimes red). Vegetatively, almost always recognizable by the upcurving (often almost openly U-shaped) round green petiole somewhat thicker in its upper half. The inflorescence is always an axillary fascicle. Some species have inconspicuous latex (at least some of the time). One (*H. scandens*) is a liana.

P: sombredito, yutubanco

Olacaceae
(Calyces Fused to Fruit or Not Enlarged)

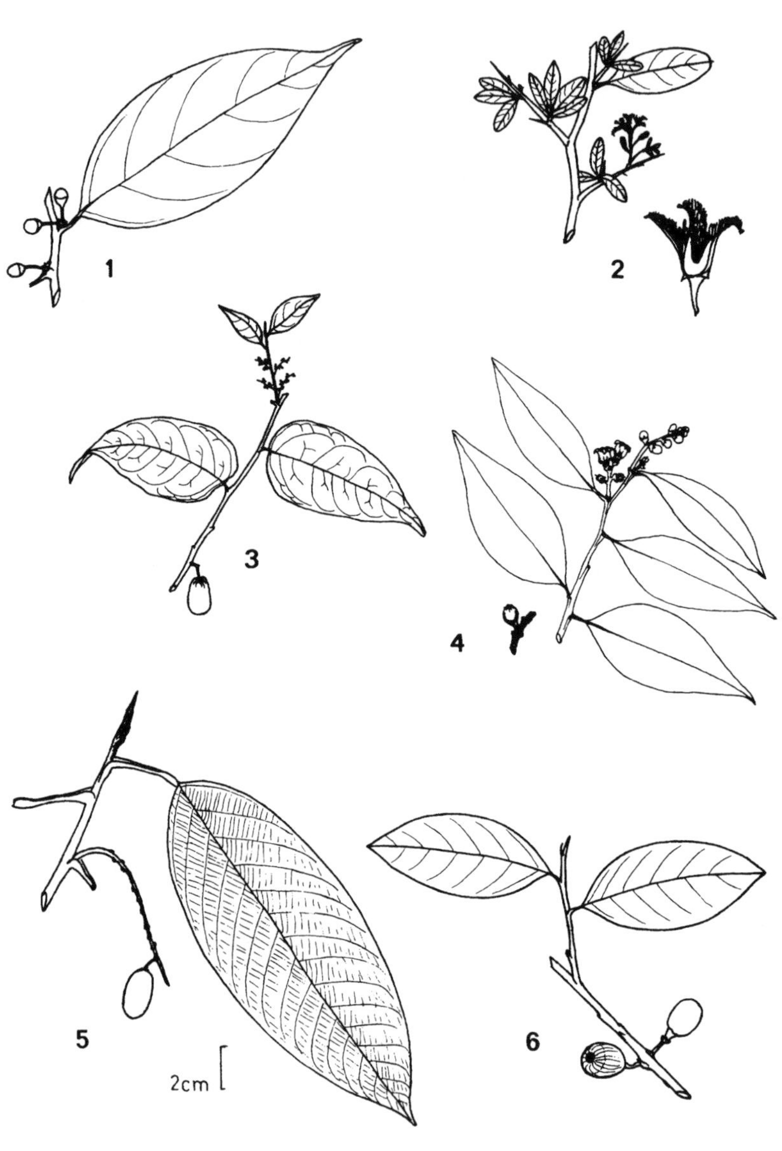

1 - *Tetrastylidium*

2 - *Ximenia*

3 - *Dulacia*

4 - *Schoepfia*

5 - *Minquartia*

6 - *Cathedra*

Olacaceae (Calyces Enlarge in Fruit) and Oleaceae

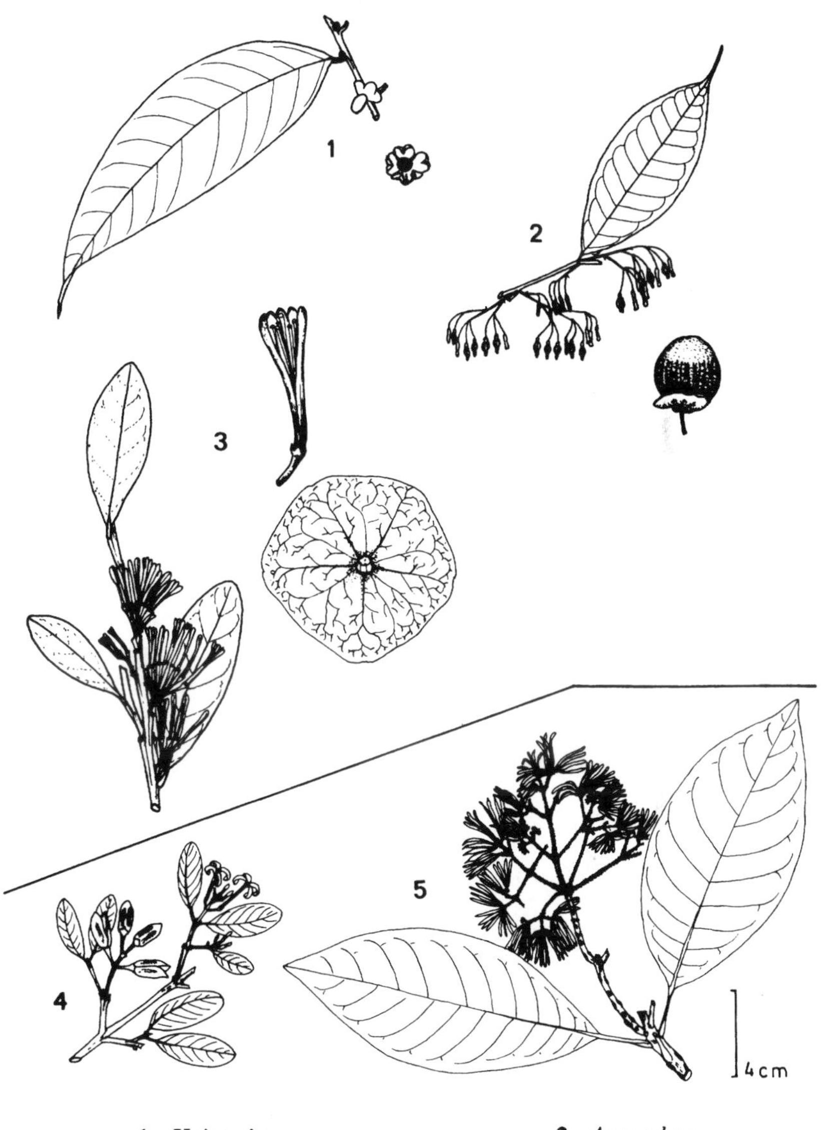

1 - *Heisteria* 2 - *Aptandra*

3 - *Chaunochiton*

4 - *Schrebera* (Oleaceae) 5 - *Chionanthus* (Oleaceae)

Chaunochiton (5 spp.) — Usually large trees with a fruit very similar to *Heisteria* but the expanded calyx brownish and much larger (usually 6–12 cm across), the fruits apparently wind-dispersed by the dry calyx. The flowers, very different from *Heisteria,* are relatively long and narrow, often showy and conspicuous (cf., *Psittacanthus* [Loranthaceae]). Formerly thought disjunct between the Guayana Shield and Costa Rica but recently discovered in Chocó. Leaves rather small, drying olive below, blackish above, the secondary veins below faint and plane or subplane.

Aptandra (3 spp., plus 2 in Africa) — Trees differing from *Heisteria* in larger fruit and in the fruiting calyx green, thicker than *Heisteria* and eventually forming a cupule around basal half of mature fruit. Inflorescence clearly branched, an obviously pedunculate axillary umbel (or bearing several such rather ill-defined umbels) (unique); the tiny flower very characteristic, in bud tubular with a central constriction and apical bulb, the staminal tube falling off soon after anthesis. Leaves with inconspicuous subplane secondary veins, usually coriaceous, medium small.

2. Disk or Calyx Not Enlarged or Enlarged and Fused to Fruit; When Enlarged, Greenish, and Forming a Cupule Covering All or Half of Fruit — Leaves typically conspicuously olive and/or with finely parallel tertiary venation.

Tetrastylidium (2 spp.) — Leaves drying olive, tertiary venation faint and inconspicuous but definitely parallel and +/- perpendicular to midrib, apex acuminate with a tiny mucro at extreme tip. Flowers in axillary fascicles; pedicels ebracteolate, the petals narrow and valvate with the lobes puberulous inside; fruit globose with the expanded patelliform disk fused to basal part.

Cathedra (5 spp.) — Vegetatively and in flower similar to *Heisteria* but the petiole usually shorter and less defined, the fasciculate flowers having somewhat longer narrow valvate petals; leaves drying uniformly olive to dark olive; fruit very characteristic, ellipsoid with at least basal half surrounded by the cupulate expanded disk, base subtended by both the cupulate calyx and a calyxlike collar of fused bracts at base of actual calyx.

Dulacia (13 spp.) (incl. *Liriosma*) — Leaves sometimes conspicuously glaucous-waxy, usually narrowly ovate, the petiole shorter and less developed than in *Heisteria.* Inflorescence short, few-flowered but definitely zigzag racemose. Petals narrow and valvate, the narrow buds tapering to pointed apex; fruit very distinctive, cylindric-ellipsoid, apparently from an inferior ovary on account of the enclosing fused calyx.

Schoepfia (19 spp., plus 4 in Asia) — Small tree or shrub of dry areas. Vegetatively similar to *Ximenia* but lacking spines. The urceolate flower is very distinctive (cf., *Vaccinium*), the inflorescence is a short few-flowered axillary spike with the flowers clustered toward tip (sometimes in part reduced to axillary fascicles). Leaves smallish, drying olive to blackish. The ovary is inferior (unique) but subtended by calyxlike epicalyx. The fruit is small, ellipsoid berrylike, its origin from an inferior ovary obscured by the subtending calyxlike basal collar of fused bracteoles.

Ximenia (7 spp., plus 1 in Africa) — Small tree or shrub of dry areas (in our area only northern Colombia). Branch spines present (unique). Leaves clustered at tips of lateral branches, drying dark olive to blackish, usually minutely retuse at apex; flowers densely villous inside.

Minquartia (1–2 sp.) — Vegetatively unmistakable with the oblong leaves having: 1) very conspicuously finely parallel tertiary venation +/- perpendicular to the numerous straight secondary veins and 2) a long petiole with a flexed swollen apical pulvinar region. The unique inflorescence is rather long and spicate or spicate-racemose; the wood is extremely hard and durable. Trunk sometimes fenestrated; also characterized by tiny blackish dots in slash.

C: acapu, guayacán negro; P: huacapú

Extralimital genera — (All Amazonian and potentially in extra-Brazilian Amazonia)

Brachynema (1 sp.) — Glandular dentate leaves and flexed thickened petiolar apex (and thickened base), thus totally unlike other Olacaceae vegetatively and looking instead like *Conceveiba* or other euphorbs; cauliflorous with tubular corolla; Southern and southeastern Amazonian Brazil.

Curupira (1 sp.) — Large tree; inflorescence like *Aptandra* (an umbel) but with long exserted stamens; leaves unique in being 3-veined from base; upper Amazonian Brazil.

Douradoa — Close to *Curupira* (and resembling *Aptandra* in the flowers in pedunculate umbels) but nonexserted stamens and only sub-3-veined leaves; Para and Amapa, Brazil.

Ptychopetalum (2 spp., plus 2 in Africa) — Very like *Dulacia* but 7–10 stamens, completely separating petals at anthesis, and the calyx not enlarged and enclosing fruit; Amazonian Brazil and Guianas.

Oleaceae

Poorly represented in our area by a single genus of mostly montane trees, easily recognized by the opposite entire leaves with distinctly swollen petiole bases; this is our only genus with opposite estipulate leaves having subwoody swollen (Sapotaceae-like) petiole bases, the cuneate leaf base and rather oblong leaf shape are also distinctive. More or less conspicuous axillary buds are usually visible in the leaf axils. The branchlets almost always have at least a few small round raised lenticels. In flower, completely unmistakable on account of the long linear or sublinear petals (fused only at base) and presence of *only 2 stamens,* the petals are pink and ca. 1 cm long in commonest species, smaller and white in most others. The fruits of *Chionanthus* are single-seeded and ellipsoid; *Schrebera* has a 2-valved capsule.

Chionanthus (incl. *Linociera*) (ca. 4 spp., plus 1 Asia) — Mostly in dry Andean forests, but one species occurs in Amazonian tahuampa. The commonest species blooms precociously while deciduous and has conspicuous pink flowers with 1 cm long narrow petals, the tahuampa species has small white flowers in a few-flowered axillary inflorescence, the other species are intermediate.

Schrebera (1 sp., plus 27 Old World) — Tree of dry inter-Andean valleys, the peculiar flower white with narrow brown petals. Differs from *Chionanthus* in 2-valved dehiscent fruit.

There are several additional genera in Central America and the Antilles, one of these (*Menodora*) with an amphitropical range disjunction (and also in Africa). *Fraxinus* (trees) and *Jasminum* (vines), both with opposite compound leaves, are frequently cultivated in our area (a few *Jasminum* species have simple leaves).

Onagraceae

Consisting largely of North American herbs, this is one of the most intensively studied plant families. It is poorly represented in our area by four genera. Only one occurs in the lowlands, semiaquatic *Ludwigia,* mostly herbs (occasionally distinctly woody), and easily recognized by the solitary axillary mostly 4-parted flowers with conspicuous early-caducous yellow petals and a more or less cylindric, usually ribbed inferior ovary topped by the reflexed calyx lobes. The other three exclusively Andean genera — *Fuchsia* plus a few weedy outliers of *Oenothera* and *Epilobium* — have uniformly 4-parted flowers with 8 stamens and an elongate narrow hypanthial tube prolonged above the ovary.

Onagraceae

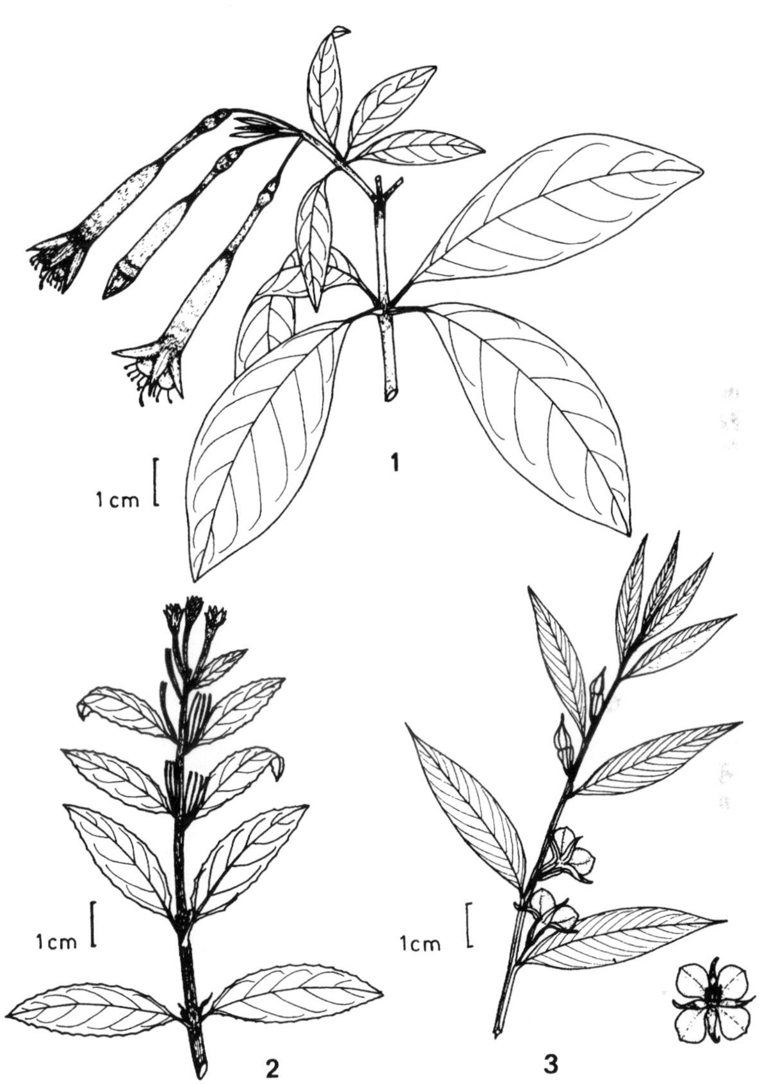

1 - *Fuchsia*

2 - *Epilobium*. **3** - *Ludwigia*

Oenothera (123 spp., mostly N. Am.) — In our area represented only by a few high-Andean herbs with alternate leaves and more or less angled stems. The 4-parted flowers have a long narrow floral tube, its basal part narrower than the more or less cylindrical ovary below it. The fruits of our species are cylindrical capsules with small noncomose seeds.

Epilobium (185 spp., mostly n. temperate and New Zealand) — Only a few species of high-Andean weedy herbs in our area, these differing from *Oenothera* in the extremely narrow elongate capsule with small seeds each having a conspicuous terminal tuft of very long silky hairs. The 4-parted flowers are smaller than our *Oenothera* species and pinkish, differing from *Oenothera* in the linear cylindric ovary as narrow as lower corolla tube.

Fuchsia (104 spp., plus 1 in New Zealand) — The northern Andean cloud forests are the distributional center for this very distinctive genus. Most species are shrubs (even small trees) but many are woody lianas, often more or less hemiepiphytic; both shrubs and lianas tend to have a characteristic reddish exfoliating bark. The leaves, usually rather small, are opposite or whorled in most species and the petioles are frequently red when fresh. The minority of species that have entire leaves could be confused with Rubiaceae but the inconspicuous stipules (or their scars) are usually separate rather than interpetiolarly fused and red petioles are lacking in similar rubs; under a lens the upper surface has cystoliths, unlike Rubiaceae. The hummingbird-pollinated flowers tend to be borne year round and are absolutely distinctive with their pendent tubular red flowers with exserted anthers and the long hypanthium topped by 4 red triangular sepals and 4 narrower more or less erect red petals (a few Peruvian species apetalous). The more or less cylindric red or dark purple berry, without any hint of apical sepals, is also distinctive.

Ludwigia (47 spp., plus 33 Old World and N. Am.) — Mostly erect herbs of wet places (sometimes floating aquatics), but occasionally becoming small trees. A very distinctive genus on account of the solitary axillary flowers with 4(–6) conspicuous early-caducous yellow (white in a few floating and extralimital species) petals, and an angled or ribbed, cylindric (to obconic) ovary topped by 4(–6) triangular reflexed sepals. The stem is usually angled or ribbed and the majority of species have alternate leaves. The subtree species have exfoliating reddish bark similar to that of *Fuchsia* and grow only in open swampy areas.

Opiliaceae

Trees with simple, alternate, entire, olive-drying leaves, mostly occurring in dry forest. Represented in our area by a single genus, *Agonandra*. Rather nondescript, vegetatively, but usually characterized by the leaf blade decurrent on the poorly differentiated petiole and the rather few, irregular, often poorly defined secondary veins; moist-forest species have the tertiary veins finely parallel and perpendicular to midvein. Inflorescence an axillary spike or narrow raceme of tiny greenish flowers. Fruit globose to ellipsoid, single-seeded, 2–3 cm long. Very similar to (and closely related to) Olacaceae which it resembles in the olive-colored fresh twigs and olive-drying leaves; especially like *Dulacia* in the poorly defined petiole; but that genus has much longer flowers in shorter inflorescences, does not occur in dry forest, and has no species with the finely parallel tertiary veins of moist-forest Opiliaceae.

Agonandra (10 spp.)

Oxalidaceae

In our area only herbs with compound leaves having leaflets with legumelike (but very short) pulvinuli (facilitating nyctinastic movement as in legumes; some *Biophytum* even sensitive to touch). One genus (*Oxalis*) has 3-foliolate leaves (often with apical notch); one has even-pinnate leaves clustered at tip of slender erect stem (*Biophytum*); and one has odd-pinnate leaves in sessile rosettes and with leaflets notched at both base and apex (*Hypseocharis*). Flowers have 5 free sepals and petals, 10(–15 in *Hypseocharis*) stamens, and 5 free styles (1 in *Hypseocharis;* carpels subapocarpous in *Biophytum*). Besides the native herbs, there are 2 rather widely cultivated paleotropical trees of the genus *Averrhoa* (carambola) in our area.

Hypseocharis (8 spp.) — Puna herbs with large thick taproot and sessile rosette of odd-pinnate leaves with sessile leaflets notched at both base and apex. Flowers white. Technically differs from *Biophytum* in 15 stamens and fused style.

P: anaruku

Biophytum (70 spp., incl. Old World) — Curious parasol-shaped, small forest-understory herb with slender erect stem and terminal rosette of even-pinnately compound leaves, the sessile leaflets asymmetric-based, more or less parallelogram-shaped. Flowers white to pink or lilac, usually in dense several-flowered cluster at end of long peduncle.

Opiliaceae, Oxalidaceae, and Papaveraceae

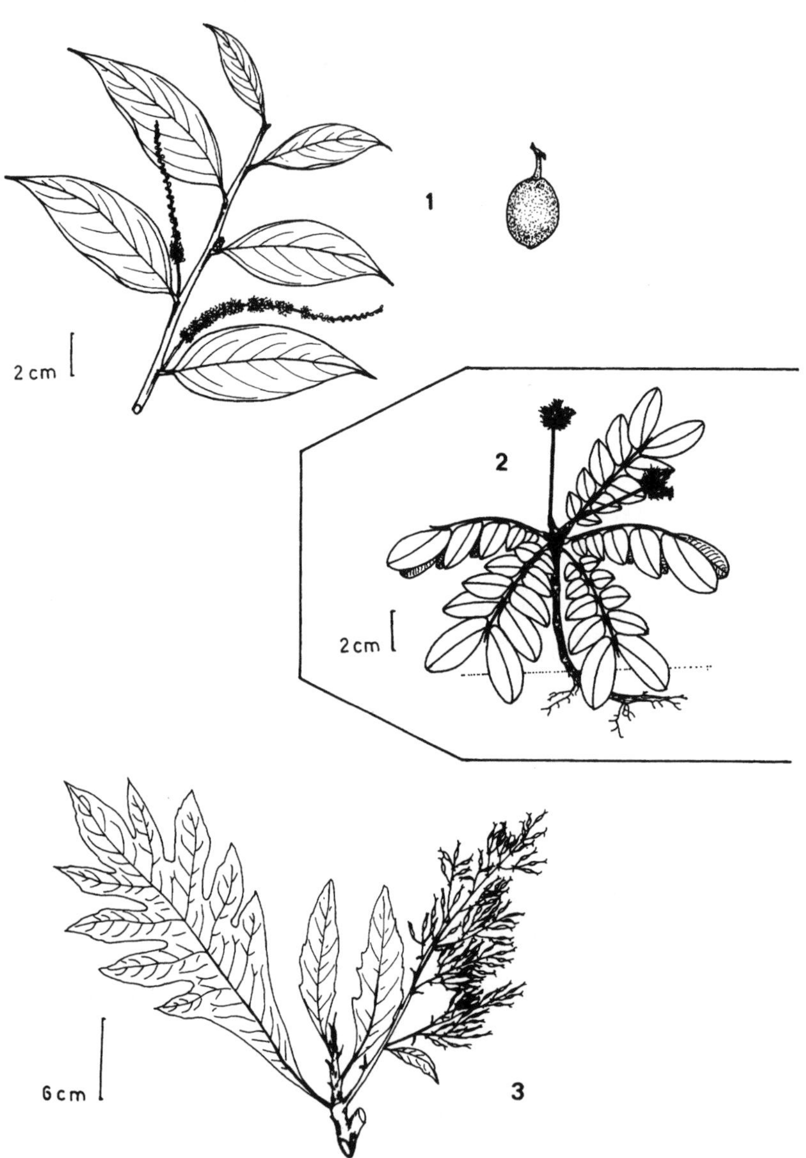

1 - *Agonandra* (Opiliaceae)

2 - *Biophytum* (Oxalidaceae)

3 - *Bocconia* (Papaveraceae)

Oxalis (800 spp., incl. N. Am. and Old World) — Mostly upland Andean and cloud-forest herbs but a few weedy species in lowlands. The 3-foliolate leaves often with broadly obovate leaflets having notched apex but sometimes merely ovate and acute and in a few rock-growing species the petiole succulent and leaflets reduced or even absent; stems often succulent; cloud-forest species often tenuous climbers. The variously 5-angled cylindric fruits are taxonomically important but poorly collected. Nearly all our species have yellow flowers (a few with red striations; very few have lilac or lavender flowers. One species ("oca": *O. tuberosa*) has an edible root.

PAPAVERACEAE

Mostly north temperate herbs, recognizable by orange latex, variously pinnatifid or palmatifid leaves, and conspicuous flowers. Our only native representative, *Bocconia,* decidedly atypical for the family, is a tree with apetalous flowers. Both it and the introduced weed *Argemone* can easily be recognized by the typical orange latex of the family and by the very characteristic irregularly pinnatifidly incised leaves.

Bocconia (10 spp., perhaps also Asia, depending on your taxonomy) — A soft-wooded tree of montane cloud forests, especially in disturbed sites as along roadsides. Easily recognized by the large pinnatifidly lobed leaves (one extralimital species is merely irregularly serrate). The inflorescence is a large panicle, the fruits fusiform, attenuate at both base and apex, rather translucently orangish-salmon in color, and with two valves which fall away to leave a ringlike replum. Flowers apetalous and exactly like the fruits except smaller and the two elongate stigmas better defined.

Argemone (10 spp., all native to N. Am.) — Two species of weedy herbs are naturalized in our area, mostly in dry inter-Andean valleys. They are spiny and very distinctive with large white to light yellow flowers, and sharply pinnatifid leaves having the irregular teeth spine-tipped as well as spines scattered on the lamina; even the ellipsoid-cylindric capsule is spiny.

PASSIFLORACEAE

Mostly vines and lianas with simple (very rarely palmately compound) alternate leaves, occasionally becoming small trees. Vegetatively characterized by usually unbranched and coiling axillary tendrils, by the usual presence of petiolar glands (when absent, usually with conspicuous glands on leaf lamina), and by the plethora of unusual leaf shapes. The leaves are usually 3-veined or palmately veined (sometimes pinnately veined or rarely subpeltate) and may be serrate or entire, ovate to round or linear or transversely almost linear (much wider than long), undivided to deeply palmately lobed, frequently oblong with a unique truncate or concave apex (a shape unique to Passifloraceae). (The amazing variation in leaf shape is probably related to the unusual coevolutionary relationship with specific *Heliconius* butterfly predators.) The flowers, usually borne singly or paired in leaf axils, are also highly distinctive, usually with several whorls of filaments forming a corona, and the ovary (in *Passiflora,* the main genus) borne on a gynophore and crowned by 3 (4 in *Dilkea* and *Ancistrothyrsus*) conspicuous diverging styles tipped by thick round stigmas (suggesting a crucifixion scene, hence the name "passion"). In *Passiflora* the usually indehiscent fruit is characterized by the remains of the gynophore stalk (except *P. apoda*) and has seeds in a mass surrounded by fleshy gelatinous arils. Liana species tend to have irregular somewhat lobed or angled stems with anomalous secondary phloem (cf., bignons in section but irregular). The tendrils may be fine and coiling or hooked and thicker at apex (*Ancistrothyrsus*). Erect species have pinnately veined leaves and are difficult to recognize when sterile. Tree *Passifloras,* mostly in middle-elevation forests, have rather broad succulent leaves with abaxial glands at base of midrib; erect *Dilkea* species, usually rather wandlike, have narrowly obovate or oblanceolate leaves with prominulous venation, that are often clustered, and rather yellowish below.

Passiflora (400 spp., plus ca. 20 in Old World) — The overwhelming bulk of neotropical species, easy to recognize by the usually conspicuous 5-parted flowers with coronal filaments and stalked ovary. This is the only genus in our area with petiolar glands; when these are absent there are usually glands in axils of basal lateral vein pair and or over the lamina. Most *Passiflora* species have undivided tendrils (except *P. tryphostemmatoides* group with small entire thin leaves). The fruit pulp of many species is edible, especially in "jugos".

E: granadilla, bombillo (*P. foetida*), badea (*P. quadrangularis*); P: granadilla; C, E, P: maracuyá; (*P. edulis*); curuba; (*P. mollissima*)

Dilkea (4 spp.) — Canopy lianas (especially of poor-soil forests) or erect wandlike treelets, differing from *Passiflora* in lacking the gynophore and in 4-merous flowers. A useful vegetative character is that the non coiling tendril tends to be short and usually trifid-tipped; the obovate coriaceous leaves with *prominulous* venation are also unlike any *Passiflora.*

Passifloraceae
(Except *Passiflora*)

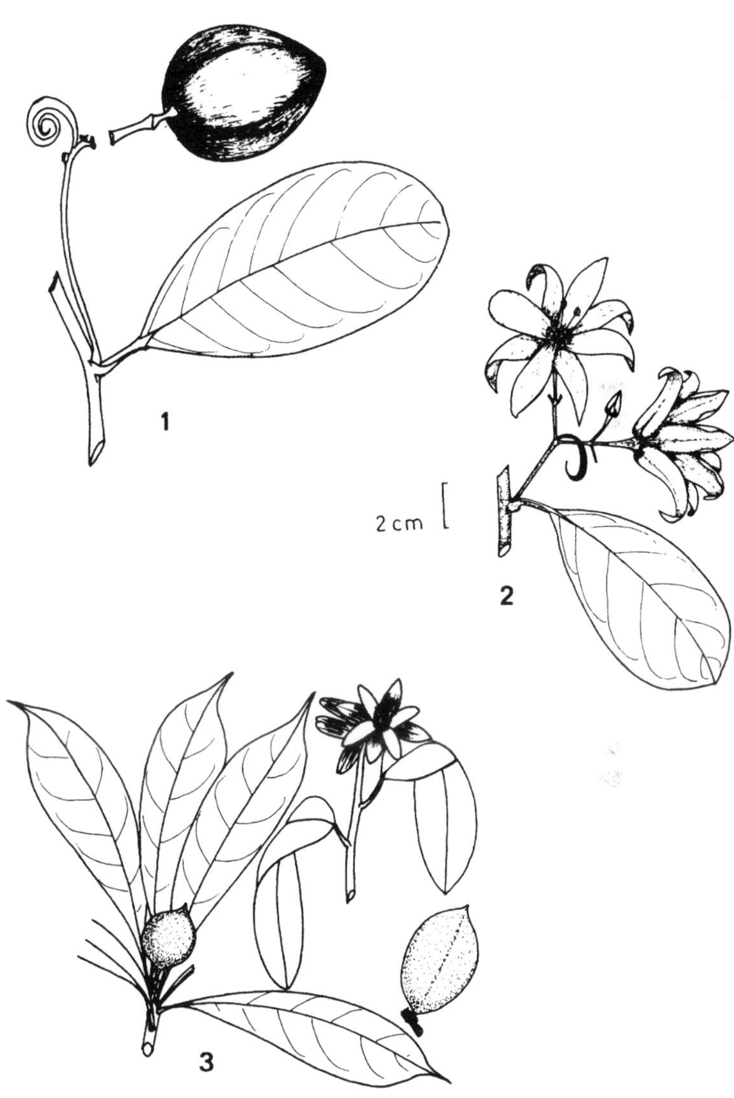

1 - *Ancistrothyrsus*

2 - *Ancistrothyrsus*

3 - *Dilkea*

Passifloraceae
(*Passiflora*)

Passiflora

Ancistrothyrsus (2 spp.) — Canopy lianas sometimes included in Flacourtiaceae. Differs from *Passiflora* in having 8 stamens, in the 4-parted fruit with a rather woody exterior and in having seeds not surrounded by pulpy arils. Vegetatively characterized by obovate leaf shape and especially by the "tendril" which is actually the inflorescence rachis plus a thicker apical hook extending beyond the flower attachment.

(*Tetrastylis*) (1 or 3 spp.) — The Central American/Chocó species are not related to the coastal Brazil type and now usually included in *Passiflora* where they are anomalous in the 4 style branches and the seeds individually enclosed in nongelatinous red arils. Vegetatively, they are typical *Passifloras* with large petiolar glands and lobed leaves.

The only other neotropical genus is *Mitostemma* of coastal Brazil.

PHYTOLACCACEAE

A close-knit and easy to recognize family, unfortunately heterogeneous in the kind of technical characters of carpel number and fusion and ovule placement on which armchair phylogenists love to segregate families. All Phytolaccaceae have entire alternate leaves that dry olive or blackish. The herbaceous phytolacs are especially closely related and may be conveniently seen as a reduction series from *Phytolacca,* from which *Rivina, Hilleria,* and *Schindleria* are indistinguishable in flower except on technical grounds. From the fleshy several-seeded berry of *Phytolacca* would be successively derived *Rivina* with a single-seeded berry, *Schindleria* and *Hilleria* with the "berry" reduced to a single dry seed subtended by the papery calyx lobes (these basally fused in the latter), *Petiveria* with the inflorescence reduced to a spike and the single-seeded dry fruit elongate and having apical bristles, and finally *Microtea* (intermediate with Chenopodiaceae) with minute flowers on a subspicate inflorescence and tiny round echinate fruits. The woody genera are also part of a clearly natural group contrary to the current tendency to split them into many segregate families. All are very easy to recognize to family, vegetatively, on account of the shared gestalt imparted by more or less succulent leaves, drying membranaceous with a characteristic olive or black color, and with sometimes longish but poorly defined petioles, the tendency to spiny habit (*Seguieria, Achatocarpus,* several extralimital genera). *Trichostigma* is essentially *Phytolacca* converted into a liana; *Achatocarpus* is *Phytolacca* converted into a spiny wind-pollinated, dry-area adapted-shrub or tree. The other three genera have wind-dispersed fruits, *Ledenbergia* via the 4 expanded calyx lobes (cf., *Petrea,* but the lobes much smaller and more delicate), *Gallesia* and *Seguieria* via a samaroid wing similar to that of *Machaerium.*

1. Weedy Herbs or Subshrubs with Berries or Desmochoric Fruits, Plant Never Spiny

1A. Fleshy berries

Phytolacca (35 spp., incl. Old World) — Succulent mostly weedy herbs. Unique in 5–16 carpels (very obvious in +/- segmented fruit).

E: jaboncillo; P: airambo

Rivina (3 spp.) — Like *Phytolacca* but berry single-seeded and red.

1B. Dry fruits

Hilleria (5 spp.) — Like *Phytolacca* and *Rivina* but the 3 sepals connate at base and fruit tiny, round, dry, and consisting essentially of a single seed subtended by the suberect and somewhat enlarged papery calyx lobes; flowers drying blackish.

Petiveria (1 sp.) — Inflorescence spicate with widely separated flowers, the exozoochoric fruit long and narrow with uncinate bristles at top; plant sometimes with noticeable garlic odor.

C: anamu; P: mucura

Schindleria (6 spp.) — In flower looks exactly like a *Phytolacca* but usually a subshrub (i.e., larger than *Hilleria, Rivina,* and the herbaceous species of *Phytolacca*) and with more stamens than *Rivina.*

Microtea (10 spp.) — Reduced chenopod-like flowers on a subspicate inflorescence; fruit small subglobose, echinate, exozoochoric.

2. Trees or Woody (Sometimes Spiny) Lianas, Sometimes Reduced to Large Spiny Shrubs in Dry Areas

2A. Samaroid fruits

Gallesia (2 spp.) — Large forest tree with strong garlic odor, mostly on good soils in areas with distinct dry season. Inflorescence paniculate; fruit a one-winged samara.

P: ajosquiro

Seguieria (6 spp.) — Spiny liana with olive-drying leaves; the short spines in pairs at nodes; inflorescence and fruit very like *Gallesia,* but habit completely different.

Ledenbergia (*Flueckigera*) (3 spp.) — Tree with the small berrylike fruit subtended by conspicuous large, expanded papery sepals and presumably wind-dispersed.

Phytolaccaceae
(Herbs and Subshrubs)

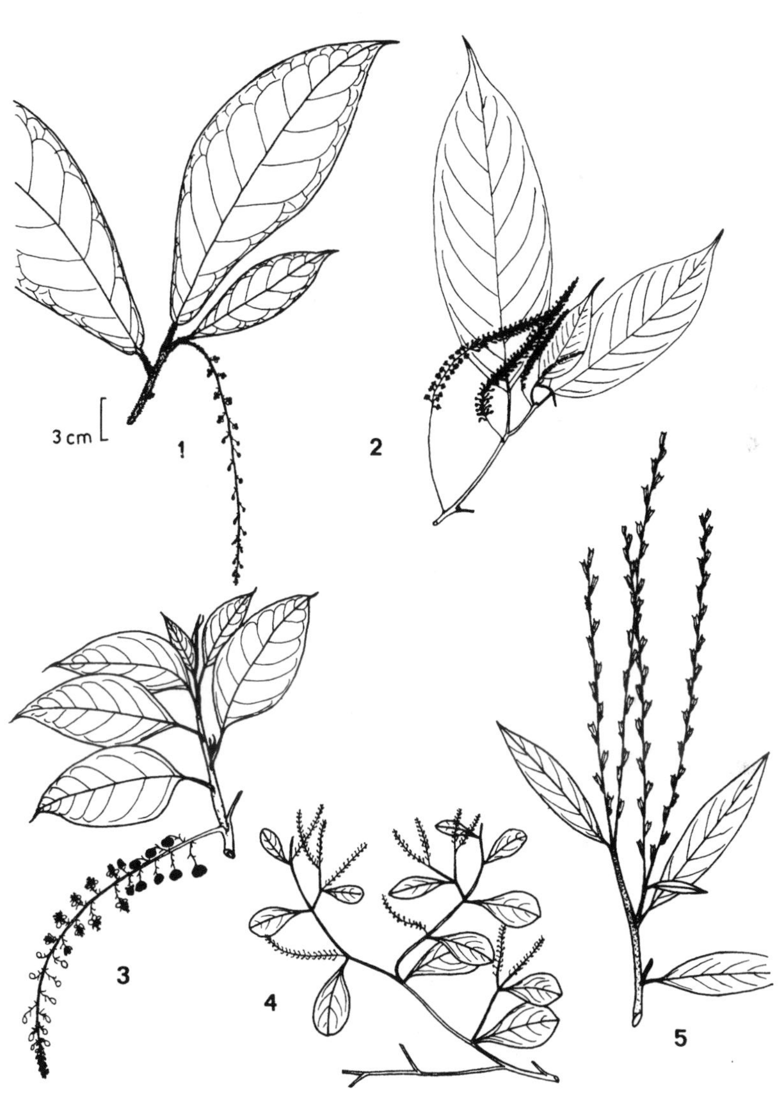

1 - *Hilleria* 2 - *Schindleria*

3 - *Phytolacca* 4 - *Microtea* 5 - *Petiveria*

Phytolaccaceae
(Trees and Lianas)

1 - *Ledenbergia* **2** - *Trichostigma*

3 - *Achatocarpus*

4 - *Seguieria* **5** - *Gallesia*

2B. Fleshy berry-fruits

Trichostigma (4 spp.) — Nonspiny liana; inflorescence a pendent axillary raceme of small fragrant white (reportedly to brown) flowers with broad spreading sepals. Fruits fleshy black berries. Essentially a scandent version of *Phytolacca,* technically characterized by 8 stamens.

Achatocarpus (10 spp.) — More or less spiny, often multistemmed tree or shrub of dry areas, dioecious, typically flowering while deciduous during dry season; fruit a berry; flowers reduced, apetalous, and presumably wind-pollinated. This genus is commonly segregated as a distinct family on account of technical floral characters with zero justification — the cited characters are all predictably correlated with the switch to wind-pollination.

(*Phytolacca*) — Several species of *Phytolacca* are small to medium-sized trees and one extralimital (Argentina) one may have a very thick trunk and be 30 m tall. The racemose inflorescences and berry-fruits of tree *Phytolaccas* are completely unlike other phytolac tree genera.

Extralimital genera:

Agdestis — Four expanded calyx lobes in fruit as in *Ledenbergia* but a semi-inferior ovary. Central America and West Indies to Brazil.

Stegnosperma — Fruit a capsule with arillate seed(s); spiny shrub or liana of Central America and West Indies.

Phaulothamnus — A monotypic narrow-leaved version of *Stegnosperma* from northern Mexico and southern Texas.

Anisomeria and ***Ercilla*** — Chilean genera very close to *Phytolacca,* the latter erroneously reported from Peru on the basis of a mislabeled specimen.

PIPERACEAE

A large but very natural family, mostly of shrubs, herbs and climbers although a few species become stilt-rooted trees well over 10 cm diameter. Important vegetative characteristics are (in woody or subwoody species) the characteristic swollen nodes (often with sheathing petiole base), the usually rather peppery Ranalean odor, and the simple entire (a few extralimital species have irregularly incised margins) leaves often with noticeably asymmetric bases. In flower or fruit, Piperaceae are absolutely unmistakable because of their characteristic densely spicate inflorescence of tiny reduced flowers completely lacking petals or sepals (a few species (*Piper* subgenus *Ottonia*) have the individual flowers pedicellate). The two main genera are succulent, herbaceous, mostly epiphytic *Peperomia* and usually woody (at least subshrubby) terrestrial (sometimes lianescent) *Piper*. The other genera are all related to *Piper* from which they differ in having axillary (rather than leaf-opposed) inflorescences. *Trianaeopiper*, usually herbaceous, is essentially *Piper* with an axillary inflorescence (this rarely once-branched). *Pothomorphe* is characterized by having the inflorescence branched with the spikes in an umbellate cluster and by having broad palmately veined leaves (these peltate in the common lowland species). *Sarcorhachis* is essentially a liana *Piper* with an axillary inflorescence.

Peperomia (1000 spp., incl. Old World) — Succulent mostly epiphytic herbs, the inflorescence branched (often verticillately) or not. The leaves may be alternate, opposite, or whorled, whereas, all other genera are uniformly alternate-leaved.

E: sarpullido (*P. pellucida*); P: congonilla

Piper (2000 spp., incl. Old World) — The great majority of species are shrubs with conspicuously swollen nodes (often with sheathing petiole bases), often asymmetric leaf bases, and rather peppery Ranalean odors. The inflorescence is always leaf-opposed and almost always elongate and densely spicate. One potential segregate genus (the subshrub *Pleiostachyopiper*) has the inflorescence shortened and thickened into a fleshy globose pseudoberry; another (*Ottonia*, mostly vines) has the individual flowers and fruits pedicellate). Leaves and habit are quite variable; some species have peltate leaves, others broadly cordate and palmately veined, others narrrowly lanceolate or oblanceolate; some leaves are cuneate-based, some cordate, some cordate on one side and attenuate on the other. In stature the plants may vary from the typical small shrubs and subshrubs to trees at least 10 m tall (the trees are all stilt-rooted) or thick-stemmed woody lianas (always with anomalous stem anatomy). Fruits are dispersed largely by the bat genus *Carollia*.

C, E, P: cordoncillo

Pothomorphe (2 spp.) — Two common weedy species, a lowland one with peltate leaves and an upland one that is merely cordate. Differs from *Piper* in the branched axillary inflorescence which consists of a single peduncle bearing an umbel of apically clustered spikes. A new typification of *Lepianthus* allows retention of *Pothomorphe* as the correct name for this taxon.

E, P: santa maría

Piperaceae

1 - *Pleiostachyopiper* **2** - *Trianaeopiper*

3 - *Piper* **4 & 5** - *Peperomia*

6 - *Pothomorpha* **7** - *Sarcorhachis*

Sarcorhachis (4 spp.) — Rain-forest or cloud-forest woody lianas with cordate, broadly ovate, palmately veined, and rather succulent leaves. Differs from *Piper* in the rather succulent axillary inflorescences in which the pistils are partially immersed.

Trianaeopiper (18 spp.) — One of the few genera endemic to the Chocó region, with a couple of species reaching into eastern Panama and several south into wet coastal Ecuador. Virtually identical to *Piper* except for the axillary inflorescences. Most species are herbs and none is more than a suffrutescent shrub. A few species may have the inflorescence once-branched.

PLANTAGINACEAE

Essentially a wind-pollinated derivative of Scrophulariaceae, in our area almost exclusively high-Andean. Ours are all herbs with parallel-veined, mostly linear or narrowly oblanceolate leaves in a basal rosette. The small 4-merous flowers have reduced narrow tannish-scarious petals and a small cupular calyx and are mostly sessile on an erect spicate inflorescence, the floriflorous part of inflorescence usually elongate but sometimes reduced to subglobose flower cluster and in a few species to a single flower. The anthers usually conspicuous and long-exserted at anthesis.

Plantago (265 spp., mostly in Temperate Zone) — All our species are rosette herbs with linear (mostly) to oblanceolate (to obovate or rhombic in some weedy species) parallel-veined leaves, exclusively high-Andean except for occasional weeds (*P. major*). The leaves may be either entire or irregularly shallowly and distantly denticulate and the petiole is always poorly differentiated. Inflorescence very characteristic, long pedunculate with the sessile flowers densely arranged in an elongate spike, or floriferous part contracted and subglobose (rarely reduced to single flower: *P. tubulosa*). Technically characterized by the fruit a circumscissile capsule with 2–many seeds.

Bougueria (1 sp.) — A reduced derivative of *Plantago* restricted to the southern puna, and differing in the single-seeded indehiscent fruit. The inflorescence is of the short ovoid-globose type and the linear leaves <4 cm long.

Plantaginaceae, Plumbaginaceae, and Polemoniaceae

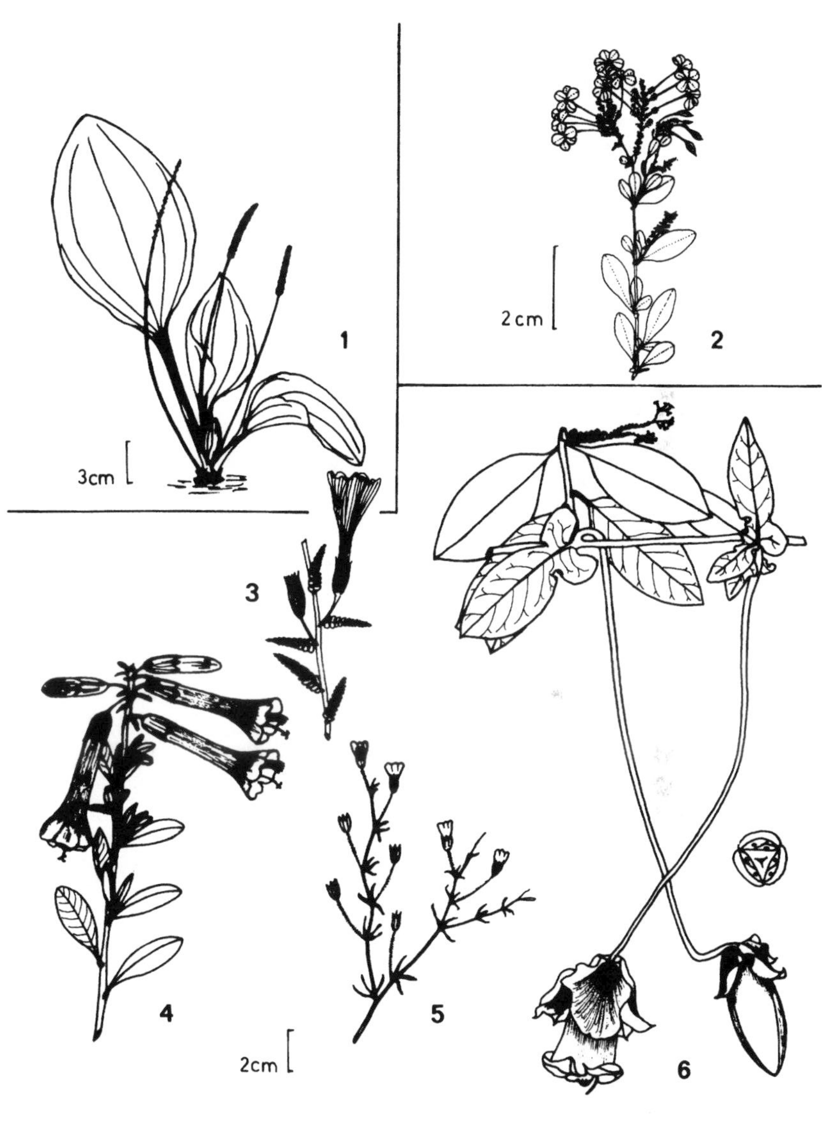

1 - *Plantago*
(Plantaginaceae)

2 - *Plumbago*
(Plumbaginaceae)

3 - *Huthia*

4 - *Cantua* **5** - *Gilia* **6** - *Cobaea*

PLUMBAGINACEAE

A predominantly north temperate, especially Mediterranean, family with only two species of *Plumbago* native in our region (and one of these may be naturalized). Our species are also unusual in being vines, occurring in disturbed lowland dry forests and at middle elevations and are very common in parts of coastal Ecuador and adjacent northwestern Peru. One of them is also in the Peruvian coastal lomas. *Plumbago* is easily recognized by the tubular calyx with conspicuous stalked glands. Vegetatively, it is recognized by the alternate leaves with the blade attenuate onto a poorly differentiated petiole which is more or less expanded and clasping at base (in *P. coerulea* the petiole base is expanded into a pair of clasping auricles).

Plumbago (20 spp., mostly paleotropical)

POLEMONIACEAE

Poorly represented in our area (and in South America), only in the Andes (one species reaches the Peruvian lomas), by two genera of shrubs (*Cantua, Huthia*), one of tendrillate vines (*Cobaea*), plus a few species of the North American herb genera *Gilia, Phlox,* and *Loeselia.* All our taxa are more or less glandular-pubescent at least on the inflorescence and when fertile all can be recognized by the conspicuously 3-parted stigma and/or 3-parted capsule. *Cobaea* is the only vine in our area with alternate, pinnately compound leaves terminating in a branching tendril. The shrubs have conspicuous tubular flowers and alternate leaves, always with the node prominently raised below each leaf base; the herbs have the leaves linear or linearly pinnatifid, except *Loeselia,* distinctive in sharply serrate acanthlike inflorescence bracts.

Cantua (11 spp.) — Andean shrubs or small trees with the +/- elliptic leaves usually more or less crenate or serrate (entire-leaved species have the leaves small and with sharp apicule and tend to have spinescent short-shoot branches). All species have the leaves arising from raised woody projections, these sometimes almost spinescent; at least the inflorescence and young branches are more or less viscid-pubescent. Flowers conspicuous, tubular, always with exserted anthers, varying from greenish or white to bright red; calyx rather large and cupular, 5-dentate to subspathaceous. Seeds small and broadly winged. One species (*C. buxifolia*) is the national flower of Peru.

Huthia (2 spp.) — Shrubs endemic to the dry western Andean slopes of southern Peru. Close to *Cantua* but with distinctive long narrow strikingly crenate-pinnatifid leaves and the blue flower more openly campanulate and the anthers not exserted. Twigs with conspicuous projections subtending leaves as in *Cantua.* Seeds more narrowly winged than *Cantua.*

Cobaea (18 spp.) — Cloud-forest vines or lianas with alternate, pinnately compound leaves terminating in much-branched tendrils (from the modified terminal leaflets). The large campanulate flowers are greenish or maroon, borne pendent on long pedicels, and apparently bat-pollinated. Capsule 3-parted, with several winged seeds.

Gilia (120 spp., mostly N. Am.) — Only a few species of this mostly western North American genus reach South America. Ours are viscid glandular-pubescent erect herbs with linear or pinnately narrowly dissected leaves and pink to purple flowers.

Phlox (66 spp., mostly N. Am.) — Our only species is a tiny annual herb of dry upland regions with narrow opposite leaves and inconspicuous solitary white to bluish flowers.

Loeselia (17 spp., incl. N. Am.) — Our only species is an opposite-leaved weed of open areas in middle-elevation northern Andean cloud forest. Looks more like an acanth than Polemoniaceae with the small bluish sessile flowers in small headlike clusters subtended by conspicuous sharply serrate leaflike inflorescence bracts. Differs from acanths in the sharply serrate leaf margin and from *Hyptis* in the round stem.

POLYGALACEAE

A predominantly woody tropical family, in our area mostly woody lianas, although herbaceous species of *Polygala* are better known to temperate zone botanists. Four of our six genera are exclusively lianas, *Monnina* is usually shrubby but also includes small trees and a few canopy climbers, and *Polygala* varies from small herbs to canopy trees (the latter distinctive in opposite leaves with a uniformly cylindric legumelike petiole). The lianas have uniformly alternate, entire coriaceous leaves with rather short petioles and usually prominulous venation; they have a characteristic stem section with irregularly concentric (usually asymmetrically so) circles of anomalous vascular tissue. Polygalac lianas might be confused with liana convolvs which have similar stem sections but the latter differ in longer petioles; *Lehretia* (Icacinaceae) is also similar but has larger leaves with more ascending secondary veins, which dry a distinctive grayish color and usually have a more acuminate apex. *Moutabea* is distinctive in having inconspicuous short spines on the twigs and also in the thick-coriaceous yellowish-olive-drying leaves with the venation (even the secondaries) immersed. *Securidaca* has greenish - olive sensitive twigs that make tendril-like twists (cf., Hippocrateaceae). Relatively nondescript *Diclidanthera* usually has the midvein drying distinctly yellowish or reddish. *Polygala* has a striking wintergreen odor in the roots and hints of a similar "medicinal" odor are sometimes found in other genera.

Most Polygalaceae have very characteristic pealike flowers with the two lateral sepals enlarged and petaloid (cf., the wings of legume flower) and the lower of the three petals enlarged, saccate, and enclosing the stamens (cf., the legume keel). The two exceptions, *Moutabea* and *Diclidanthera* have white flowers with the basal part of the five subequal strap-shaped petals fused into a narrow tube. The different genera have very different fruits, varying from a capsule (*Bredemeyera, Polygala*) to small, fleshy, berrylike drupes (most *Monnina*), large drupes (*Moutabea*) or single-winged samaras (*Securidaca*).

1. Lianas — (The first two genera have papilionaceous flowers, the next two narrowly tubular white flowers.)

Securidaca (80 spp., incl. Old World) — Canopy lianas, especially common in seasonal forest. Vegetatively characterized by the greenish-olive twigs which sometimes twist into tendril-like loops. Flowers papilionaceous, mostly magenta; fruit a very distinctive single-winged samara (in one water-dispersed species the body thicker and wing vestigial).

P: gallito

Bredemeyera (20 spp.) — Lianas, mostly in seasonally dry forest. Vegetatively characterized by the close-together, parallel, and prominulous secondary and intersecondary veins. Flower papilionaceous, white or greenish-white; fruit a very distinctive flat oblanceolate capsule with apical notch, very like single segment of *Hippocratea* fruit but smaller and dehiscing to release single seed with pilose rather than winged margin.

Diclidanthera (6 spp.) — Lianas, mostly of seasonally inundated tahuampa forest. Vegetatively rather nondescript but leaves dry a characteristic yellowish-olive (when older) or reddish (when younger), the midvein, especially, drying distinctively yellowish or reddish. Flowers white and narrowly tubular, the sepals strap-shaped like the petals (but much shorter) and similarly basally fused. Fruit a globose drupe <2 cm in diameter and black or purplish-black at maturity.

Polygalaceae
(Lianas)

1 - *Diclidanthera*

2 - *Securidaca*

3 - *Bredemeyera*

4 - *Moutabea*

Polygalaceae
(Herbs and Trees)

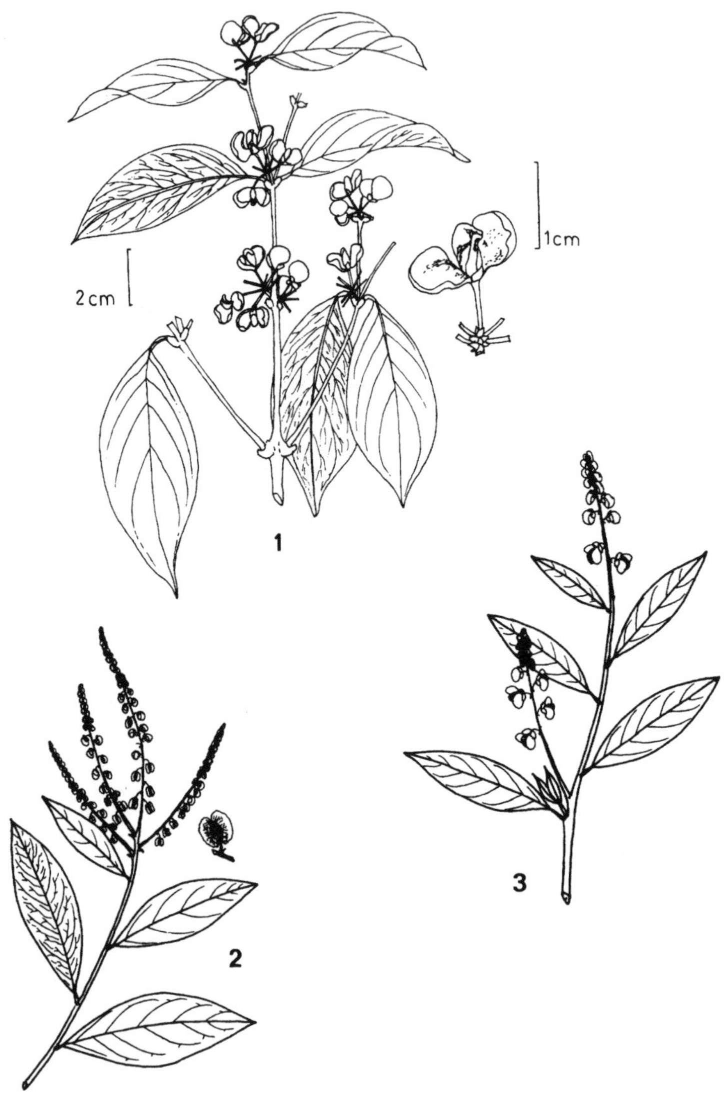

1 - *Polygala*

2 - *Monnina*

3 - *Polygala*

Moutabea (8 spp.) — Canopy lianas. Vegetatively distinctive in the scattered inconspicuous small spines on branchlets and especially the rather narrow, thick-coriaceous, yellowish-olive drying leaves with the venation (even the secondaries) immersed and inconspicuous. Flowers as in *Diclidanthera* but smaller; fruit similar but usually larger (>2 cm diameter) and sometimes orangish at maturity.

(*Monnina*) — (Some cloud-forest species of *Monnina* are scandent).

2. Erect Herbs, Shrubs, or Trees — (All with papilionaceous flowers)

Monnina (150 spp.) — Mostly shrubs or small trees of middle and upper Andean forests; a few species are more or less epiphytic and/or scandent. Leaves uniformly alternate and typically rather succulent-membranaceous. Flowers (borne all year long) blue or bluish-purple and papilionaceous. Fruit a small, fleshy, black berrylike drupe or drier and somewhat winged.

Polygala (500 spp., incl. Old World and N. Am.) — Mostly tiny herbs of open savanna areas on poor soil, a few species weedy. In addition a few species are shrubby and one is a canopy tree (with opposite leaves!). The pink or white papilionaceous flowers are very distinctive, as is the wintergreen odor of the roots. The small round fruits are capsular.

These are the only South American genera of the family although there are several others in the Antilles.

Polygonaceae

Usually easy to recognize by the uniformly alternate, entire (to irregularly crenate) simple leaves with a conspicuous ochrea (a stipule-derived ring of tissue sheathing twig around the node), this unique to the family (but sometimes reduced to nodal ring). The somewhat Moraceae-like (but usually with tip curved away from petiole base) conical stipule at branchlet apex prior to formation of the ochrea is also typical and especially useful in a few *Coccoloba* species where the ochrea is not very conspicuous. Most arborescent taxa have rather close-together, Dilleniaceae-like (but the margin entire) secondary venation, the petiole typically has a flattened base with wings connecting to ochreal margin, and in both herbs and trees the twig (or stem) is often hollow (in *Triplaris* usually inhabited by stinging ants). Frequently dioecious with the inflorescence a spike or narrow raceme, the flowers (except cultivated *Antigonon*) small and reduced, typically trimerous with 3 petals and 3 sepals (5 tepals in *Coccoloba),* 6–9 stamens and a 3-angled ovary. Fruit always

1-seeded, usually 3-winged or small and 3-angled and variously enclosed by expanded perianth segments.

The largest genus *Coccoloba* includes both trees (often multitrunked and often with very hard wood) and lianas (often with somewhat 2-parted flattened stems); three of the other genera in our area are trees (or shrubs), two are weedy herbs and two are climbing (including mostly cultivated *Antigonon* with tendrils). Two of the tree genera (closely related *Triplaris* and *Ruprechtia*) have very characteristic wind-dispersed fruits with the calyx forming cup and the 3 calyx lobes greatly elongated into conspicuous reddish wings; *Coccoloba* has a narrowly racemose inflorescence with very characteristic round ball-bearing fruits that always fall off the dried specimens.

1. Trees or Treelets

Triplaris (25 spp.) — Trees, usually with hollow stems inhabited by stinging ants; typically in riverine or secondary habitats in moist forest. Leaves larger than *Ruprechtia,* more or less oblong and with numerous rather close-together secondary veins. Dioecious; inflorescence spicate (or with spicate lateral branches), usually long and conspicuous, female flowers borne one to node, male flowers sessile or subsessile and with perianth segments connate over halfway. Fruit very characteristic, the 3 perianth segments red and much elongated, forming 3-winged samara.

E: fernansánchez; P: tangarana

Ruprechtia (17 spp.) — Mostly shrubs, sometimes trees, usually in dry deciduous-forest habitats. Very like *Triplaris* but usually lacking hollow stems and our species always with smaller leaves. In flower differs from *Triplaris* in usually shorter inflorescence with pedicellate male flowers having perianth segments fused about one-third of their length and female flowers 2–3 together on short lateral branches.

Symmeria (1 sp.) — Tree of seasonally inundated tahuampa forest. Very characteristic in +/- oblong obtuse, glabrous, subcordate leaf with long narrowly subwinged petiole. No other polygonac tree combines a subcordate leaf base with long distinctly subwinged petiole. In flower characterized by the paniculate inflorescence (the straight lateral branches with tiny sessile flowers). Fruit sharply 3-angled and trigonal-pyramidal.

Coccoloba (140 spp.) — Small to large trees or canopy lianas, the latter often with a flattened somewhat 2-parted stem. Ochrea not always persistent; when caducous leaving at least a prominent ring around node; at branch apices (i.e., prior to forming ochrea) the stipule narrowly conical and rather moraclike but usually somewhat flattened and/or the apex curved away from base of terminal petiole. Leaves usually coriaceous with more

Polygonaceae
(Trees)

1 - *Ruprechtia* **2** - *Coccoloba*

3 - *Triplaris* **4** - *Symmeria*

Polygonaceae
(Herbs or Scandent)

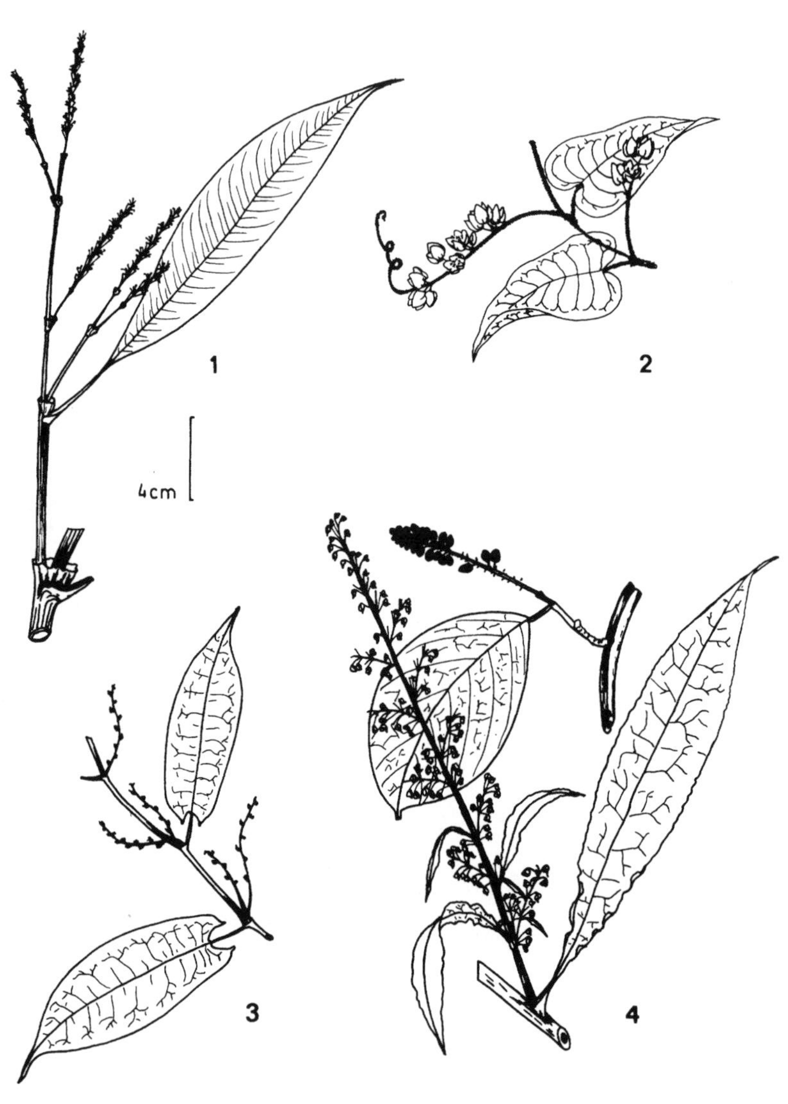

1 - *Polygonum* **2** - *Antigonon*

3 - *Coccoloba*

4 - *Muehlenbeckia* **5** - *Rumex*

or less strongly prominulous tertiary venation at least below, usually more broadly elliptic or oblong-elliptic than *Triplaris* and *Ruprechtia* and with fewer more widely spaced secondary veins. Petiole may be rather flattened, but unlike *Symmeria* lacks distinct marginate wing. Inflorescence spicate or very narrowly racemose, or with several straight spicate lateral branches along a straight central rachis. Flowers tiny, usually rather widely spaced and with 5 perianth segments (unlike *Triplaris* and *Ruprechtia*). Fruit small round ball-bearing-like (rarely somewhat acutish at apex), always falling off the dried specimens.

C: uvita de playa (*C. uvifera*); E: tangarana

2. Climbers or Prostrate Subshrubs — (One species a phyllodal shrub)

Muehlenbeckia (15 spp., incl. Old World) — Mostly south temperate; in our area exclusively Andean and always at least subwoody. Mostly high-climbing vines or lianas of montane cloud forests characterized by sagittately angled more or less cordate leaf base (common species also with conspicuous apical apicule). Also +/- prostrate puna and paramo subshrubs with tiny obovate leaves. One shrub species of disturbed northern Andean cloud forest completely leafless and reduced to jointed stem of flattened phyllodes (these without ochrea but with conspicuous annular scar). Technically differs from *Polygonum* in being dioecious and the perianth becoming fleshy in fruit.

Antigonon (8 spp.) — In our area only in cultivation or semicultivation. Vegetatively distinctive in the triangular leaf with a more or less subcrenulate margin and in having the apical coiling part of the inflorescence-derived tendril arising from a distinctively zigzag base (originally the apically branching tendril terminating an inflorescence [cf., Sapindaceae]). Flowers large (for family), pink, very conspicuous.

(*Coccoloba*) — Many *Coccoloba* species, all in lowland forest, are lianas. All have medium to large entire coriaceous leaves with more or less prominulously raised venation at least below, the base obtuse to cordate (but not at all sagitately angled), and the inflorescence simply spicate.

3. Weedy or Semiaquatic Herbs

Polygonum (300 spp., incl. Temperate Zone and Old World) — Semiaquatic usually rather succulent herbs. Leaves entire, usually narrow, the petiole usually short and/or poorly differentiated from lamina base and/or +/- winged. Inflorescence usually a spike or very narrow raceme (or reduced to axillary fascicle), the pink or white flowers either borne singly or sessile (or both). Fruit small, the calyx usually more or less accrescent but not winged as in *Rumex*.

E: yaco

Rumex (200 spp., incl. Temperate Zone and Old World) — In our area, exclusively high-Andean. Stem characteristically striate-ridged or angled. Leaves narrowly oblong, differing from our *Polygonum* in a long petiole and +/- crenate or crisped margin. Flowers greenish and pedicellate, borne in clusters along a usually more or less narrowly paniculate inflorescence. Fruits trigonal, differing from *Polygonum* in the accrescent calyx being distinctly 3-winged.

Extralimital ***Podopterus*** is a distinctive shrub of Central American dry areas with branch spines, obovate leaves in short-shoot clusters, and fascicles of peculiar trigonally 3-winged fruits borne while leafless. Several other essentially North American genera reach northern Central America.

PORTULACACEAE

An entirely herbaceous family with usually more or less succulent, always entire leaves. In our area mostly in dry inter-Andean valleys, except for *Montia,* a tiny mat-forming semiaquatic paramo herb, and a few weeds. The succulent obovate to linear either alternate or opposite leaves with poorly defined petioles are generally distinctive as are the densely pilose leaf axils (and sometimes stem) of the common genus *Portulaca;* could be confused with some species of Aizoaceae but the (usually 2) sepals and (usually 5) petals are both present and well-differentiated. All but one of our Aizoaceae have opposite leaves, which are limited in Portulaccaceae to semiaquatic *Montia* and species of *Portulaca* which are easily recognized by their strikingly pilose stems and/or leaf axils.

Portulaca (200 spp., incl. Old World and n. temperate) — mostly prostrate succulent herbs, more or less densely pilose in the leaf axils and around base of the mostly solitary sessile flowers (except in the commonest species, the yellowish-flowered weed, *P. oleracea*). Leaves linear to oblanceolate and often opposite. Technically differs from other genera in the circumscissile capsule.

Talinum (50 spp., incl. Old World) — Differs from *Portulaca* in the erect habit, in having the succulent obovate leaves larger, in lacking pilose axils or stems, and in having a diffusely paniculate inflorescence. The sepals are caducous and the flowers are usually magenta.

Calandrinia (150 spp., incl. Old World) — Like *Talinum* but sepals persistent; the inflorescence usually reduced, often to single flowers, and the plants frequently prostrate and with tiny leaves.

Portulacaceae and Primulaceae

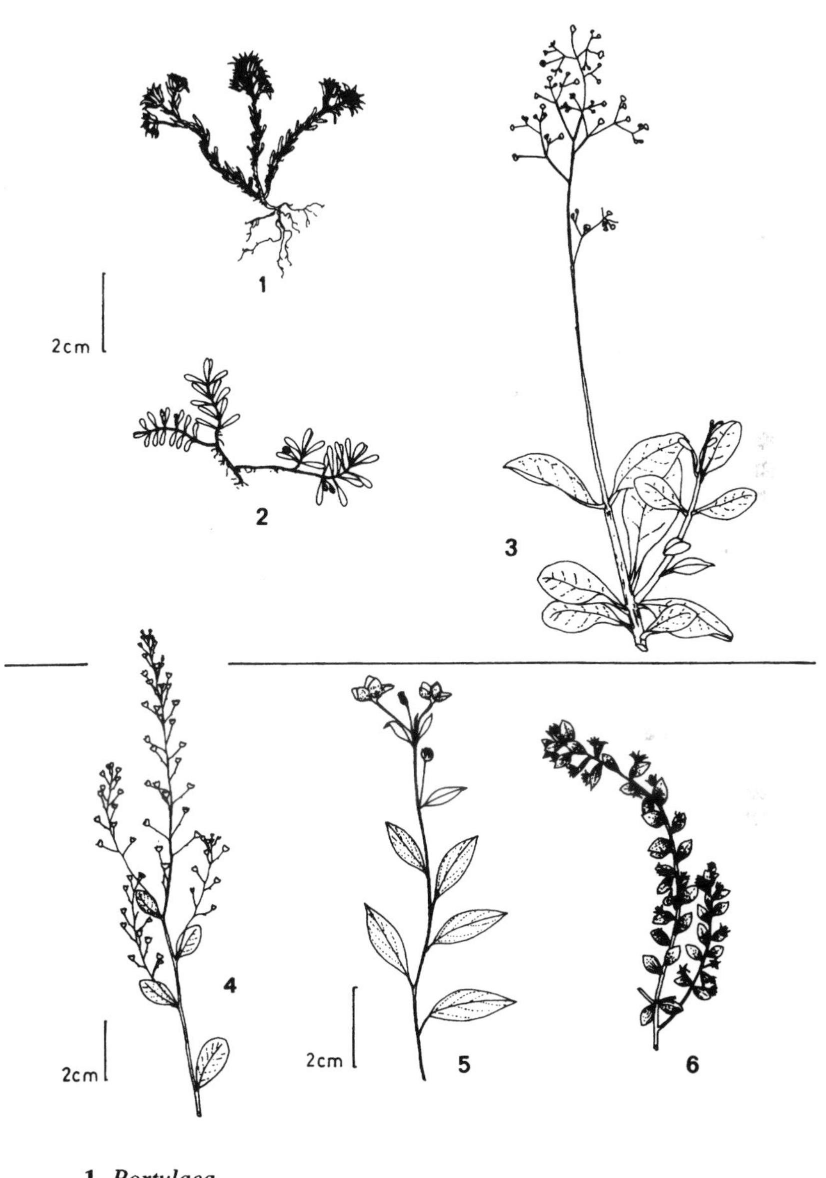

1 - *Portulaca*

2 - *Montia*

3 - *Talinum*

4 - *Samolus* **5** - *Lysimachia* **6** - *Anagallis*

Montia (15 spp., incl. N. Am.) — Tiny, nondescript, +/- semiaquatic mat-forming paramo herb with narrow opposite leaves and tiny white flowers. Technically distinguished from *Calandrinia* by having the petals fused at base.

PRIMULACEAE

Very poorly represented in tropical America, ours all more or less succulent herbs, often growing near water. Characterized by sharply angled (usually sharply tetragonal) stems and entire to more or less serrulate simple leaves. Two genera have the leaves opposite and 2 alternate; all have the small white or blue flowers borne solitary in the leaf axils, usually on long slender pedicels, except *Samolus* which looks like a white-flowered version of *Talinum* (but the inflorescence racemose). Might be thought of as the herb equivalent of Myrsinaceae on the basis of shared ovary characters and fused petals.

Anagallis (3 spp., plus 29 in Old World) — Succulent sprawling weedy herb with opposite sessile ovate leaves and square stem, the solitary axillary flowers blue with orange centers, borne on long nodding pedicels, the 5 subulate sepals nearly linear.

Samolus (10–15 spp., incl. Old World) — Our only species a widespread but not very common weed of wet places, strongly resembling *Talinum* but with strongly angled stem and white flowers in an open racemose inflorescence. The obovate, rather succulent, alternate leaves with a poorly defined winged petiole are exactly like *Talinum.* The technical differentiating character is a "semi-inferior" ovary.

Centunculus (1 sp., often lumped with *Anagallis*)— Tiny weedy herb with opposite suborbicular leaves, strongly resembling a caryophyllac but with tetragonal stem and the greenish to bluish petals not bifid.

Lysimachia (200 spp., incl. Old World) — A large north temperate genus with perhaps only a single species in our area, *L. andina* of permanently moist places in the western Peruvian Andes. Herb with alternate leaves, the stems strongly angled; flowers solitary white, on slender pedicels in upper leaf axils.

Proteaceae

1 - *Panopsis*

2 - *Oreocallis*

3 - *Lomatia*

4 - *Roupala*

5 - *Euplassa*

Proteaceae

A quintessentially Southern Hemisphere family best developed in Australia and South Africa. In our area all the species are trees or shrubby treelets, typically occurring on exposed eroded windswept ridgetops at the edge of montane or premontane cloud forest, but also scattered in lowland forest. All species have coriaceous leaves, typically with crenate-serrate margins, and the majority have long petioles, more or less thickened at base, and/or conspicuous rufescent pubescence on the inflorescence and petioles. The only genus widespread in lowland forest (*Roupala*) is very easy to recognize by the odor of the wood and slash which is exactly that of poor quality or slightly spoiled canned beef; the other lowland tree genus (*Panopsis*), lacks this odor but has a characteristic mottled wood somewhat resembling the scales of a snake (= palo culebra). *Roupala* is characterized by pinnately compound juvenile leaves and simple mature leaves, even in the same species.

The inflorescence is very characteristic with pairs of flowers arranged along a narrow raceme (or the shared pedicel reduced and the inflorescence spikelike); the narrow flowers have 4 narrow valvate perianth parts each with a subsessile anther cupped in its slightly swollen apex. All our taxa except *Panopsis* have a follicular fruit that dehisces along one side to release winged seeds; *Panopsis* has a globose drupe usually ca. 3–4 cm in diameter.

Oreocallis (2 spp., also 3 in New Zealand and Australia) — Wandlike shrubs or treelets of exposed ridgetops at the edges of Andean cloud forests. Leaves completely entire, obovate, usually more or less glaucous below (the midvein and petiole also often somewhat rufous-pubescent), lamina base more or less decurrent onto petiole. The largest flowers of any area proteac, at least 3 cm long, conspicuous, varying from white (usually with pink shading) to bright rose-red. Fruit larger and more woody than other genera with follicles 4–5 cm long, in addition with a very long apical beak ca. 3 cm long.

P: cucharilla

Lomatia (3 spp., also 9 in Australia) — Our only species a small tree of Andean upland scrub forest, very easy to recognize by the crenate-serrate leaves with a nearly truncate base and very distinctive triangular-ovate shape. Inflorescence few-flowered, axillary, the flowers white to cream, <1 cm long, unique among area proteacs in the strongly rufous-pilose perianth segments (the rachis similarly pubescent). Follicle narrower and more cylindrical than in other genera, 2–3 cm long, not at all beaked.

Roupala (52 spp.) — By far the commonest genus of proteac tree, especially common in windswept premontane cloud forest but also found in lowland forest and especially in open savanna. Vegetatively unmistakable in the canned beef vegetative odor. Most species have pinnately compound juvenile leaves (the leaflets strongly asymmetric with at least the more strongly curved abaxial margin coarsely crenate-serrate). The mature leaves tend to be rhombic, with strongly ascending secondary venation, and typically have very long petioles, the petiole sometimes as long as the lamina. Lamina base usually more or less cuneate and attenuate onto petiole, unlike *Lomatia*. Flowers short, <1 cm long, forming bottle-brush inflorescence, with the shared pedicels of each flower pair usually very short or lacking. Fruit a small asymmetric follicle 2–3 cm long with one margin nearly straight and the other strongly curved, the apex usually subapiculate.

P: gaucho caspi

Panopsis (11 spp.) — Trees, mostly of rather low-elevation cloud forest, the leaves uniformly coriaceous and simple, nearly always narrowly oblong in outline, strongly and intricately prominulous-reticulate above and below, and with a short petiole. Frequently more or less opposite or subopposite; drying blackish and glossy. Wood with a mottled color resembling snakeskin. Fruit, nonfollicular, the large, globose indehiscent drupe usually 3–4 cm across, and with a single very large seed.

C: sombrerillo; P: palo culebra

Euplassa (20 spp.) — Andean foothill (and lowland Amazonian) trees with pinnately compound leaves similar in texture and dark-drying color to simple-leaved *Panopsis*, and essentially a pinnate-leaved version of that genus. Leaves strongly resembling *Cupania* (Sapindaceae) in the subopposite leaflets with short petiolules with thickened subwoody bases and in the aborted rachis apex, but differing from *Cupania* in a glossier surface, more intricately prominulous veinlet reticulation, and the usually rufescent rachis and young twigs. Raceme with small flowers ca. 1 cm long.

(*Grevillea*) — This Australian genus is widely cultivated in drier areas. Easily recognized by the deeply irregularly pinnatifid leaves.

Pyrolaceae

Achlorophyllous saprophytic herbs, in our area represented by only a single species, *Monotropa uniflora*, of Colombian montane oak forests. The whole plant is reddish and the stem bears reduced scalelike leaves with a single terminal urceolate flower. (Figure 1).

Quiinaceae

Lowland moist- and wet-forest midcanopy trees with opposite (or whorled) usually glabrous leaves and a usually conspicuous pair of interpetiolar stipules. The leaves are usually simple but pinnately compound in two genera (*Touroulia* and extralimital *Froesia*). The interpetiolar stipules (often rather long and subfoliaceous, usually persistent) are reminiscent of Rubiaceae but completely separate (rare in Rubiaceae) and the leaf (or leaflet) margins tend to be serrate or serrulate and have a strong tendency for characteristic intersecondary veins. The pinnately compound genera are unique in the close-together secondary veins ending in spinose marginal teeth (and in *Touroulia* in usually having the leaflets not completely differentiated from rachis). The inflorescences are mostly axillary and racemose or spicate, sometimes branching, with tiny multistaminate flowers giving rise to round or ellipsoid fleshy fruits, usually characterized by a longitudinally costate or striate surface, at least when dry.

Quiina (35 spp.) — Small to midcanopy trees. Stipules of the commonest species are quite inconspicuous. The only Quiinaceae genus in our area with opposite simple leaves. Fruit smaller than *Lacunaria*, usually ellipsoid, sometimes globose, 2–4 seeded, not always prominently costate.

Lacunaria (11 spp.) — Midcanopy trees. Differs from *Quiina* in having whorled leaves (usually in threes). Fruits usually depressed-globose, large and multi(= > 6)-seeded).

Touroulia (4 spp.) — Small trees. Leaves pinnately compound with at least some of the asymmetrical leaflet bases not completely differentiated from the rachis and decurrent on it. Fruits globose, longitudinally striate.

Froesia (3 spp.) — Mostly in lower Amazonia but reaching Brazil/Peru border. Characterized by pinnate leaves like *Touroulia* but apocarpous fruits.

Rafflesiaceae

Parasitic plants with vegetative parts thalloid and immersed inside host plant. Our two genera visible only as the tiny funguslike cupular flowers emerging in rows of enations along trunk of host plant. Our genera known to be parasitic on Flacourtiaceae and Leguminosae.

Quiinaceae and Ranunculaceae

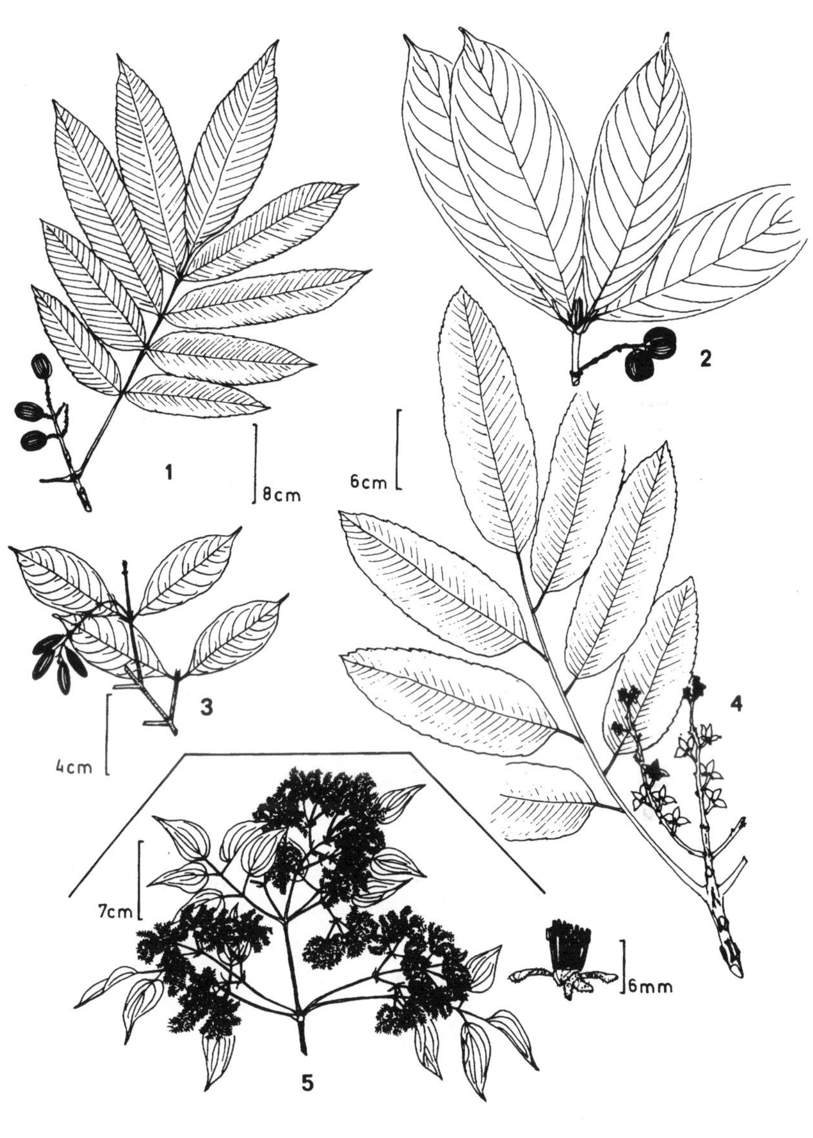

1 - *Touroulia*

2 - *Lacunaria*

3 - *Quiina*

4 - *Froesia*

5 - *Clematis* (Ranunculaceae)

Apodanthes (2 spp.) — Parasitic on *Casearia* and *Xylosma*. (Fig. 1).

Pilostyles (12 sp.) — Parasitic on *Bauhinia* (and perhaps other legumes, including *Calliandra* in Brazil).

RANUNCULACEAE

The only woody Ranunculaceae in our area is the liana genus *Clematis*, easily recognized by its opposite compound leaves with sometimes twining petiole. Herb genera represented in our area (all restricted to high altitudes except the floating aquatic *Ranunculus flagelliformis*) include *Thalictrum* (apetalous and with inconspicuous greenish sepals; leaves 2–4-pinnate), *Anemone* (incl. *Capethia*) (apetalous puna plants with white petaloid sepals and leaves divided or entire (*Capethia*), *Caltha* (low herb of wet *Distichia* cushions; yellow-flowered with broadly ovate, deeply cordate, long-petiolate leaves), *Ranunculus* (typical buttercups with 5 yellow petals and usually palmately divided leaves; mostly small and semiaquatic or weedy), and two closely related quintessentially high-Andean genera *Krapfia* and *Laccopetalum*, essentially overgrown buttercups. *Krapfia*, differing from *Ranunculus* in larger flowers with fleshier receptacle and the petals not caducous, has many species, most with deeply palmately dissected leaves, usually with yellow or orange flowers, but greenish (and 6–8 cm across!) and with entire or toothed leaves in *K. raimondii* and relatives (these intermediate with *Laccopetalum*). *Laccopetalum* has huge greenish fleshy flowers 10–15 cm across!

Clematis (250 spp., incl. N. Am. and Old World) — Lianas commonest in middle-elevation cloud forests but also descending to lower elevations mostly in disturbed wet forest. Distinctive in opposite pinnately compound (or 3-foliolate) leaves with twining petioles. The leaflets 3-veined and usually somewhat dentate, the petiole base more or less connected across the thickened node as ridge (or decurrent and this V-shaped). The branchlets always conspicuously longitudinally striate-ridged. Also easy to recognize vegetatively by the dark brown or blackish fibrous stem with deep alternating longitudinal ridges and grooves. Flowers with petal-like white sepals; more conspicuous in fruit from the numerous very elongate plumose whitish styles (which eventually function in wind dispersal).

RHAMNACEAE

A frequently thorny family, in our area with three genera of lianas (one spiny), three of essentially leafless spiny dry-area shrubs (two with photosynthetic branches), and four of trees (one usually with paired spines resembling modified stipules, another sometimes with axillary branch-spines. Most Rhamnaceae (except the distinctive essentially leafless taxa) can be recognized by the characteristic venation of the uniformly simple leaves, the secondary veins unusually close-together and parallel and either conspicuously strongly ascending or distinctly straight or both; the tertiary venation tends to be more or less parallel and perpendicular to the secondaries. Genera lacking the typical venation have the leaves more or less strongly 3-veined (*Ampelozizyphus, Zizyphus*, a few *Colubrina*), sometimes with glands at base of lamina. The commonest liana genus (*Gouania*) has a unique tendril coiled in one plane exactly like a butterfly's tongue. Most genera have alternate leaves but *Rhamnidium* and some *Colubrina* have opposite leaves; most *Colubrina* species (and all those with opposite leaves) have conspicuous glands at base of lamina (unlike Euphorbiaceae where these are on petiole apex [or in axils of main lateral vein pair below]). The flowers are always tiny and individually inconspicuous, usually 5-merous, with a well-developed disk, and typically in small axillary dichotomously branched umbelliform cymes. The fruit is usually a small berrylike drupe (with the round disk obvious at its base), but a 3-valved capsule in *Colubrina* and 3-winged (fragmenting into three 2-winged segments) in *Gouania.*

1. LIANAS

Gouania (5 spp., plus 15 in Old World) — Canopy and second-growth lianas of lowland forest, easy to recognize on account of the unique butterfly-tongue tendril borne in axil of terminal leaf at apex of short branch. Leaves ovate, usually +/- serrate and/or with gland pair at extreme base of lamina on upper side (or both), typical rhamnaceous ascending parallel secondary veins. Inflorescence an axillary spike or apparently a terminal panicle with several of these together in the axils of suppressed leaves. Fruit strongly trigonal, 3-winged, fragmenting into three 2-winged cocci.

Ampelozizyphus (1 sp.) — Canopy liana of lowland Amazonian moist forest, especially in seasonally inundated forests. The distinctive leaf coriaceous, oblong-elliptic to oblong-ovate in shape, and strongly 3–(5)-veined to apex, drying olive with main veins light tan; reminiscent of menisperm but with no hint of pulvinular flexion. Fruits globose, 2 cm in diameter.

Rhamnaceae
(Lianas)

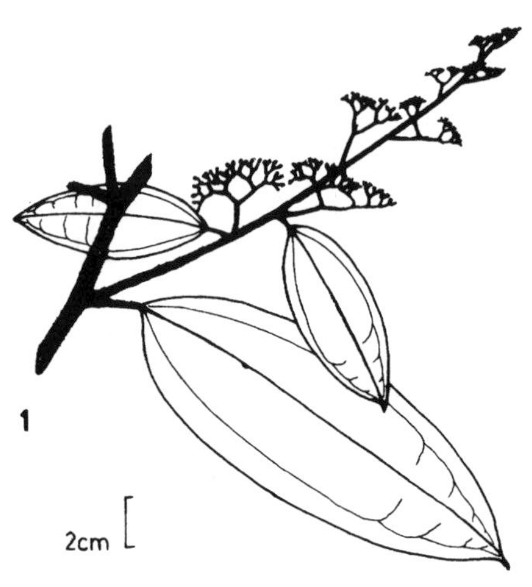

1 - *Ampelozizyphus*

2 - *Sagaretia*

3 - *Gouania*

Rhamnaceae
(Trees and Shrubs)

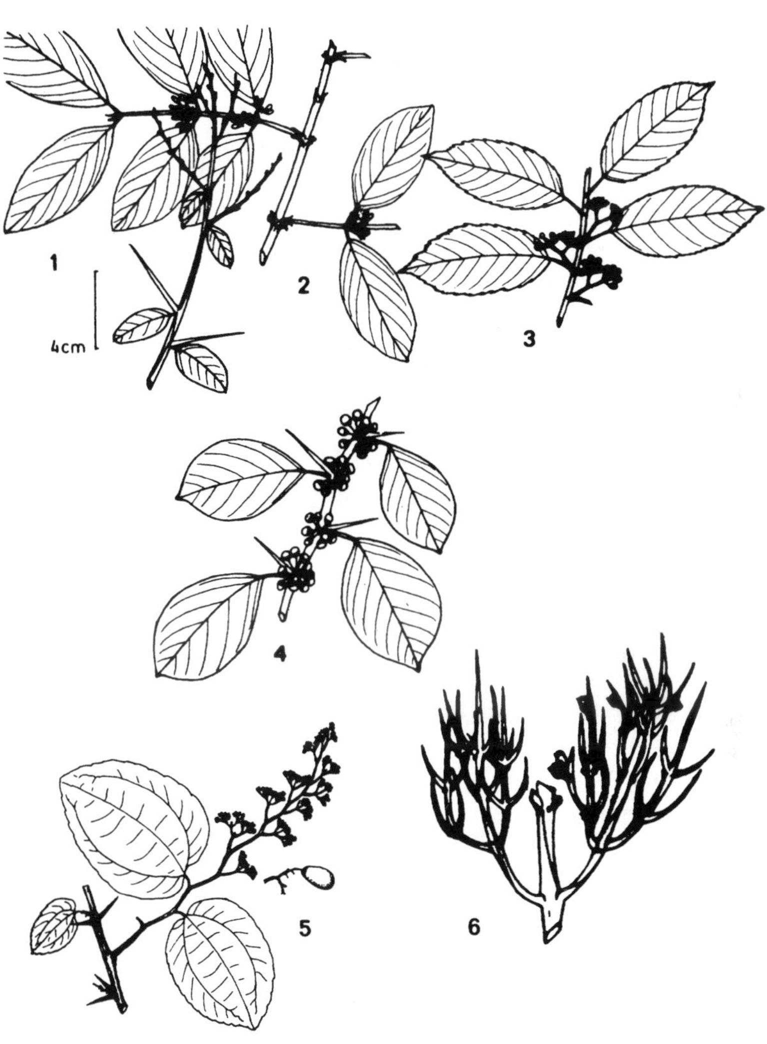

1 - *Scutia* **2** - *Rhamnidium* **3** - *Rhamnus*

4 - *Colubrina*

5 - *Zizyphus* **6** - *Colletia*

Sagaretia (1 sp., plus 30 spp. in Old World) — Dry-area liana with finely serrate ovate leaves and axillary branch spines; the leaves similar to *Gouania* but tendrils lacking. Inflorescence pyramidally paniculate with straight central axis and several sessile-flowered lateral branches; fruits 3-angled capsules (like immature *Gouania* before the wings develop).

2. Thorny Dry-Area Shrubs, Leafless or with Small, Entire, Usually Early-Caducous Leaves — (The first two genera usually have photosynthetic stems and thorns, the third has axillary spines).

Scutia (4 spp., plus 5 in Old World) — Dry desertlike areas below 1500 meters. Stems and branch-thorns photosynthetic, the latter paired and almost at right angles to main stem. Branchlets strongly angled (even in leafy extralimital taxa). Inflorescence apparently spicate in axils of the opposite leafless branch spines.

Colletia (17 spp.) — The high-altitude (> 3000 m) equivalent of *Scutia,* differing in the leafless opposite branch spines strongly ascending and often themselves branched.

Condalia (18 spp.) — Leaves small and obovate, persistent, in short-shoot clusters in axils of nonphotosynthetic *alternate* spines or along the spines. Drupe small, black, ellipsoid, edible.

3. Trees

Zizyphus (29 spp., plus 70 in Old World) — Mostly thorny dry-forest trees (often multitrunked), but one species (*Z. cinnamonum)* a nonspiny canopy tree of moist lowland forest. Leaves always strongly 3-veined from base. Dry-area species have serrate leaves and paired spines resembling modified stipules; *Z. cinnamonum* has coriaceous, rather oblong, entire leaves 3-veined to apex, with asymmetric base, glands in vein axils *above,* and a short petiole. Flowers in dichotomous umbel-like axillary cymes; fruit a small ellipsoid drupe to 2 cm long.

Rhamnidium (12 sp.) — Opposite-leaved canopy trees, mostly on good soils (especially in seasonally dry forest), easily distinguished by the rather close, straight, parallel, not strongly ascending secondary veins, pale leaf undersurface, and strongly parallel tertiary veins, perpendicular to midvein and prominulous below. Fruits ellipsoid, ca. 1 cm long, black at maturity.

Colubrina (20 spp., plus 11 in Old World) — In our area, with uniformly entire leaves, sometimes with axillary spines. Usually with pair of large glands (more or less in basal auricles) at base of lamina, these always present when leaves opposite. Flowers in fascicles or short dichotomous axillary cymes; fruit a 3-parted capsule.

Rhamnus (21 spp., plus 140 in N. Am. and Old World) — Canopy trees of montane Andean forest with typical Rhamnaceae leaf venation (except secondary veins not very strongly ascending), the leaf margin usually rather finely and distantly crenate-serrulate (occasionally subentire in one species). Fruit a small berrylike drupe, round to +/- 3-angled, with persistent disk at base.

There are several other genera of Rhamnaceae in the Neotropics, especially in dry areas of Mexico, the Antilles and also in Chile, the chaco, and even one in the caatinga. *Hovenia* of coastal Brazil and the Parana Valley has 3-veined leaves like *Zizyphus,* but in fruit has fleshy edible inflorescence branches.

RHIZOPHORACEAE

Best known for its mangrove taxa, Rhizophoraceae is actually a rather heterogeneous family vegetatively. Three of the neotropical genera are easy to recognize individually to genus (and thus family), but a fourth is vegetatively similar to Myrtaceae or *Mollinedia* of the Monimiaceae. All species are trees (except the shrubby mangrove *Rhizophora mangle*) but leaves can be opposite or alternate, entire or serrate, 3-veined or pinnate, conspicuously pubescent or glabrous. The best known genus is the well-known mangrove, *Rhizophora,* the only neotropical mangrove to have extensive stilt roots (several others have pneumatophores), also characterized by succulent opposite leaves and caducous interpetiolar stipules that form a cone sheathing the terminal bud. One other opposite-leaved genus, *Cassipourea,* occurs in the Neotropics; except for frequently having obscurely and remotely denticulate margins (usually at least vaguely serrulate with a bit of imagination), it vegetatively rather resembles Myrtaceae from which it differs rather subtlely by having fewer secondary veins (these always noticeably brochidodromous usually well before margin) and narrowly triangular, usually sericeous Rubiaceae-like stipules (caducous to leave a stipule scar). A third neotropical genus, *Sterigmapetalum,* has tannish-puberulous (especially the petiole), obovate leaves mostly in whorls of 4 with caducous, narrowly triangular, pubescent stipules between the leaves of a whorl; it has very numerous straight secondary veins, an individual leaf looking more like many sapotacs than the Rubiaceae with which it would otherwise be confused. The uninitiated would never guess that the final genus in our area, *Anisophyllea,* with alternate strongly 3-veined leaves, is a Rhizophoraceae. *Polygonanthus,* another Amazonian genus that does not reach our area, has alternate pinnately veined leaves and rather large-bodied 4-winged fruit.

Rhizophoraceae

1 - *Cassipourea* **2** - *Anisophyllea*

3 - *Sterigmapetalum* **4** - *Rhizophora*

1. Opposite Leaves

Rhizophora (3 spp., plus 4 in Old World) — One of the best known and most characteristic of all neotropical plants, the "red mangrove" (actually three more or less sympatric species) is the only mangrove with stilt roots. It is also unusual in the viviparous, elongate fruit that begins to grow into a seedling before falling from the parent tree. Another typical feature is the prominent (though caducous) interpetiolar stipules that form a Moraceae-like cap over the terminal bud.

C, E, P: mangle rojo

Cassipourea (15 spp., plus 65 in Old World) — Although frequently found in back mangroves or freshwater riverine habitats, *Cassipourea* is most typical of upland forest where it would easily be confused with Myrtaceae except for the frequently remotely denticulate (usually at least vaguely serrulate) leaves with only 3–4 far apart secondary veins, these typically brochidodromous *rather far from the margin.* The very narrowly triangular, often sericeous, caducous terminal stipule (or its scar) could lead to confusion with Rubiaceae when the serrulate leaf margin is not obvious; but no similar rubiac has so few and far apart secondary veins as the common *Cassipourea* species. *Mollinedia* might be confused with remotely denticulate individuals but lacks a stipule or stipule scar. Flowers axillary, single or in clusters, the petals (rarely seen) tend to be stalked. The characteristic small ellipsoid short-pedicelled fruit, surmounted by the persistent style and with the lower fourth or third enclosed by the characteristic sharply 5-dentate cupular calyx, is also distinctive.

Sterigmapetalum (3 spp.) — Trees typically restricted to poor-soil habitats. Easy to distinguish from Rubiaceae (with which it might be confounded on account of the narrow intrapetiolar stipules) by its whorled leaves, distinctively tan-pubescent (especially on the more or less sericeous petioles) and with straight, rather close-together secondary veins (like many species of *Pouteria*). The (rarely collected) inflorescence is a flat-topped terminal panicle with flowers and fruits similar to *Cassipourea.*

2. Alternate Leaves

Anisophyllea (1 sp., plus 30 in Old World) — A tree of poor sandy soils, unique among our Rhizophoraceae species in the alternate leaves. The distinctive leaves are coriaceous and 3–5-plinerved. The ellipsoid 2 cm long fruit is obviously from an inferior ovary.

(*Polygonanthus*) (2 spp.) — A central Amazonian genus that does not reach our area; distinctive in its single-seeded 4-winged somewhat *Petrea*-like fruits.

Rosaceae

Except for a few species of *Prunus,* the woody Rosaceae of our area are entirely restricted to the Andes. Again with the exception of some *Prunus,* the native Rosaceae taxa (including the herbs) all have serrate leaf (or leaflet) margins, except *Margyricarpus* with linear leaflets. The woody taxa have alternate coriaceous simple leaves, except pinnately compound *Polylepis* and *Margyricarpus* which are distinctive in the expanded, more or less sheathing petiole bases. The combination of alternate coriaceous leaves and serrate margins in Andean forests is +/- definitive for Rosaceae (see also *Ilex,* with green inner bark, *Symplocos* more festooned-brochidodromous, *Meliosma,* with woody thickened petiole bases, and *Escallonia,* which differs in more finely serrate, more or less resinous leaves with a characteristic undersurface from the immersed tertiary veins), as are pinnately compound leaves with sheathing bases. The herbs (including scrambling thorny *Rubus*) all have membranaceous compound leaves (either pinnate or palmate), either alternate or clustered in basal rosettes; like *Polylepis* and *Margyricarpus,* they have expanded sheathing petiole bases. The flowers of our Rosaceae are characterized by many stamens, 5 sepals, and 5 petals (or apetalous). The fruits are varied, with the trees having berrylike drupes (*Prunus*), pomes (*Hesperomeles*), apocarpous follicles (*Quillaja, Kageneckia*), or small dry fruits enclosed by the sepals (*Polylepis*); the herbs often have retrorsely barbed exozoochoric fruits (*Acaena, Geum*) or variously compound fleshy multiple fruits (*Rubus, Fragaria, Duchesnea*).

1. Trees and Shrubs

1A. Woody genera with pinnately compound (or 3-foliolate) leaves

Margyricarpus (10 spp.) — Low densely branched shrubs of dry Andean steppes, with small pinnately compound leaves with linear leaflets and conspicuously clasping petiole base. Apetalous flowers solitary, sessile, inconspicuous in leaf axils. Fruit a small white berry.

P: menbrillo

Polylepis (15 spp.) — Easily the most characteristic high-Andean tree genus, typically forming almost pure forests at altitudes that should support puna. The thin, exfoliating reddish bark and thick trunks with twisting branches are very characteristic; a few species are shrubby, especially in dry areas. This is our only tree rosac with compound leaves and the pinnately compound leaves with expanded sheathing petiole bases are unique to Rosaceae. The pendent bracteate spikes of greenish flowers suggest wind-pollination.

P: quinuar, queuna

Rosaceae
(Trees and Shrubs)

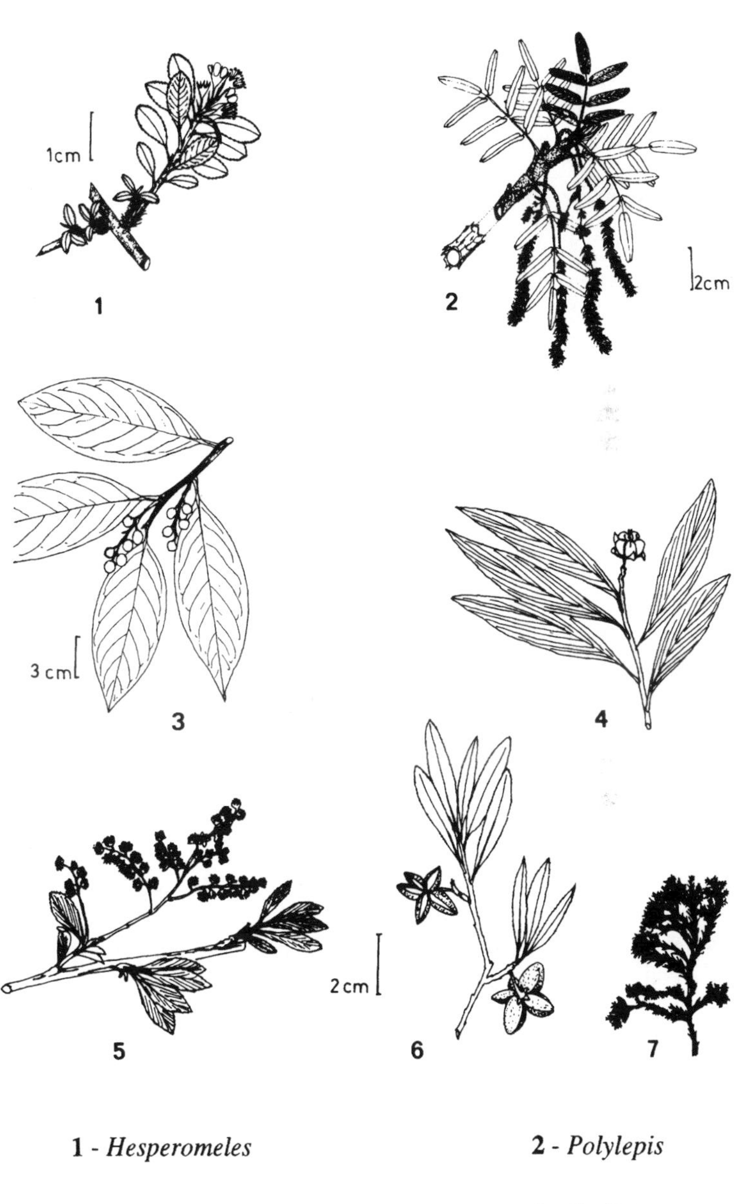

1 - *Hesperomeles* **2** - *Polylepis*

3 - *Prunus* **4** - *Quillaja*

5 - *Holodiscus* **6** - *Kageneckia* **7** - *Margyricarpus*

1B. Woody genera with simple leaves

Hesperomeles (10 spp.) — High-elevation trees and shrubs, often with some branches spine-tipped; one of the important elements of many high-Andean forests, especially in drier areas. Leaves coriaceous and serrate; more irregularly serrate than *Kageneckia* and usually broader relative to the length. Unlike *Kageneckia* in lacking well-developed short-shoots and in the terminal inflorescence distinctly corymbose. Fruit like a miniature apple (i.e., a pome, unique among native taxa).

C: mortino, cerote

Holodiscus (8 spp., incl. N. Am.) — A mostly Central American genus reaching the Colombian Andes. Leaves clustered on short shoots; white-tomentose below unlike *Hesperomeles* and *Kageneckia.*

Kageneckia (3 spp.) — Trees of dry inter-Andean valleys and remnant forest patches on western slope of Peruvian Andes. Leaves often borne more or less clustered on short-shoot branches, elliptic to usually narrowly obovate, finely and evenly serrate, usually more or less resinous. Flowers cream or greenish, borne one or few together at branch terminals. Fruit of apocarpous follicles, like miniature *Sterculia.*

P: lloque

Quillaja (3 spp.) — A basically south temperate genus; our species known only from dry inter-Andean valleys of Cuzco Department and perhaps not native. Very distinctive in the essentially entire narrow coriaceous leaves with strongly ascending close-together secondary veins (almost appearing parallel-veined). Fruit apocarpous.

P: quillai

Prunus (430 spp., incl. N. Am. and Old World) — The only Rosaceae tree in our area to reach the lowlands, but much better represented in Andean forests. Most species (including all the lowland ones) have entire leaves and are vegetatively rather nondescript except for the usual presence of pair of conspicuous ocellate glands near base of lamina (but well away from midvein) below; also characterized by the coriaceous texture, yellowish-drying undersurface with contrasting dark reddish main veins and tendency to dry dark with a distinctly reddish tint above. When serrate, with rather sharp widely separated, often +/- irregular teeth. Inflorescence an axillary raceme of small white flowers. Fruit a small round drupe, usually circumnavigated by an inconspicuous median longitudinal line or constriction.

2. Herbs — (Including clambering *Rubus;* all with mostly compound leaves)

Rubus (250 spp., incl. Old World) — Mostly restricted to disturbed cloud forest (plus a few dwarf species in paramo). Arching canes or clambering vines, sometimes +/- prostrate, usually with curved prickles on

Rosaceae
(Herbs and Vines)

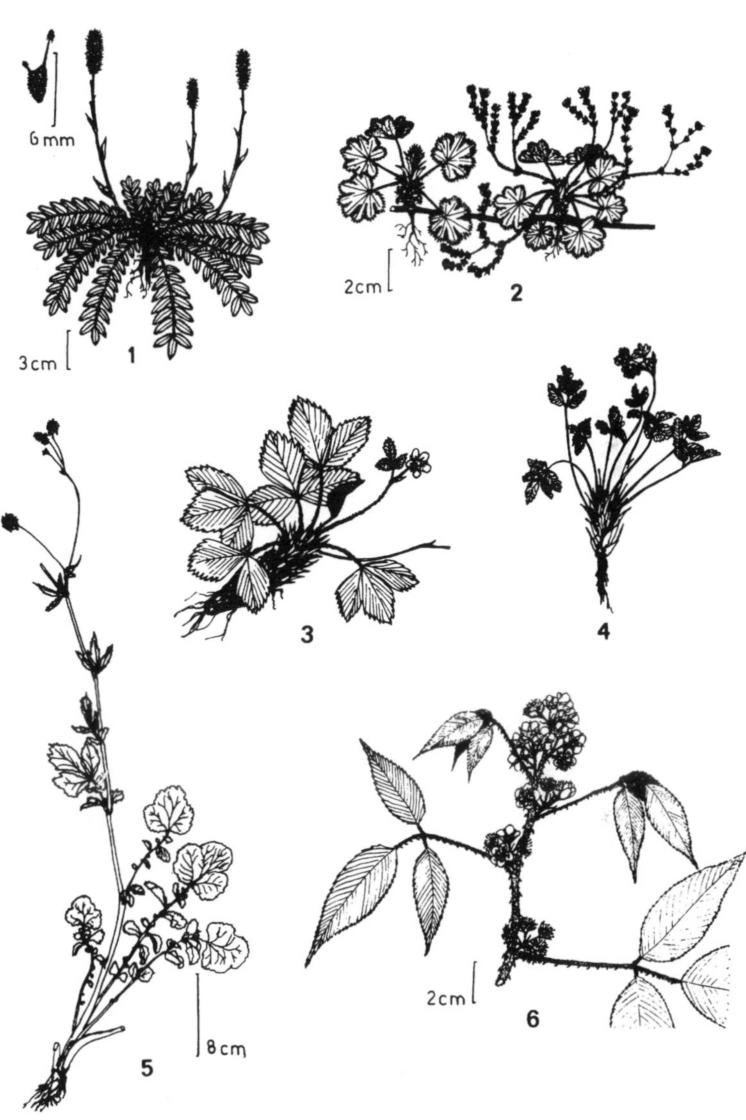

1 - *Acaena* **2** - *Alchemilla*

3 - *Fragaria* **4** - *Potentilla*

5 - *Geum* **6** - *Rubus*

stems and petioles. Leaves pinnately or palmately compound, or (usually) 3-foliolate. Fruit an edible blackberry or raspberry.

C, E, P: mora

Geum (40 spp., incl. n. temperate) — Erect herbs of high-Andean puna. Leaves pinnately compound with very characteristic division into dramatically different-sized, alternately large and small leaflets. Fruit with hooked spines (cf., *Acaena*).

Fragaria (15 spp., incl. n. temperate) — Prostrate stoloniferous herbs of disturbed cloud forest, creeping by stolons, each plant forming rosette of 3-foliolate leaves. Flowers white; fruit an edible strawberry.

Duchesnea (1 sp., introduced from Old World) — A prostrate 3-foliolate weedy herb of disturbed cloud forest. Essentially the same as *Fragaria* but differing in yellow flowers and the strawberry-like fruit with less succulent receptacle.

Potentilla (500 spp., incl. Old World) — Prostrate high-Andean paramo herbs with pinnately or digitately compound leaves. Pinnately compound leaves very like *Geum* but with the evenly incised-margined leaflets all the same size in each leaf or decreasing gradually toward leaf base.

Alchemilla (incl. *Lachemilla*) (250 spp., incl. Old World) — Mostly prostrate mat-forming high-Andean herbs. Leaves mostly palmately compound or lobed (a few species pinnate), with conspicuously sheathing petiole base. Flowers small and greenish and with small bractlets interspersed among the calyces, borne in sessile clusters on conspicuously bracteate more or less interupted inflorescences. *Lachemilla,* with 2(–4) stamens and introrse anthers (vs. 4(–5) stamens and extrorse anthers) is sometimes segregated.

Acaena (100 spp., incl. N. Am. and Old World) — High-Andean puna and paramo herbs with pinnately compound leaves with even-sized leaflets or the leaflets gradually smaller toward base. Inflorescence erect, spicate or a capitate flower cluster at end of long peduncle, the flowers reduced and greenish, perhaps wind-pollinated. Fruits spiny, exozoochoric.

RUBIACEAE

One of the largest and most prevalent neotropical families, and extremely easy to recognize to family on account of its entire opposite leaves and interpetiolar (= between the opposite petiole bases) stipules. (A very few rubiacs have intrapetiolar stipules: conspicuously so in *Capirona* and most *Elaeagia,* less obvious in *Isertia* and *Condaminea.*) While a few other families have inconspicuous interpetiolar stipules, only Quiinaceae and *Cassipourea* (Rhizophoraceae) (both usually with more or less serrate or serrulate leaf margins) have these developed as in typical Rubiaceae. Since Rubiaceae are so common, the first thing to check for in any plant with opposite simple leaves is whether it has interpetiolar stipules. Unfortunately, the stipules are often caducous: look for its scar connecting the opposite petioles and/or whether it is still present at the twig apex. (Acanthaceae can have similar lines but with swelling above, rather than below, nodes.) A few other hints: Rubiaceae never have serrate or even serrulate leaf margins although very rarely a species may have the leaves deeply lobed (*Pentagonia*) or with a few large irregular teeth near the apex (*Simira*). The ovary is obviously inferior in all but a couple of mostly extralimital genera (*Pagamea, Henriquezia, Platycarpum,* the latter two semi-inferior) and the corolla always tubular (campanulate in *Henriquezia* and *Platycarpum*); no other family of the lowland Neotropics (except totally distinctive Compositae) has opposite leaves and sympetalous flowers with an inferior ovary. The bark is sometimes papery and peeling to leave a smooth trunk (see below), sometimes breaking into thin fibrous vertical strips; when outer bark smooth and nondescript the inner bark is typically soft, undifferentiated and white, oxidizing darker.

Although easy to recognize to family, many genera of Rubiaceae are difficult to distinguish when sterile; stipules are often very useful in generic delimitation. The traditional subfamilial division is based on the technical character of whether there is one ovule (= Rubioideae) or several ovules (= Cinchonoideae) in each locule; tribal division emphasizes such traits as whether the seeds and/or ovules are pendulous or ascending, basically or laterally attached, horizontally or vertically arranged. Nevertheless, the relatively few vine genera (Group I) are easy to identify as are most of the herbs (Group III), except the weedy poorly defined genera of Spermacoceae. The trees and shrubs (Group II) can be divided into three (four, if woody epiphytes are considered separately) main alliances: 1) genera with dehiscent fruits and mostly winged wind-dispersed seeds — including most of the canopy rubiac species and all genera with conspicuously expanded calyx lobes (twenty-nine genera including the epiphytes; these further subdivided by whether they have semisuperior ovaries [tribe Henriquezieae: two genera], valvate corolla lobes [*Cinchona* and relatives: twelve genera] or imbricate and contorted corolla lobes [*Coutarea* and relatives plus *Rondeletia* and *Elaeagia* and relatives: ten genera]); 2) genera with large indehiscent fleshy fruits (probably mammal-dispersed) which are mostly subcanopy trees, often with largish coriaceous leaves [eleven genera:

Tribe Gardenieae]; 3) genera with small berrylike fruits which are mostly understory shrubs. The latter group includes two main suballiances: *Psychotria* and its allies which have 1–2-seeded fruits (eleven genera) and *Hamelia* and *Bertiera* and their allies which have several ovules per locule and several-seeded fruits (nine genera). There are also a group of miscellaneous, mostly individually distinctive genera with indehiscent fruits (splitting into 2 cocci in *Machaonia*), mostly intermediate in size between the two main types. One completely atypical epiphytic herb very common in the Chocó area appears to have alternate leaves (*Didymochlamys*).

Rubiaceae stipules are very important in generic recognition. They are of at least ten more or less distinctive types (Fig. 237). In a few genera they are intrapetiolar, although this is not always constant — *Capirona* (entire), *Elaeagia* (secretory), *Platycarpum* (enclosing terminal bud), *Condaminea* and *Isertia* (bifid); in others foliaceous (e.g., *Posoqueria, Remijia, Ladenbergia,* and the epiphytic taxa). Some genera have narrowly triangular stipules, with or without a central thickening (e.g., *Pentagonia, Randia, Cephaelis, Sommera, Warscewiczia, Wittmackanthus, Bothriospora, Guettarda, Hippotis, Genipa, Duroia, Amaouia, Bathysa*), others broadly triangular (e.g., *Tocoyena, Appunia, Coussarea, Chomelia, Borojoa, Coussarea*), some are intermediate (e.g., *Alibertia*) while others have the narrowly triangular shape accentuated (*Simira, Chimarrhis*). Several genera have conspicuously acuminate stipules: +/- broadly triangular with a long acumen (especially in *Faramea* but also in *Bertiera, Ixora,* and *Gonzalagunia,* to a lesser extent in genera like *Calycophyllum, Manettia, Hamelia*). *Psychotria* is variable, sometimes only terminal, but typically with 2-aristate or bifid stipules, similar to *Palicourea* (plus herbaceous *Oldenlandia* and *Coccocypselum*), while closely related *Rudgea* has conspicuously fimbriate ones. The herbaceous genera of tribe Spermacoceae plus *Pagamea, Coccocypselum* and one *Psychotria* (*P. ipecacuanha*) have the stipule fused into a long sheath with an aristate margin.

Some Useful Characters:

Spines: *Uncaria* (liana), *Machaonia, Randia, Chomelia.*

Hawkmoth flowers: *Cosmibuena, Hillia, Ladenbergia* (some), *Randia* (some), *Posoqueria, Tocoyena, Hippotis* (some), *Isertia* (trees only), *Amphidaysa* (herb), *Kotchubaea, Exostema.*

Epiphytes: *Cosmibuena, Hillia, Psychotria* (few species), *Didymochlamys, Manettia* (few species).

Thin papery outer bark peeling to show strikingly smooth inner bark: *Calycophyllum, Wittmackanthus, Capirona, Chomelia* (one species), *Guettarda* (?).

Fenestrated trunks: *Macrocnemum* (some); *Alseis* (some); *Amaioua* (most); *Remijia* (some).

1. Vines or Lianas — (Prostrate stoloniferous plants are treated below as herbs rather than as vines.)

Several Rubiaceae genera of diverse tribal affinities are always lianas or vines; a few other genera have some climbing species. *Uncaria, Emmeorrhiza,* and *Manettia* have dry dehiscent fruits, *Paederia* a dry indehiscent fruit, the other vine genera berries (*Sabicea, Malanea, Chiococca*) or larger, fleshy fruits (*Randia*).

Uncaria (2 spp., plus ca. 50 in Old World) — Large woody liana typical of old second growth; nodes usually with paired recurved spines; inflorescence umbellate.

E, P: uña de gato

Manettia (130 spp.) — Slender vines with rather small membranaceous glabrous leaves; flower usually red, tubular, hummingbird-pollinated (sometimes bluish, yellowish or whitish); inflorescence openly racemose.

Paederia (1 spp., plus 50 in Old World) — Vines of middle-elevation forests, leaves reputedly with an unpleasant odor. Leaves membranaceous. Similar to *Manettia* but the flower white to purplish, the inflorescence paniculate, and the distinctive flattened fruit one-seeded and wind-dispersed after the exocarp flakes off.

Emmeorrhiza (1 sp.) — Cloud-forest vine, occasionally becoming a liana. Very distinctive in the stipules fused into a sheath with long aristae projecting from its margin (tribe Spermacoceae). This is the only truly scandent Spermacoceae, also differing from other genera of the tribe in the open paniculately branching inflorescence with lax terminal umbels, the flowers tiny and white.

Malanea (20 spp.) — Usually scandent, but some species also shrubby. Leaves *Chomelia*-like with whitish-puberulous main veins below. Inflorescence a rather tenuous panicle to almost spicate, the flowers minute; supposedly differs from *Chomelia* in axillary rather than terminal inflorescence placement. Fruit apparently, a ''one-seeded''berry, rather cylindric (actually, a pyrene with chambers). Stipule distinctly foliaceous, more or less obtusely triangular and pubescent, at least in center.

Rubiaceae Stipules

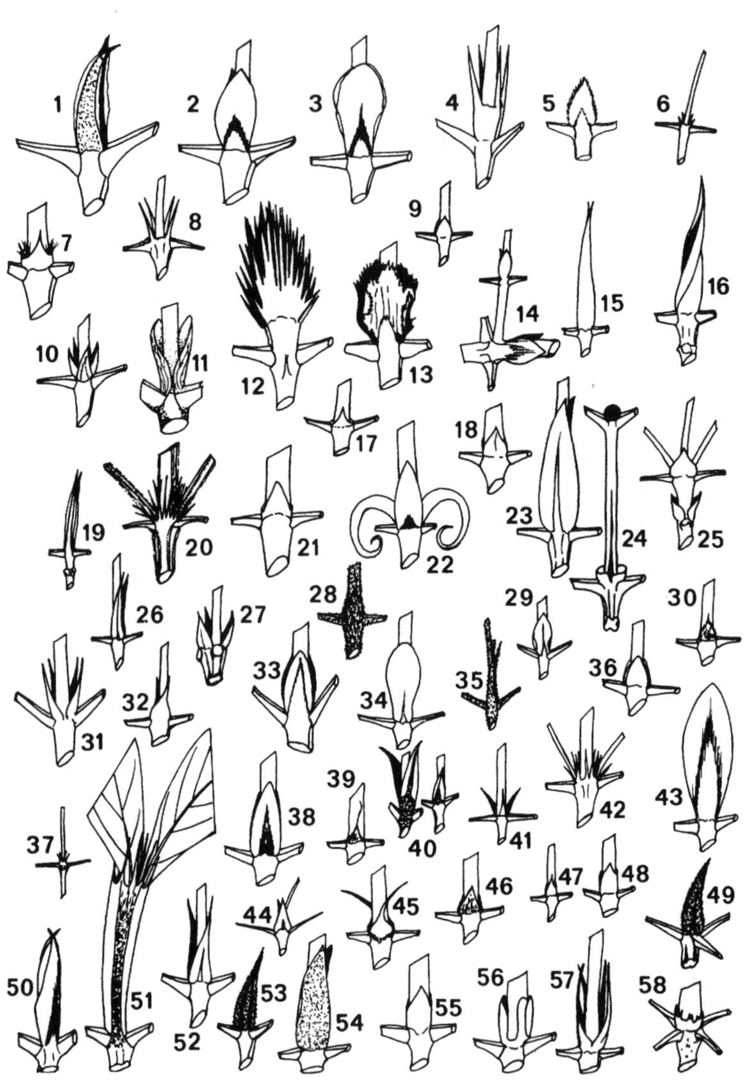

Figure 237 Legend

1 - *Pentagonia* **3** - *Remijia* **5** - *Psychotria stenostachya*
2 - *Posoqueria* **4** - *Psychotria racemosa* **6** - *Psychotria patens*

7 - *Calycophyllum* **8** - *Coccocypselum* **9** - *Cephaelis*

10 - *Randia* **12** - *Rudgea* **14** - *Sabicea* **16** - *Chimarrhis*
11 - *Capirona* **13** - *Rudgea* **15** - *Simira*

17 - *Appunia* **18** - *Coussarea*

19 - *Sommera* **21** - *Tocoyena* **23** - *Warscewiczia* **25** - *Wittmackanthus*
20 - *Spermacoce* **22** - *Uncaria* **24** - *Elaeagia*

26 - *Bothryospora* **27** - *Ixora* **28** - *Bertiera* **29** - *Malanea* **30** - *Manettia*

31 - *Isertia* **33** - *Kotchubaea* **35** - *Malanea*
32 - *Ixora* **34** - *Macrocnemum* **36** - *Manettia*

37 - *Oldenlandia* **39** - *Gonzalagunia* **41** - *Hamelia* **43** - *Hillia*
38 - *Ladenbergia* **40** - *Guettarda* **42** - *Hemidiodia*

44 - *Geophila* **45** - *Faramea* **46** - *Chomelia* **47 & 48** - *Alibertia* **49** - *Bathysa*

50 - *Hippotis* **52** - *Genipa* **54** - *Duroia* **56** - *Palicourea condensata* **58** - *Palicourea triphylla*
51 - *Pagamea* **53** - *Duroia* **55** - *Borojoa* **57** - *Palicourea stenostachya*

Rubiaceae
(Scandent)

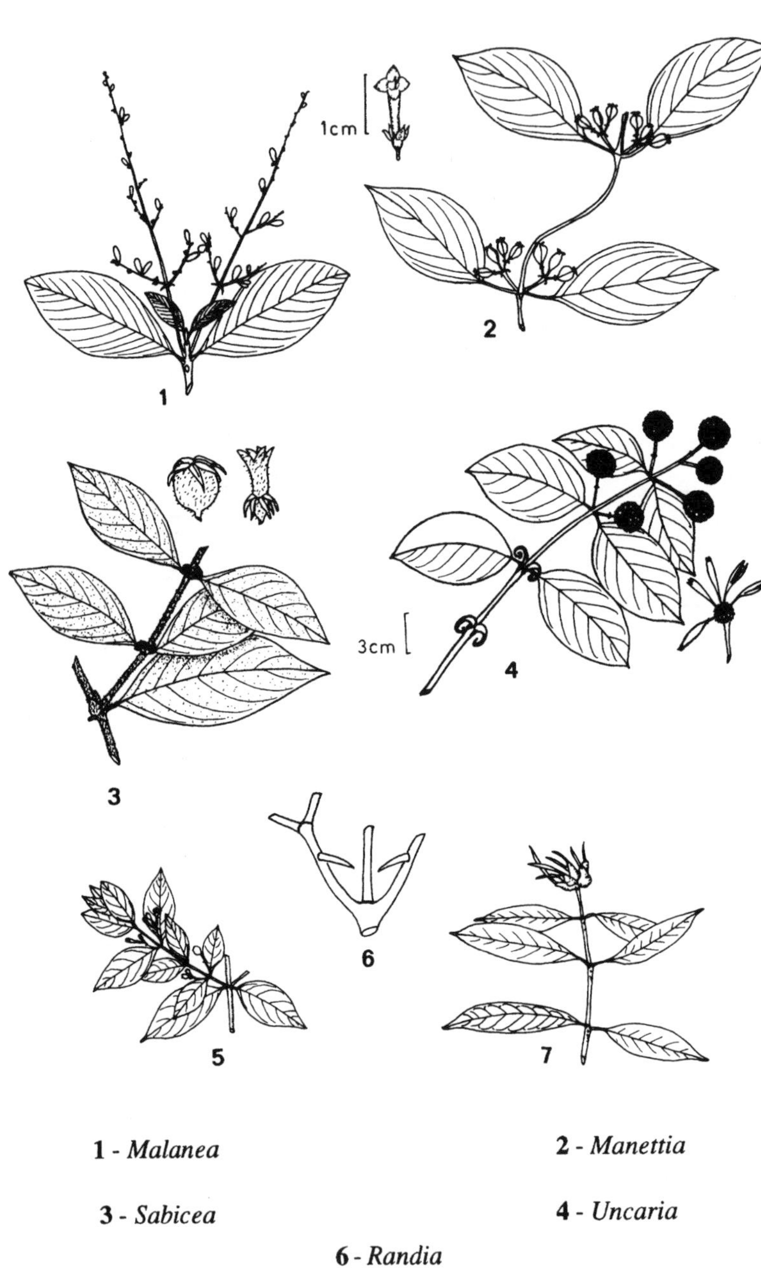

1 - *Malanea*

2 - *Manettia*

3 - *Sabicea*

4 - *Uncaria*

6 - *Randia*

5 - *Chiococca*

7 - *Schradera*

Rubiaceae
(Trees with Narrow Capsular Fruits: Epiphytic and Hawkmoth-Pollinated or Terrestrial and with Expanded Calyx)

1 - *Calycophyllum* **2** - *Wittmackanthus*

3 - *Cosmibuena* **4** - *Pogonopus*

5 - *Hillia*

6 - *Warscewiczia* **7** - *Hillia*

Sabicea (40 spp., plus nearly 100 in Old World, mostly in Africa) — Woody or subwoody vines with pubescent leaves and axillary inflorescences; flowers small, white; fruit a small red berry, turning black when ripe; common in second growth.

Schradera (15 spp.) — Hemiepiphytic liana; leaves rather succulent (looks more like Guttiferae than Rubiaceae); inflorescence capitate with two subtending fleshy bracts (cf., *Cephaelis*).

Chiococca (15 spp.) — Wiry vines with rather small stiff glabrous leaves; flowers small, bell-shaped, in short axillary racemes; fruit flattened, usually white.

(*Randia*) — A few species of *Randia* are lianas; they are characterized by a unique spine arrangement: see figure.

(*Chomelia*) — A few *Chomelia* species are lianas, usually with spines and vegetatively very similar to *Uncaria.*

2. Trees or Shrubs

2A. Trees or shrubs (sometimes epiphytic) with the fruit dry and capsular — Usually elongate and two-valved, many-seeded, usually with small winged seeds (ovules more than one per cell in flower = subfamily Cinchonoideae)

2Aa. Epiphytic trees or shrubs with succulent-coriaceous leaves — (Hawkmoth-pollinated with long tubular white flowers; fruits long and narrow.)

Cosmibuena (12 spp.) — Leaves mostly obovate, long-petiolate. The long white hawkmoth-pollinated flowers are pedicellate and borne several together, unlike *Hillia.* Seeds with an irregular wing surrounding body.

Hillia (incl. *Ravnia*) (20 spp.) — Similar to *Cosmibuena* but leaves more succulent and with shorter thicker petioles, the flowers sessile and borne singly (the base enclosed by the large foliaceous stipules), and seeds with tuft of hairs at one end. Central American segregate *Ravnia* is red-flowered and hummingbird-pollinated.

(*Psychotria*) — A few species of *Psychotria* are epiphytic shrubs; they have the standard small flowers and small berry-fruits of the genus; see also herbaceous *Didymochlamys* with alternate leaves.

2Ab. Terestrial trees, often large canopy species, sometimes shrubby — Almost all of the large-tree rubiacs have capsular fruits with wind-dispersed seeds (these mostly vertically arranged in the capsule = tribes Cinchoneae and Henriquezieae). Expanded conspicuously colored calyx lobes are a prevalent theme in this group.

(i) **Expanded bright-colored bractlike calyx lobes present (usually +/- persistent as dry bracts in fruit)** — In first three genera outer bark typically papery and peeling to leave very smooth trunk.

Calycophyllum (6 spp.) — Canopy trees, mostly in successional forests on alluvial soil. Most species with an expanded white calyx lobe; strikingly thin papery peeling bark on trunk; inflorescence flat-topped.

P: capirona

Wittmackanthus (incl. *Pallasia*)(1 sp.) — Canopy trees. Leaves with unusually long slender petioles; stipules small, triangular. Expanded calyx lobe and flowers light magenta, the flowering tree spectacular. Inflorescence elongate and spicate (unlike relatives). Strikingly thin, peeling, reddish, papery bark on trunk.

Capirona (incl. *Loretoa*) (2 spp.) — Large Amazonian trees with extremely smooth red bark. Vegetatively very different from relatives in the persistent large leaflike *intrapetiolar* stipules and noticeably large leaves. Flowers reddish and larger than in related genera; inflorescence bracts conspicuous, similar to reduced stipules, the expanded calyx lobe large and pinkish. Fruits cylindrical, rather woody, with an expanded terminal ring from the persistent calyx.

P: capirona

Pogonopus (3 spp.) — Small trees, especially in drier areas. Leaves membranaceous, either gradually tapering to long petiole (cf., *Hamelia*) or densely puberulous (or both); stipule with broadly triangular base and narrow acumen. Expanded calyx lobe red; similar to *Warscewiczia* (with red "bracts" but inflorescence paniculate and actual flowers also red. Fruits cylindrical and flat-topped except for rim formed by calyx lobes, the seeds not winged.

Warscewiczia (4 spp.) — Subcanopy trees of wet and moist forest. Leaves long, many-veined, usually drying dark; stipules thin, caducous, narrowly triangular and twisted, the middle portion +/- appressed-pubescent. Expanded calyx lobe (when present) red (rarely white extralimitally); inflorescence an elongate horizontal spray with conspicuous +/- horizontal lateral red "bracts"; actual flowers small and yellow; seeds not winged.

C: pina de gallo

(***Rondeletia***) — Some species have expanded white calyx lobes; most are shrubby and have the fruits +/- globose.

(ii) Genera without expanded calyx lobes

(a) Genera with broadly campanulate rather zygomorphic bignonlike flowers, a semisuperior ovary (unique but *Pagamea* completely superior), fruits flat and very broadly oblong, and leaves usually whorled (= tribe Henriquezieae).

Henriquezia (7 spp.) — Guayana Shield trees, barely reaching Amazonian Colombia. Vegetatively distinctive in leaves coriaceous, glabrous, and always verticillate, the stipules narrow and twice as many per node as leaves; also tending to have glands below petiole base. Fruit much larger than *Platycarpum.*

Platycarpum (10 spp.) — Guayana Shield trees barely reaching Amazonian Colombia; recently discovered on white sand in Peru. Distinctive in large (but early-caducous) intrapetiolar stipules that completely enclose terminal bud (cf., *Cecropia*). Petioles short, usually pubescent, with glands at base. Flowers smaller than *Henriquezia.*

(b) Flowers small or +/- salverform, ovary inferior and corolla lobes valvate in bud (= parallel and appressed along edges; also true of *Pogonopus* with expanded calyx lobes); fruits tiny or cylindrical.

Rustia (12 spp.) — Shrubs and small trees, the only common species restricted to mangroves. Leaves rather long and narrow, pellucid-punctate (unique, but only obvious in mangrove species); stipules narrowly triangular, caducous to leave scar. Inflorescence terminal, paniculate, anthers thick and opening by terminal pores (unique); corolla lobes valvate in bud. Fruit short, obovoid, somewhat flattened, the top +/- truncate; seeds minute, winged but barely visible macroscopically.

Chimarrhis (15 spp.) — Very large canopy trees of moist and wet forest. Leaves long-petiolate, obovate, glabrous except for nerve axils below; stipules caducous, lanceolate, usually somewhat twisted, sometimes sericeous. Flowers very small (3 mm long), whitish; inflorescence corymbose-paniculate, usually somewhat flat-topped, many-flowered, axillary. Fruit tiny, round or obovoid, splitting in half to release tiny more or less winged seeds.

P: pampa remo caspi, hierno prueba (= son-in-law's test), pablo manchana

Dioicodendron (2 spp.) — Andean trees. A segregate from *Chimarrhis* with similar tiny flowers but more pubescent vegetatively, the flowers dioecious and 4-merous rather than 5-merous, and with persistent subfoliaceous triangular-ovate stipules.

Rubiaceae
(Trees with Long, Narrow Capsular Fruits: Calyx Lobes Not Expanded)

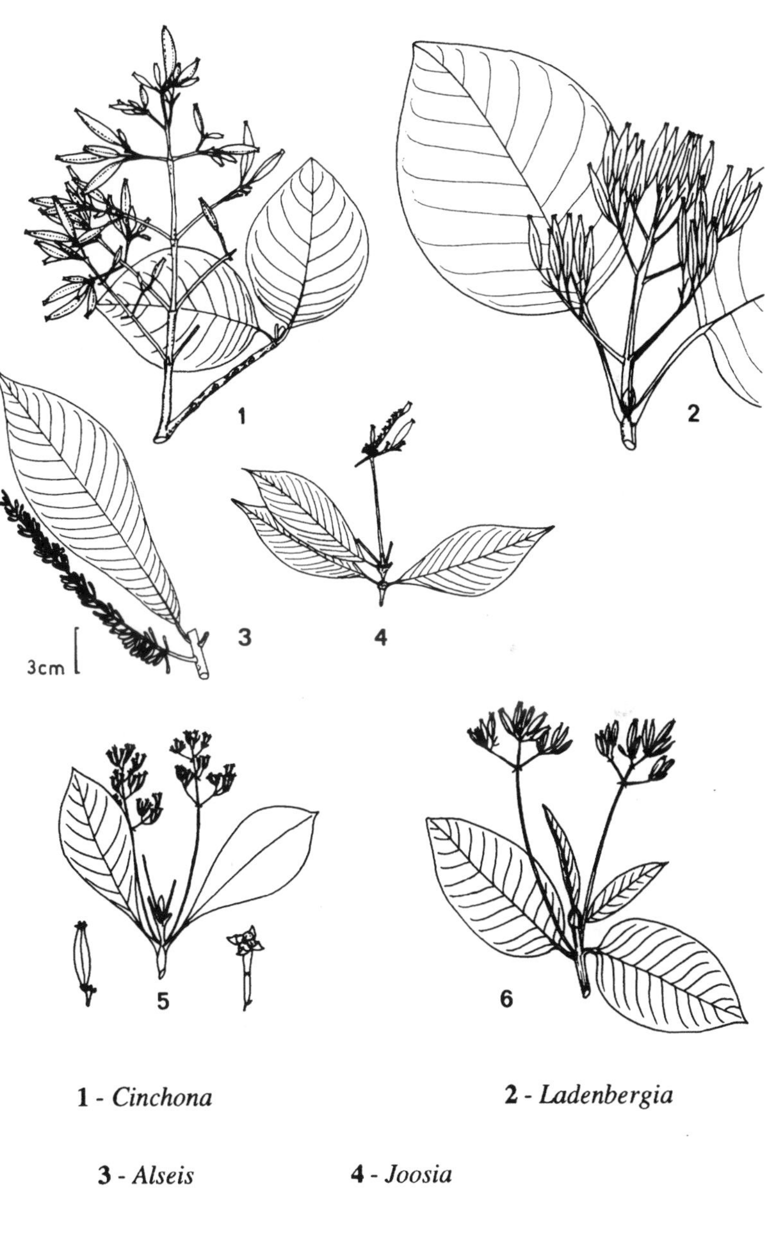

1 - *Cinchona* **2** - *Ladenbergia*

3 - *Alseis* **4** - *Joosia*

5 - *Macrocnemum* **6** - *Remijia*

Rubiaceae
(Trees with Short, Nonflattened Capsular Fruits: Calyx Lobes Not Expanded)

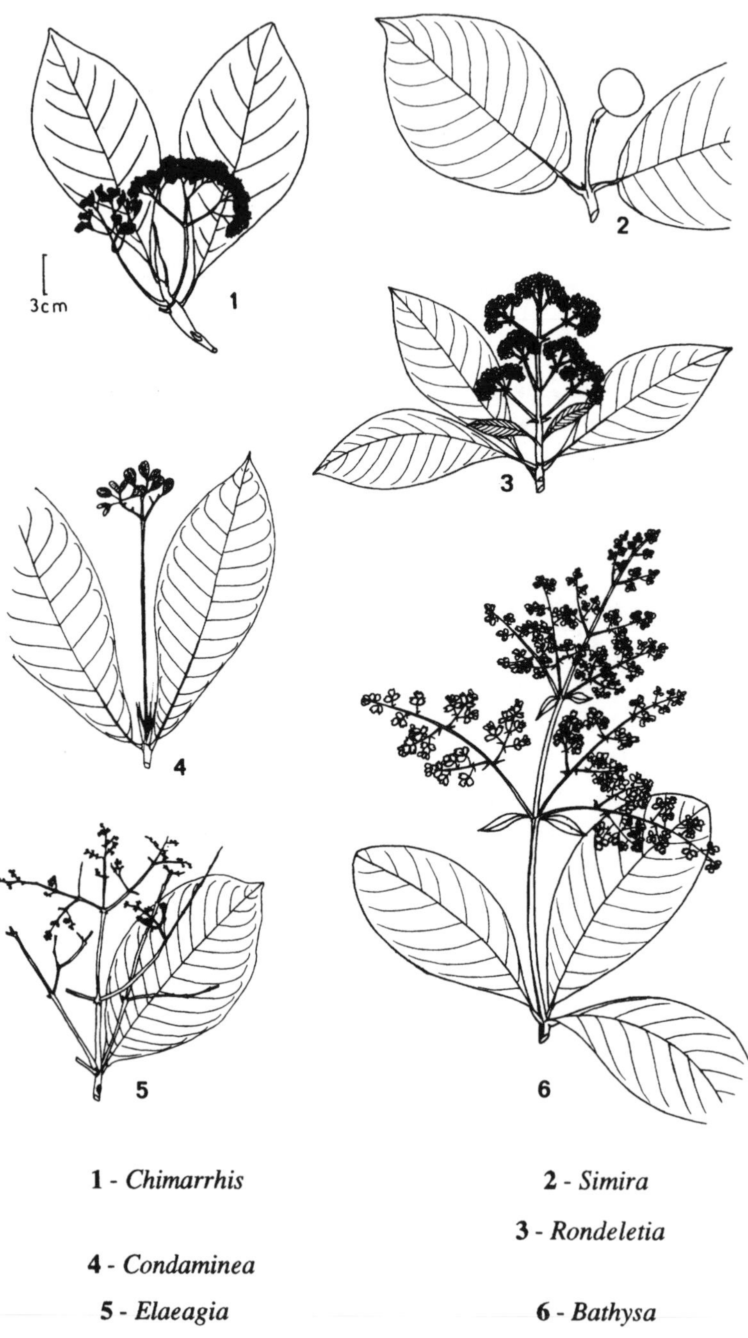

1 - *Chimarrhis* **2** - *Simira*

3 - *Rondeletia*

4 - *Condaminea*

5 - *Elaeagia* **6** - *Bathysa*

Condaminea (3 spp.) — Small often shrubby trees mostly in moist inter-Andean valley second growth. Vegetatively distinctive in the 4 long (> 3 cm) persistent stipules (technically intrapetiolar but so deeply bifid that this not apparent) and the large obovate leaves with many veins and tendency to short-petiolate cordate base. Flowers rather large and fleshy, with valvate lobes, the tube 2 cm long; inflorescence terminal, pyramidal, rather open and few-flowered with clusters of several flowers or buds on long peduncles.

Alseis (15 spp.) — Canopy trees, especially in seasonal forest. Leaves rather large, oblanceolate or narrowly obovate, usually with very many secondary veins, usually membranaceous and often pubescent; stipule caducous. Very characteristic spicate terminal inflorescence with small white flowers having exserted anthers. Fruits less than 2 cm long, narrowly cylindric, rather densely arranged along the length of spike.

Cinchona (40 spp.) — Midcanopy to canopy trees mostly in Andean cloud forests. Like *Ladenbergia* but with dense hairs inside corolla lobes and flowers usually smaller; capsule differs in splitting from below toward apex and broader relative to length.

Joosia (7 spp.) — Small trees; very *Guettarda*-like vegetatively and inflorescence also with one-sided branches; flowers white, the corolla lobes bifid at apex and with conspicuously scalloped margins; distinctive in very narrow dry fruits that split into 4 coiled valves to release the narrow, winged seeds.

Macrocnemum (20 spp.) — Medium-sized to canopy trees of moist and wet forests. Flowers small (*Lantana*-like), magenta. Capsule narrowly cylindric, ca. 2–3 cm long, 2-valved, loculicidal, characteristically splitting along outside margin.

P: rumo remo caspi

Ladenbergia (30 spp.) — Shrubs to canopy trees, especially in cloud forests. Leaves typically large, broadly oblong and with truncate or subcordate base and long petiole; stipules caducous, triangular, with a central keel. Flowers white, narrowly long-tubular, generally lacking dense hairs inside corolla lobes (unlike *Cinchona*). Capsule long and narrow (longer than *Cinchona*), septicidal, splitting from apex.

C: tanacillo

Remijia (35 spp.) — Shrubs of poor-soil areas. Related to *Ladenbergia* but with shorter, relatively broader fruit and axillary inflorescence. Stipule foliaceous, obtuse, caducous, usually pubescent at least inside. Leaves often resinous.

Rubiaceae

(Trees with Oblong, Usually Flattened Capsular Fruits; Calyx Lobes Not Expanded)

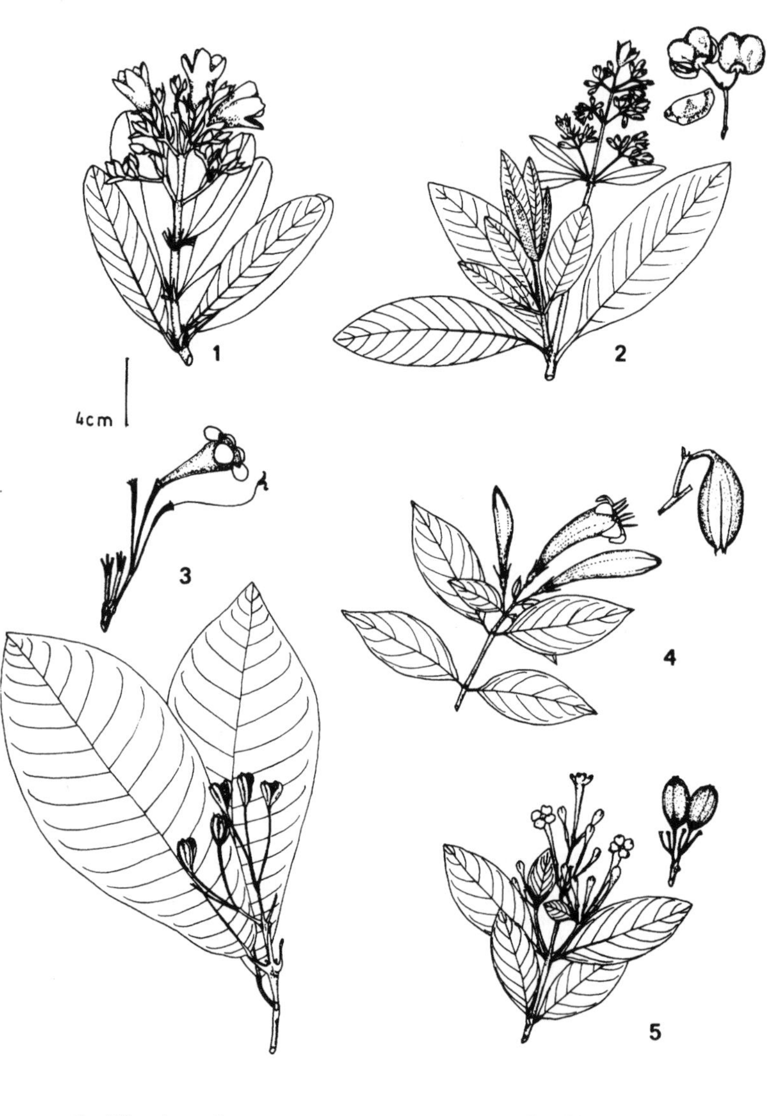

1 - *Henriquezia*

2 - *Platycarpum*

4 - *Coutarea*

3 - *Macbrideana*

5 - *Ferdinandusa*

Heterophyllaea (incl. *Lecanosperma*) (1 sp.) — Small straggly shrubs of dry southern Andean scrub, barely reaching Peru. Very distinctive in the small, narrow *Lycium*-like leaves. The solitary flowers are fairly elongate-tubular. Fruits very atypical for dehiscent rubiacs in being very small and ellipsoid in shape (cf., *Psychotria*), but dehisce into 2 valves to release minute subwinged seeds.

(c) Flowers small to large and often infundibuliform, ovary inferior, the corolla lobes imbricate (overlapping) or contorted (wrinkled) in bud; stamens often more or less exserted; fruits variously shaped but not narrowly cylindrical nor flat and broadly oblong.

Ferdinandusa (25 spp.) — Midcanopy trees, mostly on poor sandy soils. Vegetatively characterized by the glabrous (in our area) coriaceous leaves and the rather thick, narrowly triangular, caducous terminal stipules that are twisted together. The narrowly tubular flowers have distinctively notched petals unlike other Cinchoneae, except *Joosia,* (cf., *Guettarda*). Capsule cylindric; seeds obviously winged.

Coutarea (7 spp.) — Small to medium-sized trees, mostly in dry forest. Large infundibuliform-campanulate lavender to white corolla (cf., Bignoniaceae); stamens barely exserted; fruits strongly flattened, obovoid, covered with paler lenticels; seeds winged.

Macbrideina (1 sp.) — Large tree of rich-soil forest. Probably related to *Condaminea* and with an identical (though slightly larger) obovate 2-valved fruit, but the very different large infundibuliform flowers with imbricate lobes place it in a different tribe. Leaves large, membranaceous with axillary tufts of hairs below; stipules foliaceous and intrapetiolar, early-caducous and not leaving obvious scar. Also very unusual in thin infundibuliform calyx > 3 cm long and with deep obtuse lobes (cf., *Coutarea*).

Exostema (35 spp.) — Shrubs to large trees, mostly in Central America and the West Indies. In flower easy to recognize by the conspicuously long-exserted stamens and narrowly tubular white corolla also with long narrow lobes. Stipules broadly triangular, rather thin. Fruit smallish, ca. 1.5 cm long, each valve splitting, the seeds winged.

Simira (incl. *Sickingia*) (35 spp.) — Midcanopy to canopy trees, especially common in dry forest. Unique in wood of some species (including centers of twigs) turning pinkish or red-violet when exposed to air. Vegetatively distinctive in long, narrowly triangular terminal stipule and usually truncate leaf bases; rather large leaves with many secondary veins. Fruit large, round, usually splitting into 4 valves, the seeds winged, horizontally stacked into fruit.

Elaeagia (10 spp.) — Small to midcanopy trees, mostly in cloud forests. Vegetatively distinctive in resinous secretions especially on young stipules and leaves, sometimes covering terminal stipule and forming bead-like ball; stipules unusual in being large and intrapetiolar but the apex usually deciduous to leave a kind of split cup. Leaves sometimes very large. Inflorescence openly paniculate with very small flowers, the corolla tube less than 3 mm long and villous inside, the lobes contorted in bud. Fruits similar to *Bathysa;* tiny (ca. 5 mm long), round, the 2 valves with bifid apices, not splitting completely, the persistent calyx forming rim well below apex.

C: barniz de Pasto

Pimentelia and ***Stilpnophyllon*** (1 sp. each) — Trees of middle-elevation Andean forests, distinctive in the largish, obovate leaves, resinous as in *Elaeagia,* but with normal interpetiolar stipules and probably closer to *Remijia.* Poorly known and perhaps congeneric.

Bathysa (10 spp.) — Trees, mostly in middle-elevation forest. Related to *Elaeagia* but with interpetiolar stipules and the largish leaves conspicuously pubescent, rather than glabrous and +/- resinous. Like *Rondeletia* in pubescent (though often larger) leaves but the small capsule is longer than wide and split only part way at apex; also differs in exserted stamens and shorter broader corolla tube; calyx lobes never expanded.

Rondeletia (incl. *Arachnothrix*) (150 spp.) — Shrubs and small trees, commonest at middle elevations. Leaves usually similar to *Guettarda* or the whole lower surface white; stipules narrowly triangular, sometimes acuminate, always puberulous. Capsule round, the valves not woody, split to base, the seeds winged or unwinged. Corolla with narrowly cylindrical tube and reflexed lobes (cf., *Lantana;* looks like *Guettarda* but inflorescence branches not 1-sided). Inflorescence paniculate or spiciform. The segregate *Arachnothrix* differs in densely stellate (arachnoid) pubescence, 4-merous flowers and nonwinged seeds.

Schizocalyx (2 spp.) — Endemic to our area but not very distinct. Very like (and, perhaps, not distinct from) *Bathysa* with pubescent leaves and smallish flowers rather aggregated at branch ends of paniculate inflorescence. Defined by the calyx irregularly 2–3-lobed, with enlarged lobes. Fruits small and splitting in half as in *Rondeletia.*

2B. Trees and shrubs with fruits indehiscent, fleshy and berrylike — Fleshy-fruited rubiacs include three tribes of subfamily Cinchonoideae (= ovules more than 1 per cell) plus all Rubioideae (= ovules and seeds solitary in locules) except *Machaonia* (spiny shrub with mericarpous fruit; see above). I have divided the fleshy-fruited woody rubiacs into 3 main

Rubiaceae
(Trees with Large Indehiscent Fleshy Fruits and Hawkmoth Flowers)

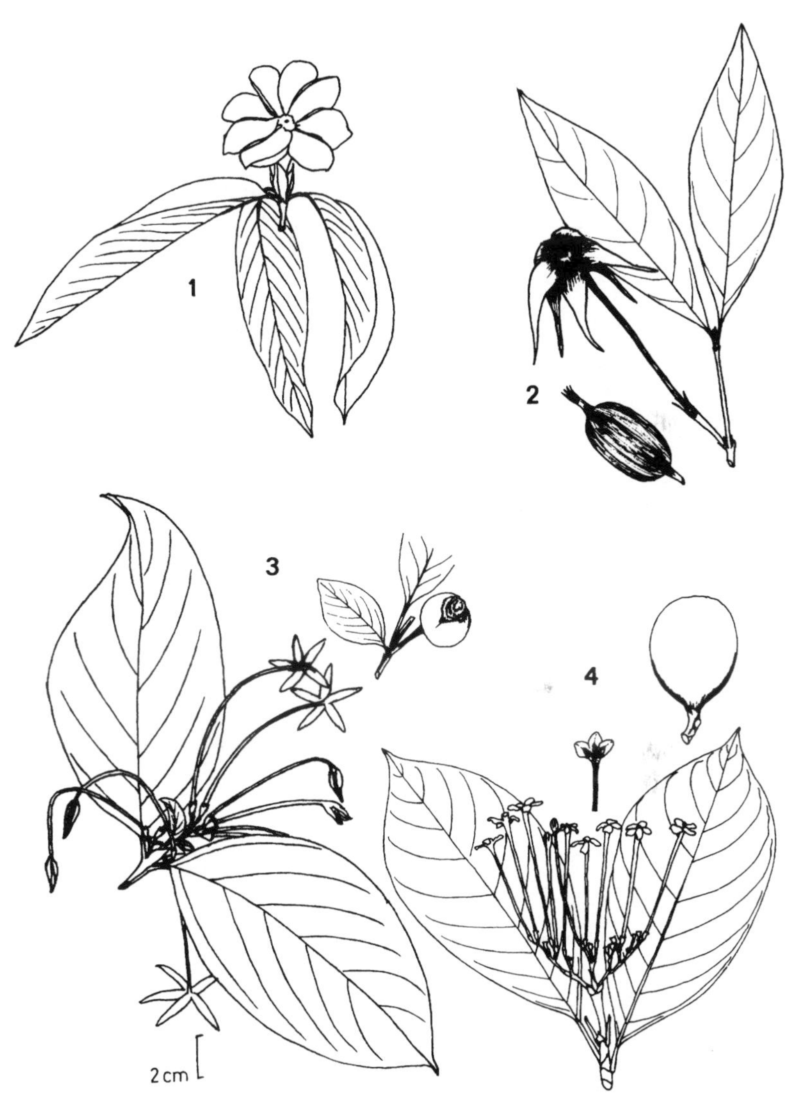

1 - *Kotchubaea* **2** - *Randia*

3 - *Posoqueria* **4** - *Tocoyena*

groups plus a number of miscellaneous genera: 1) subcanopy trees with large many-seeded fruits (mostly mammal-dispersed), 2) shrubs with small 1–2-seeded, bird-dispersed berries, and 3) shrubs and treelets with several to many-seeded, bird-dispersed berries.

2Ba. Subcanopy trees (*Genipa* and *Botryarrhena,* at least, may be canopy species) **or spiny shrubs; fleshy, many-seeded, indehiscent, usually large (mostly > 2 cm diameter) and ellipsoid (mammal-dispersed) fruits** — Usually dioecious, the unisexual flowers with corolla lobes contorted in bud. The leaves are typically rather large and more or less coriaceous. (The next eleven genera belong to Tribe Gardenieae = large many-seeded fruits, mostly dioecious, with largish, often 6-parted, white flowers, the corolla lobes contorted in bud.)

Randia (incl. *Rosenbergiodendron*) (200–300 spp., incl. Old World, the latter often generically segregated) — Shrubs and small trees (also a few lianas), usually more or less spiny (typically with whorl of spines at tip of short-shoot branch), usually growing with leaves on subwoody "short-shoots" along main branches; flowers usually solitary, sessile, white and fragrant, often elongate and hawkmoth-pollinated; calyx lobes often elongate and foliaceous.

C: kachku

Posoqueria (15 spp.) — Midcanopy trees of moist and wet forest. Leaves coriaceous; stipules thick-foliaceous in aspect, triangular to +/- oblong, subpersistent and tending to wither in place. Flowers long-tubular (8–35 cm) white, hawkmoth-pollinated. Like *Tocoyena* but smoother-surfaced yellow fruit and corolla lobes in bud bent toward one side.

Tocoyena (20 spp.) — Very like *Posoqueria* but dries black, the fruit rougher-surfaced, and flowers more slender and the terminal swelling formed by the corolla lobes in bud not bent to one side.

P: huitillo

Genipa (6 spp.) — Canopy trees. Large obovate leaves (drying black in commonest species); stipules triangular and subfoliaceous, thicker at base (except broadly obovate in *G. williamsii*). Very characteristic thick cream to tannish-yellow corolla with the tube at least 1 cm long and the anthers inserted between the bases of the corolla lobes; in most species sericeous outside. Fruit large (4–9 cm), round, glabrous (but with a roughish surface), oxidizes blue-black when cut and stains skin the same color. This is the source of the dark skin paint used by the Chocó and other Indians.

C: jagua; E: jagua; P: jagua, huito

Rubiaceae
(Trees with Large Indehiscent Fleshy Fruits and Small to Medium Flowers)

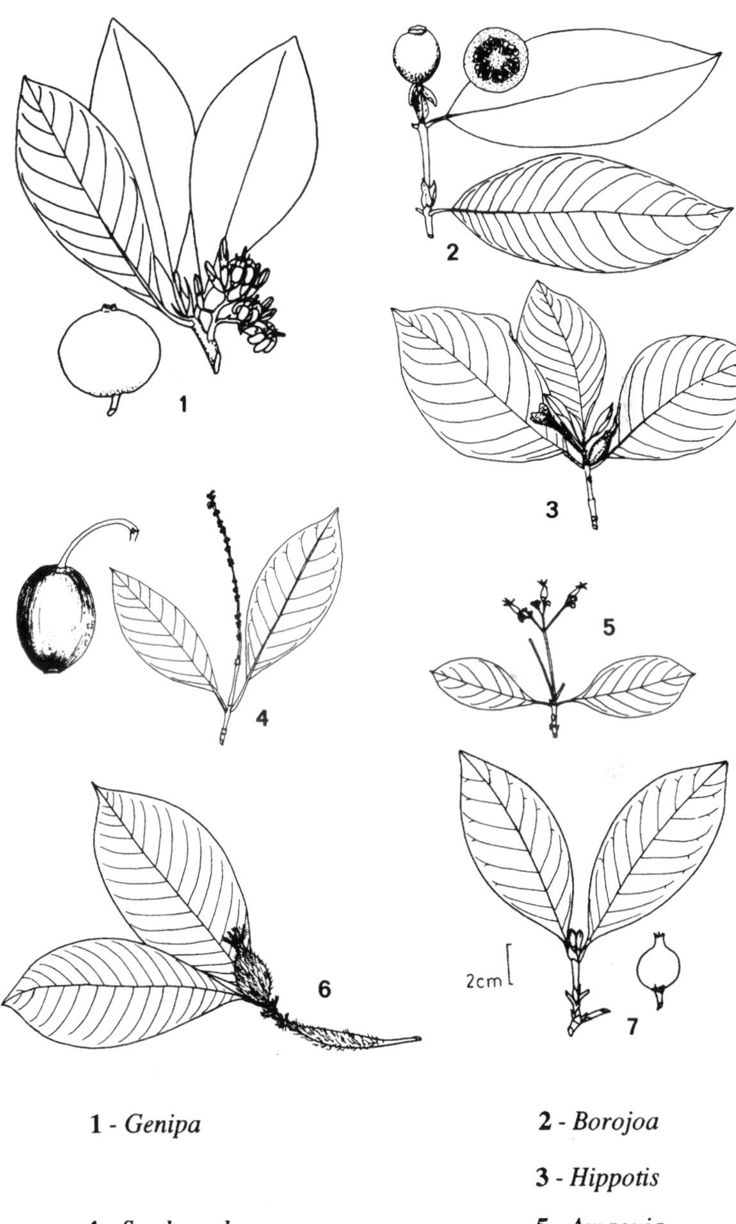

1 - *Genipa*

2 - *Borojoa*

3 - *Hippotis*

4 - *Stachyarrhena*

5 - *Amaouia*

6 - *Duroia*

7 - *Alibertia*

Borojoa (6 spp.) — Similar to *Genipa* in large usually blackish-drying leaves but the sessile flowers unisexual, more numerous, smaller and white, subtended by several whorls of bracts formed from the persistent uppermost stipules (the defining character of genus). Stipules acute to acutish, triangular or narrowly triangular, rather undifferentiated in texture. Fruit even larger than *Genipa* (to 8 cm diameter), and with a much thicker pericarp; also differing in a smoother surface and being subtended by the persistent bracts. This is the source of the famed "borojo" refresco of Chocó.

C: borojo

Duroia (20 spp.) — Small to midcanopy trees. Similar to *Genipa* in large fruit, but distinctive in caducous stipules forming conical more or less appressed-tan-pubescent cap over terminal bud as in *Amaouia;* branching and conspicuous annular stipule scars also as in *Amaioua* but twig, inflorescence, and calyx pubescence usually hispid rather than sericeous. Fruit solitary and much larger than in *Amaouia,* typically densely stiff-hairy. The commonest Amazonian species has branchlets with hollow swollen internodes which house ants.

C: borojo de monte; E: tuba abillu; P: huitillo

Stachyarrhena (8 spp.) — Small dioecious trees, mostly of wet forests. Largish coriaceous leaves. Fruit green, resembling a small calabash, pendent on an elongate pedicel (other large-fruited rubiacs have more or less sessile fruits); male inflorescence elongate and spikelike, totally unlike female, but resembling *Botryarrhena.*

Botryarrhena (1 sp.) — Large canopy tree. Leaves large, obovate, long cuneate, coriaceous, glabrous, dark-drying; stipule broadly triangular, forming stipular "cup". Fruits on short terminal raceme, smaller than in most relatives, ca. 2 cm across. Male flowers in elongate spikelike inflorescence very like *Stachyarrhena.*

Alibertia (35 spp.) — Small trees or shrubs with, typically, rather small leaves and usually with distinctive acute to strongly acuminate triangular (rarely truncate) stipules (often fused into tube at base and at least the base usually persistent) and brown twigs. Fruit smaller than in most relatives, round, sessile, yellow, ca. 2–2.5 cm in diameter, apex with a conspicuous cylinder formed by the persistent calyx tube. Flowers sessile, white, terminal, several to solitary, rather fleshy, with ca. 5 long, pointed lobes; similar to *Randia,* but the plant nonspiny.

Sphinctanthus (3 spp.) — Ours a tahuampa tree with black-drying leaves and *Alibertia*-like stipules fused into tube. Differs from *Alibertia* in the perfect flower and from smaller-flowered *Randia* in lacking spines and in the terminal flowers (or fruit).

Amaioua (25 spp.) — Subcanopy dioecious trees with fibrous, vertically ridged bark and sometimes fenestrated trunk. Pagoda-style branching with the leaves tending to cluster at the ends of the short erect or upturned branches. Perhaps best vegetative character is the characteristic sericeous pubescence of the young twigs and stipules; stipules narrowly triangular, caducous to leave scar, forming a characteristic conical cap (cf., Moraceae and *Duroia*) over terminal bud. Fruit oblong, 1–1.5 cm long, wrinkled on drying, tannish-sericeous, several together unlike *Duroia*. Inflorescence rather few-flowered, the white flowers clustered at end of peduncle, the corolla tube narrower toward apex.

Kotchubaea (10 spp.) — Related to *Amaioua* and *Duroia*, differing in the much longer (tube > 5 cm long) flower with 8–10 long narrow lobes. Stipules differ in being persistent and not sericeous nor forming cap over terminal bud, broadly or narrowly triangular, always with a noticeably thickened central area; forming secretory area and young growth sometimes +/- resinous (cf., *Elaeagia*).

2Bb. Shrubs or small trees with 1–2-seeded berrylike fruits (mostly *Psychotria*), *Palicourea* and their relatives

Psychotria (700 spp., plus 700 in Old World) — The most prevalent genus of Rubiaceae shrubs and one of the most speciose of all neotropical genera: e.g., over 100 species in Panama alone. Characterized by tiny greenish or purplish inflorescence branches. Fruit a small fleshy berry almost always 2-seeded (5-seeded in a few species like *P. racemosa*). Vegetatively, usually characterized by bilobed, persistent stipules (also in some *Isertia* and some herbs) or by triangular caducous stipules, but the stipules of other species highly variable.

E: cafecillo; P: yagé (*P. viridis*)

Cephaelis (180 spp., incl. Old World but the latter not closely related) — In its broad sense, polyphyletic, and thus often combined with *Psychotria*. Like *Psychotria* but with capitate inflorescences subtended by large, often brightly colored bracts. The two genera are usually separable in Central America but intergrade in Amazonia and are lumped under *Psychotria* in the *Flora de Venezuela*.

Stachyococcus (1 sp.) — Amazonian shrub. Like *Psychotria* but with a distinctive spicate terminal inflorescence to 12 cm long (cf., *Botryarrhena*) and with the flowers in bracteate clusters along it. Fruits larger than *Psychotria* but smaller than *Botryarrhena*.

Palicourea (200 spp.) — Shrubs and small trees, replacing *Psychotria* as dominant rubiac genus in montane forests. Like *Psychotria* but the corolla usually longer (more than 7 mm), yellow or purple with a basal

Rubiaceae
(Shrubs with 1–2-Seeded Berries)

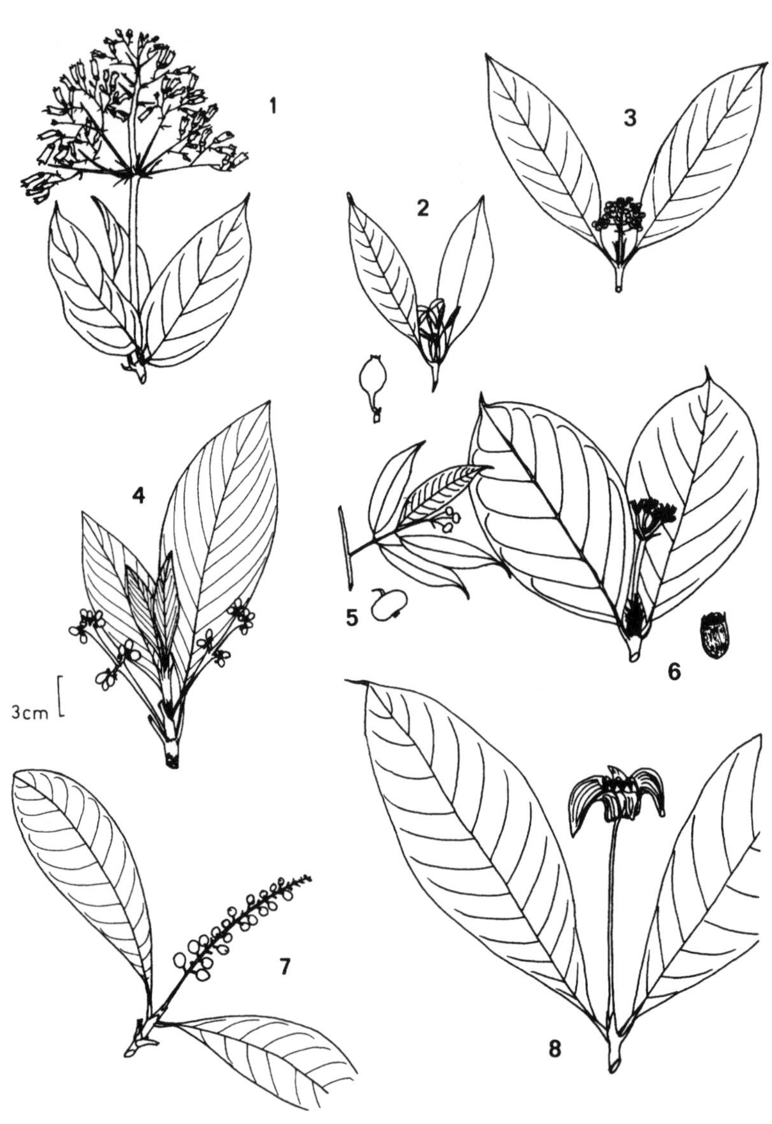

1 - *Palicourea* **2** - *Coussarea* **3** - *Psychotria*

4 - *Pagamea* **5** - *Faramea* **6** - *Rudgea*

7 - *Stachyococcus* **8** - *Cephaelis*

swelling, contrasting with the brightly colored inflorescence branches; stipules connate and persistent (in *Psychotria,* caducous or bilobed).

E: café de monte

Rudgea (150 spp.) — Very like *Psychotria* but with the stipules irregularly fringed (pectinate, but the often glandular teeth sometimes caducous). Leaves usually with shorter petioles (typically subsessile) and tending to be more coriaceous. Fruits usually white; only one seed often present in fruit.

Pagamea (24 spp.) — Small trees, mostly of the Guayana Shield; in our area on poor sandy soil. Unique in a completely superior ovary. Vegetatively extremely easy to recognize by the stipules united into a long cylindrical sheath with 4–8-setae (this persistent only at upper node). Flowers tiny and *Psychotria*-like, clustered along narrow, spiciform inflorescence branches. Fruit round with base inclosed by the cupular calyx.

Faramea (125 spp.) — Small trees with glabrous often very coriaceous sometimes 3-veined leaves. Stipules usually with a broad base and long narrow acumen typically crossed at branch tip. Similar to *Psychotria* but flowers larger and more salverform, with a narrow tube and long-pointed lobes, and usually light blue, (sometimes white). Fruits slightly larger than *Psychotria,* one-seeded, fleshy, also distinctive in being usually broader than long, seeds horizontal (ovary 1-celled).

E: jazmin

Coussarea (150 spp.) — Shrubs and small trees very *Psychotria*-like but the leaves usually more coriaceous and the white flowers usually larger. Technically, more like *Faramea* in 1-celled ovary (but with connate ovules) and one-seeded fruit but seeds vertical and fruits ellipsoid and longer than wide; seed coat thinner than in *Psychotria.* Stipules coriaceous, broad, usually blunt, forming a well-developed "cupule" at each node.

Guettarda (60 spp., plus 20 in New Caledonia) — Mostly canopy trees, especially in cloud forest. Leaves always somewhat puberulous; distinctive in the frequently rather close secondary veins and clearly parallel tertiary venation below. Inflorescence branches 1-sided; flowers similar to *Lantana,* white with wavy-edged lobes. Fruit finely puberulous.

Antirhea (incl. *Pittonionitis*) (40 spp., incl. Old World) — Mostly Antillean, in our area only northern Colombian moist forests. Trees often precociously blooming while deciduous; numerous small white flowers in large paniculate inflorescence, longer than wide, like *Guettarda* but glabrous; leaves puberulous, especially along veins.

Ixora (400 spp., mostly Old World) — Flowers with much longer more slender floral tube than in *Psychotria,* at least the pedicels and sometimes the floral tube red (technical character is lobes contorted in bud rather than valvate as in *Psychotria* and relatives). Fruits broader than long (cf., *Faramea*). Commonest species have distinctly long, narrow, subsessile leaves; stipules similar to *Faramea,* triangular with narrow acumen. Several cultivated ornamentals.

Chione (4 spp.) — Shrubs or trees, mostly Central American and West Indian. Characterized by rather flat-topped inflorescences; flowers rather resembling *Macrocnemum* in shape, but white and the stamens inserted at base of corolla; fruit usually elongate and spindle-shaped, and with a single developed seed; (technical characters of genus are pendulous ovules and several-chambered pyrene).

(*Coffea*) — Cultivated shrub, sometimes escaping or persisting after cultivation. Flowers axillary, sessile. Differs from *Psychotria* and relatives by the contorted rather than valvate petals in bud.

(*Malanea*) — Most species are scandent at least part of the time (see above) but several are more frequently shrubby. Characterized by *Chomelia*-like leaves with whitish-puberulous main veins below.

2Bc. Shrubs or small trees with several-seeded fruits — (Although this group of shrubs are supposed to be closer to the dehiscent-fruited genera (= subfamily Cinchonoideae) they look much more like *Psychotria* and its relatives.)

Gonzalagunia (35 spp.) — Small shrubs, mostly in second growth. Inflorescence a very characteristic long, pendent, spike of small white flowers. Fruit a small, usually white (sometimes blue), berry. Stipules with +/- broad base and long narrow acumen; leaves pubescent at least along veins below and sometimes the whole undersurface densely white-pubescent.

Hoffmannia (100 spp.) — Subshrubs, mostly of wet forest. Looking like a +/- herbaceous *Psychotria* with reduced mostly cauliflorous (or, in part, axillary) inflorescences; leaves often rather large and somewhat succulent, the stipules very reduced; tiny flowers and fleshy red berries typically in large part from low on stem, the ovary 2–3-celled.

Bertiera (2–3 spp., plus 25 in Old World, mostly Africa) — Shrubs; look very much like *Psychotria* but flower slightly longer, ovary and fruit 4–5-celled; inflorescence branches typically 1-sided. Vegetatively characterized by puberulous branches and the long thin acumen on the otherwise small, nondescript stipule.

Rubiaceae
(Shrubs with Several-Seeded Berries)

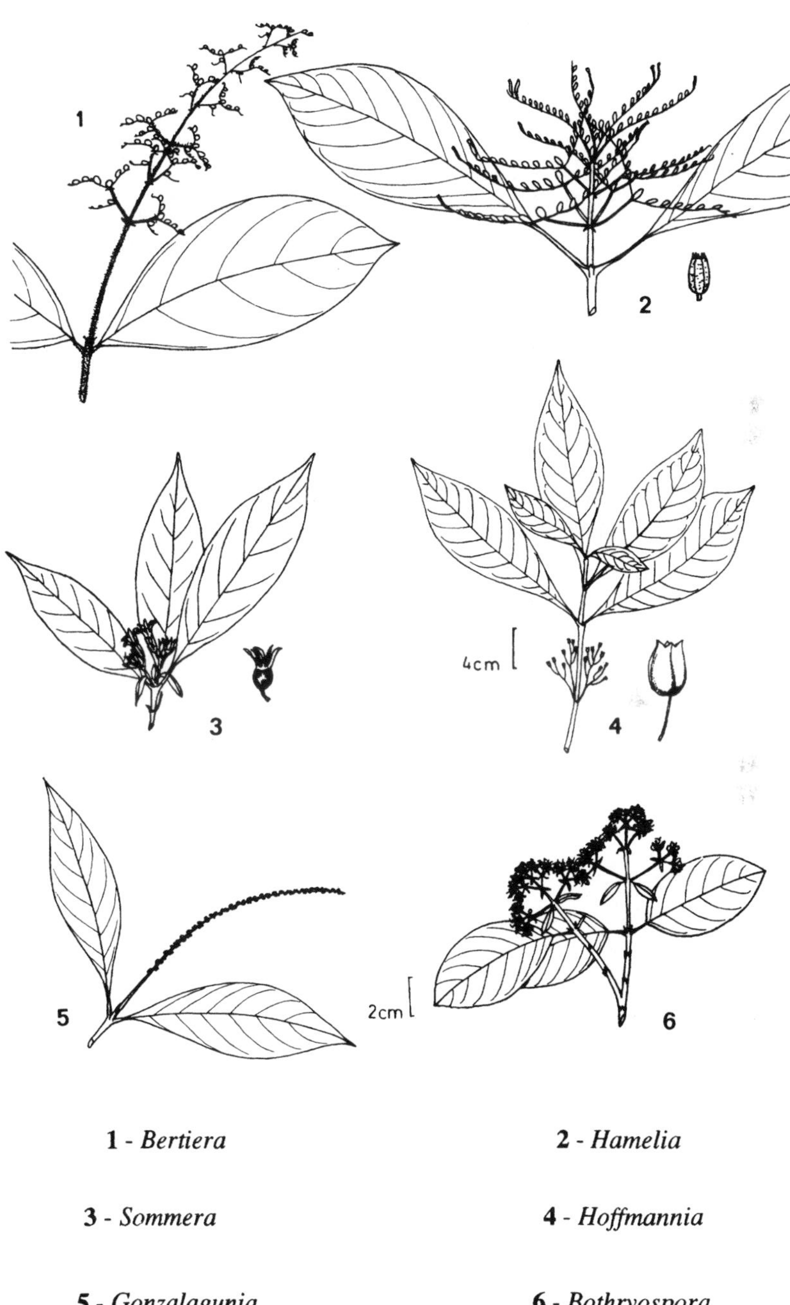

1 - *Bertiera* **2** - *Hamelia*

3 - *Sommera* **4** - *Hoffmannia*

5 - *Gonzalagunia* **6** - *Bothryospora*

Bothriospora (1 sp.) — Small, much-branched tree common in tahuampa forest. Very *Psychotria*-like but with more numerous flowers and slightly larger orangish-yellow fruits with numerous seeds. Leaves rather small; stipule lanceolate, caducous to leave a usually whitish interpetiolar line.

Retiniphyllum (20 spp.) — Shrub of poor-soil sandy areas, mostly on Guayana Shield. Distinctive in terminal racemose to spicate inflorescence with narrow-tubed, narrow-lobed flowers. Leaves coriaceous, usually with tendency for intersecondaries; stipules short, united to form a resin-secreting cup; leaves and branches +/- resinous. Fruit somewhat intermediate between *Psychotria* relatives and the large-fruited Gardenieae, a small round berrylike drupe with 5 seeds (showing as ridges on surface when dry).

Hamelia (16 spp.) — Shrubs; very much like *Psychotria* but flower larger, orange-red, and narrowly tubular with small lobes to campanulate and yellow. Fruit distinctive, rather elongate and cylindrical with a flattish apex and the calyx teeth persistent as rim around apex. Inflorescence branches with flowers on one side. Leaves often in whorls of 3, thin, and with unusually long petioles. Differs from *Hoffmannia* mostly in the terminal (rather than axillary and cauliflorous) inflorescence.

Raritebe (1 sp.) — Wet-forest shrubs, always noticeably puberulous on young twigs and petioles. Similar to *Bertiera* (or superficially to *Psychotria*) but flower longer and narrower with corolla lobes valvate in bud. Stipules narrowly triangular and puberulous.

Sommera (12 spp.) — Shrubs. Distinctive in calyx lobes elongate and +/- foliaceous; leaves somewhat sericeous-pubescent below, at least on the veins, the intervenal areas with a very minutely and closely lined surface (cf., *Hippotis*). The commonest species in forests seasonally inundated by white-water rivers. Stipules triangular, rather thin, somewhat pubescent at least in center, caducous. Flowers similar to *Coussarea* except for the long calyx teeth.

Hippotis (11 spp.) — Shrub with large membranaceous leaves, puberulous and with conspicuously prominulous and parallel tertiary venation below. Inflorescence axillary, few-flowered, the rather long (4 cm long) white or red hairy flowers with large pubescent, conspicuously spathaceous to irregularly parted calyces. Fruit ellipsoid, to 3 cm long, pubescent, somewhat ribbed, the calyx persistent.

P: sol caspi

(*Isertia*) — One common species resembles *Palicourea* but with dense hairs in mouth of corolla, several-celled ovary and several-seeded

Rubiaceae
(Miscellaneous Distinctive Taxa of Shrubs and Small Trees with Indehiscent Fleshy Fruits)

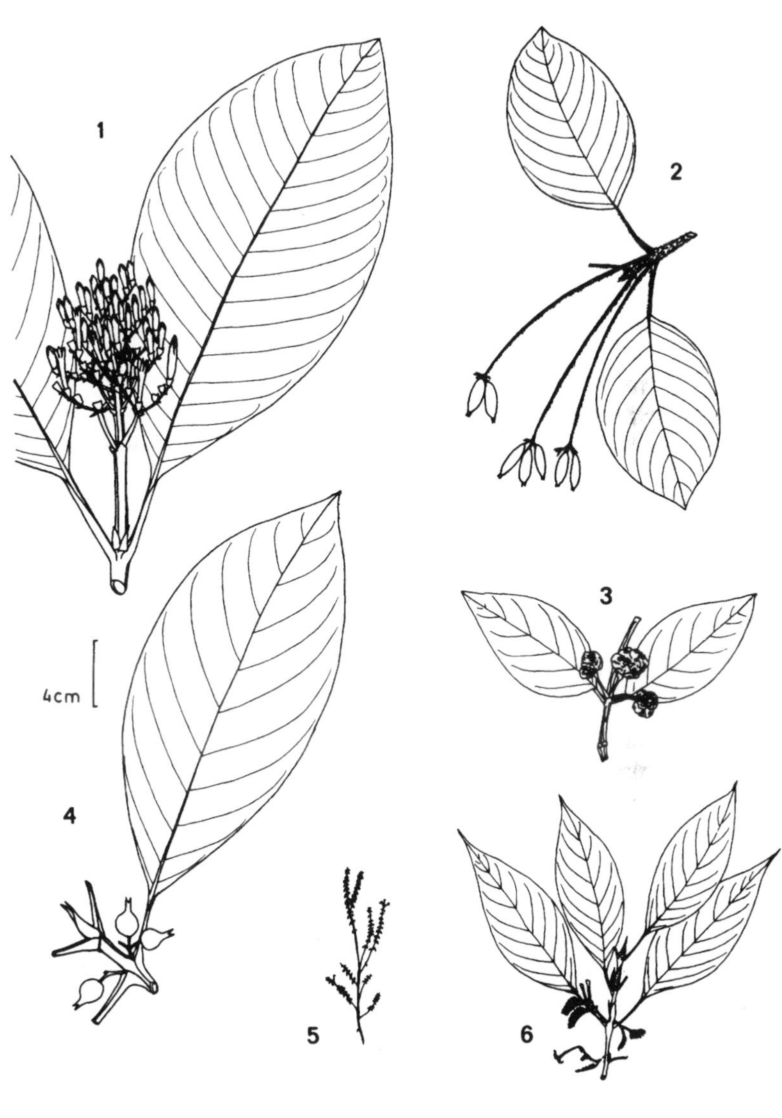

1 - *Isertia*

2 - *Chomelia*

3 - *Morinda*

4 - *Pentagonia*

5 - *Arcyctophyllum*

6 - *Guettarda*

fruits. The shrubby *Isertia* has tubular orange flowers ca. 3 cm long and an openly paniculate inflorescence with red-orange branches.

2Bd. Miscellaneous, mostly individually distinctive genera of shrubs and trees, mostly with indehiscent fleshy fruits — The fruits of the following genera are either very atypical for family (dry in *Allenanthus* and *Machaonia,* syncarpous in *Morinda*) or fleshy and berrylike but generally smaller than in Gardenieae (Group 2Ba) and larger than in *Psychotria* alliance (Group 2Bb). Two of these genera have spines (unique in erect rubiacs except for *Randia*), one has intrapetiolar stipules, and two have the largest and smallest leaves, respectively, of the family.

Allenanthus (2 spp.) — Small tree of seasonal forests reported only from Costa Rica and Panama but likely in northern Colombia. Unique, small, flat-winged (presumably wind-dispersed) fruits in large terminal inflorescence. Except for the winged fruits looks much like *Chione.*

Machaonia (30 spp.) — Mostly Antillean; in our area dry forest or coastal strand. Shrubs, usually with small leaves and spines at some of the nodes. The unusual dry fruit separates into two, 1-seeded, elongate halves (cocci). Flowers small and arranged in a paniculate inflorescence.

Chomelia (incl. *Anisomeris*)(50 spp., plus 370 in Old World) — Often spiny trees; looks like a small-fruited (fruit to 5 mm long) narrow-flowered (but not dioecious) *Randia,* although actually closer to *Psychotria* and allies; differs from *Randia* in pedunculate rather than sessile flowers and fruits; flowers white, with very narrow tube and reflexed lobes (lepidoptera-pollinated), calyx with elongate +/- foliaceous lobes. Narrow corolla lobes = typical *Chomelia;* crisped lobes = *Anisomeris.*

E: segala

Morinda (incl. *Appunia*) (80 spp., incl. Old World) — Shrubs, especially along coast. Very characteristic fruit more than 1 cm broad, syncarpous (rather resembling a succulent *Annona*); flowers in dense heads which become fleshy in fruit; leaves dry blackish.

Pentagonia (20 spp.) — Usually unbranched pachycaul treelets; leaves very large, usually 50 cm or more long, sometimes pinnately lobed (!); the venation frequently striate-parallel; flowers sessile, axillary, large and fleshy.

Isertia (25 spp.) — Trees (sometimes shrubs), mostly in wet-forest second growth. Flowers distinctive, tubular, white or red, >3 cm long and with dense hairs at mouth; leaf undersides white or silvery in white-flowered species. Stipules +/- intrapetiolar (but in most species bifid and this not obvious).

C: jaboncillo

Arcyctophyllum (30 spp.) — Low, dense shrubs or subshrubs of high-altitude paramos and puna. Very distinctive in tiny ericoid leaves.

3. Herbs — Only a few neotropical genera of Rubiaceae are herbaceous although some species of such shrub genera as *Psychotria* and especially *Hoffmannia* may be subshrubs at best. Some of the genera treated here as herbs tend to be prostrate and stoloniferous, often forming large patches of ground cover.

3A. Erect herbs or epiphytes

Didymochlamys (2 spp.) — Common in Chocó pluvial forest. Small and epiphytic; vegetatively, not at all rubiaceous-looking; leaves narrow and apparently alternate; inflorescence capitate, subtended by two large green bracts (cf., *Cephaelis*).

Amphidaysa (3 spp.) — Chocó area herb or subshrub with unbranched stem and terminal cluster of large, long-petiolate, obovate or oblanceolate leaves; the very distinctive persistent stipules are 3-aristate and 2–3 cm long. The narrow white flowers are sessile and axillary with a tube 2–3 cm long and long linear calyx lobes.

Declieuxia (40 spp.) — Savannah habitats. Sometimes subshrubby. Rather like a small-leaved reduced *Psychotria* with a rather flat-topped terminal inflorescence except that the fruit is dry and splits into two cocci (schizocarp, cf., *Machaonia*); leaves sessile.

3B. Prostrate stoloniferous creeping herbs, often forming mats on ground

Geophila (30 spp., incl. Old World) — Leaves cordate; fruits red berries.

Coccocypselum (6 spp.) — Leaves hairy, not cordate; fruits bright blue berries.

Nertera (6 spp., incl. Old World) — Leaves small, very thin, less than 1.5 cm long; stipules fused to petiole to make sheath (as in *Spermacoce*); fruit red and juicy.

Corynula (2 spp.) — Andes, especially in drier areas. Leaves and stems pilose. Fruit not fleshy, deeply costate, splitting into 2 cocci.

Galium (incl. *Relbunium*)(400 spp., mostly n. temperate) — In our area, mostly montane. Fruits usually dry; stipules as large as, and resembling, leaves, the leaves, thus, apparently whorled.

Rubiaceae
(Herbs)

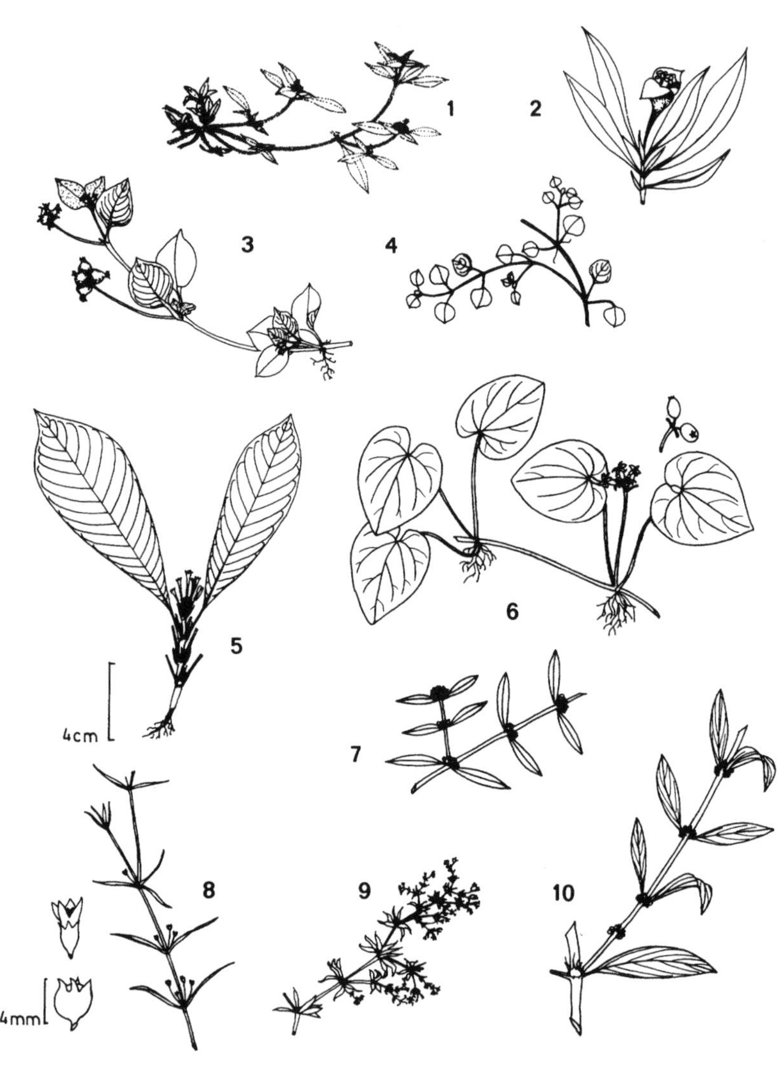

1 - *Richardia* **2** - *Didymochlamys*
3 - *Coccocypselum* **4** - *Nertera*
5 - *Amphidaysa* **6** - *Geophila*
7 - *Borreria*
8 - *Oldenlandia* **9** - *Galium* **10** - *Hemidiodia*

Oldenlandia (300 spp., of Old World) — Adventive in our area. Leaves small, or narrow, mostly less than 5 mm wide; fruits dry; similar to Spermacoceae in general aspect but inflorescence not congested.

3C. Spermacoceae

The next seven genera (plus scandent *Emmeorrhiza*) all belong to tribe Spermacoceae, easily recognized by clusters of sessile axillary flowers and stipules connate and fused with petioles to form a sheath. These genera are mostly very poorly differentiated and several can be told apart only by technical characters of the fruit. All are weeds.

Richardia (10 spp.) — The most distinctive genus of Spermacoceae on account of its exclusively terminal inflorescences, capitate and subtended by leafy bracts; fruit splitting in half.

Mitracarpus (40 spp., incl. in Africa) — Flowers all axillary, fruit dehiscence circumscissile.

Hemidiodia (1 sp.) — Flowers all axillary; fruit opening only at base, otherwise similar to *Diodia* (and sometimes lumped with it).

Borreria (150 spp., incl. Old World) — Flowers terminal and axillary, both fruit-cells opening at apex. Sometimes included in *Spermacoce.*

Spermacoce (6 spp., plus ca. 90 in Old World) — Flowers terminal and axillary, only one (the larger) fruit-cell dehiscent.

Diodia (15 spp.) — Fruit indehiscent, but the mericarps separating as in *Richardia.*

RUTACEAE

Generally vegetatively recognizable by the glandular punctate leaves, typically with a pungent, often citruslike, odor. Otherwise, extremely variable vegetatively and ranging from herbs (*Moniera*) to large commercial timber trees (e.g., *Euxylophia* and extralimital *Balfourodendron*). The largest and most widespread genus, *Zanthoxylum* is usually characterized by spiny trunks and/or leaves; several other large-tree genera may have raised lenticels on the trunk. The leaf characters that usually differentiate genera and even families are of minimal use in generic definition in rutacs where intrafamilial leaf variation may be greater and perhaps less taxonomically consistent than in any other family; leaves range from simple to unifoliolate, palmately compound, pinnately compound, or bipinnate, and may be opposite or alternate, even in the same genus. Nearly every compound-leafed genus also has unifoliolate species. The flowers vary from small and rather nondescript (most taxa) to variously elongate with white or red petals (in approximate order of increasing petal size: *Adiscanthus, Angostura, Raputia, Galipea,* (*Rauia*), *Leptothyrsa, Ticorea, Ravenia, Naudinia, Erythrochiton*). The fruits may be small, round berries (*Murraya, Amyris*), large and fleshy with edible pulp (*Casimiroa, Hortia*), winged and wind-dispersed (*Spathelia*), or variously dehiscent, occasionally as a capsule to release winged seeds (*Dictyoloma*), or usually as more or less distinct apocarpous follicles. The dehiscent follicular fruits are of several types including the tiny follicles of most *Zanthoxylum* species, slightly larger follicles, usually with seashell-like curved ridges (*Pilocarpus, Angostura, Leptothyrsa, Adiscanthus, Ticorea*), somewhat larger appressed-together follicles (*Erythrochiton, Galipea, Raputia, Ravenia, Rauia*), to woody and partially fused follicles with a subapical "horn" (*Metrodorea, Esenbeckia*).

1. Leaves Bipinnate

Dictyoloma (2 spp.) — A small tree common in second growth, especially in seasonally dry areas of Amazonia. The large bipinnate leaves are quite unlike any other rutac, although the marginally punctate individual leaflets are not unusual. The large flat-topped openly paniculate inflorescence is very characteristic. The flowers are small and nondescript, the fruit a strongly 5-ridged capsule with small winged seeds. Superficially *Dictyoloma* rather resembles *Jacaranda* (Bignoniaceae), *Dilodendron* (Sapindaceae) or *Schizolobium* (Leguminosae), other members of the same habitat in which it is found; none of these has punctate leaflets or remotely similar inflorescences.

P: huaman saman

2. Leaves Pinnate

Zanthoxylum (ca. 100 spp., plus over 100 in Old World) (incl. *Fagara*) — Small to large trees especially prevalent in drier areas and in late second-growth situations. The largest rutac genus. Most species have thick-based spines on the trunk (in our area the only pinnately compound genus with such spines) and many have spines on the leaves and branchlets, also unique; montane species tend to lack the spines. The leaflets of most species are more or less crenate or serrulate, although some species are entire; especially in dry areas the leaflets are often small and the rachis narrowly winged. The leaflets of nearly all species are pellucid-punctate at least along the margin. The inflorescence is paniculate (but often with rather spicate branches) and the flowers minute; the dry apocarpous fruit is composed of (1–)5 small single-seeded follicles. *Fagara* is a segregate sometimes used for the species with 2 whorls of perianth segments.

C: tachuelo, rudo; E: tachuelo, lagarto, sasafras; P: hualaja

Rutaceae
(Leaves Pinnate or Bipinnate)

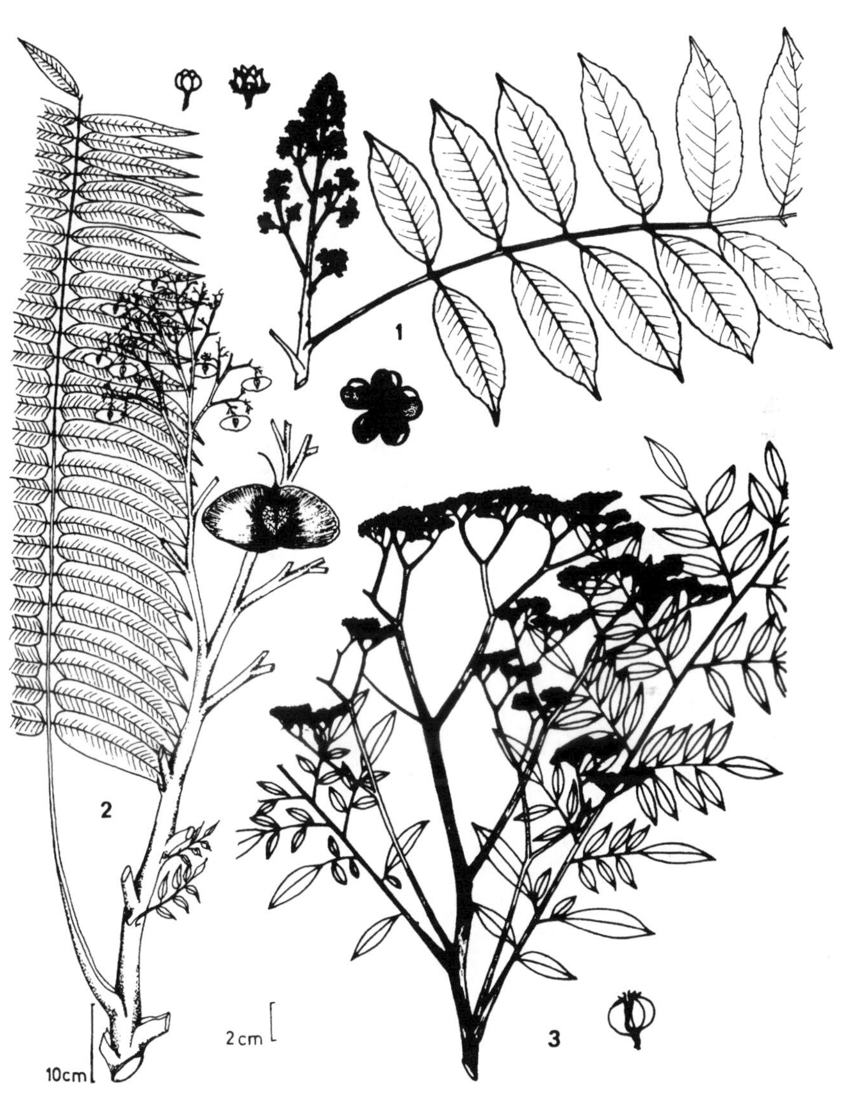

1 - *Zanthoxylum*

2 - *Spathelia*

3 - *Dictyoloma*

Rutaceae
(Leaves 3-Foliolate to Palmately Compound)

1 - *Amyris* **2** - *Metrodorea*

3 - *Moniera* **4** - *Angostura*

5 - *Galipea*

Spathelia (20 spp.) — Mostly Antillean and upland Guayanan; in our area, known only from a single collection from white-sand soil in Amazonian Peru. A slender unbranched 30 m tree with a terminal tuft of very large multifoliolate leaves (far more leaflets than any other simple-pinnate rutac and rather resembling *Talisia*) and a large paniculate terminal inflorescence. The oblong leaflets have truncate bases. The distinctive 2-winged fruits look almost exactly like those of *Terminalia* (Combretaceae).

(Murraya) — Native to Australasia. One shrubby species (*M. paniculata*) is widely cultivated and sometimes escaped. Characterized by the rather glossy, small, obovate, alternate leaflets, the fleshy red berries and the relatively long-petaled white flowers.

E: mirto

Amyris (30 spp.) — Dry-area trees. Leaves varying from 1-foliolate (with conspicuously jointed petioles) to 3-foliolate and pinnate; in our area, opposite when pinnate and sometimes when 3-foliolate. The leaflets all characterized by thin-coriaceous texture, more or less triangular to rhombic shape, and strongly ascending not very accentuated secondary veins. Flowers small and white; fruits small, round, and indehiscent.

3. Leaves 3-Foliolate to Palmately Compound — (Some genera also with 1-foliolate species)

Casimiroa (6 spp.) — Mostly Central American, perhaps only cultivated in our area. Leaves mostly 5-foliolate, often pubescent, often a mixture of opposite and alternate on the same plant. Characteristic fruit large, round, fleshy, and edible.

Moniera (2 spp.) — Very distinctive in herbaceous or subshrub habit and tenuously membranaceous 3-foliolate leaves. Vegetatively, more suggestive of Oxalidaceae than Rutaceae except for the punctations. The tiny white flowers are in small, 1–2-branched, one-sided inflorescences very distinctive in their conspicuous foliaceous bracts.

Esenbeckia (38 spp.) — Trees and shrubs, in our area with mostly simple or 1-foliolate leaves (3-foliolate only in one densely pubescent inter-Andean species), but palmately 5-foliolate in some extralimital species. The leaves have different-length petioles with a noticeable apical flexion (sometimes clearly jointed), the slender longer petioles different than in simple-leaved *Pilocarpus* species. Flowers small as in *Pilocarpus*, but inflorescence paniculate and fruiting follicles much larger and woodier, partly fused and each with a conelike subapical projection.

Metrodorea (5 spp.) — A tree of strongly seasonal forests, only reaching Madre de Dios in our area. Closely related to *Essenbeckia* but very distinctive in the opposite 3-foliolate leaves with a unique conspicuously hollowed guttifer-like sheathing petiole base with a dorsal flange. Flowers small (= Pilocarpinae), the fruits as in *Esenbeckia,* largish, with 5 partially united woody follicles each with a subapical projection.

Angostura (38 spp.) (incl. *Cusparia* and *Rauia*) — Shrubs or small trees. Like *Pilocarpus* in fruit but the buds and valvate white petals longer (> 1 cm). Leaves alternate, mostly 1-foliolate (always with longish petioles and conspicuously jointed petiole apex) but some species mostly 3-foliolate or palmately 5–7-foliolate. The 1-foliolate species can be differentiated from similar *Esenbeckia* species by the more coriaceous leaf and thicker more robust petiole; the 3–5-foliolate species have glabrous or glabrescent leaves and generally longer petioles than similar species of other genera; there is also a distinctive tendency for leaflike inflorescence bracts.

Galipea (13 spp.) — Shrubs or small trees, mostly in disturbed areas. In our area, with uniformly 3-foliolate alternate leaves and a distinct tendency for the petiole to be winged or margined. Flowers long and slender, as in *Angostura* but narrower. Fruits similar to *Esenbeckia* but less woody and the carpels lacking subapical projection (+/- intermediate between *Pilocarpus* and *Esenbeckia*). Vegetatively, rather reminiscent of *Allophylus* (Sapindaceae) except for the punctations.

Raputia (10 spp.) — Small trees, mostly of poor-soil areas, with 3–5-foliolate, long petiolate, uniformly alternate, coriaceous leaves, and a slight tendency for the petiole base to be enlarged and sheathing (cf., *Metrodorea*). Fruit as in *Galipea,* more or less intermediate between *Pilocarpus* and *Essenbeckia.* Flowers very like *Angostura* but the tube more strongly curved, especially in bud.
C: iguano

Ticorea (3 spp.) — Shrubs or treelets with long-petioled, alternate, 3-foliolate leaves, the large leaflets of our species more conspicuously and densely punctate than any other area rutac. Inflorescence long-pedunculate with flowers only at the few-branched apex; the flowers long (ca. 3 cm) narrow and white. Fruits completely apocarpous with curved ridges, similar to *Pilocarpus* but somewhat larger and usually with 5 developed follicles. Essentially combines the flower of *Leptothyrsa* with the inflorescence of *Adiscanthus.* A unifoliolate species from Panama, also densely punctate, may also reach our area; it is very distinctive in the junction of the greenish petiole and petiolule drying a very contrasting brown.

(*Erythrochiton*) — Mostly unifoliolate (see below), but one species in our area has 3-foliolate leaves.

(*Pilocarpus*) — Elsewhere often pinnate-leaved; one of our species is unifoliolate (see below) and one 3-foliolate. The 3-foliolate species is a Colombian dry-forest plant with a mixture of opposite and alternate leaves and characteristic coriaceous and blunt-tipped leaflets.

4. Leaves Simple to Unifoliolate

Hortia (8 spp.) — Large trees with rather thick branches and narrowly flat-topped growth-form; mostly in wet forests. Characterized especially by the large flat-toped paniculate inflorescence of reddish flowers and by the rather large indehiscent fleshy fruit. The simple leaves are narrowly obovate, long cuneate, and clustered at branch apices.

Euxylophora (1 sp.) — Large Amazonian timber tree, recently discovered in Peru. Leaves clustered toward apex, completely simple, ovate-oblong, distinctly *pale below,* the petiole often subwoody. Inflorescence terminal, more or less flat-topped. Combines the flowers of *Angostura* (white, valvate, thickish petals) with a fruit similar to *Esenbeckia* or *Erythrochiton.*

Erythrochiton (9 spp.) — Pachycaul treelets mostly on rather fertile soil. Mostly with large simple leaves but one species 3-foliolate. Characterized by the long (largest in area rutacs) red or white flowers with large acute calyx lobes (unlike closely related *Ravenia*) and the largish subwoody follicular fruits (cf., *Galipea* but larger) subtended by the large calyx lobes.

Leptothyrsa (1 sp.) — A locally common pachycaul treelet of Amazonian sandy-soil areas. The narrow long-pedunculate, racemelike terminal inflorescence has only abbreviated short-shoot side-branches. Flowers narrow, ca. 2 cm long, similar to *Galipea* and *Ticorea.* Fruits with individual follicles as in *Pilocarpus* but the surface more intricately reticulate than transversely ridged. Leaves clustered in terminal tuft and intermixed with bractlike reduced leaves, oblanceolate, very gradually tapering to elongate-cuneate base, extreme base subwoody.

Adiscanthus (1 sp.) — A large-leaved pachycaul treelet growing in same sandy-soil habitats as vegetatively similar *Leptothyrsa,* but the flowers thicker, shorter, dark red or maroon in color and in cluster at peduncle apex; fruits finely transversely ridged. Leaves more obviously punctate than *Leptothyrsa* and lacking the bractlike reduced leaves.

(*Angostura*) — Over half the species 1-foliolate, always with longish petiole and conspicuously jointed petiole apex, these differentiable from similar *Esenbeckia* species by the more coriaceous leaf and thicker more robust petiole, often with lenticels.

Rutaceae
(Leaves Simple: Pachycaul Treelets or Fleshy Indehiscent Fruit)

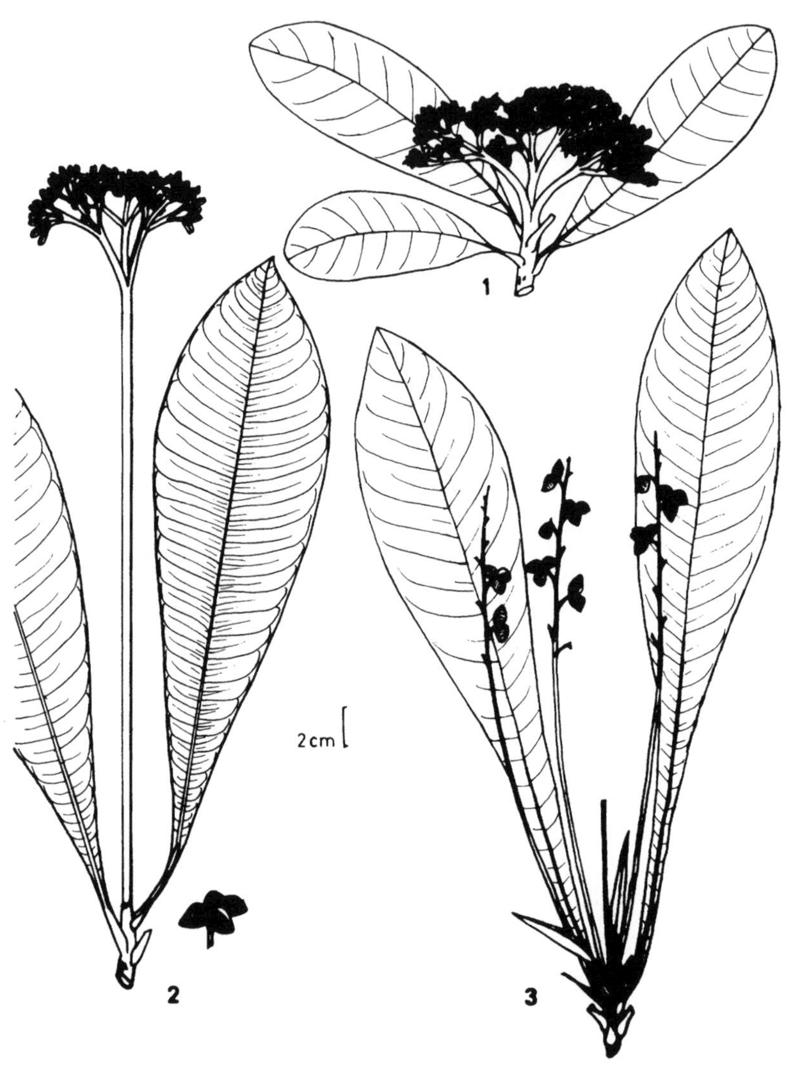

1 - *Hortia*

2 - *Adiscanthus*

3 - *Leptothyrsa*

Rutaceae
(Leaves Simple or Unifoliolate: Follicular Fruit)

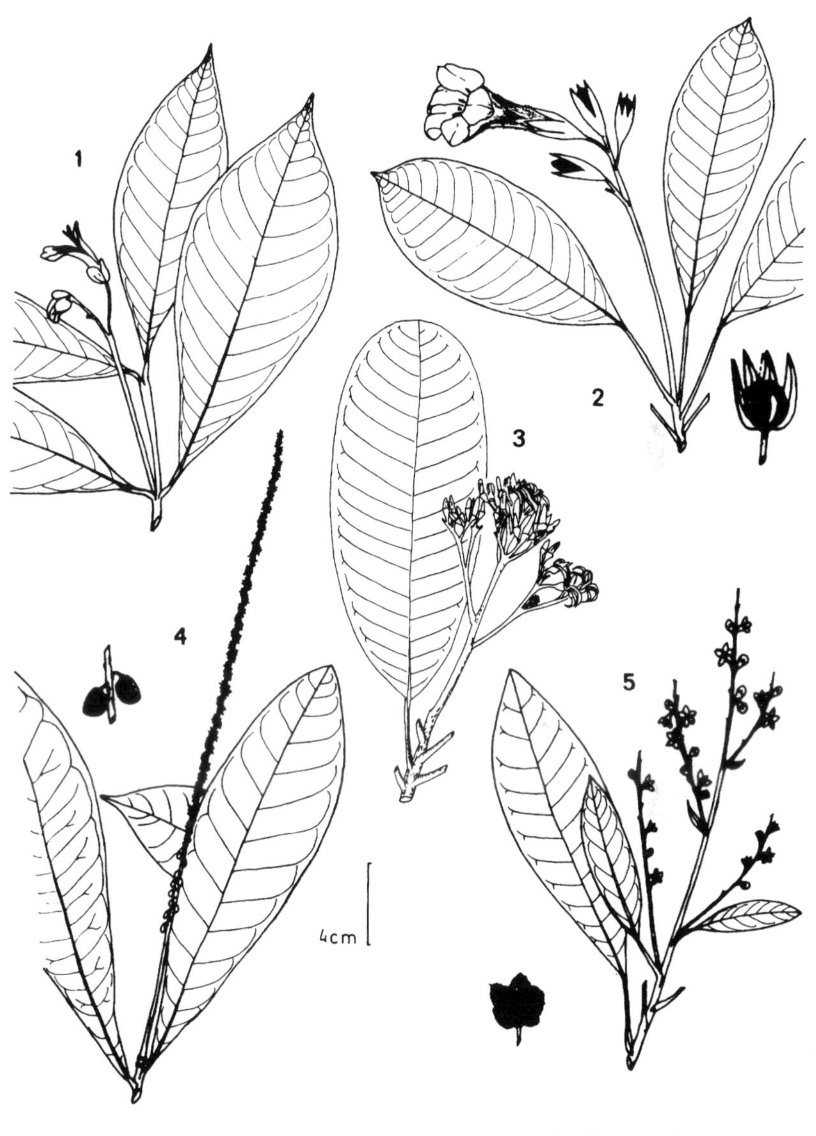

1 - *Ravenia*

2 - *Erythrochiton*

3 - *Euxylophora*

4 - *Pilocarpus*

5 - *Esenbeckia*

Ravenia (18 spp.) — Shrubs or small trees, the opposite simple leaves with sheathing guttifer-like bases (but unlike *Metrodorea* in lacking the expanded dorsal margin). Flowers strongly bilabiate, rather large (ca. 3 cm long), greenish (Peru), or bright red (Colombia). Inflorescence few-flowered, the calyx lobes large and imbricate, blunt-tipped unlike *Erythrochiton* with which it should possibly be merged.

Naudinia (1 sp.) — Endemic to Colombia. Large unifoliolate leaves with well-developed petioles; inflorescence few-flowered, the corolla long, red, and tubular. Essentially a hummingbird-pollinated derivative of *Angostura.*

Pilocarpus (13 spp.) — Only two species in our area, one 3-foliolate (see above) and the other with more or less clustered short-petioled simple (not at all unifoliolate) leaves. The main differentiating characters are small flowers (with globose buds: Pilocarpinae) on a perfectly racemose inflorescence and the smallish, completely apocarpous fruit with seashell-like ridges on the individual follicles.

(*Citrus*) — Native to tropical Asia; widely cultivated and perhaps occasionally escaped in our area (but in subtropical South America one species is completely naturalized and common even in undisturbed forest). Distinguished by the usually spiny branches, winged petiole, largish multistaminate flower, and typical citrus fruit.

(*Esenbeckia* and *Amyris*) — Both have unifoliolate species in our area (see 3-foliolate genera above).

There are many extralimital genera of rutacs, most of them small and several monotypic. At least one, *Spiranthera* (similar to *Galipea* but with longer flowers and the 3-foliolate leaves usually somewhat whitish below), reaches lowland Amazonian Venezuela and is likely to occur in our area.

Sabiaceae

Two genera of small to large alternate-leaved trees. *Ophiocaryon* has mostly pinnately compound leaves and *Meliosma* (in our area) simple leaves. *Meliosma,* a characteristic element of montane forests, is vegetatively identifiable by a usually serrate leaf margin and/or a thickened petiole base (cf., Sapotaceae but lacking milky latex). *Ophiocaryon* could easily be confused with *Talisia* or other Sapindaceae from which it differs in the smoother-surfaced, more strongly coriaceous, uniformly entire-margined leaflets on a rather flat-topped rachis; the leaves have a variable number and

Sabiaceae

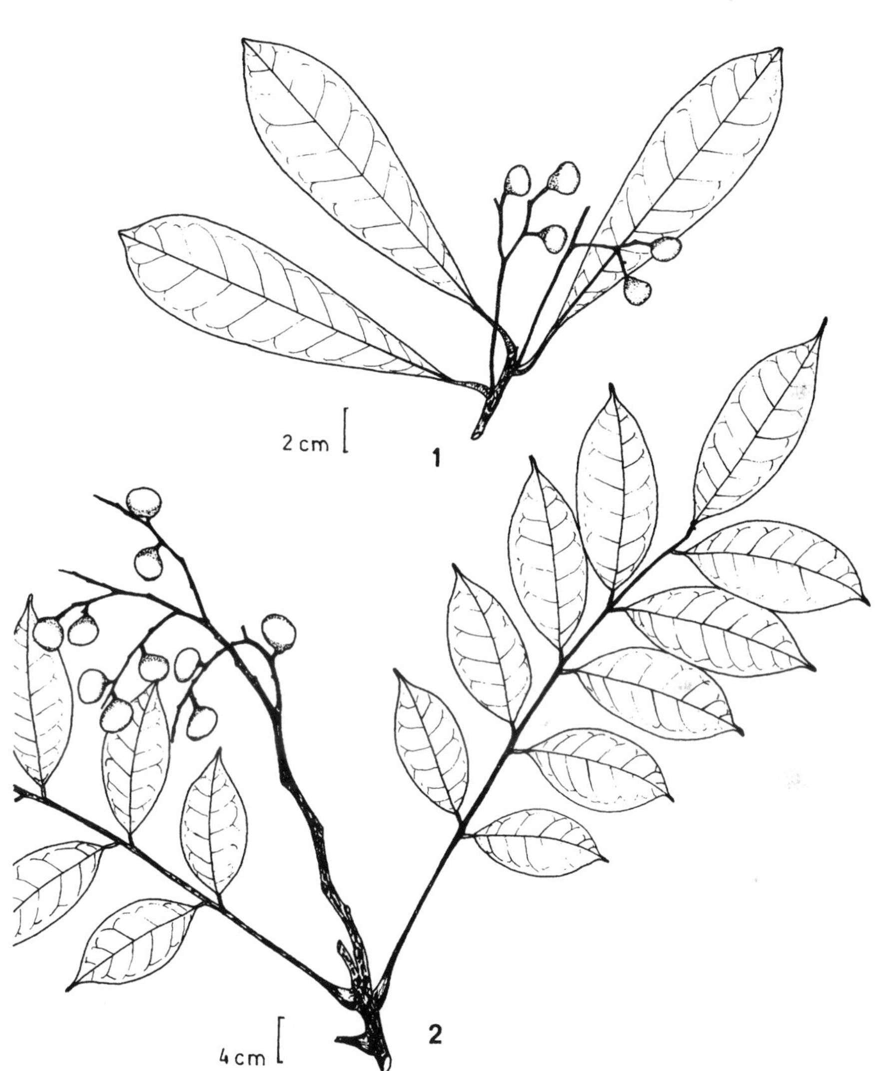

1 - *Meliosma*

2 - *Ophiocaryon*

arrangement of leaflets and can be even-pinnate and odd-pinnate (even in part simple) on the same plant. The tiny whitish or reddish flowers are borne in panicles and are noteworthy for having only two fertile stamens. The fruit is a very characteristic single-seeded, round or obovate, but distinctly asymmetric drupe with an extremely hard woody endocarp, usually black when fresh, often with a median keel.

Meliosma (50 spp., plus 15 in Asia) — Mostly montane. The petals are obtuse. The best differentiating feature from *Saurauia* is the swollen (typically subwoody) petiole base. Lacks the stellate hairs of *Clethra.* Entire-leaved species differ from Sapotaceae with similar woody petiole base in lacking latex.

Ophiocaryon (7 spp.) — Understory trees or treelets of lowland Amazonia. The very narrow petals are long acuminate. Leaflets more strongly coriaceous and with smoother surfaces than in similar Sapindaceae.

P: sacha uvos

SALICACEAE

The single South American species is very characteristic in the linear, finely serrate leaves with weakly developed secondary veins. It is mostly in the Andean region but descends to lowland Amazonia along the banks of white-water rivers where it can form pure, early-successional stands on sandbars. The inflorescence is a dense, spicate ament ca. 2–3 cm long, elongating in fruit, the fruits fusiform capsules splitting in half to release tiny seeds embedded in cotton.

Salix (19 spp., plus 400 or more in N. Temperate Zone)
P: sauce

SANTALACEAE

Poorly represented in our area by two genera of puna herbs and two small genera of trees, all with the coriaceous leaves having obscure or not visible secondary and tertiary veins (at least below) and drying distinctively olive (at least above). Twig pubescence, if present, more or less rufescent. One tree genus (*Acanthosyris*) spiny and occurring in lowland dry forest, the other (*Cervantesia*) a canopy tree of montane Andean forests, with characteristic oblong leaves (vegetatively reminiscent of a *Myrsine* but with the leaves more tomentose below and strongly discolorous).

Quinchamalium (25 spp.) — Puna herbs with thick taproot, linear +/- ericoid leaves and terminal cluster of sessile, narrowly tubular flowers.

Salicaceae and Santalaceae

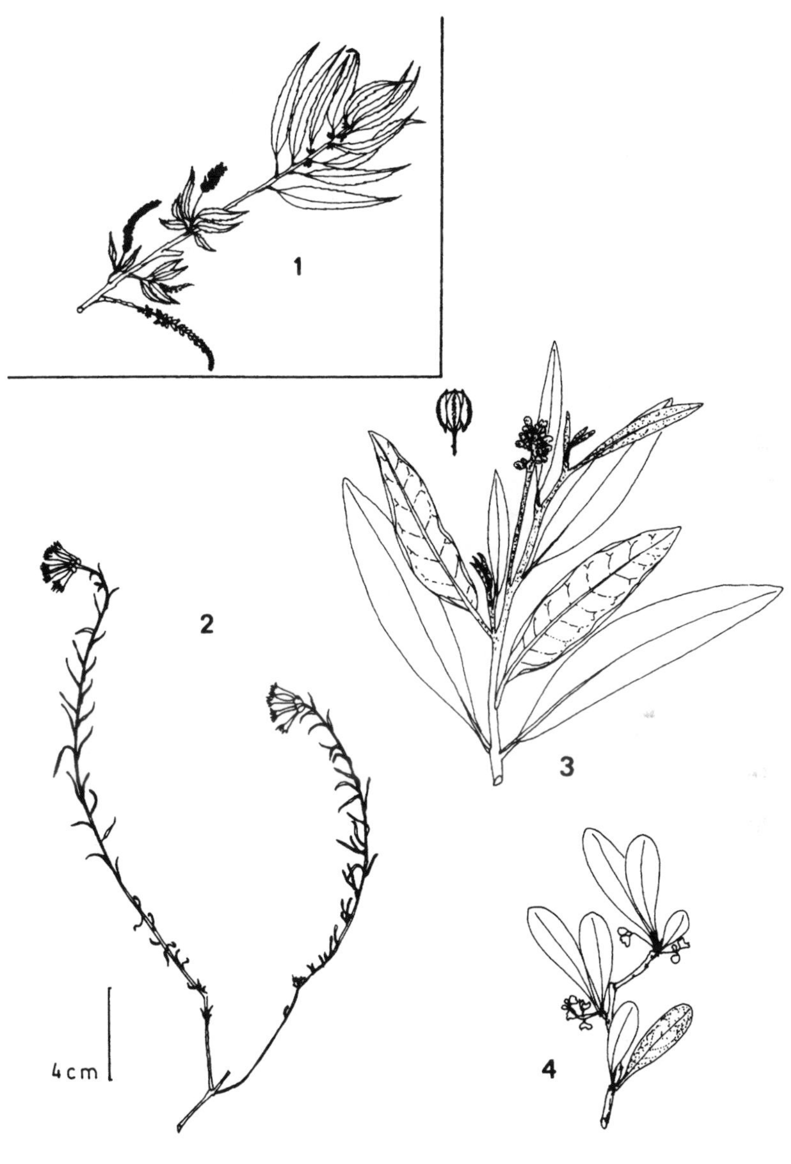

1 - *Salix* (Salicaceae)

2 - *Quinchamalium*

3 - *Cervantesia*

4 - *Acanthosyris*

Arjona (10 spp.) — Mostly Chilean, our species a high-puna herb of wet *Distichia* turberas. Like *Quinchamalium* but the ericoid leaves less linear and gradually broadened to clasping base.

Cervantesia (2 spp.) — Cloud-forest canopy trees. Leaves narrowly oblong, tomentose below and strongly discolorous, the midvein below, petiole, and twigs more or less rufescent, secondary and tertiary veins not evident below and immersed and inconspicuous above. Fruit round, ca. 1 cm in diameter, with the single large round seed covered by 5 valves that split apart incompletely from both base and apex.

P: olivo

Acanthosyris (3 spp.) — More or less spiny trees of dry thorn-scrub forest with coriaceous olive-drying entire leaves, usually narrow, and in part in short-shoot clusters. Our species (which may be an *Opilia!*) easy to confuse in vegetative condition with *Schoepfia* (Olacaceae) but differs in having the leaves mostly in short-shoot clusters.

(Extralimital monotypic *Jodina* is a Chaco element with utterly distinctive, thick-coriaceous, sharply diamond-shaped leaves with spiny apical and lateral points; there are several other genera in Chile and southern Argentina.)

SAPINDACEAE

A prominent component of the compound-leaved "rosid" families which are often very difficult to distinguish from each other when sterile. The leaves are always alternate except in Central American *Matayba apetala*. Although most genera are trees (and two high-altitude genera are shrubby), the majority of species are lianas. This is the second most important neotropical liana family and the lianas are almost always easy to recognize. Sapindaceae climbers have forking (noncoiling except the tips) tendrils that represent modified inflorescences (actual inflorescences are often borne laterally on the tendrils), a small amount of milky latex in the stems, and many of them have peculiar triangularly compound stems; their leaves, often with winged rachises, vary from odd-pinnate to 3-foliolate to variously biternate or bipinnate (one species is even palmately 5-foliolate). This is the only family with lianas with serrate leaves that are more complexly compound than palmately 3–5-foliolate and includes the only compound-leaved lianas with milky latex.

Sapindaceae trees can usually be recognized by the strong tendency to have the leaflets alternate (sometimes subopposite but almost never uniformly opposite) along the rachis and by the naked tip of the leaf rachis (resembling an inactive terminal growth tip of a *Guarea* leaf but usually subtended by only a single subterminal leaflet); even when most leaflets are essentially opposite, the "terminal" leaflet is typically paired with the aborted rachis apex. Another useful vegetative character is the tendency to serrate leaflets in several of the common genera (a very rare character in pinnately compound-leaved neotropical trees — see also Burseraceae, but most serrate-leafleted burseracs have milky latex, and tree sapindacs do not). Species with nonserrate leaflets often have strikingly asymmetrical leaflet bases, characteristically thickened woody petiolule bases, and/or prominently reticulate leaflet surfaces. A winged or margined leaf rachis with only subopposite leaflets is a combination unique to the common genus *Sapindus*. The three-foliolate genus *Allophylus* can usually be recognized by its cuneate, serrate, often pubescent, and/or membranaceous leaflets with typical close-together secondary veins (and which have well-developed petiolules as compared to sometimes similar *Anthodiscus*); a few species of *Allophylus* have simple leaves but their compound-leaved affinities are indicated by the flexed apex of the rather long petiole. The two simple-leaved Andean shrub genera of the family both have numerous close-together secondary veins and a more or less winged petiole; in *Dodonaea*, also characterized by whole plant being rather resinous, the narrow oblanceolate leaflets being long-cuneate onto the poorly demarcated petiole; but in *Llagunoa* the more or less truncate bases of the thickish, closely serrate, oblong-ovate leaves are clearly demarcated.

Sapindaceae flowers (in our area) are all small (mostly tiny) and often densely packed in spicate-racemose or paniculate inflorescences. Although superficially nondescript, with a lens many of them are remarkably complex, having floral asymmetry (on a microscale), and complicated patterns of staminodia, petal scales, and pubescence. The fruits are usually (2–)3-parted capsules with arillate seeds but are sometimes indehiscent and drupaceous (*Talisia, Melicoccus, Sapindus*), and may be variously winged with the carpels separating into 3 single-seeded samaras (*Serjania, Thinouia, Urvillea, Toulicia, Dodonaea)*.

1. Lianas with Tendrils — Six genera, each with a distinctive fruit but very similar flowers. Two genera — *Paullinia* and *Serjania* are very large and complex and, vegetatively, span most of the range of variation of the other vine genera. *Paullinia*, especially, may vary from 3-foliolate to triternate or multifoliolately ternate-pinnate. *Urvillea* and *Thinouia* are uniformly 3-foliolate. *Cardiospermum* is more herbaceous than the other genera and always multifoliolate.

Serjania (215 spp., incl. few in southwestern USA) — Fruit composed of 3 apically fused samaras; leaves typically divided into more leaflets than in *Paullinia*, usually at least triternate. Inflorescence usually racemiform, never umbellate.

Sapindaceae
(Lianas)

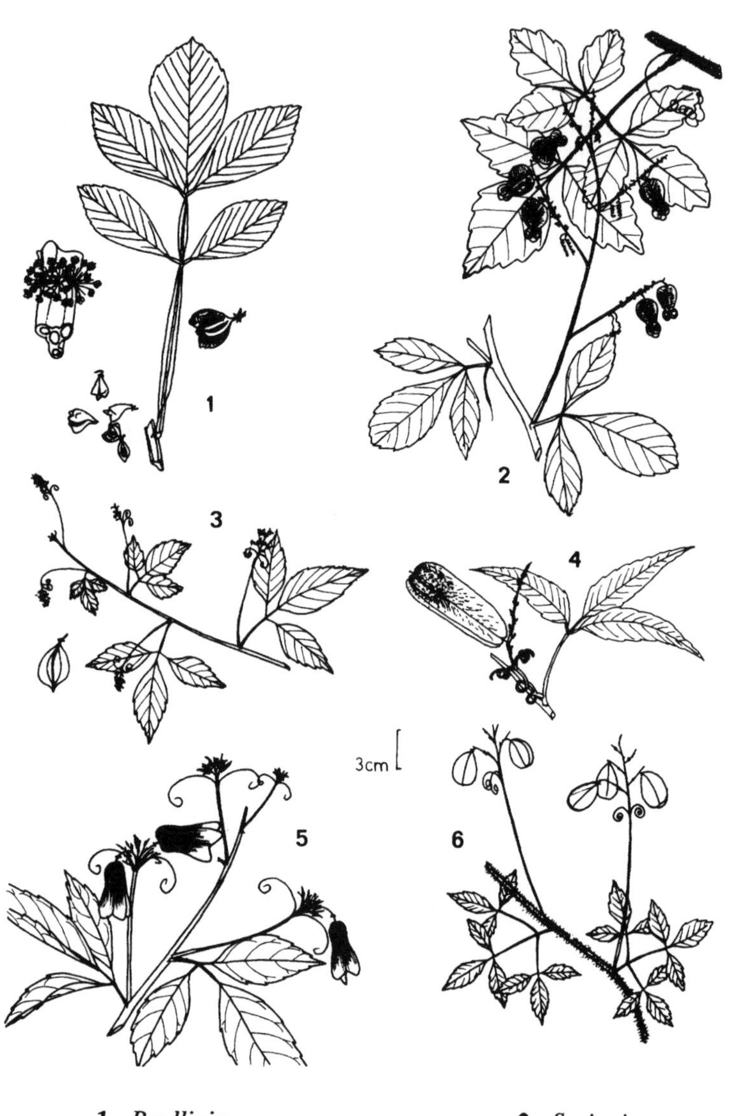

1 - *Paullinia* **2** - *Serjania*

3 - *Urvillea* **4** - *Lophostigma*

5 - *Thinouia* **6** - *Cardiospermum*

Lophostigma (2 spp.) — Essentially a 3-foliolate, Andean foothill *Serjania,* but the three oblong samaras fused along entire length.

Thinouia (12 spp.) — Fruit similar to that of *Serjania* but composed of 3 *basally* fused samaras; leaves uniformly 3-foliolate. Inflorescence umbellate.

Urvillea (1 sp.) — Fruit composed of 3 hemielliptic samaras fused along entire length; leaves 3-foliolate, more or less pubescent.

Cardiospermum (12 spp., incl. few in Old World) — Herbaceous vines (unique in family); leaves membranaceous and many-foliolate. Fruit a membranaceous, 3-celled, inflated, usually balloonlike capsule.

Paullinia (180 spp.) — Fruit a capsule with white-arillate black seeds. Leaves 3-foliolate or pinnate to triternate or ternate-pinnate. Often very difficult to distinguish from *Serjania* except in fruit; in flower the ovary is round and without the thinner, more or less winged, apex of *Serjania.* Simply pinnate leaves are common in *Paullinia* but never occur in other genera. *Paullinia,* rich in alkaloids, provides the guarana of Amazonian Brazil. Several about-to-be-proposed segregates should be rejected.

Allosanthus (1 sp.) — A very rare and poorly collected lowland forest liana known from Amazonian Peru (and perhaps Costa Rica?); vegetatively, with 3-foliolate leaves very similar to *Thinouia* but the inflorescence an axillary fascicle of spikes or spikelike racemes. Fruit unknown.

2. Trees or Shrubs

2A. Simple leaves; high-altitude shrubs or shrubby trees

Llagunoa (3 spp.) — Andes, mostly in drier areas. Leaves very distinctive, serrate, rather thick, the petiole well-defined, more or less margined and not at all flexed at apex. Inflorescence few-flowered and axillary, the wierd flowers with sepals fused into a conspicuous semicircle subtending rest of flower. Fruit a thin 3-lobed capsule. (A Chilean species has 3-foliolate leaves.)

Dodonaea (60 spp., mostly in Old World) — A common shrub of middle-elevation dry areas, characterized by the rather resinous nature of the whole plant. Leaves entire, long and narrow, long-cuneate onto the poorly differentiated petiole; very different from other Sapindaceae (and likely derived from the expanded petiole of a compound-leaved ancestor). Fruit with three membranaceous wings, eventually breaking into 3 separate samaras.

C: hayuelo

(Allophylus) — A few species of this usually 3-foliolate genus have simple leaves. They can be distinguished by the elliptic leaves with cuneate

Sapindaceae
(Trees: Simple, 3-Foliolate, and Bipinnate Leaves)

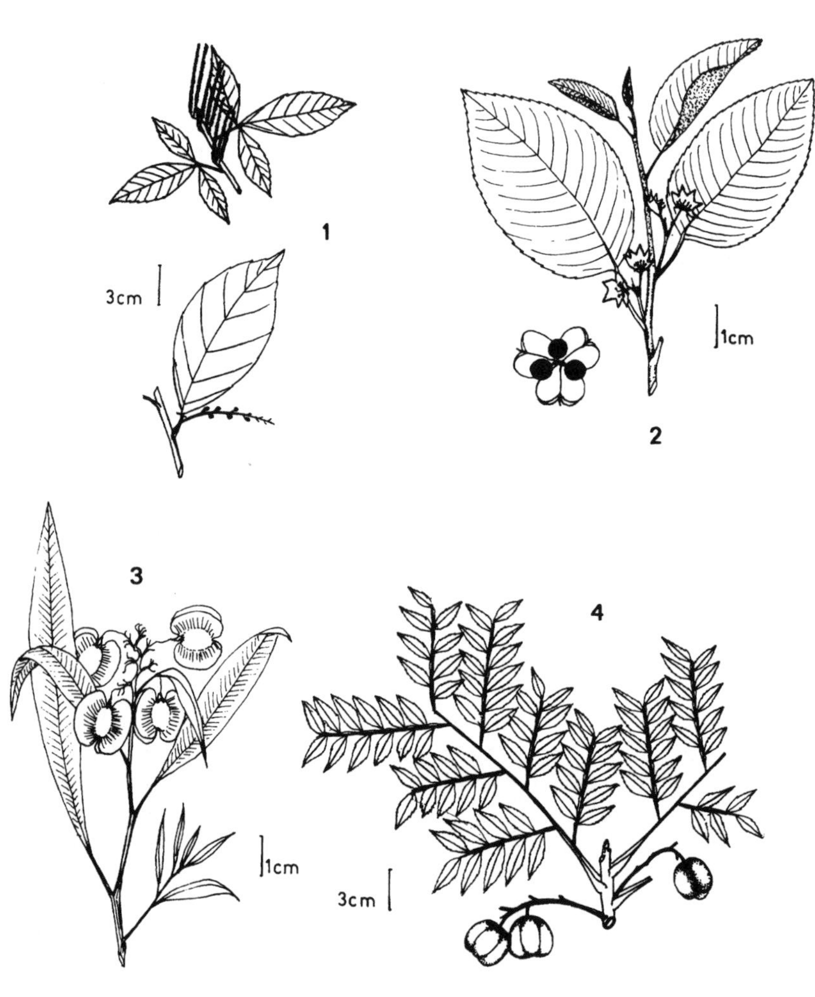

1 - *Allophylus*

2 - *Llagunoa*

3 - *Dodonaea*

4 - *Dilodendron*

bases and especially the noticeably flexed petiole apices, reflecting their compound-leaved affinities. The racemose inflorescence (these sometimes compounded into panicles) and ellipsoid indehiscent fruits are also characteristic.

2B. Bipinnate leaves

Dilodendron (incl. *Dipterodendron*) (3 spp.) — Very distinctive and very different from other genera of Sapindaceae in the bipinnate leaves; differs from bipinnate genera of other families in having very many small and conspicuously serrate leaflets. Blooms precociously while leafless.

2C. Trifoliolate leaves; mostly small trees and shrubs

Allophylus (taxonomy unclear, many spp., pantropical) — Very characteristic in the racemose inflorescence (sometimes compounded into a panicle-like structure) and the small ellipsoid indehiscent fruits. The leaflets are usually membranaceous or chartaceous, typically serrate or serrulate, and have close-together secondary veins and usually cuneate bases.

2D. Simply pinnate leaves — Usually trees and usually even-pinnate, typically with alternate leaflets, almost always terminated by the aborted rachis apex. The first three genera below (*Talisia, Melicoccus, Sapindus*) have indehiscent drupaceous fruits, *Toulicia* has dry 3-winged fruits, and the rest have 3-valved (*Cupania, Matayba*) or 2-valved (*Vouarana, Pseudima*) capsules with usually arillate seeds.

Talisia (50 spp.) — A large and rather variable genus. Frequently unbranched pachycaul treelets with the inflorescences more or less terminal and many-foliolate leaflets; sometimes larger trees and sometimes with fewer leaflets. Leaflet bases often strikingly asymmetric, the petiole (except the narrow apex) usually swollen and woody, leathery, often drying dark. Fruit ovoid to globose, one-seeded, not dehiscent.

C: mamón de mico; E: paragüita

Melicoccus (2 spp.) — The only species in our area (only in northern Colombia but widely cultivated elsewhere) has uniformly 4-foliolate leaves (that tend to have opposite leaflets), usually with subwinged or margined rachis (thus vegetatively, rather like a few-foliolate *Sapindus*). Fruits and flowers like *Talisia* but the anthers extrorse, the fruit typically more globose, fewer leaflets.

Sapindus (1 neotropical sp., plus 1 sp. north temperate and several in Old World) — Very characteristic in the leaves with a usually conspicuously winged rachis (unique in confamilial tree genera) and more or less alternate or subopposite leaflets. Somewhat resembles some *Inga* species vegetatively but lacks the glands between the leaflets. Fruit round, glossy-yellowish, with a reduced aborted carpel forming a distinctive basal bulge.

Sapindaceae
(Trees: Simple-Pinnate; Indehiscent Fruit)

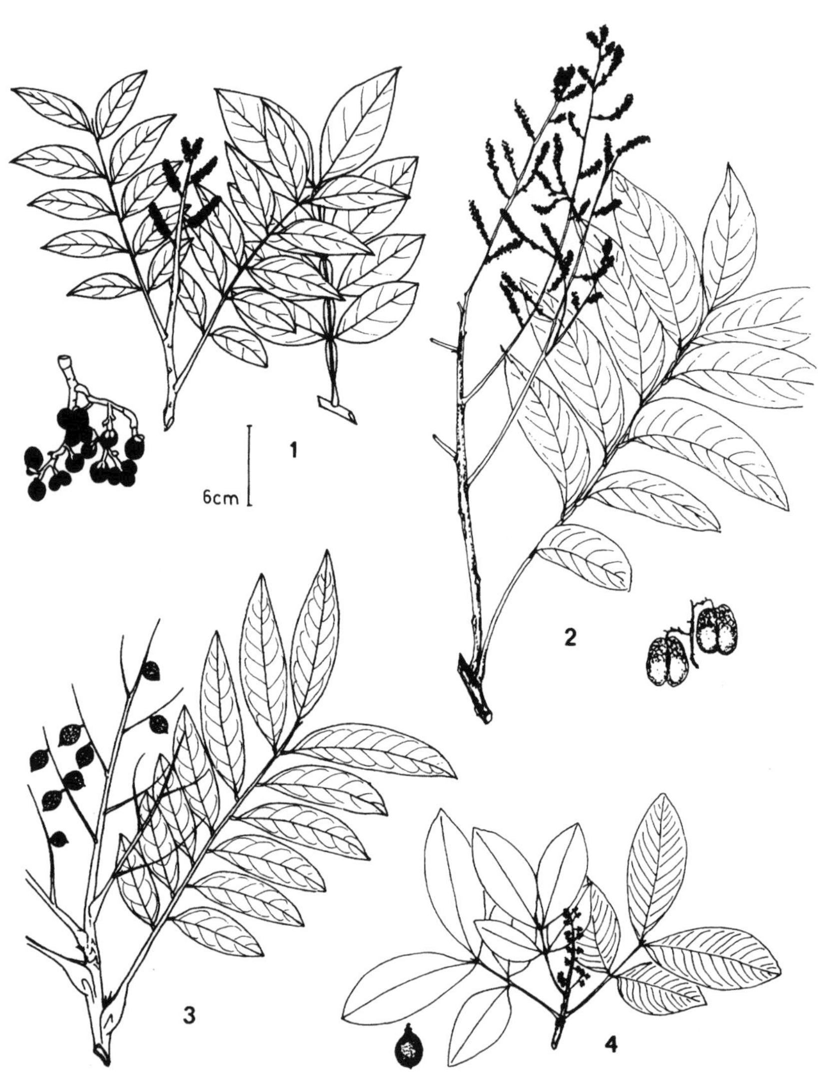

1 - *Sapindus* **2** - *Toulicia*

3 - *Talisia* **4** - *Melicoccus*

The fruits of *S. saponaria,* the commonest species, are rich in saponins and used as a soap substitute.

E: jaboncillo; P: choloque

Toulicia (14 spp.) — Trees with large multifoliolate pinnate-compound leaves like those of *Talisia,* but the fruits more similar to those of *Serjania,* 3-winged and fragmenting into 3 samaras. Petiole bases and branchlets usually with raised whitish lenticels.

Cupania (55 spp.) — Characterized by leaves with rather few usually conspicuously serrate (or at least serrulate), often noticeably pubescent leaflets. The main technical character for distinguishing it from closely *Matayba* is distinct (rather than fused) sepals. The capsule is 3-carpellate (or 2-carpellate by abortion) and easy to confuse with some species of *Trichilia.*

C: mestizo; E: huapina, sabroso

Matayba (50 spp.) — Very similar to *Cupania* from which it is technically separated by the united sepals. Can usually be differentiated by the entire more or less glabrous leaflets and the typically more stipitate fruit.

P: pinsha ñahui (= toucan leaf)

Vouarana (1 sp.) — Large tree similar to *Cupania* but with many leaflets (cf., *Talisia*) and the capsule 2-locular and rather flattened. Technically differentiated by the sepals petaloid rather than coriaceous and the seed exarillate. Known from the Guianas and Costa Rica, so probably also occurs in our area.

Pseudima (3 spp.) — Shrubs or small trees; inflorescence a terminal panicle. Capsule deeply 2-lobed with red woody valves, the seeds black with a white basal aril. Leaves multifoliolate like *Talisia* but with lepidote glands underneath; usually with raised whitish lenticels on petiole bases and branches; the leaflets with less asymmetric bases than in most other multifoliolate genera and always drying greenish.

Several extralimital genera, mostly trees with pinnate leaves, include several small technical segregates from *Matayba* and:

Diatenopteryx — Subequatorial dry areas, with bialate *Terminalia*-like fruits dehiscing into 2 samaras; essentially a 2-carpellate version of *Thouinidium* and perhaps not adequately separated.

Porocystis — With thin inflated fruits somewhat reminiscent of those of *Cardiospermum.*

Tripterodendron — Coastal Brazil, close to *Dilodendron,* but with 3-pinnate leaves with small evenly serrate leaflets.

Sapindaceae
(Trees: Simple-Pinnate; Dehiscent Fruit)

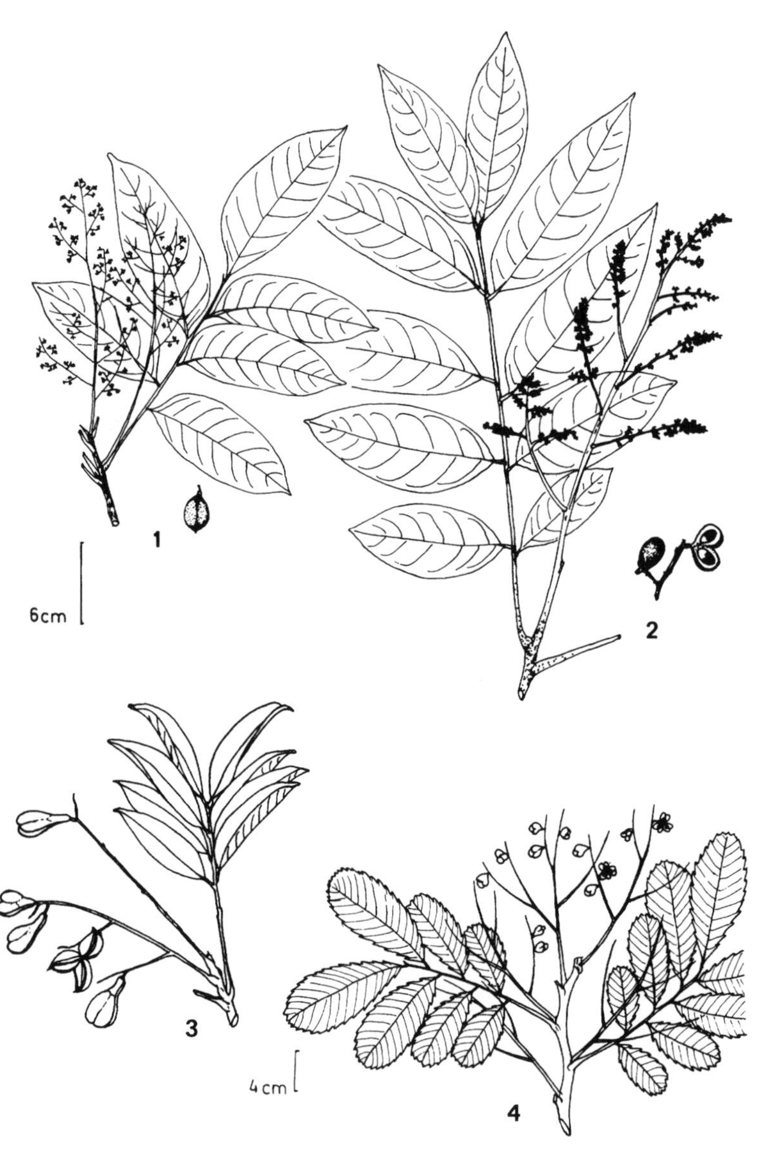

1 - *Vouarana* **2** - *Pseudima*

3 - *Matayba* **4** - *Cupania*

Magonia — A common cerrado tree, with large woody dehiscent fruits with winged seeds.

Averrhoideum — With small asymmetrically 2-parted fruits very like those of *Rourea* (Connaraceae) and conspicuously serrate leaflets.

Exothea — Central American and West Indian, with 2–4(–6)-foliolate leaves with mostly opposite leaflets and a fruit like *Talisia* but smaller.

Diplokeleba — Paraguay and adjacent areas, with smallish, long-petiolulate leaves and pyramidal inflorescence and dehiscent, rather elongate, narrowly ovoid fruit like a miniature *Cedrela,* the 3 subwoody valves falling to expose pendent, winged seeds resembling *Fraxinus* fruits.

Ungnadia — Mexico and southwest USA, with fruit like overgrown *Cupania* and leaves with terminal leaflet and short-petiolulate serrate leaflets.

Thouinia — Central American and West Indian, vegetatively 3-foliolate like *Allophylus,* but fruits with elongate lateral wings and splitting into three samaras.

Thouinidium — West Indies and Central American dry areas, with fruits like *Thouinia* but pinnate leaves with narrow inconspicuously serrate leaflets.

Bridgesia — Chilean, with simple, deeply serrate, basally somewhat 3-lobed leaves.

Valenzuelia— Chilean, with simple opposite *Buxus*-like leaves.

Sapotaceae

A diverse and ecologically important family of lowland forest trees, easily recognized by the combination of milky latex and alternate (often spiral) leaves either with the lower part of petiole thickened (= pop-bottle-shaped petiole) or with conspicuous +/- parallel intersecondary veins (venation even becoming *Clusia*-like in *Micropholis, Diploon,* most *Manilkara* and some *Chrysophyllum*). The latex usually comes out in individual droplets from the trunk slash and may be visible only in the petiole (or sometimes not at all during strong droughts). Another useful vegetative character is the 2-branched malpighiaceous hairs, typically appressed and usually with one arm shorter than the other, which often give a sericeous aspect to twigs, petiole and leaf undersides; these are usually persistent at least on the petiole. The leaves are always entire and tend to have parallel, rather close-together secondary veins with the tertiary veins often also parallel, and oriented either parallel to the secondaries or perpendicular to the midvein or obliquely perpendicular to the secondaries. Two genera (*Ecclinusa* and *Chromolucuma* with very characteristic large coriaceous leaves and parallel tertiary veins obliquely perpendicular to the secondaries) have stipules. The usually species-specific bark is highly variable between species and genera, but is most commonly (especially in *Pouteria*) reddish and fissured, scaling, or loosely fibrous with the base of the trunk more or less fluted. The commonest *Pradosia* has smooth, mottled-insculpted, greenish and whitish bark. *Manilkara* always has deeply vertically ridged bark in mature trees.

Sapotaceae flowers and fruits are remarkably homogeneous. The rather small greenish to whitish or tannish (dark red in *Pradosia*) flowers are always in axillary or ramiflorous fascicles with the petals fused into a short tube at base; they are open (rotate) in *Manilkara, Diploon, Elaeoluma,* and *Pradosia,* usually more or less urceolate in other genera. The corolla lobes are trifid in *Manilkara* and many *Sideroxylon.* The 4–6(–12) stamens are fused to the corolla opposite the lobes, usually alternating with staminodia in the sinuses; staminodes are lacking in *Chrysophyllum, Ecclinusa, Pradosia, Diploon, Elaeoluma* (usually), and a few *Pouteria.* The fleshy fruits, always indehiscent, are usually round and borne individually on short pedicels (or sessile) in leaf axils or along the branches. The seeds are especially distinctive with a shiny dark brown surface contrasting with a large conspicuous light-colored scar which may be basal (*Manilkara, Sideroxylon,* and *Diploon*) or occupy one whole side of the seed. There are two main seed types, laterally compressed with narrow scar and foliaceous cotyledons (*Chrysophyllum, Sarcaulus, Manilkara,* very few white-sand *Pouteria*) and ellipsoid with broad scar and thick cotyledons (most genera).

The generic taxonomy of Sapotaceae has been very chaotic with one specialist recognizing many segregate genera based mostly on floral characters and another basing an entirely nonoverlapping set of segregate genera mostly on fruit characters. The current specialist follows the more traditional broad generic groupings with about half of the neotropical species assigned to a vegetatively rather heterogeneous *Pouteria.* The genera can be conveniently arranged according to whether they have the typical swollen petiole base and/or narrow neck (last five genera) or not (first six genera), except that in *Manilkara* and *Pradosia* this is variable.

Sapotaceae
(Petiole Base Usually Not Enlarged; Spiny or Clustered Leaves)

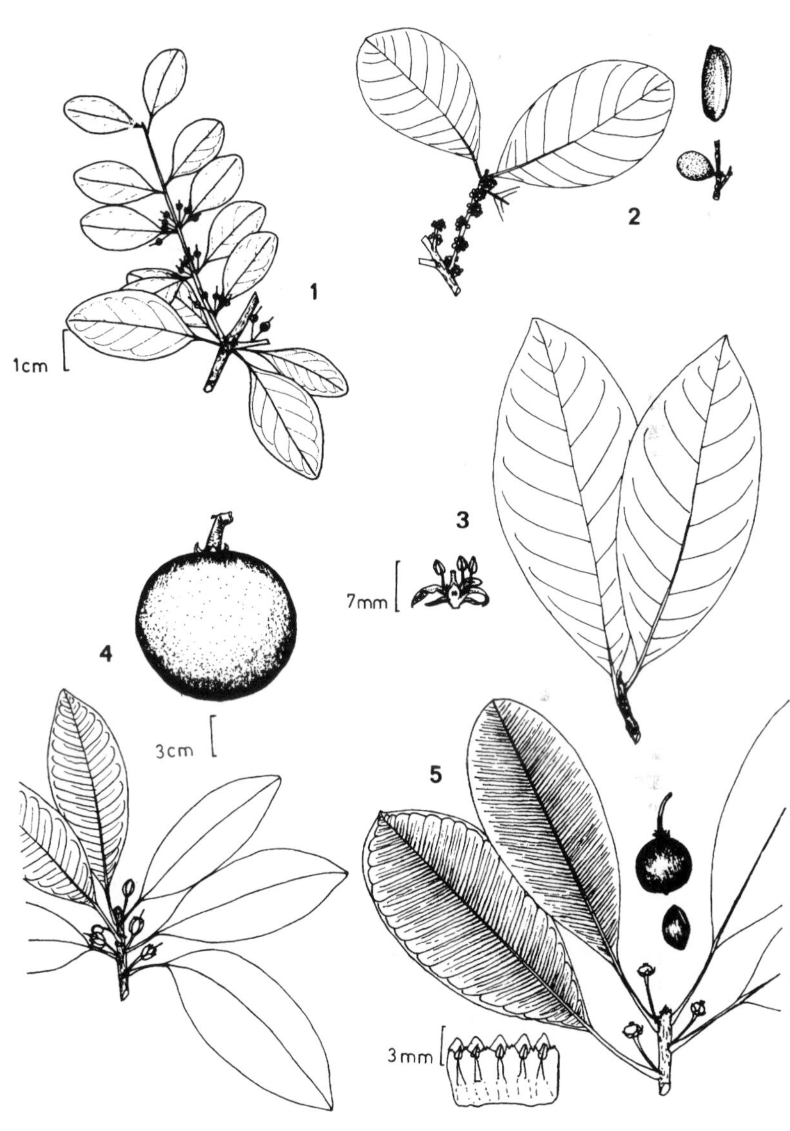

1 - *Sideroxylon*

2 - *Pradosia*

3 - *Elaeoluma*

4 - *Manilkara (M. zapota)*

5 - *Manilkara*

Sapotaceae
(Petiole Base Usually Not Enlarged; 2-Ranked Leaves with Finely Parallel Tertiary Venation)

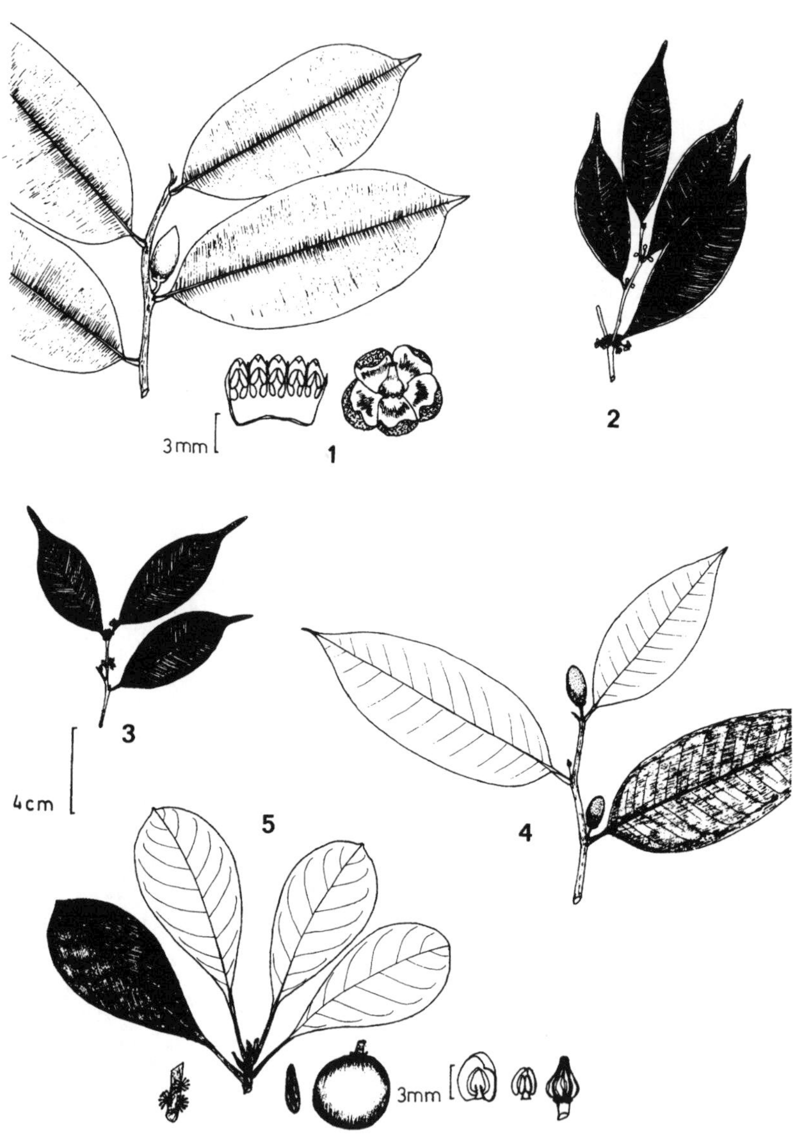

1 - *Micropholis* (*M.sanctae-rosae*) **2** - *Diploon*

3 - *Micropholis* (*M.venosa*) **4** - *Chrysophyllum* (*C. argenteum*)

5 - *Chrysophyllum* (*C. prieurei*)

1. Petiole Bases Usually Not Noticeably Enlarged; Leaves Usually Clustered at Branch Apices and/or with Conspicuously Parallel Intersecondary and Tertiary Veins — Stipules absent (except sometimes in *Manilkara*)

Manilkara (30 spp., plus 32 Old World) — Canopy and emergent trees vegetatively characterized by spirally arranged leaves clustered at branch apices and usually with *Clusia*-like venation. Unusual among taxa with parallel tertiary/intersecondary veins in the petiole base sometimes slightly swollen. Shoot apex often with varnishlike substance; small stipules sometimes present (unique except very different *Ecclinusa*). Flower very distinctive, the unique calyx of 2 whorls of usually 3 sepals each, the outer valvate (unique), the open corolla with lobes 3-parted to near base (except *M. chicle* of northern Colombia dry area), sometimes the lateral lobes again divided to give up to 30 segments. Seed laterally compressed with *Chrysophyllum*-like embryo having foliaceous cotyledons, differing from *Chrysophyllum* in basiventral scar.

C: trapichero; P: quinilla

Sideroxylon (incl. *Bumelia*) (50 spp., plus 20 in Africa) — Ours small shrubby dry-forest trees with spine-tipped branches (unique). Leaves small, like reduced *Chrysophyllum.* Petals usually 3-parted as in *Manilkara* but lateral lobes smaller; basiventral seed scar is also like *Manilkara* but the calyx is very different with 5 imbricate sepals. Fruit 1(–2) seeded, with seeds having either embryo type (cotyledons either foliaceous or thick and fleshy).

Micropholis (38 spp.) — Lowland forest canopy trees with very characteristic *Clusia*-like venation, the secondary veins reduced and closely paralleled by the tertiary and intersecondary venation, differing from *Manilkara* in the distichous leaf arrangement. Flowers like *Pouteria* with as many staminodes as corolla lobes; seeds of *Chrysophyllum*-type.

P: balata

Diploon (1 sp.) — A monotypic genus (*D. cuspidatum*) widespread in Amazonia. The leaves have *Clusia*-like venation and are very similar to *Micropholis* but thinner and with a stronger marginal vein and unique blackish-lined venation when dried. The technical characters are the unique unilocular ovary with 2 ovules and the basal seed scar.

Chrysophyllum (43 spp.) — The second most important Sapotaceae genus, often confused with *Pouteria* but differing in seeds with foliaceous cotyledons and copious endosperm and in completely lacking staminodia. A few mostly Guayana-area *Pouteria* species (only two of which reach our area, on white sand) have a *Chrysophyllum*-type seed rather than the typical *Pouteria* one. The leaves usually have more evenly parallel tertiary venation than *Pouteria,* unlike that genus sometimes with the tertiary venation perpendicular to the midvein and sometimes with intersecondaries; species with tertiary venation perpendicular to the secondary veins cannot be reliably distinguished from *Pouteria* but some have distichous leaves unlike any *Pouteria.*

C: caimito; E: balata; P: quinilla, caimito (*C. caimito*)

Pradosia (23 spp.) — Large trees, especially in dry forests where typically with smooth insculpted bark. The leaves are unusual in being often opposite or whorled and in having the midrib and usually the secondaries impressed on upper surface. Otherwise most species vegetatively essentially like *Chrysophyllum* with the tertiary venation either more or less parallel (and perpendicular to midvein or oblique to secondaries) or with intersecondaries present; species with tertiary venation oblique to secondaries that could be confused with *Pouteria* have shorter, better-defined petioles than in similar species of that genus; *P. verticillata* looks like *Ecclinusa* except for the more or less opposite leaves and lack of stipules. Petiole base variable but usually somewhat swollen at least in species with large leaves. The corolla is open (rotate) and staminodes are absent. In flower, differing from *Elaeoluma* in being cauliflorous or ramiflorous rather than having axillary flowers; in fruit, by having a drupe instead of a berry.

2. Petiole Base Usually Conspicuously Enlarged and Apex Constricted (Pop-Bottle-Shaped); Leaves Spirally or Distichously Arranged, Never with *Clusia*-Type Venation — Intersecondaries and strongly parallel tertiary venation relatively rare except when stipules present.

2A. The next two genera are unique in having stipules.

Ecclinusa (11 spp.) — Trees of poor-soil areas, vegetatively distinctive in having stipules which leave conspicuous scars after falling. The leaves, like those of some white-sand *Chrysophyllum* species, are always very coriaceous, usually (except one tiny-leaved shrub of white-sand savannas) rather large, and with conspicuously closely parallel tertiary venation perpendicular to the secondary veins. The flowers are always sessile and lack staminodes.

Chromolucuma (2 spp.) — In our area, only known from Magdalena Valley. Like *Ecclinusa* in having large coriaceous leaves with stipules but differing in pedicellate flowers and the leaves having a short incomplete intersecondary, as well as the parallel tertiary veins, oblique to the secondaries.

Sapotaceae
(Petiole Base Enlarged)

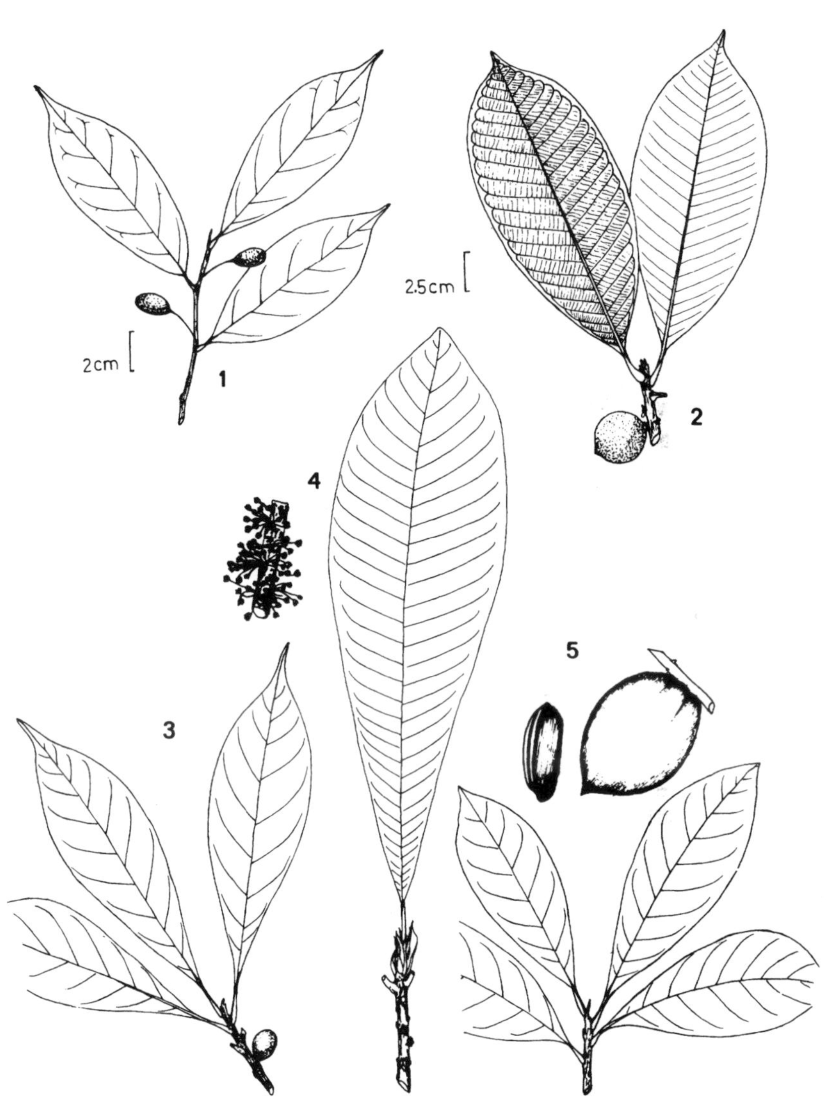

1 - *Sarcaulus*　　**2** - *Ecclinusa*

4 - *Chromolucuma*

3 - *Pouteria*　　**5** - *Pouteria*

2B. Stipules absent

Pouteria (188 spp., plus 150 in Australasia) — The main genus of Sapotaceae and, as thus circumscribed, highly polymorphic. All species have spirally arranged leaves and the characteristic pop-bottle-shaped petiole; most have reticulate tertiary venation. The few species with intersecondary veins (*P. cuspidata, P. eugeniifolia*) have smooth narrow leaves with the secondaries reduced and plane below; those with parallel tertiary veins perpendicular to the midvein (e.g., *P. baehniana, P. buenaventurensis, P. subrotata*) are distinctively broad and coriaceous. The usually urceolate flowers are characterized by having as many staminodes as corolla lobes (the small section with open rotate flowers has large leaves with obliquely parallel tertiary veins). Except one small group (*P. laevigata* and *P. oblanceolata* in our area) the more or less ellipsoid seeds have broad seed scars, large thick cotyledons and lack endosperm, unlike *Chrysophyllum.*

C: caimito, caimito silvador, caimito trapichero; E: zapote de monte, zapote silvestre, guapapango (*P. gigantea*), cauje (*P. caimito*), mamey colorado (*P. sapota*); P: quinilla, caimito

Sarcaulus (6 spp.) — Mostly Amazonian lowland trees, closely related to *Pouteria.* Differing vegetatively from *Pouteria* mainly in the distichous leaves, smaller and narrower than in most other Sapotaceae and unlike other small-leaved taxa in the relatively few strongly brochidodromous secondary veins and prominulous reticulate tertiary venation below with the undersurface drying a distinctive, slightly pinkish color. Also characterized by thick fleshy corolla and staminodes and the unique stamens which are inflexed against style.

Elaeoluma (4 spp.) — Trees and shrubs mostly of the Guayana Shield area, our only species (*E. glabrescens*) restricted to black-water-inundated forest. The narrow, extremely coriaceous, spirally arranged leaves are punctate (unique) and also distinctive in the reduced plane secondary veins (but without *Clusia*-like venation). The distinctive fruit is a 1-seeded berry ca. 2 cm long; the open flowers lack staminodes.

Saxifragaceae

Only five rather divergent genera of this overwhelmingly Laurasian family are represented in our area, three (*Saxifraga, Ribes, Escallonia*) restricted to the high Andes, with the other two (*Hydrangea, Phyllonoma*) mostly occurring in middle-elevation cloud forests. Our genera include one of herbs (*Saxifraga*), one of shrubs (*Ribes*), one of lianas (*Hydrangea*), and two of trees (to shrubs)(*Escallonia, Phyllonoma*). They are placed in three different families by Cronquist and some authors assign all five to different families. The family is relatively unspecialized in floral characters with (4)–5 free petals and sepals, usually twice as many stamens, and several only basally united pistils, differing from Rosaceae in rather weak characters such as usually estipulate leaves and more abundant endosperm.

Given the above taxonomic preamble, it is most expedient to recognize each genus separately rather than worrying about family characters. *Hydrangea* has large opposite, often serrate or rufescent-pilose leaves with expanded petiole bases connected across the nodes to form a prominent interpetiolar line. *Ribes* has distinctive more or less palmately veined doubly serrate leaves, mostly born on short-shoots. *Escallonia* is large and variable but apparently always has the alternate leaves finely serrate (sometimes this hardly visible to naked eye), usually more or less resinous (often gland-dotted), and the undersurface with weakly demarcated secondary veins and inconspicuous immersed tertiaries making a characteristic faint often blackish tracery on leaf undersurface. *Phyllonoma* is vegetatively distinct in the rather small, narrowly ovate to narrowly elliptic leaves with a very long caudate acumen. Even the herb genus *Saxifraga* is distinct: a cushion-plant with deeply narrowly 3–(5)-lobed leaf apices.

Saxifraga (ca. 2 spp. in Andes, ca. 370 n. temperate) — A cushion-forming more or less glandular-pilose herb occurring on moist rocks and canyon walls of the highest Andes (mostly above 3500 m). Leaves very characteristic, broadened apically and deeply narrowly 3-lobed or 5-lobed.

Hydrangea (80 spp., incl. Old World) — Large, more or less hemi-epiphytic liana of lowland and middle-elevation cloud forest; vegetatively characterized by opposite, large, usually more or less serrate, sometimes rufous-pilose leaves with expanded petiole bases connecting across node in prominent line (and sometimes with ochrea-like area below this). Species with entire glabrous leaves have distinctive chambered domatia in axils of lateral nerves below. The terminal umbellate-cymose flat-topped, multiflowered inflorescence is also distinctive. The individual flowers are very small, but massed they are showy and there are frequently some much larger, brightly colored, sterile flowers around the periphery of inflorescence. The fruit is a small, cup-shaped, flat-topped capsule.

Ribes (150 spp., incl. n. temperate) — Shrubs of high-Andean forest and paramo. Leaves usually palmately veined and +/- 3-lobed, also more finely serrate, mostly borne on bracteate short-shoots (a few paramo species have cuneate bases very indistinctly 3-veined well above base). Inflorescence racemose (sometimes rather reduced), often conspicuously bracteate, the flowers small, white to pink. Fruit a berry.

Saxifragaceae

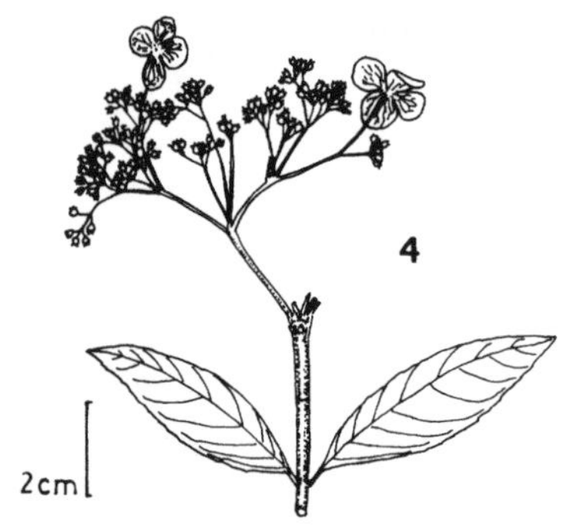

1 - *Escallonia* **2** - *Ribes* **3** - *Saxifraga*

4 - *Hydrangea* **5** - *Phyllonoma*

Scrophulariaceae

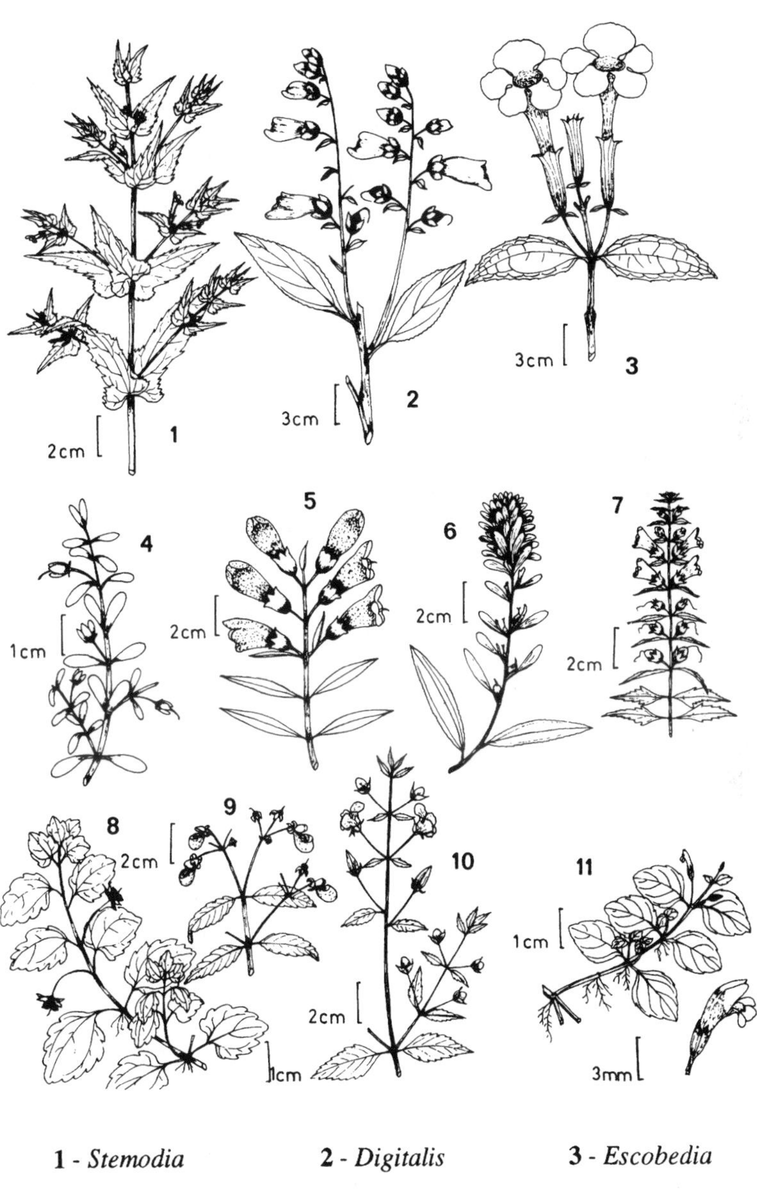

1 - *Stemodia* **2** - *Digitalis* **3** - *Escobedia*

4 - *Bacopa* **5** - *Agalinis* **6** - *Castilleja* **7** - *Lamourouxia*

8 - *Veronica* **9** - *Calceolaria* **10** - *Alonsoa* **11** - *Lindernia*

Escallonia (60 spp.) — Trees or shrubs; one of the most ecologically important arborescent genera of high-Andean forests, often dominant in the forests just below the *Polylepis* zone. Twigs often conspicuously irregularly angled and/or with exfoliating bark; some species (especially the small-leaved ones) with conspicuous short-shoots. Leaves alternate, always +/- finely serrate (not always readily apparent to naked eye), sometimes gland-dotted below and often more or less resinous, the cuneate decurrent leaf base usually not well differentiated from petiole apex. Leaf undersurface distinctive, the poorly demarcated secondary veins and immersed tertiary venation together making a characteristic macroscopically visible fine tracing. Inflorescence usually racemose or more or less paniculate, sometimes variously reduced even to single flowers, the flower with flat-topped inferior ovary, the 5 strap-shaped white petals arising from near its rim. Fruit a capsule with the ovary apex splitting.

C: tobo, tibar

Phyllonoma (8 spp.) — Small, middle-elevation, cloud-forest trees, the alternate leaves smallish, narrowly ovate to narrowly elliptic, very distinct in the very long caudate acumen. Margin serrate or not, the twigs often somewhat zigzag when leaves entire (*P. ruscifolia*); secondary veins making nearly 90 degree angle with midvein. The unbelievable inflorescence consists of a few small greenish 5-parted flowers arising from leaf midvein on upper surface near apex of leaf! Fruit a small berry.

SCROPHULARIACEAE

A nearly completely herbaceous family with tubular bilabiate flowers, mostly 4 (sometimes 2) stamens, mostly opposite, usually serrate, leaves, and often 4-angled stems. Essentially the herbaceous counterpart of the Bignoniaceae. Only *Basistemon* and *Galvezia* are distinctly shrubby. *Basistemon,* with bilabiate bluish flowers borne solitary or in axillary clusters, has finely serrate subsessile leaves, tetragonal stems with winged angles, and usually slender spine-tipped branches (dry-area species which resemble *Duranta* [Verbenaceae]); one Peruvian species is a forest-understory shrub to 2 m tall with much larger leaves distinctive in being borne on raised woody nodes. *Galvezia* occurs in the dry coastal region of Peru and southern Ecuador, and is a hummingbird-pollinated shrub ca. 1 m tall with small, entire, ovate, opposite leaves rather densely borne along the branches, and tubular red flowers ca. 2 cm long. *Ourisia,* the high-altitude puna/paramo equivalent of *Galvezia* with small tubular mostly red (white in *O. muscosa*) flowers, is prostrate and mat-forming but may be +/- woody at base. In addition to the numerous genera of erect or prostrate herbs, there are several reduced +/- aquatic genera (*Limosella, Micranthemum,* some *Bacopa*) and introduced *Maurandya* is a vine.

Of the herbs, at least several ubiquitous, highly speciose, Andean genera deserve mention: *Calceolaria* (with characteristic slipper-shaped yellow flowers), *Bartsia,* (paramo herbs with sessile narrow strongly bullate leaves and usually pink or purplish flowers in dense raceme), *Alonsoa* (4-angled stems and orange flowers, the corolla tube flat and strongly zygomorphic with "upper" lobe much larger), *Escobedia* (large scabrous leaves and long white salverform corolla), *Leucocarpus* (erect herb with sessile cordate-clasping lanceolate leaves and a contracted axillary inflorescence conspicuous in fruit from the white globose berries), and *Capraria* and *Scoparia* (common erect low-elevation weeds with small short-tubed flowers, the former (frequently with 5 stamens) with alternate, the latter with whorled, leaves). For an excellent treatment of this family see *Flora of Ecuador 21.*

SIMAROUBACEAE

Trees with alternate pinnately compound (except *Castela* and *Suriana*) leaves, usually characterized by bitter-tasting bark. Vegetatively or in flower, most likely to be confused with Sapindaceae, especially in the species with alternate leaflet arrangement, but almost all species have terminal leaflets and all lack the aborted terminal rachis-projection of that family. The coriaceous, +/- oblong leaflets with smooth surfaces and barely prominulous secondary venation that characterize several simaroub genera are unlike Sapindaceae, where coriaceous leaflets have more strongly raised secondary venation and/or intricately reticulate tertiary venation, but might be confused with *Ophiocaryon* (Sabiaceae). The flowers of simaroubs are usually tiny and inconspicuous, when larger (*Simaba, Quassia*) they have narrow strap-shaped petals. In our area, the fruits (except *Picramnia* and anomalous *Suriana*) are apocarpous single-seeded drupes (occasionally dry or flattened) sometimes on a swollen receptacle. *Picramnia* has legumelike cylindrical pulvinuli but lacks the basal pulvinus of that family. Vegetatively anomalous genera in our area, all shrubs or treelets, include *Picrolemma,* with hollow, ant-housing twigs and very thin leaflets, *Castela,* a thorny shrub with tiny simple leaves, *Suriana,* a beach shrub with terminal clusters of narrow, grayish leaves, and *Quassia,* with conspicuously winged rachis and petiole.

Simaroubaceae

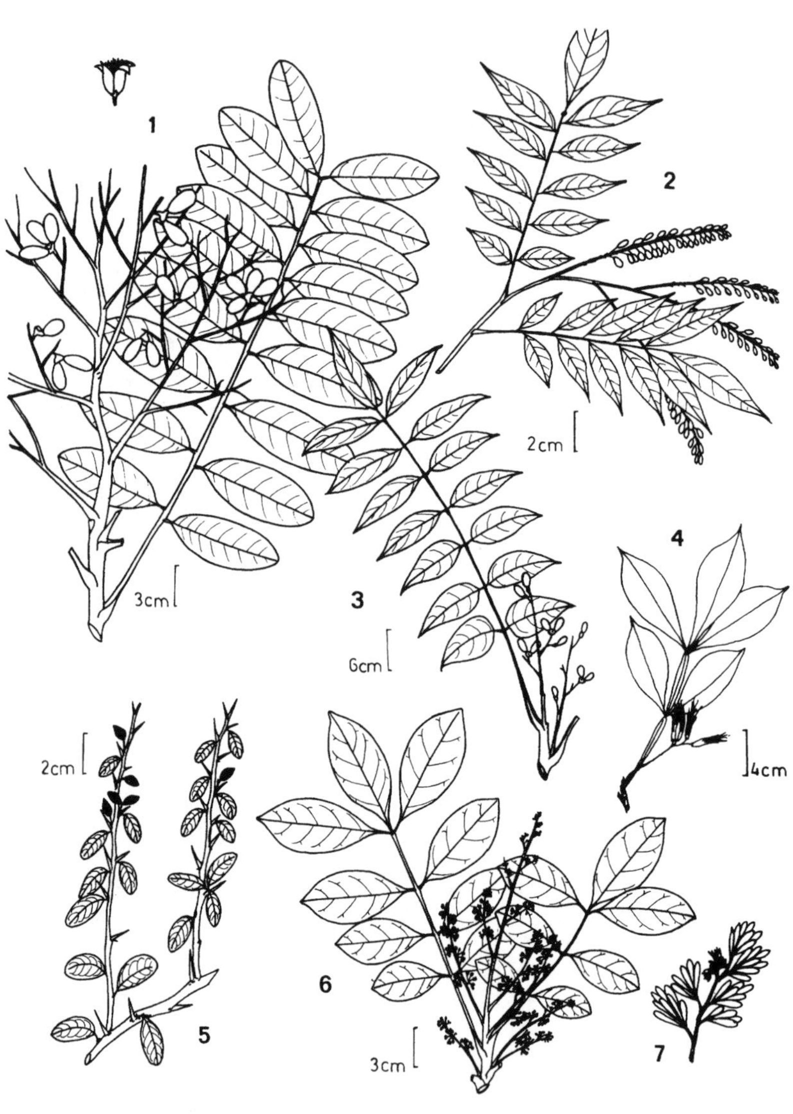

1 - *Simarouba* **2** - *Picramnia*

3 - *Picrolemma* **4** - *Quassia*

5 - *Castela* **6** - *Simaba* **7** - *Suriana*

1. Shrubs with Simple Leaves

Suriana (1 sp.) — A mostly Antillean shrub of sandy beaches, barely reaching the Caribbean coast of Colombia. Easily recognized by the succulent, narrowly oblanceolate, grayish leaves clustered at the tips of the densely arranged branches. Anomalous in the family and sometimes segregated as Surianaceae.

Castela (12 spp.) (incl. leafless Mexican and southwest USA *Holacantha*) — Mostly Central American and Antillean. A characteristic spiny, tiny-leaved shrub of dry thorn-scrub vegetation, the few millimeters long, oblong leaves borne 2–3 per node in short-shoot clusters, usually subtended by thorns. Flowers and fruits orange to red-orange and in reduced axillary inflorescences. Distinctly bitter taste.

C: jaruwa

2. Trees or Treelets with Pinnately Compound Leaves

Picramnia (40 spp.) — Shrubs to small (rarely medium-sized) trees, very characteristic vegetatively in the alternate asymmetrically ovate (often +/- rhombic and frequently dark-drying) leaflets with cylindrical legumelike pulvinuli; more easily confusable with legumes than with other simaroubacs, but differing from former in lack of vegetative odor and the combination of oblique shape and conspicuously alternate arrangement of the leaflets, usually progressively smaller toward base. Frequently lacks bitter taste. Leaflets sometimes conspicuously puberulous (unusual in our members of family). Inflorescence very characteristic, a pendent spike or raceme of inconspicuous unisexual flowers, producing small ellipsoid 1(–2)-seeded red, orange, or black berries.

E: cafetillo

Simarouba (5 spp.) — Canopy trees characterized by oblong, round-tipped, very coriaceous alternate leaflets with the secondary venation immersed, barely visible, not differentiated from intersecondaries, numerous and parallel, nearly perpendicular to midvein. Other unusual features are a frequently reddish rachis and often revolute margin. Only slightly bitter.

C: garzo, purga; E: guitaro, cedro blanco; P: marupá

Simaba (24 spp.) — Trees or treelets related to *Simarouba*, but differing in perfect flowers, pubescent petals and fused stigmas. Differs from Sapindaceae in coriaceous texture with merely prominulous secondary veins. The commonest species, *S. cedron*, is a small +/- pachycaul tree with multifoliolate leaves bearing alternate, oblong, subsessile, glandular-tipped subcoriaceous leaflets with brochidodromous venation; other species have fewer, less oblong leaflets with fewer, more ascending immersed secondary veins. The fruit of *S. cedron* is ellipsoid and much larger (5–8 cm long) than in other area taxa (1 cm long in *S. guianensis*, flattened in *S. orinocensis*), and the greenish flowers are 2–3 cm long. Only slightly bitter.

C: amargo; P: marupá

Picrolemma (2 spp.) (incl. *Cedronia*) — Understory Amazonian treelet or small tree, very distinctive in thin-textured, +/- crenate, acuminate, dark-drying leaflets and in hollow ant-containing stems. Flowers in erect, open, sparsely branched terminal panicles, orange (or cream) and ca. 5 mm long. Fruit over 2 cm long, conspicuously apocarpous, typically bright orange. Very bitter taste.

C: cojón de toro

Quassia (1 sp., plus 2 in Old World) — In our area, only in northern Colombia. Shrub or treelet very characteristic in the 5-foliolate leaves with strongly winged rachis. Similar to *Sapindus saponaria* in the winged rachis but that species lacks a terminal leaflet. Large red flowers with narrow petals ca. 3 cm long are very distinctive as is the swollen fruiting receptacle (cf., *Ouratea*). Distinctly bitter taste.

Alvaradoa (5 spp.) — Perhaps not in our area; mostly Antillean and Central American but disjunct to northwestern Argentina and Bolivia. Leaves legumelike with thin, oblong, round-tipped alternate leaflets, rather small and widely spaced; pulvinuli as in legumes but petiole base not pulvinate. Fruit a very distinctive flat, narrowly ovoid, villous, wind-dispersed samara, arranged in dense raceme.

Extralimital genera include simple-leaved ***Recchia*** (Mexican) and ***Picrasma*** (8 spp., incl. 2 in Asia), a dry-area (including Antilles) amphitropical disjunct, unique in serrate-undulate leaflet margins.

SOLANACEAE

A large and habitally diverse family, theoretically always with alternate leaves (though sometimes with a usually much smaller "minor" leaf more or less across from the regular one), entire to irregularly lobed or very characteristically remotely shallow-dentate (but never serrate). Vegetatively, entire-leaved species most likely to be confused with totally unrelated members of Caryophyllidae, which differ in branches with conspicuous concentric rings of anomalous growth. Flowers mostly 5-merous, more or less regular, sympetalous, usually with 5 (except *Brunfelsia, Schwenckia, Witheringia, Browallia*) stamens (unlike the 4-staminate related families Scrophulariaceae, Bignoniaceae, Verbenaceae, Acanthaceae, etc.), the anthers with terminal pores in *Solanum* and its closest relatives. The majority of genera rankly aromatic with the typical unpleasant tomato-plant odor of the family. *Solanum,* by far the biggest genus, usually has prickles (including on the trunks of several Andean species that are large trees). Stellate to variously dendroid trichomes are typical of several genera. The great majority of species have fleshy (to rather dry) berry-fruits, subtended or enclosed by the expanded variously lobed calyx, but some of the herbs (*Schwenckia, Browallia, Datura, Nicotiana*) have capsules.

Familial subdivision based on habit is particularly difficult in this family since many genera are a mixture of habits (predominantly shrubby) and none consists entirely (or even mostly) of large trees or free-climbing lianas. Scandent taxa (except *Solanum* species: usually spiny and/or rank-smelling) are usually more or less hemiepiphytic. Moreover generic limits are in a state of flux in some groups. Therefore, the following habit-based outline is less satisfactory than for most families. Nevertheless, except for *Solanum,* partition into four groups, one of largely hemiepiphytic climbers, one of membranaceous- and/or large-leaved shrubs and smallish trees, one of small-leaved dry-area shrubs usually with branch spines, and one of herbs is usually feasible.

1. Hemiepiphytes or Woody Epiphytes, Mostly Climbing

1A. The next four genera, are characterized by elliptic to obovate, +/- coriaceous, clustered leaves. — A natural group differentiated from each other mostly by adaptation to different pollinators: *Markea,* the variable basal group, probably mostly bees; *Trianaea* and *Solandra,* probably bats; *Juanulloa,* hummingbirds.

Solandra (8 spp.) — Large cloud-forest hemiepiphytic climbers (also occasionally in lowland wet forest). Flowers very large (>10 cm long), yellow or white with purple center, long basal tube and broadly campanulate upper tube, calyx thick and strongly triangular-toothed. Fruit, subtended by splitting calyx, can be very large (to 8 cm across). Essentially an overgrown *Schultesianthus*-type *Markea.*

Markea (18 spp.) — The main group of hemiepiphytic Solanaceae climbers. Leaves coriaceous, sometimes glandular-punctate. Flowers variable, from small greenish and campanulate (*Ectozoma, Hawkesiophyton*) to tubular and red (hummingbird-pollinated), to white or yellowish and tubular campanulate (*Schultesianthus*: bee-pollinated) to mottled-brownish and tan and openly campanulate (bat-pollinated).

Juanulloa (10 spp.) — Canopy liana mostly in moist lowland forest, but also at higher altitudes in cloud forest. Leaves or branchlets usually with vestiges of branched trichomes. Essentially a *Markea* with hummingbird-pollinated tubular orange flowers (the apex actually contracted), the corolla also distinctive in being scurfy-puberulous outside; calyx rather fleshy and deeply 5-toothed.

Solanaceae
(Epiphytic Shrubs or Hemiepiphytic Climbers)

1 - *Juanulloa* **2** - *Lycianthes*

3 - *Trianaea*

4 - *Solandra* **5** - *Markea*

Trianaea (4 spp.) — Andean cloud-forest hemiepiphytic climbers, with large narrowly obovate glabrous leaves. Differs from *Markea* in the very large (6–10 cm long) openly campanulate (without long basal tube) corolla, greenish or greenish with purplish marking and with triangular lobes; calyx very large, campanulate; flower pendent, very long-pedicellate.

1B. The next two genera (plus miscellaneous species of *Solanum*) have very different smaller, openly campanulate flowers with small calyces.

Lycianthes (188 spp., plus 12 Asia) — About half woody climbers (frequently hemiepiphytic) and half terrestrial herbs (and one species a spinescent shrub). Essentially *Solanum* but with the calyx either truncate or 10-dentate (not 5-dentate); and the inflorescence axillary. Leaves always entire (often otherwise in *Solanum),* often with small asymmetric leaf opposite large one.

Salpichroa (17 spp.) — Usually small hemiepiphytic cloud forest vines, very distinctive in the small broadly ovate, cordate leaves, appearing opposite, the "minor" leaves similar in size and shape to their pair. Corolla narrowly tubular, with narrowly triangular lobes, pendent, yellow and to 10 cm long with exserted anthers, or inconspicuous, greenish and ca. 2 cm long; calyx lobes narrowly triangular and deeply split. Drying blackish.

(*Solanum*) — A few species are tenuous hemiepiphytic climbers, mostly with pinnately compound leaves (some are also nonepiphytic lianas).

2. Mostly Small, Erect, Usually Soft-Wooded Trees (Occasionally Large Trees in *Solanum*)

Solanum (800 spp., plus 600 N. Am. and Old World) — Very large genus highly variable in habit (includes herbs, shrubs, small trees, large trees (these with smooth columnar spiny trunks) and lianas [often spiny]), but characterized by typical broadly open-campanulate corolla with 5 anthers connivent into a +/- conical central tube and opening by apical pores. Often spiny, usually +/- pubescent and usually with variously branched stellate or dendroid hairs. Leaves frequently irregularly broadly shallowly toothed, frequently spiny on midrib or main vines below, one large section (including the potato) with pinnately compound leaves. Nondescript species, with none of these features and simple rather than branched hairs, usually have small "minor" leaves opposite the regular leaves. Calyx always 5-toothed (unlike *Lycianthes*). Extra-axillary (often pseudoterminal) inflorescence different from *Lycianthes, Witheringia,* and *Capsicum.*

E: palo de ajo (*S.* aff. *arboreum*); berenjena (*S. candidum*), chitchiva (*S. americanum*); friega plato (*S. ochraceo-ferrugineum*); zorillo (*S. umbellatum*); huevo de tigre (*S. coconilla*); P: siucahuito (large purple flowers); coconillo (small flowers); cocona (*S. sessiliflorum*); vacachucho or tintona (*S. mammosum*)

Solanaceae
(Erect Trees and Shrubs with Rotate Flowers)

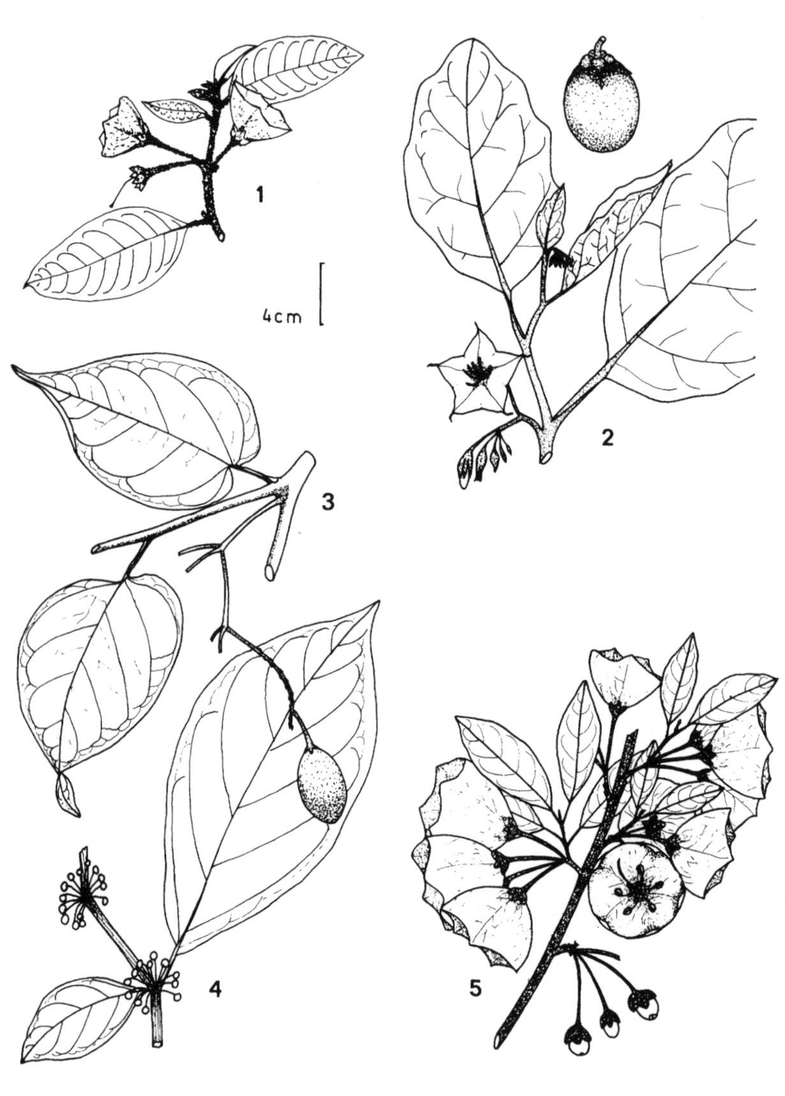

1 - *Saracha*

2 - *Solanum*

3 - *Cyphomandra*

4 - *Witheringia*

5 - *Saracha*

Cyphomandra (30 spp.) — Very close to *Solanum,* differing most obviously in the strongly dichotomous branching with pendent inflorescence arising from the dichotomy; leaves always rank-smelling, usually broadly asymmetrically ovate and cordate, the juvenile leaves sometimes deeply pinnately lobed and totally different. Most species are small trees with large tomato-like fruits, but one is a herb.

Dunalia (generic limits unclear: 7–30 spp.) — One species a small cloud-forest tree to 4 m tall with large densely pubescent leaves (cf., *Acnistus*), the flowers (longer than *Acnistus*) yellow to orangish, tubular and densely pubescent, in axillary fascicles. The other common species a more or less spiny-branched shrub of dry mattoral with dark purple flowers and small obovate leaves in short-shoot clusters.

Brunfelsia (25 spp.) — Small trees with large attractive lavender to lilac salverform flowers with large lobes and sharply 5-toothed cupular calyx, in fruit calyx becoming subwoody, splitting irregularly, and more or less enclosing fruit. Vegetatively nondescript, but the leaves usually thin-coriaceous, always entire and with short petiole, typically evenly elliptic, and borne on angled twigs.

Brugmansia (8 spp.) — Very characteristic, small, shrubby cloud-forest trees with large thin leaves, huge pendent flowers (>15 cm long), openly campanulate above basal tube, with very large inflated calyx splitting irregularly or subspathaceously. Fruit not spiny (unlike related herbaceous *Datura* in which formerly included), a woody-fleshy, indehiscent "capsule".

E: campana

Acnistus (1–many spp., depending on taxonomy) — Mostly drier areas, especially where foggy or misty. Our species a ramiflorous rather thick-branched tree with dense fascicles of small white flowers with exserted anthers and large pubescent leaves. Fruits numerous, orange.

E: cojojo, guitite

Cestrum (175 spp.) — Vegetatively rather nondescript except for the frequently apparent Solanaceae odor. Leaves usually more or less oblanceolate or narrowly elliptic; in Andean species often conspicuously pubescent with branched trichomes, but most lowland species glabrous or glabrescent. The flowers are extremely distinctive; however, narrowly tubular with short narrow corolla lobes, white to cream, very fragrant and opening at night, usually mostly axillary (or ramiflorous) and borne on short racemes. The fruit usually a slightly elongate purple berry, the base enclosed by the calyx.

P: hierba santa

Solanaceae
(Erect Trees and Shrubs with Tubular Flowers)

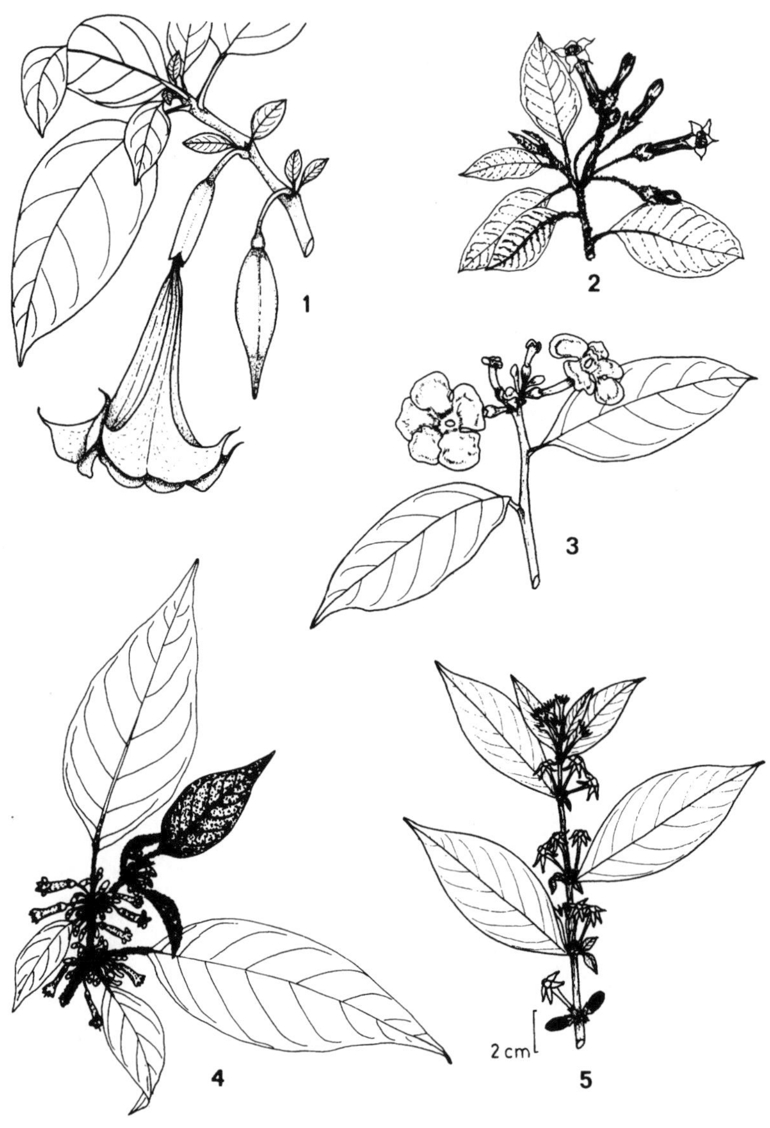

1 - *Brugmansia*

2 - *Iochroma*

3 - *Brunfelsia*

4 - *Dunalia*

5 - *Cestrum*

Sessea (14 spp.) — Paramo and high-montane forest shrubs similar to *Cestrum* (and perhaps its ancestor), but differing in openly paniculate inflorescence and capsular fruit.

Iochroma (20 spp.) — Shrubby to medium-sized cloud-forest trees with small to medium membranaceous, always puberulous leaves. Flower hummingbird-pollinated and very characteristic, the corolla tubular (usually with subexserted anthers) and bright blue, red, or deep purple in color, with very shallow triangular lobes; calyx cupular and more or less truncate, sometimes large and inflated.

Jaltomata (incl. *Hebecladus*) (12 spp.) — Mostly pubescent paramo and puna shrubs or subshrubs (one species a cloud-forest vine and a few herbs), the leaves mostly opposite and subequal. Calyx with broadly ovate lobes, split to near base; inflorescence characteristically pedunculate-umbellate. Two flower types: 1) corolla white with purple center or greenish-cream, short and broadly tubular with narrowly triangular +/- erect lobes; anthers usually exserted (= *Hebecladus*) and 2) flowers open, like *Solanum* but with longitudinal anther dehiscence (= *Jaltomata sensu stricto*). Calyx enlarging in fruit (but not closed over it).
E: jaltomate

Deprea (2 spp.) — Cloud-forest shrub 1–2 m tall, intermediate between *Physalis* and *Witheringia;* corolla distinctly green to greenish-white; calyx large, more or less enveloping fruit but not completely closed at apex. Very close to *Jaltomata* and the generic differentiation is unclear.

Saracha (3 spp.) — Andean cloud-forest trees 4–8 m tall, with smallish rather succulent, obtuse, elliptic leaves, stellate-pubescent at least on midvein. Flowers yellow to purple, pendent, at least 2 cm long, unlobed; calyx short, very broadly campanulate, 5-toothed.

(*Witheringia*) — Some species are more or less shrubby.

(*Capsicum*) — Some species are more or less shrubby.

3. Usually More or Less Spinescent Small-Leaved Shrubs, Mostly of Matorral

Grabowskia (6 spp.) — Spiny desert shrub with +/- zigzag twigs, with spines on the angles and slender branches (or clusters of obovate leaves) in spine axil these with inflorescence at tip. Flowers white and small with long-exserted anthers; berries red.

Solanaceae
(Dry-Area Shrubs; Mostly +/- Spinescent and Small-Leaved)

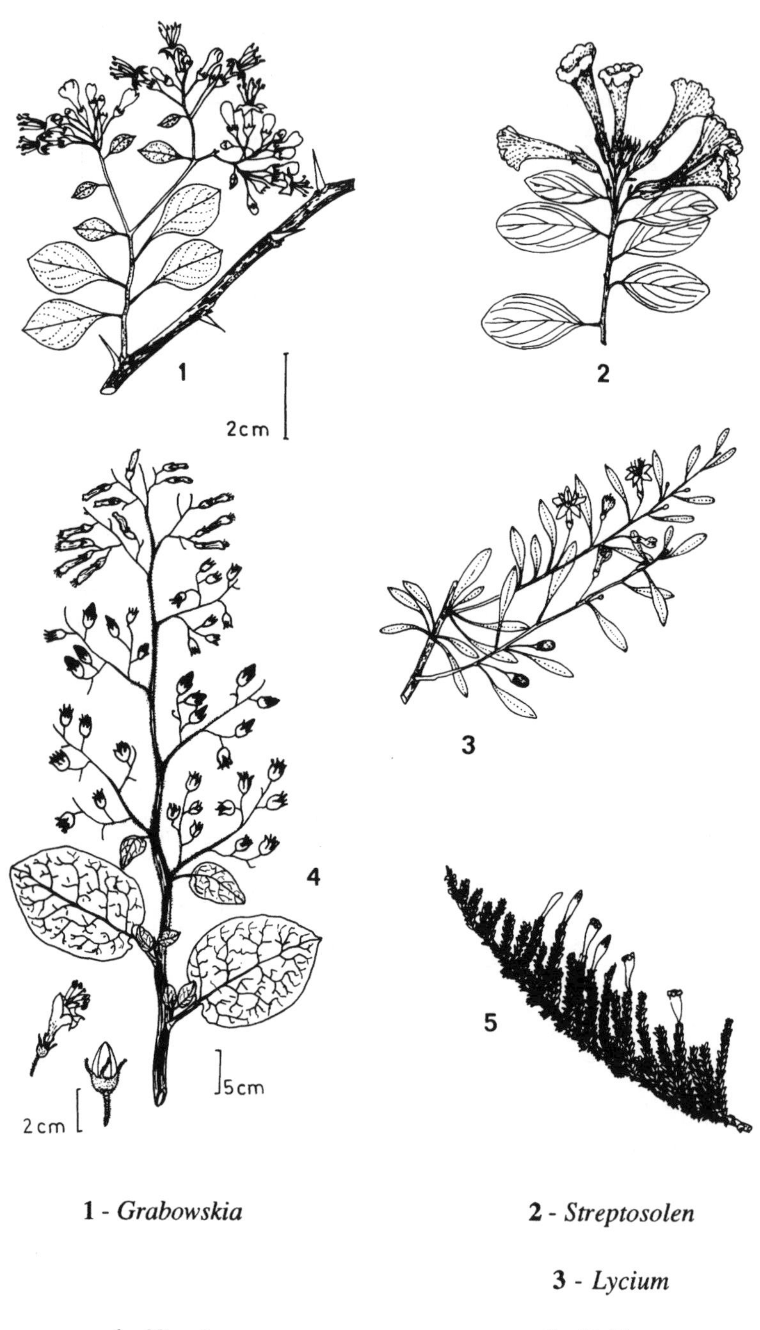

1 - *Grabowskia*

2 - *Streptosolen*

3 - *Lycium*

4 - *Nicotiana*

5 - *Fabiana*

Fabiana (25 spp.) — Mostly temperate South America; in our area, resinous shrubs or subshrubs of the dry puna and Andean steppes of southernmost Peru, with very small (<3 mm long), sublinear, dense, ericoid leaves. Flowers narrowly tubular, yellow, or purplish-green.

Lycium (15 spp.) — Spiny shrub of xerophytic thorn-scrub with slender spine-tipped branches and small oblanceolate leaves in short-shoot clusters. Flowers small, single from axils, corolla purple with exserted anthers, one species with narrow petals, the other with longer narrowly tubular corolla.

(*Dunalia*) — One species is a more or less spiny-branched shrub with dark purple flowers and obovate leaves in short-shoot clusters.

Streptosolen (1 sp.) — Nonspiny shrub or subshrub to 1.5 m tall; endemic to middle elevations of Huancabamba region, but widely cultivated. Flowers distinctive in bright orange color, very narrowly infundibuliform; leaves small, elliptic, strongly ascending veins, +/- sub-bullate and viscid-puberulent.

4. Terrestrial Herbs (Sometimes Rather Coarse and Subwoody, Especially, *Nicotiana*) and Rain-Forest and Cloud-Forest Subshrubs with Large or Membranaceous Leaves

4A. The first three genera have only 4 (or 2) fertile stamens

Schwenckia (14 spp.) — Small savannah or dry-area herb; corolla only 1 cm long, tubular with minute lobes, calyx cupular and 5-dentate; stamens reduced to 2(or 4). Fruit small and capsular.

Salpiglossis (18 spp.) — Mostly Chilean; the Peruvian species in dry steppes of western Andean slopes or loma formations. Small mostly annual herbs, more or less tenuous and distinctly viscid. Flowers large and narrowly infundibuliform or smaller and uniformly tubular. Fertile stamens two or four.

Browallia (2 spp.) — Weedy herbs with distinctly bilabiate blue salverform flower with broad lobes and yellow center. Only 4 stamens.

(*Witheringia*) — A few species have 4-parted flowers.

4B. The next ten genera have 5 fertile stamens.

Witheringia (20 spp.) — Understory subshrubs, usually with small leaf opposite main one. Very similar to *Solanum* but anthers dehisce longitudinally and inflorescences axillary fasicles. Some species have 4-parted flowers (unique in *Solanum*-like genera), and apiculate anthers (almost unique in family) are common.

Solanaceae
(Herbs)

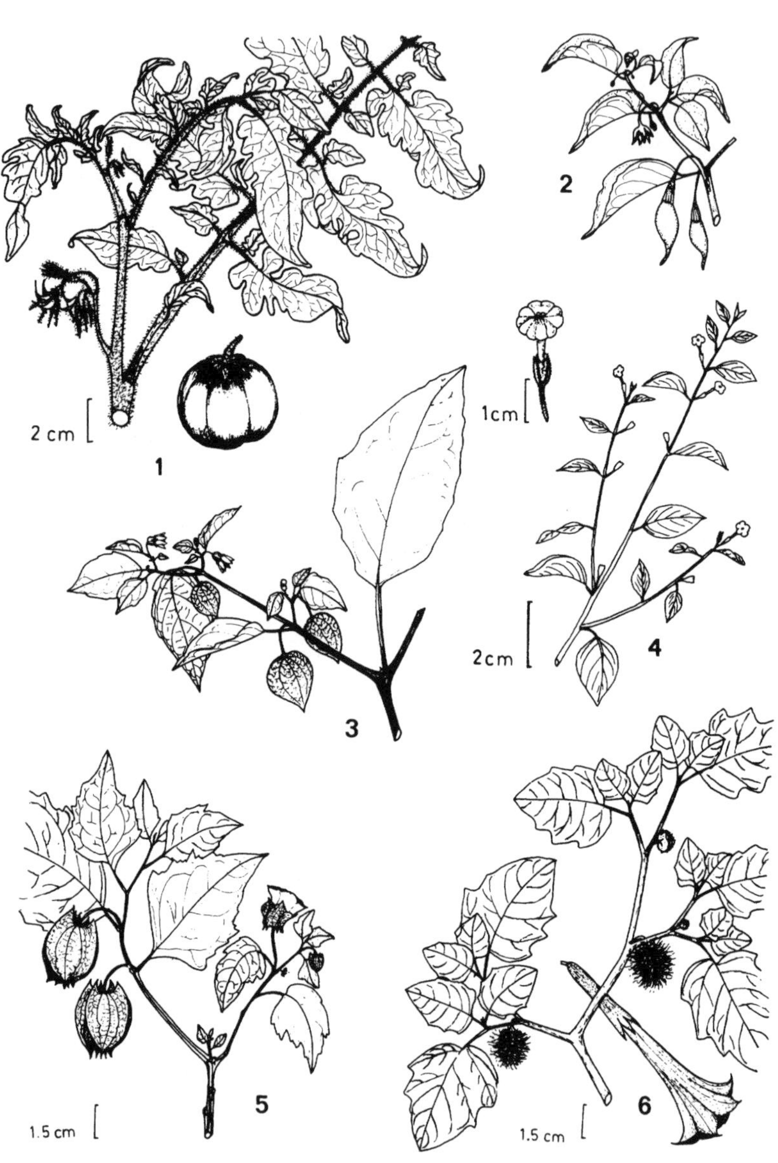

1 - *Lycopersicon* **2** - *Capsicum*

3 - *Physalis* **4** - *Browallia*

5 - *Nicandra* **6** - *Datura*

Cuatresia (6 spp.) — A dubious segregate of *Witheringia* characterized by fewer flowers per node and the reduced "opposite" leaf smaller and round.

Physalis (97 spp., plus 3 Old World) — Mostly weedy herbs with more or less triangular leaves. Corolla broadly campanulate, yellowish, the margin completely unlobed. Fruit very characteristic with calyx inflated and enveloping the fleshy berry.

Nicandra (1 sp.) — A blue-flowered version of *Physalis,* also differing in the nearly dry berry.

Lycopersicon (8 spp.) — Sprawling herbs, easy to recognize by pinnately compound leaves with alternating large and small leaflets, and the yellow flowers with the anthers conically appressed as in *Solanum.* Fruit orange to red.

Capsicum (10 spp.) — Looks like herbaceous or subshrubby *Solanum* with very small usually white flowers (one species lilac), but anthers longitudinally dehiscent and flowers axillary. Calyx small and not enlarging in fruit (unlike *Jaltomata).* Berry red, orange, yellow, or white (but never black, unlike *Jaltomata* and the most similar *Solanum* species), mostly piquant-spicy in taste.

E: ají

Nierembergia (35 spp.) — Mostly south temperate; ours tiny paramo herbs with white (or violet-lined) flowers, the corolla 2-parted, the top openly campanulate and *Solanum*-like, below this a narrow tube.

Jaborosa (20 spp.) — Low +/- creeping white-flowered rosette herb of very dry puna.

Datura (8 spp., also 2 Old World) — Weedy herbs of dry places, with very narrowly infundibuliform white to +/- purplish corolla with non-lobed margin except for apiculations at tip of the 5 veins. Leaves large, ovate, irregularly dentate. Fruit a spiny, incompletely 4-valved capsule.

Exodeconus (incl. *Cacabus*)(6 spp.) — Prostrate viscid herb of sandy washes and lomas of coastal Peru. Leaves +/- triangular (like lambsquarters, *Chenopodium*), with long petiole poorly differentiated at apex. Calyx inflated in fruit (cf., *Nicotiana*); corolla narrowly tubular-infundibuliform, white to yellow or blue.

Nicotiana (41 spp.) — Coarse, mostly viscid, pubescent herbs, mostly of inter-Andean valleys, usually with large membranaceous leaves. Calyx largish, typically +/- inflated, frequently 5-ridged; corolla tubular, the anthers often exserted.

Staphyleaceae

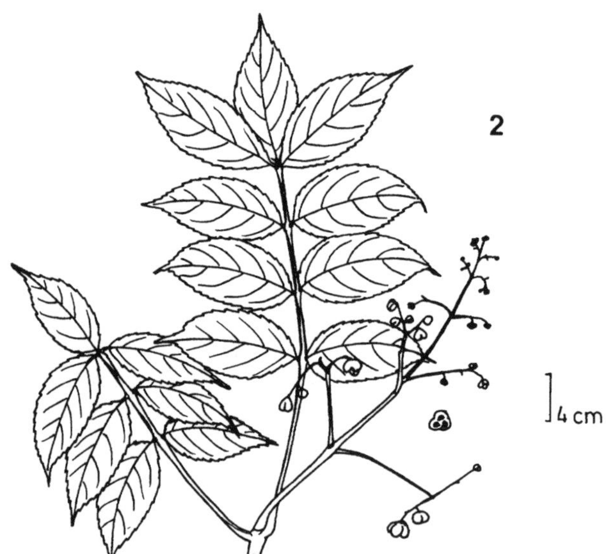

1 - *Huertea*

2 - *Turpinia*

(Solanum) — The herb species often have compound leaves and/or red-orange fruits. The inflorescence is always extra-axillary, unlike *Capsicum* and *Lycianthes.*

(Lycianthes) — The herb species all have ten calyx teeth and are often conspicuously pubescent.

There are many other neotropical genera, especially in dry, more or less, subtropical areas.

STAPHYLEACEAE

Two genera, each with probably a single species in our area. Both appear to be restricted to mesic forests on rich soil, but range altitudinally from the lowlands to montane cloud forest. Characterized by pinnately compound leaves with serrate or serrulate margins. *Turpinia* has opposite leaves characterized by a conspicuously jointed rachis, glabrous leaflets with closely finely serrate or serrulate margins, and a prominent interpetiolar line. *Huertea* has alternate multifoliolate leaves with very distinctive membranaceous glabrous or glabrescent leaflets having strongly asymmetric bases and a crenulate-serrate margin with glandular teeth. Both have openly paniculate inflorescences with tiny nondescript, 5-parted, whitish flowers and roundish indehiscent fruits ca. 1 cm in diameter and turning blackish-purple at maturity.

Huertea (4 spp.) — A large canopy or emergent tree with vertically ridged bark.

P: cedro masha

Turpinia (perhaps 30 spp., mostly in Old World) — Subcanopy tree.

E: cuero de puerco; P: cedro masha

STERCULIACEAE

Mostly trees and shrubs, with one important liana genus (*Byttneria,* our only Malvalean genus of lianas). Easy to recognize to order Malvales by the palmately 3-veined leaves, usually with stellate trichomes, and the typical swollen pulvinus at petiole apex but difficult to recognize to family in the absence of flowering or fruiting material (other than by first recognizing the genus!). All of the tree genera are entire-leaved (or palmately lobed or compound) except *Guazuma,* whereas, nearly all Tiliaceae are +/- serrate (except a few species of *Mollia* (which are lepidote), *Apeiba* (with longer petioles and more pronounced pulvinar thickening than comparable Sterculiaceae), and four genera very rare in our area: *Pentaplaris, Asterophora, Lueheopsis, Mortoniodendron*). *Guazuma* has more jaggedly serrate margins than do Tiliaceae. *Sterculia* is distinctive among Malvalean families in the terminally clustered leaves of different sizes and with dramatically different petiole lengths; *Pterygota,* vegetatively rather similar, has nonclustered nearly glabrescent leaves and slender petiole with reduced pulvinar region; *Theobroma* is vegetatively distinctive in the oblong short-petioled leaves (the entire petiole pulvinar in many species), often densely white-stellate below, but glabrescent *T. cacao* and relatives are very difficult to distinguish from some *Quararibea* species. Most of the Malvalean shrubs, all with serrate leaves, belong to Sterculiaceae; in addition to having very different fruits, the two Tiliaceae shrub genera *Corchorus* and *Triumfetta* can be differentiated from Sterculiaceae shrubs by the former's more crenate marginal serrations (except *C. siliquosus*: very small and subsessile) and the latter by the tendency to 3-lobed leaves.

The floral character that differentiates Sterculiaceae from the rest of the Malvales is the combination of 2-thecate anthers and filaments fused into a tube surrounding ovary; 2-thecate anthers are shared with Elaeocarpaceae and Tiliaceae, a staminal tube with Bombacaceae and Malvaceae, but the combination is unique to Sterculiaceae. In addition most Sterculiaceae have the stamens reduced to a single whorl of 5 with an outer whorl of 5 staminodes, these often highly elaborated. Most genera have dry, dehiscent fruits — *Sterculia:* apocarpous with 5 follicles; *Helicteres:* a spiral capsule on long gynophore; *Pterygota:* single follicle with winged seeds; *Ayenia* and *Byttneria*: round, fragmenting, 5-parted, spiny capsule; *Melochia* and *Waltheria:* with small 2-valved capsular cocci enclosed by dry calyx. *Theobroma* and its close relative *Herrania* have large berrylike oblong fruits (usually with a +/- hard shell); *Guazuma* is somewhat intermediate with indehiscent fruits, in one species wind-dispersed via long bristles, the others becoming fleshy and mammal dispersed.

1. Trees

Guazuma (3 spp.) — Common second-growth tree, especially in seasonally dry areas; also a canopy species of mature moist forest on good soils. This is the only serrate-leaved tree stercul, the leaves oblong to oblong-ovate with an asymmetrically truncate base, generally more irregularly serrate than in similar Tiliaceae. Flowers numerous and rather conspicuous, borne in more or less cymose axillary panicles, the narrow petals magenta in one species (*G. crinita*) and yellow in the others. Fruit "subspiny", one species (*G. crinita*) wind-dispersed via long bristles (cf., *Heliocarpus*), the others mammal-dispersed and short-tuberculate, resembling a *Helosis* inflorescence.

C: guacimo; E: guasmo; P: bolaina (*G. crinita*)

Sterculiaceae
(Trees with Follicles)

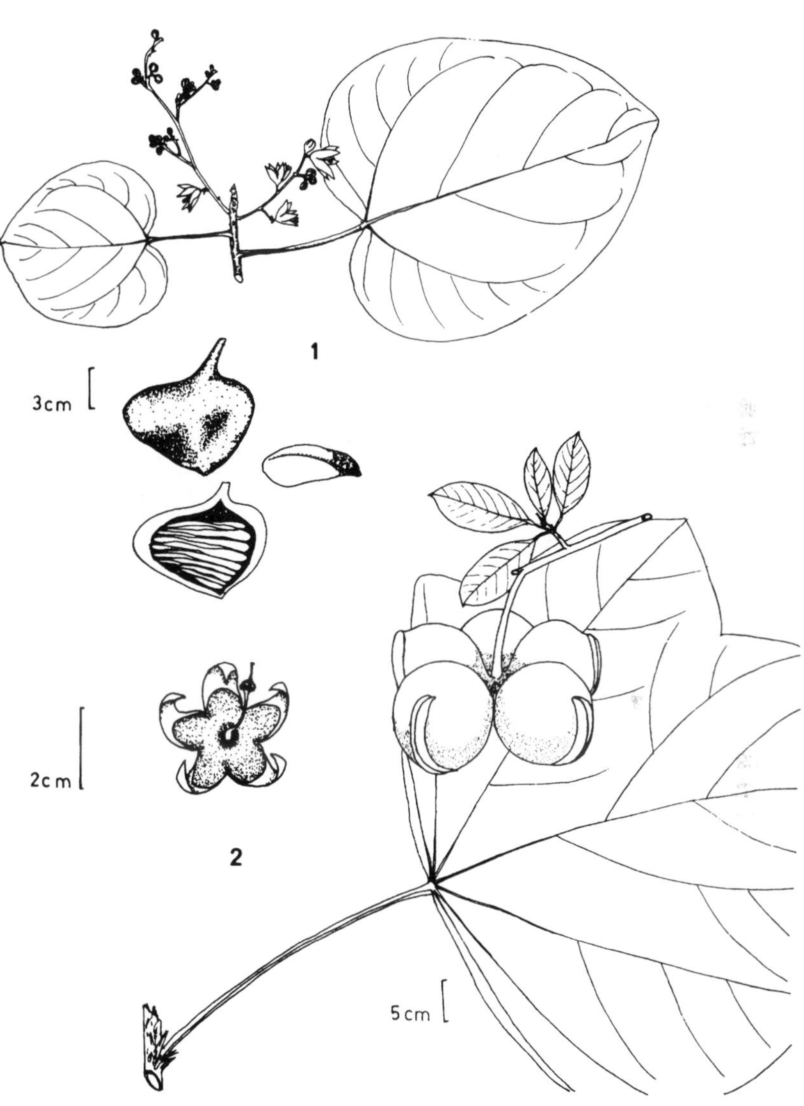

1 - *Pterygota*

2 - *Sterculia*

Sterculiaceae
(Trees with Indehiscent Fruits)

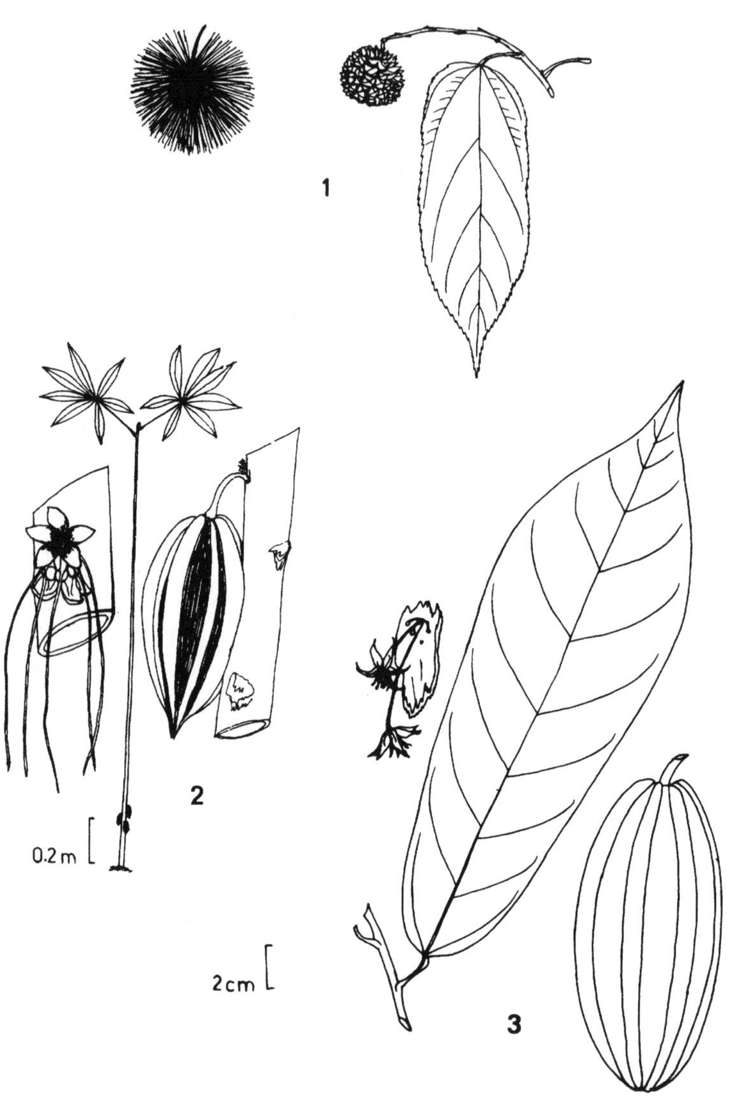

1 - *Gauzuma*

2 - *Herrania*

3 - *Theobroma*

Theobroma (22 spp.) — Middle-sized lowland forest trees mostly on relatively good soil. Leaves always oblong and with uniform-length short petioles, at least third of petiole (and often entire petiole) pulvinately thickened; leaves of most species densely stellate-pubescent and whitish below (unlike any *Quararibea*). Flowers and fruits ramiflorous and cauliflorous. Flowers small and with conspicuous staminodes, the petals with a short spatulate appendage. Fruit ellipsoid to oblong, rather large, with a fibrous exocarp and central pulp in which seeds are embedded. The +/- glabrous species, especially *T. cacao,* are very difficult to distinguish from *Quararibea,* differing most prominently in branching, with the main shoot successively replaced by a lateral branch at each node and the first branching of lateral branches bifurcate (Fig. 4).

C: cacao de monte, bacao; E, P: cacao, cacahuilla

Herrania (17 spp.) — Mostly small pachycaul trees or treelets of moist-forest understory. A segregate of *Theobroma* and with a similarly edible fruit, differing especially in the large palmately compound leaves, usually in a terminal cluster; the cauliflorous (often at extreme base of trunk) maroon flowers with remarkably elongate petal-appendages are also distinctive.

E: cacao de monte, cacao silvestre

Sterculia (300 spp., incl. Old World) — Large canopy trees of mature lowland forest, especially on good soil. Leaves often palmately lobed (especially in juveniles, even when mature foliage unlobed), clustered at branchlet apices, of markedly different sizes and petiole lengths in each cluster, with well-developed pulvinus or flexion at end of petiole. Fruit very distinctive, apocarpous, with 5 separate follicles, each lined with urticating trichomes and containing several unwinged seeds.

C: teta vieja, camajoru (*S. apetala*); P: huarmi caspi (= vagina tree, from the suggestive follicles)

Pterygota (3 spp., plus several in Africa) — Large, straight, late second-growth trees of relatively rich soils, one species with huge buttresses, one with none). Leaves large and evenly ovate with truncate or subcordate base (cf., *Bixa*), rather glabrescent (usually with small flat stellate hairs at base) and with a rather slender petiole with inconspicuous apical pulvinar region. Unlike similar *Sterculia* species in the leaves not in clusters of conspicuously different sizes nor with conspicuously different petiole lengths. Flowers greenish and inconspicuous. Fruits large, like one follicle of *Sterculia* but more irregular and with large winged seeds.

C: master

Sterculiaceae
(Shrubs and Lianas)

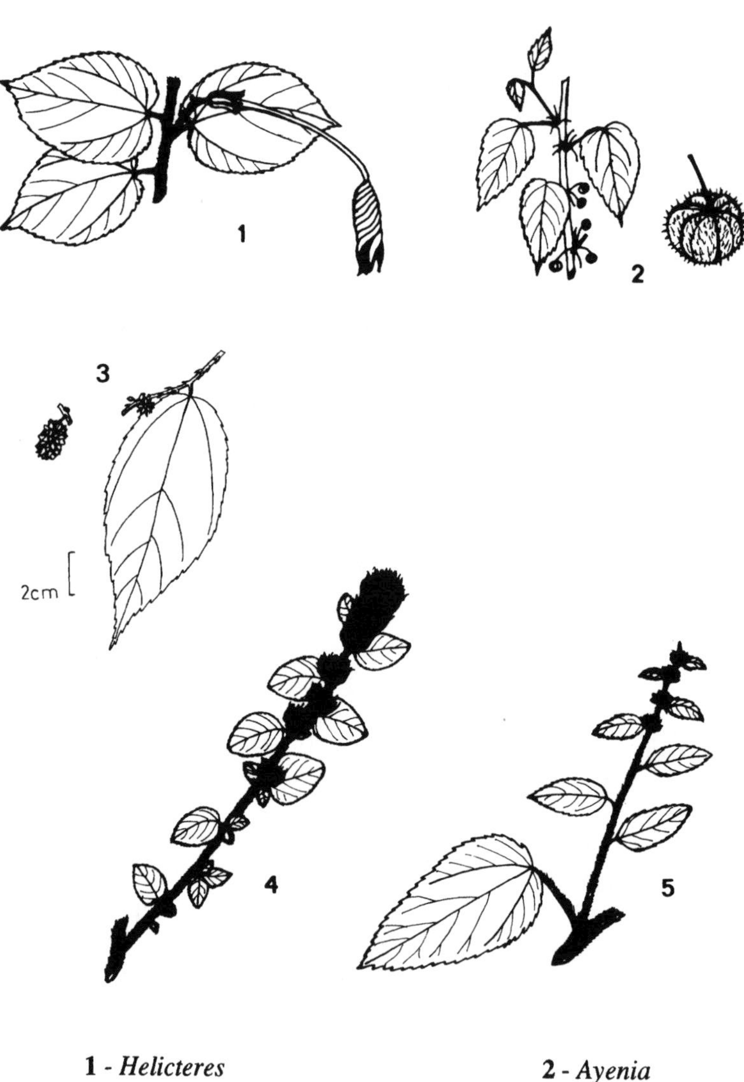

1 - *Helicteres* **2** - *Ayenia*

3 - *Byttneria*

4 - *Melochia* **5** - *Waltheria*

2. Shrubs

Helicteres (60 spp., incl. Old World) — Shrubs of dry lowland areas with asymmetrically cordate +/- oblong-ovate serrate leaves and often short petioles. Flowers axillary, with tubular calyx, short red petals, and a long-exserted staminal tube with ovary (borne on long gynophore) at its apex. Fruits very distinctive, strongly spirally twisted and on long gynophore.

Melochia (41 spp., plus 13 Old World) — Often weedy shrubs, characterized by the small white flowers in dense clusters, either fasciculate in axils or on peduncles as long as the leaves and the thin ovate calyx lobes (persisting in fruit). Leaves serrate, usually distinctively triangular-ovate (except when distinctly plicate in *M. crenata*) and with strongly ascending parallel secondary veins (cf., Rhamnaceae). Fruit small, dry, separating into 5 dehiscent cocci.

Waltheria (50 spp., plus 4 Old World) — Shrubs of dry areas, essentially a yellow-flowered version of *Melochia,* also differing from *Melochia* in either having shorter petiole or more strongly white-tomentose leaf undersurface. The inflorescences differ from *Melochia* in having several dense flower clusters each and in more conspicuous bracts and smaller calyx lobes. Technically differs in 1-celled (rather than 5-celled ovary and the fruit a single 2-valved capsule.

Ayenia (68 spp.) — Shrub or subshrub with uniformly serrate leaves usually oblong or oblongish, and either subsessile or with very short petiole or very tenuous in texture. One species (*A. stipulacea*) looks almost more like *Theobroma cacao* than *Ayenia* but has somewhat serrate margin). Flowers small, borne in a small diffuse few-flowered axillary inflorescence. The small round tuberculate capsule is a smaller version of *Byttneria.*

(*Byttneria*) — A few species of this predominantly scandent genus are shrubs, especially in dry areas, usually distinct in spiny and/or strongly angled branchlets.

3. Lianas

Byttneria (79 spp., plus 52 Old World) — Mostly lianas with characteristic 3-veined leaves and Malvalean pulvinus (our only Malvalean liana genus). Most species have spiny and/or angled stem; some have entire leaves, (either ovate or *Theobroma*-like, +/- rhombic with cuneate base and ascendingly 3-veined from above base), others are serrate. The majority of species have some kind of glandular area on or near base of midvein below or in axils of the main lateral veins. Fruit a characteristic globose-echinate capsule, splitting into 5 cocci.

E: zarza

Styracaceae and Symplocaceae

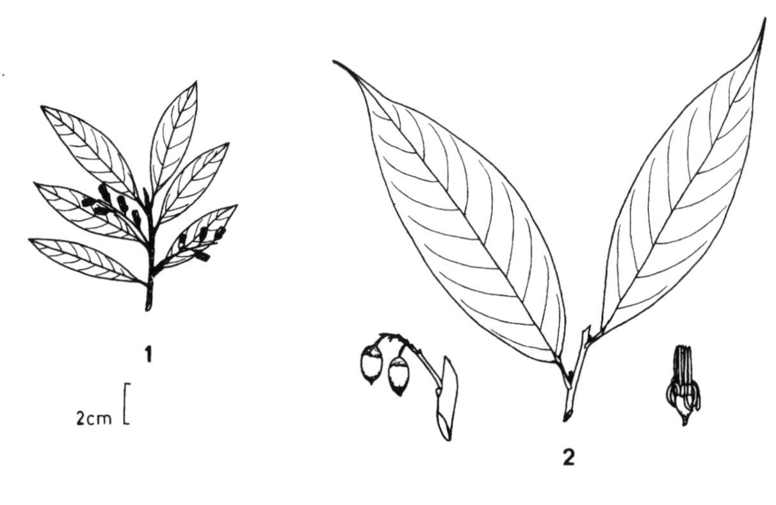

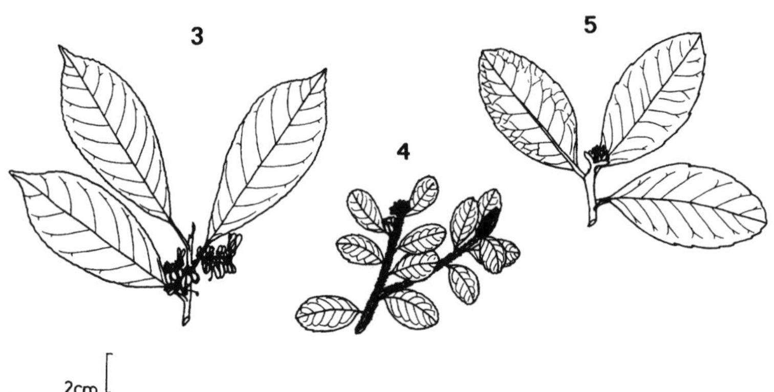

1 - *Pamphilia* **2** - *Styrax*

3, 4 & 5 - *Symplocos*

STYRACACEAE

Canopy and understory trees mostly in montane forests but occasionally in lowlands, especially in tahuampa (*S. tessmannii*) or poor-soil areas. Characterized by the densely white-stellate or lepidote leaf undersurface, sometimes also rufescent with reddish-stellate hairs, especially on twigs. The leaves are always entire (occasionally somewhat erose), the petiole not at all decurrent onto twig (unlike similarly white-stellate solanacs) The flowers are distinctive with a more or less truncate, whitish- or tannish-stellate, or lepidote calyx and valvate white petals. Although some species of Solanaceae can be similarly white below, they never have the rufous-stellate twig pubescence and are herbaceous and/or spiny and/or have asymmetric leaf bases or somewhat lobed leaf margins.

Styrax (130 spp., including N. Temperate Zone)

Pamphilia (4 spp.) — One Peru record: differs from *Styrax* in five instead of ten stamens.

SYMPLOCACEAE

Shrubs and trees mostly of montane forests characterized in our area by always at least slightly toothed festooned-brochidodromous veined leaves. The leaves are characteristically loosely and rather irregularly reticulate below, the venation generally not prominulous and the surface more or less smooth between the secondary veins. Extremely similar vegetatively to *Ilex* but lacks that genus' green inner bark layer. Inflorescence axillary, occasionally racemose but usually contracted and more or less fasciculate, the calyx lobes orbicular and overlapping (very like Sapotaceae), the petals usually narrowly obovate, the stamens numerous and usually with connate filaments; quite unlike Sapotaceae and other relatives in the inferior ovary. Fruit cylindric with a more or less truncate apex.

Symplocos (160 spp., plus 140 Old World)

THEACEAE

Mostly medium to large trees of middle-elevation cloud forests, without latex (in ours) or vegetative odor; also small sometimes dominant trees of poor-soil areas of Guayana, either on white-sand savannas or on the tepuis. A good familial character (though shared with other Thealean families like Marcgraviaceae) is the rolled young leaf at branch apex. Leaves of Theaceae, always alternate, are characteristically markedly asymmetric (most *Gordonia* and *Freziera*), at least basally, typically coriaceous and oblanceolately tapering to sessile or subsessile base, often spirally clustered at branch apices (*Ternstroemia, Bonnetia, Gordonia*), the secondary venation (and even midvein in *Pelliciera* and often *Bonnetia*) immersed and not apparent; *Freziera* is the opposite extreme with unusually numerous, nearly parallel secondary (and intersecondary) veins and the leaves distichous and evenly spaced along twig, also distinguished by the strongly grooved upper surface of petiole. The only family with which the typical Theaceae leaf could be confused is Myrsinaceae which either has less coriaceous nonasymmetric leaves with longer petioles or (*Grammadenia*) the leaves linear-punctate. Coriaceous-leaved taxa, including many of those with more or less symmetric leaves (*Ternstroemia*), mostly punctate, typically with scattered blackish glands; the punctations are dark even in bud, whereas, Myrsinaceae punctations are always pellucid in bud and usually more elongate. Even when petiole relatively long and conspicuous (*Ternstroemia,* some *Freziera*), often poorly differentiated from lamina base. Theaceae leaves may be serrate or not in the same genus (and in *Gordonia* even in same species).

When fertile, most taxa (including all montane ones) recognizable by the solitary or fasciculate flowers with strongly overlapping imbricate sepals, borne in the leaf axils or ramiflorous below the leaves. Only the Guayana Shield genera *Bonnetia* and *Pentamerista* usually have pedunculate inflorescences. The flowers usually have numerous yellow stamens and the petals are mostly white and sometimes showy (*Gordonia, Pelliciera, Bonnetia*) (sometimes pink [rarely yellow] in *Bonnettia,* mostly in species with branched inflorescence). *Gordonia* and *Bonnettia* have capsules with winged seeds; *Freziera* a usually fleshy, dark purple berry; *Ternstroemia* and *Symplococarpon* drier, larger yellow to green to purple-black berries (perhaps mammal-dispersed), *Pelliciera* a very large single-seeded fruit (dispersed by sea water).

Freziera (55 spp.) — An important genus of Andean cloud-forest tree, especially on exposed ridges or in disturbed forest. Characterized by the usually distinctly asymmetric leaf base, strongly dorsally grooved petiole, and the close-together secondary and intersecondary veins making a near right angle (75–90 degrees) with midvein; margins usually serrate, some species with characteristic sericeous pubescence. Flowers very small, often sessile or subsessile, with urceolate white corolla. Fruit a dark purple berry, subtended by the imbricate sepals.

C: zapata

Theaceae
(Small to Medium Flowers)

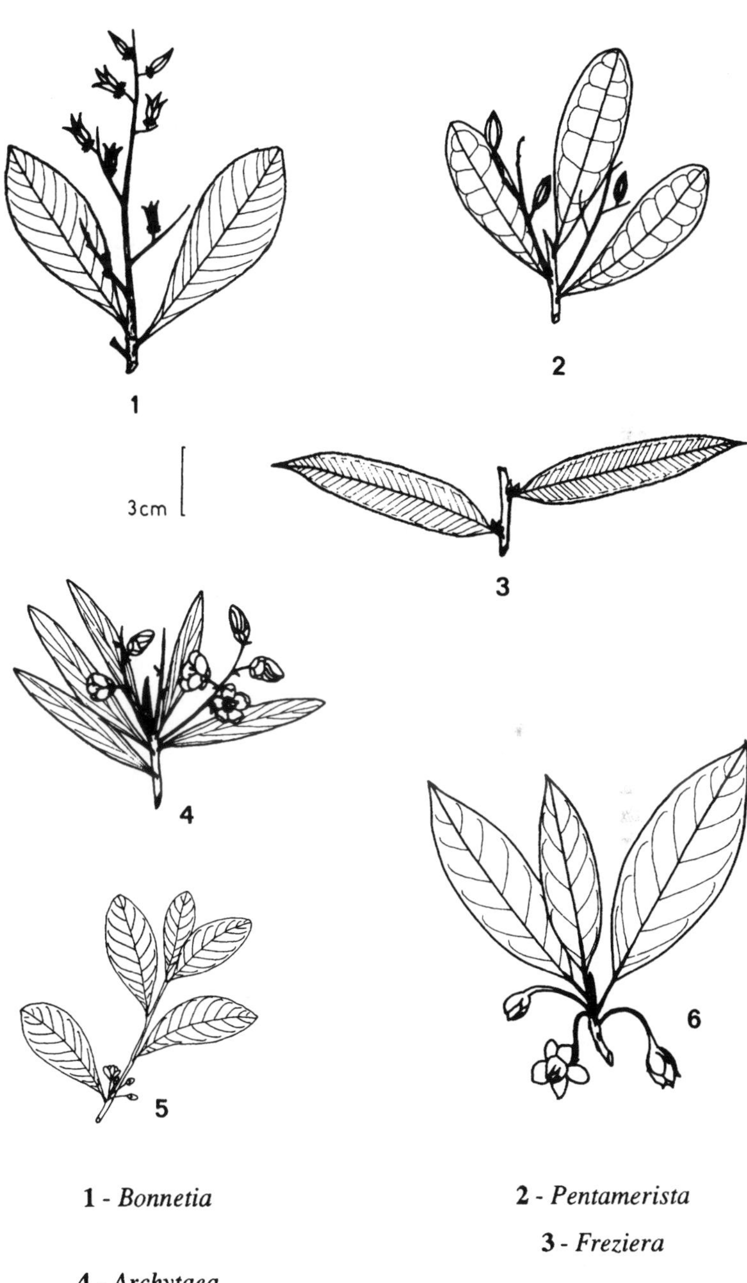

1 - *Bonnetia*

2 - *Pentamerista*

3 - *Freziera*

4 - *Archytaea*

5 - *Symplococarpon*

6 - *Ternstroemia*

Gordonia (incl. *Laplacea*)(4 spp., plus 1 N. Am. and ca. 60 Old World) — Cloud-forest canopy trees with more or less oblanceolate, subsessile, usually +/- asymmetric, and coriaceous leaves having suppressed secondary veins; could be confused only with *Myrsine* (Myrsinaceae) which has less coriaceous, more symmetric leaves with well-developed petioles. Flowers large, white; fruit a 5–(9)-valved woody capsule with winged seeds.

Symplococarpon (1 sp.) — Cloud-forest canopy trees with leaves broader and more symmetric than in *Gordonia,* with conspicuously sunken midrib and entire to serrate margins. The main generic character is the inferior ovary (= intermediate between Theaceae and Symplocaceae). Flowers whitish, rather small and with relatively small calyx; fruit ovoid-ellipsoid, indehiscent, purple-black at maturity. Vegetatively and in fruit, very similar to *Symplocos* (Symplocaceae) from which it differs in usually pubescent basifixed anthers with elongate connectives (vs. versatile and without extended connectives in *Symplocos*) and in 2 styles separate to base (instead of several fused and with a single capitate stigma).

Ternstroemia (100 spp., incl. Old World) — Trees of both Andean cloud forests and lowland forests, especially on poor soil. Our only coriaceous-leaved theac with well-developed petiole. Leaves usually with only faint serrations, often only toward apex, sometimes entire. Most species with conspicuously dark-punctate leaves, but more coriaceous than similar Myrsinaceae and the punctations less elongate and nonpellucid (i.e., dark) even in juvenile leaves (vs. glands pellucid, at least in bud in myrsinacs). Flowers pedicellate, borne singly or in pairs, axillary. Fruit conical-ovoid, subtended by the large triangular calyx teeth.

Archytaea (3 spp.) — Small trees of Guayana poor-soil area, especially white sand savannas. Leaves linear to oblong-obovate, coriaceous, reticulate-veined and essentially entire, clustered at branch apices. Inflorescence a distinctive umbel-like pedunculate cyme, the several flowers white to pink, differing from *Bonnetia* in stamens fasciculate. Fruit a 5-locular capsule distinctive in opening from base.

Pelliciera (1 sp.) — Mangrove tree with enlarged fluted trunk base. Leaves oblanceolate, thick-coriaceous, sessile, clustered at branch apex, both secondary veins and midvein completely immersed and not evident. Flower sessile and terminal, with 5 narrow petals >6 cm long and a characteristic long central staminal column. Fruit large, woody, single-seeded, compressed ovoid with a terminal apiculation, strongly longitudinally ridged.

C: pinuelo, mangle pinuelo

Theaceae (Large Solitary Flowers), Theophrastaceae, and Thymelaeaceae

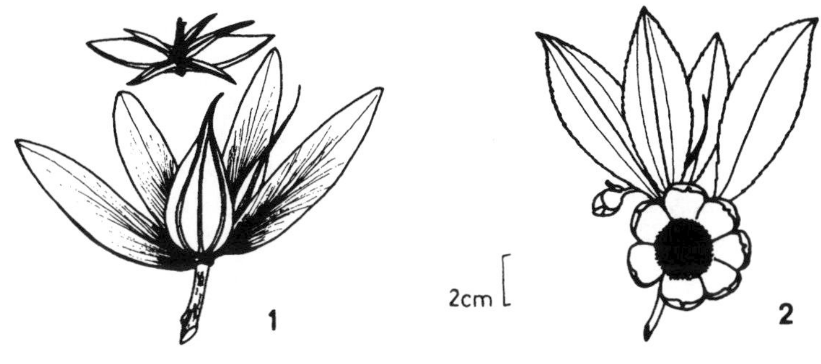

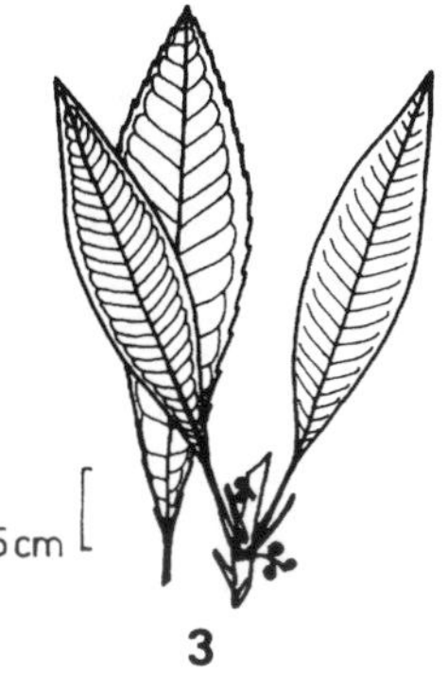

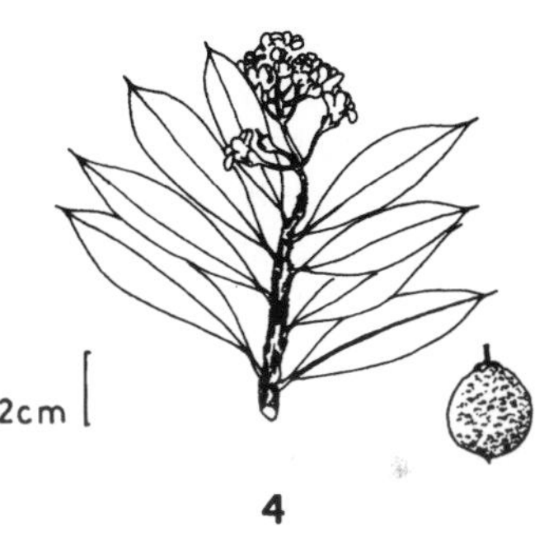

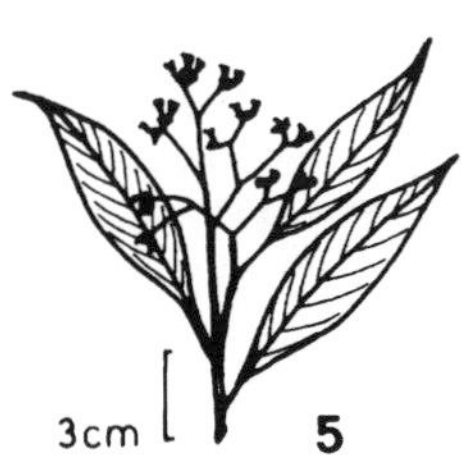

6

1 - *Pelliciera* (Theaceae) **2** - *Gordonia* (Theaceae)

3 - *Clavija* (Theophrastaceae) **4** - *Jacquinia* (Theophrastaceae)

5 - *Daphnopsis* (Thymelaeaceae) **6** - *Schoenobiblus* (Thymelaeaceae)

Bonnetia (28 spp.) — Small trees ecologically very important in Guayana-region poor-soil areas, especially the Venezuelan tepuis. Also occurring on lowland sandy savannas with one species disjunct in the savannas around Tarapoto, Peru. Leaves oblong-obovate, coriaceous, entire (or inconspicuously finely serrulate toward apex), clustered at branch apex, the secondary veins very inconspicuous when fresh and often +/- parallel with immersed midvein. Inflorescence often paniculate (unique in area Theaceae, but flowers may also be solitary or in 3-flowered cymes), the flowers white or pink (rarely yellow) with free stamens. Fruit a 3-valved capsule, the linear seeds usually winged at each end.

Neotatea (3 spp.) — Small Guayana area trees with thick branches, unique in area Theaceae in having white latex (thus, intermediate with Guttiferae). Leaves fleshy, asymmetric, sessile, clustered near apex of branches, with parallel secondary veins. Flowers large, solitary, pink, with free stamens. Fruit a 3-locular capsule with hairy, winged seeds.

Pentamerista (1 sp.) — Small trees of lowland white-sand savannas. Known from Venezuelan side of Rio Atabapo and surely also in Amazonian Colombia. Asymmetric coriaceous oblanceolate sessile leaves (similar to many *Gordonia* but with prominent midrib). Also differs from *Gordonia* in several-flowered, racemose inflorescence, with few yellowish-green flowers, and in the fruit an indehiscent berry.

The only other neotropical genera are monotypic *Acopanaea* of Guayanan Venezuela and *Cleyera* (8 spp.) of Central America and the Antilles.

THEOPHRASTACEAE

Two, vegetatively very different, but individually distinctive genera in our area, both with coriaceous leaves and similar small orange flowers with 5 basally fused petals, five fertile stamens arising from the petals, and an outer whorl of 5 petal-like staminodia. One genus (*Jacquinia*) is a shrub or small tree restricted to dry forest and with small entire, sharply apiculate, very coriaceous leaves; the other (*Clavija*) is an understory pachycaul treelet of moist and wet forest with large to very large oblanceolate leaves (sometimes with sharply remotely serrate margins. Very close to Myrsinaceae but differing in lacking resin ducts and pellucid glands, in sclerenchyma strands (not visible with naked eye) beneath leaf epidermis, and in having staminodes. This may be the only family in our area with small orange flowers, although some Antillean *Jacquinia* species have whitish flowers (and Central American *Deherainia* has large green flowers!).

Clavija (55 spp.) — Pachycaul dioecious treelets (one species <0.5 m high) with oblanceolate coriaceous leaves having intersecondaries parallel to the numerous secondary veins, sometimes with remotely serrate margins. Petiole usually long, apically subwinged, and poorly differentiated from long-cuneate lamina base, more or less swollen and black-drying at base. Can be differentiated from large-leaved *Weigeltia* group of *Cybianthus* by more coriaceous leaves, the marginal serrations (when present), and the uniformly cauliflorous spicate or narrowly racemose inflorescences of bright orange flowers. When sterile might be confused with pachycaul (or juvenile) *Gustavia,* but that genus lacks parallel intersecondary veins and has the petiole base not enlarged, or somewhat flattened, but never black-drying.

Jacquinia (50 poorly differentiated spp.) — One of the characteristic taxa of very dry forests, especially on limestone and near the coast. The very sclerophyllous spine-tipped leaves, with revolute margins and the immersed secondary veins not evident, are evergreen, even when all other species are deciduous (although a few understory species may loose their leaves during the *rainy* season). Fruits round, green to greenish-orange, ca. 2–3 cm across.

C, E: barbasco; P: lishina

THYMELAEACEAE

Mostly small to medium-sized dioecious trees with nondescript alternate entire leaves. (In addition a liana with subopposite leaves occurs in Amazonian Brazil and might reach our area.) Luckily there is one outstanding and unmistakable vegetative character: The thick homogeneous bark is extremely strong and any part of it strips as a unit from twig (or trunk) base to apex. This is the only family with twig bark that is both strong and of a thick homogeneous (i.e., non-layered) texture.

Inflorescence either open, flat-topped and strictly dichotomously branching or reduced to a raceme or (most commonly) a cluster of flowers at the end of short axillary peduncle or further reduced to sessile flowers borne singly or in fascicles; frequently more or less ramiflorous. The flowers (always smallish and white to greenish or yellowish in our area) are mostly 4-parted and lack petals but the sepals are petaloid and the calyx base often forms narrow tube. The anthers, often sessile and usually bright orange, are inserted near mouth of calyx tube and, if in a single whorl appear to arise from base of "petal". The fruit in our taxa is an ellipsoid, single-seeded drupe, in many taxa forming a fruiting cluster that strongly resembles the cluster of monocarps of many Annonaceae. There are only two genera known from our area, *Daphnopsis* with 8 stamens and *Schoenobiblus* with 4; I am unable to separate them, vegetatively, except by knowing the species.

Daphnopsis (46 spp.) — Usually small trees occurring in both low and high-elevation forests (and in extralimital subtropical forests). Distinguished from *Schoenobiblus* by the 8 stamens. Also tends to have smaller flowers or, in large-flowered species, the tepal lobes small in relation to the tube.

Schoenobiblus (7 spp.) — Entirely lowland tropical in distribution. Differs from *Daphnopsis* by having only 4 stamens. Also, more or less distinctive in tendencies to have longer narrower petals, long-exserted anthers, and sericeous inflorescence pubescence.

There are several additional genera in the West Indies (*Lagetta, Linodendron*) and temperate South America (*Drapetes, Ovidia*) as well as two others in Amazonia which may occur in our region. The Amazonian genera are: *Lophostoma,* a liana with subopposite leaves with *Calophyllum*-like venation, narrowly tubular flowers, and white foliaceous inflorescence bracts, and *Lasiadenia,* a shrub of white sand beaches of the upper Rio Negro, with rather thin, small, narrowly ovate leaves and elongate, narrowly tubular flowers, sessile at end of peduncle.

TILIACEAE

Mostly canopy trees, especially well represented in late second growth (and including two genera of weedy subshrubs). Like all Malvales characterized by alternate leaves, palmately 3-veined at base, with stellate (or lepidote) trichomes, and the petiole apex more or less swollen and pulvinar. When sterile can be distinguished from Sterculiaceae and simple-leaved Bombacaceae only by recognizing the genus. In general serrate-leaved Malvalean trees belong to Tiliaceae (except *Guazuma* with more jaggedly irregular serrations than in any Tiliaceae). Entire-leaved tiliacs (except several genera very rare in our area) have the lower leaf surface densely canescent, a character combination not found in simple-leaved Bombacaceae and only in a few *Theobroma* species in Sterculiaceae. Entire-leaved *Apeiba* differ from these *Theobromas* in a longer more slender petiole with conspicuous apical pulvinar thickening; *Lueheopsis* usually has at least a few apical denticulations but otherwise may not be distinguishable, vegetatively, from *Theobroma.*

Tiliaceae is florally the least specialized portion of the Malvalean complex on account of its multiple stamens with free filaments (shared with Elaeocarpaceae); also separated from multistaminate Sterculiaceae by the filaments in a single whorl and from Bombacaceae by bilocular anthers. The fruit of many genera is very distinctive; indeed Tiliaceae may have more fruit diversity than almost any other family, including species with fruits dehiscent (with seeds arillate, or winged and wind-dispersed) or variously indehiscent (often samaroid or otherwise wind-dispersed; even exozoochoric in one genus). However, large drupelike berries, the predominant fruit of simple-leaved bombacs and some sterculiacs (*Theobroma, Herrania*) are lacking in tiliacs.

Some genera are vegetatively distinctive. *Mollia* is mostly distinguished by lepidote scales instead of stellate trichomes; *Mortoniodendron* by an asymmetric base less 3-veined than in other genera; *Neotessmannia* by the extremely asymmetrically cordate base; *Heliocarpus* (and shrubby *Triumfetta*) by the tendency to be 3-lobed; *Luehea* (serrate) and *Lueheopsis* (virtually entire) by the unusually short thick petioles without an apical pulvinar differentiation and especially densely canescent leaf undersurface; *Goethalsia* (remotely toothed); *Trichospermum* (more finely remotely toothed) and *Asterophora* (entire) by narrowly oblong shape with the main lateral vein pair reaching far into the leaf apex.

Flower color can be a useful generic character with yellow flowers characteristic of the weedy shrubs (*Triumfetta, Corchorus*) and of *Apeiba, Vasivaea,* and *Christiana;* small whitish to cream or +/- greenish flowers of mostly Central American *Goethalsia, Mortoniodendron,* and *Heliocarpus;* larger white flowers of *Luehea* (paniculate) and *Mollia* (fasciculate); magenta flowers of *Lueheopsis* and *Trichospermum.* Wind-dispersed genera include *Heliocarpus* (plumed marginal spines around small body), *Trichospermum* (seeds like fruits of *Heliocarpus*), *Luehea* and *Lueheopsis* (winged seeds), *Goethalsia* (longitudinally 3-winged fruit fragmenting into 3 flat oblong samaras), *Pentaplaris* (5 expanded calyx lobes). *Mollia* has small thin subwinged seeds but these probably mostly water-dispersed. *Mortoniodendron* (and perhaps *Asterophora*) has a capsule with arillate seeds. *Apeiba* (and perhaps *Vasivaea*) are indehiscent and presumably mammal-dispersed. *Triumfetta* is exozoochoric.

1. Weedy Shrubs or Subshrubs —(See also similar genera of Sterculiaceae)

Corchorus (10 spp., plus ca. 80 in Old World) — Weedy herbs or subshrubs. Leaves narrower than *Triumfetta* and more sharply serrate; stipule linear and persistent. Flowers yellow, axillary, solitary or in fascicles of two to three. Fruit long and very narrow, splitting in half.

E: espada

Tiliaceae
(Shrubs and Trees with Samaroid Fruits)

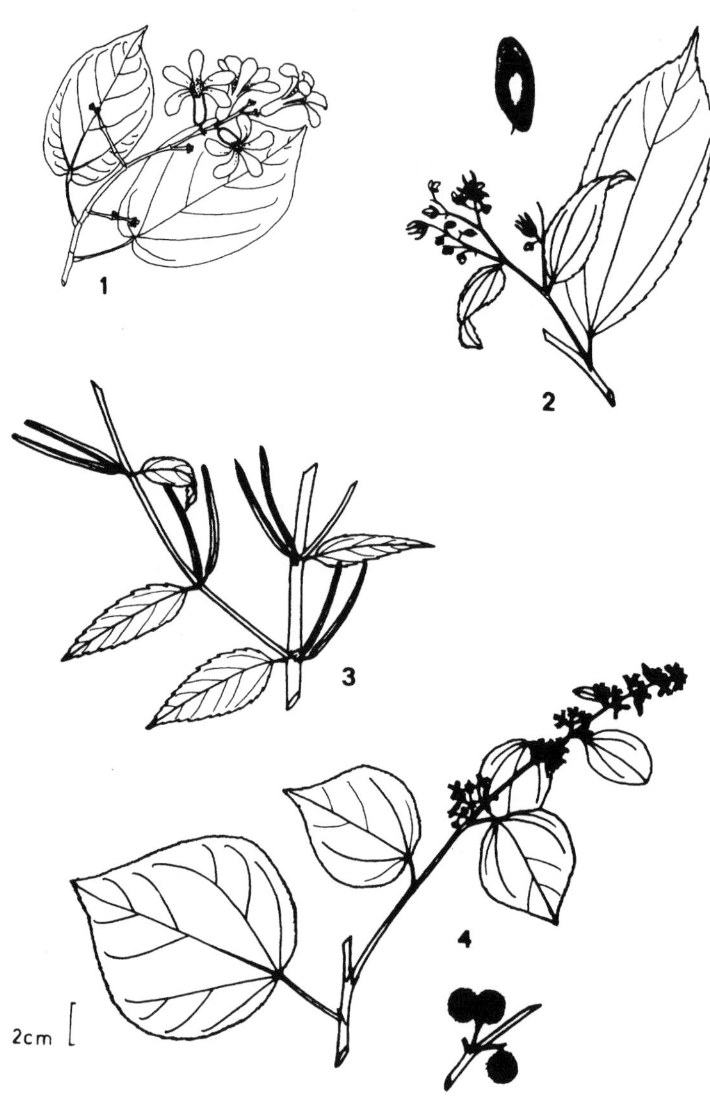

1 - *Pentaplaris*

2 - *Goethalsia*

3 - *Corchorus*

4 - *Triumfetta*

Tiliaceae
(Trees: Nonsamaroid, Wind- [or Water-]Dispersed; Mostly Capsular)

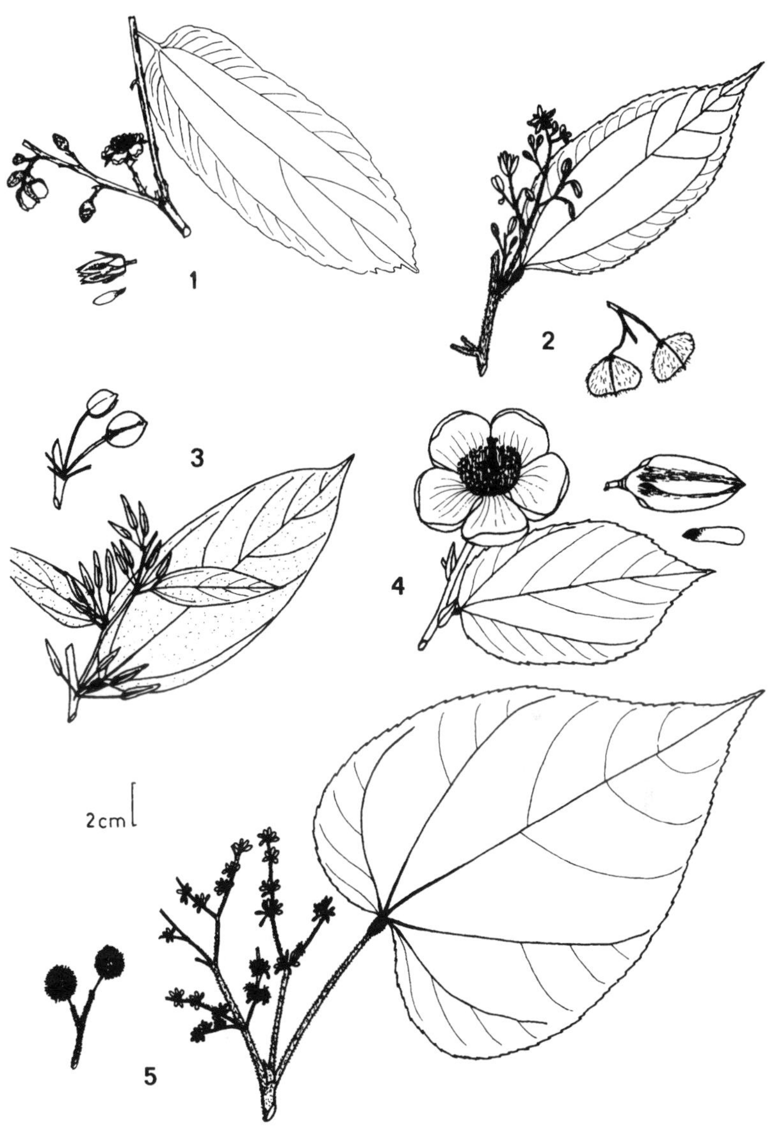

1 - *Lueheopsis* 2 - *Trichospermum*

3 - *Mollia* 4 - *Luehea*

5 - *Heliocarpus*

Triumfetta (50 spp., plus ca. 100 in Old World) — Weedy shrubs or subshrubs. Leaves more broadly ovate and more densely villous than *Corchorus*, usually obtusely sub-3-lobed, closely but rather bluntly serrate, the basal teeth glandular and larger. Inflorescence an axillary cluster of flowers or the subtending leaves suppressed and a narrow terminal panicle, the flowers pale yellow with narrow petals. Fruits round, spiny, exozoochoric.

2. Trees with Wind-Dispersed Seeds —(Probably water-dispersed in *Mollia* and *Lueheopsis*)

Heliocarpus (10 spp.) — Large, late second-growth tree, especially in cloud forest. Leaves thin, broadly ovate, +/- remotely serrate, sometimes +/- 3-lobed. Inflorescence a terminal panicle, the small flowers greenish to whitish. Fruit very characteristic, pinkish, the small round body surrounded by long soft "spines". (See also *Guazuma crinita* of Sterculiaceae.)

Goethalsia (1 sp.) — Large wet-forest tree of southern Central America, recently discovered in northern Colombia; often recognizable by orange juvenile leaves. Leaves shallowly remotely toothed with strongly ascending lateral veins (cf., *Trichospermum*). Inflorescence paniculate, the small flowers cream or yellowish. Fruit very typical, oblong and vertically 3-winged, fragmenting into 3 separate samaras, each with central seed surrounded by oblong wing.

Trichospermum (3 spp., plus 20 in Malaysia) — Tall trees of late second growth. Leaves narrowly oblong, very finely and remotely toothed; 3-veined to near apex, the base of main veins below more or less bearded with longish trichomes. Inflorescence an axillary panicle of lavender flowers. Fruit very characteristic, flat, heart-shaped, splitting in half to release small seeds with fringe of long stiff hairs (cf., fruit of *Heliocarpus*).

C: aliso; P: atadijo blanco

Pentaplaris (1 sp.) — Large wet-forest tree of southern Central America, recently discovered in northern Colombia. Leaves broadly oblong-ovate and entire, the petiole longer and more slender than in other area Tiliaceae, the apex pulvinate; stipules persistent. Remarkable fruits with large 5-lobed expanded calyx (similar to *Petrea*), presumably wind-dispersed.

Luehea (17 spp.) — Commonest in seasonal forest. Leaves always serrate and white or tan below, broadly ovate to oblong-ovate with truncate base and usually obtuse apex, the stipules often more or less persistent; petiole short, the whole petiole +/- "pulvinate". Flowers rather large, white (in our area: sometimes pink elsewhere); fruit an oblong, woody, pentagonal, 5-valved capsule with small winged seeds.

C: guacimo colorado; P: bolaina

Tiliaceae
(Trees: Follicular or Mammal-Dispersed and Indehiscent or Capsular)

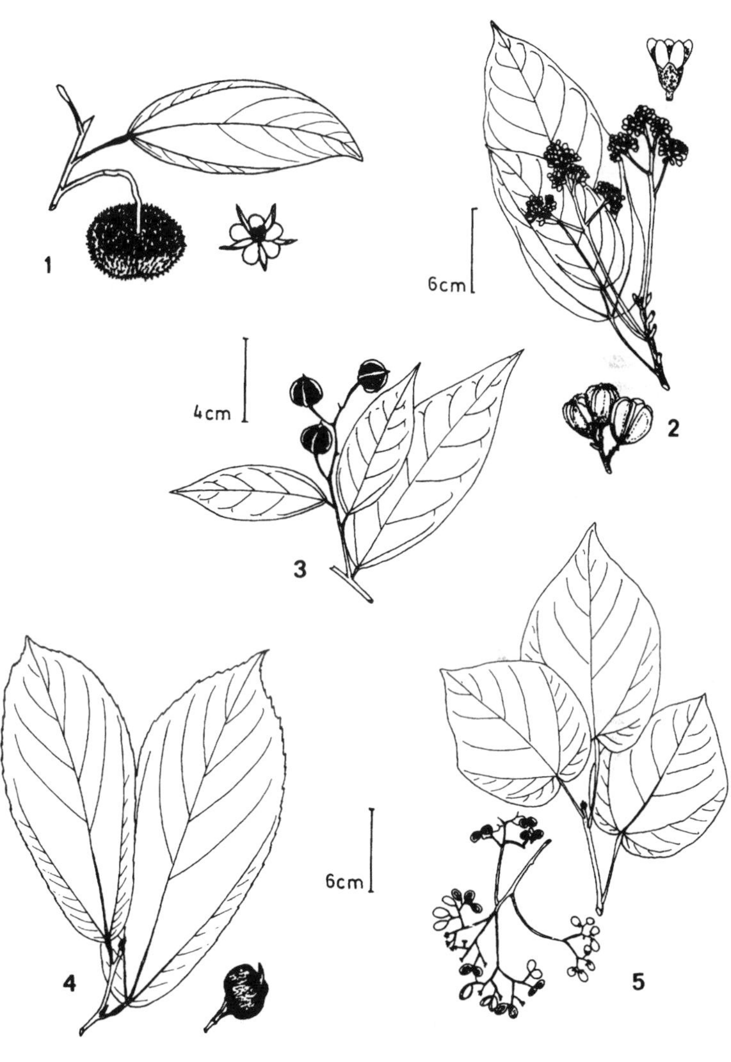

1 - *Apeiba* **2** - *Asterophora*

3 - *Mortoniodendron*

4 - *Vasivaea* **5** - *Christiana*

Lueheopsis (7 spp.) — Large tree of swamp forests. Very close to *Luehea.* Leaves similarly white below but always entire and only upper half of petiole pulvinar. Fruit like *Luehea* but seeds unwinged and the flowers magenta.

P: sapote de pantano

Mollia (11 spp.) — Trees mostly in seasonally inundated forests. Leaves mostly entire or very finely serrate, usually lepidote rather than stellate-tomentose. Inflorescence a few axillary white flowers with long very narrow petals. Fruits roundish or somewhat laterally compressed, splitting in half, the thin vertically stacked seeds not quite winged.

P: coto bara

3. Trees with Indehiscent Apparently Mammal-Dispersed Fruits

Apeiba (6 spp.) — Trees of second growth and mature lowland forest. Leaves entire (when finely puberulous) or closely serrate (when coarsely pubescent). Inflorescence few-flowered, the flowers yellow. Fruit very distinctive, more or less globose but dorsoventrally compressed, indehiscent, the surface spiny or warty.

C: peinemono; C, E, P: peine de mono; P: maquisapa ñaccha

Vasivaea (2 spp.) — Amazonian trees, recently discovered in northern Colombia. Leaves finely inconspicuously serrate, oblong, 3-veined into upper third. Quite similar to *Luehea seemannii* but more finely toothed and less tan below, the petiole more slender, and the flowers smaller, yellow, and in +/- sessile clusters. Fruits large (ca. 3 cm across), subwoody, squarish, broader than long, with a conspicuous apical projection.

4. Trees with Radially Segmenting Capsules or Apocarpous Follicles

Asterophora (2 spp.) — Leaves glabrate, entire, oblong. Fruit 5-angled, dorsoventrally flattened, fragmenting into 5 carpels. Type from western Ecuador; also recently discovered in Amazonian Peru.

Mortoniodendron (10 spp.) — Medium to large trees, mostly Central American but reaching Magdalena Valley. Leaves entire, +/- oblong, the base notably asymmetric and not very 3-veined; petiole often short and entirely pulvinar. Inflorescence few-flowered, axillary or terminal, the flowers small, white or cream. Fruit wider than long, usually more or less pentagonal, with raised angles, the surface characteristically finely wrinkled, fragmenting into (3–)5 segments to reveal shiny black seeds subtended by orange arils.

Christiana (2 spp., plus several in Africa) — Mostly southern Amazonian Brazil. Leaves large, broadly ovate, densely pubescent. Flowers small, yellow, in panicle. Fruit like miniature *Sterculia*, the follicles small, brown-pubescent, splitting in half at dehiscence.

Neotessmannia (1 sp.) — Tree known only from the type from swamp forest in Amazonian Peru. Leaves broadly oblong, remotely denticulate, the base strongly and very asymmetrically cordate, densely yellowish-tomentose below. Flowers solitary, yellow, unique in family in inferior ovary.

Extralimital genera include ***Berrya*** (incl. *Carpodiptera*) with 3 species in Cuba and nuclear Central America and several in Old World which has broadly ovate membranaceous leaves and a large open panicle of small violet flowers and ***Hydrogaster*** (1 sp.) of coastal Brazil.

TRIGONIACEAE

As treated in the recent *Flora Neotropica* monograph, the neotropical representation of the family consists of the single genus *Trigonia*. The *Trigonia* species are lianas (sometimes shrubs or treelets outside our area) with opposite leaves and a strongly malpighiaceous vegetative aspect. They are vegetatively further characterized by Rubiaceae-like caducous interpetiolar stipules (sometimes fused at base and leaving a conspicuous scar when caducous) and the usually white-sericeous (at least puberulous) leaf undersurface but with simple (usually long, spider-webby, and matted) rather than malpighiaceous hairs. Trigoniaceae may be thought of as somewhat intermediate between Malpighiaceae (from which they differ in solitary style and a more or less pealike white flower [2-petaled keel, 2 wings, spurred standard, fused filaments] without calyx glands) and Vochysiaceae (which differ in tree habit, more strongly zygomorphic flowers, and having only 1–2 fertile stamens). The fruit of *Trigonia* is unique: a 3-valved capsule with round seeds covered by long silky hairs. The typically very narrowly paniculate inflorescence and uniformly small flower size are also typical.

Trigonia (24 spp.)

The only extralimital neotropical genus of the family is *Trigoniodendron*, a large tree recently discovered in Espirito Santo, Brazil.

Trigoniaceae, Tropaeolaceae, and Turneraceae

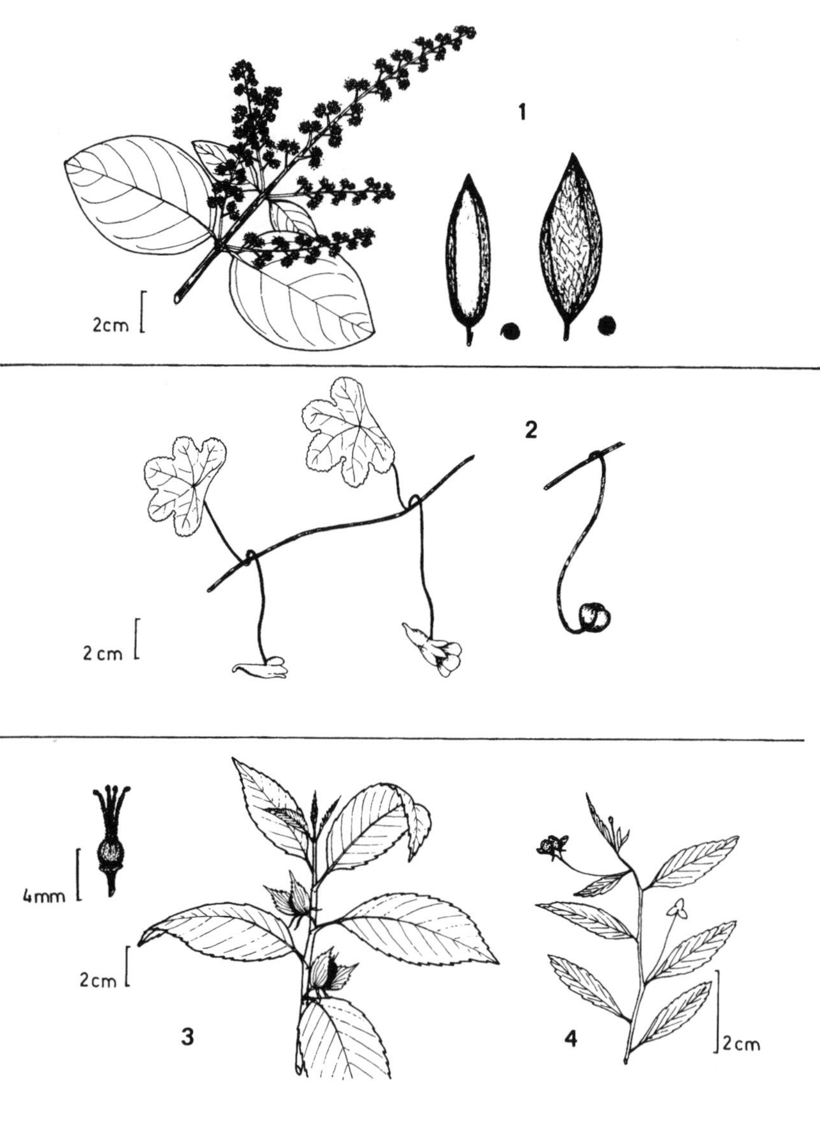

1 - *Trigonia*

2 - *Tropaeolum*

3 - *Turnera*

4 - *Piriqueta*

TROPAEOLACEAE

Tenuous cloud-forest vines, mostly of disturbed areas, very characteristic vegetatively in the peltate, variously lobed leaves with a tendency to twining petioles. The red or red-orange flowers are also unique, borne singly and pendent on slender pedicels from the leaf axils with the sepals fused to form a long conical spur from which the tips of the usually fringed petals are exserted. The capsular fruit is 3-lobed and rather euphorblike, fragmenting into 3 separate carpels at maturity.

Tropaeolum (86 spp.) — Our only genus, mostly in Andean cloud forests, with a few species ranging north to Mexico or east to coastal Brazil.

Two other small genera occur in Patagonia.

TURNERACEAE

A small family, poorly represented in our area. Mostly herbs and subshrubs but a few species small trees to 6 m tall. Leaves always alternate, serrate (mostly crenate to remotely subserrate), with cuneate base and ascending pinnate venation, usually rather small; differs from *Sida* or other Malvales in lacking 3-veined base. Flowers always yellow or orangish, solitary in the leaf axils, with 5 sepals, petals (these fused near base) and stamens. The feathery stigma is the most distinctive feature of the flower, each of the 3 styles topped by a strikingly divided feathery stigma. Fruit a short round 3-veined capsule with thin valves and small arillate seeds.

Turnera (60 spp., plus 1 in Africa) — Weedy herbs to shrubs or small trees (to 6 m in *T. hindsiana*). Leaves broader than *Piriqueta*, elliptic to +/- obovate. Flowers lack corona and the three styles are simple.

E: damiana

Piriqueta (20 spp., incl. Old World) — Coarse annual herb of dry savannahs (e.g., Tarapoto, Peru). Vegetatively distinctive in family in stellate pubescence and narrower, oblanceolate leaves. Flowers with corona and divided styles (each with a feathery apex).

Several other genera occur amphitropically.

ULMACEAE

Mostly medium-sized to large trees (one *Ampelocera* species is a small shrub and the commonest *Celtis* is a spiny liana) with simple leaves having +/- asymmetric bases. The leaves are usually alternate (except *Lozanella*), usually 3-veined (the lateral veins tending to arise below lamina base), usually serrate and/or asperous, the petioles of equal lengths, without a pulvinar apex (unlike Malvales). Although some species lack some of these characters, they are definitive for Ulmaceae when found together. Flowers tiny and inconspicuous, apetalous, typically greenish, usually with masculine and feminine flowers borne separately in small axillary inflorescences (sometimes reduced to fascicles or subsessile). Fruit single-seeded, usually yellow or orange, a small, fleshy drupe (*Celtis, Ampelocera*), berrylike (*Lozanella, Trema*) or with a pair of very unequal wings (*Phyllostylon*). The trees can often be recognized by the rather small narrow buttresses, typically with longitudinal greenish striations showing through the bark.

Three genera have conspicuously 3-veined leaves — *Lozanella,* distinctive in being opposite, *Trema* in teeth (when present) fine and close-together, or when entire with inconspicuous not very prominulous tertiary venation, *Celtis* in frequent presence of spines, sometimes liana habit, the teeth (when present) rather coarse and irregular and when margin entire, the tertiary venation distinctively prominulous. *Ampelocera* usually has entire leaves with distinctly asymmetric sub-3-veined bases, if not entire, the teeth widely scattered. *Phyllostylon* has pinnate venation (as do most of the extralimital genera).

Phyllostylon (3 spp.) — In our area, only in northern Colombia where recently discovered. Large tree of very dry scrub forests with characteristic somewhat fluted trunk and tendency to multiple trunks. Leaves rather small and obtuse with irregular teeth mostly toward apex, pinnately veined, base not asymmetric. Fruit very unlike other South American Ulmaceae in being a samara, similar to *Acer* or *Securidaca* but with a peculiar tiny second wing at base of the large elongate obvious one; wing asperous.

Lozanella (2 spp.) — Andean cloud forests. Unique in family in opposite leaves; very much like *Trema* except the opposite leaves. Leaves always finely serrate, rough-surfaced (at least above), strongly 3-veined and with a prominent interpetiolar line.

Trema (30 spp., mostly Old World) — Second-growth trees with +/- contracted axillary inflorescences and tiny orange berries. Only two species in our area, one serrate and asperous, the other entire with +/- smooth surface.

C: surrumbomo, tortolero; E: sapán, sapán de paloma; P: atadijo

Ulmaceae

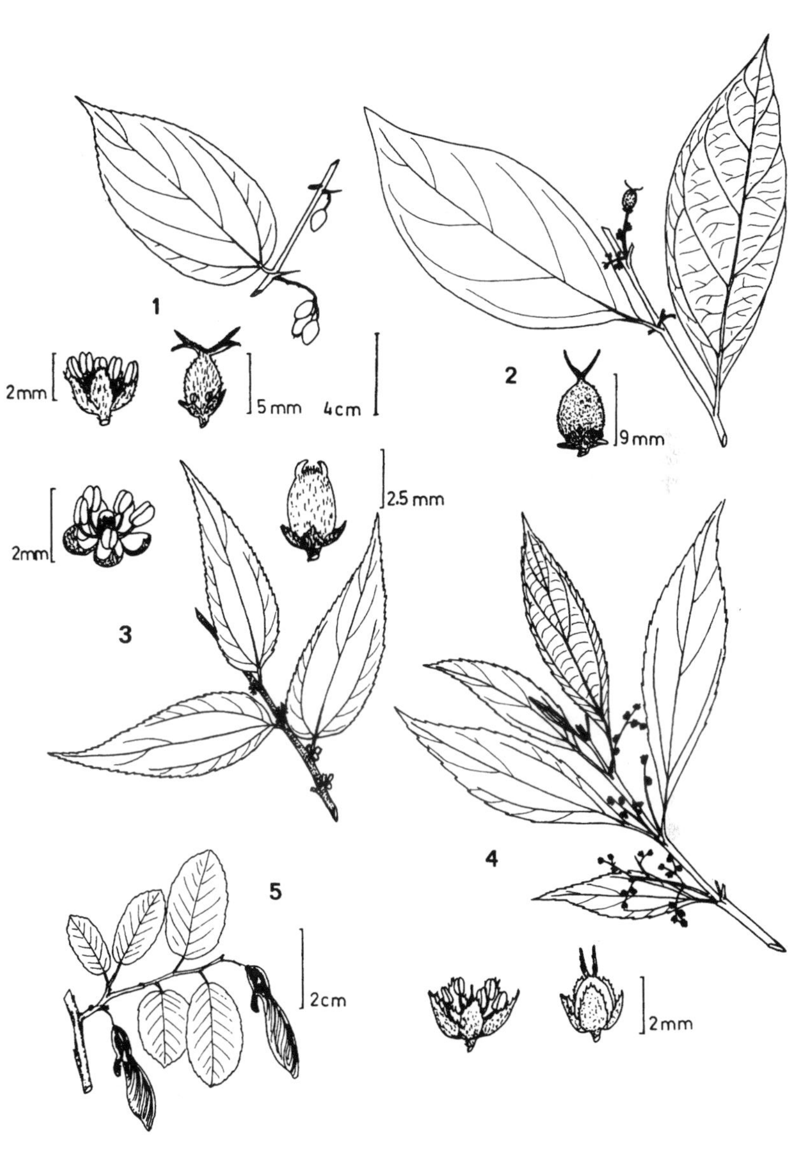

1 - *Celtis*

2 - *Ampelocera*

3 - *Trema*

4 - *Lozanella*

5 - *Phyllostylon*

Celtis (incl. *Sparrea*) (80 spp., mostly Old World and n. temperate) — Two common species in our area, one a spiny liana, the other a large tree of rich alluvial soils; also an unarmed tree of northern Colombian seasonal forests and several more or less spiny-Andean and extralimital taxa. Leaves strongly 3-veined, irregularly toothed at least toward apex (completely entire in *C. schippii*, recognizable by the noticeably asymmetric base).

E: palo blanco; gallinazo, tillo blanco (*C. schippii*)

Ampelocera (9 spp.) — Mostly large moist- and wet-forest trees (one undescribed species a small shrub), poorly known because of the small flowers, and mostly extremely short flowering periods. Leaves almost entire (sometimes with few scattered dentations) but ulmaceous in the distinctly asymmetric base, usually more or less inconspicuously 3-veined, at least on one side; tertiary venation more or less parallel and perpendicular to midvein. Fruits round or somewhat compressed, very characteristic in usually rough surface and asymmetric position of the persistent stigma, the ventral fruit margin longer than dorsal margin. Although notoriously nondescript, characterized by prominent stipule scars and a very typical appressed blunt axillary bud.

P: ají caspi, palo ají, yutobanco

Extralimital genera (All Central American): *Aphananthe*, (*Celtis*-like but pinnately veined), *Chaetoptelea* and *Ulmus*, with (very similar pinnately veined elm leaves). There is also a newly discovered genus from Central America, characterized especially by conspicuous stipules when young, which is related to Ulmaceae but has been described as the distinct family, Ticodendraceae.

UMBELLIFERAE

A large family of mostly north temperate herbs; in our area, almost entirely restricted to the high Andes. The family is the herbaceous counterpart of Araliaceae and sometimes included in that family. Umbelliferae are very distinctive in the characteristic umbellate inflorescences of small flowers, the distinctive pungent vegetative aroma, hollow stems, and the usually much dissected leaves with sheathing petiole bases. While there are at least 48 genera in the Neotropics including 16 native or naturalized in the tropical Andes, only two, both anomalous in the family, occur in the tropical lowlands of our region. The largest area genus, *Hydrocotyle* (52 spp., plus 50 in Old World), mostly Andean but with several cloud-forest species and a few weeds reaching the lowland tropics, is distinctive in its round undivided leaves and usually creeping habit. Cosmopolitan *Eryngium*, unusual in the family in the congested nonumbellate inflorescence and also distinct in the usually +/- spiny- or ciliate-margined leaves, has a few endemic Andean species as well as the widespread tropical lowland weed *E. foetidum.* There are also three genera in the coastal lomas: *Domeykoa* (4 spp.) is an endemic loma genus, *Eremocharis* (9 spp.) occurs in the lomas as well as the dry western Peruvian Andean slopes, and weedy *Spananthe* (1 sp.) has an endemic loma variety. Of the Andean genera, *Azorella* (70 spp.) is a characteristic high-Andean (and Patagonian) cushion-plant and *Niphogeton* (16 spp.) and *Bowlesia* (11 spp.) represent autochthonous Andean elements although both reach Central America, while predominantly Central American *Arracacia* (25 spp.) has a secondary radiation in Peru. Also noteworthy is cloud-forest *Oreomyrrhis* (4 spp., plus 40 north temperate) our only subwoody genus.

URTICACEAE

Mostly shrubs or small (occasionally to 10 m in *Urera*) trees but some *Urera* are lianas, *Pilea* is often epiphytic, and *Urtica, Pilea, Fleurya,* and *Parietaria* are completely herbaceous. Characterized by simple strongly 3-veined serrate (rarely entire: *Pouzolzia, Parietaria;* laciniately lobed in *Urera laciniata*) leaves, usually with cystoliths in upper surface and/or with stinging hairs (*Urtica, Urera, Laportea*). Mostly alternate, when opposite the leaves often strongly anisophyllous with one member of each leaf pair much smaller (most *Pilea*). Stipules small but usually present and very conspicuous in *Pouzolzia* and most *Pilea.* Can be told from related Ulmaceae by the cystoliths, from Moraceae by lacking milky latex and a conical terminal stipule and from Malvales by lacking a swollen pulvinus at petiole apex.

Flowers always very small and greenish, brownish, or whitish with perianth reduced or lacking; unisexual, the plants either monoecious or dioecious. Fruit usually a small achene or fleshy and berrylike (*Urera*).

1. LEAVES OPPOSITE (HERBS)

Urtica (50 spp., incl. n. temperate) — Exclusively montane in our area. Along with some Loasaceae, our only opposite-leaved genus with stinging hairs. Leaves more jaggedly serrate than in other Urticaceae taxa. Inflorescence axillary, the small green flowers in short spikes or few-branched with spicate branches.

Pilea (400 spp., incl. Old World) — Mostly succulent terrestrial herbs of cloud forests, some epiphytic, a few weedy. Often with conspicuously anisophyllous leaves, these rather bluntly serrate to crenate or even entire. Inflorescence various but usually not a sessile fascicle. Stipule intrapetiolar and often conspicuous.

Urticaceae
(Herbs)

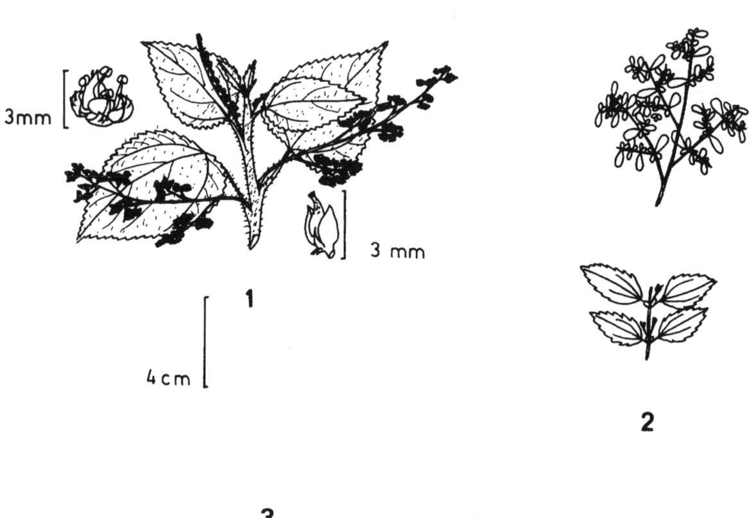

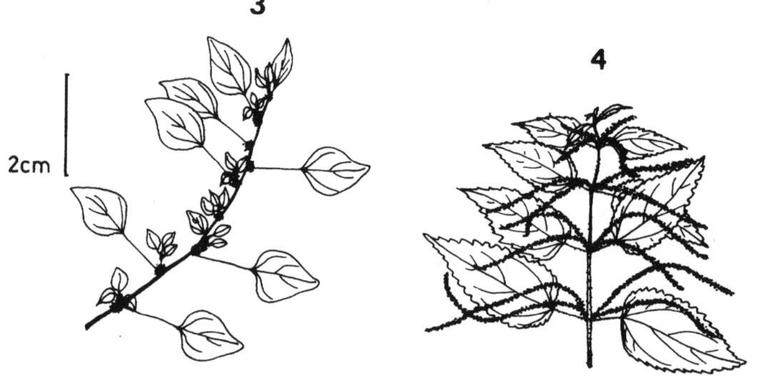

1 - *Laportea* **2** - *Pilea*

3 - *Parietaria* **4** - *Urtica*

Urticaceae
(Trees, Shrubs, and Lianas)

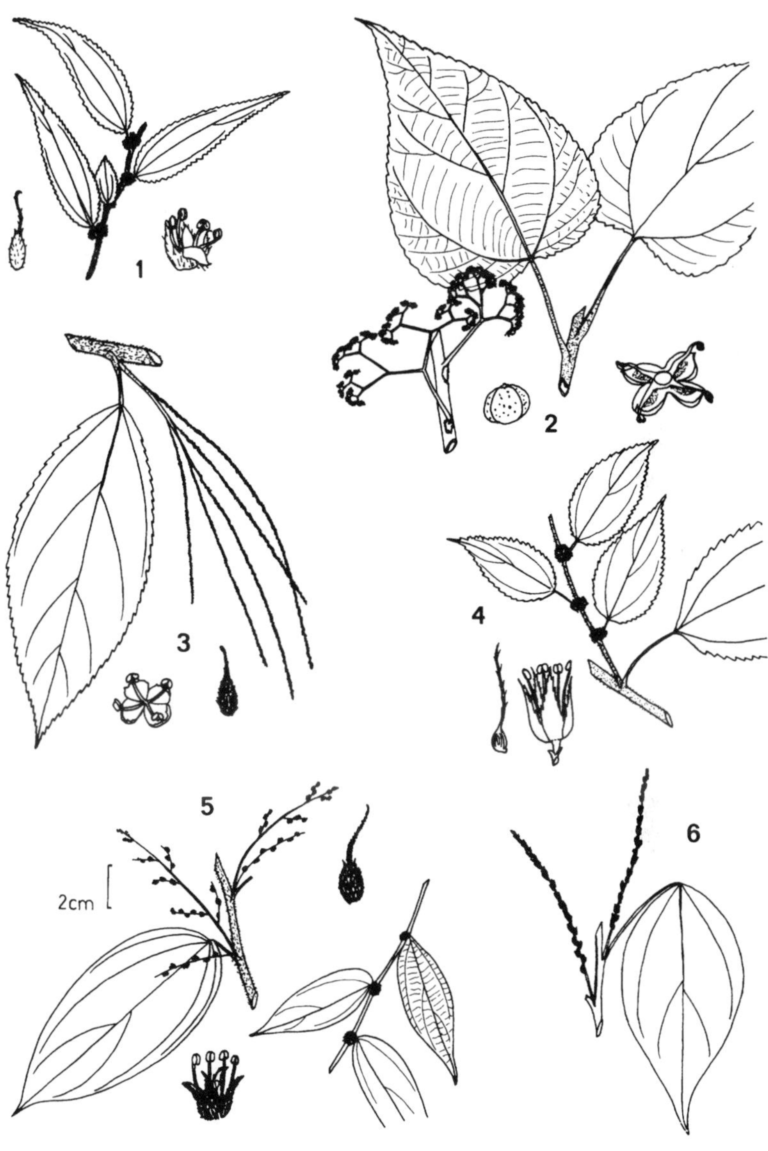

1 - *Phenax* **2** - *Urera*

3 - *Myriocarpa* **4** - *Boehmeria*

5 - *Pouzolzia* **6** - *Myriocarpa*

(***Boehmeria***) — A few species (e.g., *B. cylindrica*) have opposite leaves, differing from *Pilea* in shrubby habit and sessile axillary inflorescence (but leaves of flowering shoots sometimes suppressed and these appearing spicate.).

2. Leaves Alternate

2A. Woody trees, shrubs, or lianas

Myriocarpa (15 spp.) — Small trees mostly along streams in moist and wet forest. Leaves larger than in most other genera. Inflorescences very distinctive, the tiny flowers in long, slender, pendent, whitish spikes (or these clustered), female flowers completely lacking perianth (unique except for *Phenax*).

E: ortiguilla macho

Urera (35 spp., also in Africa and Hawaii) — Mostly small trees and treelets with stinging hairs; occasionally to 10 m (*U. capitata*) and sometimes scandent. Sometimes with small stinging spines on stems or petioles. Mostly cauliflorous, the inflorescence paniculate, +/- corymbose. Fruit a small berry, usually orange or whitish, the inflorescence often contrastingly colored.

E: ortiga, ortiguilla de tigre, crespón (*U. caracasana*); P: ishanga (*U. caracasana*), mara mara (*U. baccifera, U. laciniata*).

Pouzolzia (50 spp., incl. Old World) — Shrubs or lianas of lowland moist and wet forest. Leaves usually entire (serrate in +/- scandent herbaceous vine *P. longipes*). Stipules larger and more conspicuous than in *Phenax* and *Boehmeria.* Inflorescence an axillary or ramiflorous glomerule, usually with fewer flowers than in related genera. Commonest Amazonian species (*P. formicaria,* a tahuampa forest liana) has leaves of fertile branchlets suppressed, these thus resembling long spikes.

Boehmeria (100 spp., incl. Old World) — Mostly cloud-forest shrubs (to 5 m tall), sometimes scandent (*B. anomala, B. bullata*). Leaves serrate, usually alternate, with large and small leaves at adjacent nodes, often +/- bullate. Inflorescence sessile, ramiflorous and axillary, glomerulate (sometimes the leaves of flowering branches suppressed and the inflorescence thus appearing spicate).

Phenax (25 spp., plus 2 in Madagascar) — Very similar to *Boehmeria* with serrate leaves and flowers in axillary and ramiflorous glomerules, but differing in female flowers completely without perianth (if the fruiting inflorescence squeezed, the naked seeds fall out whereas in *Boehmeria* these are covered by perianth remains). Leaves tend to be more membranaceous than in *Boehmeria.*

Hemistylus (4 spp.) — Shrubs or treelets with a peculiar disjunct distribution from Colombia and Venezuela to Guatemala. The alternate, long-petiolate, 3-veined leaves resembling *Myriocarpa*, but the large, densely leafy-bracted panicle distinctive.

2B. Herbs (with alternate leaves)

Laportea (incl. *Fleurya*) (23 spp., incl. n. temperate) — Only one species in our area, a lowland moist-forest weed, the leaves serrate, usually with stinging hairs. Inflorescence an open more or less pyramidal panicle.

E: ortiguilla; P: ortiga

Parietaria (30 spp., incl. Old World) — Tiny tenuous prostrate herb of moist places; also in Peruvian lomas. Leaves small, entire, nonstinging. Inflorescence few-flowered, glomerulate in leaf axils.

Valerianaceae

A mostly herbaceous, basically Laurasian family, in our area restricted to the uplands, as paramo or puna herbs or as cloud-forest vines. Leaves always opposite (or in basal rosettes in high-altitude genera), often compound. Easily recognized by the strong odor (of valerian), especially from the roots and even when dry and the tendency to dry blackish. The largest neotropical genus is *Valeriana* which ranges from simple-leaved herbs to a common cloud-forest species of scrambling herbaceous vine with 3-foliolate leaves (*V. scandens*) to distinctly woody climbers with simple or pinnate leaves. The vine species are similar to comps in having the leaves 3-veined from above the base and often slightly and irregularly shallowly toothed; they are characterized by having the bases of opposite petioles interconnected to form a kind of ochrea-like nodal sheath and by the tendency to dry blackish; a number have hollow twigs. Even the fruits are similar to single composite achenes (even to the pappus). *Astrephia* (monotypic), a succulent clambering herb, has bicompound leaves and a tiny-flowered inflorescence which rather resembles an umbellifer except for the opposite leaves.

The other genera are high-Andean rosette plants with sessile inflorescences — *Anetiastrum, Belonanthus, Phyllactis,* and *Stangea.*

Valeriana (200 spp., mostly Old World) — The relatively few woody lianas (all in montane cloud forest) can be vegetatively recognized by the rather thick smooth bark with raised corky lenticel-like projections, and by the softly flexible stem.

Valerianaceae and Verbenaceae (Lianas and Compound-Leaved Trees)

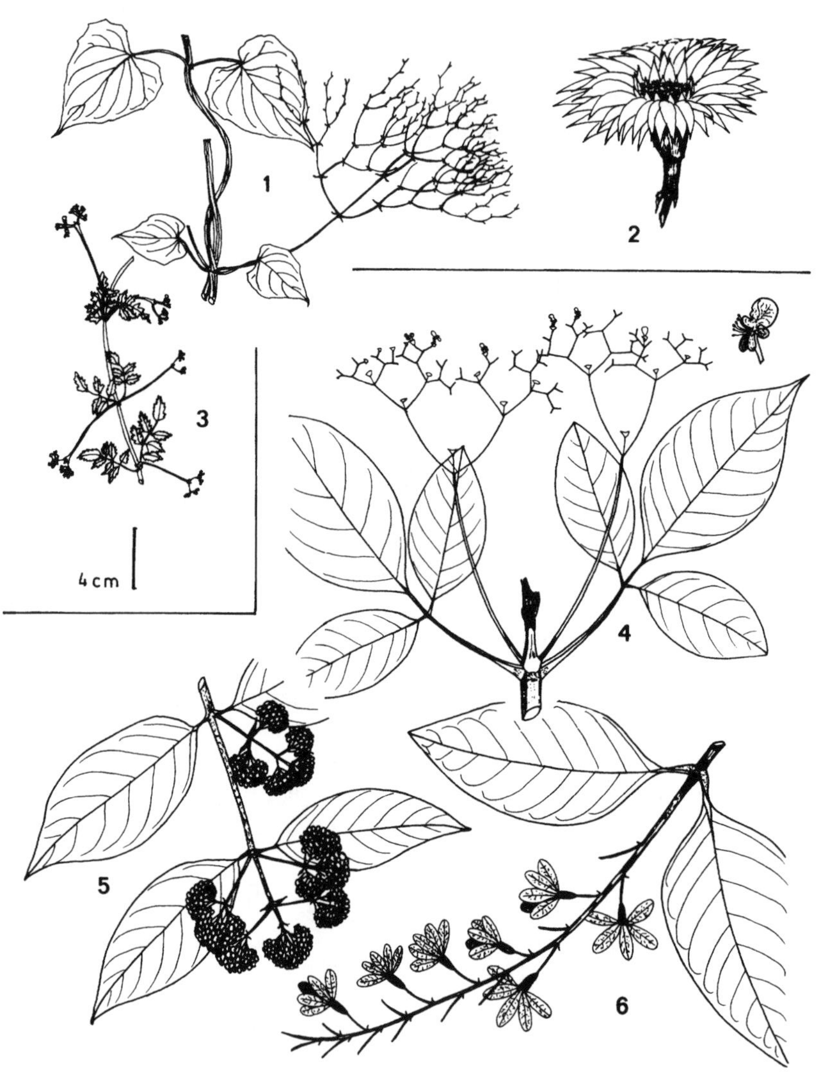

1 - *Valeriana* (Valerianaceae)

2 - *Phyllactis* (Valerianaceae)

3 - *Astrephia* (Valerianaceae)

4 - *Vitex*

5 - *Aegiphila*

6 - *Petrea*

Verbenaceae

A diverse family, in our area including genera of herbs, shrubs, trees, and lianas, all characterized by opposite (or whorled) leaves (alternate in *Amasonia* but may not reach area), usually simple but palmately compound in *Vitex*. The herb genera are more or less intermediate with Labiatae and resemble Labiatae in aromatic leaves and usually sharply tetragonal stems, differing in addition to the technical characters (terminal style and unlobed or slightly lobed ovary) in the inflorescence (except *Priva*) densely spicate or capitate (shared with *Hyptis* of our Labiatae). The woody taxa mostly are nonaromatic and with subtetragonal or irregularly angled twigs; they include one liana genus (*Petrea*) characterized by the asperous leaves (plus a number of nondescript scandent *Aegiphila* species), four tree genera — *Avicennia* (in mangroves), *Citharexylum* (white flowers, often in long spicate inflorescence; base of lamina with pair of elongate leaves where it joins petiole or leaves coriaceous and with few sharp teeth); *Vitex* (palmately compound leaves); *Cornutia* (blue flowers with only 2 stamens in large panicle; branchlets sharply tetragonal, leaves densely simple pubescent) — and seven shrub/small tree genera (*Duranta* (axillary spines); *Callicarpa* (floccose-stellate pubescence); *Aegiphila* (vegetatively nondescript; inflorescence often large and paniculate; flowers smallish, salverform, white or greenish-cream); *Clerodendrum* (flowers larger than *Aegiphila* and inflorescence fewer flowered; node slightly expanded to form almost stalked petiole attachment); *Lippia* (leaves aromatic and serrate; flowers white and usually in head); *Aloysia* (like *Lippia* but inflorescence an elongate spike); *Amasonia* (alternate leaves). Most of the trees have the base of the more or less fleshy fruit enclosed by expanded cupular calyx.

1. Lianas

Petrea (30 spp.) — Lowland forest lianas easy to recognize by the asperous opposite leaves. Flowers showy and distinctive with the openly infundibuliform dark blue or violet deeply 5-lobed corolla subtended by similar shaped, lighter blue or purplish calyx, the calyx persistent and dry in fruit, forming a wind-dispersed 5-winged samaroid.

(*Aegiphila*) — A few species are lianas, usually with membranaceous entire leaves and sometimes conspicuously pubescent but generally vegetatively very nondescript.

Verbenaceae
(Trees: Simple Leaves)

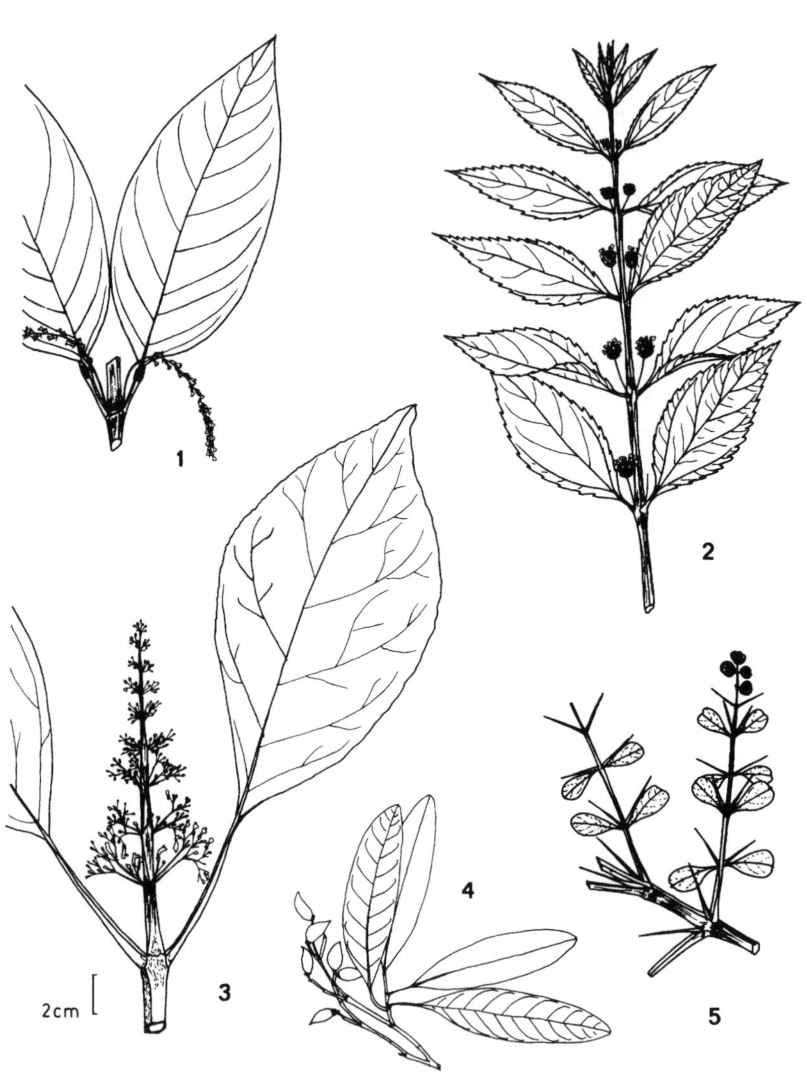

1 - *Citharexylum* **2** - *Lippia*

3 - *Cornutia* **4** - *Avicennia* **5** - *Duranta*

2. Trees and Shrubs — (Arranged more or less in order of decreasing stature).

Vitex (56 spp., plus 200 Old World) — Canopy to emergent trees, especially prevalent in seasonally dry-forest. The fibrous-papery bark is distinctive as are the usually conspicuous, blue strongly zygomorphic flowers. This is the only Verbenaceae with compound leaves (occasionally 1-foliolate), and is more likely to be confused with *Tabebuia* of the Bignoniaceae than with other verbenacs. *Vitex* differs from *Tabebuia* in the leaflet bases tapering into indistinct petiolule and from almost all *Tabebuia* species in our area by pubescence of simple trichomes.

C: trúntago; E: pechiche, guayacán pechiche; P: huingo

Avicennia (5 spp., plus 10 Old World) — Mangrove trees with pencil-like pneumatophores (thinner than *Laguncularia*). Leaves coriaceous, entire, pale below, the petiole bases connected by interpetiolar line or ridge. Petioles lacking salt-excreting glands of *Laguncularia.* Flowers sessile in reduced panicles or these further reduced to small spikes, the corolla 4-parted, white. Fruit irregularly oblong-ellipsoid, one-seeded, grayish-tomentose when young.

C, E: mangle negro

Cornutia (12 spp.) — Soft-wooded trees of wet- and moist-forest second growth with tetragonal stem and aromatic foliage like the herb genera. Leaves large and entire, usually simple pubescent. Inflorescence a large terminal panicle, the blue flowers zygomorphic as in *Vitex* but leaves simple; only 2 fertile stamens (unique in tree verbenacs).

Citharexylum (112 spp.) — Two rather different growth-forms: 1) trees of lowland second growth with tetragonal stems and large entire leaves usually very distinctive in pair of narrow glands at base of attenuate lamina and 2) Andean trees and shrubs, especially in drier areas, with coriaceous leaves usually with few sharp teeth near apex and with the twigs irregularly or not at all tetragonal. The lowland trees have long pendent spicate or narrowly racemose inflorescences, the upland taxa have these reduced sometimes to single axillary flowers. The flowers of both types are short-salverform and white.

Duranta (36 spp.) — Spiny shrubs, mostly of disturbed cloud forest or dry Andean shrub forest. Distinctive in the axillary spines and blue flowers on open pendent terminal spikes or narrow racemes. Leaves entire to serrate, often rather small; stems irregularly tetragonal. Fruit round, yellow to orangish, +/- fleshy, subtended by or enclosed in calyx.

Aegiphila (160 spp.) — Mostly shrubs and small trees of lowland moist-forest; also a few liana species. Vegetatively rather nondescript, the leaves always entire, more or less membranaceous. Inflorescence usually

Verbenaceae
(Shrubs)

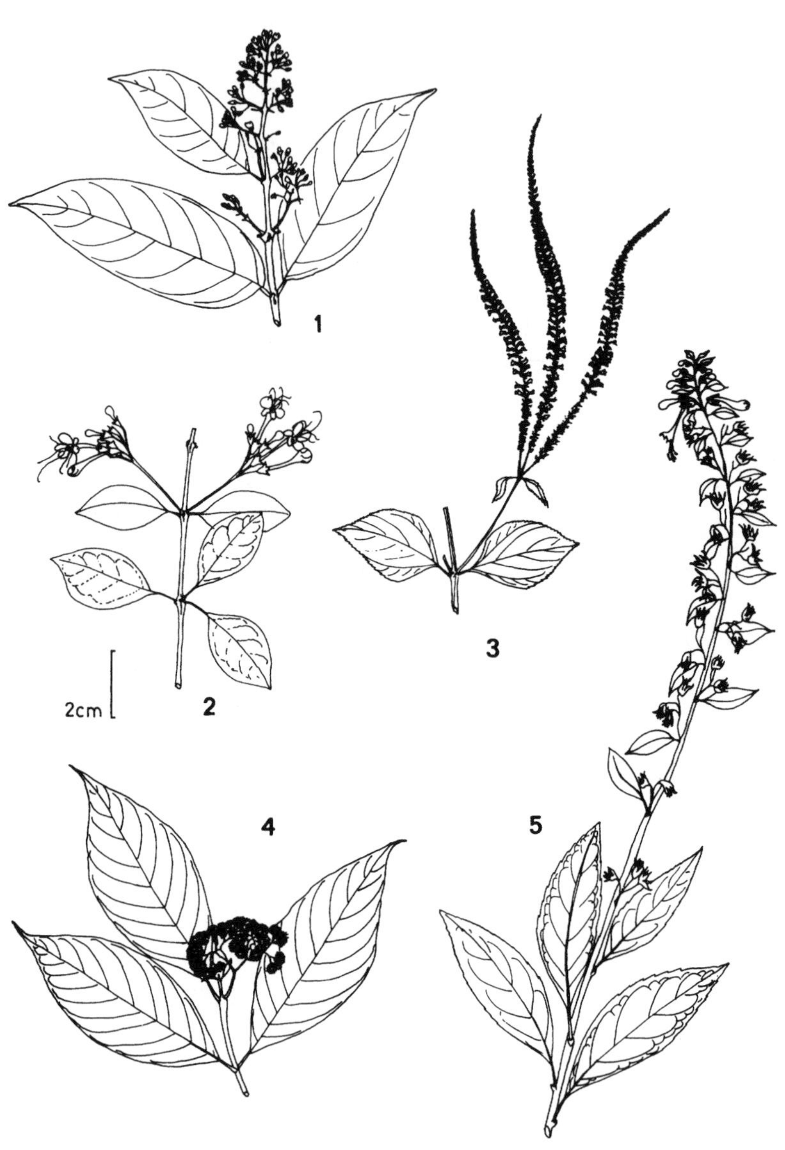

1 - *Aegiphila*

2 - *Clerodendron* **3** - *Aloysia*

4 - *Callicarpa* **5** - *Amasonia*

multiflowered and pyramidal-paniculate (reduced to few axillary flowers in several montane or cloud-forest taxa), the flowers salverform to short-salverform, white, 4-parted (except 5-parted in the cloud-forest taxa with reduced inflorescences). Calyx enlarged to form cup around base of fruit in all taxa.

C: queso fresco; E: lulo (*A. alba*)

Clerodendrum (26 spp., plus 375 in Old World) — Mostly dry-area shrubs (except for one understory species of lowland Amazonia). Vegetatively characterized by entire, usually small leaves on more or less flattened 2-angled twigs, the petiole attachments usually distinctly raised, sometimes even subspiny. Inflorescence more open with fewer flowers than *Aegiphila,* the individual flowers larger and long pedicellate, with long-exserted anthers.

Callicarpa (25 spp., plus 120 in Old World) — Shrub or small tree of weedy second growth, especially behind beaches. Vegetatively characterized by the floccose-stellate indumentum and large membranaceous serrate leaves. Inflorescence a many-flowered, flat-topped axillary cyme with small white or greenish flowers and fruit a conspicuous purple (turning blackish) berry (without subtending calyx).

Lippia (176 spp., plus 40 in Old World) — Much more prevalent in the cerrado; in our area, mostly trees of disturbed Andean cloud forests, characterized by the serrate aromatic leaves, and variously capitate inflorescences. Branchlets irregularly angled but rarely sharply and evenly tetragonal. Some species have the inflorescence heads arranged into panicles; those with simple axillary heads are very like *Lepechinia* or *Hyptis* (Labiatae) but can be distinguished from the most similar tree mints by leaves with more ascending secondary veins.

Aloysia (40 spp.) — Dry-area shrubs with square twigs and aromatic foliage. Differs from *Lippia* in the elongate spikes of small white flowers (calyx ca. 2 mm long and deeply 5-toothed). Leaves often entire.

Amasonia (8 spp.) — Erect woody subshrub, unique in family in alternate leaves. The genus is mostly found in the cerrado and there is a single old record from our area (Tarapoto). Except for the alternate leaves, looks more like Acanthaceae than Verbenaceae, with large foliaceous red inflorescence bracts and a narrowly tubular yellow corolla with long exserted anthers and broadly campanulate deeply toothed red calyx. The stem is angled but not tetragonal and the leaves of our species rather large, membranaceous and remotely toothed.

3. Herbs — (All with square stems and sessile or subsessile flowers in a spike [or this reduced to capitate head])

Lantana (150 spp., plus few in Africa) — Weedy shrubs or subshrubs, sometimes subscandent, often with small prickles on branch angles; leaves rather evenly crenate-serrate. The epitome of the Verbenaceae flower plan, the short-salverform butterfly-pollinated flowers brightly colored and clustered into dense heads. Fruit a fleshy black berry.

E: cinco negritos (*L. camara*); P: pampa orégano (*L. camara*)

Verbena (incl. *Glandularia*) (200 spp., plus 50 N. Am. and Old World) — Leaves narrower and more jaggedly toothed than in our *Lantana* (in *Glandularia* laciniate into linear segments). Inflorescence narrowly spicate, less congested than in *Lantana.* Fruit almost as in Labiatae, splitting into 4 nutlets.

E: verbena

Hierobotana (1 sp.) — More or less prostrate high-Andean herb of sandy places. Leaves grayish, linear-lobed, sometimes appearing to form whorls of linear leaves. Inflorescence more or less spicate. Differs from *Verbena* in having only 2 fertile stamens.

Phyla (10 spp., plus few N. Am.) — Trailing herbs or subshrubs, mostly in wet places. Essentially a reduced herbaceous *Lippia,* differing in malpighiaceous trichomes and the denser more cylindrical axillary spikes that elongate in fruit.

Stachytarpheta (106 spp., plus few in N. Am.) — Weedy herbs with evenly serrate leaves, the lamina decurrent on the poorly defined petiole. Inflorescences very characteristic, densely narrowly spicate and erect with the calyces appressed against rachis. Florally characterized by only 2 fertile stamens (unique except for *Cornutia* and *Hierobotana*) and the blue to deep violet color of the strongly zygomorphic rapidly caducous corolla.

Bouchea (16 spp.) — Weedy herbs. Same as *Stachytarpheta* but 4 stamens and the flowers more scattered along inflorescence and calyx not appressed against rachis. Leaves more membranaceous and coarsely serrate than *Stachytarpheta.*

Pitraea (incl. *Castelia*) (1 sp.) — South temperate herb reaching our area only in southernmost coastal Peru. Intermediate between *Stachytarpheta* (which it resembles in the individual flowers) and *Priva* (with which it shares a fruit enclosed in the expanded [but not sticky and exozoochorous] calyx); the inflorescence also intermediate with the flowers rather sparsely clustered along the spike.

Verbenaceae
(Herbs)

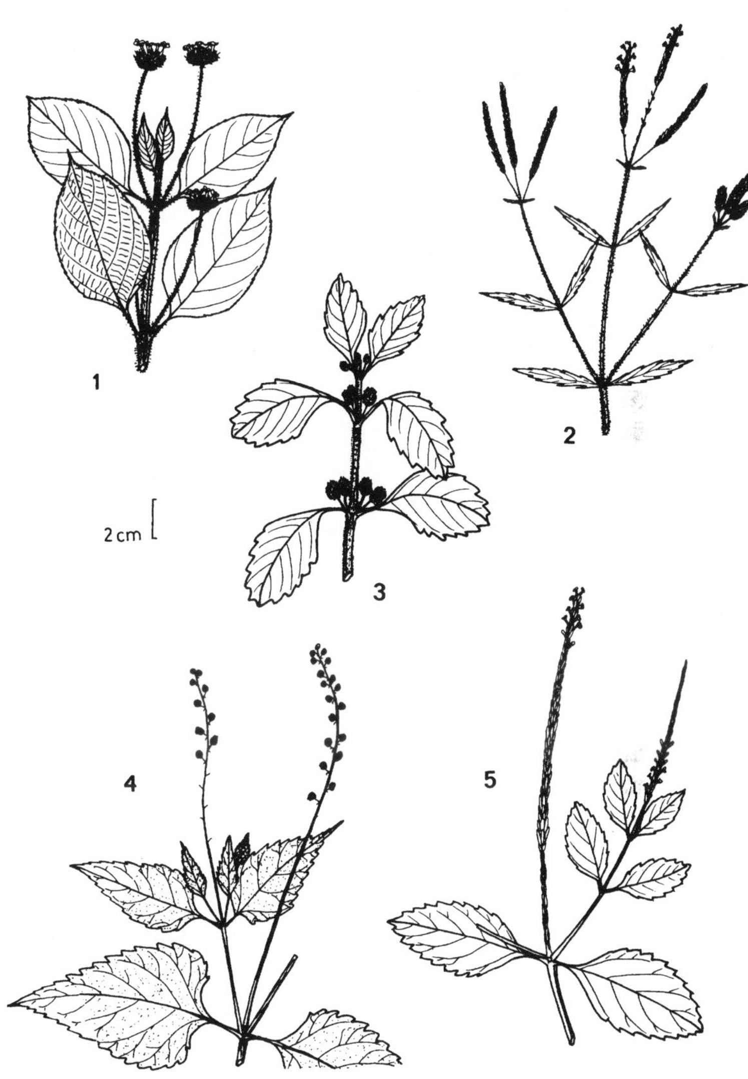

1 - *Lantana* **2** - *Verbena*

3 - *Phyla*

4 - *Priva* **5** - *Stachytarpheta*

Priva (11 spp., also 10 in Old World) — Weeds with exozoochoric fruit; the calyx sticky from minute uncinate trichomes and enclosing fruit. Leaves rather tenuous, triangular-ovate, serrate. Inflorescence spicate but with the inconspicuous pale bluish to pinkish flowers widely separated.

E: cadillo

(*Lippia*) — Some *Lippia* species are subshrubs, differing from *Lantana* in toothed rather than truncate calyx.

VIOLACEAE

This family is one of the best examples of how temperate-zone biases color botanical perception. In reality, Violaceae is a predominantly woody family of trees (even large ones), shrubs, and woody lianas with one shrubby to herbaceous genus (*Hybanthus*) and one largely herbaceous genus (*Viola*) reaching the Temperate Zone. As is so often the case, the virtual vacuum of the Temperate Zone allowed that single herbaceous genus to explode and, in fact, *Viola* constitutes half the species of the family. The persuasive influence of our familiarity with that single genus of this rather large and diversified family is obvious to anyone who has tried to convince a tropical neophyte that a tree like *Leonia* with tiny inconspicuous actinomorphic flowers and large woody indehiscent fruits is a far more typical violac.

On technical grounds Violaceae are taxonomically close to Flacourtiaceae, from which they differ chiefly in fewer stamens ([3–]5 rather than usually 10 or more). The similarity is obvious, vegetatively, as these are the two main components of the rather nondescript category of "alternate, serrate-leaved small trees". Both families also have stipules, some taxa with glandular-punctate leaves, and predominantly 3-carpellate fruits. Vegetatively, many Violaceae cannot be distinguished from nondescript flacourts like *Casearia* without first recognizing the genus. By far the commonest Violaceae genus in lowland neotropical forests is *Rinorea* and the common species of *Rinorea* all are shrubs or small trees with opposite serrate leaves, an almost unique character combination in lowland forest (shared only with a few Hippocrateaceae). *Gloeospermum* has similar but uniformly alternate *Casearia*-looking leaves; in Amazonian Peru, at least, the leaves of the commonest species tend to dry rather light green with a paler central area. *Leonia glycicarpa* is one of the commonest (but vegetatively least distinctive) Amazonian trees; its nondescript entire-margined coriaceous leaves have no obvious distinguishing characteristics, although the presence of cauliflorous fruits (or their scars) on the trunk may be useful.

In flower, the genera related to *Viola* are easy to recognize by their solitary zygomorphic axillary flowers with enlarged lower petals; the enlarged petal is spurred in *Viola, Anchietea,* and *Noisettia,* but only saccate in *Hybanthus. Corynostylis,* which has large, spectacularly spurred, white flowers, and *Anchietea,* with flowers like *Viola,* are unique in our area in being woody lianas. Even the supposedly actinomorphic-flowered genera like *Paypayrola, Gloeospermum* and *Rinorea* usually have a slightly violet-like floral asymmetry when viewed head on. Based on floral characters, a progression may be envisioned from primitive woody, actinomorphic-flowered genera like *Leonia, Fusispermum,* and *Rinorea,* to woody but subzygomorphic-flowered genera like *Paypayrola* and *Amphirrhox,* to strongly zygomorphic (but with the enlarged petal merely saccate) shrubby to subshrubby *Hybanthus* to more or less herbaceous (to subshrubby) genera with strongly zygomorphic flowers with the enlarged petal spurred (*Viola, Anchietea, Noisettia,* these differentiated largely by fruit features).

1. Strongly Zygomorphic Flowers; Mostly Herbs or Woody Lianas, Sometimes Shrubs or Subshrubs

Corynostylis (4 spp.) — Woody lianas, mostly found in seasonally inundated forests or sometimes dry forests. Vegetatively distinguishable by the slightly but distinctively raised petiole attachment of the nondescript elliptic alternate leaves. The conspicuous flowers are large, white, strongly zygomorphic, and completely unmistakable; the fruits are also very distinctive, round with 3 woody valves and thin-winged seeds packed inside.

Viola (400 spp., mostly n. temperate) — Entirely upland herbs in our area. The violet flowers are exceedingly characteristic but the diversity of leaf types striking for one accustomed to temperate zone violets; many high-Andean *Viola* species are rosette herbs and the leaves may be linear, narrowly triangular, round, elliptic, oblanceolate, entire, or variously serrate, *et cetera.*

Noisettia (1 Amazonian sp.) — Essentially a subshrubby lowland *Viola;* the flowers are yellow, whereas most Andean *Viola* species have lavender or white flowers.

Anchietea (8 spp.) — Woody lianas, mostly in cloud forest; flowers like *Viola,* but wind-dispersed, the 3 fruit valves larger and thin and papery, usually enclosing conspicuously winged seeds as well.

Hybanthus (150 spp., incl. Old World) — Mostly subshrubs to often weedy herbs but a few species are woody shrubs. The axillary flowers are recognizably similar to *Viola* but the enlarged lower petal is merely saccate rather than spurred. Augspurger has made elegant phenological studies of a Panamanian species with "big bang" flowering.

2. Weakly or Not at All Zygomorphic Flowers; Trees and Large Woody Shrubs

Rinorea (48 spp., plus ca. 110 in Old World) — The commonest neotropical Violaceae genus. Leaves always +/- serrate. There are (different) extremely common small-tree *Rinorea* species with opposite leaves in most Amazonian forests; less common species with alternate leaves are hard to recognize as *Rinorea* when sterile (interestingly, the African species all have alternate leaves). The flowers are always in a raceme, small, cream, and rather urceolate in shape. The conspicuously 3-valved fruit is elastically dehiscent throwing out the autochorous seeds as the drying valve contracts behind them.

P: cafecillo, yutubanco

Gloeospermum (12 spp.) — The other large lowland neotropical violac genus; vegetatively very similar to *Casearia* with alternate serrate leaves. All *Gloeospermum* species have noticeably membranaceous distichous leaves, while *Rinorea* and *Rinoreocarpus* are never distichous. In flower characterized by the reduced few-flowered unbranched or barely branched axillary inflorescence and tiny inconspicuous white flowers. The fruit is a small, round, indehiscent, axillary berry.

E: guayabito

Rinoreocarpus (1 sp.) — The alternate (but not distichous) leaves and flowers are similar to *Gloeospermum* (although the inflorescence is more prominently branched), the red-orange flowers longer than in *Rinorea* while the dehiscent fruits are like those of *Rinorea* (but with more seeds).

Paypayrola (8 spp.) — Mostly smallish trees. The yellowish flowers are a bit longer than in *Rinorea* and slightly zygomorphic; unlike *Rinorea* they are more or less sessile on the usually few-flowered inflorescence. The fruit is like an overgrown, nonelastic *Rinorea* one with more seeds and less deeply sulcate carpel margins. The technical defining character of the genus is the completely united filaments.

Amphirrhox (3 spp.) — The commonest species is a shrub of seasonally inundated forest; others may be small forest trees. Similar to *Paypayrola* in its rather elongate subsessile flowers with narrow petals, differing technically in the filaments united only at base (also rather similar to *Diclidanthera* of the Polygalaceae). The inflorescence usually consists of an elongate peduncle, sometimes forked apically, with a terminal cluster (or clusters) of a few subsessile flowers; the fruits are intermediate between *Rinorea* and *Paypayrola,* small like *Rinorea* and similarly 3-angled but usually incompletely dehiscent and with more numerous seeds.

Violaceae
(Trees)

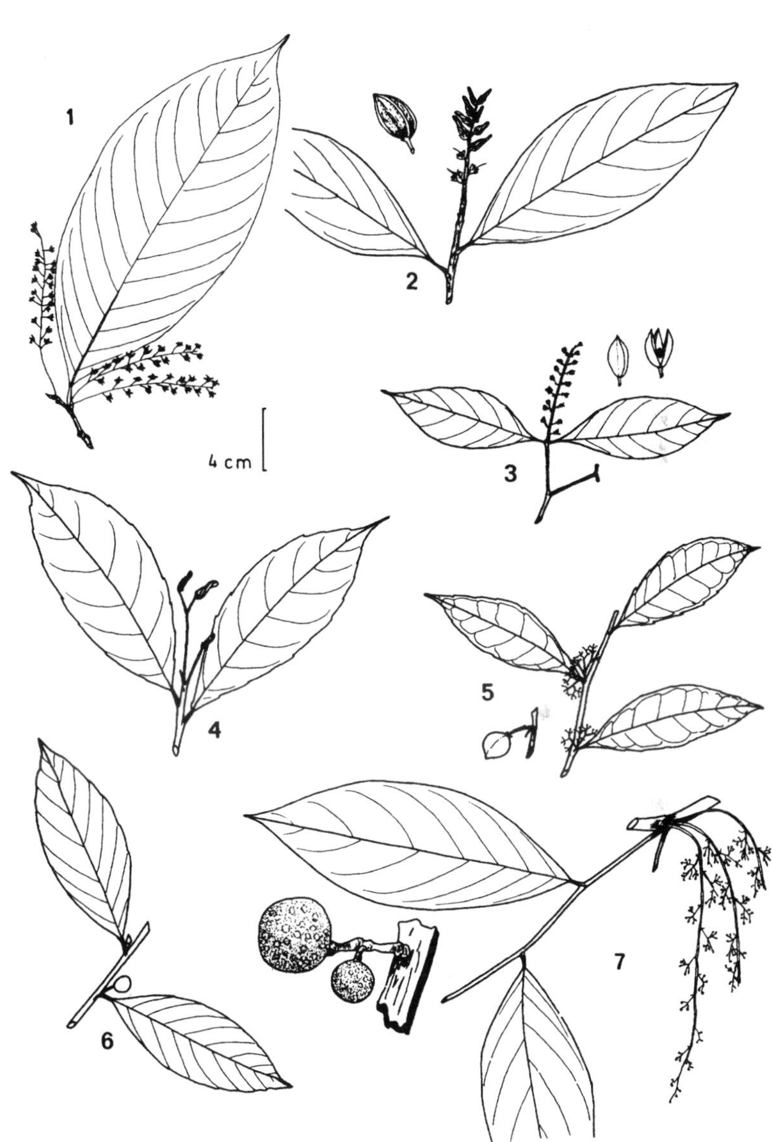

1 - *Fusispermum* **2** - *Paypayrola*

3 - *Rinorea*

4 - *Amphirrhox* **5** - *Leonia* (*L. cymosa*)

6 - *Gloeospermum* **7 & 8** - *Leonia* (*L. glycycarpa*)

Violaceae (Lianas and Herbs) and Vitaceae

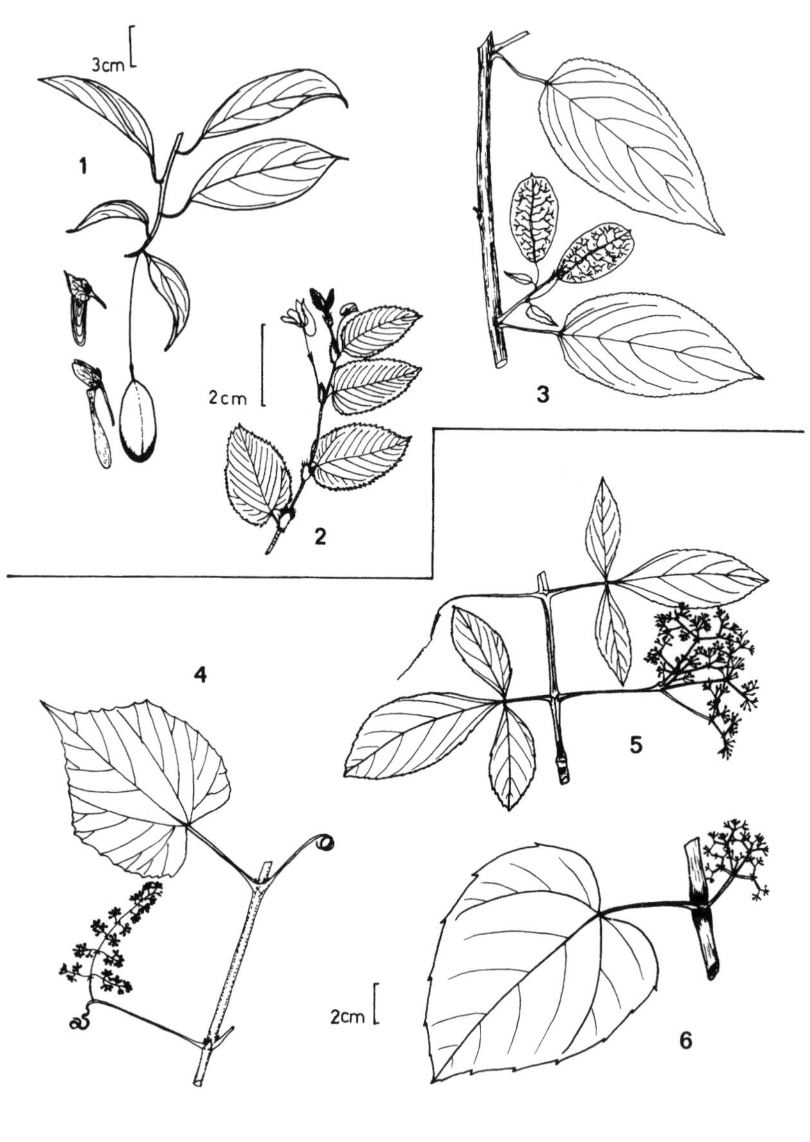

1 - *Corynostylis*
2 - *Viola*
3 - *Anchietea*
4 - *Vitis*
5 - *Cissus* (*C. elata*)
6 - *Cissus* (*C. sicyoides*)

Fusispermum (3 spp.) — Trees. In flower, similar to *Rinorea* in the small-flowered racemose inflorescence, but differs in being large trees and having tiny 3-valved nonexplosive fruits like those of most of the herbaceous genera.

Leonia (5 spp.) — Trees, commonly ramiflorous or cauliflorous. The leaves tend to be sclerophyllous and in the commonest species (*L. glycicarpa*) entire. Flowers are small, nondescript, completely actinomorphic and borne in more or less branched (i.e., paniculate) inflorescences; one species (*L. triandra,* a 25 m tree of western Colombia) has only 3 stamens (unique in the family). The indehiscent globose fruits can be as much as 5–6 cm in diameter in cauliflorous *L. glycicarpa.*

P: tamara, nina caspi

Extralimital genera:

Orthion — A Central American tree genus closely related to *Amphirrhox* but differing in an open conspicuously branching terminal inflorescence and more sharply serrate alternate leaves.

Schweiggeria — Reputedly bitypic with one species in Mexico and one in Brazil, but is not known to me.

VITACEAE

Tendrillate lianas with alternate, palmately veined or 3(–5)-foliolate leaves, differentiated from vegetatively similar Cucurbitaceae and Passifloraceae by having the tendril arising on opposite side of node from leaf. The distinctive nodes are usually jointed and/or swollen and the stems are usually rather soft and flexible, often somewhat flattened (very flat in some Asian taxa), and typically with fibrous-peeling reddish bark; frequently long pendent stemlike adventitious roots formed in *Cissus.* All species have small 4–5-parted flowers with conspicuous intrastaminal disk and borne on inflorescences arising opposite the leaves. The fruit is a berry, nearly always turning black or dark purple at maturity.

Cissus (350 spp., incl. Old World) — Lowland forest lianas, mostly 3(–5)-foliolate, the simple-leaved species without prominent lateral corners (+/- lobes) on leaf. Flower parts in fours with free petals; inflorescence flat-topped.

E: mano de sapo (compound leaves), bejuco de agua (simple leaves); P: sapohuasca

Vitis (1 sp., plus ca. 60 in N. Am. and Old World) — Mostly n. temperate, only one species reaching South America; that species vegetatively differentiated from simple-leaved *Cissus* by having the leaf with two distinct corners (or sometimes distinctly sub-3-lobed). Flower differs from *Cissus* in the parts in fives and the petals apically united into a calypterate cap; inflorescence pyramidal.

E: bejuco de agua

Several other Laurasian genera reach northern Central America or the Antilles.

VOCHYSIACEAE

Usually large trees, the bark of twigs often exfoliating. Unmistakable in flower (very zygomorphic with only 1–2 fertile anthers) or fruit (a 3-parted capsule with winged seeds or indehiscent and with several characteristic wings). Leaves always opposite (or whorled) and entire, usually with numerous secondary veins and a submarginal collecting vein. Mostly easy to recognize, even when sterile, by the conspicuous glands formed by the bases of the caducous stipules on either side of the petiole base; these are always present in *Qualea* (and *Ruizterania*), usually in *Erisma*, but not in *Vochysia*. When stipule-derived glands are not present, the leaves are commonly whorled (*Vochysia*) and/or the small subulate stipules are distinctively thick-based. Stellate indumentum (sometimes only on the calyces or inflorescence) is unique to *Erisma* which is also the only genus with an inferior ovary and an indehiscent fruit; the other genera have simple or malpighiaceous trichomes. *Vochysia* often has whorled leaves, the other genera uniformly opposite ones. The segregate genus *Ruizterania* is often separated from *Qualea;* like some species of *Qualea sensu stricto,* it has *Clusia*-like leaf venation.

Erisma (16 spp.) — The inferior ovary is unique; also distinctive in the single broad variously colored petal. Inflorescence usually paniculate; fruit one-celled, indehiscent, usually winged (common tahuampa species *E. calcaratum* has unwinged fruit). Vegetatively, the stellate indumentum is unique in family; the stipules are subulate with thickened bases, the apex is sometimes caducous leaving a glandular base (cf., *Qualea*) in commonest upper Amazonian species. When the stipule remains intact, the genus may resemble Rubiaceae but the stipules are smaller and remote from each other on opposite sides of the node.

P: cacahuillo (*E. calcaratum*)

Vochysiaceae

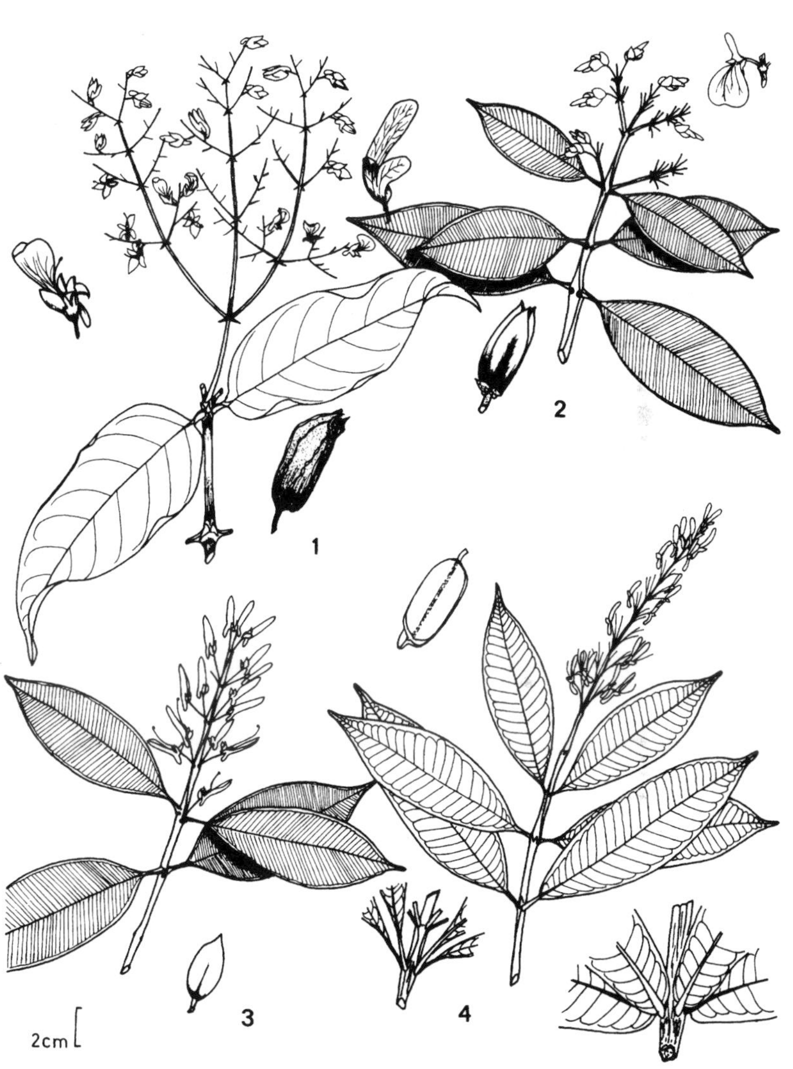

1 - *Erisma* **2** - *Qualea*

3 - *Ruizterania* **4** - *Vochysia*

Qualea (59 spp.) — Single broad petal, usually white or blue; fruit 3-celled with winged seeds; leaves often with parallel secondary and tertiary venation. Vegetatively distinctive in always having a pair of glands derived from the stipule bases at the base of each petiole.

P: yesca caspi (*Q. paraense*) (= many ant nests)

(***Ruizterania***) — A segregate from *Qualea,* differing in buds with a spur; spurred sepal 3–4 times longer than the others; pubescent thecae. Leaves always with parallel secondary and tertiary venation (cf., *Clusia*).

P: mauba

Vochysia (97 spp.) — Petals narrow, yellow; inflorescence racemose or racemiform; fruit 3-celled with winged seeds. Leaves usually whorled, typically with many rigidly parallel secondary veins and submarginal collecting vein. Some species have a characteristic flat-topped growth-form with the numerous erect spikes of yellow flowers completely covering the canopy surface at anthesis.

C: sorogá; P: quillosisa, sacha casho (*V. venulosa*)

Extralimital genera:

Salvertia (1 sp.) — Brazilian, like *Vochysia* but with 5 petals.

Euphronia (1 sp.) (from Trigoniaceae) — A small Guayana area tree mostly on low-elevation rock outcrops, characterized by alternate leaves, strongly white below.

Callisthene (8 spp.) — Brazilian dry areas, like *Qualea* but the fruit with a central column.

WINTERACEAE

Easy to recognize by the strong primitive odor and coriaceous leaves pale to whitish below and with completely suppressed undersurface secondary and tertiary venation; the laminar margin is inrolled at least at base. A small conical terminal stipule (cf., Moraceae) is present but does not leave a prominent scar (unlike Magnoliaceae). The flowers are white with numerous rather narrow petals and obviously apocarpous carpels. The fruit is Annonaceae-like with several substipitate apocarpous carpels on slender pedicels; differs from annonacs in the more or less branched several-flowered inflorescence with slender pedicels.

Drimys (perhaps only a single polymorphic neotropical species, 70 spp. in Old World) — A prominent element of high-altitude Andean cloud forest. The leaves have a peppery taste and are sometimes used as a condiment.

C: ají, canelo

Winteraceae and Zygophyllaceae

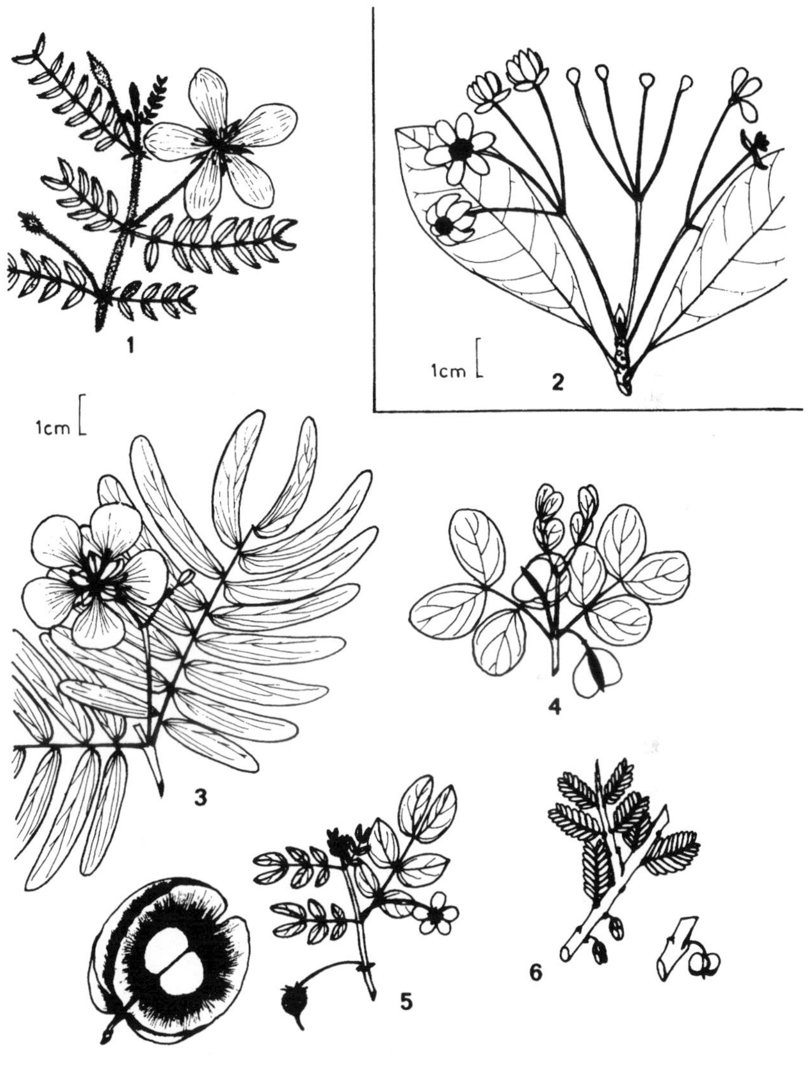

1 - *Tribulus* (Zygophyllaceae) **2** - *Drimys* (Winteraceae)

3 - *Bulnesia* (Zygophyllaceae) **4** - *Guaiacum* (Zygophyllaceae)

5 - *Kallstroemia* (Zygophyllaceae) **6** - *Porliera* (Zygophyllaceae)

Zygophyllaceae

Mostly dry-area trees (in arid areas sometimes reduced to shrubs); also two genera of rather weedy prostrate herbs. Extremely easy to recognize to family by the mostly conspicuously jointed branchlets and uniformly opposite even-pinnate (except *Fagonia:* digitately 3(–5)-foliolate) leaves with entire, opposite, asymmetric-based, sessile leaflets (often very small in desert taxa) which often curve toward rachis apex. Stipules usually present on young growth and often persistent. Flowers borne solitary or in small axillary cymes, with 5 sepals and usually conspicuous petals (usually yellow or orangish-yellow; pink in *Fagonia,* blue in *Guaiacum,* white in *Porliera*); stamens usually 10 and ovary with 5 fused carpels and subsessile stigmas. Fruit usually longitudinally 5-lobed or 5-winged (*Bulnesia*), or irregularly reduced to 2–4 lobes or wings by abortion; spiny and exozoochorous in *Tribulus.*

1. Prostrate Herbs with Solitary Yellow Flowers

Tribulus (30 Old World spp., 3 naturalized) — Prostrate weeds with conspicuously jointed stems and numerous, narrowly oblong asymmetric-based leaflets; stipules persistent. Fruits 5-parted with the segments spiny-surfaced.

Kallstroemia (16 spp.) — Very like *Tribulus* but the leaves with fewer (mostly 4–6) broader leaflets and the beaked fruit nonspiny, and consisting of 8–12 cocci that separate at maturity.

2. Trees, Shrubs, or Subshrubs

Bulnesia (7 spp.) — In northern Colombia a dominant large dry-forest tree with conspicuous orangish-yellow flowers and distinctly falcately curved, narrowly oblong leaflets; the Peruvian species a wandlike shrub with microphyllous leaflets (these deciduous most of year) and restricted to desert washes around Ica. Fruit broadly longitudinally 5-winged.

Guaiacum (5 spp.) — Large hard-wooded dry-forest tree with ca. 6-foliolate leaves having essentially sessile, broadly elliptic leaflets with +/- 3-veined asymmetric base; stipules caducous but basal leaflet pair usually very near base of petiole. Fruit a longitudinally lobed or thick-"winged" capsule, much broader than long, in our species strongly flattened and usually 2-winged by abortion.

C: guayacán

Larrea (5 spp.) — Resiniferous desert shrubs with persistent stipules, differing from all other area zygophyllacs in the evergreen leaves only 2-foliolate. Flowers yellow; fruit fragmenting into 5 villous nutlets.

In our area, found only in the lower Andean desert slopes of southern coastal Peru; of great phytogeographic interest as a potential remnant from a once more continuous distribution of this now amphitropical ecologically important desert genus.

Porliera (3 spp.) — Our species with thick more or less spine-tipped branches and very small leaves with microphyllous sessile parallelogram-shaped leaflets clustered on short shoots. Flowers white (fading cream), sessile and solitary. Fruit a small irregularly 3–4-parted capsule.

Fagonia (12 spp., plus 38 in Old World) — Dichotomously branched desert subshrub with strongly jointed stems, in our area only on the west-facing desert slopes of the Peruvian Andes. Characterized by persistent, more or less spiny stipules and digitately 3(–5)-foliolate leaves with narrow sessile, acute-tipped leaflets. Flowers pink; fruit broadly angled-conical and 5-lobed, separating from base into 5 thin-walled cocci.

INDEX: COMMON NAMES

D

E

P

INDEX: SCIENTIFIC NAMES

B

C

D

O

Q